A RETURN TO GLORY

THE UNTOLD STORY
OF HONOR, DISHONOR, AND TRIUMPH
AT THE
UNITED STATES MILITARY ACADEMY
1950-53

Bill McWilliams

A publication celebrating
the Bicentennial of the United States Military Academy

A Return to Glory:
The Untold Story of
Honor, Dishonor, and Triumph
at the United States Military Academy
1950-53

Copyright 2000
by
Bill McWilliams

ISBN: 1-890306-22-3

Library of Congress Catalog Card Number: 00132973

Jacket front illustration by Frank Wright.

Warwick House Publishers
720 Court Street
Lynchburg, Virginia 24504

For Veronica

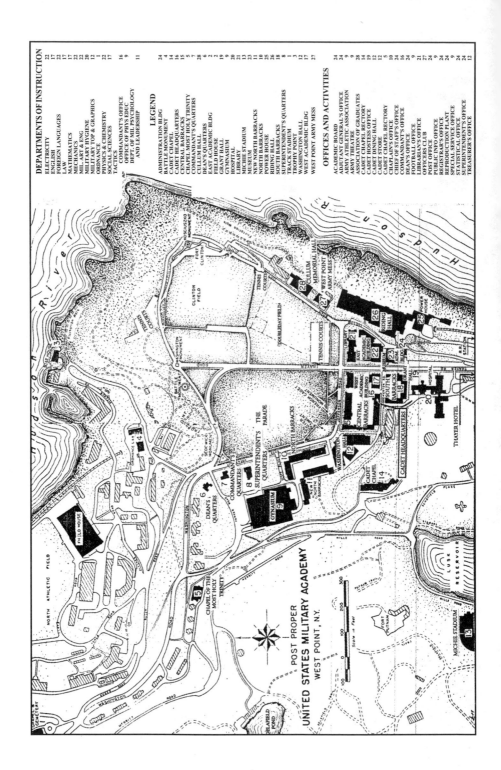

TABLE OF CONTENTS

PART TWO – TOWARD THE FUTURE

PART THREE – TRIUMPH

FOREWORD

The four national military academies, commonly known as West Point, Annapolis, the Air Force Academy, and the Coast Guard Academy belong to the American people. From the American people, from all walks of life, come the young men and women who in four years at the academies are educated and trained to deliberately high ideals and standards to become professional, career officers in our nation's Armed Forces. In *A Return to Glory: The Untold Story of Honor, Dishonor, and Triumph at the United States Military Academy, 1950-53*, Bill McWilliams brings us an extraordinary period history, and true story, of America's oldest national military academy, its cadets, graduates, and the citizen soldiers and airmen its graduates fought beside and led during the Korean War.

Most books about the Military Academy have been devoted to its founding and earlier eras in its nearly 200 years' existence, histories of the institution over extended periods, biographies of its most famous graduates and West Point's influence in their lives, unusual events, or stories of particular graduation classes in war. *A Return to Glory* is decidedly different. Seven years in research and writing, it is a comprehensive, in-depth, wartime history, written to interest readers of high school through retirement age, especially those unfamiliar with the Academy, its mission, and life in our nation's military services.

The prologue recreates the era of the Korean War and brings us into the story's setting. Then, as we follow the lives of the people who lived the events, the narrative transports us through a wide variety of scenes: from hometowns to West Point; to offices and cadet barracks; to conference rooms and academic classrooms; to football practice fields and stadiums; to the parade ground – the Plain; to military training grounds for future officers; to the Department of the Army at the Pentagon in Washington, D.C.; into the White House; through the skies over the continental United States, Korea, the Sea of Japan, the Japanese main island of Honshu; to air and ground battlefields; and back home.

Integrity, truth, honor, ethics, duty, leadership, command, competition, victory and defeat on athletic fields, in the nation's ground and air battles, and in war, are at the heart of the story. These subjects are

timeless, for they are, and will remain, always at the center of education and training for officer candidates and professional officer corps
in America's Armed Forces.

In reading of the honor incident of 1951, and its tragic aftermath,
we learn for the first time why and how two young men, sophomores
at the Academy, courageously accepted the responsibilities and lifelong burdens of exposing organized cheating in academics, an activity
which had, for some years, been eroding the foundation on which the
Academy education rested. We learn why and how the Academy and
the Department of the Army made decisions which later became controversial, in what, to that time, was an event unprecedented in Academy history. As we absorb the impacts of the search for evidence in the
undercover investigation, the formal investigations which followed,
and later the August 3, 1951 public announcements, we also read for
the first time how the Academy sought to avoid a repeat of the incident, while we receive a capsule history of the evolving honor code
and honor system, learn the code's and system's purposes, how both
were taught and passed from class to class in the Corps of Cadets, and
how they were applied in building the foundation expressed in the
Academy motto, "Duty, Honor, Country" – the foundation on which
the cadets' education rests.

In reading of cadet life during the era, we are taken through the
hectic first day at West Point in July of 1951, on through plebe year,
and the transition to cadet leaders, then finally commissioned officers.
We learn the effects of the war on the Academy and the cadets' training, while gaining insight to the cadet education: academics, military
and leadership training, physical education, athletics, and character
development, the moral underpinning of respected, successful leaders.

In reliving the Army football team's precipitous fall and rise from
the ashes of the 1951 honor incident, we read a brief history of Academy football; biographical sketches of its most famous architects; the
changing nature of the game; and the links between athletics, character, leadership, and a life in service – while being inspired by the Corps
of Cadets, their delightful good humor, and their Army teams of 1951-
'53, in one of the great untold sports stories of the 20th century.

For those who have never fought in war, or witnessed firsthand
war's devastation as seen and felt by the junior officer and America's

citizen soldiers – the GIs, the stories of Lieutenants David R. Hughes, Richard G. Inman, Richard T. Shea, Jr., and the men they fought beside and led, are powerful examples of the responsibilities they carried in war. The descriptions of battles in which they fought, and men's reactions to war's fury and chaos, are vivid and realistic. Lieutenant Hughes and K Company of the 7th Cavalry Regiment, 1st Cavalry Division, in far North Korea near the Yalu River, and on hills 339 and 347, in the fall of 1950 and '51 – and the men of the 7th Division's 17th and 32d Infantry Regiments on Pork Chop Hill in July of 1953, three weeks before the armistice, give us clear pictures of war fought on the Korean peninsula. The experiences and writings of Lieutenants Shea and Inman, and the courage and loving devotion of their wives and families, bring balance and poignant reality to lives of service, which tragically, all too often require the ultimate sacrifice.

From Korea, three years after the nation's youngest service, the Air Force, became an independent service, and before the birth of the Air Force Academy in 1954, comes the story of Lieutenant William J. "Pat" Ryan, his family, and the 93rd Bomb Squadron's B-29 missions in May, July and August of 1953. In those missions we see the realities of war fought from the air, and the hazards of missions intended to keep the hard won peace.

A Return to Glory, while telling us of the United States Military Academy at the midpoint of the world's most violent century in history, speaks to us today – and will tomorrow. For in the telling, the work is replete with implied lessons useful in educating the whole person and in developing character and leadership. As the Academy nears its 200th birthday, during the period commemorating the 50th anniversary of the Korean War, this book stands as a memorable tribute to those who fought and gave their lives in that conflict, a celebration of all that is good and right about West Point – and all Americans who lend their sons and daughters in defense of our nation, democracy, and free peoples everywhere.

John A. Wickham, Jr.
General, United States Army (Retired)
Chief of Staff, 1983-87

WHAT OTHERS HAVE SAID...

In *A Return to Glory*, Bill McWilliams brilliantly relates the enduring personal values of "Duty, Honor, Country" instilled in cadets by the US Military Academy, to the extreme character-testing demands of combat and American wars.

> David R. Hughes
> Colonel
> United States Army (Ret)
> Class of 1950, USMA

A superb, true story about the dark side of national unpreparedness and the deeper meanings of honor, character, devotion, and self-sacrifice in the early days of the Cold War – America's citizen soldiers and airmen abruptly thrust into "The Land of the Morning Calm," Korea, fighting and dying to halt the march of communism and saving a young Republic and its fledgling democracy.

> General Ronald R. Fogleman
> United States Air Force (Ret)
> Chief of Staff, 1994-97

An uplifting story of service above and beyond the call of duty – America's soldiers and airmen in Korea, their victories, defeats, and awe-inspiring triumphs over fear and the deadly chaos of war.

> Robert E. Miller
> Formerly: Private, A Company
> 17th Infantry Regiment, 7th Division

...a magnificent, important tale...[has] probably captured the spirit and soul of West Point better than any book I've ever read about the Academy.... [The] narrative is just plain *gifted*... descriptions of football games are magic... games and players literally come alive... pure and simple, the best sports writing I ever read....

> Benjamin F. Schemmer
> Editor-in-Chief
> *Strategic Review*

A Return to Glory is a must-have and a must-read. Read it to your children and to your grandchildren after them. The story treats of drama and the purpose of life through a professional examination of events at our United States Military Academy in the early 1950s. You will feel thrilled reading this book. You will appreciate why West Point is the treasure of our nation, The Corps of our core.

> The Rev. David R. Graham
> Adwaitha Hermitage
> Bellevue, WA, USA

Honor is the centerpiece in the United States Military Academy's motto "Duty, Honor, Country." In the Cadet Prayer are the words: "...Make us to choose the harder right instead of the easier wrong...." West Point strives to develop the self-discipline in each cadet to always "...choose the harder right." *A Return to Glory* tells an inspiring true story about the Academy cadets' honor code and system, while explaining their fundamental purposes – and honor's value to the officer corps of America's Armed Forces.

> John C. Bard
> Brigadier General
> United States Army (Ret)
> Commandant of Cadets, USMA,
> 1977-79

Unvarnished anecdotes, vignettes, mini-biographies, and the real story of the men and events, some well known, others virtually unknown, give readers a true insight into a critical time in the Academy's history. By focusing on the real meaning and purposes of the honor code and honor system, the reader has a far deeper understanding of honor, and how military leaders are developed in the crucible called West Point – while made ready to be entrusted with the leadership of our country's Armed Forces.

> William E. Roth
> Colonel
> United States Air Force (Ret)
> USMA Class of 1955

A wonderful, true story of love, family, devotion, sacrifice, tragedy, and remembrance in wartime. Extensively researched and told with great care, sensitivity, and compassion.

> Mary Jo Inman Vermillion
> Vincennes, Indiana

An authentic portrayal of the responsibilities, joys, rewards, tragedies, and wrenching emotional impacts that come with leading and commanding soldiers and airmen in war.

> William J. Ryan
> Colonel
> United States Air Force (Ret)
> Class of 1951, USMA

A Return to Glory vividly recreates the July 1953 last battle for Pork Chop Hill. Readers everywhere will find themselves in the trenches and bunkers, alongside the men of the 7th Infantry Division, who gave everything they had in five days of bitter fighting. An inspiring story of courage and devotion.

> James E. McKenzie
> Formerly: Private First Class
> Medic, Medical Company
> 17th Infantry Regiment, 7th Division

A Return to Glory accurately tells a true story of the infantryman's war in Korea, as seen through the eyes of the men of K Company, 7th Cavalry Regiment, 1st Cavalry Division. It's an unforgettable story of courage, valor, honor, and victory in war.

> John R. Flynn
> Lieutenant Colonel
> United States Army (Ret)
> Class of 1944, USMA
> Formerly: Company Commander,
> K Company, 7th Cavalry Regiment,
> 1st Cavalry Division, 1950-51

A Return to Glory tells it as it happened in a first-time, in-depth look at Coach Earl H. "Red" Blaik and his Army teams during the rebuilding years of 1951-53. It is a gripping story of how one of the great coaches in the history of American collegiate football was imbued with an unshakable determination to rebuild West Point football, which started Army's resurgence in pride and excellence. In this work there is another story, no less powerful – the soul and extraordinary spirit of West Point's Corps of Cadets.

> Godwin Ordway III
> Lieutenant Colonel
> United States Army (Ret)
> Class of 1955
> U.S. Military Academy
> Army Assistant Football Coach, 1957

A Return to Glory brings readers into football team meetings and dressing rooms, onto practice fields and playing fields of great stadiums – beside linemen and backs, and recreates an exciting, never-before-told, true story of defeat, victory, and honor on Army's "...fields of friendly strife."

> Freddie A. D. Attaya
> Class of 1954
> United States Military Academy

A Return to Glory is a *must read* for all Americans and the people of the United Nations whose courageous young men and women served and fought in Korea. The story taps the whole range of powerful, human emotions the War evoked, while telling an inspirational story of honor, valor, and sacrifice.

> Dale W. Cain, Sr.
> Formerly: Corporal, A Company
> 17th Infantry Regiment, 7th Division
> Two-Tour Veteran of the Korean War

Memorable, realistic, and sensitive descriptions of battles, which tell of war as it really was on the Korean peninsula – a wonderful tribute to the soldiers of the 7th Infantry Division, its 17th and 32d Regiments, and all those who fought the war never to be forgotten.

David R. Willcox
Formerly: First Lieutenant, A Company,
17th Infantry Regiment, 7th Division

The Cold War was a struggle between the young professionals on both sides who did their best – some giving their all – for their country's best interests as they saw them.

Perhaps we, in the West, prevailed because our young men and women believed in something greater than materialism. The graduates of our military academies emerged from West Point, Annapolis, and the Air Force Academy with a dedication to "Duty, Honor, Country" that the political winds and the fires of combat could not extinguish.

Bill McWilliams was there, amidst the storms of the early 1950s. His *Return to Glory* captures the defense of America's moral citadel at West Point as few writers could do. I enjoyed it.

Thomas C. Reed
Secretary of the Air Force
1976-77

ACKNOWLEDGMENTS

History is fascinating, a great teacher. To research history and bring it to life on the pages of a book, is even more fascinating, a marvelous experience.

When this work began, there was an idea, a concept fed by vague memories of events and people at the United States Military Academy – stories never told. The reach back to those events and people began in 1992, with memories of a football game, memories stirred by a talk I was asked to give ten years earlier at a leadership school. I grew up in a sports loving family, wanted to write, and believed I could write an interesting, exciting article about the game which kept cycling through my thoughts: Army-Duke at New York City's Polo Grounds on October 17, 1953.

I researched the contest and found far more than I had considered. Few readers would be interested in a collegiate football game played forty-seven years ago. What publisher would print such an article? In what magazine? Nevertheless, I couldn't quit. Something far more powerful was behind that single event, and wouldn't let go. What was it and why? Digging for the story behind the game turned up answers far more intriguing, a tumultuous, relatively untouched era in American history, and in the history of the United States Military Academy.

The football game had a setting on history's stage, the fifties, and reasons for sticking so firmly in my memory. Ah yes, West Point's cheating scandal of 1951. That was it, at least so I thought. I had been at the Academy exactly one month when the story broke. A young, wet-behind-the-ears freshman, a plebe, I was a small town boy completely awed by the Academy and the people who were its stewards. Because of that difficult first summer as a new cadet, not yet accepted as a plebe, I remembered little about the scandal, except the words "the ninety," and the Army football team's undoing as a result of the scandal. So I set out to learn what happened at West Point in the spring and summer of 1951. In digging for answers I found a whole new world, and wonderful people and events I'd somehow missed. And I found the Korean War, which was the defining event of that era, and both the central event and the backdrop for this story.

America was at war – again – far away, in a century of wars, and West Point had been producing most of America's great captains of military history since before the Civil War. The Academy was about to celebrate its Sesquicentennial when the cheating scandal occurred. The real journey of discovery was beginning. The article needed to be a book.

Honor; ethics; country; competition "on the fields of friendly strife," on military training fields, and on deadly battlefields; learning the art of war; duty; service; sacrifice; love; family; courage; devotion; heroism; "gallantry above and beyond the call of duty;" the words came tumbling out. The great intangibles are what this is about. Remembering, rediscovering, renewing, reinvigorating, studying again, relearning, learning anew, and most of all, becoming acquainted with wonderful people who lived the events – and especially those who didn't live through the events, but gave their all. These are the rewards, the gifts given me.

The real journey had to continue, the true story told, and thanks said to all who made it possible.

Not any of this work, absolutely none of it, could have occurred without the kind, considerate, generous assistance and encouragement of many people, to whom I shall forever be grateful. They brought great joy, and turned my "work into play and duty into privilege."

Pat Schrader and the nonfiction working group of California's Ventura County Writers' Club, were my first teachers in the earliest stages of the manuscript. Pat, the chairwoman of the working group, George Denardo, Carol Heckman, Ruth Carlson, Dick Hathcock, Mason Rose, and Pepper Heimowitz all listened carefully and patiently to my readings, and gave invaluable comments, suggestions, and encouragement to resurrect the joy of writing from my distant past, and set the project on the right path.

Shirley Powell, Gerry Mulvey, Sue Zylka-Thomas, Audrey Yanes, Lisa Angle, Mary Ann Trevathan, Sharon Sutliff, and Joan McCray of the San Luis Obispo, California NightWriters Club were equally patient and kind, listening, questioning, critiquing, and encouraging as I read proposed material and began the serious, in-depth research necessary for a work of this magnitude. Shirley, who chaired the all-genre working group, and Gerry Mulvey, were the bright shining stars who challenged my imagination and volunteered all manner of good ideas as the project concept and outline began maturing.

Military historian Colonel (Retired) George S. Pappas, class of 1944, provided invaluable counsel, guidance, assistance, and interview time before in-depth research began, pointing me toward sources of previously completed research and works in progress, as well as providing me his personal recollections of key events in this story. His most excellent history of the Academy, *To the Point*, was a veritable fountain of historical golden nuggets pertinent to this work, as was his knowledge and participation in the Academy Sesquicentennial celebration. His contributions were complemented by interviews with Brigadier General (Retired) Joseph F. H. Cutrona, an Academy public information officer during the era of this story, and the late Brigadier General (Retired) Jefferson J. Irvin, a member of the Collins Board, which investigated the cheating incident of 1951.

In the fall of 1994 I went to West Point to begin in-depth research. There I met Suzanne Christoff, Associate Director of the Military Academy Archives, and assistant archivists Judith Sibley and Susan Walker; later in the project came Alicia Mauldin, also an assistant archivist; Alan Aimone, Director of the Special Collections Division of the Academy Library; Charlyn Richardson and Deborah McKeon-Pogue, Library Technicians; Dr. Stephen B. Grove, Academy Historian; Colonel Anthony E. Hartle, Deputy Head and Professor, Department of English; and Harry Kubasek of the Army Football Office, football film archivist for the Army Athletic Association. Over the years since that week at West Point, I sent each of them – particularly Suzanne Christoff and Alan Aimone – innumerable written and telephonic requests for materials to help answer the never ending questions ongoing research unearthed. Their courteous, enthusiastic responsiveness was inspirational, simply through the manner of their giving and support. From them came gracious, untiring assistance, hundreds of pages of archived and special collections document copies and videos, and suggested alternate sources for materials not found at West Point. Their help was a testimonial to patience, and laid the foundation for this extensive nonfiction work. It's impossible to thank them enough.

Colonel (Retired) Morris J. Herbert, Susan Keterrer, and Leslie Rose provided a wealth of information from the Association of Graduates' *Register of Graduates and Former Cadets,* and their computer data files. Julian M. Olejniczak, Editor in Chief of the Association's magazine, *Assembly,* provided enormous encouragement by publishing two two-part serial articles based on the manuscript, and positive feedback from both articles.

Research sources included archives and libraries across the country. As a result there were numerous others who provided outstanding service, photographs, and important factual information underpinning this story. Rebecca Lentz Collier and Wilbert Mahoney in the National Archives found hundreds of Academy and Army staff papers, division and regimental command reports crucial to understanding the complete story of the 1951 cheating scandal at the Academy, and the last battle for Korea's Pork Chop Hill. Samuel W. Rushay, archivist at the Harry S. Truman Library; Brian Spangle at the Knox County Library in Vincennes, Indiana; Patti L. Houghton, Special Collections at Dartmouth University Library; Robert Schmidt, archivist at Miami University (Ohio); and Floyd M. Geery, museum technician at the Fort Bliss, Texas Museum are just a few of those who helped.

This story was given life by those who lived it, or were close to the people who "had gone before." They were kind, thoughtful, and generous in sharing their time and recollections in telephone or personal interviews, letters, faxes, e-mails, and photographs. There were over seventy who agreed to assist, and they were wonderful, unforgettable. They are all listed in the bibliography, in the section titled "Author's Interviews." Every single one provided recollections and information that brought substance, enthusiasm, and excitement to the work, and breathed life into the story.

Particularly, I want to acknowledge three beautiful ladies and some magnificent, former soldiers and airmen of the Korean War, who shared memories that often were bittersweet, if not terribly painful. All were inspirations. Joyce E. Himka, the former Joyce Riemann Shea; the late Barbara Colby, formerly Barbara Kipp Inman; and Mary Jo Inman Vermillion, sister of Richard G. Inman, told of husbands and a brother who gave "their last full measure of devotion" on the hills of Korea. William J. "Pat" Ryan, West Point class of 1951, gave not only his recollections of a turbulent time when he was First Captain at West Point, but told of his life-long love story about a lovely lady named Joanne and his combat experiences while serving in three wars.

ACKNOWLEDGMENTS

David R. Hughes, class of 1950, was the representative story from his class in Korea. He, too, was an inspiration and a genuine, unsung hero. Not only was he enormously encouraging while sharing his life and Korean War combat experiences, he enthusiastically supported this endeavor with positive feedback, outstanding ideas and suggestions. But he did more. His deep respect and affection for the American soldier, the GIs, the citizen soldiers, shined like a beacon, turning me toward the men who, for half a century, have carried in their memories countless, untold stories of courage, devotion, and heroism.

There were MacPherson Conner, class of 1952, platoon leader, 2d Platoon, E Company, 2d Battalion of the 32d Infantry Regiment, and David R. Willcox, platoon leader, 2d Platoon, A Company, 1st Battalion, 17th Regiment. Magnificent men, who, on Pork Chop Hill in July of 1953, were the connections between two incredible Academy heroes, Dick Inman and Dick Shea. Backing up Mac and David in telling of Dick Inman, Dick Shea, and Pork Chop Hill were James E. McKenzie, Lee Johnson, Robert E. Miller, Dale W. Cain, Sr., Emmett "Johnny" Gladwell, Charles W. Brooks, Robert Northcutt, and Dan Peters – courageous, devoted soldiers and never-to-be-forgotten men, who gave so much in Korea and gladly gave their recollections of "the forgotten war."

Complementing soldiers' recollections of Dick Shea and Dick Inman in Korea were 1952 Academy graduates who knew the two men well in their cadet days: Ronald M. Obach, Louis M. Davis, Thomas E. Courant, John R. Aker, George B. Bartel, Richard C. Coleman, and Walter G. Parks. My thanks to John R. Witherell, from the Sesquicentennial class, for connecting me with men who knew Dick Shea and Dick Inman.

Elizabeth Irving Maish and Colonel (Retired) Frederick F. Irving, class of 1951, provided personal recollections with interviews, correspondence, and articles about their father and mother, Major General Frederick A. Irving and his lovely wife, Vivian Dowe Irving. William Van D. Ochs, Jr., class of 1945, who performed the duties of an aide to both General Irving and Major General Bryant E. Moore, General Irving's predecessor, added much to the superintendent's story and pointed me toward a number of important research finds.

Edward M. Moses, Benjamin F. Schemmer, Emery S. Wetzel, Jr., Willis C. Tomsen, William R. Schulz III, John C. Bard, and F. Clif Berry, all from the class of 1954, and Willard L. Robinson and Preston

S. Harvill, Jr., class of 1955, brought wonderful cadet humor to life with their recollections of Army-Duke and Army-Navy pregame pranks and the gathering momentum within the Corps of Cadets to rebuild Army football. Retired General John Galvin, class of 1954, graciously gave his approval to use his delightful cartoons which kept us all laughing during our cadet days. Ronald E. Button, also '54, loaned important official pamphlets, correspondence, records and class yearbooks, which were of great value in conducting research.

Three marvelous friends gave invaluable assistance. William K. "Bill" Stockdale, class of 1951, patiently educated me on Cadet Honor Committee procedures of the era. Dr. John Krobock, class of 1953, and Roy Thorsen, class of 1955, thoughtfully critiqued four of the most complex and difficult chapters to research and write.

And to a courageous, small group of Academy graduates, most of whom asked not to be given credit in this work, I write this. You were the men who took upon yourselves the burden honor demanded, when others chose not to shoulder the inordinately heavy responsibility that integrity and truth impose. You were fallible, as we all were, but you are true, unsung heroes. You chose "the harder right instead of the easier wrong." Included among them are Richard J. Miller, class of 1952, Daniel J. Myers, Jr., class of 1951, who provided recollections about the honor code and honor system, as well as events leading to the formal investigation into the honor incident; and John H. Craigie, class of 1951, who was a close tie between two unsung cadet heroes, men who asked their true names not be divulged.

There is one more, whose real name is also withheld at his request. He regrettably had to resign following the cheating scandal of 1951, yet nearly half a century later courageously recounted events that caused great pain in his life and what he unflinchingly labeled as "family disgrace." He did far more than redeem himself, and honor himself, with a life of service to his country as a commissioned officer in the United States Army. He served as an Army aviator in Vietnam, the nation's longest war.

Next and for always – the men of the Black Knights of the Hudson, the Army football teams of 1951-1953. Their wonderful interviews and recollections were inspirational and encouraging beyond measure. They loved the game they played, the honor they gave back to Army football, and the fond memories and success it brought them. They were all special, extraordinary, and each gave a dimension to the story impossible to find elsewhere.

Peter Joel Vann was my hero on the Army football team when I was a cadet. Regrettably, I didn't seize the opportunity to meet or know Pete while we were cadets. Now, as fate would have it, after all these years I finally met Pete Vann. He's been not only a joy to meet, but a marvelous friend who has encouraged me over and over again through these years of trial, error, and hard work. A special thanks to Pete Vann for the interviews, background, photographs, cartoons, articles and other information about the game of football as it was played in that era, his comments, suggestions, ideas, and yes – he and Freddie A. D. Attaya were my football coaches throughout the project.

To Freddie A. D. Attaya, Robert M. Mischak, Gerald A. Lodge, Lowell E. Sisson, Norman F. Stephen, Thomas J. Bell, Ralph J. Chesnauskas, Godwin Ordway III, Patrick N. Uebel, Howard G. Glock, Robert G. Farris, John E. Krause, John R. Krobock, Frank S. Wilkerson Jr., Donald G. Fuqua, Alfred E. Paulekas, Paul A. Lasley, Ernest F. Condina, Leroy T. Lunn, Joseph D. Lapchick Jr., Joseph P. Franklin, and Peter J. Vann – thanks one and all. You taught me to hear, sing, and whistle that joyous song again, *On Brave Old Army Team....*

Supporting the Black Knights was Harry Kubasek, Army football's film archivist and game film recorder. He provided invaluable video copies from games of that era to assist in reeducating me on the thrills as well as the technical aspects of football as played during those four tumultuous years.

Throughout this entire project, F. Clif Berry, former editor of *Air Force* magazine whom I mentioned previously, was an invaluable advisor and counselor throughout the project, patiently helping me negotiate the twists, turns, and vagaries of writing and the publishing industry.

Finally came the manuscript readers, all seventy-three of them, nearly all of whom I interviewed. Thirteen were asked to read and critique the entire manuscript. They graciously granted their time and energies, provided corrections of fact as well as numerous thoughtful suggestions and recommendations to improve the work. Special thanks to Lieutenant Colonel Conrad C. Crane, class of 1974, Associate Professor of History at the Military Academy; Colonel Anthony E. Hartle, class of 1964,

Professor of English, whose extensive research into the 1951 cheating incident became the basis for deeper inquiry; and Warwick House Publishing's editor, Joyce Maddox, who worked so hard to purge the manuscript of errors – and improve it. They were untiring in efforts to ensure the historical accuracy so important to this work.

And to the love of my life Veronica, who has stood beside me throughout it all, encouraging me, giving me counsel and advice, and urging me on when I faltered. West Point was where we met and fell in love. This work is a gift to which you have given meaning.

* * *

AUTHOR'S EXPLANATION

Readers will see a number of cadets' names, and later an officer's name, in italics, primarily in chapters 1 through 3, 20 and the epilogue. The italicized names are aliases. Three of them – *Brian C. Nolan*, *Michael G. Arrison*, and *George L. Hendricks* – were interviewed by the author with the stipulation their true identities not be divulged. Thus, with their approvals, they were given fictitious names.

All other italicized names are aliases as well, mostly ex-cadets not interviewed but whose real names appeared in official documents sourced for this work. Their real names couldn't be divulged for reasons of privacy.

In two instances dialog in the work represents a synthesis and reconstruction of typical events and language of the era rather than specific recollections or documentation of events. Language, expressions, and content of dialog in those instances were based on the author's experiences and research confirming their typical use in daily life at the Military Academy and within the Army and Air Force. The first instance is in chapter 7, which contains radio calls and conversations between the crew of a military aircraft and flight, ground and air traffic control, conversation which, though researched, wasn't documented in whole, word for word. The second is in chapter 9, which describes the first day of a new cadet at West Point.

Bill McWilliams
June 2000

PROLOGUE

This is a true story. The events occurred a half century ago, most at a far different place, a place steeped in history and tradition. Many who lived the events are gone, but memories of them linger. What they sought to pass to others continues to inspire through the passing decades.

The United States Military Academy – West Point – though accredited and recognized for the quality of its education, is more than a college or university. The oldest of America's national military academies, established in 1802, with a Corps of Cadets now numbering four thousand, its purpose is to educate and train young men and women for a life of service as commissioned officers in the United States Army. To instill strength of character and leadership excellence in graduates are enduring goals and sources of pride in the Academy education. West Point's motto, "Duty, Honor, Country," adopted in 1888, is the cornerstone of Academy training.

Today, as throughout the nearly two centuries of Academy history, members of the Corps of Cadets come from every state in the nation, and America's territories. Additionally, as a result of government to government agreements, each undergraduate class includes a small number of cadets from other nations. Nevertheless, it can be said that the United States Military Academy is truly the peoples' Academy, brought to life by America's founding fathers and sustained by its people to help ensure the nation's security.

But mid-year of this century, when this story begins, was another, far different time. Though West Point's purpose, goals, and motto were virtually the same as today, its curriculum and training regimen were not, and the Corps of Cadets numbered less than twenty-five hundred, all men. Because there wasn't yet an Air Force Academy, approximately thirty percent of each graduating class received commissions in the nation's youngest armed service, the Air Force.

The more than one million of us who were to graduate from high school in 1950 were fortunate. It was a time of unbridled optimism and growing prosperity.

America was at peace. The Great Depression of the thirties was becoming a distant, less painful memory. World War II, the most destructive war in human history, and the defining event in our lives, ended in August 1945. The United States emerged from the War the most powerful nation in the Western world, its industries untouched by the terrible fighting. Converted from wartime production, factories were once more turning America into the land of plenty. The economy was strong. The future seemed boundless, filled with images of material goods that would make life easier.

Television, rapidly spreading across the forty-eight states, was still black and white. A license for the first color telecast went to the Columbia Broadcasting System in 1949, but a court injunction delayed the first broadcast until 1950. The automobile industry sent over six million cars and trucks off assembly lines - a new record.

It was the beginning of the computer age. A government program during World War II had developed the Electronic Numerical Integrator and Computer, or ENIAC, which was dedicated in February of 1946. The awkward machine was primarily for the military. The first working model weighed thirty tons, included eighteen thousand vacuum tubes, and covered fifteen thousand square feet of floor space - the equivalent of a building half the length and almost half the width of a football field.

Labor organizations pressed for the fourth round of wage increases since the end of the war. President Truman signed a bill raising the minimum wage from forty cents to seventy-five cents an hour in certain industries engaged in interstate commerce.

The discovery of cortisone was announced. The hormone promised to bring relief to sufferers of rheumatoid arthritis, the most painful type of arthritis.

The year in publishing was marked by an increasing popularity of books on religion, studies of world affairs, and books that discussed the plight of black Americans. Among those published in 1949 was Nelson Algren's *The Man With the Golden Arm*, the powerful story of Frankie Machine, a Chicago poker dealer whose heroin addiction keeps him from escaping the slums and ultimately leads him to suicide; *The Greatest Story Ever Told* by Fulton Oursler; and *A Guide to Confident Living* by Norman Vincent Peale.

For a change, American men cheered Parisian fashion designers as they decreed daring décolletage for evening wear and bikini bathing suits for the beach.

Collegiate and professional sports had also recovered from the dark days of World War II when nearly all able bodied athletes were in the armed forces. In 1949 the National Basketball Association, the NBA, was founded, and baseball's two major league champions staged breathtaking shows for their fans as both the American and National League pennants were clinched on the last day of the season. In the 46th annual World Series the New York Yankees defeated the Brooklyn Dodgers four games to one.

In golf, top money winners were Sam Snead, $31,594, and Mildred "Babe" Didrikson Zaharias, $4,650. In boxing, heavyweight champion Joe Louis officially retired in March. In June, Ezzard Charles defeated "Jersey Joe" Walcott to become the new heavyweight champion of the world. Kentucky defeated Oklahoma State 46 to 36 to win the National Collegiate Athletic Association basketball championship. Notre Dame's Fighting Irish once more captured college football's mythical national championship in the fall of 1949, and Frank Leahy, their head coach, was being compared to his legendary football coach and teacher at Notre Dame, Knute Rockne.

At West Point that fall, Earl Henry "Red" Blaik, the Army coach and Leahy's post-World War II rival, led his Black Knights of the Hudson to a 9-0 record, ending with a 38-0 thrashing of Navy. Army received the Lambert Trophy, symbolic of Eastern Champions, and a No. 4 national ranking. In the professional game, still several years from the explosive growth it would enjoy, the Philadelphia Eagles won the National Football League championship by defeating the Los Angeles Rams 14 to 0.

At the end of the War, hundreds of thousands of former GIs had taken advantage of the government's "G.I. Bill" and entered colleges and universities to obtain educations they would have otherwise been unable to afford. Large numbers were graduating and enrollments would soon begin declining.

Sounds of the era reflected the mood. Music was slow, dreamy, and romantic. *Some Enchanted Evening*, sung by Perry Como, was the number one song in 1949. The hit tune was from the smash Broadway musical, *South Pacific*, based on James A. Michener's Pulitzer Prize winning novel, *Tales of the South Pacific,* with its World War II story setting.

America indeed was enjoying peace and prosperity, and in no mood for war. But there were other forces and events at work which would change our lives for years to come. We were entering the mid-year of the century of the totalitarian, at the dawn of the Cold War and Nuclear Age. The Union of Soviet Socialist Republics, a reluctant but necessary ally of the West during World War II, had emerged from that conflict a powerful, new totalitarian state.

But in the high, quiet, rural San Luis Valley of Colorado that fall, the gathering storm of international confrontation seemed distant, remote, of little concern to high school seniors in a small town.

Thoughts were of buddies, girlfriends, football, music, the senior prom, and trout fishing. When will dad break loose the new, maroon Fraser automobile so I can drive it? Since I can't get an appointment to West Point this year, because of my age, should I go to college for a year while trying for an appointment? If so, where? These were my interests and burning questions of life in the fall of senior year.

We were back in a small town once more after five months in Denver, a city of three hundred twenty-five thousand, which seemed a metropolis to me. We lived in a two story duplex four blocks from the high school in Del Norte, population two thousand. Del Norte is at the western edge of the San Luis Valley in south central Colorado, on the Rio Grande River, which courses east through the relatively flat, fifty mile wide Valley toward Alamosa, before turning south to run through New Mexico. I was happier in small towns. Life was uncomplicated, fun, idyllic.

In March 1949 Dad had resigned from his high school coaching job in Los Alamos, New Mexico, where the atomic bomb had been secretly developed during World War II. We had moved to Los Alamos in 1946 from San Juan, Texas, in the lower Rio Grande Valley, near the Mexican border.

I was a foot-dragging, stubborn, argumentative fifteen year old when we left Los Alamos. I had to leave all my friends after more than two and a half years in high school at a place I considered next to heaven. My sister Mary Kay, who was born in Los Alamos, was less than three when we left.

Even as excitement over senior year at Del Norte grew, West Point was already much on my mind, and I had begun in earnest to learn how

I might obtain an appointment. Seeds for an Academy education and the military life had been planted in me early by my maternal grandmother, before the War. But World War II, our brief stay in Los Alamos, and a much-admired cousin, who in 1949 was flying fighter planes in the Air Force, had weighed far more heavily than family encouragement in determining the path my life would eventually take.

Senior year was all I hoped for, and more. In late May and early June 1950, high school and college graduations were in full swing across the land, May 23 for my high school class. West Point's class of that year graduated 670 men on June 6, the sixth anniversary of Allied landings on the beaches of Normandy, France, during World War II.

I was sixteen, still too young for an appointment to West Point when the Academy's class of 1954 entered on July 5. The Korean War had begun ten days earlier, and it was already clear a desperate fight was in progress. America had again been surprised by an attack, this time in a strange land in Northeast Asia, half a world away; and again, unprepared for war.

The last of thirty thousand occupying American soldiers had been withdrawn from South Korea the summer of 1949. Only five hundred members of the U.S. armed forces remained behind to provide advice and assistance to the South Korean constabulary, a one hundred thousand-man fledgling army ill-equipped and not sufficiently trained to face the North Korean onslaught that came the following June.

On June 30, 1950, five days after the Korean War began, President Truman authorized the return of American ground forces to defend South Korea. Overshadowed that day, and virtually unnoticed, was the president's decision to send the first seventeen American advisors to another nation in Southeast Asia – Vietnam.

I would have to wait a year to become a member of the United States Corps of Cadets. The year 1950 would not be pleasant for West Point and its graduates; 1951 would be worse.

"Not in the clamor of the crowded street,
Not in the shouts and plaudits of the throng,
But in ourselves, are triumph and defeat."

—Henry Wadsworth Longfellow

PART ONE

THE DESCENT

CHAPTER 1

DISASTER IN A FABLED PLACE

Friday morning, August 3, 1951, began earlier than normal for Major General Frederick A. Irving, the Academy superintendent. En route from the superintendent's quarters to his office in the administration building, he detoured by the South Gymnasium to speak with Colonel Billy Leer. Colonel Leer, the senior staff officer responsible for West Point's public information programs, was already at his desk. Major Joseph F. H. Cutrona, public information officer and Captain George Pappas, public information officer and Sesquicentennial project officer, were also in the office when the superintendent arrived.

General Irving opened the door, and holding onto the door knob, leaned in to speak. "I received a call from Max Taylor late last night. Drew Pearson has been making inquiries." Drew Pearson, the controversial, muckraking journalist for the *Washington Post*, was well known for his unrelenting, aggressive, sensational style of reporting. General Irving continued, "We'll have to announce this afternoon at 1300 hours."

With the one o'clock public announcements that day, one month after the class of 1955 entered the Academy, news of honor violations involving approximately ninety cadets exploded from behind West Point's gray granite walls. Word spread rapidly after months of rumor and speculation inside the Academy.

For the Academy, its Department of Athletics, the Corps of Cadets, Academy graduates, the Army football team, and the United States Army, the announcements triggered a precipitous, disillusioning plunge into public controversy, and a storm of criticism. The news came less than six months before West Point's Sesquicentennial, while the Academy was planning to celebrate the achievements of its founders and graduates, many of whom were national heroes as well as prominent figures in American history.

The United States Military Academy, a national institution of higher learning, could proudly trace its initial advocacy and beginnings to George Washington, "the father of our country," the central figure in both the American Revolution and the establishment of the nation's government. Among the more prominent graduates in its colorful history were Ulysses S. Grant and Robert E. Lee, who led The Blue and The Gray respectively in the great national convulsion, the Civil War, the bitterest, most divisive war in our history. Then came John J. Pershing, who successfully led the American Expeditionary Forces through the ghastly, horror filled trenches and "no man's lands" in Europe during World War I.

More recently, there were other West Point graduates who covered themselves with glory in victories on far-flung battlefields, and had become household names. They were World War II leaders and heroes – men such as Eisenhower, MacArthur, Bradley, Arnold, Patton, and hundreds of less well known graduates, many of whom had given their lives in that war.

Perhaps worse, the August 3 announcement of extensive honor violations at West Point came in the midst of another war that began less than five years after World War II ended. The Academy's graduates had once again been called to duty on battlefields, this time on the Korean Peninsula, half a world away, in Northeast Asia. Though the costs were terrible, the war gave the Academy's mission a renewed sense of realism and urgency. August 3, 1951, seemed an inglorious end to what had been a long period of glory at a fabled place.

On the day of the announcements, most of the men found guilty of honor violations were at Camp Buckner, ten miles from the Academy. Many were football players on Coach Earl Blaik's nationally ranked 1950 team, including first classmen, new seniors, in the class of 1952. Earlier in the summer, after the formal investigation into the honor violations had been completed, the first classmen had been removed from previously intended summer training assignments that would have placed them in leadership roles in the Corps of Cadets. They had not yet been told the investigation board, later called the Collins Board, had found them guilty of violating the cadet honor code. But the duties they had been given at Camp Buckner seemed menial "make work," although they were supporting training of the new third class, the sophomores who were the class of 1954. The first classmen's abrupt, late June reassignment into an organization separated from the rest of their classmates didn't bode well for their futures at West Point.

Now, immediately after the announcement, they were told to pack their belongings to move the next morning. Under orders, some escorted by cadet guards, those at Camp Buckner were taken by military trucks back to the Academy and placed "on restriction" in the north area of cadet barracks.

When the announcements came, they hadn't been told they were to be discharged from the Academy. Nearly all learned of the final decision by radio, television, or from their classmates who had heard the news.

Although their honor training as cadets had repeatedly made clear that penalties for honor violations were severe, the two months of waiting, uncertainty, rumors, hopes, and wishes to the contrary had convinced many that their sheer numbers would result in a more lenient response from the Academy and the Army. Their hopes were unceremoniously dashed by the radio and television broadcasts.

Further, though none were named in the carefully worded announcements, the statements had been painfully clear in describing the underlying causes for their discharges – the cadet honor code had been violated. They had cheated in academics. Surprise, embarrassment, frustration, and anger, mixed with feelings of guilt, were the emotions surging through their ranks.

While the public announcements flashed over the wires, or were handed to the press, a brief memorandum from the Commandant of Cadets, Colonel John Waters, to all cadets undergoing training at West Point and Camp Buckner, wended its way through cadet mail. It confirmed rumors circulating for four weeks. Stapled to the memorandum were copies of the one o'clock announcements released from the Academy and the Department of the Army in Washington, D.C.

HEADQUARTERS UNITED STATES CORPS OF CADETS
West Point, New York

3 August 1951

MEMORANDUM TO: Members of the Corps of Cadets

1. As you no doubt have heard, the following releases were made simultaneously in Washington, D.C., and at West Point.

2. Heretofore when Honor cases occurred you have kept the details properly within the Corps. Again I am sure you will live up to this dignified and respected practice.

3. The magnitude of this case will cause many inquiries and much conversation. Should you be called upon to furnish information relative to this case, the Public Information Officer at West Point is the agency to whom you should respectfully refer all inquiries.

4. When the case is completed, the facts will be presented to the Corps.

> JOHN K. WATERS
> Colonel, Armor
> Commandant of Cadets

The meanings of the press releases were clear. The announcement at West Point explained the talk of "something big" which began circulating in our class, the class of '55, within days after we entered "Beast Barracks" in early July.

PUBLIC INFORMATION OFFICE   8/3/51
UNITED STATES MILITARY ACADEMY
WEST POINT, NEW YORK
                                    FOR IMMEDIATE RELEASE

BREACH OF THE WEST POINT HONOR CODE ANNOUNCED

WEST POINT, NY, Aug 3 — Major General Frederick A. Irving, Superintendent, United States Military Academy, announced today that there has been a serious breach of the West Point Code of Honor involving approximately ninety members of the Corps of Cadets. Final action is being taken under the direction of the Superintendent with a view to discharging the cadets involved.

The Cadet Honor Code is the basis of the traditional high standard of personal integrity and character which has been the hallmark of West Point throughout its existence.

In this instance the infractions consisted of receiving improper outside assistance in academic work. In accordance with the accepted Code of Honor, a cadet is not permitted to seek or to accept improper assistance in any of the tests presented to him in class.

The Honor Code was established by and is administered largely by the cadets. They themselves apply the principles to all cadets regardless of their position or of their standing in the Corps. In the group being discharged are cadets who have been prominent in various activities including varsity football.

In making the announcement General Irving said, 'This action, which has been taken with the approval of the Department of the Army, is the source of deepest regret to me, as well as to all graduates and cadets of the Military Academy. Admittedly the action being taken is stern and uncompromising, but after weighing all factors most carefully I and the responsible heads of the Army are convinced that there could be no compromise solution that would preserve the vital honor system of West Point, which is the very heart of the Academy.

From the day a cadet enters West Point, the ideals which are exemplified in the Honor Code are impressed on him in all of his activities.

The ethics of honor are above and apart from regulations. A violation of regulations does not necessarily involve the integrity of the individual. But when a cadet does not accept the tenets which are the foundation of the integrity demanded of an officer, he surrenders his right to the trust imposed upon him by the nation. During the one hundred and fifty years in which the motto of the Military Academy has been 'Duty, Honor, Country', these words have guided the Army in war and in peace. These three words mean one thing — integrity. The code behind the meaning is above any one individual.

If we are to retain the faith of parents whose sons the officers of the Army must lead in battle, then we must preserve in those leaders the very highest standards not only of courage but of integrity.'

Since the cases involved the Cadet Honor Code, the names of the discharged cadets will not be released so that no undue criticism of the individuals concerned will result.

Among the "approximately ninety" were thirty-seven football players, including Coach Earl "Red" Blaik's son, Bob. He would have been a first classman, a senior that fall, the starting quarterback on Army's team, and the Army team would have once more been in contention for collegiate football's mythical national championship.

The August 3 announcement was the second "official" communication to the men of the United States Corps of Cadets saying there had been extensive honor violations. The first had been in late May, less than a week before the graduation of the class of 1951, when the cadet First Captain, William J. "Pat" Ryan, his voice wavering with emotion, read a short paragraph to the Corps at noon meal in Washington Hall. He told them an investigation was in progress into allegations of "a serious violation of the honor code by many cadets." He said no effort would be spared to drive every violator out of the Corps; "...to us, an honor violator is the same as dead."

Between the end of May, when the Corps was first told of the investigation, and August 3, the cadet grapevine and rumor mill accelerated into overdrive. Unfortunately, neither were fueled with complete or accurate accounts of what had been going on behind closed doors since May 29, when the formal investigation began. Other than Pat Ryan's announcement to the Corps at the end of May, there had been no mention of events that followed. And most assuredly, except for a few officers and cadets at West Point, absolutely no one knew what had occurred in the preceding seven months and what cadets would later come to know as the Collins Board.

The Collins Board, a board of inquiry which began taking testimony on May 29, and all other proceedings associated with the honor incident of 1951, were held behind closed doors, and board members were prohibited by law and regulation from discussing the proceedings with anyone except those responsible for advising them. Records of the proceedings were not releasable. Thus, facts and information about what was occurring were not available at West Point or to the public. But that didn't stop talk, and the talk inevitably spread far beyond the confines of the Academy.

Inside the Academy, before the public announcement, the grape-vine trafficked in bits and pieces of incomplete and erroneous information, assumptions, speculation, gossip, rumors, innuendo, half-truths, and outright fabrications of what was really in progress, and how it all began. Such talk increased dramatically when the announcement came. Now a growing public furor was added to all that was going on inside the Academy.

The events leading to the Collins Board began years earlier. No one will ever know who started the organized cheating at West Point that led to the honor incident of 1951, or precisely how and when it began. The Collins Board concluded "that there had been since approximately 1946-1947 a conspiracy within the Corps of Cadets to cheat in academics." In September, another board of officers at West Point examining the causes of organized cheating concluded "a small, closely confined ring must have existed among members of the varsity football squad...as early as the academic year 1949-50, and...may have existed as early as the fall of 1944."

SPREADING SHOCK WAVES

The Korean War was in its fourteenth month when the news release of August 3 became public knowledge. Suddenly, the savage fighting in Korea was less important. Flashed over wire services to hundreds of newspapers, radio and television stations, the announcements of events at West Point shocked the nation. The flood of news reports spread a fire storm of controversy, controversy which had been simmering inside the Academy for months. Revelations of cheating in academics by large numbers of Army's nationally ranked football team became the number one news story, sweeping the Korean War off the front pages of newspapers all over the country, and out of lead stories on radio and television.

Honor, tradition, service, and achievement provided the backdrop for events surrounding the cadets' impending dismissals. In light of the Academy's reputation, its role in national affairs, and the Army football team's seven year run of glory on the nation's collegiate gridirons, those events, and the cadets' dismissals, were to trigger a national scandal.

"WEST POINT OUSTS 90 CADETS FOR CHEATING IN CLASSROOM; FOOTBALL PLAYERS INVOLVED," said the *New*

York Times. "Corps Is Bitter at Guilty Men And Blow to Academy Honor." " 'MILD' DISCHARGES GIVEN." "President Expresses Concern – Collins [Army Chief of Staff] Says Infractions Stemmed From Varsity." "SECRECY ON CADETS AROUSES PROTESTS." "Confidence of People in the Army at Stake." From Denver, Colorado's *Rocky Mountain News,* "WEST POINT KICKS OUT 90 CADETS IN EXAMINATION CHEATING." The Seattle, Washington, *Post Intelligencer,* "ARMY FOOTBALL PERILED AS 90 CADETS EX-PELLED FOR CHEATING." In the Los Angeles, California, *Examiner* it was "POINT EXPULSIONS BLAMED ON FOOTBALL OVEREMPHASIS." In Houston, Texas, the *Press* front page covered the reaction of the cadets soon to resign from the Academy. "IT'S 'RAW DEAL', SAID FIRED CADETS." The *Daily Star* headline in the small town of Nowata, Oklahoma, told of reactions in the nation's capital. "WEST POINT DISMISSALS BRING IMMEDIATE CLAMOR FOR INVESTIGATION OF SCHOOL, FOOTBALL OPERATIONS."

Within the week, articles or editorial comment appeared in major news magazines such as *Time, Life, Newsweek,* and *U.S. News and World Report.* Consistent with its press release of August 3, the Academy never divulged the names of cadets found guilty of honor violations, but some identified themselves to newsmen, and their views of the scandal began to be told with powerful impact. The *Time* reporter said the handling of the incident had been botched, and implied the dismissed cadets had received unfair treatment considering evidence also pointed to already graduated classes. *Life* and *Newsweek* placed the incident in the context of scandals "already rampant in the nation." Both magazines decried the scandal's corrosive effects on the nation's moral climate.

The news stories brought a flood of telephone calls, letters, and messages to members of Congress, officials in President Truman's administration, General Irving, and Coach Blaik. On the floors of the Senate and House of Representatives there were emotional speeches, letters read into the Congressional record, and calls for Congressional investigations. Senators J. William Fullbright of Arkansas and William Benton of Illinois called for abolishing football at West Point and Annapolis. Fullbright, according to one reporter, had been a college football player. Benton, before becoming a senator, was instrumental in discontinuing football at the University of Chicago, which had previously been a collegiate football power.

Senator Harry F. Byrd of Virginia, who opposed any form of honorable discharge for the accused cadets, said, "These acts have struck a blow at the morals of the youth of the country which will last for a long time." Michigan's Congressman Charles E. Potter pictured the ninety as "victims of athletic commercialism." Soon after the story broke, one football player who would leave the Academy said he received offers from four or five colleges to play for them.

General J. Lawton Collins, Army Chief of Staff and an April 1917 West Point graduate, in a briefing to members of Congress shortly before the press release, revealed cheating originated in the football team and had probably been ongoing in prior graduating classes. If so, those in prior classes were now commissioned officers, some possibly serving on Korea's battlefields. Some may have been killed in action. His comments, taken from testimony of cadets questioned in the investigation, were reinforced by similar statements made to the press by cadets who would be resigning in a few days. West Point's graduates, including some of its gridiron greats, had been paying a heavy price in Korea since the war began on June 25, 1950.

Such news was heart rending for parents of Academy graduates who were serving in Korea, and staggering for some whose sons had perished in the conflict. Later, General Irving, responding to letters from distraught parents, had the unenviable task of trying to relieve anguish that would, sadly, last a lifetime.

The resignations which began eleven days after the public announcement were the results of six weeks of a preliminary, undercover investigation, an eleven day formal investigation by a board of three senior officers – the Collins Board, a review of the formal investigation board and the superintendent's recommendations by a second board, and two subsequent "screening boards," each composed of three and four senior officers respectively. All Collins and screening board members were serving on the Academy faculty or staff at the time the boards were conducted.

A noted jurist, Judge Learned Hand, who had retired from the United States Second Circuit Court of Appeals, led the board which reviewed the proceedings of the formal investigation, and General Irving's recommendations to the Secretary of the Army. With Judge Hand on the review board were two retired Army general officers. One was a retired Academy graduate, Major General Robert M. Danford, who, as a lieutenant colonel, had served as a commandant of cadets after World War I ended, under one of West Point's most distinguished alumni and youngest ever superintendent, then Brigadier General Douglas MacArthur. Later, upon retirement, General Danford had served as president of the Academy's alumni organization, the Association of Graduates. Retired Lieutenant General Troy H. Middleton, Jr. was the other member of the Hand Board. He was president of Louisiana State University when asked to serve on the Hand Board, and had distinguished himself both as an educator and as an Army corps commander in Allied campaigns in Europe against Nazi Germany during World War II.

Assisting Judge Hand, at his request, were Major General E. M. Brannon, the Army's Judge Advocate General – the Army's senior military law officer, and Deputy Inspector General of the Army, Brigadier General F. B. Prickett. They were present throughout the review board's deliberations.

The Collins Board, the formal board of inquiry, delivered an interim report to the Academy superintendent on June 4, one day before the class of 1951 graduated from West Point. The final report was delivered to General Irving on June 11. One additional session was held June 24, and a supplementary report added to the Collins Board record. The Hand Board convened at West Point on July 23 and completed its work on July 25, nine days before the public announcement of August 3.

The two "screening boards" convened on August 4 and 9 respectively, and concluded their work approximately August 15. They were to hear matters of explanation, extenuation, mitigation, and "further inquire into any additional breaches of the cadet honor code brought to their attention." On August 13, while the screening boards were in session, General Irving convened the Bartlett Board. Named for its chairman, the board of three colonels, all Academy graduates, faculty and staff members, began the first of more than eighty interviews, totaling 381 officers and cadets. Thus started the search for causes of the 1951 honor incident, as the Academy officially called it, and actions needed to prevent a reoccurrence. After the public announcement and throughout the period the Bartlett Board was in session, shock waves generated by the scandal continued to spread.

Though the Academy steadfastly refused to disclose the names of the cadets involved, some who disclosed themselves to the press began telling their stories, taking public issue with the honor code, and procedures used in the investigation and dismissals. They echoed their own testimony to the boards of officers, testimony summarized by the Army Chief of Staff to members of Congress, that the cheating had been in progress for years.

Within days, parents of some of "the ninety" began a steady drumbeat of criticism of the dismissals. They strongly defended their sons and expressed anger over the treatment the discharged cadets received. Others sought their son's reinstatement or various forms of redress. Some condemned the honor code. The group eventually banded together, calling themselves "Parents Committee of 90 Dismissed Cadets," and petitioned Congress and the president to investigate the dismissals.

Members of West Point's Association of Graduates, though not officially connected to the Academy, waged emotion-charged debates among themselves concerning football's importance to the Academy education, Coach Blaik's role in the incident, the Academy's handling of the affair, and whether the Association should conduct its own investigation into the scandal.

And so continued the public turmoil, in the midst of a war President Truman first labeled "a police action," a war now often called "the forgotten war."

The cheating scandal of 1951 was the defining event, and a disaster, in the bleakest year in West Point's illustrious history. It was also the time for a new beginning. The incident, a scandal the likes of which West Point had never seen, was a catalyst for wide ranging examinations of virtually every aspect of the Academy education and cadet life. Shortly after the Bartlett Board convened, and rebuilding of the Academy and its football team began, sensationalized press coverage tapered off, and the scandal faded from public view. It had become old news.

A DARK BEGINNING FOR A NEW YEAR

Though organized cheating had been in progress, probably for several years, its existence had been successfully concealed. Shortly after the beginning of the second semester of the 1950-51 academic year, in January 1951, that would all change. But not before the Corps of Cadets was once more sharply reminded of the Korean War.

On January 16, Major General Bryant E. Moore, West Point class of August 1917, and the Academy superintendent, received a brief telephone call from the Army Adjutant General. Earl Blaik was sitting in General Moore's office. The two men were discussing the Academy and Athletic Association budgets when the call came. Blaik remembered the superintendent's laconic reply, "....Yes, I can be ready.....You mean five today?....No, 5:00 a.m. tomorrow at La Guardia...." After he hung up he continued the budget discussion with Blaik.

An hour later General Moore was at home, packing. He called Earl Blaik and asked him to come to their quarters and "sit around" while he prepared to leave early the morning of January 17 to take command of the IX Corps in Korea. Lieutenant General Matthew B. Ridgway, class of April 1917, had become commander of Eighth Army in Korea when Lieutenant General Walton H. Walker (1912) was killed in a jeep accident in December. Ridgway had asked the Army Deputy Chief of Staff for Personnel to release General Moore from his Academy assignment.

The evening after the Academy superintendent received the call, he gave his scheduled Infantry Branch lecture to the class of 1951. The next morning, in the darkness outside the superintendent's quarters, forty to fifty members of the staff and faculty gathered quietly to wish him well. Bill Ochs ('45), who was temporarily performing duties of an aide on the superintendent's staff, was there and remembers the send-off as poignant and subdued. There was deep respect and admiration for this man who was once more answering his country's call to war, leaving behind his lovely, devoted wife Peggy, and the beauty, charm, and marvelous, placid setting of academia. The evening after, General Moore was in San Francisco boarding a plane for Japan.

General Moore had served in both the Pacific and European theaters during World War II. He commanded the 164th Infantry Regiment on the island of Guadalcanal in the early, tough island hopping campaigns in the Pacific. In Europe he had won General Ridgway's respect as an outstandingly aggressive and smart commander of the 8th Division, which for a time was in Ridgway's XVIII Airborne Corps.

20

In Korea, Ridgway expected his corps commanders to follow his example. He wanted them out of their Command Posts and up to the regimental or battalion CP's - wherever the action was hottest.

On February 24, near the town of Yoju, on the Han River, General Moore died. He "had been all over the place" that day, in keeping with his strong, effective leadership. Two fords across the Han, both important to his Corps, had been lost due to high water, and he was checking progress toward recovering their use. Author Clay Blair, in his book, *The Forgotten War*, recounted what happened.

Major General Bryant E. Moore
Superintendent
January 1949 - January 1951
Died in Korea February 1951
while commanding the IX Corps
(Photo courtesy USMA Archives.)

At 10:30 a.m....while coming in to land at Yoju, Bryant's helicopter hit a cable spanning the...River. The helicopter spun out of control and crashed into the icy water. GIs fished Moore and the pilot from the river. Shocked and cold, Moore stumbled into Hank Meyer's 24th Division artillery CP. There he dried out, got a fresh uniform, and made some phone calls. At 11:00 a.m., while sitting in a chair, Moore died of a heart attack.

War had come to the Academy and the Corps of Cadets in a far more personal way. The man who left little more than a month ago, and had served as their superintendent, was gone. On April 3, at the West Point cemetery, in a solemn military funeral, General Moore was laid to rest alongside hundreds of the Academy's sons. A classmate wrote in the October *Assembly* of that year:

...the Army did all that it could to honor a man who had so well exemplified the best traditions of West Point. The Chief of Staff and others such as General Mark W. Clark attended the services and followed the remains, on foot, to their last resting place amid the beauty of West Point which Bryant loved above all else.

The four cadet companies in the funeral cortege that day were A, B, C, and D Companies, 1st Battalion, 2d Regiment, United States Corps of Cadets. Six cadets from among three of those companies were to resign from the Academy in the late summer, found guilty of violating the honor code.

Also on April 3, on the athletic fields, across Thayer Road from The Plain, the cadet parade ground, spring football practice was in progress. Earl Blaik looked forward to the 1951 season, believing he might once more have a football team equal to his great Davis-Blanchard teams of 1944-46.

One day before General Moore's funeral, Colonel Paul Harkins, the Commandant of Cadets, agreed to an unexpected and most unusual appointment in his office, requested by a second classman, a junior, Richard J. Miller. Dick Miller was an honor representative from Company K, 1st Regiment.

RIGHT MAN FOR THE TIMES

On February 1, two weeks after General Moore's short-notice departure for Korea, Major General Frederick A. Irving began his third assignment at the Academy. He was the forty-first officer to serve as superintendent at West Point.

Born the youngest of four children in Taunton, Massachusetts, September 3, 1894, he entered West Point in 1913 and graduated a second lieutenant of infantry with the class of April, 1917. His first assignment was the 35th Infantry Regiment in Nogales, Arizona. Later that year he commanded a company of mounted troops and spent several months chasing the Mexican patriot-bandit, Pancho Villa, back and forth across the Rio Grande River, which marked the border between Texas and Mexico.

Frederick Irving, too, was a combat veteran in two wars. After his assignments on the Mexican border, he transferred for duty with the 15th Machine Gun Battalion at Camp Forrest, Georgia. In April, 1918, he boarded a troop ship with his battalion and sailed for France where he fought in the bloody trenches of World War I. In France he was twice hit by enemy fire – wounded in action – received the Purple

Heart, and the Silver Star for gallantry during the St. Mihiel offensive in September 1918.

In 1944, during World War II, he commanded the 24th Infantry Division in the bitter New Guinea campaign, where jungle covered mountains, torrential rains, and disease imposed additionally harsh demands on the men carrying the fight. His reputation as a quiet, physically tough leader who set an outstanding example, and trained his division "until they were ready to drop," paid big dividends in one of the less known and more difficult campaigns in the Pacific. His division, after its landing at Tanahmera Bay on Dutch New Guinea, was given credit for more than four thousand Japanese killed in the twelve mile drive to capture the airfield at Hollandia, while suffering 143 American casualties, including forty-one killed. General MacArthur was quoted as saying Irving's plan for the Hollandia landing "was brilliant."

General Irving commanded the 24th Division in the early stages of the landings on Leyte Island in the Philippines, and later the 38th Division toward the end of the campaign on Luzon.

Frederick Irving was a quiet man, not disposed to brash or extroverted behavior. He was modest, courteous, selfless, sincere and thoughtful. His personality reached easily through the wall of formality felt by nearly all cadets at West Point, through the images and auras the cadets attributed to nearly all the Academy's senior officers, and especially the lofty status and professional distance normally attributed to superintendents. But there was far more to Frederick Irving than his official side, his pleasant demeanor, strong professional traits, and resulting accomplishments in his nearly thirty-four years of service.

Major General Frederick A. Irving
Superintendent
February 1951 - August 1954
(Photo courtesy USMA Archives.)

23

General Irving was slightly over five feet, eleven inches tall with a slender, well conditioned, athletic build. He was an intercollegiate boxer in his cadet days, in an era when boxing was more brutal and physically demanding. Throughout his career in the Army he kept a rigorous discipline in physical fitness, and particularly enjoyed playing tennis. He played tennis beginning in the late twenties and was Post Champion at Fort Benning, Georgia, in 1935 and 1936, and at West Point, 1938-40.

His eyes were hazel and deep set. When engaged in conversation, immediate, steady, penetrating eye contact was his habit. His voice was distinct, clear, his speech precise, but not flowery. There was a quietness about him that deceptively portrayed an excess of seriousness. In the 1917 *Howitzer*, the cadet yearbook, were the brief, perceptive words, "when you analyze his face, the only reactions you get are 'grit' and 'fight' of the bulldog type."

Frederick Irving's quiet demeanor masked a reflective nature. He was a thinker, a planner, and possessed a calming, even temperament. Complementing these traits were a strong will, and an impeccable, tough, even handed sense of fairness. His contemporaries admired his great tact, and saw in him an unyielding moral courage, a willingness to make tough decisions irrespective of personal and professional consequences. Yet he was warm and charming in small group settings. And invariably, in the many formal speeches he was called upon to give, he exhibited a delightful, light sense of humor, which clearly showed he wasn't a man who either took himself, or the good natured pranks of the men in the Corps, too seriously.

Perhaps most important, in his home, the superintendent's quarters at West Point, with his family, the warmth and love he showed his wife, Vivian, and his children, reflected an uncompromising consistency in behavior, personality, and character.

Although not chosen to be superintendent because of the events soon to follow, Frederick Irving was the right man for West Point in the tumultuous year of 1951.

"SIR, SOMETHING'S WRONG IN THE HONOR SYSTEM..."

In the three and a half weeks before learning of General Moore's death in Korea, General Irving acquainted himself with the people who would serve in his command at the Academy. Getting acquainted at West Point came easily for him. In the final two years of a four year assignment at West Point he was Commandant of Cadets at the Academy in 1941-42, when the Corps' World War II expansion began. He was Commandant when Earl Blaik, an Academy graduate who had resigned his commission in 1922, returned to West Point to become Army's first civilian head football coach. In another assignment at West Point in the 1920s, Irving was on the Commandant's staff and taught military law.

Among the men General Irving met in early February of 1951, were Colonel Paul D. Harkins ('29), the Commandant, and Lieutenant Colonel Arthur S. Collins, Jr. ('38), the First Regimental Tactical Officer, who served under the supervision of Colonel Harkins. Harkins was nearing the end of his assignment at West Point and expected to leave the Academy in the summer, after the class of 1951 graduated. Collins, too, had served in New Guinea during World War II, but not in General Irving's division. General Irving also met William J. "Pat" Ryan, the senior ranking cadet in the class of 1951, and the Cadet Brigade Commander – the "First Captain" in cadet language.

Within days after General Irving arrived at West Point, in a private meeting with the Commandant, Harkins told him that as early as January Art Collins and Pat Ryan had both begun developing feelings "there was something wrong in the honor system." Collins talked frequently with cadets – many cadets in all classes. He had told Harkins and other senior officers, including Colonel John Waters, the Assistant Commandant, of his gut feeling, a feeling "he couldn't put his finger on." He put his concerns in writing to Harkins with a memorandum suggesting Harkins talk with Pat Ryan.

Independently, Pat Ryan had developed similar concerns. He talked with numerous cadets as well, but there were other, more precise reasons for Pat's feelings that things were not right in the honor system.

As First Captain, Pat was an *ex officio* member of the Honor Committee, which was composed of one first classman from each of the twenty-four companies in the Corps of Cadets. Although not the chairman, he was an active non-voting member, as well as the most senior cadet officer on the Committee. He was thoroughly knowledgeable of its activities, including honor hearing results, what was presented in hearings, and whether or not investigations of alleged honor violations resulted in hearings. After evidence was presented at hearings, he participated in discussions prior to the panel's voting, but consistent with existing procedures, couldn't vote.

Beginning in the fall of 1950, Pat noticed some puzzling "not guilty" verdicts in honor hearings. A guilty verdict required a unanimous vote of the twelve committee members chosen for a hearing panel. Voting was by secret ballot, and one not guilty vote resulted in a not guilty decision. Remembering cases he had seen, the guilty verdicts, and the circumstances

William J. "Pat" Ryan
Cadet First Captain
Class of 1951
(Photo courtesy USMA Archives.)

of each, to Pat Ryan there appeared a growing number of instances of straight forward, "open and shut" cases rendered not guilty.

One such case occurred early in the second semester. A third classman – a sophomore – in the class of 1953, slipped into the office of The Officer in Charge, the OC to cadets, a duty officer assigned each twenty-four hour period to oversee the cadets' day to day administration of the Corps. The OC was out of the room. The cadet removed a delinquency report – a demerit slip – from the box in which the reports were collected, took the report back to his room, and tore it up. He did this because, had the report been processed, he would have an excess of demerits and be unable to go on spring leave.

The Honor Committee's hearing panel divided on the issue, nine to three, choosing not to call the incident stealing. To Pat Ryan, the committee verdict clearly defied logic, and increased his concern for the honor system. A cadet had willfully and deliberately stolen and destroyed an official document, and by so doing, changed his records of military conduct and overall cadet performance. Yet, he was not found guilty. Pat, whose responsibility required him to work closely with the Commandant, told Harkins about the hearing panel's decision, adding to Harkins' belief all was not well.

Harkins had seen similar actions by the Cadet Honor Committee when he first arrived at West Point to be Assistant Commandant in 1946, but he had worked with succeeding classes to strengthen the Corps' and its Committees' support of the honor system.

There were other cases. Oddly, most seemed to involve football players. Pat remembered his experiences on the plebe football team in 1946, his first plebe year, when he came to West Point from Amherst College, and believed himself well prepared for the rigors of Army football and the Academy's tough academic standards. He had to work hard in academics in spite of his prior college experience, and had repeatedly seen Army football players receive grades on examinations that seemed strangely inconsistent with both their academic backgrounds prior to West Point, and their daily classroom performance at the Academy.

Later, Pat received a concussion in plebe football practice, an injury that eventually ended his gridiron career. Later another concussion in boxing resulted in a brief period of amnesia – a threat to his physical qualifications for a flying career in the Air Force – and led to a repeat of his plebe year. Then, in the three years as an underclassman, as he concentrated on ensuring he would graduate and become a pilot in the Air Force, the curious academic performance by some of Army's football players faded from his thoughts.

Now, Pat Ryan was once more seeing questionable academic performance by football players. He wondered, Could there be plants or sympathizers on the Honor Committee who were deliberately voting against guilty verdicts? He discussed his feelings with the Honor Committee Chairman, Stan Umstead and his company F-2 honor representative, Wayne Dozier. Pat asked if they had similar feelings. They had, yet they too were stymied. There seemed no evidence, no proof they were right, and they hadn't voiced their feelings to anyone else.

There were no members of the football team on the 1951 Honor Committee. Pat remained troubled, and his belief increased that something was amiss. In late January he again broached the subject with Colonel Harkins, and restated his concerns – about the same time Lieutenant Colonel Art Collins was preparing his memorandum to the Commandant suggesting that they discuss their observations with Ryan. The First Captain recommended to Harkins that there be a "cadets only" meeting with the entire Corps to address the subject of honor.

The Commandant agreed, and the evening of February 19, after informing the new superintendent as to the background and purpose, and obtaining his approval, convened the Corps. Cadets in the lower three classes met with Pat Ryan in the theater, and Pat asked the cadet 2d Regiment Commander, Gordon Danforth, to meet with the first class in the Electricity Lecture Hall in the east academic building.

Both talked about the importance of discipline, the honor code, and the cadets' responsibilities for maintaining the code. They spoke of serious concerns for the honor code, the honor system's integrity, and encouraged cadets to take action if there were problems.

Pat Ryan's talk with the lower three classes was pointed – a no nonsense, straight from the shoulder call to defend the code, and the system which gave life and practical application to honor's meaning. He told them, "I'm convinced we've got members of the Honor Committee that are intentionally acquitting certain cadets because they are in a certain group. I'm fed up with it, and I'm going to nail their fannies as soon as I can do it." He carefully avoided naming or labeling the group. He didn't have evidence to confirm his aroused suspicions.

The meeting was unusual, clearly out of the ordinary. Not only were Pat's words plain spoken and blunt, but the session had not been designated a period of honor training – it wasn't in the semester's master academic training schedule.

There is little doubt the February 19 session with the cadet First Captain and the 2d Regiment Commander caused a stir among members of all four classes. Pat Ryan began receiving comments indicating the meeting had struck a chord which was resonating in the Corps. However, the meeting apparently had little or no effect upon the men already involved in activities they knew were violating one of the honor code's three basic tenets, "A cadet does not cheat."

If the men who were cheating considered the meeting a warning of an aggressive attempt to expose their activities, their behavior changed imperceptibly. There was no slackening. As the semester rolled toward written general reviews – final exams – the number of men who became enmeshed in organized cheating continued to increase.

However, in Company K-1, within days of the two meetings with the Corps, the first evidence of a break appeared. The glimpse into an unhappy future was not immediately apparent, came slowly, insidiously, because two men were unknowingly being confronted with activities that had been kept under wraps and carefully controlled for years. It was a different kind of trouble, not easy to accept or believe, and it was offered by a well-meaning friend who kept his activities hidden from everyone, including his own roommates who were involved in similar activities not in keeping with the honor code.

Two yearling roommates – sophomores – class of 1953, both in academic trouble and both intercollegiate athletes, began noticing uncharacteristically strong academic performance by their mutual friend, classmate, and company mate who was on the Academy fencing team, and was himself in academic difficulty.

The two yearlings at first didn't know what to make of their classmate's sudden, noticeable improvements in the classroom. They were all in higher numbered academic sections together, which in the Academy's education system meant they had low grade point averages. The two men first responded with good natured jabs and questions, to which their classmate replied in an offhanded, noncommittal way. The slow awakening dragged on for weeks. The two yearling roommates had great difficulty recognizing and accepting what was increasingly evident. They didn't want to believe, but by late March there could be no doubt.

Finally, on April 1, the first clear indicator of impending disaster appeared. The initial revelation was a profound shock, brought to Colonel Harkins the next morning by Dick Miller, the second class honor representative in Company K-1. Harkins didn't want to believe what he heard.

"THERE IS A RING OF CADETS ... EXCHANGING UNAUTHORIZED INFORMATION IN ACADEMICS..."

Dick Miller returned to his room after supper in Washington Hall that first evening in April, and was preparing for call to quarters, the cadet study period, which was to begin at seven fifteen in the evening. *Brian C. Nolan*, a third classman in his company with whom Dick had developed an easy rapport, came into his room and asked for a private

29

meeting later that evening "in the sinks" – the basement area in cadet barracks. *Brian* let him know the meeting was to discuss honor. There was a serious urgency in his voice and request.

Nolan, in coming to see Dick Miller, had made a carefully thought out decision to approach Miller with the information he knew he should give to an honor representative. Miller was easy to talk to, quiet though not reticent, a respected second classman who accorded dignity and understanding to underclassmen. He was a cadet who could quickly gain the confidence and trust of other men. While training and correcting underclassmen, he was firm, yet exceedingly fair, lacking in harshness and any tendency toward sarcastic, unkind criticism. He was an excellent example of what many underclassmen believed a leader should be.

There was more to Dick Miller. He deeply admired his parents. His mother taught school. His dad was a citizen soldier who had returned from World War II and taught Dick the broader, deeper meanings of honor. His father had fought on the Bataan Peninsula in the Philippine Islands in the early days of World War II. Captured by the Japanese Imperial Army, his dad had survived the infamous Bataan Death March and three and a half brutal years in Japanese prisoner of war camps – and a perilous journey on an unmarked Japanese ship, to internment in Japan until the war's end.

Nolan and his two roommates, who knew of *Nolan's* decision to go to Dick Miller, as well as the reasons for the decision, all respected the second classman, soon to be first class honor representative – the K Company Honor Committee representative when the class of '51 graduated. They were confident Dick Miller would do what was right, and act responsibly – honorably – with the information given him. Miller was not prepared for what he learned that night, any more than Colonel Paul Harkins was the next morning when Miller repeated the story to the Commandant.

Dick Miller listened thoughtfully, and with increasing dismay, as *Nolan*, a member of the cadet intercollegiate swim team, told him of a classmate, and member of K Company, who, over a period of time had apparently been providing unauthorized academic information obtained from other cadets who were operating as an organized ring. *Nolan's* recognition had come gradually in a series of contacts which at first raised no suspicions. When his friend continued to provide him information that appeared the next day on daily quizzes, he became increasingly suspicious, then alarmed. One of *Nolan's* roommates had had similar experiences with the third classman, in which light-hearted teasing initially played a role. The cadet's actions appeared humorous at first, then curious and out of character with his academic performance. Over a period of time he became more open, pointed, and specific in the information he was providing to *Nolan* and his roommate.

The classmate, *George L. Hendricks*, was the son of an Academy graduate, a general officer stationed with American occupation forces in Germany. More importantly, *Hendricks* was a friend who had dated *Nolan's* younger sister, Betty, and visited his home while they were both on leave. *Hendricks* knew *Nolan* was having serious difficulty in mathematics, and at first suggested he could help *Nolan*. The offer was curious because *Hendricks* was also having serious problems in academics, struggling to remain proficient.

A puzzled *Nolan* didn't take the offer seriously, but that wasn't the end of it. *Hendricks* was sincere in his wish to help, and apparently assumed *Nolan* to be desperate enough to accept assistance not in keeping with the honor code. Several times *Hendricks* broached the subject, initially with no specifics behind his offer. Finally, over a period of time, the startling evidence of organized cheating emerged. *Nolan* could help himself by accepting the assistance offered him, which was obviously coming from other cadets who exchanged information useful in quizzes and daily recitations.

Nolan began asking questions. One day, when he was returning from intercollegiate swim practice, he met *Hendricks* in a hallway. *Hendricks* assured *Nolan* of security in accepting the help, telling him there were a large number of cadets involved, football players, even first classmen – seniors – who were Honor Committee members.

The decidedly more open explanation of *Hendricks'* academic assistance clearly indicated *Hendricks* was receiving some form of unauthorized assistance, contrary to the honor code, and the academic help *Nolan* was receiving, if he continued to accept it, knowing it was unauthorized, would be in violation of the honor code as well.

Dick Miller returned to his room that evening to encounter the usual ribbing by his roommates about "honor reps who hold secret meetings in the sinks, wearing bathrobes and hoods." It was light-hearted banter, as most cadets well understood the responsibilities Honor Committee members bore. His roommates didn't attempt to pry information from him about his meeting in the sinks.

And true to his responsibilities as an honor representative, Dick Miller said nothing to his roommates about *Nolan's* conversation with him. But he continued thinking of what he had been told, its implications, and how he should handle the stunning disclosures. He didn't sleep well that night.

By the next morning he decided he must go directly to the Commandant of Cadets and repeat *Nolan's* disclosure. But first, Dick Miller needed to tell his company tactical officer, Lieutenant Colonel Kenneth C. Robertson, of his intent to see the Commandant, and why. Robertson agreed he should take the information directly to Harkins. The implications of what he had been told were far too great, too serious to do otherwise.

A second classman's request to speak directly to the Commandant was unusual, but not nearly so startling as the news Dick Miller presented. The Assistant Commandant, Colonel John K. Waters, class of 1931, who would be replacing Harkins in two months, was called into the Commandant's office to hear Miller's account. As the story unfolded, Colonel Harkins realized he faced a serious and potentially explosive problem. The Academy, proud of the honor code and the character it sought to instill in its graduates, was apparently harboring an organized cheating ring that could undermine the very foundations of the institution.

The Commandant asked Miller if there was any evidence to support what *Nolan* told him. He replied, "Not at this time, Sir, but I believe we can get evidence." After Dick Miller answered questions from the two senior officers, Colonel Harkins thanked and dismissed him. He asked Miller not to disclose to anyone what he had heard from *Nolan*, or mention his conversation with the Commandant and the Assistant Commandant. After Dick left Harkins' office, the two officers agreed a cheating ring probably existed at West Point. But the absence of firm evidence to support *Nolan's* revelations raised many questions, and knotty problems.

How many cadets were involved? *Nolan* said he had been told the ring was extensive and was in both regiments of the Corps of Cadets.

If the number was large, if the ring reached into both regiments, what needed to be done to ensure it was broken, and its members identified? How could evidence be obtained? If there was an investigation, who would participate? Could the Honor Committee conduct the investigation, and hearings? If so, how soon?

The calendar mattered. Graduation for the class of 1951 was barely two months away. Not only would the entire class depart the Academy, but their status would change. They would no longer be cadets, but commissioned officers. Complicated legal considerations would follow should evidence implicate any of them. Within days after graduation, nearly all remaining members of the Corps would also be gone from the Academy, either in summer military training or on leave.

The class of '52 honor representatives weren't yet prepared to assume the responsibilities of the Cadet Honor Committee. More to the point, if men in the class of 1951 were involved in cheating, as *Nolan's* information suggested, members of the junior class, '52, couldn't investigate or sit in judgment of men in the class of '51. And perhaps more troubling was information suggesting members of the Cadet Honor Committee – class of '51 – might be involved.

Honor representatives from the class of 1951 were the responsible cadet leaders in maintaining the honor system, and needed to be involved in any investigation and hearings. Considering the information given him, should Harkins gamble on having the Cadet Honor Committee investigate?

Adhering to the concepts of class seniority, and the cadet chain of command, Harkins took steps to ensure Daniel J. Myers, Jr., the class of '51 Company K-1 honor representative, knew of *Brian Nolan's* disclosures. Under existing Honor Committee procedures, Dan Myers was responsible for preliminary investigation of any allegations of honor code violations in his cadet company. Myers had spent two years at Virginia Military Institute, and was firmly committed to the honor code and system. Harkins then sent Myers to Pat Ryan to tell him of *Nolan's* disclosures. Pat Ryan came to the Commandant's office, wondering what might be done. Harkins called John Waters into his office and the three discussed alternatives.

33

COURAGE AND ACTION

Paul Donal Harkins' response to what he had been told by Dick Miller in the spring of 1951 was consistent with his background and experience. Born on May 15, 1904, in Boston, Massachusetts, he was the second of five children of Edward and May Kelly Harkins. His father had been a newspaper editor and drama critic in Boston for nearly fifty years. The elder Harkins' professional interest resulted in young Paul's well-rounded foundation in the many cultural events presented in Boston. His parents' philosophy and guidance instilled in him the qualities of courtesy, self-discipline, and hard work, qualities that remained with him the rest of his life.

"Only soldiers and millionaires could ride horses," he mused in 1922. His remark told of his love for riding, and after he had worked hard to obtain an appointment to West Point, he entered with the class of 1929 from the Massachusetts National Guard on his twenty-first birthday. Among other accomplishments as a cadet, he became a member of the Corps of Cadets' color guard which carried the national colors during parades and other ceremonies, and a star polo player and captain of the team. After graduation he was commissioned a second lieutenant in the cavalry. His first assignment was in the 7th Cavalry Regiment at Fort Bliss in El Paso, Texas.

At Fort Bliss he met the love of his life, Elizabeth Mae Conner. They married in 1933, before his next assignment, Fort Riley, Kansas. At Fort Riley for six years, he was first a student, and then an instructor in the Army's equitation school. In 1939, he received an assignment to command F Troop, 3d Cavalry Regiment in Fort Myer, Virginia. These were the dying days of the Army's horse cavalry and its aura of heroic romanticism. Here at Fort Myer began a close professional relationship with the regiment's commander, then Colonel George S. Patton, Jr., which was to continue until General Patton's death following World War II.

In the months leading to America's entry into World War II, Paul Harkins participated in the Louisiana and Pine Camp maneuvers, served briefly with the 1st Cavalry Brigade at Fort Bliss, and then in January 1942, joined Patton's 2d Armored Division at Fort Benning, Georgia. In August, he became Patton's Deputy Chief of the Western Task Force, preparing for the invasion of North Africa.

As he left Washington for Norfolk, Virginia on October 22, 1942, he said to Betty, "Honey, if I'm not home for dinner tonight, I'll write you a letter. You take off for Texas." The invasion force sailed the next day, and he didn't come home for dinner for three long years.

Paul Harkins landed in North Africa with the assault landing on Fedhala Beach on November 8, 1942. He was one of eight staff officers that General George S. Patton, Jr., the tough, outspoken, flamboyant, hard driving tactical genius, took with him to Sicily, England, and finally, to the continent of Europe as the war progressed.

As Patton's deputy chief of staff for operations, he was the senior officer responsible for planning and coordinating training and combat operations for the units General Patton commanded. He also came ashore on D-day in the Allied invasion of Sicily, and went into France six days after the June 6, 1944 invasion, as Patton's Third Army prepared for the Allied breakout and dash across France from the Normandy beachhead.

Harkins was an enormously successful staff officer, noted for ensuring his boss' combat commands operated as Patton planned and directed. Harkins' abilities as an extension of his commander were never more evident than in December, 1944, when Patton's Third Army, attacking

Colonel Paul D. Harkins
Commandant of Cadets
(Photo courtesy USMA Archives.)

east through France, toward Nazi Germany, was ordered to disengage from the enemy and turn to march one hundred miles north and break the back of the German counter offensive historians call "The Battle of the Bulge." The Third Army's performance that December was a feat of arms never to be forgotten in the annals of American military history.

Following General Patton's death after the war, Paul Harkins escorted Mrs. Patton back to the United States, arriving on Christmas Day 1945. Two months later he began a five-year assignment at West Point, first as Assistant Commandant, then Commandant of Cadets.

Now, in the spring of 1951, the words beneath Paul Harkins picture in the 1929 *Howitzer*, his cadet yearbook, seemed prophetic.

...True courage of action and thought, and never diminishing appreciation of honor make his character outstanding. Outspoken and frank in his opinions, he has most acute discrimination between right and wrong...

"I ... AM DETERMINED TO BREAK IT UP ... EVEN IF I ... USE RATHER DEVIOUS MEANS TO OBTAIN EVIDENCE."

Brian Nolan's disclosures to Dick Miller shocked Harkins. However, in spite of their unpleasant portent, *Nolan's* words appeared to confirm feelings dogging the Cadet First Captain, Pat Ryan, and Lieutenant Colonel Art Collins, the First Regimental Tactical Officer. Harkins decided he must act. He had to obtain hard evidence, and quickly.

These weren't comforting thoughts for a man nearing the end of a successful tour of duty he had enjoyed immensely. In early June, after the class of '51 graduated, Harkins would be leaving West Point. He looked forward to a new and prestigious assignment on the Army staff in Washington, D.C., as the Chief, Plans Division, under Lieutenant General Maxwell D. Taylor, the Deputy Chief of Staff for Operations and Administration, and one of the Army's fast rising stars. Taylor had also served as superintendent at West Point from September 1945 until January 1949.

For Harkins, his time as Commandant had been fascinating. He knew he was working with some of the brightest young men in the country. He also knew in the four years these men were at West Point, they had to learn to shoulder the heavy responsibilities war would likely impose on them immediately after they were commissioned second lieutenants. He had to help them learn. The Army and the nation would hold them to a higher standard. Much would be demanded of them, and their lives would, of necessity, be lives of sacrifice. "You...have to be the father... mother... sister and brother... to 2500 sons," he said years later.

The Korean War, though increasingly bloody and controversial, had reinvigorated the sense of mission at the Academy, a mission tempered by the daily reminder of the war's awful price – an ever lengthening casualty list. Harkins' mission as Commandant was to give cadets military training, including physical education, to prepare them for leading young soldiers in battle, providing discipline, and leadership by example, while teaching all the facets of leadership graduates needed to be career Army officers. They had to be ready and confident enough that soldiers, mostly young draftees, would respond to the newly commissioned officers' training, trust their judgment, and follow them under difficult, sometimes chaotic and terrifying circumstances.

Now, in what should be the warm glow of personal and professional satisfaction as Harkins was ending his assignment at West Point, he was hearing of activities among the cadets that were wrong headed and totally inimical to all the Academy stood for. He didn't know the extent of the cheating ring, but the mere suggestion of organized cheating in the Corps was ugly, repugnant, and could have dire consequences. What's more, such activities might have begun on his watch. He wondered if this meant the end of his career, and several times in the coming weeks, as more and more evidence was uncovered, he openly expressed concerns for his future in the Army.

Nevertheless, Paul Harkins wasted little time with thoughts of professional self-preservation when the first revelations were brought by Dick Miller. As a staff officer under George Patton, he had learned to be cautious, deliberate, and thorough, but not timid about doing what he knew to be right, and what his boss ordered him to do. If there was a cheating ring at West Point, he would break it in the weeks remaining before leaving for Washington. If he couldn't get enough evidence to break it, there was the Assistant Commandant, John K. Waters. Harkins would ensure John Waters was fully involved in the decisions that lay ahead, that Waters could carry the work to a successful conclusion, whatever and whenever the conclusion might be. Thus from the outset, Colonel John Waters, who was the Cadet First Captain in the West Point class of 1931, and soon to become the new Commandant of Cadets, immersed himself in the details of the preliminary investigation and all that followed.

There was another serious consideration in Harkins' mind, his boss and commander, the superintendent, and the chain of command. A good staff officer, or a good commander, doesn't make decisions or take actions that get his boss in trouble. The disclosure of organized cheating at West Point was fraught with perils, although they were perils different from those of the battlefield. The hazards of professional soldiering in peacetime are far more subtle, sometimes insidious, and usually don't involve casualties. But the effects can be the same for a commander who errs in peacetime as one who makes devastating, costly, casualty producing mistakes, and suffers defeat in war. Relief from command. Firing. Transfer to a less prestigious post or less responsible duty, administrative reprimand, forced retirement, or in the extreme, prosecution under the Articles of War – all of which could be the beginning of the end, or simply the end of a career.

In the days after the first startling April disclosures, Paul Harkins, whose routine included almost daily, private meetings with General Irving in the superintendent's office, kept Irving informed of activities related to what *Nolan* had first relayed to him. Harkins' reputation as an outstanding staff officer was hard earned under George S. Patton, Jr. "Keep your commander informed," "tell him the bad news as well as the good," and "don't surprise him" were operating principles of any good staff officer. He wasn't going to destroy his reputation on this matter by failing to keep General Irving informed, and he certainly wasn't going to surprise his boss with a disaster which he contributed to with his own mistakes.

The superintendent, who had been the commandant nine years earlier, knew the importance of constantly feeling the pulse of the Corps of Cadets, knowing what was going on among these bright, energetic, young men. So did Harkins. "Know your men" was both a principle of leadership, and a responsibility of command. Both Harkins and Irving were serving as commanders as well as educators and administrators in one of the most respected educational institutions in the land. Harkins was also doubling as a member of the superintendent's staff. They needed to consult with one another as to the best means of obtaining evidence that would destroy the cheating ring. And consistent with all the rules and mores of military behavior, the commander, the boss, gave his final OK to what a staff officer or subordinate commander recommended. The commander made the decision, but he, too, kept the next man in the chain of command informed, and General Irving's boss on the Army staff was Maxwell Taylor, West Point class of June 1922.

* * *

For Daniel J. Myers, class of 1951, graduation was just over two months away when he first heard of *Brian Nolan's* disclosure of a cheating ring's existence at West Point. Dan had been elected his class' honor representative from Company K-1 in the spring of 1949, his yearling – sophomore – year. He had understudied the class of '50 K-1 representative for a year before becoming a first classman. Now, as a first classman, a senior, he was the Company K-1 Honor Committee representative.

All the Corps' twenty-four elected Honor Committee members were first classmen, and, consistent with the first class' leadership role guiding the three underclasses, they were responsible for ensuring the honor system worked as it should. That didn't mean they were responsible for honor at West Point. Honor, the honor code, and the honor system were the collective and individual responsibilities of every cadet. The officers who taught and led the Academy's future graduates were responsible for living the example of honor, fostering and upholding individual and institutional behaviors in which honor could thrive.

However, in one strikingly important way, members of the Cadet Honor Committee carried far heavier responsibilities as first classmen than any other men in the Corps. They had to sit in judgment of cadets, including their classmates, who were alleged to have violated the honor code. Honor Committee officers – such as Chairman, Vice Chairman, and Secretary were elected by the Committee, and each elected officer and each member carried specific responsibilities.

If there were allegations of an honor violation within a company, the Honor Committee member from that company had the unpleasant duty of first conducting a preliminary investigation to determine if a violation might have been committed. If evidence was sufficient to warrant a full investigation, with a possible hearing, Honor Committee members from outside the company in which alleged violations occurred would be appointed to complete the investigation, and if evidence warranted, take the matter to a hearing. A panel of twelve Honor Committee members would be convened by the Chairman, and a hearing followed, conducted with procedures similar to a court of law. The Secretary of the Honor Committee selected hearing panel members.

Honor hearing verdicts were by secret, written ballot, and guilty findings had to be unanimous. A guilty finding carried the penalty of dismissal from the Academy. The resulting discharge from the Army was a matter of record, characterized as less than honorable – neither honorable nor dishonorable – a record that could have a lifelong, adverse effect if not appealed and removed. Since 1946, there had been forty-four individual honor cases resulting in less than honorable discharges. But never in the Academy's one hundred forty-nine year history had there been an instance in which a large number of young men had apparently organized, or conspired, to violate the honor code. Cheating of any kind was inexcusable, unforgivable in the minds of all but a few at West Point. Organized cheating was unthinkable, totally unacceptable, a preposterous notion.

Dan Myers was a cadet officer, a lieutenant, a platoon leader in K-1, and took his responsibilities seriously. He was direct, strong willed, and didn't hesitate to say what was on his mind. He was not someone to shrink from responsibility, no matter how unpleasant, and he admired men who believed as he did. Such behavior was a mark of courage. He was particularly proud of his role as the company honor representative. It was excellent training for an officer candidate, because it prepared him for similar responsibilities he would face in the Army. And, like most cadets, he regarded the honor code as an ideal to be lived every day, a standard which marked the Academy a proud institution of special significance. He had also attended Virginia Military Institute two years before receiving his appointment to West Point. At VMI, the cadet honor code had prepared him well to understand the meaning of honor at the United States Military Academy.

But Dan Myers wasn't entirely prepared for what he was about to begin. By the time it was over the first week of June, he felt his involvement with the cheating scandal of 1951 had lasted far too long. Years later, in his memory, his work began early in the winter and dragged on interminably, while his academic standing suffered, and essentially alone, he sought in vain the evidence Colonel Harkins repeatedly demanded.

* * *

Paul Harkins wanted the Cadet Honor Committee to live up to its responsibilities, in spite of information given *Brian Nolan* by *George Hendricks* suggesting members of the Committee were involved in the cheating ring. Harkins wanted Dan Myers to conduct a preliminary investigation, gather evidence, work with the Committee, and thus lay the groundwork for honor hearings. But neither firm nor corroborating evidence could be found in two weeks of intense effort by Dan Myers.

Dan met frequently with Harkins during those two weeks. Additionally, he met several times with selected members of the Honor Committee whom he and Harkins believed were above reproach in supporting the code. Under the ever present insistence of the Commandant, he tried every conceivable avenue he could, plus every avenue Harkins directed him to explore, seeking hard evidence of the ring's existence. Harkins pushed Dan Myers hard, drove him, always insisting, demanding. Harkins had to have firm evidence with which to move against the ring.

From the special, small-group Honor Committee meetings came information suggesting other Committee members also believed there was trouble afoot – but none had the concrete evidence Dan Myers was seeking. While he believed there was headway being made in his preliminary investigation, some of the men who attended the meetings apparently didn't keep the subject of the meetings confidential. Word began to slowly spread that the K-1 Honor Committee Representative was "out to get men he believes are cheating in Academics," and undoubtedly some members of the cheating ring learned of Dan Myers' inquiries.

Myers also went to professors in the Academic Department, with the idea of gathering evidence through classroom observations, using both instructors' monitoring and observation of cadets, as well as hidden cameras which might be provided by the Signal Corps and the Army's investigative detachments assigned to West Point. The ideas earned little interest and no cooperation.

The discussions with Harkins also included the idea of Dan's approaching *Brian Nolan*, urging him to "join the ring" to obtain evidence. Myers recalled asking *Brian*, and remembers his urgings were to no avail. Harkins, who was bedeviled with a major problem – not wanting to expose his or Myers' efforts to find evidence – wasn't throwing his full weight behind an investigation in which there was no hard evidence suggesting precisely where to look, or who, specifically, to pursue. Thus, after two weeks of intense efforts by Dan Myers, the only evidence available remained *Nolan's* disclosures of what he had heard from *Hendricks*, and another report by one of *Nolan's* two roommates who was likewise tangled in academic difficulties.

During the same period in March, when *Hendricks* had offered to assist *Brian Nolan*, he had also created suspicion in *Brian's* roommate, first with his academic success in class, and later with an offer of assistance. After *Brian* had told Dick Miller of the cheating ring, the roommate told Dick Miller of the offer he received.

Circumstances were similar to those encountered by *Brian*. Both *Hendricks'* and *Nolan's* roommates were near the bottom of their class in academics, and *Nolan's* roommate noticed *George Hendricks* had obtained what seemed an inordinately high test grade. The roommate accosted *Hendricks* in a hallway after class and asked him good-naturedly, "How did you do that? How did you get such a high test grade?" *George Hendricks* responded by simply shrugging his shoulders and smiling.

Later, in another of several similar encounters, *Hendricks* handed him a set of problems on a piece of paper. "Go work these, and you'll do better," he said. The next day, when *Brian Nolan's* roommate went to class, he discovered on the daily quiz the same problems that *Hendricks* had given him. When *Brian's* roommate returned to his room in cadet barracks this time, he was becoming suspicious. He said to *Nolan* and his other roommate, "You don't suppose he's cheating, do you?"

But the testimony of *Nolan* and his roommate would still be insufficient to bring allegations against anyone but the cadet who had offered *Brian Nolan* unauthorized information in academics. Harkins continued exploring alternative means of gathering evidence to learn where the information was coming from, who were the other cadets supplying it, discussing the alternatives with John Waters, Pat Ryan, the cadet First Captain, and Dan Myers. The Commandant several times considered enlisting the Signal Corps detachment at West Point, with the support of the Army's investigative services, to install listening and recording devices in cadet rooms. There was also the possibility of gathering evidence through the Academic Department by systemati-

42

cally observing and examining the performance of cadets in daily recitations and tests, and using hidden cameras.

But discussions about how to gather evidence always ended with the same conclusion. There had to be specific information to target a cadet, or any group of cadets for evidence. There had to be "probable cause" to devote the time and energies of investigative agencies, and the only information available was coming from one source. Thus far *Hendricks* was the only cadet for whom there was "probable cause" for an official investigation. And if *Hendricks* wouldn't provide information, or didn't have concrete evidence against other members of the ring, the investigation would not only be stymied, but in all probability the ring would eventually be alerted there was an investigation in progress.

Unless other men came forward with information, there remained but one method of gathering the needed evidence. The cheating ring had to be penetrated and exposed. As an investigative technique, covert investigations were consistently the most effective means of breaking a conspiracy of any kind.

An undercover investigation was the answer. And it had to come from within the Corps. West Point was not like a college or university. There could be no contrived student transfers directly into the three upperclasses. Every cadet entered the Academy as a plebe. It would take years to develop evidence unless someone, or some group of cadets would agree to become part of the cheating ring.

Paul Harkins made his decision. He would "fight fire with fire." He would "use devious means if necessary...to obtain evidence." After consulting with General Irving, he asked that *Brian Nolan* come to his office. There, the Commandant personally confronted *Brian* with the proposal he accept *Hendricks'* continuing offers of help and thus join the cheating ring, emphasizing the purpose was to gather conclusive evidence to break it.

ON THE SPRING PRACTICE "FIELDS OF FRIENDLY STRIFE"

In the first two weeks of April 1951, while Paul Harkins pondered *Brian Nolan's* shocking disclosures, and Dan Myers sought to find hard evidence a cheating ring existed, Army's spring football practice was in high gear on the athletic fields across Thayer Road from the Plain.

Earl Henry "Red" Blaik, mastermind of The Rabble's preceding seven years of glory on the nation's collegiate gridirons, could survey the talent before him with silent enthusiasm.

He seldom outwardly displayed optimism, but he could now say to himself, Maybe, just maybe, here are a group of men who can match the great Davis-Blanchard teams, if they play at their best. The talented players and the team they had become were good reasons for such optimism in spite of the intense competitiveness and unpredictability of major college football. Army's Black Knights of the Hudson were firmly planted among the nation's powerhouses in major college football. Their success was no accident, and Earl Blaik knew that better than anyone.

Winning was the product of dedication, organization, planning, endlessly hard work, attention to detail, picking the right people to do the job, leadership, knowing your opponent, and much more. He drove himself mercilessly to build winning teams, and he had a presence, an aloof commanding presence that complemented all his many leadership and management skills. Blaik was able to draw from his players the best they had to give. "You have to pay the price," he would say. They did, and most gave everything they had – and more.

To nearly all his players he was "The Colonel," not "Coach." It was "Yes, Sir" and "No, Sir." He was not a "hail fellow well met." He maintained a professional distance from both his players and his assistant coaches. Conversations with his players were mostly with team captains, field captains, and quarterbacks, except when he was showing a player what he needed to do to correct a mistake. Compliments were rare. He was on a pedestal, and that's the way he wanted it. This gave him a mystique which multiplied the force of all his other talents.

It was a prestigious reward to most Army football players to be recruited by Blaik's coaching staff. Not only did many of his players hold him in awe, he had become one of the nation's deans of the college football coaching profession, which had brought him the status of a celebrity. In his seventeen years at his alma mater, seven as an assistant during the late '20s and early '30s, during the days of "iron man football," when players commonly played both offense and defense, he had absorbed and added to the traditions of football at West Point.

He also crafted and refined a set of beliefs, a logic path, and rationale for his beliefs about the game, and a philosophy about what the game meant to the men of the United States Military Academy, and to the Academy as an institution.

He believed winning football was a powerful builder of battlefield leadership. What's more, football developed a most important ingredient of such leadership, physical courage. The Army football team was the "supreme rallying point" for the Corps of Cadets and the Academy. The Army team was the United States Army's team. He frequently reminded his players they represented not only West Point, but the United States Army. And since the United States Army was the nation's Army, they must conduct themselves as representatives of the nation. He was not alone in what he believed about West Point and the game of football.

"I want an officer for a secret and dangerous mission. I want a West Point football player," was an oft-repeated, World War II quote attributed to General of the Army George C. Marshall, who led America's Armed Forces in the victory over the Axis powers during that conflict. The quote affirmed and reinforced the value of Army football.

The week of the 1949 Army-Navy game, General of the Army Douglas MacArthur, in a telegram read to the Corps of Cadets, had said,

FROM THE FAR EAST I SEND YOU ONE SINGLE THOUGHT, ONE SOLE IDEA, WRITTEN IN RED ON EVERY BEACHHEAD FROM AUSTRALIA TO TOKYO. THERE IS NO SUBSTITUTE FOR VICTORY.

Army beat Navy that year to cap another undefeated, untied season.

Along with General Marshall's words, General MacArthur's telegram became part of the legend and tradition of Army football. MacArthur's words seemed to accord winning football a hallowed importance in leadership training, and victory in battle. But there was more to General MacArthur's telegram than his words. It had a special meaning to Earl Blaik and his football team.

Earl Blaik was a cadet at West Point in the class of 1920, when Brigadier General MacArthur was the youngest superintendent in the history of the Academy. When Cadet Blaik was a first classman, there was a serious hazing incident in which a cadet committed suicide. There was intense public and Congressional interest in the circumstances of the tragedy. The War Department conducted an investigation. As part of efforts to understand the causes and avoid similar tragedies, Blaik was asked to lead a committee of seven first classmen to recommend improvements in fourth class customs. General MacArthur was impressed with Cadet Blaik's work.

Blaik's work on the committee was during the second of his two years as a cadet in a curriculum shortened by war, but he became known to MacArthur for other reasons. At the end of his second and final year, Blaik received the coveted Athletic Association Saber for best athlete in the Corps of Cadets. He was a varsity baseball player, as MacArthur had been when he was a cadet.

But Blaik's success as an athlete was mostly from his play on Army's football and basketball teams, primarily football. He was a two-year letterman on the football team, although in 1918, before MacArthur became superintendent, the cadets played only one game, because the wartime need for junior officers brought chaotic changes and curtailed the season. One result of the accelerated, stripped down, wartime curriculum was the early graduation of the class of 1920.

General MacArthur, as superintendent, was a staunch advocate of cadet intercollegiate athletics. He frequented team practice sessions and contests. Years later, Blaik often reminisced about his cadet days and his personal, athletic field encounters with one of West Point's most renowned graduates.

MacArthur also knew the physical and emotional demands of battlefield leadership, and was convinced all cadets needed both physical conditioning and competitive athletics to prepare them for leadership responsibilities. Not only must officers be physically fit, they must be able to train their soldiers to meet the brutal demands of battle.

One of his many reforms was institution of the intramural athletic program at West Point. Cadets who did not compete in intercollegiate athletics would participate in intramurals every season of the academic year. His intramural program became a model for colleges and universities throughout the country.

It was during his work as superintendent that MacArthur spoke words later chiseled in stone above the south entrance to the cadet gymnasium. "ON THE FIELDS OF FRIENDLY STRIFE ARE SOWN

THE SEEDS THAT, ON OTHER FIELDS, ON OTHER DAYS, WILL BEAR THE FRUITS OF VICTORY." These words, too, gave prosaic eloquence to the relationship between victory on the battlefield and the athletic dimension of a West Point education.

The young superintendent's term at West Point was fortuitous for Earl Blaik. It was the beginning of General MacArthur's supportive relationship with the man who later, as head coach, led Army to greater football glory.

Now, in this spring practice of 1951, the talent Earl Blaik saw on the field held out the promise of another golden year for Army football. He could look forward to twenty-five returning lettermen in the fall unless academics overwhelmed any of his charges, causing them to be "found" – separated from the Academy because they were unable to meet West Point's high academic standards. Among the lettermen were two All-Americans from the 1950 team, which had won eight and lost one – a stunning, season ending 14 to 2 upset at the hands of arch rival Navy.

The 1950 Army team had been in the chase for the mythical national championship right up to the end of the season. The upset by Navy ended a twenty-eight game streak without an Army loss, including two ties, a streak which began near the end of the 1947 season. In spite of defeat by a Navy team that had won only two games during their 1950 gridiron campaign, the Cadets were ranked number two in the nation in the season's final Associated Press poll.

And it was also during these fateful first two weeks of April at West Point that another dimension of Earl Blaik's relationship with General MacArthur was played out, as one of the Academy's, and the nation's, great military heroes came crashing down. His admiration of the General extended beyond MacArthur, the man and national military hero, into the realm of politics, a part of Earl Blaik's life that of necessity also remained behind closed doors. In prior correspondence, Blaik had been quietly, actively encouraging MacArthur's political ambitions, when, on April 11, President Harry Truman abruptly relieved the General of his commands in the Far East.

Although there were deeper, more complex reasons for his relief, the proximate cause of MacArthur's firing was a dispute with his Commander-in-Chief, President Truman, in which he had taken public issue with the president's foreign policies governing the conduct of the Korean War.

When news of his unceremonious relief reached Earl Blaik, the Army coach was shocked. MacArthur, over the years, had been an unseen but deeply admired booster of the Black Knights, with his inspiring telegrams and letters to the team and the Corps. The day the word flashed around the world, Earl Blaik called the Army team together on the practice field to relay the unhappy news. There was anger and bitterness in his voice. It was one of the very few times he ever gave in to expressing his opinions to his team about national or Academy issues. That day he cabled MacArthur:

AMERICAN PUBLIC STUNNED. TIME IS OF THE ESSENCE TO OFFSET ADMINISTRATION HATCHETMEN. MY AFFECTION AND DEVOTION TO YOU.

MacArthur flew home to the United States and received a triumphant hero's welcome in San Francisco. After addressing a joint session of Congress, and receiving a tumultuous, emotional ticker tape parade in New York City on April 19, he moved into the Waldorf-Astoria Towers and began a nationwide speaking tour that ultimately proved disappointing to him. When he returned to New York City, he was entering the twilight of his storied life.

Though Douglas MacArthur's military career had come to an embarrassing end, when he came home to a hero's welcome from the nation he had served for forty-eight years, it was to prove a blessing in disguise for Earl Blaik.

A DUTY TO PROTECT HONOR: THE LIFELONG DILEMMA BEGINS

Harkins' proposal that *Brian Nolan* accept *George Hendricks'* offers of unauthorized assistance and work his way into the cheating ring hit *Brian* hard, surprised and stunned him. He felt cornered, trapped. The enormity of his dilemma was immediately obvious. Friendship, loyalty and honor go together, don't they? He didn't remember the words in *Bugle Notes*, the small book of plebe knowledge given him that first hectic summer day in 1949 when he entered West Point. The words had remained virtually unchanged since 1923. "Honor is bigger than any one man, or group of men." Nor did he remember the words in the first of the "General Principles" upon which *Bugle Notes* said the honor code was founded. "...The Code demands courageous and fearless honesty in setting forth the truth, regardless of consequences." The thought must have been somewhere in the recesses of his mind, reaching for his conscience, the gauge for his sense of right and wrong.

But *Nolan* definitely remembered the oft repeated principle, "Every man is honor bound to report any breach of honor which comes to his attention." This was the most difficult and demanding tenet of the honor system, also the most misunderstood and controversial. It asked much of any man, and what it asked seemed in stark contrast with what most of us had learned throughout our lives.

Don't tattle. Don't tell on friends, playmates, neighbors, classmates, or teammates. And never, never tell on family members. This unyielding use of honor by the honor system, to tell of others' violations seemed a holdover from the ancient tyrannies of the European monarchies, what early Americans had fled, the trappings of emperors' regimes or ruthless dictatorships similar to the ones against which America had fought in two world wars. Snitching, that's what it was, a practice open to all manner of devious motives and abuses. It was the stuff of a brutal, all seeing internal security mechanism, a system used by such notorious organizations as Adolph Hitler's Gestapo to destroy opposition and terrorize people into submission.

On the other hand, not to disclose honor violations seemed not unlike the code binding organized crime families. "Protect 'the family' at all costs. One who discloses is worse than a 'rat,' 'mole,' or 'snitch,' deserves to die, and becomes a target for a 'contract,' a 'hit,' a 'whack'." A disciplined, ruthlessly enforced code of silence can mask anything, no matter how evil.

But cheating in academics was surely not an evil. It wasn't even against the law. Besides, cheating occurs in many high schools, colleges, and universities.

Yet, the principle that honor required reporting others' honor violations, even self-reporting unintentional honor violations, and all the contradictory history and implications it carried, was central to the Academy's honor code. It made honor work, an attainable ideal, a sought after standard, and ultimately a builder of trust, confidence, a sense of security, and pride. In the years ahead this principle made honor an irresistible force for doing what was right, identifying and revealing corruption, and violations of public trust, to say nothing of building trust and confidence in men who would lead young soldiers or airmen in battle. The principle, in fact, was built on trust of a man's motives in reporting violations. Motives of the men reporting what they believed to be honor violations should not be questioned, not unless facts emerging from the investigation of an alleged violation seemed to say otherwise.

Brian Nolan wanted to help. He wanted to do what was right, be honorable, continue to live the code he was upholding when he first disclosed *Hendricks'* participation in a cheating ring. Yet, here was the Commandant, a senior officer whose mere presence stirred awe in most cadets, in the privacy of his office, asking *Brian Nolan* to do something he had never done in his life, with perhaps far reaching consequences for the Academy and the men involved in cheating. A cadet doesn't say no to a colonel, especially if the cadet is a third classman, and the colonel is the Commandant of Cadets. And where were all the other cadets who must know about the cheating ring? Why was this for *Brian Nolan* to do alone?

It was difficult enough to name a friend and classmate as apparently involved in major violations of the honor code. *Brian* had several days earlier been asked to complete a written statement, and he remembered adding a comment to his statement: "I regret doing this and I shall remember it the rest of my life." His sense of justice told him he would probably have to confront the man he was accusing. Confronting a friend would be tough, but he knew what he must do with regard to one man. It was right, and he could do it.

But he was being asked to knowingly accept unauthorized assistance, penetrate the ring his friend had told him about, and gather irrefutable evidence against *Hendricks* – and how many other cadets? How many were there? *Hendricks* had told him "it's large," and implied "everyone was in the ring.... How did he – *Nolan* – think Army had the football team they did?" But *Nolan* had no idea how large. How could *Brian* confront a large number of cadets, if it's as *Hendricks* had told him. If his role in entering and exposing the ring were compromised, Harkins' proposal could have devastating effects on *Brian Nolan*, perhaps for the rest of his life. And should he successfully enter the ring, he would be pretending, deceiving his class-mates and friends, attempting to convince them of his sincerity in participating in their endeavor to beat the academic system, gradu-ate, and become officers. He had his own frustrating troubles with academics, and he might feel better later for his disclosures about *Hendricks*. But this was different. Completely different. He was nine-teen years old, and this was a staggering burden he was being asked to undertake.

Brian Nolan was understandably reluctant to agree to Harkins' re-quest, and wanted to discuss the proposal with his family. Harkins knew he was placing a grave responsibility on *Brian Nolan's* shoulders, and agreed to the consultation.

ADVICE AND DECISION

The young third classman telephoned an uncle in New York City, a man he admired, and a successful corporate lawyer. He asked his uncle to come visit him at West Point that weekend. *Brian* told him he had "encountered somewhat of a problem." He remembers the day of the visit, as did *George Hendricks*. *Brian* and *George* were walking to-ward the Hotel Thayer together, when *Brian* met his uncle, and the two family members excused themselves and walked up Mills Road to-ward picturesque Michie Stadium and Lusk Reservoir. The Reservoir is across the street from the Stadium – a place filled with the ghostly echoes of thousands of Saturday afternoon Army football faithful, who revel in the gridiron exploits of the Black Knights of the Hudson. Stand-ing on the sidewalk next to the waist high granite wall surrounding the Reservoir, visitors can get a marvelous view of the New York country-side, looking past the bell tower of the Cadet Protestant Chapel, across the Hudson River.

There, across from Michie Stadium, *Brian Nolan* outlined without dramatics the problem he confronted. He told his uncle he was pre-pared to leave the Academy, if need be. He didn't want to do what the Commandant was asking. Though he had passed the hurdle of plebe year, it would be no supreme sacrifice on his part. He was happy with the friends he had made at West Point, and believed he was well liked. Yet academics were hard for him. He felt his talents lay in the liberal arts, not science and math.

His uncle paused after listening, then said, "You've done nothing wrong. Don't leave under fire. It will involve you. Do as the Colonel says." Respecting his uncle deeply, *Brian Nolan* accepted his advice and reported back to Colonel Harkins.

Approximately two weeks later, on May 2, 1951, Paul Harkins wrote in a memorandum for record, "Cadet *Nolan* did [successfully join the ring]...it took a great deal of moral courage...for when he entered the ring he found several of his friends were already in it."

A SPREADING UNEASY FEELING OF SOMETHING SERIOUSLY WRONG...

General Irving, Colonel Harkins, Colonel Waters, Lieutenant Colonel Collins, Pat Ryan, and the cadets working to penetrate the cheating ring weren't the only men at the Academy who believed something was seriously wrong within the Corps of Cadets and the honor system. Colonel Lawrence E. Schick, professor and head of the Department of Military Topography and Graphics, was so troubled about what he was seeing and hearing that, independently, on May 1, the day before Colonel Harkins began drafting his memorandum for record, Colonel Schick set down his own thoughts in a memo for record. He spoke to the Dean, Brigadier General Jones, about some of its contents, but didn't send it to the Dean or anyone else. The memo makes clear he had been speaking with other Academy officers about their observations and opinions.

> There are currents of discussion among officers of the post to the effect that unhealthy attitudes are developing within the Corps on the subject of honor and duty. There are some who claim that there is a diminution of loyalty to and enthusiasm for The Corps among its own members. There are comments to the effect that 'over organization' and distortions in the sense of duty and responsibility are destroying the camaraderie and morale that were formerly such highly developed and prized attributes of the Corps. There are criticisms of the Aptitude system to the effect that it creates nervousness, self-consciousness, and unnatural apprehensions among cadets. It is declared by some that our present systems are arresting maturity and

CHAPTER 1: DISASTER IN A FABLED PLACE

promoting 'guard house' frames of mind. On the other hand, some proclaim that the 'chain of command' system creates a magnified sense of responsibility and stifles buoyancy of spirit.

Since these declarations are made by responsible persons in a spirit of grave concern I think they warrant probing. I realize that criticisms are omnipresent and that 'barracks chatter' is commonly an expression of intolerance for existing conditions and policies. But the comments I refer to are not idle complaint but considered expressions of concern.

Due to lack of specific contact with internal affairs of The Corps I am unable to challenge the comments factually. I find that for me personally to obtain or maintain an intimate picture of affairs is almost impossible. I also discover that my plight in this regard is shared by others of the Academic Board. This being the case I think there is a danger of the Board becoming engulfed by local developments or at least surrounded by unanswerable challenges instead of being in the van with its finger on the pulse of affairs at all times. If this condition should actually come to pass it would be unfortunate to the extreme. The Board would prima facie stand convicted of failing in its duty to lead and control. Certainly the moral welfare of The Corps is fully as important as its physical or mental development. We must never be guilty of hypocritical complacency in respect to this vital part of cadet life.

I therefore suggest that the General Committee (or the Board) conduct a series of hearings wherein those immediately concerned with the direction of cadet affairs expound and be queried about their activities. The purpose initially would be to obtain a clear picture of cadet environment and reaction. Possibly certain cadets themselves could appear, but only as a last resort in the event facts cannot be obtained otherwise. Those directing cadet activities who should be heard are The Commandant, a few Company Tacs, the Head of the Department of MP&L [Military Psychology and Leadership], the Psychologist, the Director of Athletics, the Director of Physical Education, the Cadet Chaplain, the Catholic Chaplain, the Surgeon.

I am loathe to advocate that the Board pattern an investigative procedure after that currently in vogue in The Congress. But I am apprehensive that the Board is not fully cognizant of some current developments. They should probe them and either give them a clean bill of health or take remedial action if needed. I suggest this proposal be discussed by the General Committee at the next opportunity.

Colonel Lawrence Schick's memo mentioned "the Academic Board" several times. He was referring to the most powerful, perennially standing board of officers at West Point. Composed of professors who were academic department heads, the Commandant, the Dean, and the superintendent, plus the Academy Adjutant General, who was a non-voting member, the Academic Board was heavily involved in every aspect of the Academy's mission of education and training.

The "Tacs" were commissioned officers, company tactical officers, one assigned to each of the twenty-four cadet companies in the Corps. The Tacs were responsible for providing leadership example, administration of discipline, instruction, guidance and counsel to the cadets and the cadet chain of command. They were the men who would, in theory, know most about what was going on within their companies.

When Colonel Schick drafted his May 1 memo, and offered his thoughts to the Dean, the Academic Board had no inkling of the trouble that lay ahead. A few tactical officers had similar thoughts based on more direct observations within their cadet companies. One was Lieutenant Colonel Raymond B. Marlin, the company B-1 tactical officer.

Since the spring of 1950 Marlin observed a widening split within B-1 between cadets on the football team and those who weren't. There were specific incidents within his company which highlighted the split. One involved the president of the class of 1952, a cadet who eventually became the captain-elect of the football team.

When Marlin arrived at the Academy in the summer of 1948 he noted that first classmen who were football players, and on the Beast Barracks detail, sought out new cadets who were soon to be plebe football players, and seemed to give them favorable treatment. It was a pattern of behavior that later caused him to inquire of other tactical officers if they had observed the same tendencies. Several had. Finally, Marlin took his observations, and the incidents he encountered in B-1 to his boss, Lieutenant Colonel Art Collins. Marlin would soon learn his discomfort over the attitudes of several cadets in B-1 toward duty and honor were well founded.

ANOTHER MAN: A SECOND DILEMMA

When *Brian Nolan* agreed to attempt penetrating the cheating ring, not only was he uncertain of what he would encounter and where it would lead, he and the senior officers directing him had little to tell them how the ring operated, and how extensive it was, except what *Hendricks* told *Nolan*. He did learn soon after he got inside the ring that the primary means of passing information was verbally, although earlier, when *Hendricks* was urging *Nolan* to accept assistance, he had given *Brian* a copy of an upcoming physics examination, known as a written general review. *Brian* also learned the operation was large, as *Hendricks* had told him. There were ring members in the 2d Regiment.

Nolan's wish not to bear his undercover duties alone, the need for corroborating testimony if there eventually was a formal investigation, and the confirmation of 2d Regiment cheating activities, resulted in a decision to seek cadet help in the 2d Regiment. Approaches were made to two of *Brian Nolan's* classmates. One attempt failed. The other was successful.

Michael G. Arrison was on the swim team with *Nolan*. He wasn't a recruited athlete as *Brian* had been. He was a "walk on." His plebe year in Company H-2 had been a nightmare. Trouble with upperclassmen dogged *Mike* at every step in those eleven months as a plebe, a long, painful time he never forgot, a period in his life that heaped misery onto the harsh treatment he was to face after the spring of 1951.

Mike Arrison's home town was Williamsport, in central Pennsylvania. As a boy, going to West Point became his dream. He remembered drawing on notebook paper, sketching the Academy's coat of arms, an intricate, richly patriotic emblem emblazoned with West Point's motto, "Duty, Honor, Country."

As it was to *Brian Nolan*, the cadet honor code was important to *Michael Arrison*. "A cadet will not lie, cheat, or steal." *Arrison* believed in it. He cared. Honorable was the right way to be. He had discussed honor with *Brian* on several occasions soon after Christmas of 1950. They talked about what they had been hearing, exchanges they had observed between cadets which caused them both to wonder if something was going on that shouldn't. And they talked about their academic struggles. Like *Brian Nolan*, *Mike Arrison* was barely holding on in academics. His troubles were chemistry and English.

There was another similarity between the cadet lives of *Mike* and *Brian*. Both were encountering classmates offering questionable assistance to solve their academic problems. In *Brian's* case, the offer was coming from *George Hendricks*, a classmate, a friend, a member of the Academy's fencing team, who was having his own academic problems. But *Hendricks* didn't room with *Brian Nolan*. He roomed with two other third classmen, one a football player. *Mike Arrison's* circumstance was different, closer, more awkward. The offer of help was coming from one of his roommates, *Theodore K. Strong*.

Ted Strong excelled in the classroom. He was a "hive," cadet jargon for a man who excelled in academics, and because of his academic performance and willingness to help his classmates, *Strong* volunteered to assist them as a tutor, an academic coach. To unsuspecting cadets, his offers of assistance to *Arrison* wouldn't seem unusual. But to *Mike*, who had been hearing discussions of honor and swim team members' observations about unusual behaviors involving honor, there was something in his roommate's words, and the inflection of his voice that caused *Mike* to wonder. *Ted Strong* was persistent, and the tone of his offers clearly different.

"*Mike*, you don't have to struggle with academics. I can get 'the poop' for you."

"Getting the poop" was a common term in cadet life and language. Organization meetings, briefings, and lectures in military training, classroom academic lectures, the cadet daily bulletin, announcements at meals in Washington Hall – made from "the poop deck," and the blizzard of letters, memorandums, regulations, regulation changes, and other papers in cadet and Army life, all filled with information, directions, prohibitions, duty rosters, and orders contained "the poop," "the word." "The poop" passed through the cadet chain of command told cadets what to do, when, where to go, and the uniform to wear. It was not uncommon to hear a cadet company commander begin a meeting with his company, "OK, here's the poop." But getting the poop had never meant, and didn't mean ...here's advance information, wrongly obtained, with which you can get better grades tomorrow... – except, that is, for the men engaged in organized cheating during those years. They had learned the subtleties of the ring's language, and its invitations. And, at first, they learned to be selective about who they invited to join.

Mike Arrison, increasingly troubled, went to see a first classman he admired and respected, Jack Craigie, captain of the swim team. Jack was in company L-2, knew *Brian Nolan* and *Mike Arrison* well, and had often given *Mike* counsel and encouragement as *Arrison* struggled with academics and the aftermath of a tortuous plebe year. "Sandy" Vandenberg, one of Jack's two roommates, remembered vividly the day *Mike* came to see Jack Craigie. Jack was not in the room, but *Mike* wanted to see him in private. Sandy suggested *Mike* wait as Jack was expected soon. *Mike* sat quietly, deep in thought. When Jack returned, Sandy Vandenberg considerately left the two alone and went to a classmate's room down the hall.

Mike went into detail telling his swim team captain of *Ted Strong's* repeated offers of assistance, assistance clearly implying the help was coming from other cadets. Without hesitation, Jack Craigie told him he should go directly to Colonel Harkins.

"Don't go to your honor representative, and don't tell anyone else. Go straight to the Commandant."

Jack Craigie and Sandy Vandenberg had feelings not unlike those of Pat Ryan, although their suspicions didn't include football players. Their concerns were based partly on an experience with the Honor Committee, when they had been called to testify in a hearing. The hearing panel's questions delved more into their respect for the accused cadet, whether they "liked him," rather than the facts Craigie and Vandenberg could offer.

When *Mike* left the room, Sandy Vandenberg returned. When he walked in, he saw an expression on Jack's face he never forgot. Jack was sitting, staring straight ahead, his face pale, ashen. Sandy, puzzled, asked, "What's the matter?"

Jack looked at him, still pale, and began, "You won't believe what just happened."

*   *   *

Brian Nolan gave *Mike Arrison* more to consider. He confided to *Mike* he was participating in the search for evidence of possibly extensive cheating in academics, and he needed help. *Mike* listened to *Brian's* appeal for help. He told *Brian* of his roommate's entreaties, and his suspicions of his roommate giving unauthorized assistance in academics. He remembered that during an earlier visit to *Nolan's* room, *Brian* had talked of *George Hendricks'* repeated offers of assistance, and hints the assistance was coming from a cheating ring. Now *Mike* was aware *Brian* had made good on his decision. He had decided Colonel Harkins must be told.

Mike Arrison would do the same. *Mike* had become convinced his roommate was involved in the "something that was going on with honor." He talked with no one else about what he believed he must do. He made his decision alone. He went directly to Colonel Harkins, told him of *Strong's* offer to get the poop. He told Harkins he wanted to help in stopping the cheating ring. The Commandant listened, then, as he had asked of *Brian Nolan*, he asked *Arrison* to accept *Ted Strong's* offer of assistance, and thereby get inside the ring – penetrate it. And, as he had with *Brian Nolan*, the Commandant also assured *Mike Arrison* the "full weight of the Commandant's office will back you up."

Brian Nolan now had an ally. *Mike Arrison* joined the cheating ring. He later confirmed his roommate's participation in activities that violated the cadet honor code. And he learned much more he wished he'd never seen or heard.

Mike Arrison's life would soon change. Though his plebe year was unhappy, his first year at West Point was only the beginning. He was completely unprepared for the turmoil and emotional pounding he was to undergo in the years ahead.

He was not alone. Few men who left West Point under the cloud of scandal in 1951 would be tested so severely as *Michael Arrison* and *Brian Nolan*, who had shouldered the most difficult, wrenching duties anyone could undertake in upholding the virtue of integrity. Completing a difficult mission and duty well performed are not always rewarding.

SPREADING THE NET

Dan Myers' and Dick Miller's responsibilities in conducting the undercover investigation were to collect information – evidence – gathered by *Nolan* and, to some extent by *Arrison*, and pass it to Harkins and John Waters. Care had to be taken to avoid exposing *Nolan's* more frequent contacts with company honor representatives, so methods were devised to pass information. *Arrison*, who lived in the north area of cadet barracks, worked mostly on his own, bringing information directly to Colonel Harkins, but occasionally giving it to Myers or Miller. He had to be careful about too many trips to the Commandant's office. It was unusual for third classmen to visit Harkins' office.

Harkins established discrete contacts in the Academic Department and told them of the investigation in progress. He needed assistance from the Academic Department in the event instructors observed questionable activities, even cheating, by *Nolan* and *Arrison*. Provisions were made for instructors to give a "stay back" – no recitation, and hence no grade to the two men – on days in which *Nolan* and *Arrison* engaged in classroom behavior that might appear to violate the honor

Daniel J. Myers, Jr.
Class of 1951
Company K-1
Honor Representative

(Photos courtesy
USMA Archives.)

Richard J. Miller
Class of 1952
Company K-1
Honor Representative

code. It was also agreed that their grade point averages wouldn't fall below the averages the two cadets had when the undercover investigation began.

As for *Nolan* and *Arrison*, when they penetrated the ring, they had to be careful of their behavior to avoid stirring suspicions. Too many questions, and carelessness in passing or receiving academic assistance, could arouse suspicions, expose their roles, and collapse the investigation.

From mid-April through May, Dan Myers dutifully collected most of the information obtained by *Brian Nolan*, including many names of suspected ring members, and some gathered by *Mike Arrison*. Dick Miller assisted, but his role was less extensive. Except for copies of examinations ring members were passing among themselves, and to *Nolan*, Myers took notes on scraps of paper and kept all the paper evidence in the right pocket of his cadet shortcoat, a gray wool, thigh length winter overcoat.

He remembers being frustrated and worried because Colonel Harkins, for more than two weeks, declined to accept the accumulating scraps of paper. As the New York spring began to warm, wear of the short overcoat became less frequent. Scraps of evidence more often remained behind in Dan's overcoat pocket, which hung in its proper place in his barracks room. Myers was increasingly concerned he might accidentally lose some of the evidence being funneled through him, or someone would inadvertently come upon his growing cache of evidence.

Finally, on May 3 in Harkins' office, the Commandant took all that Dan Myers had collected and, as Myers watched, locked it in a safe. Included in the papers he gave Harkins was concrete evidence given him by *Nolan* that *George Hendricks* had been cheating and providing unauthorized academic information to other ring members. There were also lists of names given him by *Nolan* and *Arrison*, names they had picked up in conversations with *Hendricks* and *Strong*, including first classmen, one a member of the Cadet Honor Committee. Myers furnished considerably more evidence passed to him by *Nolan*, information given verbally by *Hendricks*. Once more Harkins reiterated the need for firm evidence against more than one man. The fact that someone had "heard" of other cadets' participation, or "felt" that other cadets knew, didn't constitute firm evidence.

One day earlier, Dick Miller, *Brian Nolan*, and *Mike Arrison* had reported to Harkins, to tell him what had been gathered to date, and what was to be done in the future. In the sessions with the three men, he reiterated his determination to break the ring. They asked him for his assurance the Tactical Department – the Commandant's organization – would "back them up." He firmly reiterated they had the Department's backing.

Brian Nolan was becoming more concerned about the drastic effects, the penalties, all these young men might receive as a result of his activities. This had become far larger than he had ever anticipated. There may be hundreds involved. Will all these men be dismissed from the Academy? he wondered.

Harkins responded, "Because of the large numbers that may be involved, I don't know what might happen. Some may get to stay."

Brian Nolan didn't want his friend, *George Hendricks*, to go, especially not on his account. He began to wonder if there was some way *Hendricks* could stay. *George* seemed to have known a lot about the ring. Perhaps his complete cooperation in the formal investigation could result in his being able to remain at West Point.

Harkins began his memorandum for record that day, and added entries periodically. He continued his almost daily routine of closed door sessions with General Irving, the superintendent, to keep him informed of progress. Throughout the month of May, *Nolan, Arrison*, Myers, and Miller worked to obtain more hard evidence of a cheating ring. There were periodic meetings with Harkins, each adding to the mounting evidence against *Hendricks* in particular, and *Strong*, one man in each regiment. Harkins pressed for more and firmer evidence in each meeting, and at the same time repeatedly emphasized the need to find out where these two men were getting their information.

Who was providing them the unauthorized assistance? He could not break the ring without finding out, with concrete evidence, who else might be involved. Just one more was all he needed. He also reiterated his determination to break the ring, if at all possible, before he left West Point for his new assignment on June 10. He reassured them the Tactical Department would back the men who had put themselves at risk, and that Colonel Waters was well prepared to resume work next academic semester if sufficient evidence could not be found this semester. On Wednesday and Friday, May 23 and 25 more evidence came in, but still no hard evidence against anyone other than *Hendricks* and *Strong*.

The morning of Monday, May 28, 1951, Paul Harkins received the final pieces of evidence he needed to begin an extensive, formal investigation by a board of officers – a fact finding board that could develop and assemble information sufficient to judge the cadets involved as guilty or not guilty of honor violations. That morning Dan Myers brought what he believed to be copies of several written general reviews. They had been passed by cadets in the ring to other members of the ring. A new name surfaced, and there was paper evidence of the cadet's involvement.

Dan Myers remembered the May 28 meeting in which he gave Paul Harkins the last pieces of evidence. It was also the last time he came to Harkins' office. Harkins was pleased to finally receive the evidence he needed, and was anxious to get moving on a formal investigation. Yet Dan Myers observed once more a growing sadness in Harkins. The Commandant reiterated the lament to Myers, that his – Harkins' – Army career may be over, but he would do what was necessary to stop the cheating ring. He also told Dan that as a young, newly commissioned officer, Myers might be ordered back to West Point to be a witness in the investigation. Dan Myers was ready and expected the call, but it never came.

In the nearly two months since the first disclosures by *Brian Nolan*, firm evidence had been gathered to formally investigate the activities of two cadets, and now there was paper evidence pointing to another. There was information that could implicate fourteen more, and a list of thirty-five other names collected as a result of contacts with ring members.

Additionally, there had been conversations implying honor representatives were involved, deliberate efforts to elect honor representatives to effect outcomes of future honor investigations and hearings, academic tutors passing advance information on daily recitations, tests, and final examinations. So-called "name athletes" were said to be part of the ring. Among the names were several football players, some who doubled as varsity athletes in other intercollegiate sports. The Corps Squad tables – athletes' training meal tables in the cadet dining hall –

were said to be an area where extensive advance academic information was being passed. If this were true, the athletes were passing information at the tables in spite of repeated cautions to avoid discussing examinations they had just taken.

The ring had "runners" who carried information to other companies, battalions, and between regiments. There were men who would receive questions taken out of daily recitations and tests, work out answers to the questions and pass them to ring members taking the same class later in the day, or the next day. There were estimates of "fifty to one hundred" cadets being involved, and perhaps "many more" who knew of the ring but made no move to report its existence.

Paul Harkins knew the time was ripe to begin a formal investigation. He informed General Irving of the just-uncovered, hard evidence and recommended the formal investigation proceed. The superintendent's response was brief and to the point. "Go ahead."

Later in the day Harkins met with Colonel Counts, Professor of Physics and Chemistry, and one other officer in Counts' department. Harkins asked them to look at the examination copies given him by Dan Myers. There was no doubt of the copies' authenticity. They were identical to written general reviews numbers one, two, and three, and one other physics test.

Paul Harkins prepared to convene a board of officers. He had already decided who would take on this most unpleasant duty. The three primary members of the board were all graduates in West Point's class of 1938, each a combat veteran of World War II, and each a member of Paul Harkins' Tactical Department. The four alternate board members were all Academy graduates, also combat veterans, from the classes of '35, '38, '39, and '41. Four of the seven officers were sons of Academy graduates.

Harkins notified the seven board members, explained the task they were undertaking, and delivered to the board president the evidence gathered. The officers then began consultations with the Academy's Staff Judge Advocate – the equivalent of a district attorney – and finally, late in the day the three primary members met with Dan Myers, *Brian Nolan* and *Mike Arrison,* one at a time. The board was to begin taking testimony at seven o'clock on Tuesday, May 29. As the board prepared, and began mapping a strategy for questioning the young men to first appear before them, Paul Harkins signed a brief directive:

MACC   28 May 1951

SUBJECT: Letter Orders
TO: Officers Concerned

1. A Board of Officers consisting of
                    Lt Col Arthur S Collins, Inf
                    Lt Col Jefferson J Irvin, Inf
                    Lt Col Tracy B Harrington, Armor
and the following alternate members
                    Lt Col Cornelius DeW W Lang, Arty
                    Lt Col Joseph A McChristian, Armor
                    Lt Col Birdsey L Learman, Inf
                    Lt Col Robert W Garrett, Inf
are appointed to investigate the alleged honor violations of Cadets *Strong, TK*, and *Hendricks, GL,* Third Class.

2. Report of the Board in quadruplicate will be submitted to the Commandant of Cadets at the earliest practicable date.

BY ORDER OF COLONEL HARKINS:
                    FRANK D MILLER
                    Lt Col, Infantry
                    S-1, USCC

CALM BEFORE THE STORM

Two months of apparently successful, covert investigative work had come to an end. Primarily, it had been the work of two courageous young men who, though weighted with periods of doubt, felt they were doing what was right, however reluctantly and painfully they had undertaken their duties. Another, much larger group of young men, who had carefully masked organized cheating activities and had been the recipients and custodians of what had been passed to them from class to class for an unknown period of years, were about to be exposed and judged for their actions.

A small number among them had begun to suspect they had been found out, that someone was passing information to honor representatives. *Michael G. Arrison*, Company H-2, was already in trouble, and had been since sometime in mid-May. His roommate, *Ted Strong*, apparently suspected he was passing information about the ring when *Arrison's* other roommate had been asked by Dan Myers to help in uncovering the ring. *Strong* had undoubtedly told other ring members. Tomorrow, May 29, word of the cheating ring's undoing would flash through their ranks like electricity, and the controversy that would explode in public in August would begin building in earnest inside the Academy's gray granite, ivy-covered walls.

Still, between May 28 and August 3, the gathering of evidence, the investigation board, and the review board, all the hundreds of letters, messages, memorandums, board reports, phone calls, conferences, staff meetings, briefings, staff papers – and decisions, continued behind closed doors. Command guidance required confidentiality, the silence necessary to protect the strategy and success of the investigation, the futures of both the guilty and innocent, and an emerging plan to hold off making a public announcement until all those found guilty resigned and left West Point. There was a veil of secrecy over events grinding toward a tragic conclusion. The bottle remained capped for two months, but pressures were building inside – rapidly.

*   *   *

On May 28, the day Paul Harkins signed the letter order convening the formal investigation into cheating at West Point, Coach Earl Blaik, totally unaware of events that would soon engulf his football team and his family in bitter controversy, wrote a letter to MacArthur. "...I shall write you about the football prospects after the Academic Board has applied the pruning knife to the turnouts." Turnouts were cadets who failed academic courses, and could, upon successful reexamination, and recommendation of the Academic Board, be "turned back" to the following year's class. Failure of turnout examinations resulted in a cadet being "found," that is, separated from the Academy.

Retired General MacArthur, the avid behind-the-scenes fan of Army football since his days as superintendent at West Point, was continuing in that role. Since serving as Army Chief of Staff in the early 1930s, he had asked each head football coach at West Point to keep him informed about the Army team's status and prospects for coming seasons. Due to Earl Blaik's splendid coaching record at the Academy, and the association dating to Blaik's cadet days when MacArthur was the Academy superintendent, Blaik's correspondence with him over the years had grown to a deepening friendship.

That General MacArthur now lived scarcely fifty miles from West Point would be a great help to the nationally known Army coach in the months and years ahead. Blaik would sorely need the Old Soldier's counsel and advice before the summer of 1951 was over, for the disaster about to shake the foundations of the United States Military Academy in the spring and summer of 1951 was to become a catastrophe for Earl Blaik and his family, the Army Athletic Department, and West Point's much admired football team.

CHAPTER 2

PROFILE OF A YOUTHFUL CONSPIRACY

Lieutenant Colonel Arthur S. Collins, Jr., the board president, was ready. The three officers gathered shortly before the board convened at seven o'clock in the morning, Tuesday, 29 May 1951. It was a typical New York spring day along the Hudson River. Comfortably cool in the early morning hours, warming and muggy as the sun heated the air toward mid-afternoon temperatures into the nineties. The location was a conference room near the Commandant's office on the second floor, overlooking central area of cadet barracks. The cadet guard room, center of day-to-day administration of the Corps of Cadets, was on the first floor.

Semester final examinations were coming to an end. For reasons that had been rites of spring for one hundred and forty-nine years, nearly everyone looked forward to graduation, this year the class of 1951. For the three underclasses graduation meant a completed academic year, a summer without the hectic pace of academics, new military training experiences, one more year gone in the long trek toward their graduations – and four relaxing weeks of summer leave.

June Week was beginning, always a festive time when thousands of people converge on West Point: families, friends, "drags" – cadets' dates, and fiancées of cadets soon to be commissioned second lieutenants in the Army or Air Force. Many a young lieutenant marries in the days immediately following graduation, at the "marriage factories," the cadet Protestant and Catholic Chapels. Hundreds of West Point graduates, their wives, and some family members come for class reunions, or to see a son or grandson graduate from the Academy. Among the graduates are wonderful old soldiers whose classes are celebrating fifty-year reunions, or more. These elderly gentlemen, many of whom have fought battles on and above faraway lands, truly give substance and meaning to the "Long Gray Line." Then there are thousands more visitors and tourists from all around the world who simply come to see the pageantry of June Week, graduation week at West Point.

Graduation week is normally a happy occasion, filled with excitement and eager anticipation. There are parades and other ceremonies, intercollegiate athletic contests, receptions, dances, concerts, and endless chatter about plans for the future. Throughout the week anticipation builds toward the climactic events, graduation parade on the Plain the evening prior to graduation, and the graduation ceremony, held in the Field House down below Trophy Point, near the Hudson River.

Immediately following the parade was the event plebes looked toward all year – "recognition," signifying the end of fourth class year and all the work learning to live on the lowest rung of the seniority ladder at West Point. For plebes, recognition brings a dramatic change for the better, a tremendous emotional release, a lifting of burdens, which give way to relief, and finally, elation. Plebes become young men again, shortly before they become new third classmen. They can begin to relax, feel free to talk, look around, and not be ordered around.

Emery Scott Wetzel, Jr., class of '54, well remembers his recognition that spring, not only because it was the end of a long, tough year, but he encountered a gentleman from the class of 1906 he would never forget. "Scotty" was a plebe in company F-2, the First Captain's – Pat Ryan's – company. Ryan had left instructions with the F-2 orderly room that after the plebes went through their recognition ceremony in the company area, they were to "drive around to the First Captain's room for recognition."

While "Scotty" and several of his classmates stood at a "brace," the plebe's rigid position of attention, and Ryan was talking to them, into the room walked retired General Jonathan M. Wainwright, a tall, gaunt man whose nickname to his close friends was "Skinny." He introduced himself to the cadets, then in easy conversation he pointed to Pat Ryan's desk and said, "That's where I studied." He looked around the room and pointed to a bed and said, "That's where I slept." "Scotty" Wetzel stood wide-eyed, seeing a man he remembered was a national hero when he was a boy in the early days of World War II.

General Wainwright was First Captain in his graduating class of seventy-eight men in 1906. He was the American general given command of defenders on the island of Corregidor against the onrushing Japanese Imperial Army in March of 1942, during the early, bitter days of defeat in World War II, when President Franklin Roosevelt ordered

General MacArthur to leave the island for Australia. Wainwright led his gallant soldiers in prolonging the defense of the island until, with no relief or reinforcement in sight, running out of food, water, ammunition, and medical supplies, he surrendered the garrison in May. He was interned in Japanese prison camps, as was Dick Miller's father, when American and Filipino forces were overwhelmed in March, on the Bataan Peninsula. General Wainwright remained a prisoner of war for the duration, an indescribably harsh imprisonment that broke his health. For his gallantry on Corregidor he received the Medal of Honor.

For Pat Ryan, General Wainwright was a reminder that war wasn't far away. Graduation for the class of 1951 was different from graduations of classes since World War II. The Korean War tempered excitement for the normally joyous occasion. They were the first class to graduate after the Korean War began.

Though preceding Academy classes had already received a deadly baptism of fire in Korea, there was always something different about a "war class." There is an increased importance, purpose, and urgency in a wartime graduation. Young graduates don't leave the Academy in wartime with a sense of dread or fear because of casualty lists. Instead, prevailing sentiments are exuberance and optimistic beliefs about the indestructibility of youth, imbued with a deeply ingrained sense of duty.

To get into the fight, to volunteer for units bound for combat, is doing what cadets are educated and trained to do. It's right, proper, and what the country asks. Pride and idealism, fueled by patriotism are the mixture of emotions that drive men and women to volunteer for battle. The class of '51 was experiencing emotions similar to those felt by men in the classes of 1836, 1846, 1861, April and August of 1917, and 1942. Though '51 graduates didn't express their feelings the same way their predecessors did more than a century earlier, most were anxious to "see and feel the elephant," as their predecessors would have said. The elephant was battle, war experienced by a soldier. To have "seen and felt the elephant" was an expression used by a graduate in the class of 1846, writing home during the war with Mexico, when General Winfield Scott was leading the invasion of Mexico, and the Academy class of 1846 was taking heavy casualties. The term later came into widespread use during the Civil War.

In the spring of 1951, among those educating and training the first Korean War class to graduate from West Point, were many officers and enlisted men who had fought in World War II, and a few who had been in World War I. When these veterans witnessed the pageantry of June Week that year, their feelings were different than those of the men about to be commissioned. These men who had seen combat in the World Wars knew full well what war meant, what it would bring to the young men who would go into the combat arms of the Army or Air Force, what was needed to lead, and lead well when the bullets began to fly. For all those serving at the Academy, and watching '51's graduation ceremonies, there were other feelings: satisfaction for doing their best to prepare the young graduates for what lay ahead; a tug of nostalgia in recalling their own graduation from West Point in years past; for some, a pause, a sigh of relief, and another beginning as they shifted into preparations for summer military training and entry of the newest class, '55, into the Academy. For some, such as Paul Harkins, there was the satisfaction of completing a rewarding assignment at the Academy, while for a few there was anticipation of a return to war for the second time – after a pause of less than six years. They were officers and enlisted who fought in World War II and were being reassigned to units in Korea.

But the joys and anticipation of June Week, 1951, weren't felt by the men on the Collins Board, or the cadets awaiting its judgments. Upstairs, in the conference room near the Commandant's office, the mood was decidedly grim. This was the beginning of something West Point had never experienced.

The Collins Board left its mark on the Academy, the United States Army, and the life of every single man involved in its proceedings. There were one hundred and thirty-six men who testified before the board, most in the ten days ending June 8. After an additional one day session June 25, the board reaffirmed its June 8 conclusion that ninety-four cadets were guilty of honor violations, and should be discharged.

Among the ninety-four were thirty-nine from the class of 1952, fifty-four from the class of 1953, and one from the class of 1954. The board named sixteen men in the class of 1951 as probably involved, but there wasn't evidence to sustain guilty findings.

In August, after the public announcement, the fact that not one man in the recently graduated class of 1951 was found guilty triggered anger and frustration among many of those dismissed from the classes of '52 and '53, and became another source of public controversy in the weeks, months, and years ahead.

The details of what occurred, was said, and decided in the Collins Board have been held in strict confidence, behind closed doors, and later in confidential files, in the years since the Board began. Surrounding the Board's actions has been a cloud of speculation, half truths, factual errors, rumors, and controversy for many West Point graduates of that era.

The Collins Board was a fact finding board, an administrative proceeding, not a court martial. The Board's purpose was to investigate, obtain facts, reach conclusions, and make recommendations to the Commandant and superintendent.

An honor investigation by a board of officers, without the convening of a cadet Honor Committee hearing, and a judgment of guilty or not guilty by a panel of Honor Committee members, while not the usual practice, was permissible under existing honor system procedures. Under honor system procedures, a cadet had the option of reporting an honor violation to an officer instead of his company honor representative. If he did, the matter became "official" and required an investigation by officers. Officers are sworn to uphold the law and Army regulations, and it's their duty to investigate allegations that laws or regulations have been violated. The disclosure of a cheating ring, direct to Paul Harkins, certainly made the matter "official." As evidence was developed in April and May of 1951, it became abundantly clear that a confluence of factors made investigation by the Honor Committee neither practicable nor feasible. Thus the decision was eventually made to proceed with an investigation by a board of officers.

Army regulations didn't require boards of officers to transcribe verbatim testimony in administrative proceedings. The Academy staff judge advocate, the board's advising law officer, and the board president concluded there was no need to change the practice for the one about to begin. Instead a tape recorder was used to take testimony, and Jean Reissmann, the board recorder, transcribed summaries of testimony, leaving board members' statements and questions off the written record. Each day Jean Reissmann and typists assisting the board kept pace with completed drafts of testimony.

Since the hearing was administrative, and not a court martial, there were no attorneys present. Further, this was an honor investigation conducted by a board of officers. Procedures governing honor investigations didn't include retaining attorneys to defend cadets accused of violations. The principles of the honor system said each cadet was responsible for maintaining the honor code and system, telling the truth, irrespective of the consequences to one's self. Honor was "more important than any one man or group of men."

Cadets testified under oath, one at a time. Prior to being sworn, each was read the twenty-fourth Article of War by a board member. The Article explained the cadet's rights against self-incrimination. At the end of each day, board members discussed results of testimony, and prepared for the next day's proceedings.

As the days passed, Lieutenant Colonels Jefferson J. Irvin and Tracy B. Harrington analyzed testimony taken to date, and reviewed and discussed emerging conclusions. Jeff Irvin was responsible for writing the report, completing it after the board evaluated testimony and reached agreement on its conclusions and recommendations.

The board completed an interim report the evening of June 4, intending to reach conclusions and recommendations on the future of men from the class of 1951 who might be implicated and were to graduate the next day. By the time June 4 rolled around, members of the Collins Board tentatively concluded that 107 cadets in the four classes could be proven guilty of cheating, or to have had knowledge cheating was in progress. Art Collins provided their names to Harkins in the interim report. The Board also concluded sixteen men in the class of 1951 were involved in the cheating ring, but evidence couldn't be obtained to establish their guilt.

In its first seven days, the Collins Board, the young men who participated in the undercover investigation, and everyone who testified went on a wild, emotional roller coaster ride. And the ride was just beginning. In those seven days, and in the one day session on June 25, a tragic story unfolded.

Among men involved in the ring, word that the Board might be unraveling the organized cheating created its own set of emotions: suspicion, distrust, fear, anger, frustration, embarrassment, disillusionment, anxiety, guilt, shame, and for a few, everlasting shame. Then there were the questions – "Who gave it away?" "Who squealed?"

Among their roommates, other cadet friends, teammates, and classmates who were unaware the ring existed, there were other emotions created as rumors and half truths ran rampant: shock, disbelief, sadness, disappointment, sympathy, disgust, and different forms of anger and disillusionment. What the hell is going on? Why? How could they? What about their parents and the people back home? What happened to honor? Yet, in some who kept their own counsel was sympathy, and a measure of support for men feeling the weight of the Academy's investigation.

Day One: A Quick Break Shattered

When the first day's proceedings began, Art Collins, Jeff Irvin, and Tracy Harrington hoped *Cadets Hendricks* and *Strong* would provide evidence sufficient to break the ring quickly. Much to the officers' surprise, matters didn't turn out that way. And some weeks later, the three board members would find themselves accused of misconduct. For Jeff Irvin, the allegations of misconduct became a life-long unhappy memory.

The first cadet called to testify was *George Hendricks*, class of 1953, the K-1 third classman who would become a second classman, a junior, in one week, the day the class of 1951 graduated. His name had been called at reveille formation that morning, along with *Brian Nolan, William E. DiSantis*, one of *Hendricks'* roommates, and a third cadet in K 1, *Gordon F. Seabold*, who had also been implicated during the undercover investigation.

At the same time, throughout the cadet area, in companies where men had been implicated by the undercover investigation, and were to be called to testify that day, cadet company commanders read off the names of men who were to report to the office of their company Tacs after breakfast. Thus, the men who had been implicated in cheating found themselves seated next to one another, outside their Tacs' offices, wondering what was going on. For those who were definitely involved in cheating, and knew one another as members of the ring – and nearly all did know, the call of their names at reveille, and the sudden, mandatory meetings must have been profound shocks.

73

To protect *Brian Nolan's* role in the undercover investigation, he was among the first group of cadets called to appear before the Board at seven the morning of May 29. As the four men from K-1 waited outside the conference room following breakfast that morning, *Nolan* was the only one who knew why they had been called. In spite of the circumstance, he was optimistic about the eventual outcome of the investigation. His friend, *George Hendricks*, would be able to remain a cadet at West Point. *Brian* had the Commandant's word – at least he believed he had his word.

"If your young friend comes in and tells everything he knows, he'll be able to remain a cadet."

These were Harkins' words *Brian Nolan* remembered. As the cadets waited to be called into the conference room, *Brian* couldn't tell *George Hendricks* what the Commandant had said. But he remembered clearly the meeting he had had with Paul Harkins in early May. Dick Miller came to tell him the Commandant wanted to see him. *Brian* hurried to Harkins' office, wondering why the Commandant was asking to see him again – personally.

Harkins wanted as much evidence as he could get, and as many names as he could obtain, of men who were known or believed to be violating the honor code. A stated intent to allow *Hendricks* to remain in the Corps if he told everything he knew was a way to encourage *Brian Nolan* in uncovering as much evidence as possible. Harkins worked "one on one" with each of the cadets participating in the undercover investigation. He made no such commitment to *Mike Arrison* or *George Hendricks*.

Brian Nolan and *Mike Arrison* were called before the Board that first morning, but the questions to them were perfunctory and general. Collins, Irvin, and Harrington knew who they were and what their roles had been. Board members needed to make it appear that they were accused cadets because they had been acting as ring members for six weeks or more. While gathering evidence, *Brian* had been working closely with his friend *George Hendricks*, the first cadet to testify. *Brian* remembers Art Collins asking that first morning his "impression of what's going on in the Corps of Cadets." *Nolan*, unquestionably worried and ill at ease, replied, "Every time a rock is turned over, ants run out from under it."

The evidence against *George Hendricks* was overwhelming. When he reported to the Board, was advised of his rights and sworn, the three officers confronted him with copies of examinations he procured and passed to ring members, along with names of the men who had received test copies. They carefully presented the names, disguising the Board's sources of information. *Hendricks* had no idea who gave the Board the information they had, or how it had been obtained. He only knew he couldn't hold back. He must tell the three officers everything. If he didn't he would be adding "false swearing" – lying under oath – to the list of honor violations he had committed. He completely unburdened himself, telling all he could remember.

Within minutes the three officers began believing they would easily complete their assignment. *Hendricks* admitted complicity in organized cheating and implicated fourteen other cadets. He answered questions candidly, with a series of disclosures that stunned the Board, in spite of all Harkins had provided the three officers before the Board convened.

After confronting *Hendricks* with evidence collected against him, questioning began. The questions were simple and direct, the kinds of questions asked in any investigation into wrongdoing. When? What? Where? Who? How? and Why?

Hendricks responded,

I first heard about this ring during fourth class year [1949]. Initially I had nothing to do with it. After Christmas [of that year], when [we were ordered to change rooms], I moved in with another classmate, Cadet *DiSantis*, [in K-1]. Until that time I had no trouble with academics, but I then began to have difficulty with English. Cadet *DiSantis*, [who was on the football team] said he would help me out and wanted to know if I could help get the writ [or test] questions. I agreed to go along with him, and on several occasions I got the answers to writ questions and gave them to *DiSantis*. He in turn took them to [two other classmates in company B-1, both football players, one Cadet *Thomas S. Grayson*]. At that time he said that they couldn't help me in English because I was in the first hour class. I believe they took the information to [four other men, two in G-2, and one each in F-2 and M-2, all football players].

The thing quieted down at the end of last [academic] year but when academics started again in September [1950], [two of the previously named B-1 men, including *Grayson*, and one other man in B-1], contacted me and said they needed information if they were to go proficient.

At this time it started to get so big they formed blocs in each company so, in addition to passing information, they could get enough votes to control the election of honor representatives. They felt that if they could get their man nominated to the Honor Committee they would be able to block any discharge for violation of the honor code. I believe that this applies to the second and first classes as well as to the third class.

Hendricks' testimony, at this point, not only implied a large number of honor violations but, if true, implicated a small number of honor representatives in the first, second and third classes. First classmen, the leaders in the Corps, the men responsible for fourth class honor training, men who investigate or sit in judgment of cadets reported for honor violations. Men who in one week would graduate from West Point and take the oath of regular officers in the Army or Air Force. In a few months some could lead American soldiers on the battlefields of Korea.

Second class honor representatives who had spent a full year learning by observing and doing were within a week of full assumption of leadership responsibilities the first classmen were to vacate.

And if the second and first class honor representatives were not involved, there were allegations that third classmen knowingly elected honor representatives who, as first classmen on the Honor Committee a little more than a year hence, could ensure cadets involved in cheating were not found guilty.

Art Collins, Jeff Irvin, and Tracy Harrington continued to press *George Hendricks* with questions. He divulged what he believed to be the origins of the ring, why and how it began, its overall size and methods of operation, as well as his beliefs as to why no one in the group had previously disclosed its existence:

In the upperclasses it is mainly a football group and the only outsider I know of in the first class is [a cadet in company M-

1]. The background on this thing as I know it is that it was started by football players for football players. Their classes differ so much from the rest of the Corps that they frequently take the writs after others take them. When they enter as fourth classmen, they generally work very hard, but in case they are having difficulty with academics, this system of procuring information is passed on to them and they continue what they consider to be a standard practice. I am not certain who told the third classmen to procure the information this way but I heard someone say, and I can't remember who, that it was [two second classmen, one in B-1 and another in M-2, both respected football team leaders].

The way the ring works is that they have a representative in each company and the man taking the writ the early period has to leave the writ whether he finishes or not, to bring the answers to the people who will be going to the next class. They meet in the [cadet] area, in the library, or the hallways of the academic building. If they have a little time and there is an hour between classes, they have someone go to the man's room, get the problems, work them out, and then pass them on to the other members of the ring.

There are two rings operating, one in the First Regiment, and one in the 2d Regiment. In the First Regiment, I think [Cadet *Grayson*, one of the previously mentioned football players in B-1, and another in D-1], might be the head men. [Another football player in D-1] knows that they do it, and he works the problems for them. Cadet *Corley* [in B-1] is a go-between and also works the problems. I believe that all members of the football team in my class are involved except [one in company A-1].

Here is a specific instance. On or about 24 May, I was present when [my roommate, *DiSantis*] gave [the academic coach, *Corley*, in B-1] a physics writ which had been procured. Another instance, at a hygiene lecture during the first week in May, I gave a mathematics writ to [another classmate, a previously mentioned football team member in G-2].

It is hard for me to explain this. I know it is wrong. I know we should not have done it but the group that was in it just felt

that there was safety in numbers and there were so many finally involved that no action could be taken without disgracing
the Corps of Cadets.

I have thought a lot about how to correct the situation and
what we should do to stop it. Everyone has been scared to death
because they have had an idea it is going to break.

So ended the testimony of the first West Point cadet to go before
the Collins Board on Tuesday, May 29, 1951.

Before excusing *Hendricks*, a Board member handed him a third
class roster, and Art Collins asked the cadet to mark the names of all
the men in his class he knew to be members of the ring.

But *George Hendricks* was only the first to testify on May 29. A
baffling turnabout followed as the hours passed and testimony by other
cadets continued. The pattern of testimony which came after *Hendricks'*
was unsettling, not what the Board members expected. By the end of
the first day, Collins, Irvin, and Harrington were uncertain how to proceed in the investigation. They were confused, unsure the cheating ring
existed.

The second witness was the B-1 academic coach *Evan A. Corley*,
whose name had surfaced in the last few days of the undercover investigation, along with documented evidence of his passing copies of tests
to other ring members. *Corley* was one of twenty-five cadet tutors who
volunteered to assist members of the football team. His talents and
success in academics, and willingness to help the football team in "the
war for tenths," led him into activities he hadn't remotely considered
possible at the outset. He was also a "star man," wearing a gold star
sewn on each side of the black braided collar of his tunic, stars sym-
bolizing academic standing in the top six percent of his class and above
90 percent in general order of merit. If *Corley* was as forthcoming as
Hendricks, the Collins Board would be well on its way to breaking the
ring.

But they were surprised by *Evan Corley's* testimony.

Art Collins advised *Corley* of his rights and administered the oath.
Evan Corley swore to tell the truth "so help me God." After taking his
name, rank and organization, the Board president continued.

As he would with seventeen other cadets that day, Art Collins informed *Corley*,

"Allegations have been made that numerous violations of the honor code have been in progress for some time. The violations are said to be by a group of cadets operating to pass unauthorized information in academics. Evidence suggests you were involved. Are you aware of such an organization?"

"Sir, I know of no group of cadets which operates to pass unauthorized academic information."

Art Collins then told *Corley* he knew the third classman had worked problems for *George Hendricks*, and the problems appeared on writs the following day. He went on to tell *Corley* the Board had learned he passed unauthorized academic information to his roommate, Cadet *Grayson*. *Corley* replied,

"I am an academic coach and I have coached Cadet *Hendricks* but have never worked problems for him which later appeared on writs. I have never passed unauthorized academic information to *Grayson*."

If the yearling's answers were correct, they were puzzling, disappointing, frustrating. Was Corley lying under oath? Surely not. Surely a cadet who believed in the honor code would not compound cheating by lying under oath? Cadets did not lie or knowingly make false official statements. They certainly wouldn't lie under oath. Their explanations were accepted as truth, without question. They were preparing to be officers and gentlemen. Wasn't that the way it was? Evidence found during the undercover investigation clearly indicated the third classman before them was lying.

Art Collins bore down.

"Mister *Corley*, I remind you the answers you've given us were provided under oath. Your testimony is in direct conflict with information previously obtained by the Board. If you have knowingly made inaccurate statements, you could face the charge of false swearing. You can correct your testimony. Do you want to change any of your answers?"

Corley didn't waver. "No, Sir. I desire to make no change to my answers."

Art Collins excused *Evan Corley* and called the next witness, Cadet *Thomas S. Grayson* from company B-1. There was hard evidence *Grayson* had received unauthorized assistance from *Evan Corley*.

When called into the room, *Grayson* reported to the Board, and before being told of the investigation's purposes, surprised the three officers. "I'm not going to say anything until I see Colonel Blaik. I was in trouble like this last fall."

Art Collins answered, "The Board proceedings have nothing to do with Colonel Blaik. We're investigating alleged violations of the honor system within the Corps of Cadets." Art Collins then advised him of his rights, swore him, asked him to be seated, and described the allegations. As he did with *Hendricks* and *Corley*, Collins named the cadets who had given *Grayson* unauthorized academic assistance, as well as the men to whom he had passed information, described the circumstances of the allegations, and asked him if he was aware of cheating by other cadets Collins named. Some were football players.

"Do you have any knowledge of the alleged cheating?" Collins asked.

"I have never heard of any cheating. I'll tell you anything about myself but if it's football players you're asking about, I won't squeal. I want to talk to Coach Earl Blaik and I'll say anything he tells me but I won't say anything until I talk to him."

Again, based on the prior investigation, the board was confronted by a cadet whose testimony appeared less than truthful.

Art Collins bore down:

"Mister, this Board has nothing to do with football players. We are investigating alleged honor violations. While many of the names we asked about were football players, the only reason their names were brought up was we had proof positive they were involved as violators of the honor system."

Grayson was obviously agitated and nervous as he continued his statement:

"I don't know anything about this honor violation. There are people around here who don't like football players. Colonel

Marlin, my tactical officer is one of them." He went on to name a cadet who "doesn't like football players either."

The Board questioned him more directly, attempting to turn his attention to honor violations. In response to their questions he said he had heard of the ring, although he could give no specific information on it. He had never received any papers and had never been given any answers. He had indications of what to study for some recitations. His roommate, [*Evan Corley*] often has fourteen people in his room every night he is tutoring. *Grayson* said he had been told by an upperclass-man, name not remembered, "You can beat the plebe system by falling out. You can beat the academics the same way." He restated his intent to see Coach Blaik. He would do anything Coach Blaik told him, for he knew the Coach would never let him down.

Except for two other cadets, *Corley's* and *Grayson's* refusals to cooperate with the Collins Board became the pattern of testimony by every cadet who followed them that first day. *Bill DiSantis, George Hendricks'* roommate, denied any knowledge of cheating, or that he was involved.

Ted Strong, Mike Arrison's roommate in company H-2, was one who admitted participation in cheating. *Strong*, an academic coach, had been given problems to work for a football player in G-2, whose two roommates were also football players. The three roomed near *Strong* and *Arrison*.

Strong explained his involvement to the Board and named three cadets in G-2, one definitely implicated, the two others possibly impli-cated. While tutoring one of the men, another of the football players had given him problems to work. *Strong* discovered the next day the identical problems were on the examination he took, but he didn't stop the practice of working problems for his classmate. Though he knew he should have reported the man for honor violations, and therefore was guilty of violating the honor code, he continued to participate in the group's activities. He believed one of the man's roommates knew of the cheating but didn't take advantage of it, while the second room-mate probably knew about it.

Ted Strong's and *Mike Arrison's* third roommate, an assistant man-ager on the football team, had been implicated in the preliminary investigation. Dan Myers, the K-1 Honor Committee representative had asked if he would assist in exposing the ring. He refused, saying he "might have friends involved." He testified before the board next and admitted knowledge of cheating in progress, but firmly denied participating. As with many cadets who roomed with men trafficking in advance information in academics, he could scarcely avoid knowing what was occurring almost daily in his cadet room.

The seventh man to testify on May 29 was from K-1, *Gordon Seabold*, class of '53, a member of the intercollegiate boxing team. He readily admitted his participation in cheating, corroborated the participation of *DiSantis*, one of *George Hendricks'* two K-1 roommates, and gave direct testimony implicating two more cadets. One was *Hendrick's* second roommate, the other in company B-2.

Then the Board's progress abruptly ground to a halt. Twelve cadets in succession emphatically denied any knowledge of a group cheating in academics, or that they had participated. Their denials, coupled with denials by *Corley, DiSantis*, and *Grayson*, totaled fifteen. Of the twelve later in the day, seven were from the class of '53, two from '52, and three from '51. The twelve named no other cadets as having been involved in cheating. What's more, they refuted previously firm, damaging testimony against every single man the Board was convinced had been involved, in spite of a warning to each of the twelve he was testifying under oath.

While the Board was in progress that first morning, Tom Courant, a second classman and Cadet-in-Charge of academic tutoring for the football team, was in an upper floor room in company C-1, across central area from the entrance to the Cadet Guard Room. A member of company E-2, Tom was coaching a cow football player in mechanics. The cadet, whom Tom knew well, seemed agitated, frustrated, and finally slammed his book on the desk. "What's the use? I'm going to get thrown out of this place anyway." Surprised and puzzled by the cadet's outburst, Tom asked, "What do you mean? What are you talking about?"

The distraught second classman gazed out the window across central area, and suddenly pointing, said, "See that man going into the Guard Room? That's the captain of the football team. He's been called in to testify before an investigation board." The young junior then told an astounded Tom Courant of his involvement with others in cheating.

Thus the Cadet-in-Charge of tutoring the football team began learning he had been carefully excluded from activities which would initially cause Art Collins to believe Tom Courant was the ring leader.

When Art Collins, Jeff Irvin, and Tracy Harrington summed the day's results, they had made little progress, if any, in breaking the back of the cheating ring. Of nineteen cadets who testified, only three admitted participating in cheating. Two of the three were the men named in the letter order issued by Paul Harkins when he convened the Board. They were *Hendricks* and *Strong*. The third was but one of the seventeen additional names against which considerable information had been gathered during the undercover investigation. Another, *Mike Arrison's* and *Ted Strong's* roommate, had admitted knowing of the ring but apparently hadn't participated, and denied knowledge of any participants.

Not only was the Board stymied by the testimony it had received on May 29, guidelines for the proceedings had been carefully developed to ensure cadets would not be called to testify if there wasn't firm evidence suggesting the witness' involvement in the group's activities. The Board was concerned the proceedings could appear an inquisition if cadets were called to testify without direct, firm evidence implicating them. The fact that a cadet said another was "believed" or "thought" to be involved was not sufficient to call the implicated cadet to testify.

There were other troubling aspects of the first day's testimony before the Collins Board. If it was disturbing that upwards of a hundred cadets would organize to cheat in academics, it was equally disturbing that cadets, against whom considerable evidence had been accumulated, would compound their circumstances by falsely testifying under oath. When Art Collins, Jeff Irvin, and Tracy Harrington were cadets, the Corps was half its present number. There was a sense of closeness not unlike the "band of brothers" of pre-Civil War days. Organized cheating in the Corps of Cadets would have been outlandish, lying to avoid cheating's consequences, beyond comprehension.

There was also testimony describing contention between members of the Corps of Cadets and football players. Strangely, the men who were most willing to disclose their involvement in the group weren't football players and were pointing their fingers at the football team as both the source and leaders of the cheating operation. The Board members wondered aloud if there was such strong feeling, perhaps resentment, between certain members of the Corps and the nationally ranked football team, was there some strange scheme afoot to discredit the football team? After all, Army football players had for years received lavish attention in the national press, and because of their practice and travel demands, were frequently excused from training activities and duties assigned the rest of the Corps. If there was a move to discredit the football team, it was indeed a bizarre scheme, farfetched. Were the men lying about the football team, falsely accusing the players? Or were they telling the truth?

The first day's testimony caused uncertainty and confusion among the three officers. There was serious doubt a cheating ring existed. The unusual, contradictory testimony resulted in a decision to call no witnesses on May 30, which was Memorial Day. The Board needed a breather to discuss how to proceed. As events unfolded in the ensuing twenty-four hours, the discussion proved unnecessary.

As for the cadets who had joined activities contrary to the concepts of honor at West Point, and were now alerted to the Collins Board, far too much was at stake in their lives. Pat Ryan's announcement of the inquiry during noon meal on May 29 energized the ring's grapevine, and the grapevine was now being infused with information that witnesses were bringing from their appearances before the Board.

The talk of an investigation into honor violations had a disquieting effect on men caught up in cheating. Only a few had been called to testify before noon when Ryan spoke to the Corps. Orders to appear before the Board were coming one at a time, except in companies where names of the first men to testify had been called at reveille formations. No reason had been given for instructions to report to company tactical officers right after breakfast. Now, after the noon meal, as word spread, the questions began. Who had been called for appearances before the Board? Who had reported them? For the first time, the men involved in organized cheating began feeling the heat of the investigation. There were meetings held to determine what to do in the face of the spreading inquiry. In the rest of the Corps, among men not in the ring, the cadet rumor mill began grinding furiously.

AN UNFORGETTABLE DUTY

Though May 29 didn't bring the hoped for break, Art Collins, Jeff Irvin, and Tracy Harrington would find a way to proceed. They were experienced officers who had been in some tough fights in World War II. Each knew and understood the concepts of honor from their days as cadets, and the three men, particularly Art Collins, were sensitive to the improper use of honor to get answers or enforce regulations.

Lieutenant Colonel Arthur S. Collins, Jr. was a graduate of the West Point class of 1938, as were Lieutenant Colonels Jefferson J. Irvin and Tracy B. Harrington. All were combat veterans of World War II, the "good war" of the twentieth century. Jeff Irvin and Tracy Harrington had fought in Europe against Nazi Germany's Wermacht – the German Army, and Art Collins in the Pacific against Japan's Imperial Army in the brutal island hopping campaigns led by General MacArthur. Jeff Irvin was thirty-seven years old, Tracy Harrington thirty-six.

Art Collins, two months shy of thirty-six, was confident he knew the meaning of honor at West Point, its unquestioned importance to men of character, particularly its inescapable necessity for officers leading men in war. He was a respected, highly decorated infantryman, noted for being a tough, yet compassionate regimental commander. Consistent with one of the principles of leadership taught at West Point, he looked out for the welfare of his men.

Collins was born and grew up in the Mission Hill section of Boston, Massachusetts. As a young boy he delivered papers, sold magazines and worked in the public library. He loved baseball and hockey. He was selected to attend the Boston Latin School, one of the oldest and finest high schools in the state, a school noted for preparing its students for the finest colleges in the nation.

Throughout his four years as a cadet he was never an academic star. He took pride in playing on the "goat" team in the annual "goat-engineer" football game, the traditional, full-equipment game between second classmen with high grade point averages – "hives," as they were called – and those with low grade point averages – "goats."

He participated in three intercollegiate sports at various times during his first three years at West Point: hockey, baseball and track. His

The Collins Board

Lieutenant Colonel
Arthur S. Collins, Jr.,
(Board President)
Cadet, Class of 1938

Lieutenant Colonel
Jefferson J. Irvin
Cadet, Class of 1938

Department of Tactics

Lieutenant Colonel
Tracy B. Harrington
Cadet, Class of 1938
(Photos courtesy USMA
Archives.)

classmates saw in him an Irish, fun-loving sense of humor. His nickname as a cadet was "Shono." In 1951 his close friends knew him as "Ace," as did many cadets, who would not dare call him that to his face.

In the Pacific theater he received the Silver Star for gallantry, and distinguished himself commanding the 130th Regiment of the 3d Infantry Division, in the steamy, malaria infested jungles of New Guinea and the Philippines. He had come out of the war one of the youngest colonels in the Army, and like thousands of other officers in the Armed Forces had reverted to a lower rank during the massive demobilization following the war.

In early May, during cadet drill and practice parade periods on the Plain, Art Collins had noticed something uncharacteristic of his boss, Paul Harkins. It was at a parade one afternoon that Collins noticed Colonel Harkins carrying a brown manila envelope. Art had been in the Tactical Department for three years and had never seen Paul Harkins carry a paper home. At that time officers seldom took their work home with them. But here was the Commandant with the brown manila envelope in his hand. How unusual, Art remembered thinking. Parades were in late afternoon, twice a week in each regiment, and for three weeks, every time Paul Harkins was at a parade, he had that envelope, and after he watched the parade, Art Collins saw Harkins walk from the Plain toward his quarters, carrying the envelope.

The middle of May, Harkins called Art Collins into his office. Harkins had a manila envelope in his hand. He said, "Art, I have something here that's going to shock you. You won't believe it, but I've had some knowledge for about a month now; it's in this envelope, and I have never let it out of my hands." Harkins went on to say the contents were such a shock to him and so unbelievable that he wouldn't even leave it in his desk and had kept it in his personal possession.

He told Collins,

"This pertains to a violation of the honor system, but it might be a massive violation. I'm going to appoint a board, and you're going to be the President of the board. I have orders [for my next assignment] and am scheduled to leave here in about three weeks. I want this cleared up before I go. This has happened during the time I've been here as Commandant, and it really hurts to think of that, but you investigate it and let the chips fall where they may."

Harkins' words remained with Art Collins the rest of his life. He would say 30 years later,

"...as you go through your service, you run into people... you respect and... admire... People... talk... about individuals who... cover things up...; I haven't run into that, although later events... [showed]... that it happens. This was one of those events, however, which meant... much to Colonel Harkins at that time and could have hurt... [him]... There was so little to go on. He could have... easily said, 'Let's see if we can get something more concrete,' and... passed the envelope... to the new Commandant, but he didn't. His ... 'let the chips fall where they may, get to the bottom of it, and... get it cleaned up... before [I leave,' were] inspirational."

Art Collins didn't relish the task given him at the end of May 1951, to conduct this investigation. But he would long remember the role he performed, and he would, for the rest of his life, consider the assignment a compliment to his performance as an officer.

A Reasoned Policy of No Disclosure

While the cadets' grapevine was hard at work before the end of the first day of testimony, the three officers on the Board could not and would not discuss any aspect of the investigation, except among themselves, their advisors, and the convening authority – Colonel Paul Harkins. There was to be no discussion outside the Board until all the facts were in, analyzed, conclusions reached, recommendations formulated, and decisions made. The proceedings were "restricted" information, and any release of information would be solely through the Public Information Office, at a prearranged date and time.

Honor, and accusations of its violation, required quiet, dignified deliberations. To discuss the facts of an investigation before it was complete could embarrass, if not disgrace, an innocent man. In an effort to protect men alleged to have committed honor violations, the procedure which had evolved over the years was not to discuss the facts of cases until cadets found guilty had resigned and departed the Academy. Then, and only then, were members of the Corps made aware of the case.

And while a man's name may be divulged in discussions of his case in meetings in each company area, the discussions began and ended with a request that after the meeting was over there be no further discussion of the matter, or the man's name in connection with the case. Cadets understood and respected such requests.

Under any circumstance, the Academy would not publicly release the names of cadets who resigned or were dismissed from the Academy. They deserved the opportunity to redirect their lives without fanfare or embarrassment. As with other cases of cadets' separation, discharge, or dismissal from the Academy each year, not only were names never announced, there were no public announcements as to type of discharges or numbers of cadets leaving the Academy, whether the separations were voluntary or involuntary.

An Emotional Evening: The Story of Two Meetings

George Hendricks, the lead-off witness before the Collins Board, told the three officers everything he knew or had heard about the cheating ring. They "...had the goods on him," he remembered. He couldn't do otherwise. But immediately following *Bill DiSantis'* appearance,

Hendricks realized he was in trouble of another sort – with the men in the cheating ring. *George* hadn't told *DiSantis* or anyone else in the ring he had invited *Brian Nolan* to join. When *DiSantis* and *Gordon Seabold* saw *Brian Nolan*, their K-1 classmate, waiting to testify, they were both surprised and angered – but they held back saying anything to *Hendricks* until they saw him in private after testifying.

"What the hell was *Nolan* doing in there?" *Hendricks'* roommates demanded. *Hendricks* explained why he had invited *Nolan* to join. *Brian* could be trusted, he told them. *Nolan* was his friend, and was in academic trouble. *Hendricks* had dated *Nolan's* sister while visiting at *Brian's* home. Besides, *Nolan's* brother had been reported missing in action in Korea. *Hendricks* sympathized with *Brian Nolan's* circumstance because one of *Brian's* roommates also had a brother missing in action, a brother who, they would later learn, never returned from the war.

George Hendricks wasn't aware the missing status of *Brian's* brother had quickly changed when he was found alive and well. *George* wanted to help *Brian* get out of academic trouble, and believed the Academy asked entirely too much of a man to be an intercollegiate athlete and carry the academic load and all the other demands placed on cadets. Besides, "the system wasn't fair."

Ring members were careful about whom they invited to join, until the last several months, when their numbers began multiplying. It was an exclusive group and required careful observation of each man before offering him assistance, or asking him to help. Members knew who other participants were within their own companies and classes, and believed they knew who, outside the group, supported the honor code, and thus shouldn't be invited to join. And most ring members were amazingly consistent in their knowledge of men outside of their companies who were participating in the ring's operation. Admission to the group, though not approved by vote, normally required discussion before an invitation was extended. At least that's the way it had been in the past. By the spring of 1951, the small, closely knit group of years past had lost control of who would be included, and the number of men involved had increased rapidly, especially within the class of 1953.

Early the evening of May 29 a group of approximately fourteen third classmen, including *George Hendricks* and *Evan Corley*, met in company A-1 barracks. Most had testified before the Collins Board that day. Each told what they had said to the Board. When *Hendricks* told them he had divulged everything he knew, they were furious. In spite of his protesting that the Board had extensive evidence against him, which made it impossible for him to withhold the truth, the meeting with his classmates, nearly all football players, became increasingly heated, intimidating, and threatening for both *Hendricks* and *Evan Corley*. *Corley* had denied knowing or participating in anything, while Hendricks laid before the Board all he knew.

When the meeting concluded, both men had been told in no uncertain terms they were to "take the fall" for the entire group. *Hendricks* was to go back to the Board and tell the three officers he fabricated the story of a cheating ring, that he and *Corley* were the only ones involved. *Corley* was to admit his involvement and corroborate *Hendricks'* story, nothing more.

For *Hendricks* the situation was impossible. He would have to recant the truth and give false testimony. He was convinced his alternative was physical harm by men who were quite capable of retaliating if he didn't do their bidding. His only way out was to tell the truth. He phoned Art Collins and explained what had occurred, including the fact both he and *Corley* had been given explicit instructions to take the fall for the group. He told the Board president he feared for his safety.

Art Collins decided he must respond to *George Hendricks'* call and made arrangements for *Hendricks'* protection that evening. He posted a guard outside *George's* cadet room. The guard took the name and organization of any man who came to the room, discouraging unwanted or unexpected visitors.

George Hendricks didn't ask to reappear before the Collins Board, and didn't change his testimony. Art Collins told him he needn't do any more. When *Evan Corley* asked to reappear before the Board when next it met, his testimony confirmed what *Hendricks* told Art Collins the evening of May 29. *Corley* was prepared to take the fall, willing to testify falsely again, willing to admit his participation, and deny that anyone other than *George Hendricks* and him were involved in cheating.

The meeting in Company A-1 early the evening of May 29 was the first of two emotionally charged gatherings for men in the class of 1953 now feeling the heat of the Collins Board. The second was in

Coach Blaik's film projection room in the South Gymnasium. Soon after their meeting with *George Hendricks* and *Evan Corley*, one of their classmates on the football team phoned the Army coach at home and asked him to meet with a group of players at nine forty-five. The telephone call to "the Colonel" was the second hint to Earl Blaik that day something was seriously wrong on his football team.

EARL BLAIK'S NIGHTMARE BEGINS

The first hint came the afternoon of May 29. Blaik was walking across the Plain, the cadet parade ground, toward a baseball game. Bob, his son, was on the baseball team, one of twelve football players who participated in other intercollegiate sports. As the Army coach crossed the Plain he encountered another of his football players who told him an investigation was underway concerning a breach of the honor code. The player said he thought it wasn't serious. Blaik gave little thought to the cadet's remark. He assumed the investigation was the result of an individual breach of honor.

Then came the phone call early the same evening, requesting the nine forty-five meeting. After the twelve men arrived in the projection room, Earl Blaik listened to their stories for nearly an hour, and re-membered the conversation:

> Each was anxious to unburden himself. Most of them had yet to go before the board. Of those who had, some had impli-cated themselves but had not implicated others. A few had not told the truth.
>
> I had no doubt about the seriousness of their acts, yet I knew them to be men of basic integrity. I needed no reflection to advise them.
>
> You know how we do business in the squad and at the Mili-tary Academy. Each of you should state the facts to the board without equivocation.

After the players left, Blaik sat in the projection room, where his assistant coaches and players had often gathered for happier purposes. He was filled with foreboding.

Near the gymnasium, where Blaik's office was located, were the superintendent's quarters. Within a few minutes Blaik was outside throwing pebbles at the superintendent's darkened bedroom window. Blaik had known General Irving for years, and their families were close friends. The superintendent came downstairs in his bathrobe and let Blaik in. They sat in the living room, and Blaik recounted the meeting with the twelve third classmen. From what he had learned, the circumstances presented a grave situation. They talked awhile longer, and at the end, Blaik recalled some years later,

> I said, 'If I ever gave good advice, this is it: this affair is so serious that you should remove the investigation from the board of young tactical officers and place it in the hands of the Academic Board. This may be a catastrophe, and it demands the most mature judgment. I beg you to do so.'

The next afternoon, his son Bob, and one of his roommates, the captain-elect of the football team, came to Blaik's office. The Collins Board was not taking testimony that day. Bob and his roommate told Blaik the class of 1952 was involved in the cheating, and they were as well. Blaik was shocked, hurt, angered, not wanting to believe what he had been told. He later described his anguished reaction to what his son had said that May afternoon. "My God, how could you? How could you?"

After the two left his office, he immediately went to the superintendent. He told him of his son's statement and offered Bob's resignation as proof of his sincerity. Blaik was more convinced than ever that without the judgment of the Academic Board, a board of senior professors and department heads, and historically the most powerful Board at the Academy, the investigation could have frightful results. He knew the men on the Collins Board were members of the class of 1938. To him they were too young to render the mature judgment needed. They would be responsible for committing the Academy to actions he believed could destroy it.

Further, they were members of the Tactical Department, and Earl Blaik had detected what he firmly believed was a growing rift between the Tactical Department and his Department of Athletics, particularly the Football Office and his team. In his view, Colonel Harkins, the outgoing Commandant, and the Assistant Commandant, Colonel Waters, were contributors to an increasingly strained interdepartmental relationship. At the time, however, he mentioned nothing of the rift to Irving.

General Irving, whose family had been friends of the Blaiks for years, had known of the undercover investigation from the outset, all that led up to it, had sanctioned the formal investigation, and couldn't acquiesce to Earl Blaik's request. He could neither tell Blaik the particulars of what had been found in the preliminary investigation, nor could he simply direct that the formal investigation be turned over to "more mature" men, as Blaik wished him to do. The Army football coach and his superintendent were now on a collision course. Their friendship would suffer, and sadly, never be fully repaired.

One day after disclosing to his father his involvement in cheating, Bob Blaik described to the Collins Board his father's feelings about the turn of events. "This has been a terrific blow to my father," he said. "It has broken his heart."

Then came Friday June 1, when the first class returned from its Aberdeen trip. Three first classmen on the football team came to their Coach's office seeking advice. They knew what he had told the men of the second and third classes. But to them he said, "You are to graduate next week. Some of you are to be married. Your parents are here. Either look to them for advice or make your own judgment as I shall not advise you."

With the return of the first class, and the visit to Coach Blaik by three of their number, he had what he believed to be a complete list of the ring's participants. It included members of all four classes. He wanted to testify before the Collins Board. He hoped to dissuade them from what he believed to be precipitous action.

What he might have said, no one knows. The Board didn't call him, heightening his concerns about the superintendent's decision to conduct a formal investigation.

But the Board's neglect of information he had was not because he was the football coach, or his team's and son's involvement in cheating. Neither was there a move afoot to embarrass Coach Blaik, the Department of Athletics, or its intercollegiate teams. The list of names and information he had was of no value. It could not be used as evidence, nor could it be used as a basis to call witnesses. The men who told Blaik they were involved were giving him names of other cadets they believed were involved but were not providing him specifics as to how they knew they were involved.

Unfortunately, there was much Earl Blaik didn't know about the events that had been set in motion long before the shock he received the night of May 29. He did firmly believe the cheating was in every class, and he was equally certain he knew the truth of the matter. Yet, at the time, he could not know the depth and seriousness of the testimony the Board was hearing, the procedures the Board was governed by, nor would he afterward. Once the Board completed its work, copies of the Board proceedings were closely held, and Board members could neither discuss the proceeding with him nor disclose to him what they were learning from the testimony. He was hearing what the young men involved in cheating told him in the privacy of his office, and even if they were completely forthcoming in what they disclosed to him, none of them had at their disposal all of the pieces of the dramatic testimony being given the Board.

Thus began Earl Blaik's participation in the agonizing events of the 1951 honor scandal. He was entering the most difficult game of his life, but it was no game. Behind the closed doors and veil of confidentiality shielding events that spring and summer, Earl Blaik increasingly involved himself in a futile struggle to retain the young men as cadets at West Point, arguing they should not be made to suffer the extreme measures the honor code demanded – dismissal. Instead, given their character, the quality of men he believed them to be, he concluded they should be severely disciplined, learn from their mistakes, and participate, indeed help lead the way, in important changes in the honor system that Blaik now saw as largely responsible for the men's actions. These two days were the beginning of Blaik's long nightmare, his "catastrophe," which he likened to the devastating March 1913 flood in Dayton, Ohio, when he was a boy.

There was more than a bit of irony in Earl Blaik's May 29, 1951 awakening to the stain of dishonor which had found fertile soil in his nationally admired football team. The advice he gave the men who came to see him that night helped open the floodgates for the Collins Board and break the back of the cheating ring which had spread far beyond the football team. "Each of you should state the facts to the Board without equivocation," he told them. Two days hence, one by one, the men who had given false testimony took Earl Blaik's admonition. They returned to tell the truth to the Collins Board as they understood the truth, as did most of the men who had played on his team, and who later resigned.

There were, however, some notable exceptions in the classes of 1952 and '53, men who doggedly denied participating or having any knowledge of cheating, and who, in spite of hard evidence to the contrary, refused to tell of their involvement until the final days of their time at West Point. In the class of 1951, none ever incriminated themselves, nor did they provide evidence implicating any of their classmates, or members of the underclasses.

As for Earl Blaik, the course he set for himself, and in behalf of his players, almost from the outset stirred such controversy that his role in breaking the cheating ring was totally obscured and never publicly acknowledged – not even by the Army coach.

Earl Blaik, General Irving, *Brian Nolan*, and *Michael Arrison* were in exclusive company with one another beginning in May 1951, although none knew all the others' trials – and never would. All were scarred by the honor incident of 1951, but in ways far different from the experiences of the men who resigned from West Point that summer. Each of the four was to learn that tragic surprises which besmirch the reputations of respected institutions, icons, heroes, or people presumed upright and innocent, have ugly ways of taking their own revenge on those attempting to find and correct what has gone wrong. None deserved what they endured.

And while Earl Blaik unknowingly contributed to ending the string of denials and false testimony that had been given the Collins Board on May 29, that same evening and Wednesday, May 30, others among the young men enmeshed in the investigative net began seeking advice elsewhere. They went to chaplains, friends whom they trusted as confidants, and tactical officers to reveal what had occurred and ask what they should do when confronting the three officers asking hard questions. *George Hendricks* was among those who sought counsel from the Catholic Chaplain after his appearance before the Board. He could get no relief from his growing predicament. The Chaplain didn't want to discuss the matter.

The ring's unraveling would begin on Thursday morning.

WEDNESDAY: PRESSURE BUILDS AND BACKFIRE BEGINS

The convening of the Collins Board and Pat Ryan's announcement of an investigation into honor violations stirred a flurry of questions. Rumors and speculation, ever a part of conversation, and often pain in any setting, can be destructive in a military organization. Both thrive on incomplete information, particularly when "big things" are in the offing. And now some word, however skimpy, was out, and the investigation board, or board of officers as it was called by cadets, became the impetus to multiply rumors and create endless speculation, especially among cadets who knew nothing of the cheating.

For the men who were definitely involved in cheating, knew with certainty about the cheating operation, or who had heard and were convinced cheating was in progress, the investigation board and Pat Ryan's announcement were profound shocks. After the first day's proceedings, among the men who knew they were involved, the word brought back from those who had testified triggered all manner of emotions – and questions. *George Hendricks'* and *Evan Corley's* experiences in the A-1 company area, followed by *Hendricks'* call to Collins, were just the beginning. *Hendricks* had already been angrily confronted by *DiSantis, Seabold*, and his other roommate in K-1 prior to the meeting in A-1. Four times in the first days of the Collins Board, Art Collins received phone calls from cadets asking for assistance because they "had been threatened." *Brian Nolan* was particularly agitated when he called Collins. He had been threatened, and knew he'd carried heavy responsibilities in the undercover investigation. He'd been confronted with threatening words from *DiSantis, Hendricks'* roommate.

Elsewhere men who knew or suspected they were in trouble met with other participants in the ring. What was the Board told by the men who testified? Were names given to the Board? If so, by whom? What other information did the Board know? How should they answer questions if called before the Board? Did the Board know how many were involved? Who had "ratted" or "squealed" on the ring? Wasn't everyone checked out before being invited in? If so, why would anyone turn them in? Didn't they realize how big this thing was?

Art Collins, always on the move in the cadet area, feeling the pulse of the Corps, was approached Wednesday, May 30, with numerous questions, many he couldn't answer for reasons of confidentiality. He also encountered some who had testified the previous day. One told him he wanted to come back and testify. He wanted to set the record straight. The Board's badly needed break was gathering momentum.

Late that evening, the class of 1951 formed up on Jefferson Road facing north, in four provisional companies; two at eight thirty and the second two at eight forty-five. They marched down to the West Shore Railroad Station on the banks of the Hudson River and boarded a train for Aberdeen Proving Ground in Aberdeen, Maryland. They returned to West Point by train shortly before six in the morning on Friday, June 1. Among those on the train were three first classmen who had testified before the Collins Board on Tuesday. Two were cadet company commanders from Companies B and C, 1st Regiment. The third was a cadet captain on the 1st Regiment staff, and the captain of the 1950 football team, also from C-1.

One cadet from '51 had been taken from the Aberdeen trip roster to testify before the Board. He was the Honor Committee secretary whose name repeatedly surfaced during testimony by second and third classmen. The Committee secretary's responsibilities included appointing hearing panel members as well as cadets to investigate alleged honor violations. He denied any knowledge of cheating, but during his testimony suggested that if organized cheating was in progress on the football team the people involved may have "thrown the 1950 Navy game." His remarks later came back to cause a furor for the officers on the Collins Board.

By the time the first class returned from Aberdeen early Friday morning, the Collins Board was well on its way to destroying the cheating ring at West Point, at least in the classes of 1952 and 1953. For the classes of 1951 and 1954, the outcome would be far different.

DAY THREE OF THE COLLINS BOARD: THE FIRST COLLAPSE

Testimony resumed Thursday morning while the class of '51 was in Aberdeen. The lead-off witness was a manager on the football team, a second classman from company M-1. He denied any knowledge of cheating or a cheating ring, saying he had only heard of it on Tuesday when he learned Cadet *Grayson*, a football player in B-1, had been called to appear before the Board. A few days later the same second classman voluntarily reappeared before the Board, recanted his denial, and admitted he was a participant in the ring. He was later found guilty and resigned in August.

The next witness was a fourth classman, a football player who denied knowledge of any cheating or participation in the group. He was found guilty by the Board, because of circumstantial evidence, but the case was later reviewed by the Army's staff judge advocate, the evidence found insufficient, and he was returned to duty. He resigned from the Academy more than a year and a half later, on January 20, 1953.

The third witness was *Evan Corley*, the B-1 academic tutor for the football team, who, along with *George Hendricks* from K-1, had been given stern marching orders to "take the fall" in the Tuesday evening meeting in company A-1. On Wednesday, consistent with his "instructions" he came to Art Collins and asked to reappear before the three officers.

When *Corley* reported to the Board on Thursday he was reminded he was still under oath, and began testimony as he had been instructed by ring members. He admitted his involvement in cheating, and implicated no one else other than *George Hendricks*. Art Collins knew *Corley* was again withholding the complete truth. He reminded the third classman the Board had evidence there was much more to tell – that he had changed his original statement to tell the truth, and might as well tell the whole truth. *Evan Corley* relented and began describing how he became involved in the cheating ring.

Like *Ted Strong* in H-2, who had told of similar circumstances, *Corley* recounted how a football player had brought him problems to work, and *he* discovered the next day the same problems were on his graded recitation. He continued working problems given him though they kept appearing in the next day's classroom recitations. He became suspicious, but "felt it was just a way to give football players a little extra help." Then one day he overheard two men talking about obtaining the problems from another cadet. He then knew they were cheating but decided he was in too deep to get out.

Evan Corley named ten other cadets as direct recipients of his assistance, or who had given him unauthorized information. He also told the Board he ordinarily passed information to men in his own company. After his testimony he marked a drill roll indicating the names of those he knew or believed were involved.

Nine more cadets who had denied involvement on Tuesday voluntarily reappeared before the Collins Board on Thursday, May 31, and recanted their testimony. Nearly all were football players. Each admitted participating in the ring's activities, implicated other cadets, and marked drill rolls as *Corley*, *Hendricks*, and *Seabold* had done. The nine were followed by three more second classmen, football players, each appearing before the Board for the first time. All three admitted to cheating and directly implicated other cadets, either through testimony or by marking drill rolls.

At the end of the second day of testimony the Collins Board had received thirteen additional, corroborated admissions of guilt, bringing the total to sixteen. Of the sixteen, seven were in '52 and nine were in '53. Thirteen were in the 1st Regiment, three in the 2d Regiment. Twenty-eight more cadets had been directly or indirectly implicated: eight in '51, nine in '52, ten in '53, and one in '54. In the total of forty-four cadets were twenty-five from 1st Regiment, and nineteen from 2d Regiment. There was at least one man each from eight of the twelve companies in 1st Regiment, seven of twelve companies in 2d Regiment.

There was no elation on the Collins Board at the end of the second day of testimony. Art Collins remembered years later the sadness he felt on seeing before the Board the son of the first company commander he served under when he graduated from West Point in 1938. The cadet admitted his deep involvement in cheating. His father was an Academy graduate.

The full force of what the three officers had heard began to sink in. First had come denials, evasions, and false testimony. Then on Thursday one of the men who returned to change his testimony admitted there had been a conspiracy to keep silent or deny the ring existed, more evidence of collusion. Further, there was the acknowledgment that the group carefully selected most of the men who joined in cheating.

There had been misappropriation, copying, and trafficking in term examinations, other tests, and daily graded recitation problems, as well as verbal transfer of information intended to gain better academic grades and the resulting improved overall records of cadet performance. Unauthorized information had been routinely exchanged between companies, battalions, and regiments. Academic coaches, "star men," and other cadets who weren't football players were involved, some working problems to copy and distribute to others. Exchanges took place in cadet rooms, hallways of academic buildings, in tutoring sessions at the library, in athletic locker rooms, during meals at athletic training tables, and other locations.

Though evidence suggested the football team was where the organized cheating began, and had remained confined mostly to the team, the ring had apparently grown at an accelerated rate in the last six months, spreading rapidly in the class of 1953, well beyond the football team. This was disturbing.

Perhaps most disturbing was the disclosure of deliberate attempts to elect members of the group as company honor representatives who, upon becoming first classmen little more than a year hence, would be on the Corps' Honor Committee, responsible for leadership in the operation and training for the honor system. Even worse was repeated testimony that a small number of members of the 1951 Honor Committee were believed to be involved. None of the officers on the Collins Board had expected the startling evidence or cadet behaviors they encountered those first two days.

There was no doubt the Academy's football team was deeply involved, and testimony clearly suggested organized, systematic cheating had begun on the team. Art Collins, Jeff Irvin, and Tracy Harrington had heard admissions from the team's captain-elect, two players who had been named as All-Americans, the number one and two quarterbacks, men acknowledged and admired as team leaders, and soon to be leaders in the Corps of Cadets. And the three officers were stunned to hear of three graduates, football heroes, two of whom had become heroes in Korea, one killed in action, alleged to have been participants in the group in years past. This, if true, meant the practice had indeed been passed from class to class by men who were leaders, expected, and assumed to be examples to the underclassmen in the Corps and on the football team.

The second day of testimony had seen the collapse of the cheating ring – apparently.

A NEW STRATEGY: BOUNDED BY BOARD PROCEDURES AND RULES OF EVIDENCE

Evidence unearthed on May 31, as disquieting as it was, enabled the Collins Board to change its strategy. The three officers first consolidated the lists of all cadets identified in testimony or on drill rolls, and for each man who testified, listed with him the names of others he had implicated. They cross referenced individual testimony and noted remarkable consistency in individual knowledge of who was involved, indicating witnesses knew who most of the other participants were. The Board was surprised by the high degree of correlation on lists separately prepared by cadets who had testified. The three officers were able to see clearly how many times each cadet had been named by others as direct participants, believed to be involved, or simply having knowledge of the ring. And they were able to begin charting the growth and spread of the ring, as well as how the men became involved. The Board was ready to begin systematically calling witnesses by class and cadet company.

There were other reasons for the revised strategy: the curriculum's structure, and scheduling of academic classes.

"Cadets recite every day in every class," was a long standing academic policy governing the West Point education. This meant cadets were to be graded every day in every class. West Point's instructors seldom varied from that policy. In addition to daily graded recitations there were usually monthly "writs," cadet slang for a written test. Like most colleges and universities, there were mid-term and final examinations, called WGRs, written general reviews. To ensure cadets were graded equitably, everyone taking the same class received the same recitation questions and tests.

During the academic year, the Corps of Cadets was organized into a brigade composed of two regiments. The two regiments contained three battalions of four companies each. Companies were approximately one hundred men each, a proportional number of members from the four undergraduate classes within each company.

The two regiments were divided into two groups each by the Academic Department, A and B in one regiment, C and D in the other. One group from each of the two regiments alternated weekly in first attendance of academic classes. This meant, for example, one week A and C groups would recite daily, or take writs, the day before B and D groups. The next week the reverse schedule was in effect: B and D would recite one day, A and C the next.

The cadets engaged in cheating had been passing information, usually questions and answers, to classmates scheduled from the opposite groups, within or across regiments, to recite the following day, or later the same day.

The manner in which the group passed information was in part because the Academy curriculum was virtually the same for each graduating class, and was organized in a building block fashion. For each class, the only electives were a choice of one among five foreign languages. Each cadet thus attended classroom instruction only with his classmates all four years. For a few cadets who had prior college work, and could validate courses by taking pre-course examinations, there were opportunities to take other, more advanced work – but only with their classmates. There was no course of instruction in which members of different undergraduate classes sat in the same classroom together, although it wasn't uncommon for men in the upperclasses to tutor cadets in the underclasses.

The Collins Board was operating under carefully prescribed procedures and rules of evidence. As testimony and evidence accumulated the first two days it became apparent that cheating was stratified within each of the four classes, and hard, first hand evidence of cheating could only be developed and obtained among those stratified groups. That is, a cadet could provide hard, admissible evidence only about his classmates, because there would be no passing of unauthorized information between undergraduate classes, except, perhaps through academic coaches, upperclass to lowerclass.

A witness could testify and give firm evidence he received a test copy from another classmate or gave information to another classmate, but could only testify to his "belief" or that he "had heard about" members of other classes trafficking in unauthorized information. He might have "observed papers being passed" or "heard" upper or lower classmen speak words which sounded incriminating, but under rules of evidence such testimony would not be sufficient to convict should the case go to a court martial. It would have been considered hearsay in a trial by court martial, and absent other firm, corroborating evidence, of little value.

Nor would the Collins Board call a cadet to testify simply because one man said he believed another was, or might be involved in cheating. There had to be independent testimony by two or more cadets saying they had observed behavior indicating another man was involved before the Board would consider calling him.

Additionally, procedures for recalling a witness were carefully considered. Most reappearances before the Collins Board were voluntary, as a result of an implicated cadet's decision to recant false testimony, or add to testimony previously given. Cadets were not recalled involuntarily unless later testimony injected hard evidence justifying involuntary recall.

WALL OF SILENCE: THE INVESTIGATION THWARTED AGAIN

Because graduation was so near, the Board decided to call men from the class of 1951 first, on Friday, upon their return from Aberdeen Proving Ground. Three of their number had been called before the Board the first day of testimony. If there were to be consequences for first classmen implicated in cheating, there was little time to act. Graduates in '51 were to be commissioned, sworn as officers in the Army or Air Force after they returned on Friday, prior to their graduation the following Tuesday. The change in status from cadet to officer prior to graduation would not prohibit punitive action should any be found guilty of cheating. Unfortunately, Thursday's success in breaking open the ring in the classes of 1952 and '53 did not carry over to Friday and the class of 1951.

One by one, twenty-five members of the class of 1951 denied participation in cheating, bringing the total to twenty-eight, including the three on Tuesday, May 29. Two of the twenty-five acknowledged they knew cheating was in progress at some period during their four years as cadets, but stated they didn't want to incriminate themselves, and provided no further information about their activities. Nor did they provide information about men they may have observed cheating. One volunteered seven names of classmates he believed knew unauthorized information was being passed, but also refused to incriminate himself.

All who were called to testify had been identified by underclassmen as "believed" or "heard" to be participants, or through their words, the language known to the ring, or their actions, had evidenced probable participation. But not one single first classman stated he had either given or received unauthorized information to others – or had encouraged, or otherwise suggested to underclassmen they should engage in such activities.

The Collins Board was again stymied. There was considerable testimony implicating men in the class of 1951, and one in '54, but the three officers had no admissible evidence with which to decide guilt and recommend discharge for any named in the testimony, a circumstance that later became a source of frustration and anger for men who resigned from the classes of '52 and '53. The absence of firm evidence with which to proceed against men in the class of '51 also stirred frustration among senior officers in the Army, and Academy graduates – and, though for different reasons, bitterness for Earl Blaik.

Art Collins' investigation board continued its work, taking testimony next from members of '52 and '53, and submitted an interim report, dated June 4, to Harkins and the superintendent. General Irving faced a difficult question. Did he have sufficient evidence to hold up graduation for any men in the class of 1951? There was no satisfactory answer to the question. Rules of evidence and carefully considered Board procedures ensured the proceedings didn't become an inquisition, and made any further investigation of the class of 1951 both impractical and, if pursued, probably completely unsuccessful.

The interim Board report named sixteen men in the class of 1951 whom Collins, Irvin, and Harrington concluded had cheated, but added the three officers did "not believe that cheating can be proved." In the meantime, to men in '52 and '53 who testified before the Board, it became apparent the first classmen they were certain had been involved in cheating, had knowledge of the ring's activities, or seemed to have condoned or encouraged the practice, would graduate on schedule. The final, dramatic scenes at West Point for the class of '51's role in the honor investigation were played out the morning of their graduation parade.

Three men in the class of '52, angered by the turn of events, came to Art Collins the evening before graduation parade and told him they were certain of at least two first classmen's participation in the ring. They were willing to confront one of the first classmen before the Board, in an attempt to obtain firm evidence from him, and perhaps an admission of guilt. Art Collins ordered the first classman to appear before the Board a second time. He didn't attend his graduation parade, an event all cadets look forward to for four, long, demanding years.

In the meantime the three "cows" told the first classman they intended to confront him before the Board. He pleaded with them not to go through with their plan. He had family, friends, and a fiancée at West Point for his graduation, and he was getting married after graduation. He told them if they wouldn't go through with their plan, he would write a letter to the Academy after graduation and marriage, and admit his involvement in cheating. The three men relented.

When the four appeared to testify before the Board, the underclassmen backed away from their statement to Art Collins, saying they "felt certain he [the first classman] was aware of [the] practice [of cheating] but they could not testify because they had no conclusive proof." Their refusal to testify against him scuttled the last hopes of obtaining firm evidence against anyone in the class of 1951, and wrecked the three underclassmen's credibility with the Board. The new second lieutenant never wrote the promised letter.

Later that morning, after General Irving received the last minute interim report, he asked Art Collins if there was reason to hold up the graduation of men in the class of 1951. Art Collins replied, "There isn't sufficient evidence." Earl Blaik, who argued vigorously against the Collins Board's continuation, was now certain a large number of first classmen had been involved. He feared holding up graduation would wreck the Academy, and also gave counsel to Irving – only he was far more blunt than Collins. "Better to let sixty or a hundred crooks graduate, than to hold up graduation to try to find them out."

This last scene at West Point for the class of 1951 would prove not to be the end of the drama, however. Several months after the class graduated, one man who testified on Tuesday, May 29, told Pat Ryan, then a lieutenant in Air Force pilot training, he had been involved in the cheating ring, but could not bring himself to confess his involvement to the Collins Board. Ironically, he wasn't one of the sixteen men in '51 that Art Collins, Jeff Irvin, and Tracy Harrington concluded had cheated.

The wall of silence had held firm. No investigation of a deeply rooted, long running conspiracy is perfect. There were men who cheated, or who knew of the ring, avoided the investigation net, and graduated with their classes. Each knew who he was, and would have to live his life knowing what had happened. Like the men who were to suffer the consequences of being found out, they would have to reconcile the questions raised in their minds. How would they go on with their lives knowing they had successfully evaded discovery? How would they conduct themselves with men who had been in the ring, resigned, were encountered in the years ahead, and who knew the secrets of those who avoided the consequences of dishonor?

THE OLD CORPS' SILENCE STILL AT WORK: AN IRONY

On June 1, the day twenty-five members of the class of 1951 were completing their testimony before the Collins Board, another kind of silence was coming to an end for two other members of the same class, one from company D-1, the other from L-1. Though the two first classmen resigned for "deficiencies in conduct," they had, in fact, endured "the silence," a punishment believed to have had its origin in the Corps late in the 19th century, when the honor code and system were "unofficial." During that era the honor code and system were entirely cadet administered, while Academy officers, aware and encouraging of the code's practice, turned a blind eye and were uninvolved in the code's administration.

The Vigilance Committee, the forerunner of the Honor Committee, used the silence when cadets found guilty of honor violations refused to resign or leave the Academy, and were subsequently reinstated in the Corps by Academy officers. When the honor system was finally, officially recognized in the 1920s, the silence remained in effect, and its provisions codified in honor system procedures, though there were several later attempts by members of Honor Committees to end the practice.

Prior to the start of the 1950-51 academic year, the silence was described in a set of Honor Committee Notes, approved by Paul Harkins. The provisions outlined clearly the treatment an accused cadet would receive if he was found guilty in an honor hearing but refused to resign

106

and instead elected to go before a board of officers or face a court martial, and was reinstated in the Corps of Cadets.

1. The silence lasts his entire Army career.
2. He is not permitted to wear his class ring.
3. If the silenced man goes to a hop, all other cadets leave the floor as soon as he steps on.
4. If the man goes to the Theater, the remainder of the row he sits in is left empty.
5. If he comes into the Boodlers, that is vacated.
6. He is addressed only on official business, and then as 'Mister.'
7. He will not be allowed to have roommates.
8. If possible the man will eat at a separate table in the dining hall.

The two cadets in '51 had been returned to duty in the Corps after proceedings before boards of officers, and all the provisions of the silence put into effect. They were subject to increased and frequent inspections, when their conduct was already more than seriously suspect – thus their resignations for deficiencies in conduct on June 1. During the Collins Board, one of the men was several times mentioned by testifying underclassmen as having been a member of the cheating ring

PUSH TO THE FINAL REPORT

In the days following testimony from members of the class of 1951, the Collins Board pressed hard to end its work. Summer military training, leaves, and entry of the new fourth class, 1955, were coming fast on the heels of June 5 graduation exercises. In the two or three day period after '51 graduated and departed on leave, two of the three remaining classes would also depart the Academy for summer military training or thirty days' leave. In less than three weeks the First Beast Detail, men from the new senior class of '52, responsible for leadership and the first month's training for the class of '55, were to start intense preparations for the July 3 entry of the new plebes. Thus the Collins Board concluded testimony from men in the class of 1952 must come next.

Based on previous testimony and evidence given on drill rolls, the carefully consolidated, analyzed, and cross referenced information enabled the Board to continue methodically, company by company, through the classes of '52 and '53, calling or recalling witnesses. From the class of 1952, forty-three men were called or voluntarily reappeared before the Board. Of that number, the Board was convinced thirty-nine were guilty of cheating or had first hand knowledge of cheating and had not reported the activity.

Most admitted involvement in the ring's activities, and identified classmates who gave them, or to whom they had given, improper academic assistance. Some repeated the names of men from '51 they believed knew of or were involved in the ring's activities. Twenty-seven of their number admitted cheating, and two acknowledged they knew of the cheating, but didn't report it. Seven of the twenty-nine were guilty of false swearing, in that they had not told the truth in previous appearances before the Board.

There were ten more who denied involvement or chose not to admit cheating; nevertheless, the Board concluded there was sufficient evidence to sustain a guilty finding in each case. On June 25 there would be one more day of cadet testimony, as the Board sought to obtain more direct evidence concerning the ten who avoided giving information about their participation in the ring.

Next came testimony from fifty-three men from the class of '53, in addition to those who had testified in the first day's proceedings, a total of sixty-six. The Board found fifty-four guilty of cheating or having knowledge of cheating. All fifty-four admitted their participation or knowledge. Fifteen falsely testified in previous appearances before the Board. As did the seven in '52 who had committed the same offense, the fifteen men in the class of '53 had compounded their cheating by lying, in a desperate bid to deny the existence of the ring and remain cadets at West Point.

Perhaps the most stunning testimony from those in '53 came from the two men who had been elected class honor representatives from companies B-1 and B-2. Both admitted their elections were the result of deliberate attempts to provide a measure of security for men engaged in cheating. In Company K-1, a similar attempt failed when *Gordon Seabold* failed in a bid to become the class' honor representative.

As the Collins Board completed taking testimony from the class of 1953 and began deliberating its findings, conclusions, and recommendations, numerous disturbing facts and clear patterns of cadet behavior had emerged from the investigation into the honor incident of 1951.

The last man to testify before the Board submitted its 10 June report was Tom Courant, the football team's head academic tutor, from Company E-2. When he was finally called before the Board, he didn't know what to expect. When he reported to Art Collins, he was told to have a seat. Then Collins leaned forward across the table, fixed his gaze on Tom, and said,

> "Mr. Courant, when this thing broke, I figured you were the ring leader. But now I know better. I want you to know you are the last person we talked to. So we would like to talk with you about corrective measures."

The questions to Tom Courant were first to ensure the Board understood the system of academic tutoring for the football team, and from there into Courant's thoughts about possible involvement of the Football Office in organized cheating. Tom Courant's answers were straight forward and firm.

He was certain the Football Office knew nothing of the cheating and explained his conclusions. He told the Board that ring members carefully selected who would be recipients of their assistance, and why. He said the cadets avoided men they knew supported the honor code and system. When word of the investigation spread, one of the men enmeshed in cheating told Tom he was among the men ring members carefully skirted in their activities. Tom repeated the names of several cadets whom ring members avoided. He had enormous respect for Doug Kenna, the assistant football coach responsible for the team's academic tutoring, and was certain ring members would have been careful to keep him from knowing of their activities. He was equally certain about Coach Blaik.

TURMOIL BEGINS FOR TWO AND MORE

When idols and icons fall, the falls are often hardest on those who bring them down. If the United States Military Academy was at a zenith in its prestige following World War II, and its nationally ranked football team was much admired by followers of collegiate and Army football, the crash of both would be heard across the nation, and many would be hurt.

Two young men in the class of 1953 were to quietly bear for their lifetimes the pain of duties they reluctantly performed, and the Academy had taught must be performed. Though both men were struggling hard to hold on in academics, and were being repeatedly urged to accept what they knew to be improper assistance, they refused to compromise their integrity. They did what was right, and exposed the rapidly expanding cheating ring.

Their acts required acceptance of enormous responsibilities. For their troubles they paid dearly, to people who were cold and unthinking, and the two, with a few exceptions, were left to fend for themselves, struggling, doubting, wondering if they had indeed done what they should. They were never labeled heroes, but their actions spoke of courage, a form of courage which marks true honor.

For *Brian Nolan* and *Mike Arrison*, early in June Week, the turmoil which the Academy and Army were about to enter had already begun. They had somehow been compromised, their roles in the undercover investigation exposed. Word spread quickly among the men implicated in the investigation. No one knew precisely how, but it did.

George Hendricks first briefly experienced what *Brian* and *Mike* were about to receive, the anger of men exposed by the investigation. When *Hendricks* told ring members he disclosed all he knew about the ring, he heard their threatening response, and conveyed concerns for his safety to the Board President, Art Collins. The Commandant reacted by placing a cadet guard outside *Hendricks'* door. The Commandant also restricted *Brian Nolan* to his room for his own protection for the duration of June Week. While the restriction may have been prudent because of *Brian's* role in the undercover investigation, the anger festering among the men being called before the Collins Board, and the threat conveyed to him by *Hendricks'* roommate, it seemed a strange response to all he had endured the preceding eight weeks. There was no enjoyment in June Week for *Brian Nolan*.

Before the week was out, *George Hendricks* and *Bill DiSantis* received information from members of the ring that *Brian Nolan* was one of the two men who had provided evidence of their activities.

Hendricks and *Bill DiSantis* promptly confronted *Brian* with the disclosure. *DiSantis* angrily reminded *Nolan* "...You son of a bitch! Remember what happened to Cox?" The remark referred to Cadet Richard C. Cox, who mysteriously disappeared in the winter of 1950 and was never found, despite an intense search involving participation by the Federal Bureau of Investigation. The implied threat drew an immediate response from the Commandant when the information was made known to him. *DiSantis* was taken out of cadet barracks under arrest by the FBI for several hours of questioning before he was returned.

There were indications in mid-May that *Mike Arrison* was already in trouble because his role in the undercover investigation had been unwittingly compromised by Dan Myers' attempt to enlist 2d Regiment cadets in the undercover investigation. An Honor Committee member also undoubtedly spoke with the H-2 honor representative.

Later, during the undercover investigation, the H-2 Honor Committee member stopped *Mike Arrison* in the hallway outside the third classman's room and told *Mike* he was "aware of what he was doing and supports" him. The gesture might have been well intended, but wasn't good news for *Mike*. Matters got progressively worse, as suspicion grew among his roommates and other ring members that *Mike* was supplying information to the Honor Committee about the group's activities.

Brian Nolan, in the months and years ahead, was the recipient of cutting, slashing, hurtful remarks from friends, acquaintances, people he didn't know, and even his mother, about his "tattling on" friends, classmates, the Army football team, and the Academy. A high school friend came home from Notre Dame between semesters and told *Brian* he was no longer his friend. The man had met one of the former cadets dismissed for cheating, then a student at Notre Dame, and had been told of *Brian's* role in the incident.

One of *Brian's* instructors on the faculty, also aware of his role in the investigation, made known his displeasure at what *Brian* had done, to the point that Brian had to ask for transfer to another academic section.

Worse, he was stung by what he believed to be a completely unethical turnabout by Paul Harkins, who had told *Brian* that *George*

Hendricks would be able to remain a cadet if he "came in and told everything he knew about the cheating ring." Despite the confrontation between *Brian* and *George*, after *George* had been told of *Brian's* role in the undercover investigation, the two friends temporarily reconciled. *Brian* had confidently told *George* his "tell all" testimony before the board would mean *George* could remain and graduate from West Point. Both *Brian* and *George* retained vivid, life long memories of the two friends happily walking across south area of cadet barracks, arms over each other's shoulders, excited by the revelation *George* wouldn't have to leave the Academy.

The commitment given *Brian Nolan* by Paul Harkins never came to pass, although later that summer Harkins tried vainly to obtain leniency. The authority for that kind of commitment was not his to give. The friendship between *Brian* and *George Hendricks* ended when it became clear *George* would have to resign from the Academy and receive a less than honorable discharge from the Army, a harsh penalty for any young man, but especially for the son of a proud brigadier general and Academy graduate.

George Hendricks was one of several men in the ring who, during the events of that summer, called their parents, some asking for parental intervention to perhaps avoid discharge from the Academy. As with others of "the ninety," his request was met with a less than pleasant response. More than one was virtually disowned by their parents. In *George Hendricks'* case, his father's reply was brief and simple. "You're on your own."

Brian Nolan watched in frustration and dismay as events unfolded that spring and summer. He had, as a young nineteen year old, been the first to bear the brunt of the enormous responsibility given him and *Mike Arrison* by Paul Harkins, a fact *Brian* never forgot. As the 1950 football team disintegrated that summer and many other young men left West Point, guilty of honor violations, *Brian* became aware that Earl Blaik's son was among them. These were not easy thoughts to bear, especially since Earl Blaik, in his capacity as Director of Athletics, was primarily responsible for *Brian* being successfully recruited for the Academy's swim team.

Michael Arrison's experiences were quite different from *Brian Nolan's*, and more painful. *Brian's* two roommates fully supported his role in the undercover investigation and all that followed. One provided information leading to the success of the Collins Board, and the three men remained good friends throughout their lives. *Brian* was a competitive, respected Corps squad athlete and cadet leader. What's more, the men in Company K-1 who knew of his participation in the events of 1951 didn't disparage or disclose his role, and at least tacitly supported what he had done. This was not the life *Michael Arrison* knew at West Point.

His roommate, *Ted Strong*, the fifth man to testify before the Collins Board, resigned on September 14, having admitted cheating, with corroborating testimony from classmates and ring members he had assisted as an academic coach. *Michael Arrison's* second roommate also admitted to the Board he had known of the cheating, but had not reported his observations and hadn't participated in the ring. Due to a minority report written on June 24 by the Board President, he and all men who admitted knowledge of the cheating – but testified they had not cheated – were returned to duty. *Michael Arrison's* second roommate was one of nine who graduated from the Academy, among eleven men reinstated in the Corps of Cadets following the Board's June 11 recommendation they be discharged.

By the time *Michael Arrison* returned to Company H-2 in late August to begin the new academic year, word of his role in the undercover investigation had swept through the upperclasses in the company. His predicament would only worsen should he remain in H-2. He knew he would be virtually ostracized in a company where he already had unhappy memories of an unrelentingly harsh plebe year. He asked to be transferred to Company E-1, H-2's sister company. His request was granted.

In his new company, ready for the fall semester, he was optimistic and hopeful for a new start in cadet life, but his academic troubles persisted. Tormented by what had occurred, he was unable to concentrate on academics when he was already struggling to hold on. He faltered and was turned back the following January to the class of 1954, for academic deficiencies in mechanics of solids and social sciences. When he joined his new class the summer of 1952, *Michael Arrison* was transferred back to Company H-2, where he remained until he graduated in June 1954.

Life in H-2 was less than pleasant for *Michael Arrison* in the remainder of his time at the Academy. He was subject of much the same treatment received by *Brian Nolan* years later, but for *Mike* it began immediately within the company which had lost six men to honor violations. There was an enduring coldness within the class of 1953, an unreasoning refusal to see the dutiful and honorable nature of his role in the events of 1951.

The day he walked back into H-2, he put the best face on what he hoped would be an improved circumstance. It was to no avail. When, as a new second classman, he walked into the cadet company as it was forming in north area, he extended his hand to the cadet company commander, a former classmate, expecting a welcoming handshake. Instead his former classmate refused to shake his hand, shunning him in front of the entire company. The incident was an example *Mike's* former classmates would largely imitate, and was a cruel second beginning that got worse.

Michael Arrison's military standing in H-2, unquestionably shaped by his tough plebe year and his role in the honor scandal of 1951, continued to dog him. When he became a first classman, he was the only man in the upper two classes in his company initially assigned no leadership responsibilities. He was shunted aside. His company tactical officer, soon to transfer from the Academy, compounded *Arrison's* isolation by inviting every member of the new first class to his quarters for social calls – except *Arrison. Mike* remembered the disillusionment and overwhelming sadness that gripped him when one day, standing in front of the entrance to the cadet Protestant Chapel gazing down on the cadet area and beyond the Plain and adjacent athletic fields, across the Hudson River, tears began streaming down his face.

Arrison would soon know he was to pay an even higher price for his adherence to the first two words in West Point's motto. He was learning that Duty, Honor, Country – the words inscribed on the coat of arms he had daydreamed about and idly sketched on notebook paper in high school – were excruciatingly difficult to live.

During the undercover investigation the spring of 1951 he had stumbled onto the fact a classmate in Company M-2, a friend from his home town, was also in the ring. He then had the painful duty of obtaining and providing information implicating his friend, as well as others. The two had been in different high schools in the same town, and had known each other in the same elementary school. At West Point they had become reacquainted and became friends.

When his M-2 classmate and friend resigned on August 17, he returned to their home town and disclosed *Arrison's* role in his embarrassing resignation from West Point. When *Michael Arrison* next returned home on leave, he would be greeted repeatedly with the same coldness he was encountering in cadet company H-2. Telling on friends was not acceptable, especially if the friend was a home town boy.

Not until the fall of 1953, some time after he began his fifth and last year at West Point, would *Michael Arrison* begin to heal from the deep wounds he suffered. There was a roommate, and other good men in his company, who helped, and a fine example of an officer who encouraged and inspired him and lifted his flagging spirits. Ironically, his roommate was a varsity football player – Freddie A. D. Attaya, from Picayune, Mississippi. The officer, Captain Martin D. "Tiger" Howell, was a 1949 graduate, had been a small, tough, spirited tackle on the Army football team, and in the fall of 1953, was assisting Earl Blaik in "a football miracle." Freddie Attaya was playing equally inspired football on the '53 team.

Michael Arrison would later be personally thanked for "saving the West Point honor system" by a well-known, distinguished graduate of West Point, an Air Force lieutenant general who was on the United Nations' armistice negotiating team in Korea when negotiations began. His name was Laurence C. Craigie, class of 1923, father of John H. Craigie, class of 1951, and the captain of *Michael Arrison's* swim team.

During the fateful ten days the Collins Board had been in session, a storm had been loosed inside the Academy. Aside from the shocks received by everyone directly involved in the proceedings, General Irving, the superintendent, Earl Blaik, and the senior officers on the Army staff, now knew that within the Corps of Cadets, young men believed to be among the finest in the nation, had engaged in organized cheating, and conspired to both hide and deny its existence. Now came the hard part: decisions, what to do and how to react to what was unquestionably an onrushing disaster.

THE FINAL REPORT – BUT JUST THE BEGINNING

When Art Collins, Jeff Irvin, and Tracy Harrington finished taking testimony on June 8, they spent the next two days completing the final report, including a review by the new Commandant, Colonel Waters. On June 11, Waters signed the letter of transmittal and sent the report to General Irving.

The Collins Board concluded ninety-four cadets were guilty of honor violations and recommended the cadets' discharge from the Military Academy. Among the ninety-four were fifty-four from the class of 1953, thirty-nine from '52, and one from '54. In August, after the Collins Board conducted an additional one-day session on June 25, and three more Boards – the Hand, Jones, and Barrett Boards – had completed their work, the first of eighty-three cadets began resigning. All but three of the resignations were accepted between August 17 and the end of September.

Throughout the period of the undercover investigation and the Collins Board, General Irving kept his boss, Lieutenant General Maxwell Taylor, informed on progress and results. When it became clear organized cheating involved a large number of cadets at West Point, the decisions about how to respond to such an unprecedented incident went well beyond authority vested in the superintendent.

Since its founding in 1802, the Academy had been one of the nation's premier institutions of higher learning, dedicated to providing the United States Army and the country the best trained officers it could produce. The stature and well being of the Academy, the Corps of Cadets, and the people who served to educate and train the cadets, had continually grown in real and perceived importance. Presidents, members of congresses, secretaries of War, Defense, and the Army – Army chiefs of staff, and senior Army officers, over the 149 years the Academy had existed, had become increasingly sensitized to its role in the nation's security. As the importance and prestige of the Academy grew, largely through the accomplishments of its graduates in all walks of life, but particularly in the nation's wars, the public also became aware of the Academy's importance to national security.

West Point had faced many difficult circumstances throughout its history. During the Civil War, World War I, and World War II, the nation was absorbed in extraordinary crises. During those crises, shifts in national security priorities, rapid expansions of the Army, with accompanying urgent demands for junior officers, resulted in major curriculum changes and accelerated graduations. The curriculum changes unbalanced the Academy education, and placed emphasis on immediate needs to prepare graduates to enter combat, consistently at the expense of a broad based education to prepare graduates for long term growth in leadership and management skills. Then, after each conflict, the Academy would go through a period of curriculum reform and sharp budget reductions, as the nation demobilized and shifted to a peacetime economy.

During those post war periods of demobilization, the Academy almost invariably faced an assault by various groups who considered the Academy and its mission superfluous in a new millennium of peace and prosperity. Now, in 1951, in the midst of the Korean War, the institution of the United States Army and its Military Academy were about to enter another crisis, a crisis unprecedented in the history of either.

There would be no loss of life or battle casualties in this one. No humiliating defeat on the battlefield. The Academy, which had prided itself in educating and training the best of the Army's leaders, with the foundation of that education resting squarely on the concepts of "Duty, Honor, Country," – the Academy's motto since 1888 – had confirmed widespread, organized cheating in the Corps of Cadets. Its much prized young men, who had been rigorously screened before being accepted into the Academy, and who were officer candidates expected to be the best the Army could give the country, were about to be publicly tarnished and humiliated.

Honor was the central word in West Point's motto, and honor was at the heart of the problem both the Army and the Academy now faced. Both individual and institutional integrity were at stake. Further, integrity's place in beliefs about leaders and perceptions of their effectiveness were at stake. There was absolutely no doubt the circumstance must be confronted head on. Equally, there was no doubt the public would know – and would want to know, one way or another, what happened, especially to the young men who were to suffer the consequences of their actions. The entire matter was indelicate at least, required thoughtful decisions, and most delicate handling.

From the very beginning, when General Irving first told Maxwell Taylor what the undercover investigation had disclosed, it became inevitable the Army Chief of Staff and nearly all his deputy chiefs of staff would be intimately involved in the decisions to follow. They were, and as the summer and events ground toward their agonizing conclusion, decision and controversy ultimately involved the Army Chief of Staff General J. Lawton Collins, Secretary of the Army Frank Pace, and the President of the United States Harry S. Truman.

While it was inevitable the Army staff, Department of the Army, and the president would be involved in the events of that year, their involvement, coupled with pressing, higher priority national security issues, and the lack of precedent for dealing with such circumstances also dragged out and delayed decisions for the ninety-four. There was another complicating factor in the affair – sharp differences about how matters would be resolved.

The differences not surprisingly began at West Point, and were mirrored in the Army staff, a corporate body composed of senior officers, most of whom were Academy graduates. The wounds resulting from those differences were to become deep and lasting, and scars persist to this day, fed by myths, rumors, half truths, outright fabrications, and an absence of facts about what went on behind closed doors at West Point and in the decision making machinery of the Army Staff. Most of all, however, the differences were magnified by strong personalities, strong wills, and unending questions, many never publicly answered: When did all this begin? Where in the Corps of Cadets did it begin? How and why did it start? How and why did it become so big? Who was involved and who were the leaders of the conspiracy? Why didn't anyone find out sooner or report the ring's existence sooner? Who should have known and didn't? Who was responsible? What needs to be done to avoid a repeat of the incident? What was done to avoid a repeat? This last question, years later, became particularly troubling when the Academy would again be torn by incidents of cheating.

For the young men awaiting their fate, the period between June 8 and August 3, when the surprise public announcement was made, became a long period of uncertainty. They were not to be passive. They fought for what they wanted – which, for most, was to remain as cadets at West Point, graduate, and receive commissions in the regular Army or Air Force.

There were several incidents during the Collins Board that gave hints of the attitudes, powerful emotions, fight, and tenacity of the men who were to suffer the consequences of their mistakes. Art Collins later wrote of an "air of violence" that seemed to be growing among the cadets involved in the Board proceedings. Four times during those ten days, outside the proceedings, cadets phoned Art Collins in his quarters in the evenings and told him of concerns for their personal safety. One was Dan Myers, the K-1 Honor Committee representative, who received a number of threatening phone calls. As a result of the calls, one of Dan's K-1 classmates, Ernie Condina, an Army jayvee football player, offered to be Dan's "bodyguard."

In each of the four cases, Collins went to the cadet area and took actions intended to calm their fears. The four believed themselves threatened. *George Hendricks* was one of the four.

Another cadet reported to Art Collins and told him there were rumors of an attempt to kill the Board President. A few days later, *Bill DiSantis*, the K-1 football player and *George Hendricks'* roommate, came to see Collins and asked him if he had heard any rumors that he, *DiSantis*, was threatening to kill Collins. Collins assured *DiSantis* he had heard no such rumors, and if he did, he would place no value in them.

In the intervening weeks after the Collins Board, while waiting to learn what was in store for them, some of the accused cadets took matters into their own hands, and fought the best way they knew how, without realizing their actions were working to their disfavor.

Board members suddenly found themselves on the defensive, being questioned by the FBI about information, supposedly originated by an unnamed Board member, that the football team might have "thrown the Navy game last December." The suggestion originated in the testimony of the first class Honor Committee secretary, and somehow became a more public allegation, supposedly originated by the Collins Board. Art Collins was angered by the allegation he knew had come from unspecified sources outside the Board.

There was also a deliberate attempt by a small group of cadets found guilty, to discredit the Collins Board, with accusations the men called to testify had been coerced, intimidated, tricked, or not accorded their rights during Board proceedings. The campaign was so successful that on August 7, four days after the public announcement, Collins Board members were asked to sign affidavits stating they had conducted the proceedings according to law and Army regulations.

Before the affidavits were presented to the Board, Art Collins reacted to the cadets' allegations by asking that the Screening Boards convened to hear the cadets being considered for discharge also investigate evidence the Collins Board had failed to adhere to the law and regulations governing their investigation. The success of the cadets' campaign to discredit the Collins Board was in spite of enthusiastic compliments given the Collins Board in late July by Judge Learned Hand, a distinguished, nationally acclaimed jurist asked by Army Secretary Pace to examine the Board's work.

Finally, on August 16, as resignations were beginning, one man among the eighty-three came to Art Collins and admitted, and personally apologized, for attempts to discredit the Board. He explained he sought to discredit the Board because he "would do anything to remain at West Point."

The same day, another came to Collins to say good-bye, and apologized for his behavior and the behavior of a small group of men who were "nothing more than 'bums' and shouldn't have been at West Point in the first place." He knew he had violated the honor code, and was wrong in doing so, and was leaving the Academy holding it in the highest regard.

The apologies, though well intended, could not undo the damage already done. For years afterward, rumors and accusations circulated among Academy graduates that the Collins Board had "tricked the ninety" into admitting they had cheated – an egregious myth unsubstantiated by facts.

There was a deliberate attempt by other men found guilty to implicate additional cadets. One cadet had come to Art Collins to warn him of their intent. The cadets involved were convinced, primarily by erroneous assumptions, that hundreds of cadets, perhaps as many as five hundred, either participated in or knew of the cheating. There was a misguided belief that if more were implicated, the Academy would decide dismissal of everyone implicated would be too costly, thus forcing the retention of all participants. They gave Art Collins a list of names, but refused to sign it or give written statements, saying they "couldn't testify in court the men were involved in cheating." The Board couldn't act on the list of names because there wasn't any evidence to go with it. The cadets then complained there were forty-three names given the Board, and no action was taken in response to their allegations.

Then on June 25 they gave twenty-nine names to Art Collins, which he accepted. Each of the twenty-nine allegedly were involved. During the subsequent two screening boards, convened August 4 and 9 respectively, the additional allegations were investigated. Again, there wasn't evidence to add any of the twenty-nine names to the ninety-four found guilty in June.

Finally, after giving the names to Art Collins on June 25, they compiled yet another, much longer list of members of the graduated class of '51 they believed were participants. This list they signed and gave to Earl Blaik. In August the Army coach furnished the list, with sworn testimony, to one of the two screening boards. As before, there was no evidence to go with the names. The list couldn't be used as a basis for calling cadets in to testify, and no evidence was turned up to add more names to those found guilty by the Collins Board.

Some of the soon to resign cadets expressed their belief cheating was so widespread that the Academy authorities and the Football Office must have known what was going on. To the implicated cadets, the Academy's failure to act in the face of such knowledge implied tacit approval of cheating, and proved the cadets' contention the Academy knew cheating was in progress.

There were opinions, mostly assumptions, that members of the faculty – instructors and professors – were involved, or at least knew cheating was in progress and winked at it. During classroom instruction there were indeed echoes of "get the poop" or "here's the poop." All but a few instructors were Academy graduates and the language of "getting the poop" had been part of their vocabulary as cadets. It was easy to resort to the same jargon when they returned to the Academy after assignments to regular Army units and graduate school. Some instructors, while giving lectures, would literally knock on their lecterns or desk tops, or stomp on the floor once or twice, supposedly indicating points of emphasis or answers to forthcoming quiz questions. Such signals were seldom used in the classroom, and less seldom taken seriously, although a few instructors indeed developed reputations for delivering "foot stompers" during instruction prior to daily graded recitations or quizzes.

During the Collins Board, several former West Point football players who had assisted on the Army coaching staff before and after graduation, were named by implicated cadets as having talked openly to them about "getting the poop" – implying the officers had cheated while cadets and were encouraging their young charges toward "beating the academic system." If cheating had been in progress for several years, as testimony indicated, and any of the men involved had returned to assist on Blaik's staff after graduation, there was a distinct possibility a few were encouraging cheating through their influence and verbal insinuations. The language that had evolved in the ring, the verbal cues, were subtle however, and could have been misinterpreted by the cadets who heard such remarks. Nevertheless, if the allegations were true, it would be virtually impossible to obtain evidence sufficient for administrative or disciplinary action against the officers in question.

On the Army staff, in the event sufficient evidence could be unearthed, preparations were made to prosecute officers who had graduated from the Academy, for offenses they might have committed as cadets. There had been precedents for such action – two cases wherein graduates were court-martialed, and General Brannon, the Army's Judge Advocate General advised the Secretary of the Army that precedence for such action existed.

On hearing testimony about past years of cheating, and names of graduates said to be involved, the Collins Board called five officers to testify, all of whom were former Army football players. One had been specifically named as having implied to an accused cadet that he had "gotten the poop." "How do you think I was able to graduate?" he allegedly said to his young, admiring listener. The Collins Board apparently didn't confront the officer with the allegation, but simply asked him if cheating was evident on the football team during the years he was a cadet. The answer was no, as was the answer from all the officers called before the Board, including one of Earl Blaik's most admired assistant coaches, Captain Johnny Green ('46), a small, tough, All-American guard on Blaik's 1944 and '45 national champion teams.

EARL BLAIK ENTERS "THE FIELDS OF UNFRIENDLY STRIFE"

In the meantime, Earl Blaik, severely shaken by what his football players were telling him during and after the Collins Board, was beginning his own passage through turmoil and torment to decision. He was torn by an immediate, not well founded conclusion that a formal investigation by a board of young, "less mature" tactical officers was not the way to solve the problem. He tended to believe everything he was told by his players, and wanted them to be truthful before the board of inquiry after their devastating errors in judgment. He was equally torn by his loyalty to the Academy he had come to love, the weaknesses he firmly believed existed in the Academy's education system, and the shattering knowledge that his son, unknown to him, had been involved in the ring.

He had confidence in his assistant coaches and all the other Academy graduates who had been involved in identifying and recruiting the men who had "paid the price" on the Army football team. Parents, educators, and athletic coaches had recommended these young men, praised and vouched for their good standings and achievements. What's more, his players had been through rigorous academic and physical aptitude testing to verify their capabilities to succeed at West Point, and the Academy's Academic Board and Admissions Office had been heavily involved in the final selection of men appointed to fill vacancies. They had to be good men, solid citizens, capable of succeeding in the fast pace and high standards of the Academy's demanding curriculum.

Then there was Earl Blaik's personal pride, born of confidence, a glowing record of winning seasons, and nationwide acceptance and admiration in the competitive, tumultuous world of major college football. There had to be other reasons these young men had violated the Academy's honor code. "I don't condone their actions," he said. They weren't right in what they did, but they were not entirely to blame, and they surely didn't, to Blaik's mind, deserve the harsh, public penalty which was apparently coming their way.

Blaik, ever independent, frank, unabashed, straight from the shoulder, was reaching his own conclusions about why the looming disaster had happened, and he wasn't shy in letting senior officers in the Army know, tactfully, what he concluded, and what the Army and Academy needed to do to rescue the situation.

He began his campaign to stand up for the accused cadets on June 17, with a letter he typed without benefit of a secretary. His typing was unpracticed, full of errors, the "hunt and peck" letter apparently an effort to keep the matter behind closed doors.

He wrote directly to General Maxwell Taylor, the former superintendent who had renewed Blaik's contract for another five years in 1946. In addition to renewing Blaik's contract that year, in 1947, Taylor appointed Blaik Chairman of the Athletic Board, adding to other positions he already held – Director of Athletics and head football coach. His increased responsibilities were, in part, the result of a tightened Academy budget following post World War II demobilization.

Earl Blaik's letter strongly advocated the Army not go through with the "drastic action" being recommended by General Irving – discharge of "the ninety" – and provided Taylor his rationale for supporting their retention as cadets. He reasoned that the men who had come to West Point were of sound moral character, that they had been the recipients of a system of cheating that had been "handed down to them," and that conditions existing at West Point, including weaknesses in the honor system, which needed correction, had encouraged them to continue the practices they had inherited. He hit hard on the theme "West Point is not perfect."

He argued that adherence to the honor code was motivated too much by fear of dismissal. Further, the scheduling of academic classes on alternate days was but one illustration of the way young men were being tempted to take advantage of an onerous curriculum load – especially intercollegiate athletes who had the additional demands and pressures of representing the Academy on "the fields of friendly strife." He believed administrative punishment of the cadets, and rehabilitative work in which they would teach the lessons of honor, would be far more effective means of making examples of the men involved, especially given their potential to be fine officers and leaders of men.

Attached to his letter to General Taylor was the first draft of a statement he would make to sports writers and broadcasters when the affair became public knowledge, an eventuality he would fight to avoid, yet knew he would undoubtedly confront.

To the extent possible, between June 17 and July 6, Earl Blaik continued to deepen his involvement in the sometimes heated deliberations affecting the outcome of events later that summer.

During the same period Vince Lombardi and four other assistant coaches went to Korea to conduct football clinics on coaching and officiating. The delegation of coaches had been requested of Blaik in the spring by Army Special Services, and was intended to be a morale booster for the troops. Lombardi and the four assistants conducted many clinics. They traveled near the front lines of Korea but worked mostly in secure areas. They were also sent to Japan to teach the game to the Japanese. They returned to West Point in mid-July, full of enthusiasm about what they had seen and heard, only to be greeted with the news that their prospective national championship team would soon be decimated.

On July 6 Earl Blaik sent a letter to Army Chief of Staff J. Lawton Collins similar to the one he earlier sent to General Taylor. The letter to General Collins was more polished, detailed, and thoughtful, but expressed essentially the same ideas put forth to General Taylor. The letter to the Army Chief of Staff was dated one day after Blaik had been in Washington accompanying the superintendent, General Irving, to present to Generals Taylor and Collins the Army coach's views concerning what action should be taken, views clearly in opposition to the superintendent's recommendation for discharge.

Major General Frederick A. Irving was a gracious man. He invited Earl Blaik to accompany him to Washington several times that summer, though Irving knew full well Earl Blaik didn't see eye to eye with him on what was to be done. General Irving and Earl Blaik, and their families, had been friends many years, and the superintendent held Blaik in high esteem. He also knew Earl Blaik was deeply admired by Academy graduates everywhere, and admired no less among his colleagues, competitors, sports writers, and broadcasters in the world of college football. Blaik's influence reached far beyond West Point.

Because of his respect for the Army coach, "Red" Blaik's years of hard work and sacrifice for the Academy and its athletic programs, and his untarnished reputation, Irving gave Blaik an unfettered opportunity to separately present his views to the Army's senior officers. The mutual respect the two had for one another never diminished, but their friendship suffered greatly as a result of the cadets' dishonor and the two men's disagreement over how matters should be resolved. Sadly their families were never again the friends they had been. For however Earl Blaik might wish things to be, it was Frederick A. Irving who must bear the responsibilities for decisions thrust upon him by the honor incident of 1951. And the gentlemanly, unaffected superintendent who had the mark of a quiet, tactful, gritty bulldog about him, had, on February 1 of that year, walked unknowingly into circumstances not his doing, heading pell-mell toward the most shocking event in West Point's illustrious history.

CHAPTER 3

DECISIONS AND TRAGEDY

With graduation on June 5, 1951, a quiet began settling over West Point, a quiet sorely needed by the men who had to deliberate the fate of the ninety-four cadets. General Irving was to make decisions based on the Collins Board's report and consultations with his staff, including the Academic Board. There would also be consultations, approvals, and decisions by the senior officers in the Army chain of command. The Chief of Staff, General J. Lawton Collins, Secretary of the Army Frank Pace, and ultimately President Truman would participate in decisions affecting separation of the cadets from the Academy.

The Academic Board, which undoubtedly had been kept informed of the affair beginning with the decision to convene the Collins Board, began its consultative role in earnest June 1, after the Board took testimony that day from the last of twenty-five men questioned in the class of '51. The Academic Board, established early in the Academy's history, in 1951 was composed of sixteen senior officers, most Academy graduates. All but five were permanent professors. Eleven were heads of academic departments. The five who weren't professors included General Irving, the Board President; the Dean of Academics, Brigadier General Harris Jones, who was General Irving's classmate and had been on the faculty at West Point since July of 1931; Colonel Harkins, the Commandant – followed by his replacement – Colonel Waters; Colonel Charles L. Kirkpatrick, chief surgeon and Academy Hospital Commander; and Colonel Robert S. Nourse, the Academy's Adjutant General, who served as Secretary of the Board and was a non-voting member.

The eleven permanent professors were colonels in the Army. One, Colonel Herman Beukema, who graduated in Dwight D. Eisenhower's class of 1915 and headed the Social Sciences Department, came to the faculty in August of 1928. Another, Colonel Gerald A. Counts, Physics and Chemistry, who had given support to Colonel Harkins' undercover investigation, joined the faculty in August 1931. He also served on the Board of Athletics, which Earl Blaik chaired. Colonel

Oscar J. Gatchell, Mechanics, and Colonel T. Dodson Stamps, Military Art and Engineering, both became faculty members in 1938. The other seven joined the faculty in the 1940s. Colonel Boyd W. Bartlett, Professor of Electricity, was the first, on 1 June 1942, six months after the United States entered World War II. In August, Boyd Bartlett would be appointed president of the board charged with inquiring into causes and recommending actions to avoid a repeat of the '51 incident.

After the men in '51 testified on June 1, following their return from the Army's Aberdeen, Maryland, installation, the Academic Board received a briefing on the Collins Board's plan to methodically proceed through the classes of '52 and '53, understanding that furloughs and summer military training for the two classes would continue as scheduled. By June 8, few cadets remained at West Point, all classes having departed after graduation, '52 and '53 for scheduled summer activities. Until their return to the Academy beginning June 23, General Irving, his staff, and senior officers at the Academy, would have more time to deliberate the decisions confronting them.

On June 11, Colonel Waters forwarded the Collins Board report to General Irving, who gave copies to the Academy Chief of Staff and the Academic Board for their study and recommendations to the superintendent. Waters' letter transmitting the Board's report was brief, and filled with emotion. His closing words, inferring his recommendation the cadets be discharged from the Academy, were from the Cadet Prayer, which every new fourth classman memorized as plebe knowledge, and recited each Sunday in chapel.

> ...Endow us with courage that is born of loyalty to all that is noble and worthy, that scorns to compromise with vice and injustice and knows no fear when truth or right are in jeopardy.

The superintendent met with the Board on June 15, and after inviting comment, presented his views. The Board voted their approval for what became General Irving's decision, but not unanimously. There were eleven for and four against the superintendent's decision. His decision was to press for discharge of the ninety-four, and recommend to the Army they receive honorable discharges – recommendations which were to generate sometimes heated disagreement in the weeks ahead. On June 18 he traveled by military aircraft to Washington, accompanied by Earl Blaik, and acquainted General Taylor, and General Haislip, the Vice Chief of Staff of the Army, with the contents of the Collins Board report, leaving copies with them. Earl Blaik presented his views separately and left his self-typed letter addressed to General Taylor.

On June 23, when the classes of 1952 and 1953 began returning to West Point, arrangements were completed to move most of the cadets found guilty of cheating into a single building at Camp Buckner, ten miles from the Academy. When the moves began, final decisions on their cases were still weeks away; however, indications remained that matters would not be resolved in their favor.

On June 25, a one day session before two members of the Collins Board – one an alternate member – heard testimony from twenty-one cadets. Most were men recalled in an effort to obtain additional evidence concerning the ten men in the class of 1952 who had refused to incriminate themselves in prior testimony. Although the men had chosen not to incriminate themselves, the Collins Board had obtained what it believed was sufficient evidence to find them guilty and recommend discharge in the final report on June 8. Seven of the ten later resigned.

Another curious incident occurred on June 25 involving a cadet interviewed by the Board that day. Late in the evening he reported to the Officer-in-Charge, Lieutenant Colonel Garrett, who was also an alternate member of the Collins Board, that he had been "struck in the head from behind" at night in the automobile rally port, just outside the north area of cadet barracks, near the fifty-first division. The cadet was examined by a doctor, wrote a statement about the incident, and was interviewed by Jeff Irvin as well as Colonel Garrett, who obtained a statement from the cadet's acting cadet company commander. The officers concluded no further action was necessary. The same cadet told Art Collins there was a plan afoot to "stack the list" of men the ring members were accusing as participants, to make the ring appear larger than it was.

On June 27, Irving returned to Washington and conferred with General Collins, the Chief of Staff, and General Haislip with reference to the Board report. The next day a conference was held at West

Point with General Brannon, the Army's Judge Advocate – the Army's senior military lawyer – and General Craig, the Army's Inspector General, and other senior officers. They determined that it wouldn't be necessary to resort to court-martial to separate the cadets who admitted cheating. A prior court decision in the case of a cadet who requested a court-martial, had ruled cadets were not commissioned officers, and since appointed by the president could be administratively separated by the president without court martial. The cadets found guilty would be given an opportunity to resign and receive an honorable discharge. Those cadets who elected not to resign would be brought before the Academic Board for separation under Academy regulations.

During the conference, Colonel John Waters, the new Commandant, voiced his strong objections to the cadets receiving honorable discharges. His views didn't prevail at the time, and General Irving directed the Academy Chief of Staff, Air Force Colonel John J. Morrow to prepare a letter to the Secretary of the Army setting forth proposed methods of separating the cadets.

The letter, dated July 2, 1951, was carried by General Irving to Washington the next day, and delivered to Secretary Pace. Contained in the letter was the provision that, should the press inquire into events connected with the honor violations, a public announcement would be made.

Tuesday, July 3, was the day the class of 1955 entered the Academy. The absence of the Academy superintendent on the day a new fourth class was received at West Point was indicative of the intense efforts to decide what must be done.

Again, on July 5, the superintendent conferred with the Army Chief of Staff in Washington. Earl Blaik, who accompanied the superintendent on the trip, restated his views to General Collins separately, as he had to General Taylor, recommending an alternative to discharging the cadets. The following day Blaik mailed his letter of July 6 to the Chief of Staff. The letter included the text of a statement Blaik explained he would make when he would – inevitably – face sports writers. General Irving sent a letter to Secretary Pace outlining the advantages and disadvantages of the proposed procedures for separating the cadets.

At West Point the Public Information Officer, Colonel Leer, prepared public information plans and a draft of proposed procedures, including press releases, which General Irving approved. A conference was held at West Point on July 6 and 7, with officers from the Army's Public Information Division, in which plans were discussed for Department of the Army approval.

Then, on July 9, in Washington, General Irving conferred once more with Generals Collins and Haislip, concerning a review board, later called the Hand Board. The Hand Board, composed of three distinguished, highly respected men, one a nationally known jurist, the other two retired general officers, was to review the Collins Board proceedings, and the recommendations of the superintendent, and report its findings to the Chief of Staff, General Collins, and Secretary of the Army Frank Pace. The Board was to take an outside, objective look at the Collins Board proceedings, and the superintendent's decisions and recommendations. The Board's conclusions and recommendations were to prove crucial in the decisions that followed.

On July 18, with "First Beast," the first month of new cadet training, still in progress for the class of '55, General Collins wrote separate personal and confidential letters to the three men who comprised the Hand Board. With each letter was a copy of the Collins Board report. The letters requested the Board to convene in New York City two or three days prior to their July 23 opening session at West Point. General Collins' letters contained nothing that could remotely be considered foregone conclusions regarding the Collins Board, or any of the superintendent's recommendations.

Meanwhile, the days were passing ever so slowly at Camp Buckner for the men waiting to learn their fate. At the Academy, matters involving the honor violations seemed to be rushing toward a precipice – all still behind closed doors. Tension was building, but the knowledge and emotions accompanying that knowledge, were confined to a relatively small number of people at West Point and in Washington. However, that would soon change. The rumors of "something big" continued to circulate among the men in '52 who were at West Point and Camp Buckner. The rumors spread into the new third class, '54, at Camp Buckner, and to the new cadets soon to be plebes, the class of '55.

THE CLASS OF 1954: INTO THE FIGHT EARLY – TO WIN

Competition, deeply embedded in the American spirit, is equally, if not more important, at the nation's military academies, and in the armed forces. A fundamental belief is that competition brings out the best in its participants, whether they are individuals or organizations. Douglas MacArthur's now legendary quotation, "On the fields of friendly strife...," about the relationship between athletics and victory on the battlefield, is the most notable, outward expression at West Point of the role of athletics and competition in training future military leaders.

There are yet other objectives for immersing cadets in vigorous athletics and physical education. As officers they must be able to lead by example, and train young men and women to successfully meet the physical and emotional demands of soldiering. And there is, of course, much more than athletics and physical education in the making of a future military decision maker. The discipline of the mind, ways of thinking, solving problems under the most extreme circumstances, and the technical knowledge that must be factored into such thinking – those subjects taught in classrooms and during military field training – along with leadership training and character development, are the indispensable balancing factors in the education of an officer.

In mid-July, Beast Barracks' fast pace at West Point was equaled at Camp Buckner, where the class of 1954, the new yearling class, uninformed, yet part of the rumor mill about the brewing storm at West Point, was undergoing intense summer military training. On the morning of July 16 a group of forty-eight yearlings slipped out of Buckner, on orders, for Stewart Air Force Base in Newburgh, New York, where they boarded an Air Force transport aircraft for Guantanamo Bay, Cuba – "Gitmo." There they boarded the *Whiskey*, the battleship U.S.S. *Wisconsin*, for an educational midshipman cruise. For the forty-eight, this was their second encounter with the U.S. Navy since they entered West Point as new cadets, little more than a week after the North Korean Peoples' Army (NKPA) struck south across the 38th parallel. The first encounter was the Army-Navy game in Philadelphia the previous December, where Navy won a stunning 14-2 upset to end Army's string of twenty-eight games without a defeat.

What happened on that cruise – years later described as "Mutiny on the *Whiskey*" – told of the future, and another Army-Navy game in

132

the fall of 1953. The men in '54 liked to poke good-natured fun at "the system," and, like all cadets, they relished winning.

Then, on July 18, the same day the Army Chief of Staff, General Collins, dispatched letters from Washington to members of the Hand Board, another group of yearlings at Buckner hatched a plot, seemingly unrelated to the "Mutiny on the *Whiskey*." At the time, the scheme – a cadet prank, a practical joke – which Earl Blaik would have laughingly categorized as typical cadet humor, "situational humor," was a delightful joke at the expense of "the system." Though the plan went awry, there was more to the story than any of the cadets involved knew at the time. "The Reveille Gun Caper," the brainchild of third classman Ed Moses, portended better days ahead, the rebirth of "the twelfth man," and the birth of Army football's "victory cannon" at "a game never to be forgotten," Army vs Duke, at the Polo Grounds in New York City, October 17, 1953.

THE HAND BOARD: A REVIEW OF THE COLLINS BOARD AND THE SUPERINTENDENT'S RECOMMENDATIONS

Earl Blaik had evidently planted additional seeds of caution in the minds of General Collins, as well as General Irving and other senior officers on the Army staff. Their natural inclinations toward caution had no doubt also been fueled by the cadets' burgeoning effort to discredit the Collins Board.

Then, there were the obvious differences of opinion among the senior officers over the type of discharge the cadets should receive. The recommendation from General Irving for honorable discharges was in conflict with Army policy established when Maxwell Taylor was superintendent. The new Commandant, Colonel Waters, argued strongly against honorable discharges. And the Academic Board's eleven to four vote in favor of General Irving's decision to discharge the cadets was not unanimous. There had been forty-four discharges for honor violations since the toughened policy went into effect shortly after World War II, nine since September of 1950. All forty-four received less than honorable discharges. Given the evidence of a "conspiracy of silence," at least twenty-two known instances of "false swearing" before the Collins Board, and deliberate attempts to stack the cadet Honor Committee with ring members, a reversal in Army policy seemed unlikely.

But again, Earl Blaik's arguments were probably having a telling effect, and undoubtedly obtained a sympathetic hearing from General Irving, who had been Commandant of Cadets in an era when men who resigned following honor violations received honorable discharges. But General Irving was his own man. Indeed, arguments could be persuasively made that Academy cadets were held to a higher standard, and many of the cadets found guilty by the Collins Board had been the disillusioned, trapped observers and recipients of flawed cadet leadership, and an honor system corrupted by a small number of individuals who failed to understand their roles and responsibilities as leaders in the system.

The senior officers on the Army staff, having read the Collins Board report, and in consultation with General Irving, concluded it was necessary to have an external review of both the Collins Board and the superintendent's recommendations, which he had carried to Washington in his July 2 letter. The result was the Hand Board, which convened at West Point July 23-24, 1951, and completed its report the following day.

Judge Learned Hand, a distinguished, nationally respected jurist, retired from the Federal Second Circuit Court of Appeals in New York, was named chair of the Review Board. The two additional members were retired Army general officers. Lieutenant General Troy H. Middleton, who was a corps commander during the Allied drive into Nazi Germany during World War II, was a distinguished educator and President of Louisiana State University. Major General Robert M. Danford was an Academy graduate from the class of 1904, had served as Commandant of Cadets, from 1919 to 1923, under then Brigadier General Douglas MacArthur, when MacArthur was the Academy superintendent. He was familiar with the numerous reforms MacArthur instituted while superintendent, including the formal approval of the cadet honor system. After Danford retired from the Army, he served as president of the Association of Graduates from 1945-47, spanning most of Maxwell Taylor's time as superintendent, another period of reform and reconstitution of the Academy curriculum.

When General Collins wrote the Hand Board members on July 18, mailing each a copy of the Collins Board report, he enclosed a copy of an order from the Secretary of the Army, which outlined the Board's charter:

17 July 1951
CONFIDENTIAL ORDER

A Board is hereby appointed to inquire into certain violations of the Honor System at the United States Military Academy.

The Board will consist of the following persons:
Honorable Learned Hand
Lt. General Troy H. Middleton, USA-Ret
Major General Robert M. Danford, USA-Ret

The Board will review the report of a Board of Officers convened pursuant to Letter Order, Headquarters, United States Corps of Cadets, West Point, New York, dated 28 May 1951, and the recommendations thereon of the superintendent, United States Military Academy.

The Board is authorized to call such witnesses and develop such additional evidence as it deems appropriate to determine further the nature, causes and extent of the alleged violations.

The Board will recommend to the Secretary of the Army appropriate disciplinary and corrective action.

The facilities of the offices of The Inspector General and The Judge Advocate General of the Army are hereby made available to the Board for the purposes of this inquiry.

BY ORDER OF THE SECRETARY OF THE ARMY:

M.F. HASS
Colonel, GSC
Secretary of the General Staff

The Hand Board's work began in New York City before they came to West Point. They carefully studied the Collins Board report, and were thoughtful in selecting witnesses to be interviewed. They wanted to hear testimony from men with divergent views, and heard from twenty-one witnesses. The twenty-one were: General Irving; five members of the Academic Board, including officers who unhesitatingly expressed minority opinions about the superintendent's decision; Coach Blaik, who wasn't on the Academic Board, but obviously had a huge stake, and strongly opposing views in the decisions to be made; Art Collins, Jeff Irvin, and Tracy Harrington, the three Collins Board members; seven members of the cadet Honor Committee, class of 1952, including one who requested to appear on behalf of three of his friends who had been implicated. The Hand Board also received testimony from Dick Miller, Company K-1, the honor representative who had assisted in the undercover investigation; and four of the implicated cadets, two football players and two who weren't intercollegiate athletes. The father of one of the football players was an Academy graduate and became the acknowledged leader of the "Parents' Committee of 90 Dismissed Cadets," who launched a vigorous public campaign for a Congressional investigation, and when that failed, requested a presidential investigation of the affair.

Two proposed alternative solutions emerged during witness testimony before the Hand Board, in addition to the recommended dismissals formulated by General Irving. One was suspension of the cadets for one year. The second was termed an "intramural cleanup," which would have retained the cadets and attempted to restore them to "normal status." Earl Blaik's argument for this solution had found adherents in the Academic Board, and the Hand Board report took note of the "sharp and strong" differences of opinion among Academic Board members, but pointed out the majority favored dismissal.

The Hand Board report also took note of the high regard cadets held for the honor code:

> ...The Board finds from cadets themselves that they have been thoroughly briefed and indoctrinated with the principles of the Honor Code; that they regard it the very soul of West Point, and they approve and expect the drastic penalty which the Code imposes upon those who violate it...

During the interview of Collins Board members, Judge Hand complimented the three officers on the quality and thoroughness of their work. He told them he was particularly impressed with their careful analyses of cadet testimony, and "had never seen such detailed cross referencing and analyses of witness testimony in all my years on the bench." For Jeff Irvin, Judge Hand's remarks became life long, pleasant memories, among numerous memories of a difficult duty.

When the Hand Board sent its July 25 report to Secretary of the Army Frank Pace, it contained one conclusion and two recommendations:

...The Board concludes that the Report of the Board of Officers and the recommendations of the Superintendent should be affirmed...

It is recommended that:

a. The course of action as contained in the letter of the Superintendent to the Secretary of the Army, dated 2 July 1951, be followed.

b. The Superintendent consider the advisability of giving different written general reviews to successive groups of cadets.

The Hand Board gave the Secretary of the Army the outside, objective look he needed to finally decide dismissal of the cadets was necessary, which was the recommendation made by General Irving. In discussing the alternative of dismissal, before reaching its conclusion, the report said:

...The Board believes that the action recommended by the Superintendent is the only possible one he could recommend as strictly adhering to the long established and cherished Honor Code...

The report contained other prophetic words, pointing toward the outcry that would be heard on August 3, when the surprise public announcement was released:

This solution, [dismissal], when it breaks in the newspapers, will bring an instant storm of criticism and abuse upon the Academy. Our people will not generally understand such action. Cheating is so common a practice in our schools and colleges that our people will accept it as entirely inadequate for such a tragedy. If 5 or 10 men were involved it would attract little attention, but with nearly 100 dismissed, there will be outraged public sentiment, criticism and political clamor that will not subside for months. Our people will be convinced that there must be something definitely rotten in the administration of West Point and their sympathies will be strong for the injured cadets.

FINAL DECISIONS

The behind the scenes struggle at West Point and in the Army staff was entering its final stages. Later, there would be other decisions to ensure the case of every cadet found guilty by the Collins Board received careful attention and review. The cadets were to go before one of two Screening Boards after the public announcement of August 3. In the Screening Boards, composed of three and four senior officers respectively, all assigned to the Academy, the cadets received benefit of military lawyers they could select from the Academy's law department, or civilian attorneys. They could provide explanations, introduce matters of extenuation and mitigation, and, if they wished, provide additional evidence regarding honor violations.

After the Screening Boards received additional testimony or statements from the cadets, investigated any new evidence submitted, and made their final recommendations, the cadets who admitted to cheating met with the Commandant and were given the option of resigning. If they chose to resign, their resignation, along with their case file was sent by courier for final review by the Army's Judge Advocate General. If the Judge Advocate General's review of evidence upheld the recommendation for discharge, confirmation was sent to the Academy, and each cadet was given a final interview with the superintendent before leaving the Academy.

Those cadets who admitted to cheating, and might elect not to resign, were to be administratively discharged. The superintendent had been given the authority to court-martial cadets who didn't admit to cheating, but against whom evidence had been gathered sufficient to warrant court-martial. Under any circumstance, their cases would be reviewed by the Army's Judge Advocate.

After the Hand Board met, and prior to the public announcement of August 3, Lieutenant General Maxwell Taylor, in a memorandum to General Collins, recommended that cadets who had admitted to cheat-

Lieutenant General Maxwell D. Taylor, Superintendent, USMA, September 1945 - January 1949; Army Deputy Chief of Staff for Operations and Administration (Photo courtesy USMA Archives.)

ing, and compounded their violations by lying under oath, be offered the opportunity to resign. If they refused to resign, he recommended they "be dismissed for undesirable habits or traits of character," using procedures established to bring cadets before the Academic Board at West Point. For those who admitted to cheating, and had not falsely sworn, he recommended they receive administrative punishment at the discretion of the superintendent, the latter to include disbarment from holding [cadet] rank or representing West Point before the public. Maxwell Taylor had largely agreed with the recommendation made by Earl Blaik – that the cadets be retained at West Point, except for those who had lied under oath before the Collins Board. His memorandum was silent on the type of discharge the cadets should receive.

Paul Harkins, who was in his new assignment under the supervision of General Taylor throughout the period of deliberations after the Collins Board, and was the one officer on the Army staff having detailed knowledge of what had occurred at West Point, participated in the Army staff's deliberations. He undoubtedly drafted General Taylor's recommendations to General Collins, the Army Chief of Staff, recommendations which urged leniency for the cadets who had told the truth to the Collins Board.

While the Screening Boards were in progress between August 4 and 15, General Irving continued to press for honorable discharges for the cadets. However, if the Army's senior officers agreed with the superintendent, they would have to support a reversal of the Army policy established five years earlier. Their dilemma was exacerbated after the public announcement, when some of the cadets, stung by the surprise announcement, began to fight back through statements to the press. Some criti-

Frank Pace, Jr.
Secretary of the Army
(Photo courtesy USMA Archives.)

cized the "raw deal" they had received at the hands of the Academy, and a few publicly attacked the honor system, attacks echoed by some of the "Parents Committee of the 90 Dismissed Cadets." The cadets unknowingly generated anger among some of the senior officers on the Army staff because, while the Academy and Army had steadfastly refused to disclose detailed contents of the Collins Board report, and some of the incidents that followed – information that would have been even more embarrassing to the cadets involved – the cadets were not being forthcoming about their own activities before, during, and after the Collins Board. Particularly rankling was the unpublished evidence of cadets' "false swearing" before the Board, and the schemes among a small number of them to elect ring members to the Honor Committee.

On August 6, 1951, the final decision was made to give the cadets administrative discharges. When the decision was received by General Collins from Secretary of the Army Pace on August 9, the Chief of Staff prepared a memorandum for record documenting the decision.

General Irving soon received word of the decision, and like any good, loyal soldier, never let on to anyone he had argued for honorable discharges, or that the Secretary of the Army decided against his recommendation. The responsibility of an officer is to argue a recommended decision as vigorously as facts and principles allow, but once the decision is made, and though it be contrary to what he proposed, he is to carry it out as though it were entirely his own.

There would be no discharge certificates given, and there would be no characterization of the discharges as honorable or dishonorable. The decision was a compromise, less severe than in the previous forty-four cases, but not in keeping with the superintendent's recommendation. The decision was made by Secretary Pace, after consultation and advice from the Army's senior officers and the Army's Judge Advocate General.

In the end, to ensure there wasn't a publicly apparent double standard in the administration of military justice, the Army held to its five year old policy with respect to men found guilty of honor violations at West Point. Nevertheless, the aggravated circumstances of a confirmed conspiracy among most of the accused cadets, "false swearing" by nearly twenty-five percent of them, attempts by a small group to deliberately discredit the Collins Board, and admitted deliberate elections of two of their number to influence honor hearing results, were serious indeed. Though three of those aggravated circumstances were never made public, the cadets' punishment could have been much more severe.

August 9, at Mama Leone's restaurant in New York City, Earl Blaik held a press conference before a large group of reporters, sports writers, and broadcasters. Though he said he "didn't condone what they had done," he was well prepared to defend the character and good names of the young men who would soon be resigning. His opening statement was essentially the one he enclosed in his July 6 letter to General Collins. He held the press conference in spite of a telephone request from retired Lieutenant General Robert Eichelberger, the former Academy superintendent who had brought "Red" Blaik back to West Point in 1941, to be Army's first civilian head football coach. General Collins called Eichelberger and asked him to intervene, hoping "Uncle Bobby" could dissuade his protégé Blaik, from any public statements.

Earl Blaik was no longer a colonel in the Army, as he had been during World War II. Nor was he a Federal Civil Service employee. He had served the Academy as a football coach for better than eighteen years. In his mind and heart, his teams had performed superbly in representing the Academy, the Army, and the nation. During and after World War II, the Black Knights had been one of the most effective recruiting forces in West Point's history. Now he must be a determined, single-minded, independent man, who believed he was doing right by the men who had been cadets at the United States Military Academy and had "paid the price...on the fields of friendly strife." Earl Blaik would not be turned from the course he had set.

The next day, one of the cadets who would resign on August 22 wrote General Collins and carried the letter to General Irving, requesting the superintendent transmit it to the Army Chief of Staff. The cadet

141

who signed the letter "representing The Eighty-Eight Cadets," was in the class of 1952. He asked that the letter remain confidential until it reached General Collins' hands, saying he placed no restrictions on its release after the Chief of Staff received it.

10 August 1951

General J Lawton Collins
Chief of Staff
Department of the Army
The Pentagon
Washington, D.C.

Dear Sir:

Although this action is not normal military procedure, we feel that by writing directly to you we may be able to facilitate clearing up some of the unpleasantness that has resulted from the recent investigation. None of us – that is, of the eighty eight cadets, feel that we should be allowed to remain at the Military Academy. That we violated our own Honor Code cannot be denied. However, we believe that there are extenuating circumstances that need explanation and entitle us to special consideration.

As a former Cadet, you realize that all Honor cases are confidential. Should we not have expected the same consideration for our wrong? Since the entire issue has been made public, we feel that all the criticism is unjust. How can we return home with our reputation intact? Please understand, we are not saying that we are not guilty of violating the Honor Code. We are guilty and we do not belong in the Corps of Cadets. But we do not deserve to be branded as liars, traitors, and cheats, as the newspapers called us when the news was released.

Every one of us expects to enter military service. Many would like to enter OCS, Air Cadets or ROTC. How can we continue our military service with honor? In our group are veterans. Men who have served in combat and in the occupational forces. These men have served honorably in the past and should be allowed to continue honorably in the future. Men who have never been to the Academy become officers and all we ask is an even start with them.

We deeply regret that all of the adverse criticism has appeared against the Military Academy. We did not mean to imply that the Honor System is unfair. Neither did we mean to imply that the Academy instruction is inferior. West Point is part of the American tradition. To think of disbanding the Academy is absurd. The greatest disappointment for the eighty-eight cadets, as you must fully realize, is to leave this institution.

Of all the discharged cadets, there is not one that feels bitter toward the cadet that reported his classmate and touched off the investigation. We feel no bitterness toward this cadet, because he believed, just as we believe, that the Honor System is the Corps' business, and should be dealt with by the Corps of Cadets.

It would have been easy for us, during the investigation, to hide behind legal technicalities by use of evasive answers to direct questions. But to have hid behind legal technicalities is an insult to our personal integrity. But you know, as well as we, just what the code of Honor means – even to us. We violated this honor system. We are guilty of violating the Honor Code of the Corps of Cadets, but we have not committed a sin against the American people. We were the violators but not the instigators of the violations. We inherited this method of violation from the upperclasses.

Again, we say, we deeply regret the adverse criticism against the honor code and the Academy. Even in our position, we are proud of the honor system, and the tradition of the past 150 years which surround both it and the Military Academy.

We ask, in order not to be branded for life as cheats and as liars that we be given a chance to resign, and to receive an honorable discharge.

Respectfully yours,
The Eighty-Eight Cadets*

*After the 10 August letter, five more cases were reviewed which resulted in cadets found guilty being reinstated in the Corps of Cadets. Thus 83 resigned instead of 88.

General Collins responded on August 16:

In replying to your letter of 10 August 1951, in which you say you speak for the cadets involved in honor violations at West Point, I would say at the outset that it represents one heartening aspect in this sad affair. The frank admission of guilt and wholesome spirit of repentance which I find therein confirm my first impression, that those involved in this affair, while openly admitting their violation of the Cadet Code of Honor are not fundamentally men of bad morals or character. They have, however, fallen into evil ways; they have not, in the words of the Cadet Prayer, preferred the "harder right instead of the easier wrong," and their failure has brought sorrow upon themselves and upon the Military Academy.

I am also glad to note that you recognize the inevitability of drastic action on the part of the authorities in order to restore the honor system to its indispensable place in the ethical structure of Cadet life. Indeed, many of your group have made partial amends for past offense by the straightforward way in which you have assisted the authorities in investigating this matter and in putting the situation to rights. I sincerely hope that you will all leave the Academy determined to redeem this early fault by the character of the lives which you lead hereafter.

The decision of the Department of the Army with regard to the type of discharge to be given was based upon a thorough consideration of the equities in this case. For the reasons which I have indicated above, I did not feel that the conduct of the Cadets warranted the usual discharge presently given in honor cases, that is to say, one which indicated discharge under conditions other than honorable. However, in justice to Cadets who have received this type of discharge for honor offenses in the past, one cannot justify an honorable discharge in the present case. Instead, it has been decided to allow the Cadets to resign and to receive a simple administrative discharge without specification as to circumstances. In case the Department of the Army is queried by colleges, governmental agencies and the like as to the meaning of this discharge, the reply will indicate that it was administrative in nature and not dishonorable.

General J. Lawton Collins
Army Chief of Staff
1949-1953
(Photo courtesy USMA Archives.)

I have observed your concern over difficulty in pursuing a military career in Officer Candidate Schools, The Reserve Officers' Training Corps, or in the Air Cadet Program. While I cannot speak for the other services, I will say for the Army that your separation from West Point in itself will not be grounds for disbarment from the Reserve Officers' Training Corps, Officer Candidate Schools, or other similar programs under Army jurisdiction.

Sincerely yours,
J. Lawton Collins

Neither of these two letters were ever made public, although several senior officers who struggled through the painful decisions of that summer wanted to publish the cadet's letter, hoping to put to rest all the questions surrounding the guilt or innocence of the men involved – and the oft repeated charges of a "raw deal" from the Academy and the Army. The Army's Chief of Information argued against the release, noting that press attention to the incident had begun to diminish, that the press would focus on the character of the discharge rather than the attitude evident in the cadet's letter.

At West Point, while the exchange of letters between General Collins and the cadet representing the men about to be discharged was in progress, the first formal steps toward rebuilding the Academy began. The Bartlett Board convened on August 13, to begin the search for causes of the incident, and actions needed to avoid a repeat of the tragedy. The Bartlett Board was the equivalent of an accident investigation, necessary for correcting what the Academy and the Army found to be the underlying causes of the incident.

"YOU HAVE TO PAY THE PRICE..."

It would be a long time after the public announcement of August 3, 1951, before the effects of the honor incident would begin to subside. Many idols had fallen. True, cadet lives were compartmentalized and fast paced. No less so were the lives of the people who educated and trained the Corps. Such factors helped soften the incident's impact. But the reality was that nothing could immediately bury the losses, take away the questions, or remove the rubble left behind.

On June 12, shortly after the Collins Board finalized its report, Art Collins, Jeff Irvin, and Tracy Harrington completed an analysis of extracurricular activities, including intercollegiate athletics, engaged in by the ninety-four cadets recommended for discharge. They summarized and confirmed that the numbers of cadets implicated were largest, percentage-wise, from intercollegiate athletic squads such as football, basketball, baseball and hockey. As they continued the analysis in activities other than athletics, the Board members discovered, to their complete surprise, that the highest percentage of participants came from Catholic Acolytes, the activity for which Art Collins was officer-in-charge. Jeff Irvin and Tracy Harrington couldn't resist poking good natured fun at their board president. He, along with many others, had been hoodwinked. But they also knew deep down it was no laughing matter. Perhaps for years the cheating ring had successfully disguised their work from Art Collins, Earl Blaik, twenty-one company tactical officers, untold numbers of academic instructors, and hundreds more officers and cadets.

Among the eighty-three who resigned in August were sixty varsity athletes. In the end, all eighty-three admitted cheating, including those who had steadfastly refused to admit their involvement to the Collins Board. Among the men who resigned were some of the most admired members of the Corps of Cadets. Aside from the thirty-seven football players, which included the team captain-elect and two All-Americans among twenty-three lettermen, there were seven of fourteen members of the basketball team, including the team captain. Most of the seven were also football players. There were hockey and baseball players, several of whom were also football players.

The Army football players who left that year paid dearly for their mistakes, perhaps more than others of the eighty-three. Winning in the glare of national collegiate football spotlights, amid the clamor at huge stadiums filled with thousands of cheering fans, normally gives team members lifetimes of pleasant memories. Not so for the Army team of 1950. The events of the following spring and summer tore the heart out of friendships and shared team loyalty which had grown from their mutual sacrifice and hard won accomplishments on the college grid-iron. Their fierce loyalty to one another, a cohesiveness that coaches and teams build painstakingly, piece by piece, year by year, had, in large measure, been their undoing. Their pleasant memories of the season of 1950 would always be blurred, painful, and not nearly so fulfilling as what might have been. And for those few who were left to play in the next two seasons, there were some of the same thoughts of what might have been, a certain sadness, a frustration, an ambivalence, and a legacy of adversity none had counted on.

Even more devastating, however, was the number of emerging leaders in the Corps of Cadets who were to resign – men who at young ages had the proven potential to be exceptional cadet and commissioned officers, capable of accepting and successfully managing increasingly heavy responsibilities. There were twelve who would probably have been cadet captains, all of whom never received the opportunities they had otherwise earned. They would have been one third of the Corps' ranking cadet leaders the coming year – likely including two of the three most senior cadet captains. All were in the class of 1952. One might have been the Cadet First Captain, the Brigade Commander, an *ex officio* member of the Cadet Honor Committee, and the ranking member of the Committee – Pat Ryan's successor. The other might have been a Regimental Commander. Two of the eighty-three were class presidents. Additionally, there were cadet academic tutors and men who wore gold stars on the tunics of their traditional gray uniforms, stars emblematic of academic excellence at West Point.

Worst of all, in the class of 1953 there were two elected honor representatives who in just over one year would have become members of the Cadet Honor Committee. Their nominations, in hopes of election, had been planned, and they had successfully hidden their cheating from unsuspecting classmates in their companies.

Twenty-one of twenty-four cadet companies lost one or more of their men discharged in the scandal. The "flanker" companies – the tall man companies – where most of the intercollegiate athletes lived, had been hardest hit. Company M-2 lost fourteen men; L-2 and H-2, six each; E-2, four; K-2, F-2 and C-2, three each. In the 1st Regiment, B company lost twelve; D and K, four each; A, C, E, H, I, and M, three each. The numbers said nothing of the disillusionment suffered, friendships and professional relationships torn, lives changed, betrayals felt, and trust destroyed.

But perhaps the most poignant and devastating impacts of all, though far from West Point, were felt by families of Academy graduates who had lost their lives in Korea or were still serving on Korea's battlefields. General Collins' briefing to members of Congress, in which he divulged cheating had been in progress for several years, was widely reported in the press after the public announcement and repeated frequently by some of the cadets who resigned.

General Irving received a heart-wrenching letter, dated August 6, from the father of Howard Gallaway Brown, class of 1950, who had been killed in Korea on September 22, less than four months after he graduated from West Point. Howard had been one of the men in '50 who had been rushed to his first troop duty with an Army infantry unit already engaged in Korea. Howard was in Company L-2 when he was a cadet. He was a big, broad shouldered, pleasant, easy-going southerner from Tupelo, Mississippi. He was proud of his heritage.

The letter from his father told of the anguish brought to his family by public implications that their dead son, and others like him, could have been involved in the affair. "Tupe's" father was never told that in all the thousands of words of testimony given to the boards that year his son was never mentioned as one of the men in past classes involved in cheating.

The 1951 cheating scandal had brought pain to those whose sons were still risking their lives in Korea, and more tragedy to those who had already suffered tragedy of enormous proportions.

Earl Blaik was nearing the end of the line in attempts to retain the cadets who were already resigning, or to alter the type of discharges they were receiving. In his several trips to Washington, the Army coach had met with General Maxwell Taylor; General Haislip, the Army Vice Chief of Staff; General Collins, the Army Chief of Staff, and discussed the cadets' fates in a twenty-minute session with Secretary of the Army Frank Pace.

On Monday, September 3, while resignations of the eighty-three cadets were still in progress, Earl Blaik met for thirty-five minutes with President Truman in the Oval Office of the White House. Blaik asked for the meeting through Brigadier General Harry Vaughn, the president's long time friend and military aide. The Army coach made one last ditch appeal to alter the character of discharge given the cadets. The president, as Earl Blaik remembered years later, "was a hard nosed Army rooter" and during the meeting expressed sympathy for the cadets found guilty. Blaik also remembered President Truman's expressed dissatisfaction with the Department of the Army's handling of the affair, while saying he couldn't now reverse what had already been decided. The next day, Earl Blaik wrote a letter thanking the president for the appointment:

My dear Mr. President:

You were most thoughtful and considerate in devoting so much time to my version of the West Point affair.

It has been a saddening experience as daily I learn from many of the dismissed boys who are now like derelicts going about from one college to another in search of a place that will accept them as the fine Americans I know them to be.

They have found that the lack of an honorable discharge does affect an effort to be accepted in many colleges. It is so difficult for their families and friends to accept the Army version of the discharge. Anything short of an honorable one suggests that they are blighted by their stay at West Point.

It is even more difficult for me to explain something to parents, for as a parent, I do not believe the action taken ever considered the type of boy involved or his future as a citizen.

But you were kind to listen to me Mr. President, and I shall always be grateful to you.

Respectfully,
EARL H. BLAIK

Earl Blaik's characterization of the difficulties the young men he described as "like derelicts" was somewhat overstated. Due to the nationwide publicity the resigning cadets received, many were given considerable assistance in their efforts to appeal their discharges, enroll at major colleges and universities, and for a few, to enter Reserve Officer Training Corps at their places of enrollment. For all but a few, rebuilding of their lives had begun – quickly.

In the next two months, additional decisions barred reinstatement or reappointment of cadets discharged in the 1951 incident. The last doors were closed to the eighty-three men who left West Point after resigning, but the wounds left behind, and the sadness taken with them would be slow mending – and for a very few there would never be mending. Eleven others from the ninety-four found guilty were returned to duty in the Corps of Cadets, evidence in their cases judged insufficient in subsequent reviews by the two screening boards or the Army's Judge Advocate General.

There would also be a new beginning for everyone else, including the men who left West Point that summer, and those left behind who lived through it all and had to rebuild the Academy and its football team. Though the Academy's officer corps, and the huge majority of cadets who remained at West Point were enormously proud of the Corps and its honor code and system, rebuilding began among ashes of disillusionment and defeat – while far away in a war torn land, graduates continued to perform their duty, honor their country, and give their lives.

CHAPTER 4

KOREA: ONCE MORE INTO THE FIRE

August 27, 1951. Transition week is in progress at the Military Academy, when the Corps of Cadets reassembles at West Point to begin the academic year. This year is different however. A war is in progress, along with resignations following the cheating scandal. Far, far away, near a desolate, war torn place – Chorwon, Korea – a young, infantry lieutenant writes a letter.

Truce negotiations between senior officers representing the United Nations Command, and their battlefield foes, the North Koreans and Communist Chinese, had begun on July 10. The battle lines which had been so fluid, with the two bitterly contending armies surging up and down the Korean Peninsula, were now less fluid. A pattern of stalemate had slowly emerged, yielding a form of warfare that brings its own brand of anguish, not unlike that produced by the ghastly trench warfare of World War I.

> Again from Korea. Again from a mountain top.
>
> Yesterday I took out a patrol. It was Sunday... a Sunday without services... a Sunday out in that troubled land that lies between two armies. There is no room for a church on our misty hill in this lonely land of many battles.
>
> No, the day seemed only like a wet, slick day anywhere, and I wondered, as we moved down the slopes to seek out our enemies, why the feeling of Sunday had so completely deserted me. But the ridges, and the woods, and brush, soon pulled our bodies into a shallow sort of fatigue, and thinking became tiresome. We wandered far, under the fitful skies.
>
> Then, a group of Chinese who had been waiting, opened up and shot our lead man... and we suddenly became involved in a short, sharp struggle of grenades and bullets. But we, at a disadvantage, had to pull back without our dead soldier.
>
> Yet we knew what we had to do, and soon we set out again to risk much to get to him. This time we moved – not to gain knowledge, for we knew about our foe – not for ground, for we

were turning back – not for glory, for we had been there a long, long time. We returned into a holocaust of bullets to recover the symbol of someone who had been so alive a short while before – and we returned in the hope that we, too, would be treated in the same way, were we ever there.

We set out, taut in every nerve, moving in a high-tension sort of way. I happened to look at the wet, bony wrist of someone beside me. He gripped his rifle with a chalky hand. Flesh and caution, against the savagery of bullets and sharp little fragments....

We set out... an intense group of men... under that terrible, broken sound of artillery, and the snicker of machine guns in the bushes. Then, in a final, fearful second of confusion – in a second of awful silence, one gutty private crawled up, and with the last ounce of his courage, pulled our soldier back to us.

We had succeeded. We started back, rubbery legged and very tired... feeling a little better, a little more certain there would be a tomorrow. We had done something important. We were bringing our soldier with us.

Then it was night, and the rain was soft again. We drew up on a nameless ridge and dug into the black earth to wait for the enemy, or for the dawn. The fog moved in among the trees. I sat for a long time looking at the end of the world out there to the north.

Nine months in a muddy, forgotten war where men still come forth in a blaze of courage. Where men still go out on patrol, limping from old patrols and old wars. Weary, jagged war where men go up the same hill twice, three times, four times, no less scared, no less immune but much older and much more tired. A raggedy war of worn hopes of rotation, and bright faces of green youngsters in new boots. A soldier's war of worthy men – of patient men – of grim men – of dignified men.

A sergeant sat beside me. For him, twelve months in the same company, in the same platoon, meeting the same life and death each day. Rest? Five days, he said, in Japan, three days in Seoul... and three hundred and fifty-seven days on this ridge! Now he sat looking, as I was, at the same end of the world to the north.

Nine months, and I am a Company Commander now, with the frowning weight of many men and many battles to carry. A different, older feeling than of a platoon leader. New men...I must calm them, teach them, fight them, send them home whole and proud... or broken and quiet. But get them home. Then wait for new replacements so the gap can be filled here, that gun can be operated over there.

There is much work to be done. I must put this man where he belongs, and I must send many men where no man belongs. I must work harder and laugh merrier... and answer that mother's letter to tell her of her lost son. Yes, I was there.... I heard him speak.... I saw him die. So, in many ways, I must write the epitaph to many families.

There is always that decision to make as to whether a man is malingering or sick... whether to send him out for his own sake, and for another's protection, or return him for a necessary rest. And one must never be wrong.

One must be ready and willing, always, to give his life for the least of his men. Perhaps that is the most worthwhile part of all this... the tangible sacrifice that an infantryman, a soldier, can understand.

I see these things still I am slave
When banners flaunt and bugles blow
Content to fill a soldier's grave
For reasons I shall never know

Now it is raining again. The scrawny tents on the line are dark and wet, and the enemy is restlessly probing. It will not be a quiet night.

Lt. David Hughes

Lieutenant David Ralph Hughes graduated with the Academy's class of 1950, less than three weeks before the *In Min Gun*, the North Korean Peoples' Army (NKPA) crossed the 38th parallel, to launch a massive, surprise attack against the Republic of Korea. For David Hughes it had been a long, torturous road to August 27 near Chorwon, Korea, and he traveled it with a celebrated group of men who will undoubtedly be remembered by Academy graduates for generations to come.

They were the men of '50, who covered themselves with a different form of glory, in "the forgotten war."

A FIRST FOR '50… FOR OTHERS, A SECOND OR THIRD TIME AROUND…

Tuesday morning, June 6, 1950, the Corps of Cadets, and the families and friends of the graduating class, heard an address by Secretary of the Army Frank Pace. His remarks included a brief insight into Academy history, "…every West Point class except those of 1945 through 1950 has shed its blood for America on the battlefield." The implication was unmistakable. There would be men among those six classes who would eventually be tested on battlefields somewhere in the world. When Secretary Pace made those remarks, he hadn't the slightest notion how soon those six classes would be adding to the long history of that grim statistic.

To many in America's armed forces, Korea would mean going into battle for the third time in their lives. Young soldiers and officers who survived the horrors of World War I trench and aerial warfare – with its advancing technology of machine guns, heavy artillery, tanks, the airplane and weapons it carried, and lethal gas – had become many of the non-commissioned officers and commanders in Army and Air Force combat units of World War II. Others became the senior military leaders, the master strategists and tacticians in the "great crusade," "the good war" of the twentieth century. And many who survived the devastation of the vast, mobile, killing machines of World War II would now confront another of duty's call to war.

For West Point's graduates, the graduates of all military academies and officer training courses, and the American GIs who fought in their commands, Korea was once more into the fire. Consistently, throughout the Academy's history, the classes that graduate just before the onset of war suffer disproportionately grievous losses in the battles to follow.

In the Civil War, there were two classes from the year 1861. The Civil War, the grim national convulsion in which tactics on the battlefield lagged far behind the growth of weapons technology. A war that marched long lines of thousands of men, in close order, with drum cadences and flags flying, into devastating sheets of close range cannon and rifle fire. Frequent bayonet charges by foot soldiers, followed by bloody hand to hand fighting, and cavalry charges with bugles sounding, thundering horses' hooves beneath soldiers organized in troops, squadrons, and regiments, their officers shouting commands, leading the charges with sabers or pistols drawn.

The classes of May and June 1861 totaled seventy-nine. Forty-three of their number fought at the First Battle of Bull Run, the first major battle of the Civil War, near Washington, D.C., on July 21 of the same year. Before Confederate General Robert E. Lee surrendered at Appomattox in April of 1865, twelve were killed in action or died of wounds.

In the twentieth century the classes of April and August of 1917 went into the "war to end all wars," "Black '41" into World War II, and in the summer of this year, the class of 1950.

Why did these classes suffer more than others? The explanation is brutally simple.

THE LIEUTENANTS' WARS

In war, on the battlefield, the young lieutenant leads the small units of the combat arms. In the mid-20th century, the lieutenant is the infantry, armor, heavy weapons, reconnaissance, or combat engineer platoon leader, the artillery gun section leader, a wingman in a flight of four fighters, a co-pilot or aircraft commander on a bomber.

Unless they have served as soldiers or airmen in combat before they become officers, they have no experience in war. Yet their responsibility is to lead soldiers and airmen into the deadly, terror-ridden chaos, confusion, and irrationality of war.

They seek the responsibility, sometimes with brashness and bravado, facades which bolster and shield them from the fear they all will know. They grow up fast if they survive. Thoughts of heroism and glory rapidly vanish. They quickly relearn an ancient lesson. There is ever a youthful, prewar naiveté with respect to the battlefield's realities. The battlefield, whether on the ground or in the air, is not the place to be without prior months of tough, realistic individual and unit training. In the absence of carefully prepared training, mistakes are more frequent, with deadlier consequences.

The sacrifices of the young who do not survive are silent, almost reverent, unless extraordinarily heroic, and observed by someone who

155

survives. They have made no mark in life save the joys given by their youth, the promises seen and never fulfilled. Acts of heroism are more often matters of circumstance and luck, desperate acts from individuals driven by powerful emotions, which overwhelm fear and concern for their own safety, and bring out the noblest good within them. Most have gone to war with the greatest of confidence they will come back. "It will not happen to me. Life will go on forever. The world will continue to be my oyster."

War pauses little, if at all, to honor their loss. Their pictures can be found in high school and college yearbooks, and tenderly kept family photograph albums. The grief of loved ones is private, shared only with family, friends, the lost son's or daughter's comrades in arms who write, call, or might escort their bodies home for burial, and the officers – their commanders – who had the painful duty of committing the young soldiers to combat, who understood all too well the terrible price inevitably paid, and who, when they fell, performed the sad duty of sending letters of condolence, facts, and circumstances.

Those who survive the battlefields as lieutenants, and can accept and understand the managed violence of war, exorcise the sad memories that sometimes haunt them, and become inspired – they will go on to become the military leaders of tomorrow.

The Military Academy doesn't publicize or boast of the sacrifices of its young graduates. It cannot. Such actions would be crude, callous, and supremely disrespectful. Irreverent. Yet sacrifice in the extreme is an accepted part of life called to duty in the armed forces.

West Point's class of 1950 learned early in Korea the meaning of leadership on the battlefield, and its price. But the price was probably increased by an unusual set of peacetime circumstances and decisions distant and seemingly unrelated to battlefield performance.

In the spring, well before the North Korean Peoples' Army invaded the Republic of Korea, a decision had been made by the Army to send the Academy's newest graduates directly to troop duty with Army units, without their attending a series of schools en route, such as a generalized introductory school at Fort Riley, Kansas; followed by the Infantry School in Fort Benning, Georgia; or Artillery at Fort Sill, Oklahoma; or Armor at Fort Knox, Kentucky. The new second lieutenants could continue to receive specialized schooling they might request, such as airborne (parachute) training at Fort Bragg, North Carolina, or Ranger training at Fort Benning, if the Army had need of men with those specialized skills.

The decision for troop duty right out of the Academy resulted after repeated complaints received from various branch schools that Academy graduates were not performing up to standards. The courses, in some cases, lasted nearly a year, and instructors were complaining of disciplinary problems, a general inattentiveness, and a puzzling lack of adequate performance.

In the Academy and Army staffs' subsequent discussions of the problem, the decision was made that graduates should bypass the schools. Senior officers knew West Point's entire four year curriculum was carefully structured to prepare graduates for troop leadership. They found the schools, in fact, duplicated some of the graduates' branch technical and field training, and leadership training received while at the Academy. They also concluded there was one other factor at work contributing to the problem – graduates' additional months, almost a year, in student status, with responsibilities solely as students. The prolonged student status was adding to the frustration and boredom among well-qualified young officers anxious to begin that for which they had been trained for four years. The new lieutenants needed to sink their teeth into troop-leading responsibilities without additional schooling.

Unfortunately, the decision that followed the conclusions didn't take into account the wartime circumstances the class of '50 would soon face. The last branch field and technical training of any consequence received by most of the class of '50 was the summer of 1949, and for some, the last such training had been two years prior to graduation. In the absence of recent field or technical training and a total lack, or all too brief unit training, many in the class were scrambled to their first troop duty as fillers or replacements in units about to engage, or already engaged, in fighting on the Korean peninsula.

The absence of recent training in the technical combat skills required by the various branch missions certainly complicated life for the new graduates who found themselves platoon leaders in combat. The battlefield is not a place to get a hurry-up refresher course. War's dreadful chaos causes enough errors having lethal consequences.

157

General MacArthur's reminiscences of Korea fit the circumstances of West Point's class of '50 remarkably well: "The Korean War meant entry into action 'as is.' No time out for recruiting rallies or to build up and get ready. It was move in and shoot."

So into the theater of operations came the class of '50, swelled with the pride of young officers anxious to do their duty. What they faced that summer was not at all what they expected, although the steady drumbeat of bad news had kept coming all summer, warning them of what lay ahead.

American and South Korean forces were in a desperate fight to avoid being driven from the Korean Peninsula. On September 3, Edmund Jones Lilly III was the first in '50 to be killed in action. He graduated two months and twenty-seven days earlier. He was in the 9th Infantry Regiment, 2d Infantry Division when he died. He received the Silver Star and the Purple Heart, posthumously. Before the month of September ended, six more members of the class died.

By war's end, 365 from the graduating class of 670 had served in combat in Korea. Of their number, thirty-four were killed in action, eighty-four more were wounded, and seven died in accidents or of natural causes while serving in combat. Other West Point graduates, before and after the class of '50, paid heavily in Korea, but none more dearly. The class of '50 had a special kinship with the classes of 1836, 1846, 1861, 1917, and 1941.

The five preceding West Point classes bloodied for the first time in Korea were only slightly more fortunate than '50. In the class of 1949, twenty-seven were killed, fifty-two wounded. About one-half of the West Pointers from the classes of 1949 and 1950 went to Korea as "unit fillers" to serve as platoon leaders. Casualties among junior officers from the classes of 1945-1948 were less severe, a total of sixty-one killed, 124 wounded, but still heavy compared to casualties among officers commissioned in college and university officer training programs, and officer candidate schools.

The Academy class of 1951 lost nine, with forty-seven wounded. In the class of 1952 five were killed in action, and thirteen wounded. By the time the classes of '51 and '52 had reached Korea's battlefields, "volunteers" from the Peoples' Republic of China had entered the war, and the opposing armies were in the brutal stalemate reminiscent of the trenches of World War I, while armistice negotiations were droning on at Panmunjom. The emergency that had rushed the class of 1950 into the breach in Korea was over. Fewer members of '51 and '52 were sent to the stalemated front before the armistice agreement was signed. Their casualties were less.

As football practice began late the summer of 1950, American and South Korean forces were continuing to retreat toward the southern tip of the Korean peninsula. Overwhelmed by the power of the attackers' onslaught, they had suffered a series of stinging defeats after the *In Min Gun,* the NKPA, launched their attack across the 38th Parallel in late June.

The *In Min Gun* had been taken far too lightly, almost with disdain, by many American commanders. The advanced elements of American forces rushed to Korea from Japan also painfully relearned old military lessons: occupation duties are not adequate substitutes for field training in preparing for combat, and never underestimate your enemy.

The soldiers of the NKPA were well trained, tough, disciplined, and well equipped. As a result, American and ROK soldiers coined a new term – "bugout" – GI slang for fleeing the enemy, frequently leaving weapons and equipment behind. In the first two months of the NKPA offensive there were several shameful bugouts by both ROK and U.S. units, all green combat units, most committed piecemeal, in a futile effort to stem the tide of North Korean advance.

ONE MAN'S ROAD TO KOREA

By the time David Hughes went to war in Korea in November of 1950, the conflict had gone through several distinct phases. By September the United Nations forces had pulled back into what became known as the Pusan perimeter, on the southeastern tip of the Korean peninsula. Their shortened lines of resupply and communication, coupled with better natural barriers behind which stronger defensive positions could be maintained, enabled the Americans, ROKs and their Allies to consolidate, replenish, and rebuild their strength with a growing stream of fresh combat units, equipment, and supplies.

Then, in mid-September, they went over to the offensive, executing a planned breakout from the perimeter, timed to begin shortly after the surprise landings far to the rear of the NKPA, at Inchon, on South

Korea's west coast, near Seoul. The badly overextended NKPA, lacking air and naval support, had suffered increasingly heavy losses in its bid to push the UN forces off the peninsula. Faced with MacArthur's stunning tactical surprise at Inchon, the NKPA was forced to retreat northward or be destroyed. The retreat became a near route and the UN forces were soon pursuing a shattered NKPA all the way to the Yalu River, bordering China.

"The boys will be home by Christmas" became the optimistic rallying cry. The Communist Chinese had differing views, and the rallying cry vanished in the bitter cold, eerie, flare-lit November nights in far North Korea. With unearthly sounds echoing in the cold night air, from hundreds of unseen bugles, horns, and whistles, and hundreds of thousands of screaming Chinese infantrymen, swarming from the hills into advancing, thinly stretched, road bound United Nations forces, the Korean War again became another war.

Shattered were any illusions of a quick end, with a quick UN victory. It once more was a war of withdrawal, retreat, rear guard and delaying actions, ambushes, and stinging defeats for UN forces, the vast majority of which were Americans and South Koreans.

David Hughes would see the Yalu River, but only briefly. No sooner did he see the Yalu, than he learned the cold, hard facts of life in combat no American infantryman ever wants or ever likes. His introduction to the battlefield became the fighting withdrawal, the retreat, the enemy roadblock and ambush, ferocious and confusing night attacks designed to sow fear as well as destruction. Not a good beginning.

But he would soon learn he was more fortunate than many young officers entering combat for the first time. His company commander was a patient and thoughtful trainer and teacher, a member of West Point's class of 1944, and Dave Hughes had been assigned to one of the most illustrious, storied units in all the United States Army.

* * *

David Hughes wrote many letters during his sixteen months in Northeast Asia. Most were to his mother in Denver, Colorado, where he grew up. He was close and deeply devoted to his mother. His father died when David was six years old. The family was living in Pueblo, Colorado, at the time.

His grandmother and grandfather were immigrants from Wales, in the British Isles. They had settled in Colorado in the 1880s. David's father was one of five children born to the elder Hughes, and Dave Hughes was an only son. He had three sisters, two older, and one younger. His father had been a retail food salesman who traveled up and down the "front range" on the east side of the Rocky Mountains, between Walsenburg in the south, and Boulder, north of Denver. David often traveled with his dad on sales trips, and remembered dancing on counter tops while singing *Home on the Range* for his dad's customers.

After his father's death, the family moved to Denver, where his mother could more easily find employment, and they could be near his aunt, his dad's sister-in-law, who was successful, wealthy, well connected, and influential in Colorado. His aunt and uncle had no children of their own.

David's aunt helped support him when he attended the Colorado Military School in Denver, where he fit well in the disciplined life of a young military academy student. His aunt, who also helped David's widowed mother and family through the harsh depression era, became an important influence in Dave's life.

It was as a boy in the Colorado Military School, during World War II, that he learned of West Point, and its name and allure grew in his thoughts, forming a lasting impression that eventually led to David securing a principal appointment from Senator Milliken of Colorado – with the devoted assistance and intervention of his aunt.

At West Point, David Hughes was in company F-2. He was independent, a rebel. He considered himself a maverick, proud of his Welsh heritage. The Welsh were noted for their love of the lyrical in the English language. They were poets, writers, speakers, and the preachers in the Calvinist churches in England, the clergy that rebelled against the Church of England. Precise use of the language was deeply admired among the Welsh, and in David Hughes deeply ingrained. The same Welsh poets and preachers had also been the eloquent spokesmen for the coal miners in Wales, and their unions, which fought hard for better pay and safe working conditions in the harsh life and health threatening environment of the mines. There was some of that in Dave Hughes blood – powerful feelings for the people who sweated and toiled in hard, dangerous labor. To him they were truly "the salt of the earth."

He played lacrosse. He joined the staff of *The Pointer* magazine, and the chess club. He liked chess and on the team was paired with a Filipino cadet and classmate, Fidel Ramos, who in 1992 was elected president of the Philippines. They were rated number two and three on the chess team, and prided themselves in beating Navy in 1950.

On *The Pointer* staff he not only wrote for the magazine, but became its lead photographer. Photography was a skill he acquired when he stayed with relatives in Omaha, Nebraska, one summer, and worked in their camera

David Ralph Hughes as a cadet in the graduating West Point class of 1950. (Photo courtesy USMA Archives.)

business. His company mate and close friend at West Point was Paul Gorman, *The Pointer* editor their first class year, who reached the pinnacle of his Army career as Commander-In-Chief, United States Southern Command, two years before he retired in 1985. As new first classmen, they had traveled to Europe together the summer of 1949.

Dave Hughes' last five months as a cadet at West Point were less than joyful. He was a cadet private, a "clean sleeve," one of five in his entire class. He was "slugged" and busted, given stern disciplinary action and reduced in rank when he was a first classman. His camera work had brought him grief as a cadet.

His journalistic photo hobby reached great heights of interest and excitement for him late in 1949 when he climbed atop the West Academic Building, adjacent to the central area of cadet barracks. Assembled on the concrete-covered quadrangle below was the entire plebe class ('53) in the form of a postage stamp, standing stiffly at attention, looking up at him. He snapped a picture that drew him irresistibly toward instant fame, and another kind of fame he didn't want.

Life magazine, headquartered in New York City, expressed interest in the photograph and asked to meet him. The only problem was that *Life* representatives insisted on a time of day Dave was scheduled for academics. Faced with the choice of his day in the photographic sun, or academic classes, he chose fame, and immediately became infamous with his company tactical officer when he was reported absent from class.

He met a Commandant's Board resulting in removal of his stripes, and more. He spent nearly the entire last semester of his four years at the Academy serving "special confinement" in his room.

Second Lieutenant David Hughes took leave beginning June 6, 1950, after graduation and commissioning ceremonies. His first assignment after graduation leave was Fort Riley, Kansas, directly into the 77th Infantry Rifle Company, a special training support unit. In less than three weeks his life and the lives of the men in the class of 1950 changed – forever. Secretary of the Army Frank Pace's words, intoned at Dave Hughes' graduation, became more than words. The class of '50, and the five classes before them, were to receive their baptism of fire, and soon.

After graduation, in spite of his tenure as a cadet "clean sleeve," he remained in the vicinity of the Academy and visited with George Moore, from Highland Falls. Moore, a much loved father-confessor to cadets on *The Pointer,* was the magazine's publisher, and also kept Dave and others out of trouble with the officers who reviewed material to be printed in the cadet publication. After a few days in the vicinity of the Academy, he left and began a driving tour about the country before reporting to Fort Riley.

When branch and unit assignment selection came in the spring, he had deliberately selected the separate infantry company at Fort Riley, far from Fort Benning, where he knew many of his classmates would be. Now, the ever independent Dave Hughes was anxious to get to the company.

No Ranger or Airborne schools for Dave Hughes. No "elite" units. The average GI infantry unit was what he sought, where men who were "the salt of the earth" would be found. It was an outlook entirely consistent with the heritage and beliefs that motivated him. A line infantry unit was where the Army's hard, tough, most difficult work was done, and the average GIs including draftees, did the toiling, sweating, fighting, and most of the dying. They were the men, seldom volunteers, who bore the Army's heaviest burdens and greatest risks in battle, and when committed to the fight, they largely determined who won or lost.

Dave Hughes reported to Fort Riley, to the 77th, after graduation leave. He was there only forty-five days when orders came, sending him to Korea and the 1st Cavalry Division. He was flown to Tokyo, then went by ship to the port of Inchon, South Korea, where General MacArthur's surprise landing had broken the NKPA's assault on the Pusan perimeter less than three months earlier. He traveled to an Army replacement company in Pyongyang, North Korea, the North's capital, which had been abandoned without much resistance in the NKPA's hasty flight north.

In Pyongyang he received orders sending him forward to the 1st Cavalry Division, which had joined the fight in the Pusan perimeter, acquitted itself well, and been part of the pursuit of the NKPA as it fled north. By the time he joined Company K, 3d Battalion, 7th Cavalry Regiment, the Eighth U.S. Army and its UN Allies were pressing close to the Yalu River, the border between North Korea and the Peoples' Republic of China.

7th Cavalry Regiment! George Armstrong Custer of Civil War fame – and the legendary Battle of the Little Bighorn in 1876, where Custer (class of June 1861) had been killed. At the age of twenty-three, Custer had been one of the youngest brevet brigadier generals in the Union Army, as well as one of the most dashing cavalry commanders in the Civil War. At the Little Bighorn in North Dakota, eleven years after the Civil War, his command was destroyed by the Lakota Sioux Indians, Cheyenne, and other tribes allied with them. Custer died with 247 of his troopers.

In spite of Custer's disastrous defeat, the 7th Cavalry could claim a glorious history. Aside from its many battle streamers, there were warm memories associated with Custer's time in the 7th. He had been its commander when *Gary Owen* had been adopted as the Regimental song – its anthem. The jaunty tune became the official cavalry march for the United States Army.

A young second lieutenant couldn't ask for a more colorful Regiment. Dave Hughes later believed he couldn't have been in a better infantry company than Company K. His commander was extraordinary, a man Dave Hughes would never forget – Captain John Robert

Flynn, West Point class of 1944, a veteran of World War II's European theater. John Flynn gave much of himself to his officers and men. He taught them well. He undoubtedly saved many lives with his devoted teachings about how to fight the infantryman's war in Korea – though the reality of the infantryman's war is great loss, even in victory.

In David Hughes' mind, one conclusion became a certainty. He owed much to John Flynn, perhaps his life. David's road to Korea had been long, and he had traveled far north in a strange land to join what appeared to be the tail end of a war.

War is filled with uncertainty and violent change. Today's apparent victory can be quite another matter tomorrow.

HEADED SOUTH – AGAIN

When Dave Hughes arrived in the 7th Cavalry in the first days of November, war greeted him with a scene out of Dante's *Inferno*. In the daytime a pall of smoke drifted through company areas. The night sky glowed reddish orange from the light of forest fires raging in the mountains near Unsan, North Korea. It seemed as though all of far North Korea was aflame. Talk was of the 8th Cavalry Regiment's defeat near Unsan. The 8th, one of the 7th's sister regiments in the 1st Cavalry Division, had been overrun and virtually destroyed by two full Chinese divisions of twenty thousand men who had fallen on the 1st and 2d Battalions of the 8th, and the ROK 15th Regiment, at Unsan

The attack came at dusk on November 1, simultaneously from the north, northwest, and west of Unsan.

> Blowing bugles, horns, and whistles and firing signal flares, the Chinese infantrymen, supported solely by light mortars, swarmed skillfully – and bravely – over the hills. To the ROKs and Americans, the oncoming waves of massed manpower were astonishing, terrifying, and, to those Americans who believed the war was over, utterly demoralizing.

> Most of the Chinese troops were veterans of the victorious CCF [Chinese Communist Forces] campaigns against Chiang Kai-shek's Nationalist forces. Since they had no close air support, no tanks, very little artillery, and were experienced in the tactics of guerrilla warfare, they specialized in fighting under cover of darkness.

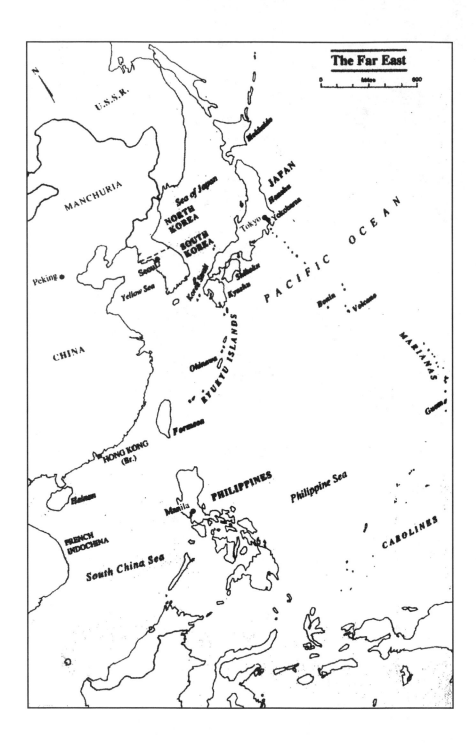

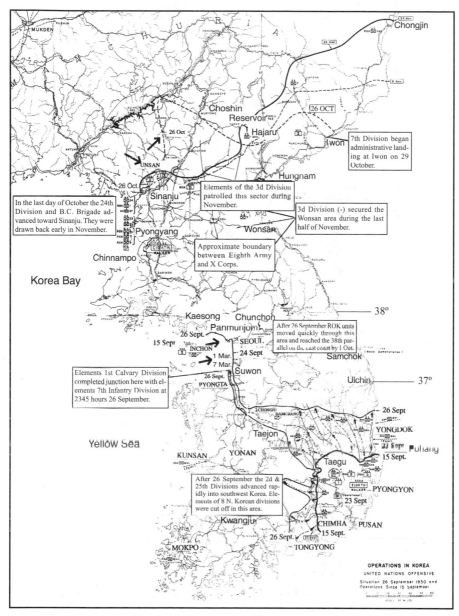

The line of the Pusan Perimeter on September 15, 1950, is indicated. Arrows mark the Inchon landing on the same date. Other arrows mark the line in North Korea, representing UN dispositions on October 26, 1950, and the location of Unsan, N.K., where the 1st Cavalry Division was located when David Hughes arrived in K Company, 3d Battalion, 7th Cavalry Regiment.

The whistles, bugles, and horns were not only signaling devices (in place of radios) but also psychological tools, designed to frighten the enemy in the dark and cause him to shoot, thereby revealing the position of men and weapons. The fighting tactics were relatively simple: frontal assaults on the revealed positions, infiltration and ambush to cut the enemy's rear, and massed manpower attacks on the open flanks of his main elements. War correspondents were to describe the attacking waves of the CCF as a 'human sea' or 'swarm of locusts.'

The UN forces at Unsan caved in under the massive weight of the CCF. Within about two hours the ROK 15th Regiment collapsed. Attached American tanks, artillery, and anti-aircraft elements began a hurried and disorderly withdrawal through Unsan to the south. At the same time the CCF drove a wedge between the loosely tied 1st and 2d Battalions of the 8th Cavalry. Both battalions gave ground, forced back on to Unsan. By ten o'clock that night, both units were out of ammunition, more or less overrun, cut off from the rear, and desperate.

The two battalions of the 8th Cavalry Regiment were cut off and surrounded, virtually destroyed. Survivors took to the hills in small bands, abandoning their equipment. There was no semblance of organization, though the commander had received orders to withdraw the remnants of the two battalions through the 3d Battalion. Withdrawal was an order there was no chance of pulling off. By this time, at about three o'clock in the morning on November 2, the CCF was swarming into the 3d Battalion, blowing bugles and horns. In the wild melee and hand-to-hand fighting the battalion commander was mortally wounded, and his executive officer took command. Many men bugged out, but others heroically banded together into tight perimeters to fight to the death in a replication of Custer's Last Stand.

Fortunately, the majority of the 1st and 2d Battalions of the 8th Cavalry eventually found their way out of the hills, including the three commanders. But the 1st Battalion took heavy casualties: 265 killed or captured out of about eight hundred men.

By daylight on November 2, it was clear to the Division Commander, Major General "Hap" Gay, the 8th Cavalry Regiment had suffered a disaster. No one knew what was left of it at Unsan, but whatever was left certainly must be rescued. Heroic measures were necessary. He committed all three battalions of the 5th Cavalry, plus the 1st and 2d Battalions from the 7th, in a desperate effort to break through to Unsan. Without sufficient artillery and close air support the rescue couldn't succeed. The casualties suffered in the 5th Cavalry were ghastly: 350, 250 of them in one battalion.

Alarmed by the heavy casualties, the Corps Commander ordered General Gay to cease the rescue attempt. The 1st Cavalry and the remaining units of I Corps were to pull back south of the Chongchon River. Gay, with heavy heart, ordered his battered division to withdraw, abandoning the remainder of the 8th to their fate. Enclaves of brave men fought on for several more days, inflicting five hundred or more casualties on the CCF. When it was all over, about six hundred of the 3d Battalion's eight hundred men were dead or captured.

This was to be one of several gut-wrenching defeats for the Eighth United States Army, the US Marines, the ROKs, and the rest of the UN forces, as the full fury of the Chinese Communist intervention began dramatically, once more altering the course of the war. The 1st Cavalry Division, I Corps, and the UN forces would have to go over to the defensive. Another long trek south was beginning.

A BATTLEFIELD EDUCATION BEGINS

Captain John Flynn, Dave Hughes' K Company Commander, called his officers together those first few days in November, as he had done and would often do in the months ahead, to give them the word. "It looks like we're in for a long pullback." Dave was a rifle platoon leader, 2d Platoon of K Company. He would spend many hours during breaks on the retreat, listening to John Flynn talking with his officers and non-commissioned officers. Bull sessions. Discussions about what was important: principles, values, how to fight, tactics, techniques, the attack, counterattacks, withdrawals, road blocks, rear guard actions, high ground and how to take hills; automatic weapons, mortar, artillery, and air support, anti-tank warfare; marching fire, the importance of discipline under fire – don't stop firing in a fire fight, every trooper keep firing – mutual support for units on the flanks or when leading assaults, patrols, night fighting, how to calm men before a fight, how to lead under fire.

It was an endless list of what John Flynn had written of in infantry doctrine and manuals at the infantry schools in Fort Benning, Georgia, where he had been an instructor before being called to war a second time. He was the consummate teacher and soldier on the battlefield, who fought beside his men as well as led them, a man his officers and noncoms could admire.

And it was during one of these sessions that John Flynn, while preparing his officers and noncoms for what lay ahead, did what every good commander should do. Flynn believed it isn't enough to train and prepare well the men in a command who are to be committed to battle, and to prepare the command's officers to be good leaders in battle. He believed it's equally important to know in advance who is to step forward and lead during a crisis, should the commander be killed or seriously wounded – wounded sufficiently to incapacitate him, or require his evacuation to an aid station. Command and leadership continuity must be in place before battle. A unit engaged in a fire fight can't afford to suddenly become leaderless. The cost can be dreadful, and extend far beyond the unit whose commander has fallen. A good commander designates a chain of succession for just such a contingency.

During the retreat, the night when Dave Hughes first heard John Flynn name Company K's chain of succession, and heard his name on the list, this was the night he suddenly realized he might one day carry the full burden of command. It was a sobering thought in circumstances already filled with uncertainty.

Dave Hughes listened intently to all John Flynn's talks, learned, and admired.

In the intervening months there were rear guard actions to allow UN forces to withdraw southward, and remain intact. The CCF leapfrogged rapidly along the hills above the withdrawing Eighth Army columns, setting up roadblocks and ambushes, harassing, launching sharp night attacks when the retreating columns drew into perimeters to defend against assaults. The Eighth continued down the western half of the Korean peninsula, past Pyongyang, past Seoul – giving up the capital city once more, moving further south.

Down the middle of the peninsula, the CCF and its reconstituting allies, the NKPA, poured slowly through the rough mountainous terrain of a growing gap between UN forces in the two halves of the peninsula.

In the eastern half, the X Corps, commanded by Major General Edward M. "Ned" Almond (Virginia Military Institute class of 1915), who also doubled as MacArthur's chief of staff, reeled backward before the Chinese onslaught. The X Corps separated into two enclaves, one containing the Army's 7th Infantry Division and the 1st U.S. Marine Division. The two divisions, which had come ashore in September's landing at Inchon, closed into a shrinking defensive perimeter around the North Korean port of Hungnam, and resisted increasingly strong attacks by the oncoming enemy.

Further south the second enclave was established around North Korea's port of Wonsan. Forming a defensive perimeter around Wonsan were elements of the Army's 3d Infantry Division, and a ROK Marine division. They gradually pulled back, shrinking the perimeter, were finally evacuated aboard U.S. Navy ships, and moved north to Hungnam to bolster the defensive perimeter there, which was under heavy pressure from the advancing Chinese and NKPA. Slowly, the forces around Hungnam contracted their defensive positions, while a massive evacuation was under way, closely supported by air and naval power. By December 24, the evacuation was complete. In all, 206,600 persons were evacuated, including 108,600 troops, of which 3,600 had been airlifted.

Nine days earlier, President Truman, concerned about a possible outbreak of general war with the Soviet Bloc, declared a national emergency.

The UN forces in the eastern half of Korea had been saved from the pursuing CCF. The X Corps disembarked near the tip of the Korean peninsula, in the South Korean port of Pusan, and moved north to link up with the retreating Eighth Army in the west. The two forces reestablished a continuous defense line from coast to coast in South Korea, south of Seoul.

The evacuation of UN forces from North Korea, by sea and air, had been the greatest rescue of ground forces since the evacuation of Dunkirk by the British in World War II. But the losses had been staggering since the CCF entered the war in late October.

On December 23, 1950, one day before the closure of the Hungnam evacuation, while UN forces were continuing to withdraw southward before the flood of CCF manpower, Lieutenant General Walton H.

"Johnnie" Walker (West Point class of 1912), the Eighth Army commander, was killed in a jeep accident north of Seoul. His replacement was Lieutenant General Matthew B. Ridgway (April 1917), reassigned from his position as the Army's Deputy Chief of Staff for Operations and Administration. On January 4, 1951, Seoul changed hands for the third time.

Ridgway confronted a daunting task. The UN forces were in retreat, dispirited, and demoralized. In the six months since the war began, the first two months had been a desperate fight to delay, hold on, buy time at terrible cost, hoping to reinforce soon enough to stem the NKPA tide. Then had come the breakout from the Pusan perimeter, fast on the heels of the Inchon landing, and the month and a half dash northward, with quick victory apparently in sight. Crushing reversal and bitter disappointment followed when the CCF entered the war in early November. Ridgway's mission was to grab hold of retreating forces, expunge the pall of defeat, revitalize their battered morale, and instill the belief they could turn the tide and win. It would not be easy, but in the space of one month after he took command, the turnaround took hold.

He started by quietly replacing corps and division commanders who were weary of the fight, men who seemed to have absorbed defeatist mentalities. Ridgway had the backing of the Army Chief of Staff, General J. Lawton Collins, and the Chairman of the Joint Chiefs, General Omar Bradley. Major General Bryant E. Moore, the Academy's superintendent was among those Ridgway asked for to help take the offensive. The new Eighth Army Commander got nearly all the replacements he asked for, and more and better weapons and equipment continued to pour onto the peninsula. In the meantime the advancing Chinese and North Koreans were extending their lines of communication and resupply, and absent effective air and naval support, subjecting themselves increasingly to destructive air and naval bombardment by American forces.

By January 25 Ridgway was ready to go on the offensive, with punishing firepower complemented with cautious, well coordinated offensive maneuvering. In the UN's first counteroffensive against the CCF and NKPA, he continued a series of hammer blows to the enemy, inflicting heavy losses. His forces absorbed a counterattack, gave ground, and attacked again. The UN kept up relentless pressure through

April 21, and continued to pound the CCF and NKPA with withering firepower.

Ten days before the UN's first counteroffensive came to a close, General Douglas MacArthur was relieved of his command by President Truman. No sooner than Ridgway got the Eighth turned around, and in an aggressive fighting stance, he was selected to fill the position vacated by MacArthur's release.

Now, in April of 1951, it would be Lieutenant General James A. Van Fleet (class of 1915) who would take command of the Eighth Army and UN field forces in Korea. His work as commander was to include twenty-two agonizing, frustrating months of alternating offensive and defensive position and attrition warfare, and finally, trench warfare, as the UN forces fought determinedly to gain ground, and leverage, for the truce negotiations which began in Kaesong on July 10.

When General Ridgway turned the Eighth Army around, the 1st Cavalry Division and its Company K, 3d Battalion, 7th Cavalry Regiment, were ready to begin its long, last, Korean War fight – to win.

WAR'S GRIM CLASSROOM

Classrooms and field training exercises can never adequately prepare a soldier for what awaits him on the battlefield. There are circumstances faced in war which are incomprehensible, unbelievable, brutal beyond description. The old adage about "hours and hours of boredom punctuated by moments of stark terror" doesn't come close to describing what men face in infantry warfare. If a man doesn't listen, doesn't learn every day he survives, his chances of survival, never mind success, only diminish.

Dave Hughes' first battlefield classroom was defeat and fighting withdrawals – retreat. Not the best beginning. But his education began the day he arrived in K Company. Teachers, and men he looked up to, were John Flynn, First Lieutenant Shanks, Master Sergeant Abeticio, who later got a rare battlefield commission – and other veterans who had "been there" before he arrived. He learned nearly everything from John Flynn and those he looked up to.

But he also had to learn much about himself and the puzzling, ugly randomness of war, small things that loomed large in the struggle to live and lead soldiers – the GIs – the way they deserved to be led.

173

Ears. Hearing. Alertness to sounds. The ability to pick up sounds, instantaneously know where the sounds and firing are coming from, instantaneously interpret them, and what type of weapons are being fired. Ears, discriminating ears, were the most crucial sensors God gave men to survive on Korea's infantry battlefields. Quick, careful, accurate hearing was absolutely essential.

Almost unconsciously, Dave Hughes developed the habit of jerking his helmet off his head when the enemy opened fire – a habit most would consider unwise, completely unsafe when bullets were flying. "Crack – Thump!" Machine gun? Where is it?

"Ka – choonk!" Incoming mortar round. A sixty millimeter? Eighty-two millimeter? Hundred-twenty millimeter? One was more deadly than the other. Each dug more deeply than the other when it hit, and the shrapnel spray patterns were different, progressively more lethal. A split second's hesitation in interpretation of a whistle overhead, and the decision could be fatal for him, and many more. The sixty millimeter is hard to beat, but smaller. The hundred twenty is a killer, but slow. The eighty-two matches a man's reaction, and is deadly.

Eyes? Yes, they were terribly important. But in Korea, as in all preceding wars, at night eyes were severely limited by darkness, the time when the CCF and NKPA far more frequently attacked. Even when flares were launched to illuminate the night battlefield, eyes could be severely confused and impaired. And when flares weren't launched, muzzle flashes, and the flashes of exploding enemy rounds induced temporary night blindness, were disorienting and confusing.

Luck? Fate? Charmed life? Guardian angel? God's will? What was it? Why do some men repeatedly walk or run through hell, barrages of exploding artillery and mortar shells, automatic weapons, rifles, and grenades, and never get hit, while others' survival is measured in seconds, minutes, or hours after their very first battle starts? It's a mystery none can decipher. To dwell on the questions and their possible answers could be both maddening and fatal.

One day a hundred-twenty millimeter mortar round impacted so close to Dave Hughes he couldn't possibly have survived the explosion. He heard it coming, and knew instantly what it was. He knew it would be close. He dove for the earth's cover, flattened himself as much as humanly possible, hugging the ground. The projectile's tail fin broke off on impact, severing the weapon's arming and fusing chain. It didn't explode, a "dud." He received scratches when the tail fin broke off, tumbled, and struck him on his arm.

In an assault on a high hill, his platoon was pinned down close to a strong Chinese defensive position, and the enemy began throwing grenades. Someone near Hughes yelled, "Grenade!" He rolled over, felt something against his leg, and looked down just in time to see the handle of a "potato masher" – the handle of a Chinese hand grenade against him. "Blam!" The handle gave him a real "Charlie-horse" in his leg, and trouble with one eye for a while, but not a severe puncture anywhere. His minor wounds brought the award of his first Purple Heart. The grenade killed the man lying next to him, an unexplainable, sickening surprise.

Enemy tactics and enemy ruses. During an assault, when allied infantrymen overran an enemy defensive position, a tactic of the Chinese soldier was to keep out of sight, well hidden in his bunker or "spider hole" or play dead. Then, after the attackers went past their positions, they would stand up and open fire from the rear.

Dave Hughes nearly lost his life one day, exactly in that manner. He was still carrying the M-2 carbine as his weapon at the time, a thirty-caliber, semiautomatic weapon carried by officers and most medics – along with a trench knife. As the platoon leader, he routinely carried the carbine, a map, and a hand held radio – a "walkie talkie," so he could, in coordination with his company commander and the other platoon leaders, exercise command and control of his unit, coordinate supporting fires – and fight. The carbine had a reputation for jamming, not a desirable trait or reputation for any weapon used in close combat. The weapon was light, more easily affected by the Korean soil's grit than other heavier weapons.

On this day they were in the assault, pushing hard to take high ground, and overrunning Chinese defensive positions. Intent on keeping the assault going, he happened to turn his head slightly, and out of the corner of his eye, caught sight of a wounded Chinese soldier rising up behind him. He wheeled to fire his carbine, and the weapon jammed after one round, which missed its target.

Time is one of several enemies in such circumstances. Slow reaction is another. Friends are training, adrenaline, which speeds reaction, and practice at clearing jammed weapons. He was able to react, clear the jam, and fire before the Chinese soldier had a chance to hit him. From fifteen feet away Dave Hughes' second attempt to save himself was successful. He killed the enemy soldier.

That did it! Another lesson. Get rid of the carbine. Its reputation and performance could no longer be ignored. The carbine had very nearly cost him his life, and would be of no value if he had to save the life of one of his troopers. He took a vertical clip loading, forty-five-caliber Thompson submachine gun from a dead Chinese soldier. Ironically, the Thompson was American made, had been shipped to the Nationalist Chinese during the 1940s, lost to the Chinese Communists, and had been used against American soldiers in Korea.

Dave Hughes was left handed. He rigged the weapon with a sling to carry it on his left shoulder, muzzle pointing forward, so he could rapidly reach down and fire from the waist. His recaptured Thompson had a heavy bolt which could crunch through lots of grit. No misfires.

Battlefield lessons must be indelibly stamped in one's memory, ready to be recalled and acted upon with great speed, under enormous pressure. In the heat of a furious fight, men quite literally at times run on automatic. No time to think, reflect, discuss, or consult. Decisions have to be instantaneous and right, under the worst of circumstances, or the entire situation can become irretrievable, with catastrophic loss of life. And the lessons can't be learned overnight. Not in one fight, not two, not three. The learning has to go on and on, layer upon layer, as experience and responsibility expand, especially if an officer charged with leading, is to be ready, and is to be accepted by his troopers as ready to assume greater leadership responsibility.

Dave Hughes "ran on automatic" many times in his months in Korea. The first independent offensive action K Company participated in was an assault his platoon led. When the operation was over, much to his astonishment he realized he had taken a hilltop almost single handed. The hill was steep on both sides. His troopers met heavy, determined resistance, and the attack bogged down. They took cover and stopped their advance. He couldn't get them to move. Execution of the assault plan, coordination of fire and movement, his troopers pinned down by fire, taking casualties, had been too much. He lost his presence of mind and stormed the crest of the hill. When he got to the top he realized he was alone, and he didn't remember what he'd done – but the fight was over. His soldiers were witnesses to his charge up the hill.

John Flynn recommended him for a decoration, the Silver Star. Dave was perplexed, ambivalent about the nomination. This was his first real fight in the assault, on the offensive. He remembered how little he really knew at the time, and how much he had to learn. Leading in combat was an art, and he hadn't practiced the art well. These were discomforting thoughts to go with a decoration.

In the spring, the Assistant Division Commander selected Dave to be his aide, and Hughes was able to rest, get cleaned up, and look civilized. He quickly got bored and unhappy. He felt he belonged in the fight, and his boss knew it, and understood. He sent David back to K Company.

Captain John Flynn received serious wounds in June of 1951, while the Collins Board and its aftermath were grinding on in confidentiality at West Point. The wounds ended his participation in the war, and earned his evacuation to the States. John didn't forget K Company. He wanted to know what was going on, how they were doing. His replacement tried many times to answer John Flynn's inquiries, but the letters went unfinished, never seemed adequate, always out of perspective – at least until March of 1952, when David Hughes was on a Japanese ship, bound for the United States.

Hughes had become the K Company commander. By June Dave was at the top of Flynn's leadership succession chain. Flynn's choice wasn't entirely a matter of skill, or the art of his leadership. Much of it had to do with luck, surviving, gaining experience while surviving, while the rest of the company's lieutenants had been killed, wounded, or rotated, and their replacements had met the same fate. Turnover among lieutenants is high in line infantry units engaged in battle. Other factors may have been John Flynn's conclusions about the qualities he had seen in his officers, qualities necessary for command in combat – plus their combat experience.

Though he had become company commander, the battlefield lessons continued for David, some far more difficult and painful than he remotely imagined. Cruel choices forced by impossible circumstances, followed by years of smothered regrets, regrets that in war had to be immediately swallowed or buried.

Once, while his depleted company was preparing to launch an assault before daylight against a heavily defended hill, he received approximately thirty replacements in the early hours after midnight. He couldn't see their faces, and if he could have, he wouldn't remember them. The acquaintance was too brief.

Dave Hughes had no choice. His officers and noncoms had to position the replacements in the company's area, immediately, to strengthen the attacking force. In the dark of night, the young soldiers had no opportunity to see or become acquainted with other men in the company, their squad leaders, platoon sergeants, or platoon leaders.

David Hughes never forgot the number of casualties his company suffered that day – twenty-two cut down. Several of those hit were among the thirty replacements he received during the night. Grievous losses. He hadn't seen their faces, and they probably never saw his. Yet he bore the responsibility, the losses, and the buried frustration, anger, and pain, along with their parents and loved ones. He swore to himself that would never happen again. He would never accept green troops in the midst of close combat, nor would he send them to units already engaged in a fight.

In another long, seven hour night fight, when K Company was holding an important hilltop outpost, enemy assaults by four companies – a battalion – began to take a heavy toll, encroaching ever closer, nearly collapsing one flank of the company's defensive perimeter. After repeated assaults, the enemy succeeded in overrunning positions on their left flank, to within thirty-five yards of the company command post. Hughes sent his First Sergeant with orders to direct the pull back of a fifty-seven millimeter gun section, which was now in an untenable position because of the advancing Chinese. A Chinese soldier came up the hill with the withdrawing K Company troopers and penetrated into the company CP. After a tussle in the CP, he jumped into the hole with Hughes, who killed the enemy soldier with his submachine gun.

In the advance up K Company's left flank, the Chinese took three prisoners. During the heat of the engagement, Dave Hughes rightly concluded he didn't dare risk his entire company, and loss of the key hill, by sending a rescue force after the three prisoners. The Chinese immediately pulled one prisoner from the area of the fighting, down the hill, not to be seen again. The other two they pushed up the hill with them during the assault. One of the captured troopers, seeing a K Company heavy machine gun kill every man in a Chinese mortar crew, and cut down the enemy attack wave, kicked his captor, jumped over the side of the steep ridge, and escaped. In the continued Chinese advance, the third American soldier was killed by fire from K Company.

Another lifetime memory, a regret, buried, to keep fighting without putting others at increased risk. Regrets and mourning take place, but these are luxuries none can afford to dwell on – not in a fight, and not while struggling to keep others alive, hold unit integrity, keep fighting against this assault wave – and be ready for the next one.

In the fall of that year, classmate Lieutenant John Ross was assigned to K Company. But in his first hard action his arm was shattered by machine gun fire, ending his brief stint in combat. Hughes remembered the odd feeling of commanding a classmate, and sending him, possibly, to his death. The responsibility and lessons of command.

HILL 339 – THE PRELUDE

In late August of 1951, while cadets found guilty of honor violations by the Collins Board were resigning at West Point, truce negotiations broke off at Kaesong, South Korea, where they had begun on July 10. Though both sides had agreed hostilities would continue while negotiations were in progress, it was obvious neither side would start an all-out offensive unless talks became deadlocked or broke off completely. But the no-man's land between the armies didn't remain inactive in the seven week period that talks continued.

Air strikes, continuous artillery bombardment, constant combat patrolling, and offensive ground operations of battalion, and occasionally regimental attacks, characterized the fighting. The attacks were to secure key terrain, bring in prisoners, and relieve enemy pressure on the UN lines. "Except for several offensives that grew out of intermittent breaking off of armistice negotiations, this was generally the pattern of operations that prevailed until the signing of the armistice [in July 1953]."

When the truce talks broke off in late August 1951, General Van Fleet decided to resume the offensive. The objectives were to drive the enemy further back from the Hwachon Reservoir, which supplied water and electrical power to Seoul, and away from the Chorwon-Seoul railroad. The offensive began first in the eastern half of the peninsula, when two divisions in X Corps, the 1st Marine Division and the Army's 2d Infantry Division, launched heavy attacks intended to secure high ridges to the west and north of the bitterly contested area called the "Punch Bowl."

Further west, in I Corps, the 1st Cavalry Division was one of four divisions, plus the British Commonwealth Brigade, that advanced on a forty-mile wide front from Kaesong to Chorwon. On September 21, in the area near Chorwon, K Company, 3d Battalion, 7th Cavalry prepared to go up more hills, "get off the road" and "take high ground," as General Matthew Ridgway had directed when he took command of the Eighth Army the previous December. General Van Fleet, who replaced Ridgway after MacArthur was fired, continued Ridgway's aggressive, methodical approach to winning battles and administering punishment to the Chinese and North Koreans. The 7th was to be part of an effort to advance north from Line Wyoming, to seize and hold Line Jamestown.

The number 339 was the hill's elevation in meters above sea level – approximately 371 yards.

The 3d Battalion of the 7th Cavalry was given the mission of taking and holding a patrol base extending from Hill 339 to Hill 343, and back over to Hill 321, a four-thousand-yard perimeter. K Company, commanded by David Hughes, received the mission of securing Hill 339 and one thousand yards of the 3d Battalion's perimeter. It was a mission not relished by Hughes because in the preceding three weeks there had been three attempts by other battalions to seize and hold 339, all ending in failure, with entire companies overrun, and heavy casualties. Nevertheless, up K Company would go on September 21.

This time the fight to gain 339 was relatively easy. The Chinese set off a red flare, and simply pulled off the hill. Five minutes after K Company took the crest of the hill, the Americans found out why the assault had been successful so quickly, and how life would be for the next week.

Dave Hughes remembered, "They suddenly began shelling us and mortaring until I thought the roof was going to come off the hill. They kept working the front slope with a battery of seventy-five millimeter and self-propelled artillery, and they shook us to pieces with more hundred-twenty millimeter mortars.... The rain of eighty-two millimeter and sixty millimeter was just incidental. The fewest incoming rounds we ever reported for twenty-four hours was 350, and we estimated twelve hundred on the second day."

Not until the second day did the men in K Company learn why the Chinese had targeted them while hardly touching the rest of the 3d Battalion perimeter. From the Observation Post on the crest of 339, observers in K Company could see more of the Chinese troop and gun positions and access routes than the Chinese could afford to have their adversaries see.

The Company dug in and began organizing to defend the hill. Below them, on the slope of the hill, were dead enemy soldiers from previous battles. The Chinese, from their positions, watched like hawks, and from the flanks of K Company could observe the rear slope of Hill 339, not an enviable circumstance for the Company. In the daylight Dave Hughes couldn't put men on the ridge top, or on the forward slope. The Chinese would immediately open fire with self-propelled artillery and dig them out of their holes. In one week, from bombardment alone, the company suffered 33 casualties from direct hits on "fox holes" with mortars, including a regular midnight dose of hundred-twenty millimeter mortars.

The first night on Hill 339, K Company had a scrap. The Chinese came across a "saddle" in the hill line, the same one they had used when they had hit and overrun C Company two weeks earlier. And they came down a road on the extreme right flank of the company. On the road they ran into an American tank, which scattered the attackers. Mortar fire kept them dispersed. On the peak of 339 the enemy plastered K Company with everything they had, and the Chinese infantry came in right under their own mortar fire on the shoulder of the hill – into the teeth of a heavy machine gun. The gunner, Sergeant Malloy, held his fire until the enemy soldiers were ten yards away. When he opened up, he stopped them cold. In the dark, confusion, and noise they never located Malloy's firing position. The assaulting Chinese force continued to crawl around and pour machine gun fire on K Company for a few more hours, then withdrew, pulling some of their dead with them.

The next morning five enemy dead were found within those ten yards in front of Sergeant Malloy's machine gun. One had his hand draped over the parapet. K Company took no casualties from enemy small arms.

181

K Company Troopers during a pause in Fighting on Hill 339, late September 1951. (Photo courtesy David R. Hughes.)

And so the cat and mouse game continued for seven days, while K Company sent out patrols from their base on Hill 339, attempting to locate enemy positions, and determine their strength.

Dave Hughes dreaded sending out patrols. He knew his men were under observation from the moment any patrol's point man began moving forward. Each patrol normally came under heavy attack by the time it had moved forward no more than six hundred yards – slightly more than a third of a mile. The Chinese had become better at cat and mouse as the days went by, and better at holding their fire, factors which increased Hughes' concern for the men sent on patrols.

In spite of his well-founded dread, on September 22, he was ordered to send a combat patrol the following day against positions Hughes knew were in front of K Company. The purpose of combat patrols was to pinpoint and determine the strength of enemy positions in front of friendly forces, whereas reconnaissance patrols were intended to avoid enemy contact while gathering tactical intelligence. Thus the mission of combat patrols nearly always implied contact with the enemy.

Lieutenant Charles Radcliffe's 1st Platoon was given the mission, a platoon depleted by losses, as was the entire company. Radcliffe, a fine young Canadian who called New Jersey his home, was K Company's most experienced platoon leader. Hughes called the men together the evening of the 22d, and briefed them. Radcliffe had designated Private Leo J. Bernal of San Bernardino, California, the point man on the patrol. As patrol members gathered for David Hughes' briefing, Bernal turned to Private Edward J. Escalante, and quietly told him of his assignment as point man. To Escalante, Leo Bernal's face and voice expressed sadness, and the thought Bernal believed he wouldn't survive this patrol.

Edward Escalante had been on patrols, but this one was different. Fear was present on any patrol, but he was more afraid than he'd ever been. He had twisted his ankle several days earlier. Weakness and the distracting pain were still there. He had asked David Hughes to let him off this mission, fearing that if the patrol got into trouble he wouldn't perform as he should, and wouldn't make it back. Dave told Escalante he couldn't spare him, his experience was needed. If a fight started he would forget his pain. Edward Escalante wasn't a soldier to dispute his company commander's decision.

When the patrol moved forward from K Company's Hill 339 positions on September 23, they threaded their way, in column, down the north slope and crossed a mine field, moving toward the hill directly in front of them. Bernal was on point and Radcliffe was not far in front of Edward Escalante, who was a rifleman in the second squad in line. The platoon sergeant was at the rear of the column.

Someone in the patrol saw a Chinese soldier at a considerable distance, crawling forward toward K Company positions atop 339, apparently intent on reporting his observations to his unit. A soldier asked Radcliffe's permission to fire on him. The platoon leader refused the request, not wanting to expose their

Pvt. Edward J. Escalante

patrol while Bernal was clearing the way ahead.

Pvt. Leo J. Bernal
(Photo by Escalante)

They started uphill. Escalante took out his camera and began taking pictures of his friends. He saw what appeared to be a clearing in the brush on the hill in front of them, and told his squad leader, who didn't seem concerned.

The patrol had gone barely two hundred yards more when the Chinese opened up with a deadly machine gun cross fire, from thick trees on the hill directly above them, and another position on their flank. The first rounds immediately killed Bernal, and mortally wounded the platoon leader, Lieutenant Radcliffe. Escalante and the entire patrol hit the dirt, hugging the ground for cover. Pinned to the earth by enemy fire, Escalante watched Radcliffe's wounded body roll downhill toward him. He could hear Radcliffe's moans, as he lay dying on the ground in front of him. Charles Radcliffe's moans seared the young private's memory, leaving him sounds he'd never forget.

Combat patrol
(Photo by Escalante)

More fire from the second enemy machine gun killed two more patrol members and wounded others. For a few confused moments the patrol was leaderless. An American light machine gun crew moved forward toward the second enemy machine gun, firing, while the gunner, "Mississippi" Jewel Collins, cradled the weapon in his arms, trying to find cover in a position from which they could continue keeping the enemy machine gun position under fire. But Jewel Collins and his assistant gunner were now trapped in an open area.

Edward Escalante's first response to the patrol's predicament was to quickly raise himself up and fire his M-1 rifle toward the second enemy machine gun, then hug the earth again, fully expecting return fire. None came. When the flanking enemy weapon had opened fire on the column, Escalante heard the distinct sound of the British-made automatic weapon, the Bren gun. Escalante described its guttural, killing noise as similar to an automobile engine that had thrown a piston. There was also the sound of another automatic weapon firing from the second position – a burp gun.

During the momentary confusion when the patrol was pinned to the earth, raked by the killing cross fire, he also heard Jewel Collins' shouts for more ammo for his machine gun. No one knows what entered Edward Escalante's mind. He never said. Perhaps it was rage. But he did forget his ankle. He had a box of thirty-caliber ammunition, as did a corporal near him. Escalante shouted, "You're the corporal! Go on up there!" The man didn't want to go. Escalante jumped to his feet, yelling, and ran forward up the column, pulling a belt of ammunition from the can he was carrying, readying it to feed the machine gun. It was well he did. Collins' last belt was almost empty. Had the gunners run out of ammunition they would have been at the mercy of the three automatic weapons firing at them.

When Ed reached the gun crew, he was greeted with a sight he'd never seen. An American soldier, a South Carolinian, Private First Class S. A. Hyatt, was on his knees, pitched forward on the side of the hill, dead.

Though shocked by the sight of Hyatt, Escalante didn't stop after he delivered the ammunition. He decided to cross to the right flank of the hill, which was an open area, and move straight up the slope – firing. He did, and was in a march, now, firing. Others in the patrol began following. Escalante was quick in reloading the eight-round clips his M-1 used. He kept moving forward, firing as rapidly as he could, a single shot each squeeze of the trigger. He moved to a position in front of the bunker, paused, and decided to take a chance on reloading again. He quietly ejected the rifle's clip, reloaded, slammed the rifle bolt forward, and discovered the clip he ejected contained two unfired rounds.

He didn't know how many Chinese soldiers were in the bunker. At that moment, one's head appeared out of the trench. Escalante fired two rounds from his rifle, and threw a hand grenade into the trench, from which enemy soldiers had been hurling grenades. He entered the bunker and found three dead Chinese soldiers.

Immediately, more grenades came in from a position about twenty yards further up the hill. He turned his attention to silencing the enemy hidden there. Another American soldier was now at Escalante's side. Edward asked for another grenade and threw it into the trench above. He recalled, "We went over the top of the trench, and found three dead Chinese soldiers. One was the bastard who'd killed the Lieutenant with the Bren gun. I killed his ass that day. He should have been in China, eating rice."

Escalante had now outdistanced his patrol and was in a dangerous position. Yet he continued to bring fire to bear on the other machine gun on the next higher ridge. Throughout the period when Escalante, almost single handed, rallied and led the patrol with his courage and aggressive actions, the platoon sergeant was able to gather the remaining patrol members, including four wounded, and begin withdrawing. He then shouted at Escalante, telling him it was "all clear." Only then did Edward Escalante leave his position. The 1st Platoon had finally successfully broken contact with the enemy, at the end of a disastrous ambush and brief, sharp fire fight – and Edward Escalante's last patrol. A few days later he was seriously wounded as K Company moved forward in an assault, in the continuing United Nations offensive. He was evacuated to the continental United States, his war ended. For his actions in front of Hill 339 on September 23, 1951, Edward Escalante received the Silver Star.

*   *   *

Every night, enemy patrols would crawl up Hill 339, plot and count K Company's positions. Every night Dave Hughes would have to calm down a squad that thought the whole Chinese Army was in front of them. But there were positive effects from all this – the men in K Company dug in tighter, kept their weapons spotless, slept during the daytime, and watched at night. The sixty millimeter mortar crew increased its responsiveness under the leadership of its black platoon leader, Lieutenant Walker, and the Company was able to strengthen its defensive positions with additional heavy and light machine guns, to totals of five and seven respectively by September 28.

Then came the 28th, the defining day for K Company's stay on Hill 339. The day was quiet, a good day for what David Hughes was sure would come, the first acid test of the unit now under his command – an all out Chinese assault, intended to dislodge the Company from 339. The bombardment commenced at eleven thirty in the evening and continued to midnight, when, for a few minutes, it stepped up to a frenzied pace, with everything the enemy had raining into K Company positions. Then came the Chinese infantry, up the hill. In a bitter, seven hour fight with a battalion strength attack that surged to within yards of the Company Command Post, the men of K Company held firm.

This was the night two companies of Chinese penetrated the Company perimeter, moved in behind the defenders, and cut communications with the 3d Battalion; when two more companies of enemy soldiers began overrunning both flanks of the Company; when three K Company troopers were captured, and two of them were marched up the hill with the attacking Chinese captors – resulting in the death of one American from the fire of his company mates; the night David Hughes killed a Chinese soldier with his submachine gun when the lone enemy jumped into the Company Command Post after trailing a withdrawing fifty-seven millimeter recoilless rifle team toward the hilltop; when everything Dave Hughes had had to learn in the preceding eleven months about defending a hilltop had to be brought to bear to repel an assault by overwhelming numbers.

By eight o'clock the next morning K Company counted seventy-seven enemy dead within their perimeter. The Company had suffered ten killed, fifteen wounded, and one captured. There were no "bugouts." Discipline held. Those who gave their lives fought to the death where they stood, in their assigned positions.

Lieutenant David R. Hughes, Commander, Company K, 3d Battalion, 7th Cavalry Regiment, the morning after the Chinese assault on Hill 339 was repulsed.

In the seven days on Hill 339, K Company, including men attached to support its operation, had taken fifty-four casualties. On September 29, the company was rotated around the Battalion perimeter, and I Company replaced them.

This was merely the prelude. Four days later, on October 3, with no replacements for their losses, K Company jumped off in the continuing Eighth Army offensive. A series of objectives, including two smaller hills, led to Hill 347, which, like several other hills in the vicinity, was nicknamed "Baldy."

There was "Old Baldy" and "Chink Baldy." There was "T-Bone Hill" and "Alligator Jaws." But "Baldy" had a particular meaning, a name coined by GIs because of the barren, hostile hill crests gouged and completely denuded of trees or foliage from constant shelling by the two opposing armies during the numerous times the hills changed hands.

There was another name given Hill 347 by the GIs who toiled up and down its battle scarred slopes - "Bloody Baldy."

Air strikes softening up Hill 347 on October 2, 1951.
(Photo courtesy Edward Escalante)

THE TAKING OF HILL 347

Action began on October 3, with the 4th Battalion, the Greek Expeditionary Force, to the right of K Company, and I Company on the left. At the end of the first day's fighting, the remainder of the late Lieutenant Radcliffe's 1st Platoon was destroyed, and two K Company officers were critically wounded. Company G took 130 casualties, including four officers, on Hill 418, and the Greek company on the right of K took 135. No units gained their objectives, and the 2d Battalion won and lost Hill 418 five times.

On October 4, the attacking UN forces again started up the outlying hills toward 347 with all the support they could muster. Again K Company was in the enemy trenches, as were the Greeks, but the tremendous mortar fire and the seemingly unlimited numbers of enemy threw out K Company and the Greeks. That was the night Dave Hughes received thirty replacements whose faces he never saw before he committed them to the early morning assault.

The next day, they tried again. The Greeks reached and held their objectives. K Company didn't until all the companies of the 3d Battalion attacked just after dark. K Company took the two smaller hills with seventeen more casualties, including the artillery and 4.2 inch mortar Forward Observers.

On October 6, the attackers, including K Company, reorganized, while the Chinese threw three thousand rounds into the 7th Cavalry Regiment's area of operations.

Having taken the two smaller hills nearer 347, and paused to reorganize, K Company was ready to advance on their primary objective. No more replacements came. The hill was bare for four hundred yards down from the peak, which meant assaulting troopers would be completely exposed in the final lunge at the crest. The Chinese were well covered, deeply dug in with interconnecting trenches, tunnels, dugouts, and bunkers. The trenches were four feet deep, displaced down the hill a few yards below the crest. Direct fire from below, even from tanks, couldn't dislodge them. They had sixty, eighty-two, and one hundred twenty millimeter mortars on the hill, and supporting artillery aimed for probable approaches to their defense network.

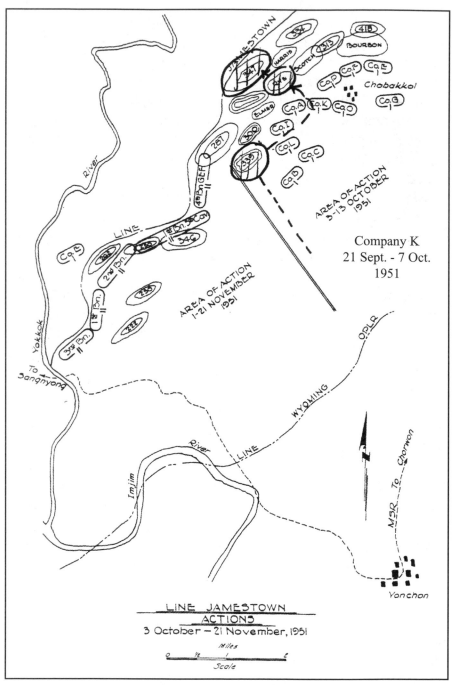

Heavy dotted line indicates K Company's route of advance toward Hill 347, September 21 - October 7, 1951.

All elements of the 7th Cavalry's 3d Battalion were committed. The assault began at ten o'clock in the morning, with L and I Companies attacking up the other side of the peak. While L Company was fighting up the hill, I Company had to turn and counterattack toward the 3d Battalion Observation Post, where senior officers in the Battalion were fighting off grenade attacks on their flank.

When the first assault by K Company began, Dave Hughes was seven hundred yards to the rear of the company's lead elements, where he could – with maps, radio, and weapon in hand – observe and control the remaining two platoons' progress, and coordinate all the supporting fires. The 2d Platoon led the attack.

The assault initially appeared successful. They reached the enemy's trenches but were forced off by intense, concentrated fire, and grenades, and pinned down, losing one officer and twenty more men. In each assault the Chinese rained grenades on their attackers, including anti-tank grenades. Sergeant Eugene F. Chyzy, the M Company machine gun section leader attached to K Company for the attack, remembered seeing three and four enemy grenades in the air at any one time during the assaults.

The second time up 347 that day, 3d Platoon attempted to assist, but the attack bogged down, and the remaining men had to pull back part way. In this attempt the officer leading the 1st Platoon was wounded. K Company fell short again, having lost another officer and more men.

In the third attempt the same thing occurred. The last officer in the Company, other than David Hughes, was wounded by a grenade, breaking the attack. Platoon Sergeant Monroe S. McKenzie, now leader of the 3d Platoon, radioed Hughes. McKenzie could see only three men left in the assault element. He asked Hughes what he should do. Hughes told him to "Hang on."

There were actually now six K Company riflemen left, still able to fight on Hill 347. What happened next was something Dave Hughes remembered little of afterward, until he began reading operations reports in Japan weeks later, to piece it all together.

On October 7, when matters seemed to have once more ground to a halt on the slopes of "Bloody Baldy," his mind working at a furious pace, he was at once reacting and acting, on automatic – putting to use

191

all he had learned from the men who had served with him, and what he had experienced in assaulting hills in the months after the 1st Cavalry Division went on the offensive again in January.

In observing the previous three assaults, he saw what had to be done differently to reach and hold the crest of 347 that day. He began moving forward, gathered the Company's headquarters element, mortar crews, and remaining Forward Observers, and told them they were to be riflemen. They, with him, were to continue forward and join the surviving K Company riflemen on the northeast slope of the hill. There would be a total of thirty men, including him, the equivalent of one, under strength platoon – less than one fourth a company's normal strength.

They loaded themselves up with ammunition and as many hand grenades as they could carry. Because he had taken the Forward Observers as riflemen, he could no longer coordinate distant fire support – artillery and mortars – for the Company. When they joined with the men pinned down on the side of the hill, he organized them for another assault.

He led Sergeant Chyzy and his M Company machine gun section to a position where they could support the fourth assault on Hill 347. He told Chyzy they were "to fire at any enemy up there [they] could hit, even if he or other GIs were up there and got counterattacked."

He returned to the riflemen and gave them orders for the assault. They were to use marching fire – fire repeatedly as they moved forward – to keep down the enemy's fire. They were to run through enemy return fire and grenades, no matter how intense, no matter the cost, until they crossed their trenches. Then they were to turn around and come back down on the trenches from above and behind.

He moved to the front of his men, waving them forward to begin the assault up the last 150 yards toward the crest of "Bloody Baldy," shouting encouragement. The time was about one o'clock in the afternoon. Three hours had elapsed since the attack on Hill 347 had begun.

His fatigue shirt and pockets were heavy with all the grenades he carried. Up they went, up a hillside being savaged by heavy mortar barrages, up through air laced repeatedly with shrapnel from exploding grenades, and small arms fire from their right flank. The men faltered, slowed, wanting to once more take cover. Hughes kept going, stayed in the lead, firing his submachine gun. Men began to notice, to see their commander, this young lieutenant, showing what seemed extraordinary courage and calmness in a storm of return fire, mortar rounds, and grenades. He was in the lead, showing the way. They had to follow. They didn't want to let him down. They couldn't leave him out front, alone.

Master Sergeant McKenzie saw two men killed by enemy anti-tank grenades. They were the Forward Observers. Hughes didn't slow his ascent. McKenzie and Corporal Robert W. Holden then saw Hughes throw his weapon down, leaving it behind. His submachine gun had quit firing, jammed or empty. He began hurling grenades - and kept going. He rushed the top, continuing to throw grenades, going straight for the bunker which had repeatedly stopped the previous attacks. To McKenzie, Hughes was "pulling his men through the fire," keeping the Chinese from standing up to fire back, and allowing the remaining riflemen to spread out and overrun the trench line. They followed, trying to keep pace.

Up he went, for the bunker, knocked it out, and continued to throw grenades into the lateral trenches, tunnels, and dugouts.

Company K's Sergeant First Class Arthur J. Shuld, Jr. was following not far behind Dave Hughes when he surged the final yards toward the top. Shuld tried to keep up, but was hit, wounded. As he was being helped back toward the aid station, he turned to see his Company Commander had made it to the top. The surviving K Company men were on the trench line with him, moving about, systematically throwing grenades and firing into the enemy positions.

When they reached the crest, they saw the L Company lead men coming up the other side, nearing the hill top.

Sergeant Chyzy, with his M Company machine gun section, also watched in admiration, as Dave Hughes led the rush toward the crest of Hill 347. He saw the men of K Company hesitate, then follow. When next he saw David Hughes, he was on top of the hill, shouting for Chyzy to bring his section up.

Chyzy would later say, "By his outstanding courage and leadership, Lt. Hughes inspired us so much that my section and myself under any conditions would stay with Lt. Hughes to the last man."

Sergeant Ray P. Moses, K Company, said "During the... push on Hill 347... I have never seen such heroism and courage as that Lt. Hughes

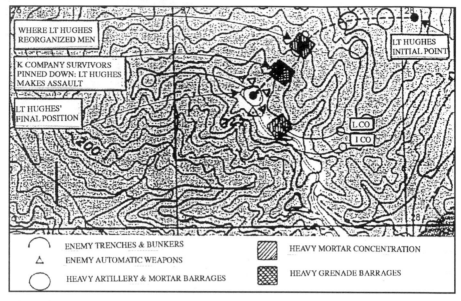

The taking of Hill 347. Lieutenant David R. Hughes' route during the fourth, and final assault, leading the rush of K Company to the Hill's crest. October 7, 1951.

had shown during the attack... It was only one of many times I have seen... his leadership and lack of fear while he was my company commander, but this was the greatest."

Corporal Holden recalled, "His cool-headedness and lack of bother for the terrible enemy fire in the attack presented an example to the men which alone held the Company together.... Upon being first up the hill and I being down behind him I saw he was fighting and killing like a mad man so we could get up there. Running in and out of the bunkers he threw everything he could, including Chinese grenades, until the hill was taken."

Tunnel by tunnel, the men of K Company rooted the enemy out – either as prisoners, or dead men.

A reinforced battalion of enemy soldiers had defended Hill 347. They were defending a Chinese division and regimental artillery command post, facts not known to K Company when the attacks began. By dark, K Company marched 192 prisoners off the hill, and counted a hundred dead enemy soldiers in the open, just within the perimeter of the Chinese trenches, which were only ten yards down slope from the peak, and two hundred yards long, ringing the hilltop. An unknown number of enemy casualties were attributed directly to David Hughes.

There were 250 enemy dead on the hill, many in their bunkers. The 3d Battalion, 7th Cavalry suffered seventy casualties. They had captured or killed all but eighty men in the enemy battalion, according to one of its soldiers, a clerk, captured later.

With all the attachments to K Company, including Sergeant Chyzy's fourteen man Company M machine gun section, which joined them on top, David Hughes had only thirty-seven men left under his command that day, only fifteen from Company K.

The fight for Hill 347 was over – for October 7, 1951.

GOING HOME

K Company was soon relieved from Hill 347 and rotated to another sector in the regimental front, where the 1st Battalion had just been overrun. The Company stayed relatively stationary on the hills for ten more days, while the 5th and 8th Cavalry Regiments moved up to take their objectives on Line Jamestown.

The last of the men who had been with the Company at the peak of the fighting were rotated out. The last K Company GIs were gone. Only Lieutenant David R. Hughes remained. He was the only officer for a brief period. A short time later he was reassigned as the assistant Regimental S-3 (Operations Officer). That was his duty when the 1st Cavalry Division was pulled out of the line into reserve, and prepared to ship out to Japan.

In the fighting near Chorwon, Korea, the 1st Cavalry Division had taken a severe pounding, suffering more casualties than in any equal period the Division was in Korea. K Company was fifth in the number of casualties by company. They had lost 167 men and six officers – and won all their battles.

David Hughes remembered first learning of West Point's cheating scandal in December when he got to Japan. He was stunned. He had been on the 1st Cavalry's front almost constantly since before the public announcement of August 3, and the story had faded from press reports by the time he was pulled off the line in October.

Not much news was forwarded to troops on the line. No radios were allowed. No *New York Times* delivered. And there aren't many free moments to pay attention, if there is any news – especially if you're an infantry platoon leader or a company commander. Information flows mostly one way from men on the line, and it's all business – up the chain of command, and seldom much more than orders come back down.

On January 11, 1952, David Hughes, for his actions on Hill 347, was nominated for the Distinguished Service Cross, the nation's second highest decoration for valor. The citation accompanying his award read in part, "First Lieutenant David Ralph Hughes...is cited for extraordinary heroism in action against an armed enemy on 7 October 1951, near Sokkogae, Korea."

David Hughes was fortunate, and he knew it. He survived unscathed. Good, brave men witnessed his act of courage on their behalf, and they had lived to tell what they saw. There were thousands more in Korea who gave their all in no less heroic acts. Sometimes their final acts were performed alone, without the benefit of witnesses. For others, no witnesses lived to tell of their courage and deeds.

But he had lived to see numerous acts of courage by men in K Company, and was determined they should be recognized. While in Japan, assigned to the 7th Regiment's headquarters, he wrote scores of recommendations for decorations. After he returned to the United States, in a bitter twist of irony, fire destroyed the headquarters building, and with it the Regiment's records of Korean operations, including all the recommendations for decorations he had written for the troopers of K Company.

In March 1952 Dave Hughes boarded a Japanese ship, the *Oturu Maru*, and sailed for home. During his recrossing of the Pacific, which lasted fifteen days, he finally paused and answered the letters Captain John Flynn had written him, inquiring of K Company and its well being. "...Here beside me I have several false starts on letters to you. But they were inadequate and out of perspective."

His answer was lengthy, completed on the only typewriter on board the ship. It was in the form of an "after action report," and clinically told of K Company's combat actions since Captain Flynn left Korea. He told of the battles for Hill 339, and "Bloody Baldy" – Hill 347, and many other details, answering Flynn's questions. He mailed the letter in Seattle, Washington, and headed home on leave.

Home. The Korean War was over for David Hughes, but it would not be his last war. For his West Point class, 1950, the Korean War wasn't yet over, and some of its members would also fight in another war. Many Academy graduates would fall in Korea, on the ground and in the air, before the Armistice was signed, though the opposing armies remained essentially static for twenty-two more months after K Company's last major battle.

Nor was Korea over for other West Point classes, and thousands more Americans who came after K Company left. There were more bloody battles all along the lines separating the two opposing armies and more fights for Hill 347 before the Armistice agreement took effect on July 27, 1953. And another, smaller hill, an Outpost a short distance to the north of 347, achieved a blood-soaked notoriety all its own in the spring and early summer of 1953 shortly before the Armistice.

Pork Chop Hill was the name.

CHAPTER 5

BOYHOOD DREAMS

The year is 1942, a late fall Saturday morning on the Rio Grande River several miles east, southeast of El Paso, Texas. A beautiful day, sunshine filled, only a few wisps of clouds in the sky. A nine-year-old boy, fishing pole in hand, walks to the left of his father along a gravel, levee-top road, with the river running silently past to their right. They are going to fish for bass, carp, and perch in the lily pad covered, still waters of a drainage canal, which is to his left, paralleling the levee.

The drone of airplane engines can occasionally be heard in the distance behind them. As the two fishermen walk and idly talk of the day's prospects for a catch, the boy turns to gaze over his left shoulder at the airplanes in the distance. He can see the silhouette of Mount Franklin, which is a short distance to the west of their home. The mountain is the highest of several peaks in a range rising behind the circling airplanes. The color of the desert mountains make the big, lumbering, four engine, propeller driven aircraft hard to see. They are B-24 heavy bombers flying in the traffic pattern at Biggs Army Airfield, beyond the northeast edge of the city. The boy has wondered admiringly about the great airborne machines while watching them as he walked home from Crockett Elementary School in the afternoons.

As he turns back to look for open water suitable for fishing in the canal, his eyes catch sight of something on the ground a few yards beyond the far bank. It's a soldier's hat, a real soldier's hat – not a store bought, play soldier's hat. The hat is a dark, olive green drill sergeant's hat, a campaign hat, lying in the Johnson grass growing around the lowest rung on an aging, rusting, broken-down barbed wire fence running parallel with the canal.

He excitedly tells his smiling father what he's seen and points to a land bridge over the canal a few yards ahead of them. The crossing covers a flume through which the canal waters ebb and flow. A breathless excitement seizes him as he hands his fishing pole to his father and dashes over the crossing to pick up his find.

And what a find it is! It truly is what he thought it to be, a wide-brimmed, felt, drill sergeant's hat, a campaign hat, complete with a leather chin strap. As the boy picks up the hat and rotates it to examine the front, he makes another exciting discovery. There, mounted on the front, just above the brim, is the small, metal, yellow and black unit emblem of the 1st Cavalry Division. It's beautiful! Untarnished. The yellow, elongated shield is edged in black and a wider black diagonal stripe runs from the upper left to the lower right of the shield. And in the yellow, upper right field of the emblem there is the famed black silhouette of a horse's head. The 1st Cavalry Division's home is at Fort Bliss, in El Paso, near Biggs Army Airfield.

There is no name inside the hat. Finders – keepers! This is a trophy of war a young boy could rightly claim. He can't wait to show his playmates and friends. Who is the owner of this prize? he wondered. Whoever he is, he's an American soldier. I wonder if he's like my uncles and cousins who have already enlisted in the Army Air Corps or Navy?

The find, the drill sergeant's hat, and the memories of that day, were part of a growing mosaic of images in his mind's eye. The images were of pictures and heroic stories, mixed with admiration and love of family members going off to war, all irresistibly urging him toward a dream not yet clear. Six more years passed before the dream took a clearer form – when his decisions and directed energies began turning the dream into reality.

During Christmas holidays of that same year in El Paso, he began imagining the dream, adding to the collection of images – memories – accumulating in his mind, yet still absent a clear meaning for his future. He is in an upper berth of a Pullman car, on a train, at night, riding through the countryside deep into Mexico, between Chihuahua and Torreon. The boy, his father and mother are on their way to visit his paternal grandparents in San Benito, Texas, near Brownsville, where he was born, eight hundred miles to the southeast of El Paso. Gas rationing is in effect in the United States. No long driving trips. Civilian travel on American trains or airlines isn't possible. The war effort has totally absorbed the country's transportation system, to move troops, military equipment, and supplies. The best way to grandmother and granddad's is by train through Mexico.

It's late. The car gently sways from side to side. The steady, rhythmic clickety-clak made by the car's wheels at each steel joint in the tracks relaxes him, urging sleep ever closer. The boy, in his pajamas, is lying on his side, propped up on his right elbow, next to the wall of the upper berth. Standing upright on the bed sheet before him are two armies of small lead soldiers, carefully deployed in battle formation, facing one another. They are Mexican soldiers of the Napoleonic era, their uniform cut distinct in spite of the lead's solid gray color. The small stand for each foot soldier and cavalryman is broad enough to keep them upright in spite of the gentle rocking of the Pullman. The boy is fighting a great, imaginary battle, a battle that will live in history, and shape the future of the United States, and the world. What more important battle can there be? It is, after all, war in the mind of a nine-year old. What he knows of war is innocent, childlike, absent comprehension of its deadly realities.

During World War II such were the beginnings of dreams for life, and dreams of young men destined for West Point and lives of service in the Armed Forces. The children of the Great Depression were growing up at a most impressionable age, in the greatest, most devastating war in human history. Patriotism was everywhere, and it was wonderful, and terribly exciting. Patriotism and love of country had suddenly bound us together after the struggle for economic survival that was the legacy of the Great Depression. Families, friends, small towns, cities, the whole nation seemed to pull together. Sacrifice came with relative ease, and little grumbling. America, indeed the whole world, was threatened by the dark evil of the Axis powers' Nazi Germany, Fascist Italy and the militaristic Empire of Japan. America's friends and allies were being brutalized long before the Japanese attack on Pearl Harbor on December 7, 1941. The attack had galvanized the entire nation. Even nine and ten-year-old boys and girls could understand why family members had to go off to war or to work in support of the war effort.

In the spring of 1950 dreams would begin to come true for the boy, who was now sixteen, still too young to enter the United States Military Academy. Others like him, with long held dreams, were seeking appointments to West Point. For many, however, experiences had been different. Dreams and wishes for the future hadn't included West Point or the armed forces. Though their ideas and plans for the future were different, circumstances and reality channeled them to the same place. There were other avenues and paths leading to the Military Academy.

EXAMPLES, IDEALS AND IMAGINATION – INSPIRATIONS FOR THE FUTURE

Far from the small town of Del Norte in Colorado's San Luis Valley, where I was nearing the end of my senior year in high school, Benjamin F. Schemmer was also finishing his senior year. He was at Phillips Academy, an exclusive boys preparatory school in Andover, Massachusetts. Through initiative and hard work, Ben received a scholarship to Phillips, but certainly he wasn't an ordinary Phillips student. His family couldn't afford the school's tuition. Their home was near the small town of Winner, South Dakota, where they had lost their family farm in the drought and devastating "dust bowl" that swept the plains of the Midwest during the Great Depression.

During World War II Ben admired his two older brothers. The oldest served as a Technical Sergeant in the Army Air Force, until forced to retire, disabled. The second older brother enlisted in the Navy and reached the rank of lieutenant commander before he was tragically killed in an accident, one week prior to his assumption of command of a destroyer escort.

After World War II, Ben spent four years at Phillips washing dishes in the kitchen which three times daily fed seven hundred boys, the great majority of whom came from wealthy families. Ben, given to hard work, wanted some direction, some meaning in his life, something more than "What car are mom and dad going to buy for me?" or "Who am I going to marry?" – the questions he heard most among the other more fortunate boys at Phillips.

For Ben Schemmer, answers to his question "What do I want to do with my life?" began to come his junior year at Phillips, when a new headmaster arrived at the Andover school. His name was John M. Kemper. He was a 1935 graduate of West Point and had served in the infantry during World War II. Ben, proud of his dad's service as a private first class during World War I, was impressed with John Kemper's example as the Phillips headmaster, particularly his integrity and sense of purpose. John Kemper became an important force in Ben's decision to seek an appointment to West Point.

201

Ben applied for both West Point and Annapolis. Early in 1950 he received a Congressional, third alternate appointment to West Point, from South Dakota. The third alternate appointment didn't assure his entry to the Academy that summer. He later was designated a qualified alternate, however, and was notified he had appointments to both Academies. He remained uncertain as to which to take, until the Korean War broke out. Soldiering. West Point. That was the place to go. In July 1950 he entered the Military Academy with the class of 1954.

Cadet Benjamin F. Schemmer
Class of 1954
(Photo courtesy USMA Archives.)

FOR WEST POINT – THE PLACE, ADMIRATION OF FAMILY AND LOVE OF THE GAME

Further south, in New York City's Bronx, at Mount St. Michael's High School, Thomas J. Bell had the previous fall completed a great senior year on the football team. Like many boys in the World War II era, during his junior high school years he was drawn to the football exploits of Army's "Doc" Blanchard and Glenn Davis, seen in the local theaters' "Movietone News" on fall, Saturday afternoons in 1944, '45, and '46.

When he entered Mount St. Michael's High School, Tom was determined to play on their football team, at the time one of the best teams in the city. Tommy wanted to be a halfback, and knew he had to be fast to be a good one. As a result he was on the track team and played football all four years. In the fall of 1949 he became a standout halfback, a record setter on a team that was a perennial powerhouse in "The Big Apple's" high school leagues. He won All-City recognition in football his junior and senior years.

Tom's parents immigrated from Ireland in 1930, during the Great Depression, two years before he was born. His dad worked at odd hours as a bartender in the Bronx, and his parents didn't own a car. Consequently they seldom attended his high school games, and while they knew Irish football, they didn't understand American football. They were baffled by its complexities. One day, Tom remembered, he came home and told his little Irish mother he had scored three touchdowns. Her reply in her soft Irish brogue was, "Is that good, Son?"

The tall, muscular, handsome redhead loved athletics, and paid less attention to his studies. His dreams were of Notre Dame, Fordham, Holy Cross, or Georgetown – mostly Notre Dame. He was getting state-wide notice as a strong, tough, hard-running ball carrier. His solid reputation as a football player spread, and in the spring of 1950 his athletic prowess paid off. Heavily scouted by several large universities, he had hopes of a scholarship offer. None came.

Subsequently, through two of Tom's high school friends and teammates, all three received invitations from Army assistant coaches to visit the Academy. Tom declined, telling his friends he didn't think he wanted "to be a cadet in a tall stiff collar." When his two friends came back from West Point that first Saturday, they were excited, and convinced him the Army coaches still wanted to see him. He should go with them next week. He did.

Like his friends, Tom was impressed with the Army coaching staff, the beautiful facilities, the grounds, and of course, the Corps of Cadets. The parades, uniforms, and the pretty girls lining the reviewing stands also caught his eye. He thought, This is for me!

Vincent T. "Vince" Lombardi was his contact for recruiting and shepherded Tom's appointment and participation in the four week "cram school" at Smith [ice hockey] Rink, where he prepared for his entrance examinations.

Vince Lombardi was also a product of "The Big Apple," Brooklyn's Cathedral Preparatory Seminary and St. Francis Prep, and Fordham University, where in the mid-1930s he had been a guard on one of college football's legendary front lines, the Fordham Rams' "Seven Blocks of Granite."

Vince Lombardi was particularly happy after the season of 1949, when Earl Blaik moved him to offensive backfield coach. The new assignment gave Vince the opportunity to see and work the entire offensive system Blaik had established at Army. When the coaching staff began their annual recruiting drive between the '49 season and '50 spring practice, Lombardi launched enthusiastically into recruiting young athletes who would one day be his backfield.

There was a kinship between Vince Lombardi and Tommy Bell. Vince was the son of Italian immigrants. His father, Harry, was the local butcher and his family had been supportive during Lombardi's five years of study for the Catholic priesthood. Because he grew up and got his start in the great immigrant-fed melting pots of New York City and northern New Jersey, his natural recruiting base was in high schools within those areas. It was no accident that some of his greatest finds for Army football were sons of immigrants.

When he came to the Bell household in the spring of 1950, Vince was an assistant coach, the varsity backfield coach at Army, under a man he referred to as "The Colonel," Earl "Red" Blaik.

Tommy Bell wasn't home when Vince Lombardi came to tell Tom's mother her son had passed his entrance exams for West Point. She told Tommy that when Vince, a mercurial, volatile, charismatic personality, excitedly told her the good news, he grabbed her in a big bear hug and they danced around the kitchen – and he said to her, "I'm going to make Tommy an All-American!"

Cadet Thomas J. Bell
Class of 1955
(Photo courtesy USMA Archives.)

Tom's appointment was as a qualified alternate, like Ben Schemmer's appointment, a means of filling appointments vacated late in the annual cycle by men who, for a variety of reasons, either elected not to attend the Academy, or failed entrance examinations. It was also through qualified alternate appointments that the Academy's Athletic Department each year recruited many of its intercollegiate athletes.

The same spring Lombardi recruited Tommy Bell, Vince and Captain Johnny Green, the defensive line coach on Earl Blaik's staff, visited another young high school football player, who later entered West Point with the class of 1954. He was Robert M. Mischak, a big, fast, single-wing tailback from Union, New Jersey. Bob was also the son of an immigrant.

Bob's father was born in the Ukraine, and came to the United States as a boy, before World War I. The elder Mischak grew up with a deep appreciation for the freedoms enjoyed in his adopted homeland, so much so that when the United States

Cadet Robert M. Mischak
Class of 1954
(Photo courtesy USMA
Archives.)

entered World War I, he falsified his age, enlisted at 16, and as a private went back to Europe with the American Expeditionary Forces to fight in "the war to end all wars." When he returned home, he went to work in the oil industry in Carteret, New Jersey. Initially dissatisfied with his line of work, he joined the U.S. Marine Corps. After a short enlistment, he returned home and started both his family and a career in Newark, then moved to Union. Bob had two sisters, one older, the other younger, who died shortly after birth. His father's legacy to Bob was pride in his country and service in the American military. He taught Bob that success required hard work and honesty.

Like other young members of first generation immigrant families in America, Bob Mischak knew economics and hard work were major considerations in thoughts of college. His parents couldn't sustain the costs of college educations for their children. Any scholarship Bob could obtain would help. He was a solid student, but not a scholar. He had deliberately taken courses to prepare him for college, but he knew football would have to be his path to a good education.

Bob Mischak admired Army football. He remembered his family living in a modest two bedroom Cape Cod style home. As a young boy on Saturday afternoons in the falls during World War II, while scraping and repainting front porch railings, or "stoop" railings as they were called, he played "stoop ball" with friends. As they used the steps of the porch to launch thrown tennis balls, they listened to Army football games on the radio. For Bob Mischak, as for Tommy Bell and most of us of that time, it was the era of "Doc" Blanchard and Glenn Davis.

The University of North Carolina, a powerhouse in college football, actively recruited Bob Mischak, but he chose West Point instead, after visiting the Academy with his father. His father met Earl Blaik and they talked of their experiences during World War I, when Blaik was a cadet at West Point, and Bob's father was a private in the Army. Bob's pride in his father, and his dad's admiration for the military were the determiners for Bob Mischak's entry into West Point. Football would become the determiner of his life after graduating from the Academy.

"THEY 'UNDERSTOOD' I WANTED TO GO TO WEST POINT…"

He came to West Point an artillery-man's son, an Army "brat," his father an Academy graduate in the class of 1931. Edward M. Moses didn't graduate from high school, most unusual for any man seeking an appointment. His route to the Academy was difficult.

Ed Moses' father and mother married right after his father graduated from West Point. Ed was born at Fort Bragg, North Carolina, his father's first assignment after graduation, and in the years before World War II they twice moved between Fort Bragg and Fort Sill, Oklahoma, home of the Army's field artillery school. During the war they lived in Bethesda, Maryland, and after the war was over they went to Germany where his father served in the occupation forces. As a freshman at an American school in Frankfurt, Germany, Ed wasn't receiving the education needed to prepare him for West Point.

It was time for some disciplined study at St. James, an Episcopal boarding school in Hagerstown, Maryland. He went back to the States for two years at St. James. A highly competitive route, a presidential appointment to West Point, appeared to be the only avenue open to Ed, and in the summer of 1949 he learned that there was to be an unusually large number of presidential appointments for the class entering in 1950. He was encouraged by the availability of more appointments. To improve his chances he elected to forego completing high school at St. James. Instead, he entered Hilder Prep School in Washington, D.C., a school that specialized in preparing young men for service academy entrance examinations.

He took the examinations in March 1950. His scores placed him fifth for the Naval Academy and forty-fourth for the Military Academy. In the meantime, his mother sought an appointment for him through Congressman Bland, who was in his final term representing Virginia's First District. Ed received Bland's principal appointment, gave up his presidential appointment to another candidate, and entered West Point in July with the class of 1954.

Cadet Edward M. Moses
Class of 1954
(Photo courtesy USMA Archives.)

There was a spark of mischief and "the live wire" smoldering in Ed Moses when he came to the Academy in July 1950. Not a troublesome or rebellious mischief, but one laced with good humor, initiative, and the ability to call up powerful emotions and "the twelfth man," the spirit of the Corps of Cadets.

"I WANT AN EDUCATION, AND TO BE AN OFFICER..."

John Chapman Bard was born in November 1929, the month after the stock market crash of '29, the famous precursor of the Great Depression. By the time World War II ended in 1945, John and three of his friends had tried unsuccessfully to enlist in all of the services. The next year he was making plans to attend the University of Michigan and its law school and learned of the GI Bill. Because his family couldn't afford a college education for him, he decided to quit high school in the eleventh grade and enlist in the Army. In 1947 he applied for and completed Officer Candidate School and was assigned to Alaska, first at Fairbanks, then to a Signal Heavy Construction Company on Adak in the Aleutian Islands

He was one of two white officers in an all black unit. The other white officer was his company commander. The older black soldiers had enlisted in the 1930s, served in World War II, and were the company's non-commissioned officers.

While on Adak Island John decided he liked the Army enough to make it his career and asked his congressman for an appointment to West Point. Since his test scores were so low he received a third alternate appointment, making him fourth in line for the one vacancy his congressman could fill in the class of 1953. But his test scores were good enough to get him to the West Point Preparatory School at Stewart Field in Newburgh, New York, and a two month crash course before the entrance examinations. He didn't get

Cadet John C. Bard
First Captain, Class of 1954
(Photo courtesy USMA
Archives.)

into the Academy that year, but after a year of college he tried again in 1950, and with high test scores and the principal appointment from his congressman entered West Point with the class of 1954.

John Bard was twenty, older than most of his classmates when he entered "Beast Barracks" a week after the Korean War began, and he had served twenty-two months as a junior officer in the Army. There was an intensity about John. Though he'd been through something akin to "Beast Barracks" when he was in OCS, he didn't let starting over again dissuade him from learning. Nor did he approach his studies in plebe year with the notion, "I've seen this before." Instead he studied as though he'd never seen it before. This was the mark of an achiever, a leader. He was to hold fast to his dream of being an officer.

A SURPRISE AFTER CHURCH, AND THE 'WALK-ON' BEGINS

It's 1948. Waterloo, Iowa. Lowell Ellis Sisson's ambitions for the future are all near term. No thoughtful consideration beyond next year – senior year at West High School in Waterloo. He loves football. His heroes are "Doc" Blanchard and Glenn Davis. His high school team completed a good season in the fall of '48, and he looked forward to his final year at West High. He had absolutely no thought of West Point or the military life. There was no military background in his family. That changed abruptly one Sunday morning after church.

A man walked up to Lowell and introduced himself as Russell Broholm. "Would you be interested in going to West Point?" he asked. Lowell Sisson first looked at him in disbelief – completely surprised by the question. Then he answered enthusiastically, "Yes!" This man whom he hadn't known, went to his high school principal first, then to their local member of Congress, Representative Gross. Congressman Gross awarded Lowell Sisson a principal appointment to West Point, to enter with the class of 1954, but not before he completed his best football season yet.

Waterloo's West High School team won the state football championship Lowell's senior year, and he was named best athlete in the graduating class. He was a halfback, tall, slender, and fast. To prepare for examinations to qualify him for appointment to the Academy, the school provided him a separate room for study. He was never approached for recruiting by a member of the Army coaching staff, apparently he was not one of approximately eight hundred boys initially identified for screening each year at the start of the recruiting cycle. He didn't go to the "cram school" in the spring at Smith Rink, as normally did about twenty-four football recruits, to study for qualifying examinations.

After entering West Point in July of 1950, Lowell Sisson "walked on" to the plebe football team. It would be a difficult up and down road, but in four years he would repeat the best athlete award in his West Point graduating class. He received the symbolic sword of athletic excellence, as did Earl "Red" Blaik when the Army coach graduated from the Academy in 1920.

Cadet Lowell E. Sisson
Class of 1954
(Photo courtesy USMA
Archives.)

"EVER SINCE THE THIRD OR FOURTH GRADE..."

The Military Academy was the only school Gerald A. Lodge gave serious consideration to from the time he was in grade school in Cleveland, Ohio. He was spurred on toward the Academy when he saw friends, two to three years older than him, obtain appointments. He was a standout student and athlete in a Cleveland public high school, sixth in a graduating class of 360. He was a fullback and linebacker on the high school football team, in a section of the country heavily recruited by major colleges and universities because of the quality of high school play. Ohio, Pennsylvania, and West Virginia, steel and coal mining states, these were important college football hunting grounds for recruiters.

But Gerry Lodge hadn't been contacted by an Army assistant coach, a recruiter, not until he applied for an appointment to West Point. His circumstance was different from Lowell Sisson's. Gerry Lodge may have been on Army's list of potential football recruits, for when he applied he was immediately contacted by Captain Johnny Green, Earl Blaik's defensive line coach. Murray Warmath also contacted Gerry. Warmath had been named Earl Blaik's First Assistant Coach before he moved on to be head coach at Mississippi State and the University of Minnesota.

Though he was sought after by universities such as Ohio State, Princeton, Michigan, Dartmouth, Harvard, and Cincinnati, Gerry stuck with his dream of West Point, and won an appointment by competitive examination in the Congressional district in which he lived. While Earl Blaik and his coaching staff may have called Gerry Lodge a recruited football player, like many other young men who come to the Academy, Gerry was simply making his dream a reality.

Cadet Gerald A. Lodge
Class of 1954
(Photo courtesy USMA Archives.)

He was "discovered" after he applied for an appointment, and was exactly the kind of man Army football recruiters hoped to find, a scholar-athlete. When Gerry went to his first football practice in the summer of 1950 he discovered, much to his amazement, eighty-eight of 110 players reporting to the varsity, junior varsity, and plebe teams, were all-state picks. Five were high school All-Americans, and sixty-five had been team captains. He knew he was in a tough league, but that didn't turn him aside.

A BROTHER'S INFLUENCE

Leroy Thomas Lunn's brother graduated from West Point in the class of 1950, and lettered three years on the Army football team. He was left offensive guard on the undefeated, untied 1949 team. This was the team that ended its season with a crushing 38-0 win over Navy, the last victory over the midshipmen Army would see until Leroy Lunn became a first classman. At the time Bob Lunn was reveling in the joys of the win over Navy, Leroy had no thought of following in his older brother's footsteps. The Military Academy wasn't of interest to him, wasn't in his plans.

Leroy wanted to be an engineer – a mechanical or civil engineer, and he wanted to go to the University of Tennessee. He was a good, solid student. But at this time of his life football was his primary interest. His dad had played guard. His brother played guard, and in high school in Spring Valley, Illinois, Leroy played guard, with some time at fullback.

When Bob Lunn came home on Christmas leave, he encouraged Leroy to consider West Point. Leroy had the scholarly ability, the drive to be a good engineer, if that's what he wanted to be, and he was a good football player. The Academy's curriculum would suit Leroy's engineering interests.

When Bob Lunn returned to the Academy to complete his last semester, he told the Army coaches about his brother, and Leroy soon received a telephone call from Captain Johnny Green. "Go to your local congressman and tell him you want an appointment to West Point."

Leroy tried and was told, "There are none available." He called Green back and repeated the results of his inquiry. "Don't worry," came the reply. Not long afterward a letter came telling him he had a principal appointment. His brother's influence on him would win out, though he had football scholarship offers from Tennessee and the University of Illinois. He entered the Academy in July with the class of 1954.

In addition to being a good student nearly always on the Dean's List at West Point, Leroy Lunn was feisty, a fighter, and he earned the nickname of "Brutus" among his friends and classmates. He was small as major college football players go, especially linemen. For four years his weight held steady at 194 pounds, and the long climb from the ashes of 1951 were to be hard for him. His scrappy toughness was respected though. His teammates elected him captain for the 1953 season.

Cadet Leroy F. Lunn
Class of 1954
(Photo courtesy USMA
Archives.)

BLANCHARD AND DAVIS, BROTHERS HE ADMIRED, AND EARL BLAIK'S ACQUAINTANCE

Picayune, Mississippi is a small town. There were sixty-seven students in Freddie Alvin Daniel Attaya's graduating class, and Fred was number twelve. There are thousands of small towns in America where high school football is king each fall. Picayune High School was no exception in the fall of 1948, Freddie Attaya's senior year.

Since elementary school, West Point had been primary in Fred's consideration of his future. So had football. As did Tommy Bell in New York City's Bronx and Bob Mischak in Union, New Jersey, Fred saw and admired "Doc" Blanchard, Glenn Davis and Army football in local movie theaters' Movietone News clips. The screen images of these two Army greats, the examples of two older brothers important in Fred's life, and a fortuitous set of circumstances brought Freddie into the Academy class of 1954.

Freddie's oldest brother was his military inspiration when Fred was a freshman in high school, and Army's "Mr. Inside and Mr. Outside" were first casting their spell over young boys interested in football. During World War II, his brother became a genuine hero to Fred. He was a waist gunner on B-17 bombers in the Eighth Air Force in Europe. The older Attaya flew eight missions before the war ended, but his third mission was a near catastrophe.

On climb out from its base in England, the airplane's electrical system malfunctioned. The aircraft commander could have aborted, but elected to continue the mission. When they got over the target, the bombs would not release due to icing in the bomb bay. The heavy load made fuel critical for the return trip, and the bombs never could be jettisoned. A decision was necessary to belly land, and the pilot successfully put the big, bomb-laden machine down in Belgium, twenty to thirty miles behind Allied lines. The entire crew survived, and Fred's brother later served out the war training gunners for bomber crews. Afterward he obtained a degree in civil engineering and a commission in the Army.

Picayune High School won the state football championship in their division in 1948. Freddie was team captain, and led in rushing, pass receiving, punt and kickoff returns. He also doubled at other backfield positions when needed. He captured the interest of football recruiters from Louisiana State and the University of Mississippi – 'Ole Miss.

Unknown to Freddie, he also captured the attention of the Professor of Military Science at 'Ole Miss, where the second of Fred's older brothers was back-up fullback on the Rebel's football team. The officer, Colonel James O. Wade, was from the West Point class of 1926. He had played on the Army football team as a cadet, knew Earl Blaik, and approached Fred's brother and inquired about the younger Attaya.

Fred's interest in the Academy immediately intensified. Though he had football scholarship offers from 'Ole Miss and LSU, he sought an appointment to the Academy through his local congressman. The congressman told him he'd receive a principal appointment, but something went wrong. The congressman said he'd "overlooked" Fred's name.

He was disappointed but soon received a call from Johnny Green. "Would you be willing to prep somewhere for a year?" Green asked. "Yes," Freddie replied, and he accepted LSU's scholarship offer. He made the LSU freshman team that fall, at defensive halfback, and scrimmaged regularly against the Tigers' varsity. LSU was number nine in the nation at the end of the '49 season and played in the Sugar Bowl January 2, against the number two Oklahoma Sooners.

In the spring of 1950, Fred dropped out of LSU. Johnny Green was still working on an appointment for him. Fred took the entrance exams and was awaiting final word on the results when the Korean War broke out. He wondered what he should do, and called Johnny Green. "Should I go to West Point or the draft?" he asked. Came Johnny Green's reply, "Stay at home and wait." A few days later Freddie Attaya was on his way to the Academy.

Cadet Freddie A.D. Attaya
Class of 1954
(Photo courtesy USMA
Archives.)

He would learn, under Earl Blaik and Vince Lombardi, that football practiced and played at West Point was sharply different from what he'd seen at LSU. He had some lessons to learn from both men, some powerful lessons, and he learned them well.

FOR LOVE OF THE GAME, AND A FATHER'S PERSUASION

Peter Joel Vann loved football. He had a brother and a sister, and was quarterback on the Hamburg, New York, high school team. He was tall and gangly, with long arms, and could throw a football with considerable accuracy, almost as far as a kickoff – fifty to sixty yards. He also played center on the 1950 high school basketball team.

But football was the game he loved. His brother played end on the high school team, and in 1949 was a senior when Pete was a junior. In the last game of the '49 season, Pete lived an unforgettable memory. He threw a brother-to-brother touchdown pass, a pass that brought local fame to both boys, and made his dad even more proud of them.

Pete's high school principal, Spencer Ravel, who had been a Naval officer, first stirred his curiosity about West Point. He asked Pete if he was interested in the service academies. There wasn't any history of military service in his family. His father made a modest living, holding three jobs, one in nearby Buffalo's Marine Midland Bank. It was unlikely Pete could go to a good college without the aid of a football scholarship.

His play as Hamburg High's quarterback drew a lot of favorable press in upstate New York, particularly in the *Buffalo Evening News,* and Earl Blaik's systematic approach to recruiting had picked up on Pete's praises in the sports pages. So had other football recruiters. By the time his senior season came to an end the fall of '50, there was interest in Pete's skills at North Carolina, North Carolina State, Duke, Cornell, Syracuse, Colgate, St. Bonaventure, and Virginia Military Institute.

But Pete, too, had remembered Army's halcyon football days and liked the appearance of the Army lettermen's jacket. He decided to write a letter to Bobby Jack Stuart, who had been a well-known halfback on Earl Blaik's undefeated 1948 team. Pete wanted to find out if he might obtain one of those good looking letterman's jackets. The letter apparently increased the interest of Army recruiters, and he received a phone call from Stuart.

Virginia Military was particularly aggressive in pursuing Peter Vann's football skills. He visited their campus, they put him in a cadet uniform, and introduced him to some attractive young ladies. He came back from the trip convinced VMI was the place he wanted to go. But his dad was convinced the Military Academy was a wiser choice, and worked patiently to urge Pete toward the same conclusion. But all that came later, after a real scare that began on December 23, 1950.

He came home that day to do some chores around the house, including clearing ice from the rain gutters along the edge of the roof overhang. He was on a ladder removing the ice when he suddenly became dizzy and passed out. Their family doctor came over to examine him, and rendered his diagnosis – spinal meningitis.

He was taken to Our Lady of Victory Hospital in Lackawana, New York, with high fever and delirium. He received the standard, intense treatment for the serious illness, which in his case included a spinal tap with a needle which seemed eighteen inches long, 144 shots of streptomycin, and twenty-one days in the hospital.

He came out of the hospital in mid-January, skinny as a rail, soon to be asked to take the entrance exams for West Point. He was a National Honor Society member, carrying a B+ average in school, but didn't do well on the entrance exams when he took them in April. His weight had climbed all the way back to 140 pounds by April, not exactly major college material. He received a call telling him he didn't do well – he'd better come to West Point to go to the "cram school" at Smith Rink and prepare to take the entrance exams again.

Pete made the trip to the Academy on the train, arriving at the West Shore railroad station, where he was met by an Army car and whisked to Smith Rink. He finished his schooling and retook the exams, while Doug Kenna and Johnny Green, two of Blaik's assistants worked with him on the practice fields. Kenna, an All-American quarterback on Army's 1944 national champion team was impressed with the power of Pete Vann's throwing arm, but had to teach him how to throw a good spiral instead of an end over end ball. Pete's fifty to sixty-yard throws not only traveled as far as a kickoff, but looked like a kickoff before Kenna showed him how to change his grip on the ball. Kenna was impressed enough to tell Blaik about Pete Vann's passing ability. Kenna was right, and in the season of '53 would be vindicated in his judgment of Pete's abilities.

Peter Vann finished his cram school, passed his exams, and was subsequently given a qualified alternate appointment arranged by Earl Blaik. He returned home just one day before obtaining his high school diploma and a B average. The color, the beauty of West Point, and his dad's persuasiveness had won out, and he reported to the Academy on July 3, 1951, to enter with the class of 1955. Later, due to academic struggles, he would be "turned back" to the class of 1956.

Cadet Peter J. Vann
Class of 1956
(Photo courtesy USMA Archives.)

HE ADMIRED AN ENGINEER, BUT WANTED TO BE A PILOT – TO EUROPE IN '44 AND BACK

William J. "Pat" Ryan, like nearly half the men in the West Point classes of 1949 through 1952, had prior military experience. He was one of the millions of young men returning from the war seeking an education they might otherwise have been unable to afford.

Pat's path to the Academy began in Paducah, Kentucky, where his family lived next door to an Army colonel and his family. The colonel, the recipient of Pat's boyish admiration, was an engineer, the commander of the Army's Corps of Engineers' District which included Paducah.

Pat was sixteen when the United States entered World War II. He wanted to be a pilot. When he turned seventeen, he applied for an appointment to the Academy, but erroneously checked the wrong block on his application form. He received an appointment to Annapolis instead. He declined attending the Naval Academy, and in April 1943, joined the Enlisted Reserve Corps. The still expanding Army Air Force called him to active duty in October of that year and the following April he arrived in England, assigned to the 8th Air Force's 398th Bomb Group.

His assigned duty was to maintain "quad 50" anti-aircraft machine guns defending the 398th's air base. He volunteered to be a supernumerary gunner, hoping to fly combat missions, but was never called to fill in for a crew member. He applied again for the Military Academy, and in June 1944 his squadron commander recommended him for one of three Academy appointments allocated to the 8th Air Force.

Pat returned to the States in late August, to the Army's Academy prep school – Amherst College, where he reported on September 4. A year of hard study prepared him to successfully pass qualifying examinations for West Point. But no appointment came. He stayed on at Amherst and the next year he received an appointment from Congressman Noble Gregory of Kentucky. He entered West Point's class of 1950 in July 1946, two months shy of his twenty-first birthday.

The road to West Point had been long for Pat Ryan, and the road to graduation would be five more years, caused by head injuries in football and boxing. The injuries nearly cost him a West Point education and his cherished dream to be a pilot, but he persisted. He was mature, determined, hard working, and never gave up on his dream. The road he traveled led him to the role of First Captain and Cadet Brigade Commander in the class of 1951, and back to the Air Force and pilot's wings after graduation.

"WEST POINT WAS A BOYHOOD DREAM INSPIRED BY MY DAD'S BROTHER"

Howard H. "Dan" Danford's dad came from a farm family. Aledo, a small town of three thousand in northwestern Illinois, was where he grew up. The home he lived in had been built by his Grandfather Danford after his retirement from the farm. His grandfather was a successful farmer, and with Dan's grandmother, raised three sons. All were to become college graduates.

Uncle Rob, an Academy graduate in the class of 1904, completed his career a major general in the Army. Dan's Uncle Fred graduated from the University of Illinois with an engineering degree and later became Head Engineer at the Texas and Pacific Railroad. Thornton Danford, Dan's father and the youngest of the three sons, was a chiropractor.

Dan Danford's Uncle Rob was the inspiration for Dan's decision to enter the Academy. Dan and his father were close, but his father wasn't interested in a West Point education for him, preferring that Howard, as his parents called him, remain close to home and raise a family. General Danford didn't have a son and Dan became the son he never had. Major General Robert M. Danford retired from the Army in 1942, having served in the field artillery most of his thirty-eight year career. His last four years of service were as the Army's Chief of Artillery.

General Danford had another distinction. He was Commandant of Cadets at West Point from 1919 to 1923. He had been selected to serve as commandant under then Brigadier General Douglas MacArthur, when MacArthur was sent by the Army to reform and revitalize the Academy after World War I, the same period when Earl Blaik was a cadet and star athlete. Lieutenant Colonel Robert Danford's assignment as Commandant of Cadets brought him great satisfaction, because he had been deeply involved in some of the major reforms that MacArthur instituted, several of which remained fundamentally unchanged over the years. There were many fine stories to tell a young Howard Danford, stories that could stir dreams and wishes.

In spite of his Uncle Rob's inspiration, Howard Danford was doubtful he could physically qualify for the Academy. His eyes were bad – vision twenty-sixty. Yet he was a good football player, enough so that he received considerable notice in local newspapers. He went to Yankton College, in his home town, where his maternal grandfather had been college president for thirty years, 1895-1925, and where his three siblings were to graduate. Dan continued to play football, while his dad collected press clippings.

Dan's father, apparently out of pride, and to let his brother know of Dan's success in football, sent some clippings on to Uncle Rob, who was, by this time, president of the Academy's Association of Graduates at West Point. Retired General Danford delivered the clippings to Earl Blaik. The result was a qualified alternate appointment from a congressman in the state of Illinois, and Howard "Dan" Danford entered the Academy with the class of 1952.

Howard Danford played football three years at West Point, both as a starter on the plebe team, and on the junior varsity two years, where

he scrimmaged nearly every week of the '49 and '50 seasons against Army's powerful varsity. He spent a lot of time with his uncle, who served on the Hand Board in July of 1951, which completed its work nine days before the shocking public announcement in August.

Dan, as he was called by his friends and classmates, also became well acquainted with Colonel Arthur S. Collins, Jr. He was the Cadet First Regiment Commander his first class year at the Academy when Art Collins was the Tactical Department's First Regiment Commander – the officer who was Dan Danford's counterpart and commander-teacher.

Cadet Howard H. Danford
Class of 1952
(Photo courtesy USMA Archives.)

A SMALL VIRGINIA FARM – AND A RACE
FROM EUROPE TO WEST POINT

Richard Thomas Shea, Jr. was Virginian through and through. Raised on a small family farm near Portsmouth, he developed a disciplined work ethic, a seriousness, and an unrelenting determination, that mixed well with the Christian virtues given him by his parents – and a pleasantness in dealing with others that told of his deep respect for every person. His route to the Military Academy class of '52 was circuitous, roundabout, somewhat like Pat Ryan's long road to '51. There was a marked difference, however. Dick Shea literally ran his way to West Point.

At a young age he emerged a leader. He was president of his class beginning in the second grade, and his classmates elected him president nine more times before he graduated from high school. In high school he was an honor roll student, senior class president, and voted "Most Likely to Succeed" and "Best All Round." He held fond memories of his high school days. He particularly remembered his lead role in the senior class play, *Spring Fever,* and his work as art editor for the school newspaper, *The Countryman.*

By the summer of 1944, Portsmouth was a center of World War II military activity. The high school student body was well represented at the launching of the aircraft carrier U.S.S. *Shangri-La* in the local shipbuilding yards. When Dick graduated from Churchland High School, he was anxious to leave and fight in the war. He enlisted in the Army immediately after graduation.

Instead of sending him overseas, the Army assigned him to the Army Specialized Training Reserve Program. The program was to prepare young enlisted soldiers for duty as military technicians. High aptitude scores, and his age, seventeen, resulted in his assignment to the ASTRP, and the role of a student at Virginia Tech, where he remained from July 1944 to March 1945. Though athletically skilled, he remained too busy at Virginia Tech to participate in athletics, taking an average of twenty-one semester hours in his eight months at Tech.

On March 31, 1945, he received an assignment to Fort Ord, California, and believed he would be fighting the Japanese in the Pacific. Boxing was as close as he came to fighting in World War II. By the time his training was complete, the war in Europe ended, and he was assigned duties in Berlin, Germany, as a member of the occupation forces. He was in the 53d Constabulary Squadron, which had been the 53d Armored Infantry Battalion, 4th Armored Division, before its role was changed to occupation duties.

It was in Berlin, Germany, where the serious side of him was much in evidence as Sergeant Shea worked his way up through the ranks to become the outstanding communications noncommissioned officer in the American Constabulary Forces. He completed the Radio Operators Course, was promoted to staff sergeant, and earned the Expert Rifleman's Badge. It was here that Dick Shea also gained recognition as an exceptional athlete. He began to run races. He trained faithfully, knowing that hard work could make him a good, competitive runner.

His early efforts weren't in vain, nor were any of his later ones. In the GI Olympics Constabulary Championship in Berlin he won the fifteen hundred and five-thousand meter runs. He went on to win the five-thousand at the European Championships and the Steeplechase at the Allied Olympic Games. The well-known track man, Nate Santwell, believed Dick should run the ten-thousand meters for the United States in the Olympics. He passed up the Olympics, however.

His intense competitiveness and soldierly qualities helped earn Dick an appointment to West Point, and he returned to the States to enter the Military Academy Prep School at Stewart Air Force Base in Newburgh, New York. During his year at the Prep School he starred as a runner on the track team. His quiet, friendly nature and strong Virginia accent gave him a charm, a charismatic presence. A house was a "hoose," a mouse a "moose," sounds and language of the Virginian that earned the nickname of "Cornpone" from his classmates.

While his classmates good naturedly called him "Cornpone," they thought enough of him to also name him their battalion athletic officer in their cadet chain of command. For the first time, in the role they gave him, he was able to combine his athletic abilities and leadership qualities. Dick's ability to lead and inspire began to shine. Each day, an hour before reveille, he and several of his classmates would quietly dress and take a morning run to keep in good physical condition and competitive trim. And it was in the Prep School that he began drawing his trademark cartoons which were to keep the United States Corps of Cadets laughing the next four years.

Also at Stewart, Dick met Joyce Elaine Riemann of New Milford, New Jersey. Romance blossomed and they soon fell in love.

Dick Shea was already running life's race well when he entered the Military Academy on July 1, 1948. In his three years as an underclassman he won many more championships and set records on the Academy track and cross country teams.

In July of 1951, as the Army and the Military Academy were struggling behind closed doors toward painful decisions necessary to resolve the most devastating incident in West Point's history, the men in the class of 1955 met Dick Shea for the first time, as he continued his run to ath-

Cadet Richard T. Shea, Jr.
Class of 1952
(Photo courtesy USMA Archives.)

letic fame. Dick was on the Beast Detail, in Fourth New Cadet Company, our cadet company commander – "Mr. Shea, Sir" to us.

IT TAKES MORE THAN A RECORD OF EXCELLENCE

Richard George Inman, class of 1952, was on our Fourth Company Beast Detail as a squad leader. Dick Inman was also on the track team with his classmate, Dick Shea, and received the coveted letterman's major "A" in both football and track.

Dick Inman was born in Indianapolis and completed his secondary education in Vincennes, Indiana. As a boy in elementary school, he exhibited excellent leadership qualities among his classmates. He was strong in scholarship as well. In his pre-high school days, he gained many honors and awards – in school, the Young Men's Christian Association, and in church.

Like many of us, when the United States entered World War II and Dick wasn't yet in high school, it was off to the vacant lots to "play war," fight great battles important to the world. Dick Inman was the oldest of four children, and during their "wars," when there were only three Inman children, he was a twelve-year old "General" in a whole neighborhood gang, with ages ranging down to seven. His brother Bob was his trusty cohort, as was his sister Mary Jo.

They marched the few blocks near home, wearing their makeshift helmets and shouldering rifles carved from wood. Their frequent mythical battles included construction of forts, trenches and other defense networks dug on vacant lots in the local neighborhood. They, and even their dogs, fought valiantly from behind mounds of dirt at home construction sites. Their extended warfare eventually caused neighbors to complain that Dick and his "army" were unnecessarily blighting the aesthetics of the area, creating eyesores.

In high school Dick Inman continued to excel in scholarship, and was active in athletics and drama. The awards and recognition he received reflected his character, the quality of his work as a young man, and the balanced, well-rounded background and preparation the Academy looked for in its candidates for admission. His school selected him outstanding freshman athlete. He won the citizenship award given by the Daughters of the American Revolution, the American Legion football award as most valuable to his team, and was accepted as a member of the National Honor Society, a membership accorded young people based on scholarship, leadership, character, and citizenship. Because of his study and work in drama he became a member of the National Thespian Society.

On Indiana's athletic fields he did equally well. His father was coach of the Vincennes High School football team, and was Dick's hero along with the legendary Carlisle Indian football player and Olympian, Jim Thorpe. His father, in addition to serving as the high school football coach, taught history and civics. In football Dick garnered All-State recognition, and was named to the all-Southern Indiana and all-Wabash Valley teams. In the State track meet, as a junior, he won fourth place in the high hurdles, and in his senior year was third in the same event – in a race that set a new state record.

In 1948, Dick Inman's reputation and record helped earn him a principal appointment to West Point, from Congressman Gerald P. Landis of Indiana's Seventh Congressional District.

There was more to Dick Inman than his record of performance as a boy in school. Persistent, determined, hard working at every task he un-

Cadet Richard G. Inman
Class of 1952
(Photo courtesy USMA Archives.)

dertook, he was not one to give up or be deterred by pain. High ideals were noticeable in his words and actions. He admired courage, ambition and honesty. Honor came above everything. It is well that Dick possessed those qualities, for adversity would be his brother after three years of Army football, one as a plebe, and two on the "Golden Mullets," the junior varsity. In the fall of 1951, his first class year, he became one of the men of whom Earl Blaik said, "They didn't know they hadn't the experience or skill to play varsity football." Like many men on the '51 team, he might never have played varsity were it not for the tragic honor incident which shook West Point that spring and summer.

BOYS WHEN THEY ENTERED – ALL HAD TO BE MEN WHEN THEY GRADUATED

Boyhood dreams change, become better defined, and mature as boys grow to be men. As we each took different paths to the Military Academy classes of 1950 through 1957, so each of us took different paths when we left West Point. With few exceptions, those who graduated and were commissioned as officers in the armed forces served the required number of years we knew would be necessary when we entered the Academy. Throughout the years following mandatory service, the paths we took diverged, as had Academy graduates' paths in previous generations. Among those who remained on active duty through the difficult years of the Cold War, some became senior officers in the Army and Air Force, in positions carrying enormous responsibility. Those who chose to leave the service became engineers, educators, business leaders in all forms of enterprises such as electronics, surface and air transportation, finance, and space exploration; physicians, lawyers, clergy, community leaders, state and local elected officials, congressmen, senators, appointed government officials – and some who lost their way.

But many held to dreams of being officers, and lives of service in the armed forces. For all who would "see and feel the elephant," – go into battle – there was a more sobering reality about dreams of West Point. We might dream and enter the Academy as boys. We could continue to dream as we grew toward manhood while at the Academy, but we must not graduate as boys.

For many, especially the war classes, graduation would thrust us immediately into leadership roles demanding maturity, skills, decisiveness, and wisdom far beyond our years. We could not be called "boys," or be "boys" one day, and the next day be transformed to men by a diploma and gold bars of a second lieutenant. Neither could we beg nor disregard the meaning of honor one day, and become honorable by virtue of a commission the next. We had to grow, step by difficult step, throughout the four years, to be men, honorable men.

The consequences of failure to prepare, learn, and grow could cost the lives of others, and ourselves. There are no guarantees in battle, and no guarantees of survival to see dreams become reality. If we prepare, learn, and grow there's nothing more than an increased probability of avoiding disastrous errors in the presence of "the elephant."

Dreams can be turned to dust by the bullet, whether it is aimed or randomly fired. The bullet knows not our dreams, who we are, what we stand for, and whether we are good or evil. These truths are the constants of men and women called to duty and a life of service in the armed forces.

CHAPTER 6

THE END OF ARMY'S GOLDEN ERA
OF FOOTBALL

On September 1, 1950, Army began practice for another sparkling football season. There was once more an air of excitement and great anticipation at the Academy in spite of the summer's news of lengthening casualty lists and defeat in Korea. Over its sixty year history at West Point, football had become an important, popular sport and a pervasive influence in Academy and cadet life. Earl Blaik's teams had once more made Army a national football power, and during the dark war years, given the game a shining new stature at the Academy. The Black Knights had become a powerful magnet in recruiting young men for West Point and a life of service – whether they were football players or not.

From his first season at West Point in 1941, when Earl Blaik took the cadets to a 5-3-1 season, after a dismal 1-7-1 Army season the previous year, the Black Knights' football fortunes steadily improved, and were given a substantial boost by the influx of talent during World War II. In 1942 they were 6-3, losing to Navy, Pennsylvania, and number six Notre Dame. In 1943 the cadets went 7-2-1, losing to national champion Notre Dame and number four Navy, and made their first Blaik era appearance in the top twenty ranked teams, at number eleven. Then, from 1944 through 1949, football glory lived at West Point, and there was little doubt it would remain through the 1950 season.

In 1946, after the national championship seasons of '44 and '45, and Army's split, first place national ranking with Notre Dame in '46 – following the "game of the century" 0-0 tie with the Fighting Irish – the Football Coaches' Association named Earl Blaik Coach of the Year. Contrary to his critics' predictions, Blaik's last Davis-Blanchard team had remained undefeated and very nearly won another national championship outright, despite the flood of football talent returning from World War II to teams representing prewar major college football powers. There was good reason the Academy and Hollywood collaborated in the making of the movie, *The Touchdown Twins*, star-

226

ring "Doc" Blanchard and Glenn Davis, the legendary "Mr. Inside and Mr. Outside."

The '46 game of the century with Notre Dame was notable for other reasons. The greatest collection of college football talent ever assembled was on the two squads: fourteen current or future All-Americans and ten future Hall of Famers. Coaches Earl Blaik and Frank Leahy would also join the Hall of Fame.

In '47, while General Maxwell Taylor was superintendent and adjustments to the reconstituted curriculum were still in progress, Blaik, his assistants, and the Army team were going through their own rebuilding period. That fall the Black Knights went 5-2-2, including Columbia's stirring, come-from-behind 21-20 upset of Army, to end the cadets' 32-game undefeated streak. In spite of what many believed would be the end of Army's reign as a football power, the cadets were number eleven in national rankings at the end of the '47 season.

Then, in 1948 came an 8-0-1 season and a number six national ranking, followed by a 9-0 season and number four in '49. Army was climbing steadily back toward the mountain top, in spite of critics' predictions the Black Knights would be eclipsed in the post war years. There were good reasons their predictions didn't ring true.

A winning tradition and a head coach who develops the habit of winning bring their own rewards. Talented athletes who love the games they play, also love to play on winning teams – and love to win. They are drawn to winning traditions and winning combinations, and Earl Blaik had it all at Army. Sports writers and coaches accorded due recognition to the men who excelled on Blaik coached teams.

During the years preceding the 1950 season, eighteen Army players were first team All-Americans. Two of those eighteen were two-time All-Americans. Felix "Doc" Blanchard and Glenn Davis were three-time All-Americans and the first Army players to win the Heisman Trophy, symbolic of the nation's outstanding college football player, in '45 and '46 respectively. In 1943 All-American Casimir Myslinski received the Knute Rockne Award for outstanding lineman of the year. Arnold Tucker, Army's All-American Quarterback on the 1946 team, won the Sullivan Award as the nation's amateur athlete, who by performance, example, and influence did the most to advance the cause of good sportsmanship. In 1947, All-American Joe Steffy, a guard and team captain, won the Outland Trophy as college football's outstanding interior lineman.

What's more, after World War II, Army's core of football talent was fed by the same stream of returning GIs that colleges and universities all over the nation were receiving. At West Point, on the post-war football teams, were former enlisted men, non-commissioned officers, and even junior officers, seeking educations and commissions as regular officers in the Army or Air Force.

In the coaching ranks of college football, a winning tradition draws other talent. Earl Blaik, who developed a formidable reputation as a football teacher, as well as a leader, and was always thoughtful in selecting his assistant coaches, was able to select from among the best of eager, ambitious assistant coaches anxious to learn and go on to become head coaches. During the war years, he assembled probably the best coaching staff the game had ever seen, and after the war there was no decrease in its skills or performance in spite of the turnover among its members.

Blaik's coaching staff became a training ground for college head coaches. His football program had a reputation similar to that of the "cradle of coaches" at Miami (Ohio) University. The University's reputation had its roots in the era when Blaik graduated from Miami, before he entered West Point. In Blaik's eighteen years as Army's head coach, thirty-two varsity assistant coaches passed through his staff, several of whom were Academy graduates. Among his assistants, nineteen became head coaches at major colleges or universities. One became a legend in the professional ranks – Vince Lombardi, the great coach of the Green Bay Packers, who, from the time he left West Point after the 1953 season until his death, repeatedly credited Earl Blaik as the man who taught him the most about coaching. In his 1963 autobiography, *Run to Daylight*, Lombardi wrote "Earl Blaik..[is]..the greatest coach I have ever known."

But most of all, Earl Blaik's detractors and critics, who pointed accusing fingers at his wartime success at Army, often cruelly calling the wartime West Point cadets "slackers" or "draft dodgers," and saying Blaik was merely the fortunate beneficiary of "those types of football players," were ignoring Blaik's prior success as a coach, and the service records of former Army players. While World War II was unquestionably a factor in Army's rise to national football prominence under Earl Blaik, there was far more to his success than the talent funneled his way by war. There was also far more to the young men who played on his teams than was implied in the harsh, biting remarks directed toward them.

Throughout the years before Brigadier General Eichelberger, the Academy superintendent, brought him back to West Point as the Academy's first civilian head coach, Blaik built an excellent reputation in the coaching profession, first as an assistant at Army from 1927 through 1933, and then as head coach at Dartmouth for seven years before returning to Army.

The Army coach's critics also failed to consider the history and development of football, and the status accorded the game in the armed forces before, during, and after World War II.

In 1914, the year World War I began in Europe, football emerged in the United States as the most popular of all crowd drawing sports. While high school football was flourishing and professional football was relatively unknown, still in its infancy, collegiate football games were already pulling huge crowds. Harvard dedicated the first of the great stadiums in 1903, with twenty-seven thousand permanent and fifteen thousand temporary seats, and in the fall of 1914 the Harvard-Yale game drew sixty-eight thousand five hundred fans in the Yale Bowl.

During the same era, football became quite profitable for colleges and universities competing in the great national arenas. Receipts from football funded construction of athletic facilities other than football stadiums, and provided funds to cover the costs of expanding athletic programs.

Football was also an enormous boon for young men who were unable to pay the cost of education beyond high school. Football scholarships were becoming "a ticket to a college education."

When America's young men returned from war to begin the 1919 football season, the game became a partner in "The Golden Age of Sports" (1919-1930). As would be the case in 1946, there was a veritable explosion of talent on the college gridirons, and in the Roaring Twenties, hippodroming – the construction of huge stadiums – resumed, spread across the country, and reached its peak.

In the nation's armed forces, competition had ever been a means to stimulate unit and individual morale as well as affect better unit performance. Football, "the manly sport," was a natural for such competition, particularly in the Army and Navy. Despite the potential for injury, the sport was seen to increase physical conditioning of the troops as well as stimulate competition among regiments or divisions, and between historic Army posts. As in other types of sports in the Army, football championships became marks of unit pride and *esprit de corps*.

When the United States began mobilizing prior to its entry into World War II, it became apparent to many colleges and universities that the manpower drain would make it virtually impossible to sustain many athletic programs, including football. There was no such thing as a college deferment from the draft, and members of coaching staffs were either being drafted or volunteering for service, as were the able bodied young men who might otherwise go to college and play football. Many of the former college coaches, athletic directors and administrators received direct commissions in the armed forces and became coaches for military teams, or conducted physical education training programs.

It was during this period that Earl Blaik, the head football coach, was recalled to active duty as an officer, a lieutenant colonel in the Army, as were many of the Academy's faculty members who were either not physically qualified for combat duty, or were considered best qualified to teach at West Point because of their academic backgrounds, rather than retrain and prepare to lead in combat duty.

As the armed forces built toward a peak strength of over twelve million men and women, the military activated hundreds of Reserve Officer Training Corps (ROTC) units at colleges and universities throughout the country. Revamped secondary school curriculums resulted in high school graduations a year earlier than normal, and physical education received extraordinary emphasis within those curriculums.

The service academies, as had been the case in the Civil War and World War I, were also subject to the dramatic changes the entire nation was undergoing. At West Point, graduations accelerated, the curriculum compressed to three years, and changed to include a drastic increase in technical and combat arms training to prepare cadets for combat as soon as possible after graduation and commissioning. A prime example was the establishment of flight training in 1942, at Stewart Army Airfield in Newburgh, twenty-five miles north of West Point. The flight qualified cadets obtained their pilot's wings when they received their Academy diplomas, to meet the needs of a rapidly expanding Army Air Force.

Within the armed forces, huge training centers were built where thousands of men and women were given basic and technical training, as well as training in Officer Candidate Schools. The shifts and changes into a wartime education system, emphasizing short term needs for trained soldiers, sailors, airmen, and the officers to lead them, had a dramatic effect on who played football and where it was played. Football literally joined the armed forces.

During the period before the NCAA Football Rules Committee met in early 1942, over 350 colleges and universities had already dropped football. Nevertheless, after the country entered the war, and on through the '42 and '43 seasons, major college and university football powers continued to dominate the top twenty teams. However, by the time 1944 rolled around, teams began appearing in the top twenty national rankings that had never been seen before – and have never been seen since. Among the top twenty at the end of the 1944 season were Randolph Field (3), Bainbridge Naval Training School (5), Iowa Pre-Flight (6), March Field (10), Norman Pre-Flight (tied at 13 with Georgia Tech), El Toro Marines (16), Great Lakes (17), Fort Pierce (18), St Mary's Pre-Flight (19), and 2d Air Force (20). This was the first year under Earl Blaik that Army won a national championship. That same year, rounding out the top 20 were Ohio State (2), Navy (4), Southern California (7), Michigan (8), Notre Dame (9), Duke (11), Tennessee (12), and Illinois (15).

In 1945, Army was again chosen by sportswriters and broadcasters as national champion, while Navy was number three. But the end of the war once again brought change to the make-up of the top twenty teams. Demobilization proceeded so rapidly after the war ended in August that not one of the armed forces' wartime training centers fielded a team ranked in the top twenty at the end of the '45 season.

In those nine years preceding 1950, Coach Earl "Red" Blaik's Army teams enjoyed five undefeated seasons, two national championships, shared a national title with Notre Dame, and five Eastern titles, symbolized by the Lambert Trophy. Blaik's record during the period was a sparkling 67-10-6, including undefeated streaks of thirty-two and twenty games. And in the six seasons beginning in 1944, his teams achieved a 49-2-4 record. The two losses and two of the four ties were in 1947.

SHADOWS IN THE GOLD

When Army's 1950 football team began practice sessions on September 1, there was indeed reason for great anticipation at West Point, in spite of the series of battlefield defeats and growing casualty lists in Korea. This was to be another great Army team, and one of the deans of college coaching in America, Earl Blaik, was preparing his magic formula for another golden season.

Throughout Blaik's years as head coach, he often asked and received Army approval to retain selected graduates as assistants on the plebe or junior varsity teams the succeeding season. John Trent, captain and end on the undefeated, untied '49 team, and All-American Arnold Galiffa, the '49 quarterback and Trent's classmate, stayed on at West Point to assist in preparations for the 1950 season. However, with the outbreak of hostilities in Korea, both were ordered to report to Fort Benning where they became platoon leaders in the 3d Infantry Division. Both left West Point before the season began.

The 3d Division was scrambling to come up to strength in men and equipment, training hard, readying for its overseas move by ship onto the Korean Peninsula. As with other divisions which preceded them into war that summer and fall, there was precious little time to prepare for what lay ahead.

TALENT AND CONFIDENCE

The Army squad of 1950 had depth and some of the finest athletes on the nation's gridirons. When practice began in September, twenty games without a defeat was an imperative driving the Army team. Team members expected to win. As Blaik said many times, "You have to pay the price." The team's attitude toward winning confirmed its willingness to pay the price. The attitude was complemented and strengthened by Blaik's thorough planning and game preparation and his disciplined organizational skills. One of his players that fall remarked about the team's detailed scouting reports and game preparation. "I not only knew where that tackle was going to scratch himself, but when he was going to scratch."

The men in '54 who made the plebe team, some of whom scrimmaged against the varsity that fall and the following spring, were awed by the football skills and fierce competitiveness of the men they worked against. And, as with every Academy class, there were some in '54 who, though they never played football, felt they were good enough to try out – have a try at "walking on."

Lowell Sisson, Company A-1, who had starred on his Waterloo, Iowa, state champion high school team, walked on. He began on the bottom rung of Army football's ladder, on the fifth team of the plebe squad. He pushed himself hard, and with his well developed football skills, by season's end worked his way to second team.

However, Lowell did pay the price on the plebe football team the fall of 1950. In a game against the Pittsburgh University freshmen, he took an elbow to the throat, causing a severe injury to his larynx, an injury which nearly ended his football and cadet careers. For several moments after the blow, he was fighting for air, trying to breathe, and for the first time in his life felt genuine fear. The next day, still unable to talk and feeling as though every breath was a battle for survival, he checked himself into the Academy hospital, where he remained for twelve weeks. His next struggle was trying to catch up and stay up with curriculum demands while he recuperated. It wasn't easy.

Ben Schemmer was less fortunate. Ben was in Company L-2, where there were several varsity players who received nationwide publicity as members of Army's undefeated '49 team. One of them encouraged him to try out for the plebe team and put in a good word for him with the Army coaches. Blaik responded by suggesting Schemmer suit up. Team practice had been in progress for some time.

Ben was tall, slender, fast, and a natural athlete, a candidate for the backfield. After warm-up exercises and drills his first day out, some of the plebes were sent to run plays against the varsity defense, and Ben Schemmer was among them. The coaches put him in at halfback and instructed him on the play they wanted him to run. It was a quick opening, straight ahead dash by Ben, with a hand-off from the quarterback. After he tore through the hole the linemen were to open, he was to cut

233

at an angle toward the sidelines, where, theoretically, down-field blockers would mow down the defensive backfield and clear his path to a practice touchdown. It didn't work out that way.

The quarterback took the snap from under the center, and handed the ball off to an excited, quick starting Ben Schemmer, and that's the last thing he remembered. Big, muscular, tough, defensive linemen met him head on at the line of scrimmage, with a crushing tackle. When he came to, he was in the Academy hospital with a broken collar bone and two broken ribs. His brief football career was over, and the 1950 Army team rolled on without Ben Schemmer.

Guided by Blaik and his hard working assistants, Army was regarded once more as a contender for the mythical national champion-

Earl "Red" Blaik, his coaching staff, and the 1950 Army Football Team – organized, disciplined, tough in the fall before the end of Army's "Golden Era of Football." The varsity is in the foreground in black jerseys, the junior varsity in the right rear wearing yellow jerseys, the plebes in the left rear in white jerseys. (Photo courtesy George Silk/Life Magazine © Time, Inc.)

ship. Famed sportswriter Grantland Rice, in *Look* magazine's annual preseason review, picked Army and Notre Dame as the top two teams in the nation, but he was hesitant to say which one would be the best. Articles about "The Rabble" in the September issues of *The Pointer* magazine exuded confidence. It would be up to Army to "clarify" Grantland Rice's uncertainty, though the two great rivals suspended their series by mutual agreement at the end of the 1947 season. In another piece the cadet sports writer said, "Army may, at some time, hit that long expected post-war slump, but, it surely won't be this year.... I doubt if that slump will ever hit as long as Army teams are Blaik coached."

But all was not well at West Point. One cadet writer in the September 29 *The Pointer* was critical of the Corps' spirit and support of the football team. He bemoaned the air of complacent overconfidence and indifference he heard among some members of the Corps. There were far more serious reasons for concern. Though Army was to roll through eight more victories before colliding with Navy, the 1950 season would be considerably different for more than one reason.

TWO MEN IN TWO DIFFERENT ROLES – UNKNOWINGLY FANNING THE FIRES

Earl Blaik and the Commandant, Paul Harkins, had been slowly, ever so slowly, developing a strained professional relationship since 1946, when Harkins arrived at West Point to become the Assistant Commandant. Each sensed characteristics they disliked in the other, and they looked for behaviors and actions in one another to justify or explain what they considered inappropriate actions. Perhaps it was personality differences, jealousy, or some form of professional rivalry in which each perceived of the other an influence contrary to their beliefs regarding their respective responsibilities at the Academy, or their contributions to the Academy mission.

Both were strong personalities, proud men, achievers in their work. It was the kind of rivalry or misunderstanding that could have damaging effects, especially in a disciplined, otherwise close-knit military organization with a clear, well-defined mission. Unknown to both, that was exactly what was happening.

Though they were both on the Academy's Athletic Board and, according to Blaik, never disagreed on athletic policy, they circled one another warily, never lashing out in the open, but contesting one another through their subordinates or the superintendent, General Moore. As Commandant, Harkins was also on the Academic Board, a far more influential board of Academy senior officers, chaired by the superintendent.

Harkins, as did other senior officers in his Tactical Department on the faculty and in the superintendent's staff, saw Blaik as too powerful. The Army coach had direct access to the superintendent, an "open door," an authority he had negotiated with General Eichelberger when he came to West Point – as he had done when he went to Dartmouth University from West Point after the 1933 season. He had the same access to the superintendent as the Commandant, the Dean, and the Academy Chief of Staff, who was responsible for coordinating and running the day-to-day administrative activities of the superintendent's staff. Not only was Blaik not subordinate to either the commandant, dean, or chief of staff, but General Taylor, the superintendent during the post-World War II draw down with its accompanying budget squeeze, had given Blaik additional responsibilities as chairman of the athletic board, when he already filled the positions of head coach and athletic director.

To add fuel to the fire, there remained a few senior officers, and no small number of graduates, who remained skeptical, if not resentful of the idea a civilian was head coach, though Blaik was an Academy graduate. Not until Blaik came to Army had anyone but active duty officers served as the head football coach.

And perhaps worse in their minds, in 1942, when the Academy began its wartime expansion from a Corps of 1900 to nearly 2500 cadets, Congress agreed with the Army's request to enact legislation enabling "qualified alternates" to be appointed to the Academy when vacancies appeared late in the annual appointment cycle. The law was badly needed because too many vacancies were going unfilled in each class, due mostly to entrance exam failures and candidates' withdrawals from appointments shortly before the new classes entered at the beginning of July. In an expanded Corps of Cadets the result would otherwise be a Corps considerably under strength from year to year.

But it also happened, because of NCAA recruiting rules governing the timing of recruiters' contacts with and commitments of recruited athletes, the law was the primary vehicle enabling Earl Blaik and his Washington athletic liaison, retired Colonel "Biff" Jones, to secure appointments to the Academy for promising athletes. In short, Earl Blaik and "Biff" Jones, the Academy graduate and former, highly successful Army coach, who had hired Blaik as his assistant in 1927, together, made Blaik the only man at West Point directly responsible for securing appointments to the Academy for many young men.

Among cadets who saw themselves as recruited athletes, it was not uncommon to find strong feelings of obligation and loyalty to Blaik, for his role in their appointments. The football players among the recruited athletes, over the years, also developed a collective name – "Blaik's Boys" – a name that, unknown to Blaik, and for a variety of reasons, was increasingly used in a derisive tone, especially among cadets who had never played the game of football, didn't understand its demands, or simply disliked the game because of what they perceived it to be.

There were other irritants that bothered Harkins, though he cautiously avoided confronting them in the open. He was unhappy about the Army coaches' seemingly endless requests for what Harkins regarded as "special privileges" for the football team, and excusals from duties and responsibilities required of everyone else in the Corps. The requests were never made through Harkins, but through tactical officers, professors or instructors in the Academic Department if the matter involved academics; or the Cadet Brigade Commander, First Captain Pat Ryan, who, along with many other responsibilities, scheduled and managed Dining Hall seating arrangements for the entire Corps, including athletic training tables.

Harkins was both the Commandant of Cadets and the officer commanding the Tactical Department, which was the link in the chain of command leading from the Army chief of staff into the Corps of Cadets. If there was anything he was sensitive to, it was just and even-handed administration of discipline and command, and he was the boss.

Like Blaik, Harkins was a consummate planner and organizer, and both came from an era in which iron-willed Army commanders were unfortunately often not open to "understanding by communication."

The Military Psychology and Leadership course, instituted as part of the post-World War II reforms at the Academy, was barely four years old at West Point, and George Patton, the man Harkins had served so loyally throughout the war was not the kind of man to encourage "understanding by communication." "Obey orders. I've laid out the plan. Perform your duties. Get the job done." These were the mission oriented watch words of a tough-minded commander.

And like Blaik, underneath Harkins' impressive exterior and commanding presence was a fierce competitor. Any loyal staff officer who served George S. Patton, Jr. for as many years as Paul Harkins did wasn't going to come away with a desire to lose on the battlefield, whether it be in war or in peace. Harkins' admiration of Patton was apparent. The controversial tactical genius of World War II had a profound, lasting influence on the Commandant, such that Harkins chaired a committee at West Point which finalized plans to memorialize Patton at the Academy.

On August 19, while second Beast Barracks was in progress, before the Corps returned to begin the academic year and football practice began, Paul Harkins participated in the dedication of the Patton Memorial, a heroic statue of the famed 3d Army commander, which stood across Jefferson Road near Doubleday Field, facing the library.

Earl Blaik was a coach whose job was to teach his players the game of football so well they would win – consistently. Harkins liked to teach young men to win in combat, too. He, like Blaik, was not accustomed to losing, and he was the cadets' real commander, including the men on Blaik's football team. Harkins knew he was responsible for training cadets for military leadership in peace and in war, and that included learning to think and make decisions under enormous pressures. If the training the Army saw fit to provide for cadets was warranted for one cadet, it was warranted for all cadets, including athletes. Earl Blaik wanted his players to perform under pressure on the practice fields and the game fields. But to do that he sought to relieve the pressures and demands of cadet life.

Unconsciously, both were engaging in a contest that should never have happened. And unfortunately, the more intense their respective efforts to hold their ground, or prevail, the more fault lines were appearing between the two departments – and between an important leadership nucleus on the football team and a growing number of cadets in the Corps. In the middle was the Academic Department, which, like the Tactical Department, had come to be considered by many of the football players as another "system" to be beat. The same attitudes could also be found among other cadets who were not football players, attitudes that, unknown by both Blaik and Harkins, were already giving way to a variety of rationalizations for behaviors the cadets knew were wrong.

Blaik, who deeply admired Douglas MacArthur, his long distance correspondent and confidant over the years, was not much different from Harkins with respect to communicating and understanding his responsibilities. Blaik, who either consciously or unconsciously imitated MacArthur, was from the same mold as Harkins. All business. No nonsense. "Here are the rules. If you don't want to obey the rules of the squad, you're off the squad." "Give 'em an order. They'll obey it." If ever there were two men who could have sharp differences for lack of communication and understanding, Earl Blaik and Paul Harkins were leading candidates.

Earl Blaik, in his tenure as a cadet after World War I, came to West Point after graduating from Miami (Ohio) University. He was thus older than most of his classmates. He was an assistant coach at Army from 1927 to 1933. Then he came back to Army as head coach in 1941.

By 1950 Blaik firmly believed he understood cadet life because he had been at the Academy a total of nineteen years, including his time as a cadet. He was head coach under four superintendents and served with five commandants, including Harkins. He knew the demands placed upon cadets by the Academy curriculum. He concluded the formal relationships between officers in the Tactical and Academic Departments and cadets were impediments to the officers' really knowing and understanding the young men in intercollegiate athletics, especially the football team. Blaik believed the more relaxed circumstances on the athletic squads enabled him to know team members better. They could confide in him and his coaches with greater ease.

He was right to some degree, but there were more complex reasons his team members tended to come to him or his assistant coaches, not the least of which was the influence the cadets were convinced he exerted at the Academy – and in their lives.

Blaik was also aware that competition for time in the cadets' training day was fierce, and the combined academic and athletic hours, plus military training, discipline, and regimentation were factors he knew weighed heavily upon West Point's intercollegiate athletes – especially the football team. The game, which was becoming more complex and intensely competitive each year, required enormous mental, emotional, and physical discipline, and commitment. For these reasons, Earl Blaik made it a point to relax his players when they came to practice, prepared for games, or got respites from the fast paced routine of cadet life.

"Leave cadet life and your troubles behind you when you come to practice and play football. Leave it all behind. Football requires total concentration – no distractions. You have to keep your mind on the game as long as you're on the practice and playing fields." While he was nearly all business, he occasionally arranged ice cream socials at the Blaiks' quarters near Lusk Reservoir, especially during Beast Barracks for each year's new crop of players.

Blaik devoutly believed football was the game most like war and he was contributing to training leaders for the battlefield. In Blaik's mind, football's need for total concentration, and tremendous mental, emotional, and physical discipline, courage, and sacrifice, justified some compromises in the rigors of cadet life. That was one way his athletes could have the time and the mind set to learn what would stand them in good stead on the playing field – and the battlefield.

Earl Blaik and Paul Harkins were coming at their respective missions from within two different worlds, two entirely different frames of reference. As far as Harkins was concerned, the only wars Blaik ever fought were the gridiron wars. He had never been on or near a battlefield. It was an outlook indicating strong, if not somewhat myopic views of Earl Blaik's role and what he contributed to the Academy and the Army. And in Blaik's mind, Harkins had never been a real soldier. "He was just a staff officer," not at all a thoughtful, realistic regard for Harkins and his World War II service.

And unknown to both, their words and actions, and those of some of the men who served under them, were increasing tensions between their two departments, and causing subtle shifts in attitudes affecting both the football team and the Corps of Cadets. Within the Corps of

Cadets, to which the football players belonged, the growing tensions were evident, but not understood as to their origin. The problem didn't stop there. Between increasing numbers of Earl Blaik's football players, and a growing number of officers in the Tactical and Academic Departments, and within the Corps of Cadets, there was a growing undercurrent of disaffection.

There was another, deeper, more corrosive element in the fissures beginning to appear between departments and among the Corps of Cadets the fall of the 1950 football season.

Organized cheating in academics was already in progress by a small group of cadets, many of whom were varsity athletes, primarily football players. What's worse the honor violations were spreading rapidly in the Corps. After the 1950 football season ended, the spread of cheating accelerated, until, by the end of May 1951, it had reached epidemic proportions. And all the while a small but growing number of men in the Cadet Corps, and the Academy's officer corps, sensed something was not right.

WAR'S SAD REMINDERS

While the West Point football team was preparing for the 1950 season, far away in Korea the United States Army and the Academy's young graduates were suffering grievous losses. Edmund J. Lilly III, a platoon leader in the 2d Infantry Division, was the first member of the class of 1950 to lose his life in combat. Killed in action September 3, he had graduated less than three months earlier.

Ted, as he was called when he was a cadet, was a "brat," his father a career Army officer. His father was in the infantry in the Philippines in 1941, and Ted graduated from an American secondary school in a class of three students. His junior high school classmates were Gail Wilson and Frank Lloyd, both of whom became his classmates when, after a year at the Citadel, Ted entered West Point. Major General Jonathan M. Wainwright, West Point class of 1906, and the senior officer in command on Corregidor Island when American forces surrendered there in May of 1942, was the guest speaker for his secondary school graduation. Ted returned to the States with his mother in May 1941 when prewar tensions in the Pacific resulted in the U.S. Government ordering military dependents home.

While in Fayetteville attending the Citadel, Ted was a member of St. John's Episcopal Church and took part in the Young People's League. At this time he became interested in a career in the ministry and had many long talks with his pastor on the subject. He eventually decided to try for West Point and entered the Academy in 1946.

The day after he graduated, on June 7, Ted married a lovely El Paso girl, Mary Alma Russ, whom he had met on a blind date while on a cadet training trip to Fort Bliss. While at Fort Sam Houston, Texas, on honeymoon leave, he became concerned about press reports of world conditions and notified his unit, the 2d Infantry Division, of his exact location. Several days later his leave was canceled and he reported to Fort Lewis, Washington. By the end of July he was in Korea.

That early September day, his platoon, in B Company, 9th Infantry Regiment, held the crest of a high hill overlooking the Naktong River front, which roughly marked the final line of defense for the Pusan Perimeter. While the remainder of the Regiment was driven from its defensive positions, Ted's platoon held fast. Why they didn't withdraw, no one knows. Before he died, he was seen walking among his men during intense automatic weapons fire and grenade explosions, encouraging them to greater efforts against overwhelming odds. He was doing what he had been taught. He was performing his duty as an infantry platoon leader, exemplifying the courage his men must have in the face of fear and the ever present urge to abandon their threatened position.

Edmund Jones Lilly III
Class of 1950
May 26, 1928 - September 3, 1950
(Photo courtesy USMA Archives.)

There were many others from the Academy classes of '45 through '50 who were to fall between June 25 and Army's first football game of the season: forty-eight in all, including two who were taken prisoner and later died in captivity.

But there was no one keeping count at West Point, certainly not with the intent of publishing the mounting toll of graduates. That would be cruel, disrespectful, irreverent. Life at West Point was not like that. The Academy graduates who gave their lives, many having just begun their tragically brief service, were professional soldiers, paying the price of what noted British soldier-scholar Sir Archibald Hackett defined as "unlimited liability." Most had graduated from the Academy before 1950, and the fast pace of life at West Point quickly blurred memories of them, easing the shock that invariably accompanied such losses.

The friendships formed among most cadets were stratified, the class seniority system was a strong inhibitor to lasting associations between members of successive graduating classes. There is always sadness when a graduate gives his life, but it comes slowly, a name at a time, and it strikes families and friends hardest. For the most part, under-classmen left behind in their cadet companies are spared the personal hurt in losing someone for whom they might have had great, yet distant respect.

But the men who became teammates on the Corps' intercollegiate athletic teams, or worked closely in other activities, were another matter. Close, lasting friendships often cut across class seniority lines. The crucibles of tough, demanding intercollegiate competition and team-work are powerful influences in forming such associations. The loss of a former teammate can be keenly felt, the result of strong bonds of friendship, deep mutual respect, fierce loyalty, confidence, and trust. The more so in sports such as intercollegiate football – where those associations are formed on an exciting, potentially gratifying national stage, leaving team members happy, lifelong memories of accomplishment and friendship.

Less than a week before the September 30 opening game against Colgate, Tom Lombardo ('45), one of two great quarterbacks, and captain on Army's 1944 national championship team, was killed in action at the Naktong River. His death came nine days after the landing at Inchon, as the 2d Infantry Division was preparing for offensive operations to widen the breach of the Pusan Perimeter. The 2d would add strength to the advance of the 25th Division and link up with forces which would land at Inchon. Tom, aide to the 2d Division Commander, Major General Laurence "Dutch" Keiser (West Point's class of April '17), was also in the 9th Infantry Regiment's sector assisting in the offensive preparations when elements of two NKPA divisions crossed the Naktong in force and attacked. Caught flat-footed and ill-deployed for defense, the 9th Infantry was almost immediately overrun. Lombardo, the Regimental operations officer, and many others were killed in the onslaught.

Undoubtedly Earl Blaik responded to Tom's loss as he always had – and would continue to do – when he learned one of his former players had made the ultimate sacrifice. When he became aware of the loss, often by phone call or letter from a team member, he would immediately express his heartfelt sympathy and condolences to the man's family.

Earl Blaik's feelings were deep and genuine. They were his "boys." He felt their loss. He had, more often than not, been directly responsible for their appointments to the Academy and had been a prime motivator toward a career in the armed forces, perhaps more than any other figure in their lives.

THE SEASON MUST GO ON

In spite of all that was going on, good or ill, that fateful fall of 1950, the Army football team roared through eight straight victories before the Navy game. The Corps of Cadets shouted its support and approval as the Black Knights rolled over Colgate in the season opener 28-0 at Michie Stadium, crushed Penn State 41-7, and stunned Big Ten champion Michigan 27-6 in New York City's Yankee Stadium. Then came a 49-0 pasting of Harvard, an easy 34-0 win over Columbia, a 28-13 defeat of always tough Pennsylvania at Franklin Field in Philadelphia, and a 51-0 thrashing of a lightly regarded University of New Mexico at Michie Stadium on November 11.

The New Mexico Lobos were played before a crowd of 30,476, a new attendance record for Michie Stadium. Among the fans that day were sixteen thousand Boy Scouts visiting West Point. And for the second time in his ten years at Army, Earl Blaik was deliberately absent from the field while his team played a football game. Anticipating a romp over New Mexico, he left First Assistant Murray Warmath in charge and traveled to Baltimore to scout Navy's performance against Tulane. The following Tuesday, college coaches ranked Army number one in the nation.

But November 1950 was an unusually rainy month at West Point. There was only one day of outdoor practice without rain. Blaik moved the team into the Field House.

The week after the New Mexico game, Army traveled to Palo Alto, California, to play Stanford. The rain and mud followed, and the Cadets, in a hard fought game, slogged their way through the weather to a 7-0 win. Army slipped to number two in the polls after the Stanford game. The stage was set for the annual showdown with Navy.

The team's trip to Palo Alto by plane was long and tiring. Twenty-three hours elapsed before beds were available for sleep. Upon their return to West Point, it took them several days to regain their vigor. As a consequence Blaik and his staff didn't put too much pressure on the team until the first two days of Navy week. Cold, rainy weather continued, as did indoor practice in the Field House.

MORE BAD NEWS

It was during Navy week preparations that the Corps of Cadets, Earl Blaik, his coaching staff, and his entire team received a harsh, emotional jolt. Blaik learned John Trent, the captain of Army's undefeated 1949 team, had been killed in action in Korea the night of November 15. He died four days after his unit disembarked at the North Korean port of Wonsan to join the Eighth Army drive toward the Yalu River. At the time, little information was available concerning his death.

Exactly what happened to Johnny Trent that night, no one knows. The 2d Battalion, 15th Infantry Regiment, the battalion to which John's platoon and company were assigned, was part of the 15th's mission to relieve elements of the 1st Marine Division in the vicinity of Wonsan.

The mission required "mopping up" protection for the port, with assistance by several Republic of Korea (ROK) battalions. Because the mountains to the west and south of Wonsan were still crawling with NKPA troops bypassed by the rapid allied advance toward the Yalu, the mopping up task proved to be an arduous and dangerous chore. The 3d Division historian wrote:

It became apparent that the Division was engaged in something considerably more extensive than had been anticipated. The strength of the NKPA remnants and guerrillas in the area seemingly had been underestimated.... Men in every [3d Division] unit learned the bitter taste of ambush, the sudden shock of receiving burp-gun fire from darkness or other concealment. They learned that there was not a moment during which they were completely safe from sneak attack, never a time when danger could not appear from any point of the compass.

It was in this dangerous operational environment the night of November 15, 1950, that "Big John," as his classmates called him, did what combat leaders are expected to do. His platoon was deployed in protective positions, on the alert. His platoon sergeant volunteered to leave their position and move about in the inky darkness to check on their men. John saw the task as his responsibility, declined his platoon sergeant's offer, and left to look to his men's welfare. A fire fight broke out. In the terrible confusion that inevitably comes with night combat, random grenade and mortar round explosions, blinding muzzle flashes from rifles, burp guns, and other automatic weapons hide soldiers from help they might normally receive in a daytime encounter. John's body was carried from the field the next morning.

John Charles Trent
Class of 1950
October 11, 1926 -
November 15, 1950
(Photo courtesy USMA Archives.)

A Marine officer, William B. Hopkins, departing with the 1st Marine Division, wrote that John Trent's body was being escorted to the rear by a weeping master sergeant with a thick southern accent, who blamed what he believed was a fouled-up fight on the incompetence or panic or even perfidy of the ROKs attached to the 2d Battalion. The angry master sergeant's words were filled with bitterness and remorse.

"I been in this man's Army mor'n half my life," Hopkins quoted the sergeant as saying, "and this is the most fucked up unit I ever seen. We burnt three machine-gun barrels last night, and nobody ain't seen...[no enemy].... Trouble is we'll ever know whether he [Trent] got kilt accidental or on purpose. Cain't nobody say there ain't some gooks among them ROKs."

John, the quiet, strong, handsome young man from Tennessee, was personable, affable, and possessed a commanding presence. He was a much admired team leader. To cadets, teammates, and graduates serving at the Academy, John Trent was a shining example of all that was good and right about West Point and Army football. Johnny, as Coach Blaik called him, was posthumously awarded the Purple Heart, and after the 1950 season came to a close, the Football Writers' Association of America named him Football's Man of the Year.

When the news of John Trent's loss reached Earl Blaik less than three months after word of Tom Lombardo's death, "The Colonel," as nearly all his players knew him, must have also remembered the angry shouts of some rabid, unsportsmanlike Notre Dame fans in New York City. During the 1944, '45, and ' 46 seasons, Army and Notre Dame's Fighting Irish were meeting in Yankee Stadium, vying for national football honors. Some of the mean-spirited "subway alumni," the ones who believed the Army – Notre Dame rivalry was healthy and sportsmanlike only as long as Notre Dame won, would scream "Slackers!" at the Army players as they left the field for the dressing room after the games, insinuating Army's football players were at West Point to avoid their obligations to serve their country in time of war.

Earl Blaik and John Trent, Collegiate Football's 1950 Man of the Year, pictured in far more memorable and happier times during the 1949 football season. (Photo courtesy USMA Archives.)

Now, in the summer and fall of 1950 West Point's young graduates, including its gridiron greats, were once more paying the price of being inexperienced, junior officers in leadership positions on battlefields far from their country's shores. Three former captains of Earl Blaik's Army football teams had been killed in action, two in Korea within a brief span of time, in a devastating period of great loss.

Colonel John K. Waters, the Assistant Commandant looked at Johnny Trent's death far differently than did Earl Blaik. He wondered aloud if John had received adequate tactical training as a cadet, because football players, during the fall training weeks, were excused from scheduled hours of drill and ceremonies, which included several hours of tactics instruction. Waters studied how many hours of drill and tactics instruction football players routinely missed, and told his boss, Colonel Harkins, of his findings. Neither took any action, but the results rankled them further in an already flaming interdepartmental prairie fire.

* * *

In spite of the November weather, and the sad news of John Trent's death, Earl Blaik believed the week's preparation for Navy was the best ever since he became head coach. But within the Corps of Cadets there was at least one who had the same nagging discomfort Earl Blaik was feeling. In the November 30 issue of *The Pointer,* Ransom E. Barber ('51), the magazine's sports editor, again warned of trouble ahead. In a piece titled "Army Sitting Ducks for the Middies," he thoughtfully analyzed the recent history of the Army-Navy series, and urged the Corps "to get up for the game and let the team know how you feel about Navy."

The same nagging feelings appeared in another way two days before the game. They were expressed by the captain of the football team and were another blow in the downward spiral pulling the football team and the Corps of Cadets in opposite directions. The team captain rose to speak to the Corps at the traditional Thursday evening rally in the cadet dining hall. Instead of dwelling on Corps and team unity and the need to pull together to win, he criticized the cadets for failing to show up at practice in the Field House, remarks which added fuel to the fire. The Assistant Commandant was particularly irritated at the remarks

248

because Blaik, as was his normal practice, had "closed" the practice sessions to everyone, including cadets, a decision Waters didn't understand. He was baffled, and remained miffed that Blaik wouldn't allow cadets to watch practices for fear they might "give something away to Navy" and then the team captain would criticize the Corps for failing to adequately support the team.

Yet, even among the national sports writing fraternity, there were precious few who believed Navy would upset highly regarded Army. The day of the game, banner headlines in the *New York Times* proclaimed Army a three-touchdown favorite. It was the kind of proclamation that would guarantee a dogfight in one of collegiate football's most storied rivalries.

After losing to Navy in his first three years as Army's head coach, Blaik's teams played the middies seven years without a loss. But there were some haunting memories for Blaik in those seven games, the kind of memories that have long been staples of the series. In 1946, the last year of the powerful Davis-Blanchard teams, heavily favored Army barely eked out a 21-18 win over a Navy team which lost seven of its prior eight games. The game ended with Navy driving inside the Army five-yard line.

Again in 1948, undefeated Army held on to gain a 21-21 tie with the middies, who lost eight games before its annual trek to Philadelphia. In 1949, "The Rabble" took their revenge by administering a 38-0 defeat to Navy, until then the most points scored against a Navy team since the cadets won 40-5 in 1903.

To a band of hungry midshipmen, the season of 1950 and the recent history of Army-Navy football, was exactly the right mixture from which they could stir a powerful, emotional brew. And they did.

Before President Truman, several cabinet members, and a crowd of more than 101,000, on a cold, dry, overcast afternoon, an inspired, furiously energized band of midshipmen administered a physical and emotional 14-2 beating to Army. It was a beating West Point cadets, Earl Blaik, and his coaching staff, would remember for years. In the first five minutes of the game, Army's safety and second platoon quarterback Gil Reich, had to leave the field because of a hard hit which left him dazed. He never returned to the game, and "didn't return to himself" until later that evening. The hard-hitting, gang-tackling middies had served notice of a long afternoon. Four times in the first quarter they stopped Army's offense without a first down, once on the Navy 21-yard line.

Early in the second quarter, Navy made the first of five pass interceptions in the game, and promptly drove to its first touchdown with a slashing running attack. Then twenty seconds before the end of the half, with Gil Reich out of the game, the fired-up Navy quarterback, Bob Zastrow, completed a 30-yard touchdown pass over the head of Army's substitute safety. The extra point gave the middies a 14-0 lead. The Navy first half offensive performance was matched by its defense, which smothered the vaunted bevy of Army ball carriers, holding them to a net three yards rushing and only one first down.

The crowd and sportswriters were expecting an Army turnaround in the second half. A change in momentum never occurred. The cadets received the second half kickoff, and in the first offensive series Navy intercepted its second pass. Army's only score came in the third quarter when the defense caught the Navy quarterback in the end zone for a safety.

Then came one of the wildest fourth quarters in the history of the great Army-Navy rivalry. In those last fifteen minutes of play three Army passes were intercepted, each team lost the ball twice on fumbles, Army blocked a Navy punt, and the middies were penalized time and time again. Army, in the final quarter, was on Navy's 21, 15, 6, and 3-yard lines. On the last play of the game, Navy end John Gurski made his second of two interceptions and ran the ball to midfield, where he fumbled.

After the game, the Army dressing room was like a morgue. Earl Blaik, talking to reporters said, "We were defeated by a great team." There were a few tears from such Army stalwarts as team captain and All-American end Dan Folberg, fullback Gil Stephenson, and quarterback Bob Blaik, Earl Blaik's son. From behind one locker door a sob could be heard. And then out of the nearly deathlike silence came, "Oh, hell, boys. It was tough, but there are plenty more years."

For all but a few of the men in the Army dressing room that day there would be no more Navy games, no more Army football, and precious little time as teammates – except at spring practice the next semester.

When the team returned to West Point on Sunday, it was cold, raining, and miserable. A long-standing tradition was that the Corps of Cadets turned out to welcome the team at the West Shore Railroad Station when they returned to the Academy from the gridiron war in Philadelphia – irrespective of weather or whether the Black Knights won or lost. Not this year. No one knew why, although years later it was said that Earl Blaik called the Cadet Guard Room from somewhere short of West Point and asked that the Corps not meet the team.

DARK ECHOES

There was more than one sad irony in the beating the Army football team took that day. In Korea, along a winding, four mile stretch of road, in hills east of the Choshin Reservoir, the last scenes of an enormous tragedy were being played out. The United States Army suffered one of the most bitter and costly defeats of the war, and survivors of Task Force Faith – units from the 32d Infantry Regiment, "The Queen's Own," from the 7th Division – were straggling southward toward the small North Korean town of Hudong, dazed, exhausted, frostbitten, and hundreds wounded.

The Task Force, named for its commander, Lieutenant Colonel Don Faith, a Fort Benning OCS graduate, had been cut off, and withstood assaults on its defensive perimeter for four days and nights. Because his units hadn't sufficient ammunition remaining, and there were hundreds of wounded requiring better care, he believed his force couldn't withstand another major attack. He decided they would fight their way south to Hudong, where he expected to be reinforced. The Chinese Communist Forces were dug in on hills overlooking virtually the entire route. They had set up road blocks, and blown two bridges.

At one o'clock in the afternoon of December 1, his column of thirty trucks began to move. The trucks were loaded with six hundred wounded, protected by all the infantrymen that could be mustered. To give the column added protection and firepower, a vehicle mounted with quad fifty-caliber machine guns and two half-tracks carrying twin forty millimeter anti-aircraft artillery pieces were attached. The quad fifty vehicle was in the center of the column, one twin forty in the lead, and the other giving rear guard support.

George Everett Foster
Class of 1950
March 4, 1928 -
December 2, 1950
(Photo courtesy USMA Archives.)

The move began with a horrible, shattering error when pre-planned Marine close air support aircraft, in their first pass, dropped napalm short, on the head of the column. The men who suffered the blow were infantrymen who were leading the column. The billowing, searing flames engulfed about a dozen Americans, among them two platoon leaders. Both were badly burned, one fatally. He was George E. Foster, West Point class of 1950.

Then, in the ensuing eighteen-hour, four-mile, fighting withdrawal, the entire force was virtually destroyed. Of the original twenty-five hundred Americans who comprised the Task Force, about one thousand were killed, left to die of wounds, or captured and placed in CCF or NKPA prisoner of war camps. Only about 385 survivors were fit for duty.

The Task Force commander, Don Faith, a paratrooper, was mortally wounded in the battle. He had been selected out of OCS to be General Matthew B. Ridgway's aide, and was at Ridgway's side throughout World War II, jumping with the 82d Airborne into Normandy. On the battlefield he was intense, fearless, relentlessly aggressive, and unforgiving of error or caution. In Korea, he was one of two battalion commanders to receive the Medal of Honor.

The tragedy of Task Force Faith the first two days of December was virtually overshadowed by what was occurring all across North Korea. United Nations forces were reeling southward, engulfed by the onslaught of a sudden, massive intervention by Chinese Communist "Volunteers," who, unknown to the world, actually began pouring across the Yalu River in October. By November 24, nine days after Johnny Trent had fallen near Wonsan, the CCF offensive was administering punishing, demoralizing blows to UN Forces. The tide of war was once more reversed.

EVAPORATED GOLD

No one at West Point knew the defeat by Navy on December 2, 1950, was the beginning of a long, agonizing drought for Army football. The upper three classes could look forward to Christmas leave in spite of a middie victory. The plebes, the young men of '54, who, like classes before them, looked forward to a Christmas at West Point without the upperclasses, had had high hopes of an Army victory. A win over Navy would have meant "falling out" instead of the usual plebe "brace" or rigid position of attention throughout all of December, not just Christmas leave. For '54, however, "gloom period" came early.

On December 27 Earl Blaik wrote a lengthy letter to General MacArthur. He was continuing a correspondence, and deepening a friendship with MacArthur that began shortly after Blaik graduated from West Point. Typically, the letter summed up Blaik's thoughts about the season just past, and reflected his consistently detached, analytical assessment of his team's prospects for the coming season, including its strengths and weaknesses.

The letter also revealed the importance he attached to the Army-Navy game, and the strong impact the December 2 loss had on a head coach accustomed to winning. He devoted an entire page to explaining the reasons Navy won. His letter also reflected the mood engendered in the team from the growing casualty lists from Korea.

He wrote of "the sad blow to read of General [Walton H.] Walker's death." General Walker (class of 1912) was the Eighth Army commander and had been killed in a jeep accident December 23.

The loss of John Trent was deeply felt, and considered by Blaik one of several key factors in the stinging defeat at the hands of the middies. He wrote:

> The news during the Navy week of [John] Trent's death in Korea relegated thoughts of the Navy game to second place...our '49 captain, [he] had been an assistant coach in early season preparation, but had been ordered along with [Arnold] Galiffa, to join the 3d Division.

His letter offered another reason for the Navy win, and some emotion that for him was unusual. He understood the importance of the Corps' and Academy officers' influence on the team, and recognized what the cadet writer for *The Pointer* had seen and heard in September and November.

> The Corps and officer personnel were cock-sure of a victory forecasted by them to scores ranging up to fifty points. Try as the coaches did to dispel this feeling, it was impossible to obtain among the players a true respect for the Navy team.

There was an edge of bitterness, an undertone of anger in these words, which were uncharacteristic of Blaik's behavior, and the way he analyzed his team's mistakes and losses. Perhaps his words were reflections of a growing frustration he couldn't understand, manage, or control. Perhaps he sensed something was wrong, puzzling. If so, he was responding the way people frequently do when they can't comprehend why plans go awry, why they fail to achieve what seems so achievable, and for Earl Blaik, the master football coach, the most tormenting event of all – defeat of a team that seemed to have everything going for it. He was pointing his finger at the cadets and officers who were overconfident, and among whom his players lived and studied. He was unconsciously reflecting the growing distance between some of the men in the Corps, and their football team. There was, indeed, something seriously wrong. And it was more than he ever dreamed.

However perceptive and successful Earl Blaik was as a coach, leader, and teacher of young men; however thoughtful he was in analyzing the game of football, victory and defeat on the gridiron, and the young men who paid the price for him, he hadn't the vaguest notion that members of his 1950 football team were among a larger number of cadets violating the cadet honor code, cheating in academics, and that the violations had been going on throughout the fall, and in lesser numbers, perhaps several years prior to the 1950 season.

No one, including Earl Blaik, was aware the loss to Navy ended Army's golden era of football. Nor did anyone have an inkling of the disaster that would bring the Academy and the Army team to their knees the following year. The unraveling began while spring football practice was in progress, but it had been a long time in the making.

CHAPTER 7

1951: WEST POINT'S YEAR OF ADVERSITY

Saturday, December 22, 1951. Except for men on the Honor Committee, and in the cadet chain of command, thoughts of the previous spring and summer's events, the war in Korea, and the losing football season – which ended with a 42-7 thrashing by Navy – had been pushed into the background. Preparations for written general reviews, the semester final examinations, and eager anticipation of the Christmas holidays were of more immediate interest beginning the first days of December. Now, as the hands on the central area clock pointed to twelve noon, the class assembly bell reverberated in the cadet area, triggering loud whoops and yells of joy, which echoed up and down barracks halls in all twenty-four cadet companies.

Christmas leave! Written general reviews were over, and cadets eligible to take leave – except for the class of 1955, our class, the plebes – were ready to storm out of barracks doors.

There it was in the cadet "Blue Book," officially titled "REGULATIONS for the UNITED STATES CORPS OF CADETS," Chapter 13, page 19 of 84 pages, plus maps, which governed every facet of cadet life at West Point:

13 – LEAVE OF ABSENCE

13.01 Types and Eligibility:

Furlough	1st, 2d and 3d Classmen (Summer)
Christmas leave	1st, 2d and 3d Classmen
Week-end leave	1st, 2d and 3d Classmen
Sick leave (Hospital)	All Classes
Skiing leave	1st, 2d and 3d Classmen
Fishing and camping leave	1st, 2d and 3d Classmen
Emergency leave	Under exceptional circumstances

There could be only two forms of leave for plebes, sick leave, in the hospital, and emergency leave, "Under exceptional circumstances."

Puzzling. Many seventeen, eighteen, or nineteen year olds who volunteered to be soldiers in the Army, or airmen in the Air Force, could take Christmas leave. But not plebes at West Point, who were the same age and training to be officers. Oh, for life beyond plebe year!

For the superstitious, thirteen is an unlucky number. What better chapter to tell plebes completing their first semester at West Point they wouldn't be granted leave while the three upperclasses fled their gray granite surroundings?

The upperclassmen would scatter to the winds on Christmas leave. Most went home, the place where we would like to go – at least those of us who had left home for the first time in our lives last summer. "Home sickness," the disease that afflicted many in Beast Barracks, and had faded in the pace and excitement of our first semester at West Point, again invaded our thoughts.

Actually, we had known before we entered Beast Barracks in July that our first Christmas as cadets would be celebrated at West Point. We kept hoping, wishing, spreading rumors there would be a change in Academy policy. No luck. But we took comfort in knowing all three upperclasses would be gone from the Academy, leaving us peace and quiet. Besides, there was the feeling in every plebe class who stayed behind at Christmas, "We can run this place better than *they* can, anyhow." What's more, we could look forward to a full schedule of social activities, and opportunities to escort young women about the Academy, just like upperclassmen.

Karl Brunstein, Oscar Raynal, Dave Pettet, and I were in Company K-1, in south area barracks, in a first floor room, an unusual privilege for plebes. I had moved from a room two divisions away, after getting the worst end of an altercation with one of three other roommates, Harry Comeskey, who was under considerable academic pressure at the time, and was "found" in academics at semester's end.

Late that fall, one afternoon right after lunch, Harry and I got into a shouting match over something neither of us would remember. Carrying the usual load of books for afternoon classes, as I was backing out of the room, dressed in class uniform and the cadet short overcoat – and pulling the door closed with my right hand – I decided to get the last word in the argument. Poor judgment, and worse tactics. Harry's reaction was instantaneous and nonverbal. My adrenaline had been spurred into circulation by the argument, and I had an unusually firm grip on the door knob. From inside the room, before I could pull the door closed, Harry grabbed its edge, and in a lightning motion jerked the door and a surprised me back into the room. Surprise was complete when I met his clenched left fist flush on my nose as I came toward the rapidly opening door. My grip on the door knob abruptly relaxed when his blow sent me reeling unceremoniously across the hall, up against the wall and down on the floor, as my books went flying. I was dazed and bleeding from the nose, but had to recover quickly to avoid being reported late to class.

I did, but the one-sided battle scars, a swollen nose and some black eyes were evident to discerning upperclass cadet officers. They told our company Tactical Officer Major Royem that it was best I move to another room. Shortly thereafter I moved in with less excitable room-mates – or perhaps roommates I wouldn't provoke to such violence.

There was Karl – the serious minded son of Russian-Jewish immi-grants, an academic genius – a star man – an excellent gymnast who, as an enlisted man, obtained a regular Army appointment to the Acad-emy; Oscar – whose home was Chihuahua, Mexico, was equally gifted, as well as a warm, gentle, devout Catholic, who years later, after serv-ing in the Mexican Army, became a Jesuit priest; and fun-loving, rabble-rousing Dave – an Army "brat" whose father was a Provost Marshall in the Army's Military Police at Fort Bliss, Texas. Dave's reputation as a "goat's goat," a man who is consistently low in academic standing, belied his intelligence. He was interested in things other than studies, such as mountain climbing, the rifle team and later, the pistol team. He also had a facility for languages. Not easy languages, but difficult to learn languages – such as German and Russian, which led to his be-coming a military intelligence officer years later.

I was glad to be on the first floor with new roommates, two divi-sions away from the third floor room where I'd lost a most embarrass-ing argument.

Across the hall from our first floor room were two yearlings, Nelson Byers and Noel Perrin, "Mr. Byers and Mr. Perrin" to us. They were from California. As the bell rang to sound the start of Christmas leave – which also signified "falling out" time for plebes, we opened the door to our room to watch the flood of upperclassmen dash to the transportation that would take them to freedom. "Nellie" Byers and Noel Perrin were leaving their room across the hall. They wore dress gray with the long, wool, cadet overcoats, carrying their B-4 bags – the military, olive drab, heavy canvas suitcases – while most of the upperclassmen wore civilian clothes and carried suitcases of various descriptions. Both turned and glanced at us, grinning broadly, as they headed down the hall toward the entrance to the division. They clearly were elated at leaving "Sing Sing on the Hudson." As he pushed open the door to the barracks stoops, "Nellie" Byers called out cheerily, "You men have a good Christmas. We'll see you New Year's."

"Yes, Sir. Merry Christmas. Have a good trip," I said, as each of my roommates voiced similar good wishes for the holidays.

PREPARATIONS FOR A JOYOUS SEASON

Cadet Guy L. McNeil, Jr. was in Company I-2. He, too, was headed home to California on Christmas leave, along with three other members of his company, two classmates, Hugh R. Wilson, Jr. and Herman Archer, and first classman Hilmar G. Manning. An Air Force C-47, a two-engine transport aircraft, the old "Gooney Bird" of World War II fame, was waiting on the ramp at Stewart Air Force Base in Newburgh, New York, twenty-five miles north of the Academy. The pilot, Colonel Guy L. McNeil, instructor pilot Major Lester G. Carlson, and crew chief Staff Sergeant Marion A. "George" Sampson, prepared for take-off that afternoon to carry the four men from I-2, Nelson Byers and Noel Perrin from K-1, Karl Glasbrenner from C-2, and twelve more cadets from nine other companies, home to California for Christmas leave. Eight men in the class of 1952, and eleven in the class of 1954 comprised the California bound revelers.

Guy McNeil's father was a senior staff officer and a widower, stationed at Hamilton Air Force Base, twenty miles north of San Francisco. He was the 4th Air Force Inspector General, and his duties required extensive travel, doing annual IG inspections and operational readiness inspections of Air Force units assigned to 4th Air Force.

His duties weren't pleasant. Not only did he travel frequently, but he and his inspection teams were also responsible for reporting any deficiencies they observed in units' mission performance, and violations of Air Force policy or regulations, duties that didn't endear him to Air Force people who lacked realistic understandings of the IGs responsibilities and obligations.

Another important responsibility Colonel McNeil bore was to ensure the IG complaint system worked as it should. In each Air Force wing or group there was at least one officer assigned to hear, investigate, and resolve complaints and grievances by enlisted men and women, government employees, and officers assigned to Air Force installations for which the 4th's commander was responsible. Resolution of complaints and grievances was an important matter in any command.

But there was satisfaction gained from his duties. Though he was responsible for inspecting Air Force units for their adherence to policies, practices, regulations, and effective mission performance, discrepancies and deficiencies found during inspections resulted in "corrective actions," and usually improvements in efficiency and mission performance by the units his teams inspected. And there was always some satisfaction in knowing complaints were thoroughly and objectively investigated and resolved.

Also under Colonel McNeil's supervision was the 4th Air Force's Safety Office, including flight safety. Safety duties carried a certain unpleasantness as well. Work in flight safety was seldom the kind of duty to bring smiles and sunny dispositions. Staff officers assigned flying safety responsibilities not only did preventative inspections of 4th Air Force units, they kept accident statistics, reviewed accident investigation reports to ensure investigations and reports were thorough and complete, and occasionally participated in accident investigations. Satisfaction in the duty could be gained only from finding and eliminating hazards that might cause an accident, or finding and correcting the causes of aircraft accidents.

Guy McNeil, Sr., and his son had been in communication during the early fall. He had found a way to bring Guy, Jr., and other West Point cadets home on Christmas leave by military aircraft – if his son could round up enough cadets to nearly fill available seats. Fly an airplane from Hamilton to Stewart, with a load of service personnel headed "space available" to the east coast for Christmas. Pick up the cadets, including his son, and bring them back to Hamilton. Then fly them back to Stewart before New Year's to ensure the cadets got back to the Academy on time.

He could then pick up a load of service men and women returning to the west coast from Christmas leave. The thought of having his son home for Christmas was undoubtedly pleasing for a man who had lost his wife and was subject to the pressures and lack of time with his family that came with being an Air Force IG. Besides, there was a war on, and it would be good to put another airplane in the air to take "the troops" home for Christmas.

There surely would be a large number of military personnel stationed in and near Hamilton Air Force Base who would be delighted to take a space available military flight to the east coast for Christmas at home. If necessary they could fly back to California commercially, should there be insufficient seating available on return military flights from the east coast.

On November 27 Guy McNeil, Sr. requested the 4th Air Force Director of Operations to schedule two C-47 airplanes for flights from Hamilton to Stewart Air Force Base, one to depart on December 20 and return December 23, the other leaving on December 30, with return January 3. The stated purpose was to "ferry West Point cadets." On the same day, at West Point, an item appeared in the Cadet Daily Bulletin:

> Any cadets desiring transportation to San Francisco, via military aircraft, who have not already done so, contact Cadet Glasbrenner, Company C-2, Room #2312, prior to 2215 [10:15] tonight. It is requested that only those cadets who intend going the entire distance apply. Gas stops and other details are not available.

Young Guy McNeil and C-2's cadet company commander, Karl Glasbrenner, worked hard to identify and contact cadets from California whose homes were in the San Francisco area, hopeful of filling as many seats as possible. They succeeded in finding the seventeen other men who would gladly travel military space available. Space available was a legitimate means of travel on leave, and a way to avoid expensive air line tickets, as well as the frustrations of transportation to and from major airports during the holiday period. The only possible drawbacks were flight on a military airplane, even on leave, required wear of the uniform, and seats on the average C-47 weren't made for comfort.

The 4th Air Force Flying Safety Officer assigned to Colonel McNeil's staff was Major Lester G. Carlson, who in July had been recalled to active duty from the Air Force Reserve. He graduated from pilot training in class 43-D during World War II and was released from active duty into reserve status in 1950. Since his return to active duty, he had been pushing hard to build up his flying time, to achieve a senior pilot rating and a green instrument card, in part because Colonel McNeil believed a senior pilot rating and a green instrument card were necessary for any officer assigned the duties of a 4th Air Force Flying Safety Officer. *Minimum Individual Training* was the term describing the training received in routine flying to maintain proficiency and build experience. Major Carlson was an instructor pilot in the C-47. He had almost two hundred hours in the airplane, nearly all in the preceding three months. Two trips across the United States and back would add considerably to his experience.

It all fit nicely. Christmas of 1951 would be grand after a trying year. The C-47 lifted off from Stewart Air Force Base the afternoon of December 22 and, after refueling and an overnight stop en route, landed at Hamilton late the next day.

PLEBE CHRISTMAS

Outwardly, there was always an air of formality at West Point, a reserve; to the uninitiated visitor, a stoicism encasing the individual personalities that we were. The plebes, subdued and inclined to remain distant, formal, and almost flawlessly courteous, after nearly six months in the fourth class system, relaxed during the Christmas holidays more than they ever had since the "plebe hike" in late August, at the end of Beast Barracks.

There were no academics, except for men deficient in academics after final exams who had to take turnout exams – exams to determine if they were to be reinstated in their class, turned back to the next class, or in the extreme, "found," ineligible to return to the Academy. The upperclasses were gone, except for a few who ended the semester deficient in conduct or academics. Though they lived in their own rooms in the company areas for a few days until they were released to go on leave or permanently depart the Academy, they kept to themselves, and formed up in front of Washington Hall for roll call and meals, rather than march from their company areas with the plebes.

But plebes were not to relax too much. The self-discipline, the formality, though less strenuous, would remain through the Christmas holidays.

"Cadets will recite every day in every class." That was the common saying in the academic department, an accepted way of academic life, a statement of academic policy, and very nearly the norm in our lives, right from the first day of our first semester at West Point.

Preparing for Christmas, to govern ourselves, and behave as future officers and gentlemen was no different. We received a three-hour course in "Customs and Courtesies" in the nine-day period before plebe Christmas. Our instructors for those three hours were from the Tactical Department and the first class, the class of '52, not the Academic Department. But the instructional principles and methods were the same.

For education and training in the classroom we read "Official & Social Courtesy, USMA," and a mimeograph on "Courtesy and Table Manners." Lectures. Films. Practice. Role play the circumstance, the hypothetical situation, or the case study. Practice. Critique, learn, do, redo, practice some more. Review and prepare for a writ – a test.

The first hour of training was "Conduct During Christmas Week," and explained how we were to conduct ourselves at hops (dances), while visiting at the Thayer Hotel, and eating at the Cadet Restaurant, which was in Grant Hall, the Cadet visitors' reception center. The subjects covered in the next hour were conduct at officers' quarters, table manners, acknowledgment of courtesies, visitors, and wearing of the uniform. The last hour was a series of skits in the Army Theater, in which selected members of our class role played situations requiring the use of accepted customs and courtesies – and we also took the usual writ. Lieutenant Senior Grade Tingle, a Navy exchange officer whose rank was the equivalent of an Army captain, was our instructor for the last hour.

We were ready. Bring on the holidays! Christmas would have been much more enjoyable if we'd gone home, but we had to make the most of circumstances.

There was a hop, either a formal hop in the evening, or a "Tea Hop" – an informal afternoon dance – every day for ten days beginning Saturday, the day Christmas leave began. Seven of the ten were formal occasions, three in Cullum Hall, overlooking the Hudson River, the rest in the East Gymnasium.

At formal hops we wore full dress gray and our "drags" – the young ladies we escorted – wore evening dresses. There was a receiving line at each one, usually composed of a senior officer and his wife, and one or two other officers and their wives, to act as chaperones after everyone had gone through the receiving line. At each hop there was a classmate from a different company designated Hop Manager, and another who, in the receiving line, acted as an "aide" would for a general officer. The receiving cadet was at the head of the line, and, as each cadet, followed by his date, passed through the line, he introduced them, one after the other, to the senior officer in the line, who in turn, introduced them to his wife.

Many plebes whose homes were distant from West Point, and whose families couldn't visit the Academy during the holidays, signed up with the Cadet Hostess, Mrs. Barth, for "blind dates." From her office upstairs in the north end of Grant Hall, Mrs. Barth canvassed colleges, universities, professional schools, the Young Women's Christian Association, churches, and other organizations to find young women interested in coming to the Academy for weekend and holiday social activities with cadets. They stayed overnight at the Thayer Hotel or in dormitory rooms at Ladycliffe College, just outside the Highland Falls entrance to the Academy. Most came under sponsorship, often with group chaperones. Many a cadet met his future wife through the Cadet Hostess.

As for me, my first date since leaving home the previous summer was through Mrs. Barth, for a formal hop that first Christmas away from home. I was uncertain what to expect, but was delighted to be somewhere in the vicinity of female company. There were always rumors and a fair number of cadet horror stories about blind dates arranged through Mrs. Barth. Not too surprisingly, such stories were usually exaggerated, designed to taunt the female-starved plebes, and any other gullible cadet who wouldn't challenge the authenticity of the tales.

Aside from hops, there were the Ice Follies on Thursday, the second day after Christmas. It was an evening of hilarity at Smith Rink, the hockey rink, where any fourth classman daring enough to put on ice skates could engage in games he signed up for in advance. There was broom hockey, an obstacle race, musical chairs, a relay race, and the southern sweepstakes, a game specifically designed for cadets who had never previously worn a pair of ice skates.

There were "open houses" for two days beginning Christmas, where parents and friends of cadets could see displays related to Tactical Department training we received, academics, and the Department of Military Psychology and Leadership. On Christmas day, between eleven and twelve noon, the open house extended to fourth class rooms, where cadets brought visitors to their rooms. There they could see how meticulously we were required to maintain rooms and what "inspection order" meant. Yep! Sure enough, on Christmas day all plebes were to have their rooms in inspection order for two hours beginning at ten in the morning, the first hour to have an inspection to ensure our rooms were in inspection order.

There was a sumptuous Christmas Eve supper, which families could attend, and Christmas dinner beginning at twelve thirty, to which cadets could also escort their families. Plebe's visits and meals with their families took the sting out of the Christmas day inspection, at least for those whose families could visit the Academy.

In addition to the specially scheduled activities, there were movies at the Army Theater, athletic facilities available in the gymnasium, including the varsity swimming pool, the Weapons Room, where there was a snack bar serving sandwiches, salads, soft drinks, tea, coffee, and other light fare. Then there were all the normal club activities which were open on a prearranged schedule, including skiing nearby.

Ah, yes! For four whole days there was no reveille: two Sundays, Christmas, and New Year's Day. Reveille on all the other days was an hour later than normal, 6:50 in the morning instead of the usual 5:50, which meant meal formations were an hour later.

We had our own chain of command, classmates who were identified in a December 19 Administrative Memorandum as "acting cadet officers." They were given the responsibilities belonging to the first class when academics were in progress.

We definitely were busy, if not always happy. The holidays passed rapidly – too rapidly. The upperclasses would soon return. Those from the west coast would be the earliest to start back to West Point.

THE RETURN FLIGHT

Sunday, December 30, 1951, a few minutes before nine o'clock in the morning, Pacific time. The C-47, tail number 44-76266, sits on the ramp at Hamilton, engines idling. A few parents and friends of the passengers are standing outside base operations, watching the aircraft preparing to taxi for takeoff.

Passengers and their baggage are already loaded, and the aircraft's one door, several feet in front of the tail on the left side of the aircraft, is open, with a removable aluminum step ladder hanging from detents on the cabin floor. The ladder will be lifted inside the airplane and the door closed after the crew chief completes his preflight duties and climbs up the ladder, carrying the chocks into the aircraft.

There has been a change of plans. Colonel McNeil is ill. He had intended to get some more "air time" and fly the cadets, including his son, back to the east coast. Major Carlson, who was the instructor pilot on the trip out to Stewart, is filling in for him, doubling as the aircraft commander as well.

The crew chief, Corporal Bertram J. Audley, stands well forward of the left wing and outside the radius of the big propeller on the port engine, where he and Major Carlson, the aircraft commander sitting in the left seat in the cockpit, can see one another clearly. There is a red, emergency fire bottle on a two-wheeled cart standing waist high next to Audley. He knows Major Carlson is talking on the radio, calling for his flight clearance, and will soon be giving him the traditional "thumbs up" with both hands to signal removal of the chocks before taking Audley on board and taxiing for takeoff. Private Audley is from Rockville Centre, Long Island, New York.

"Hamilton tower, this is Air Force 76266, standing by for ATC clearance. Over."

"Air Force 76266, Hamilton Tower, I have your clearance. Ready to copy? Over."

"Hamilton Tower, 266, ready to copy."

"ATC [Air Traffic Control] clears Air Force 76266 to Goodfellow Field, Texas, Instrument Flight Rules, Amber 1 to Bakersfield, Blue 14 Riverside, Green 5 Blythe, thence Visual Flight Rules Red 8 to Goodfellow. Maintain 8,000 feet, climb and maintain 11,000 passing Bakersfield, and 13,000 passing Riverside. Read back. Over."

The clearance was as filed by Major Carlson and his crew, and it was read back to Hamilton Tower before 266 received clearance to taxi for takeoff. The C-47 was fully loaded, near the maximum gross weight, slightly over twenty-nine thousand pounds, with twenty-four passengers, their baggage, and a crew of four, including Major Carlson, and Corporal Audley. Carlson and his crew planned the flight to Goodfellow estimating seven hours and thirty minutes en route with fuel for eight hours and forty minutes of flight. The good weather forecast for the destination and the long stretch without an overcast sky from Tucson to Goodfellow augured for a routine flight.

There was one minor hitch. Weather was moving south toward Phoenix. There were two low pressure areas aloft, well to the northwest and northeast of Phoenix, and their movements were causing changes in upper wind patterns, affecting wind speeds, directions, turbulence, and cloud cover north of their planned route of flight after passing Blythe. They would cross two mild cold fronts: one in the vicinity of Bakersfield, and another moving northwest to southeast, between Blythe and Phoenix, spawning scattered showers and light rain north of the route of flight through Arizona, and snow in the high mountains near Flagstaff.

With the shifting weather pattern, there was the possibility of sufficient changes in wind speeds and direction to reduce ground speeds along the planned route of flight. If that occurred, the margin of reserve fuel might be too close, thus causing a decision to stop short of Goodfellow and refuel. The heavily loaded aircraft, if operated at higher than planned altitudes and temperatures, could also burn fuel at a faster rate than planned.

Storm detection radar reported no thunderstorms along the route of flight and none in the weather moving slowly southeast toward the route of flight through Arizona. Nor were there thunderstorms forecasted. There were icing conditions in the clouds above the altitudes they planned to fly, but neither was that a concern. The aircraft had deicing systems, and the systems were operational. The oxygen system was fully serviced and operational, and there were thirty-three parachutes on board, a safety feature not found on commercial aircraft.

Though the weather could be a complicating factor, there are two other pilots on board to assist Major Carlson. One is First Lieutenant

Walter Boback from Little Falls, New York. He is in the 78th Air Base Group at Hamilton, and is attached to the 78th Fighter Interceptor Wing for flying. Before being recalled to active duty earlier in the year, he was a sheriff's deputy in Los Angeles. The other co-pilot is First Lieutenant John A. Harrison from Berkeley, California. He is in the Headquarters Squadron, 4th Air Force.

Five other passengers joined the nineteen cadets from the Military Academy for the flight to Goodfellow. Sergeant Jeane Garafalo, from the Women's Air Force 4727th Squadron at Hamilton, boarded for Stewart Air Force Base, planning to go home to Plainfield, New Jersey, to surprise her mother for New Year's.

Private First Class George Thomas, from the Headquarters Squadron of 4th Air Force signed on for Stewart, then it would be home to Rensselaer, New York, for a few days.

Private First Class Robert L. Baesler, Jr., who spent Christmas with his dad and mother, Lieutenant Colonel and Mrs. Robert L. Baesler, at Hamilton, needed to return for classes at the technical training school at Keesler Air Force Base, Mississippi. His father was the Director of Operations in the 78th Fighter Interceptor Wing at Hamilton.

Shore leave was up for Navy Petty Officer 3d Class Richard Mulholland from Pacific Grove, California. He and "Nellie" Byers were classmates in Carmel High School's class of 1948. He was awaiting reassignment after duty in the Mediterranean on the aircraft carrier USS *Coral Sea,* and came to Carmel on Christmas leave. Nelson Byers had enlisted in the Air Force, and Richard Mulholland the Navy after graduating from high school. When they crossed paths again that Christmas in Carmel, they spent considerable time together. Nelson told his friend about the December 30 return flight to the east coast, and arranged Richard's space available booking on 266.

Army 2d Lieutenant George Ahlgren, in the 7th Armored Division at Camp Roberts, California, caught the flight headed in the general direction of Fort Benning, Georgia, for infantry training. In 1948 George Ahlgren was on a winning American rowing crew in the London Olympics.

At ten minutes after nine, Pacific time, the C-47 lifted off at Hamilton and began the climb to its assigned altitude. On reaching Bakersfield, the crew of 266 made the first of two changes in the planned route of flight. Because Air Route Traffic Control couldn't clear the aircraft at altitudes lower than fifteen thousand feet on through Riverside from Bakersfield, Major Carlson received clearance at a lower altitude to proceed via Needles, California, then south to Blythe. The change considerably lengthened the overall route to Goodfellow, and reduced the airplane's ground speed on the leg from Needles to Blythe.

As he approached Blythe, he requested a second change in plan, that his Instrument Flight Plan be extended to Phoenix. He intended to reroute to Phoenix, land, and refuel at Williams Air Force Base approximately twenty-five miles southeast of Phoenix. They were flying in and out of cloud layers en route, and there was a thick layer of clouds below them, an overcast condition which kept the crew from seeing the ground. As he would learn, there was a ceiling of approximately six thousand feet above the earth's surface at Phoenix, with light rain showers. At flight altitude, winds made the ride bumpy but not uncomfortable. Visibility beneath the clouds was ten miles, certainly not conditions to cause concern.

He reported over Blythe at 1:38 Pacific, thirteen minutes behind his original estimate. At 1:58, Air Route Traffic Control called and asked, "Are you landing or overflying Phoenix?"

The reply, "Request land Williams Air Force Base. Estimate Phoenix at 2:20."

Air Traffic Control responded, "ATC clears Air Force 76266 to the Phoenix Range Station, descend to 9,000 until advised by Phoenix Approach Control. No delay expected. Contact Phoenix Approach Control, Baker Channel, over White Tanks for further instructions." White Tanks intersection was a navigational check point over the White Tank Mountains, along the route of flight, west of Phoenix.

A few minutes prior to 2:20, Phoenix Approach Control attempted contact with 266, making several calls on all available frequencies. At 2:20 Air Route Traffic Control instructed 266 to contact Phoenix Approach Control. The crew acknowledged, and finally, at 2:21 established contact with Phoenix Approach on "D" channel, informing the controller they were unable to transmit on "B" channel.

Approach Control asked their position. They responded they were five minutes east of White Tanks. Approach Control asked them to verify their position. They responded they were five minutes *west* of White Tanks, altitude 9,000 feet. The time was 2:22.

Phoenix Approach Control, hearing that 266 was five minutes west of White Tanks, cleared 266 to descend to 6,000 feet and report over White Tanks Intersection, then upon passing the Perryville Fan Marker, they were to descend to 2,600 feet. Two minutes later, at 2:24 they reported passing White Tanks "two minutes ago" and were estimating the Phoenix Range Station at 2:30, ten minutes later than estimated when they passed over Blythe.

Now, something more ominous was becoming a factor in the crew's beliefs about the airplane's position along the route of flight. Terrain.

In a large arc swinging from west through north, to almost due east of Phoenix and Chandler, the location of Williams Air Force Base, the terrain's elevation above sea level increases. Cedar and pine covered mountains become more numerous and reach increasingly to higher altitudes. To the west were the White Tanks. Flagstaff, at an elevation above 7,000 feet, is 139 miles north, northeast of Phoenix, which is 1,573 feet above sea level. A few miles north of Flagstaff is Humphrey Peak, 12,633 feet above sea level, the highest peak in Arizona. To the northeast of Phoenix, at distances of forty to seventy miles are six peaks, reaching above 7,000 feet. They are part of the Superstition Mountain Range. The range extends beyond seventy miles to the northeast where another group of peaks frames the Mogollon Plateau. Their highest points are nearly 8,000 feet above sea level.

Time: 2:28. "Phoenix Approach Control, 266 is descending through 8,000. Be advised our fan marker receiver is inoperative. We'll be unable to receive the Perryville Fan Marker." If they passed above the Perryville marker with an operating receiver a light would flash on the aircraft radio panel, and an aural identifying signal would be heard in their radio headset.

"Roger, 266, understand your fan marker receiver is inoperative. Maintain 6,000 feet to the range station."

"Phoenix Approach Control, 266, roger, maintain 6,000 to Phoenix."

Time: 2:34 Pacific. "Phoenix Approach Control, 266, passing through 7,000."

"Air Force 266, copy. Passing 7,000."

Silence. The airplane is now fourteen minutes past its Blythe estimate for the Phoenix Range Station, and four minutes past the revised estimate of 2:30. Winds along the route of flight are boosting ground speed to approximately 225 miles per hour.

Puzzled that the pilot still had not reported passing the Phoenix Range Station, Phoenix Approach Control began attempting contact shortly after the 2:34 call from 266.

"Air Force 266, this is Phoenix Approach Control, over." No answer.

"266, Phoenix Approach Control, do you read me, over?" Still no reply.

"Air Force 76266, Phoenix Approach Control broadcasting on all channels and 'Guard' [the emergency radio channel]. Do you hear me?"

Broadcasts continue, "Air Force 76266, this is Phoenix Approach Control, over." A pause to listen. "Air Force 76266, this is Phoenix Approach Control. Do you read me? Over." No reply. Again, "Air Force 76266, this is Phoenix Approach Control broadcasting on all frequencies. Do you hear me? Over." And so it continued at two minute intervals for a while, then at five minute intervals until 266's estimated time of fuel exhaustion, 5:50 in the afternoon, Pacific time, more than three hours after the last contact with 266.

At 5:50 Pacific, on Sunday, December 30, 1951, at Williams Air Force Base near Chandler, Arizona, the airfield tower operator began completing actions sequenced in an emergency checklist for overdue aircraft, to both notify Air Force commands and prepare air and ground searches for the C-47. First was a communications check to determine if the airplane might have landed at another airfield. The response was negative, confirming search and rescue operations would be necessary. Among the organizations notified were the major air command to which the aircraft was assigned, Continental Air Command; Headquarters, United States Air Force in Washington; the Military Air Transport Service at Scott Air Force Base in Illinois, responsible for search and rescue operations; and the Air Force Flight Safety Research Center at Norton Air Force Base, near Riverside, California. In the Pentagon in Washington, D.C., the Air Force Operations Center notified operations staff members in the Army and Navy, and the Army staff, in turn, called the Military Academy to inform senior officers of the missing aircraft.

Throughout the night Air Force telephones and teletypes in command centers were busy as people were alerted and planes prepared for airborne searches as soon as daylight and weather permitted. Messages were dispatched to all services and commands which had people on board the airplane, confirming the missing C-47. Simultaneously, men and equipment were alerted and marshaled, ready to begin a ground search, in response to an aerial sighting of possible wreckage, or reports that might come from people who believed they heard or saw the missing airplane. As plans for the search were made ready, weather forecasts indicated an airborne search probably couldn't begin until the afternoon of December 31.

It had been a grim holiday week for air travelers. A search had been in progress in Oregon and northern California for another Air Force C-47 which disappeared the day after Christmas, with eight men on board, on a flight from McChord Air Force Base near Seattle, Washington, to Travis Air Force Base, northeast of San Francisco. In New York state, Saturday night, a charter Continental Airlines C-46 with forty people on board disappeared en route to Buffalo from Pittsburgh. A single seat P-51 fighter aircraft, en route from Castle Air Force Base near Merced, California, to El Paso, Texas, disappeared after requesting a letdown from thirty-five thousand feet nearing Tucson, and was believed down the same afternoon 266 was reported overdue.

At Hamilton, where the flight of Air Force 266 originated, copies of the crew and passenger lists permitted notification of military units to which they were assigned, who, in turn, notified next of kin that the airplane was overdue. In some instances news releases to press agencies, and ensuing queries resulted in families and loved ones learning of the missing aircraft by radio or television before they received notification from the military that there were family members on the aircraft. After the next of kin were notified the airplane was overdue, the names of all aboard were released to the press.

The anguished waiting and hoping began for family members late the evening 266 vanished, while preparations for the air and ground searches proceeded through the night. Colonel Guy McNeil, an experienced senior Air Force officer and pilot who had flown his son and eighteen other cadets home for Christmas, told reporters the next morning, "I was going to fly the airplane myself. I needed the air time. But

I decided not to... I'm an old hand at this business... I don't know whether to keep on hoping or not. My head says no and my heart says yes." Colonel Guy McNeil had been ill. He decided not to fly. His troubles were just beginning.

A MASSIVE SEARCH

The greatest air search in the history of Arizona began early Monday, the last day of 1951. Searchers disregarded hazardous flying weather to make more than seventy flights into cloud shrouded mountain areas in an effort to find both the C-47 and the P-51. The two airplanes were presumed down, but not in the same general area. The P-51 pilot had asked for landing at Tucson, and was believed down to the south of Phoenix, toward Tucson.

Hopes were high that survivors would be found. An urgency fueled the searches. Temperatures were dropping as the huge band of clouds and rain, now mixed with snow, moved deep into the state.

The search for the Hamilton C-47 was first concentrated northwest of Phoenix. Toward late afternoon, as forecasted, weather began improving. As more aircraft joined the effort, the search area expanded eastward, covering twenty-four thousand square miles. During the day several persons reported hearing "a large transport type plane flying low in bad weather around Roosevelt Dam," northeast of Phoenix, about the time the aircraft became lost.

There was a flurry of excitement late in the day when a Williams-based aircraft reported seeing what was believed to be the wreckage of an airplane. The glimpse was fleeting, seen through a break in the clouds. The observer said he saw what appeared to be a plane's fuselage with wings folded back along the sides. The location was north, northeast of Phoenix, about twenty miles north of the town of Superior, in the Superstition Mountains. Clouds prevented reestablishing the airborne sighting before darkness closed in. Nevertheless, the search and rescue commander directed a ground party into the area. They moved quickly, but arrived in the vicinity after dark.

Tuesday, New Year's Day, 1952, the first day of West Point's Sesquicentennial year. Monday morning nationwide press reports of the missing C-47 had brought worrisome news to everyone at West Point, and to upperclassmen converging on the Academy, returning from

Christmas leave. General Irving, the superintendent had received copies of teletype messages sent to all military units of the people listed as passengers and crew members on board the airplane. The messages summarized the scant, known details of the flight, and listed names, ranks, and organizations of everyone on the flight. By 2:00 in the afternoon, yearlings, cows, and first classmen were filtering into their respective company orderly rooms, signing in before the 5:30 deadline marking the end of leave.

In Phoenix, two o'clock Eastern time was noon. By then the airborne search included Air Force, Navy, Coast Guard, and Civil Air Patrol aircraft from eight military and civilian airfields. Among the more than sixty airplanes participating in the search was a Hamilton-based twin-engine B-25, the same type of airplane made famous in General Jimmy Doolittle's World War II raid on Tokyo in April 1942.

The B-25, piloted by Lieutenant John F. Rich and Captain Robert J. Clounch was flying an assigned search area near Lookout Mountain and Theodore Roosevelt Lake, approximately sixty miles northeast of Phoenix, when the crew spotted what appeared to be the upturned, partially burned tail section of a C-47. The site was on the steep, southwest, rocky slope of jagged Armer Peak, below a sheer cliff that reached to a summit approximately 7,000 feet above sea level. The B-25 circled the area, radioing the location of the sighting to the search command center, and maneuvered to get a closer look at the wreckage.

Over the years there had been several crashes in the Superstition Mountains. Snow had fallen since 266 disappeared and covered most of the wreckage. There was no sign of movement. This was possibly an old wreck.

Then came the first observation toward ending the aerial search. On a low altitude fly-by of the wreckage, men on board the B-25 could read the last three digits of the aircraft tail number – 266. For Lieutenant Rich, the discovery was not what he hoped. He was the officer in the operations section at Hamilton who signed Colonel McNeil's requests to schedule the C-47s for the flights to Stewart. The 266 tail number was confirmed by the crew of another search airplane.

More disturbing, the crash scene, from the air, appeared still and foreboding. The tail section was the only large piece of wreckage visible, and there was no sign of life. A burn darkened spot on the sheer cliff above the wreckage, approximately 150 feet below the summit, gave stark evidence the airplane may have flown straight into the side of the shale rock cliff and exploded. Lieutenant Rich, on board the B-25, concluded "everyone on board went damn quick," a conclusion confirmed the next day when a ground search party reached the crash site.

Arnold Johnson, a cowboy, and ranch foreman, was the first person on the scene. His wife had spotted the wreckage through binoculars, and Johnson inched his way near the summit, and downward along the base of the cliff, to reach the site. He told the Air Force search party what he saw before he led them, struggling, three miles up a steep slope through snow and ice-covered rocks and pines to reach the wreckage. There were no survivors. Breath and life ended at 2:38 Pacific on Sunday, four minutes after the last radio transmission from 266. Watches found in the wreckage marked the time.

At West Point, as the upperclasses returned from leave on January 1, radio and television broadcasts told of the airborne sighting of the wreckage and no sign of life.

The news was shattering. The crash which took the lives of twenty-eight people, including nineteen cadets, was, to that time, the largest single air disaster in Arizona history. It was also the largest single loss of life in the 150 years of the Cadet Corps, and a stunning, tragic punctuation mark at the end of the darkest year in West Point history.

The first few days of the Sesquicentennial Year, which had been planned for years to be filled with celebration, were overwhelmed with sad, sharply contrasting thoughts, and necessary, grim tasks: nineteen young men gone forever; lives full of promise ended; classmates, roommates, and friends lost; memorial services; military funeral cadences; pallbearers; rifle volleys; the mournful sound of Taps; letters and messages of condolence; and the hundreds of details necessary to give comfort to their families and friends, and ensure the men were given honors they deserved as they were laid to rest.

The suffering at West Point was filled with irony, yet was a small fraction of the suffering borne by all involved in the flight and destruction of 266 – and the loved ones left behind. In all such accidents there are heart wrenching, seldom told stories, and unbelievable tragedies.

TRAGIC IRONIES OF AIR FORCE 266

Two investigation boards convened in the hours following the crash of Air Force 266, one the accident investigation, to find the cause, and recommend what must be done to avoid a repeat. The other was a "special investigation" to determine if there were violations of law or regulations in the scheduling or conduct of the flight. Nearly a year later another special investigation began, based on information uncovered in the months following the crash. The findings in each investigation were, for different reasons, held in closed records; nevertheless, the investigations' effects reached far, adding quietly, selectively to the suffering which inevitably follows a fatal crash.

The safety board found that the pilot, Major Carlson, for reasons never precisely known, made a navigation error. He misjudged the position of his aircraft, probably because of an on-board failure in a key navigation aid. He obviously never confirmed his position by relying on other navigation methods and aids, for which he had been trained and which were at his disposal. There were no failures in navigation aids broadcasting from the ground, and pilots flying aircraft in the area at the time of the accident verified the aids were operating. Condition of the wreckage prohibited determining positively whether or not there was a failure in on-board navigation receivers, or that the pilot erred by tuning the wrong ground station. There was no evidence of engine failure or any other mechanical failure that would have caused the crash.

Evidence was clear. Major Carlson and his crew had flown eighteen minutes past their estimated time of arrival at the Phoenix radio range station and were still flying on an easterly heading, descending slowly in clouds and rain, when the end came. Had the airplane been 200 feet higher, or a few hundred feet right of its course, it would have missed Armer Peak entirely. Instead its left wing plowed through the tops of pine trees for thirty feet before the abrupt deceleration of the wing caused the airplane to first yaw violently to the left as the wing structure failed, then cartwheel up and to the left, slamming the fuselage nearly head-on into the cliff.

The ensuing explosion shattered the aircraft into small pieces, triggered rock slides, and sent pieces of wreckage careening over the summit of Armer peak. The remainder of the wreckage was scattered over a half-mile area, having fallen among the rocks and trees below the cliff.

The special investigation board found that the scheduling and authorization for flight of 266 violated the intent of an Air Force regulation defining and describing the conditions under which an Air Force airplane could be scheduled for flight. The language of the official report was damning:

> It was determined that the flight was scheduled at the request of the Inspector General, Fourth Air Force. The Inspector General, Fourth Air Force, who was scheduled as pilot, did not go on the flight due to illness. An examination of the flight scheduling log at the Operations Section revealed that a prior flight of a C-47 aircraft assigned to the Fourth Air Force had been made from Hamilton AFB, California to Stewart AFB, NY, on 20 December 1951. The pilots for this prior mission were the Inspector General, Fourth Air Force, and the same instructor pilot on the fatal flight. While it has been stated by all personnel concerned that passengers were loaded on a non-priority space available basis, it must be recognized that on each of these flights the son of the Inspector General, Fourth Air Force, and the sons of other military officers living in the San Francisco area were passengers. It is apparent that although these missions were scheduled as training missions, there exists an apparent intent to transport Cadets of the U.S. Military Academy, West Point, to Hamilton AFB during Christmas holidays and then return them to Stewart AFB. In a signed statement the Inspector General, Fourth Air Force indicated that the carrying of passengers, and specifically the West Point cadets, was incidental to the training, and that the training was intended to qualify the instructor pilot for a green instrument card and a Senior Pilot rating in order for him to more adequately perform his duty as Flying Safety Officer. The Inspector General, Fourth Air Force further stated,

> 'My underlying thought was to transport these West Point cadets for the Christmas holidays on the West Coast. My thought was that as long as the flights were necessary for the above

stated purpose, [Minimum Individual Training], the destination of the flight was immaterial.'

It is therefore apparent that the scheduling of these flights was in violation of the intent of Air Force Regulation 60-1, which provides that under no circumstances will flights be permitted whose sole purpose is for the convenience or prestige of an individual or group of individuals.

Colonel Guy McNeil meant well, but he now faced the torments of good intentions gone horribly wrong, which must have dogged him the rest of his days. He had lost his own son while believing he was serving the interests, welfare, and morale of the other cadets and their parents, and the other service personnel who had availed themselves of either of the two C-47 missions. There were twenty-seven other lives lost, two whose parents resided at Hamilton. Guy McNeil, Sr.'s illness, which caused him not to fly on the return mission, created an unforgiving, nightmarish question, and probably an answer that would never bring him comfort. "If I had been on that flight, would I have made the same mistakes? I'm certain my experience would have prevented such errors – and the accident."

Then there was the responsibility and trust given Colonel Guy McNeil, the Inspector General. The officer filling that position was to inspect for violations of regulations, and accepted principles of good management and leadership. If he didn't already know, Guy McNeil would soon learn he had ignored his responsibilities as an officer and leader, and compromised the trust and confidence he earned through his years of service. He was relieved of his duties February 25, 1952. He paid dearly for his well intentioned actions.

There was another officer on the Fourth Air Force staff who suffered the price of failure to adhere to Air Force policy. The Deputy for Operations, responsible for approving or denying requests to schedule and fly airplanes assigned to Fourth Air Force headquarters was also relieved of his duties, to which he had been assigned November 4, less than two months before the crash. His responsibilities included review of such requests for their validity, consistent with the policies and regulations governing flying operations, and to ensure crews scheduled were properly trained and qualified to fly planned missions.

Nearly a year later another special investigation revealed that Major Lester G. Carlson, the instructor pilot on 266 had been suspended from flying status in 1948, two years before he went from active duty into the Air Force Reserves. He had had at least one episode of losing consciousness – passing out – while on active duty, but not while flying. Flight surgeons were unable to pinpoint the cause, but believed it to be a form of allergic reaction. They treated him and appealed the Surgeon General's decision to ground him.

The Air Force Surgeon General denied the appeal, but due to a breakdown in procedures as Carlson was leaving active duty, the permanent suspension wasn't entered in his flight or personnel records.

When he was recalled to active duty his flight records showed him still qualified for flying. His medical records had been retired and were never retrieved from the Air Force's archives. Major Carlson obviously knew of his questionable flying status, because he was involved in the appeal in 1948, before he left active duty in the spring of 1950, before the Korean War began. He completed a medical history questionnaire when he was recalled to active duty in July of 1951 and received a flight physical examination. He made no mention of his questionable flying status, or the error in his records, and no questions were raised by the absence of recorded flying time in his flight records from May 1948 until his separation from active duty in May of 1950.

Subsequent reinvestigation of his flying status, and possible questions about Major Carlson's condition on December 30, 1951, that he might have lost consciousness or become incapacitated during the fatal flight, didn't affect the final outcome of the accident investigation. There was no evidence he ever had a recurring loss of consciousness after his return to active duty. There had been no radio calls from 266 to indicate an emergency of any kind prior to the crash. Statements from his fellow officers at Hamilton revealed Major Carlson had no indicators of a recurring ailment. As a pilot he was considered more than competent, thorough, and detailed in his planning, and his conduct in flight was the same.

Nevertheless, a question can rightfully be pondered. If Major Carlson were not flying 266 that day because he had been disqualified from flying, would the man flying in his place have made the same fatal mistakes?

Aircraft accidents occur at the end of an unbroken chain of events. So said the theory of accident investigation. If the chain is broken, the accident will not occur. It will be avoided. Sadly, the chain of events wasn't broken on December 30, 1951, high in the mountains of Arizona. But many lives were.

There were other, equally poignant, tragedies associated with the loss of Air Force 266.

The small town of Carmel, California, suffered grievously. "Nellie" Byers and Richard Mulholland, the two high school classmates, were gone. Nelson's mother, whom he had been visiting over the holidays at her home in nearby Jacks Peak, had lost her husband, Nelson's father, the previous April. Now she had lost a son. Richard Mulholland's mother and father had lost their only son. And strangely, the two men, who had spent considerable time together during their leave, had given up part of their holiday joy to be pallbearers at the funeral of a 1950 graduate of Carmel High School, Keith MacKenzie, who lost his life in a fire on Christmas Eve. Keith and Nelson Byers had been teammates on the Carmel football team when "Nellie" was the quarterback his senior year. Keith was on leave from the Navy, in Maryland, where he was preparing to enter the United States Naval Academy the coming summer, with the class of 1956.

Noel Perrin's mother lived in Palo Alto. Noel planned to follow in his late father's footsteps, and enter the Air Force after graduating from West Point. Noel's father was a member of the Academy's class of 1930, and had become the youngest brigadier general in the Air Force, having served in every major theater of operations in World War II, before he died of a heart attack in Dayton, Ohio in 1946. In October before the crash of 266, the *Assembly*, the Academy's alumni association magazine, published a memorial article about Noel's father. Now Noel's mother had lost her only son, in addition to her husband. With full military honors, Noel was laid to rest beside his father in the Golden Gate National Cemetery.

William F. Sharp, class of 1952, cadet Company L-2, came home to visit his father and mother, retired Army Colonel and Mrs. Robert Sharp, in Los Altos, near Palo Alto. The night before he left to return to the Academy on 266, he and Alice Hanoum, who lived in San Francisco, announced their engagement. His parents, accompanied by Alice, would now accompany Bill Sharp's body to Salt Lake City for funeral services in the Yale Ward of the Church of Jesus Christ of the Latter Day Saints.

At West Point, the true enormity of 266's tragedies were never known. Nevertheless, sadness gripped the gray, wintry onset of "gloom period" like none other in recent memory. Aside from roommates, class-mates, and friends of the cadets whose lives ended so abruptly during the season of joy, officers such as the superintendent, General Irving, responsible for seeing to the letters of condolence to the families, and assisting the families in arranging for services and burial of their sons, felt the stinging losses as well.

Lieutenant Colonel Jeff Irvin, who, the summer before served on the Collins Board, and had seen the exodus of eighty-three cadets for violating the honor code, was assigned the painful duty of going to the west coast to assist bereaved families and personnel in the Sixth Army at San Francisco's Presidio, in making arrangements for funeral ser-vices and burials. He was the Academy's representative, and officer in charge of twenty-one cadets who attended funerals or escorted bodies back to West Point for burial at the Academy cemetery.

The day we learned the C-47 was overdue, shock numbed the four of us in the class of '55 who lived across the hall from Noel Perrin and Nelson Byers. Shock turned to disbelief when word came New Year's Day there appeared to be no survivors. Conversation was subdued, filled with questions. We speculated about what happened. What could have gone so wrong?

There was always a noticeable quiet when members of the Corps returned from Christmas leave and marched in the winter dark to their first New Year's meal in Washington Hall. Supper formation the evening of January 1, 1952, was nearly silent, and talk was in hushed tones at meal tables in the huge Washington Hall.

Perhaps no man was stung more by the crash of Air Force 266 than General Irving, the superintendent. The past year had been grim, harsh, filled with adversity brought by loss, tragedy, controversy, and sting-ing criticism of the Academy for its handling of the honor scandal. Now this.

The much anticipated year of the Sesquicentennial Celebration had been slow in coming. Hopes for an end to the fighting in Korea were high when armistice negotiations began July 10, while deliberations were in progress concerning the young men involved in the scandal. Now the war was grinding on, relentlessly, the front essentially stabilized. A ghastly war of position and attrition was still in progress on the Korean Peninsula as the two opposing armies, surging back and forth only a few miles, fought bitterly over every inch of terrain, trying to gain advantage for the armistice negotiations. Casualties mounted, adding to the lengthening toll of Academy graduates who were falling in that far off land.

Letters and messages of condolence poured in to the Academy the first two weeks of January. All were read or passed to the men of the Corps, from: General of the Army Dwight D. Eisenhower, Commander in Chief of the Allied Powers in Europe; Secretary of the Army Frank Pace and Chief of Staff, General J. Lawton Collins; Secretary of the Air Force Thomas K. Finletter and Chief of Staff, General Hoyt S. Vandenberg; Lieutenant General Willis D. Crittenberger, Commander, U.S. 1st Army; Vice Admiral Hill, superintendent of the United States Naval Academy; United States Merchant Marine Academy; Royal Military Academy at Sandhurst, England; Royal Military College of Canada; Military Aviation School of Venezuela; Donald W. Griffin, secretary, National Alumni Association of Princeton University; President Hunter Guthrie, for the faculty and students of Georgetown University; the Nicaraguan and Guatemalan Military Academies; and the New York Chapter of the Daughters of the American Revolution.

There were many more, but none more memorable to Bill Ochs, General Irving's aide, than the letter from former president of the United States, Herbert Hoover. It had been a dark year, indeed. And now, at the beginning of the New Year, the Sesquicentennial Year, the superintendent and his staff were immersed in preparations for mourning, working tirelessly to ensure the nineteen cadets' parents and loved ones were given solace and support in their time of greatest need. A celebration didn't seem appropriate, and Mr. Hoover's letter, perhaps, was intended as good wishes for the Sesquicentennial planned to begin with great fanfare and ceremony Saturday, January 5. To General Irving, Mr. Hoover's letter was far more than a Sesquicentennial greeting.

The Waldorf Astoria Towers
New York, New York
January 3, 1952

My Dear General:

The West Point Academy has had so many misfortunes during the past year that it seems to me the duty of its friends is to express to you their unflagging admiration and respect for the institution on this 150th Anniversary.

There is no need for me to recall its magnificent contribution to the safety and progress of our country over almost the whole of our national life. Nor does the Academy really need an expression of confidence for the future.

Yours faithfully,

General Irving was deeply touched by the former president's letter.

4 January 1952

Dear Mr. Hoover:

Seldom at any time in my career have I sensed such a feeling of frank sincerity as expressed in your most thoughtful letter of yesterday.

Your few poignant remarks encompassing West Point's past 150 years of Duty, Honor, Country – with its accompanying achievements, blessings and tragedies – and your affirmation of faith in its future are gratefully appreciated.

Respectfully,

F. A. Irving
Major General, USA
Superintendent

AN END TO THE BEGINNING

At 12:55 in the afternoon on Wednesday, January 9, 1952, the entire Corps of twenty-four cadet companies assembled in class uniforms, gray caps and gloves, and long winter overcoats, with capes buttoned down in the back. Except for the sounds of commands given by cadet officers, and the occasional rustling of the few remaining fall leaves, a reverent silence had descended over West Point. We marched in columns of platoons, three men abreast, to assembly points, and formed two unbroken lines, standing at parade rest, one pace apart, on both sides of the roads. The lines, facing toward the center of the roads, stretched from the West Shore Railroad Station, on the banks of the Hudson River, up the hill on Cullum Road to the intersection with Jefferson Road on the level of the Plain, and from there along Jefferson Road in front of the library, across Thayer Road, past central area barracks, Washington Hall, and north area barracks, to the intersection of Scott Place and Jefferson Road.

The Cadet Brigade Staff and Color Guard were assembled at the Railroad Station, facing the three hearses which would carry the caskets to the Old West Point Chapel located at the cemetery.

At 1:13 a train pulled into the station bringing the bodies of Cadet Karl F. Glasbrenner, Jr., class of 1952, Company C-2; Cadet Maurice J. Mastelotto, class of 1954, Company G-2; and Cadet Hugh R. Wilson, Jr., class of 1954, Company I-2. While the band, assembled to the right of the Brigade Staff, played solemn hymns of faith and, finally, the alma mater, from the time the train came into view until the hearses bearing the caskets disappeared around the turn onto Jefferson Road at the top of the hill, the Corps of Cadets silently welcomed back to West Point the three men who were to be buried at the Academy.

To the Corps of Cadets they represented the nineteen who perished on Air Force 266. As the train came to a stop, the Cadet Brigade Commander commanded "attention" and "present arms," and in sequence, up the road from the station to the top of the hill, cadet company commanders ordered their companies to attention. When casket bearers loaded the caskets on the three hearses, the brightly polished, olive drab official car of the West Point Provost Marshall led the hearses in a slow procession toward the Old West Point Chapel which stood on the grounds of the cemetery.

As the procession moved along its route, cadet company commanders called their companies to "attention" when the lead car came within a hundred yards of their companies' flanks. Each cadet rendered a hand salute when the first hearse came within six paces of his position, and held his salute until six paces after the last hearse passed. When the procession reached the intersection of Jefferson and Washington Roads, at the far northern corner of the Plain near the superintendent's quarters, company commanders began reassembling their companies in columns of platoons, three files in each platoon, to march silently back to the cadet area.

We returned to the normal afternoon class schedule for the second hour of academics, and life continued at West Point. Funerals were held the next day for Karl Glasbrenner and Hugh Wilson and the following day for Maurice Mastelotto, and the three men were laid to rest in the West Point cemetery.

Perhaps some of the saddest ironies in the many tragedies of Air Force 266 was the havoc wreaked on the Corps of Cadets and the classes of '52 and '54 – in a year filled with adversity. Among the eight men killed in the class of 1952 was the C-2 cadet company commander, Karl Glasbrenner, and two honor committee representatives, William F. Sharp, Company L-2, and William E. Melancon, Jr., Company F-1. Six men in Company L-2 resigned the previous summer following honor violations. L-2 had been one of the four companies hardest hit by the scandal, along with B-1, M-2, and H-2. Now, Company L-2 had lost three more men in the accident, including its honor representative, Bill Sharp. William N. Pedrick and Harry K. Roberts, Jr. in the class of 1952, both from L-2, had been killed in the crash of 266.

Company I-2, which had been one of only three cadet companies unscathed by the cheating scandal, lost four men aboard the ill-fated aircraft: Guy L. McNeil, Jr., Hugh Wilson, and Herman Archer, all from the class of 1954, and Hilmar G. Manning, class of 1952.

Company L-1, which had also been untouched by resignations resulting from the scandal, lost Kenneth J. MacArthur, class of 1954.

And in Company, K-1, where four men had resigned the previous summer for honor violations, we would never again see Nelson Byers or Noel Perrin, both from the class of 1954. We were unaware that Nelson Byers had been among forty-eight members in his class, who in July, the month our plebe class entered the Academy, sparked a small, cadet version of a good-humored *Mutiny on the Whiskey* – the Navy battleship USS *Wisconsin* somewhere in the Caribbean Sea. He had been part of a beginning call to arms against the Navy football team, and a coming together for the Corps of Cadets and Army's young, inexperienced team.

Nelson Sawyer Byers   *Two of 19 cadets killed*   *Noel Sanders Perrin*
Class of 1954   *in an aircraft accident near*   *Class of 1954*
Carmel, California   *Phoenix, Arizona,*   *Palo Alto, California*
December 30, 1951
(Photos courtesy USMA Archives.)

For the Corps of Cadets the losses for the year were staggering. In addition to normal annual attrition, which was always a matter of concern, 102 cadets were gone, never to return: eighty-three resignations in the honor scandal, and nineteen killed in the crash of 266.

But life and education at West Point, though deliberately intense and hard, had their merciful qualities. Like war, life at the Academy for many of us seemed a seven day a week push just to stay up with its demands. Sadness, unless one chose to hold on to such anguish, was quickly brushed aside by worry about falling irrecoverably behind in meeting those demands.

The lack of detailed knowledge about the depths of tragedies is often an insulator against prolonged sorrow. At the time, there was much we didn't know about the destruction of Air Force 266.

In the end, youthful exuberance and optimism were our most powerful defenses against excesses of mourning. We can never forget those who were our friends, brothers, classmates, roommates, and teammates, and gave all of themselves. We won't. And neither would they want us to remember them in perpetual, consuming grief.

Life goes on. There will be better days ahead.

PART TWO

TOWARD THE FUTURE

CHAPTER 8

REBUILDING BEGINS

Splendid accomplishments sometimes begin with seemingly random, unrelated events, and involve people who haven't the slightest notion they are doing anything but having fun. Not until later do they learn they struck sparks and ignited fires that would burn for as long as they lived. They didn't realize that within themselves were great, untapped strengths that drove them to rise above circumstance.

Ten miles from West Point, the class of 1954, the new yearlings who just returned from four weeks of summer leave following their plebe year, were at Camp Buckner for their second summer of military training. Already at Buckner when '54 returned from leave were most of the men found guilty by the Collins Board. They were housed in a separate building.

On July 16, 1951, just over a month after the Collins Board, while the men found guilty were still awaiting word of their fate, the first steps were taken in the Corps of Cadets to rebuild what had been torn down. Rebuilding began with the men in '54, cadet style, while a foundation for rebuilding was being laid at the Academy, and in General Irving's discussions with the Army's senior civilian and military leaders in Washington, D.C.

The foundation had rough edges. Rebuilding wasn't perfect, easy, or painless. Heated disagreements on the decisions to be made had flared behind closed doors for weeks, and would continue. Later, on August 3, after the honor violations became public knowledge, controversy became more intense. Many decisions remained for General Irving and the Army's senior officers, while members of Congress and the Armed Forces' Commander-in-Chief, President Truman, who was being urged to form a Presidential Commission, expressed intents to intervene and conduct their own investigations.

Rebuilding within the Corps of Cadets began at Buckner, quietly, and with good humor, three weeks before the public furor aroused by the announcement of August 3. When it began, the men in '54 knew little of the going's on at West Point.

MUTINY ON THE *WHISKEY*

The opening paragraph of movement order Number 10:

In accordance with arrangements made with the superintendent, U.S. Naval Academy, fifty (50) cadets of the class of 1954... will participate in the U.S. Naval Academy Cruise during the period 16-26 July 1951. Cadets will depart by air from Stewart Air Force Base, New York, on 16 July 1951, join the cruise at Guantanamo Bay, Cuba, and disembark at Norfolk, Virginia, back to Stewart Air Force Base.

For reasons of security, the movement order didn't name the ship on which the cadets embarked, or any other ships in the naval task force. She was the USS *Wisconsin*, the *Whiskey* to her crew and passengers, a World War II battleship which had just come out of "mothballs" and was on a shakedown cruise preparing for fleet duty in waters off the Korean Peninsula. The fifty cadets included two alternates, in the event anyone dropped out due to illness or injury.

When the order was published on July 10, it stirred great interest, rumor, and good-natured speculation among the third classmen. One question frequently asked was, "How were the cadets selected for the cruise?"

No one knew the answer with certainty, but trips away from the Military Academy were always welcome, especially after plebe year. One rumor was each man had prior military service. The cruise was to "relax" them while their classmates went through qualification training on the M-1 rifle, which they had already received. Another tale, passed partly in jest, grew from ever-present thoughts of the Army-Navy football rivalry, and the still smoldering frustration over the 14-2 beating administered by Navy the previous November. They were going on the cruise to provide replacements for the Navy football team. The "Boys from Crabtown" were returning to "Canoe U" to begin football practice.

Forty-eight yearlings assembled near Camp Buckner's staff building at nine o'clock in the morning, July 16. The uniform was "Drill C" – khakis – with a B-4 bag and a small, gray weekend bag. After the ride to Stewart in two buses, and an uneventful flight to "Gitmo" – Guantanamo – they received a surprise when they came aboard the *Whiskey*. Among the Naval Academy midshipmen on the cruise were yearlings, men from the USNA class of 1954, and Naval Academy first classmen, class of '52, learning through practicing the seaborne leadership responsibilities and tasks of ensigns, the equivalent of Army second lieutenants. There were no Military Academy first classmen accompanying the men from West Point's class of '54. West Point year-lings with prior military service, some as officers, feeling their new found freedom after being liberated from plebe year, and still smarting over the loss to Navy "on the fields of friendly strife," were in no mood for the Navy's version of seagoing "kitchen police," or KP duty.

In this instance the Navy KP was paint scraping, the removal of peeling, scarred, or chipped paint, preparatory to repainting sections of the ship's surfaces to protect against the sea's corrosive effects. The Naval Academy's yearlings obediently went to work scraping paint, with some of their first classmen in supervisory roles.

Though outranked by Navy's first classmen, the yearling cadets refused to be intimidated by a lack of seniority. There were certain types of orders given by the midshipmen that simply would not be obeyed, no matter that the middies' first classmen were understudying ensigns. Defend. Resist. Sidestep. Avoid. Refuse if necessary. Don't give up, and don't give in – especially to Navy. "On Brave old Army team..." What transpired was nothing akin to the *Caine Mutiny* or *Mutiny on the Bounty*, but the men of '54 dutifully recorded for posterity *Mu-tiny on the Whiskey*.

The cruise was an education. Cadets had the chance to study naval operations on all parts of the *Whiskey*, from engine room to gun turrets. We observed a nine-gun primary battery shoot of the big 16-inchers against a target barge twelve miles away. The target was quickly blown into history. New state-of-the-art jet aircraft flew from the aircraft carrier USS *Leyte*. The chow and rough seas took a toll; some hung over the railings and threw caution (and a few other morsels) to the winds. There were no women on the ship and the rumored "Hops" never came (except as beer ashore; see below). No one expressed a desire to transfer to the Navy.

Unlike Buckner, where the emphasis was on learning useful military skills, the first class midshipmen seemed to see the cruise as a seagoing Beast Barracks. The sleeping quarters were

Cadets in the class of 1954, sailors, and midshipmen relaxing on the deck of the USS Wisconsin *after some Navy "chow" in July 1951. (Photograph courtesy Robert J. Downey.)*

so hot the enterprising cadets slept outside under turrets and up in the crows' nests. Among the nautical activities offered was wielding wire brushes to scrape away layers of encrusted paint. The middies scraped away. But there was unanimous and vocal disinclination on the part of the cadets, even under the supposed threat from the skipper (conveyed by the first class middies) of confinement in the brig. An incident was averted when the cadets were set ashore at "Gitmo" to enjoy a brief period of liberty in the surrounding region, including an introduction to the Hatuey Indian brand of Cuban beer.

"Hell no, we won't chip paint!" Downey, Qualls, Brewster, class of 1954, on the deck of the USS Wisconsin, *in front of a liberty boat. (Photograph courtesy Robert J. Downey.)*

The Navy seemed anxious to rid themselves of these mavericks who wouldn't scrape paint. They flew the group from "Gitmo" to Norfolk and onward in severe weather to Stewart Field. We rejoined our classmates at Camp Buckner on July 26.

...Apparently by mutual consent of both services and academies, cadets in such numbers never joined another middie cruise....

West Point's class of 1954 had sent a detachment of forty-eight to meet the Navy on the high seas in the summer of 1951. None knew that in the future all but a few of the forty-eight would set sail for another port on another day: Philadelphia, November 28, 1953.

THE REVEILLE GUN CAPER

While the *Whiskey* mutiny was underway, somewhere in the Caribbean another rebellious band of cadets in '54, operating independently, decided to conduct their own brand of mutiny at Camp Buckner.

Throughout plebe year at West Point, the class of '54, like classes before it, had endured the quaint tradition of the seventy-five millimeter reveille gun, emplaced on Trophy Point, east of the Plain. The reveille gun rudely blasted everyone awake precisely at five fifty in the morning, six days a week. The roar of the gun's blank shells, which with great fury blew a large puff of gray-white smoke into the air high above the Hudson River, was convincing proof that artillery guns and cannons were designed for the sole purpose of making cadet life miserable. The loud, vibrating "Boom!" which shattered the badly needed stillness for sleep-starved cadets, unfailingly came an instant before sounds of the "Hellcats'" bugles, drums, and fifes, strategically placed inside the echo chamber that was the central area of cadet barracks. The reveille gun and the "Hellcats" routed the entire Corps out of bed for reveille. If cadets weren't normally grouchy at wake-up, the reveille gun could teach them to be grouchy the rest of their lives.

After the summer leave following plebe year, when '54 began training at Buckner, much to their chagrin they discovered another reveille gun existed. It was one more of those seventy-five millimeter "first thing in the morning, you're going to be blasted awake or else cannons." And as luck would have it, the dastardly artillery weapon was strategically placed for the strongest possible noise effect. But a small group of cadets in '54 decided things were going to be different in yearling summer at Camp Buckner. It was time to get rid of the noise, be more civilized about waking up in the morning, "kick the system in the shins," "stomp on its toes," retaliate with a little "return harassing fire," do something with that damnable artillery gun, piece, or whatever it was called.

The rebellious cadets recorded what happened in *The Reveille Gun Caper*. It was another blow for freedom, a sign of spunk and scrap in the class of '54, and occurred just three days after the *Whiskey* contingent left for their Caribbean cruise.

> ...A group of cadets conjured up a daring mission. It was to capture the French 75 mm reveille gun located on the peninsula bordering the cadet swimming area, and move the gun across Lake Popolopen to the top of an opposite mountain where it would cease its daily harassing bursts.
>
> The team divided into two groups who launched a coordinated assault in the early hours of 19 July 1951. The previous day, the first group requisitioned an engineer raft. Under cover of darkness, they moved it to a nearby position, while the second group infiltrated the peninsula, avoiding a sentinel post covering the access route. By 0130 [1:30 a.m.] hours both groups were in position. The raft was moved into shore, the chain was severed and the cannon was quietly placed on the raft. Suddenly, the unexpected occurred. The cannon shifted on the raft and the raft became unsteady. The raft tilted and the quiet night air was torn with a horrendous splash as the gun fell and settled into fifteen feet of water. Everyone froze in place. The game was up! It was every man for himself.
>
> Amid shouts of "Who goes there?" and "Halt!" the team members sprinted down the peninsula, breaking cover like a covey of quail – the guards in close pursuit. There was but one thought at the moment – get back to the sack fast!
>
> One member of the team, Ed Moses, had the misfortune in the middle of his darkened barracks to encounter the Cadet Corporal of the Guard, Joe Peisinger, who was in hot pursuit. Ed assured the guard of his unconditional surrender and delivered his 'name, rank, and serial number'.

The next day volunteers dove into the depths of the lake on a new mission – a directed salvage operation. Ropes were secured to the submerged cannon. It was unfortunately recovered undamaged and after a thorough cleaning, it resumed its harassing bursts.

Few of the team were apprehended, but for those that were their punishment was swift. However, relief was unknowingly close at hand. The Emperor of Ethiopia arrived for a state visit in the early fall and, as luck and custom would have it, granted amnesty to all.

Ed Moses received a "Delinquency Report," a "Quill slip" in cadet parlance, from the Cadet Corporal of the Guard, who was required to complete and sign an official report at the end of his twenty-four hour tour of duty stating that he "had reported every violation of regulations he had observed." Ed's punishment was indeed swift, as it was for all the other team members caught in the sinking of Camp Buckner's rev-

DELINQUENCY REPORT

Rebuilding begins, "cadet style." The perils of mixing cadet humor and military discipline. The delinquency report received his yearling (sophomore) year, by Edward M. Moses, class of 1954, for the "Reveille Gun Caper" at Camp Buckner, near West Point.

eille gun. He received twenty demerits, forty-four punishment tours, and two months' special confinement. Though the mission failed, it was well conceived, and spoke volumes about the days ahead.

As a plebe in the fall of 1950, Ed Moses tried out for the powerful, nationally ranked Army football team. He was a "walk on" and worked hard, hoping to make the team. But a week and a half later, Earl Blaik approached him, put his arm around Ed's shoulder and said, "Moses, you're just too small for Army football." Now, as a yearling at Buckner, after his daring raid on the reveille cannon, he heard about the cheating scandal. He wanted to try out for the football team again, but he couldn't. He had forty-four punishment tours to walk off, and two months of "special confinement" in his cadet room. The Emperor of Ethiopia didn't arrive early enough to grant Ed amnesty. He did play intramural football that fall as a halfback on the Company A-2 brigade football championship team. The following year he decided if he couldn't play football for Army, he would do the next best thing – be a cheerleader for the Corps of Cadets and "The Rabble."

Ed Moses had an affinity for armor – tanks, steel monsters that could move on the battlefield with great speed, shock, and firepower, and with their own deadly accurate, high velocity cannons, were holy terrors for lightly armed enemy infantry. The tank could withstand rifle shots, mortar rounds, and even most artillery shells, unless the shells were powerful antitank rounds designed to pierce heavy armor. Though his affinity was for armor, another artillery man's cannon was in Ed Moses' future at West Point, two years hence, before he would become a second lieutenant of armor in the United States Army. That cannon would become a fixture in the future of the Academy and Army football for generations to come.

Don't scrape the *Whiskey's* paint! Go! Take the cannon, and fire it after every touchdown, every extra point, and when victory's ours! Go! Hold that line! Defense! Defense! C'mon Rabble! Go! ... Go! ... Go!

The slow, rumbling cadence was beginning, though its sound wasn't yet voiced, and not yet heard. It would be heard first in the early fall of 1951, in West Point's Michie Stadium. Then the sounds and the cadence would continue and build. On a fall day in New York City's Polo Grounds, on October 17, 1953, it would reach a thunderous, deafening crescendo, and roll irresistibly through the remainder of that joyous, colorful season. For those fortunate enough to be in the Polo Grounds that October day, the sights, sounds, and emotions were to be stirring, lifelong memories.

THE SEARCH FOR CAUSES AND REMEDIES

Following the Collins Board, during General Irving's and Coach Blaik's trips to Washington in June and July, it became clear there must be a follow-on investigation to find and remedy the causes for the devastating honor incident. The Collins Board's June 8 report had been read by General Maxwell Taylor, the former superintendent assigned to West Point at the end of World War II, and General J. Lawton Collins, the Army Chief of Staff. Nearly all the policies instituted by General Taylor, when he was superintendent, were still in effect when the Collins, Jones, and Barrett Boards were finally completed in mid-August. But it was the Collins Board, with its three page discussion summarizing testimony and facts gleaned from the investigation, the board members' conclusions and recommendations, and a strong letter from the Army Chief of Staff, General Collins, that set the direction of the Bartlett Board.

The fundamental purpose of the Collins Board was fact finding. The Board members were to give the superintendent factual bases for decisions regarding the futures of the men found to be involved in organized cheating. To provide the facts meant to obtain evidence on every cadet participating – if evidence existed sufficient to meet the evidence tests of a court martial. In gathering facts and evidence, the Collins Board was able to verify the existence of organized cheating; learn generally to what extent the Corps of Cadets had been affected; where, how, and why the cheating got started; and the various ways the ring operated and spread.

In addition to concluding a total of ninety-four men in the classes of '52, '53, and '54 were guilty, and the conspiracy to cheat in academics "existed since approximately 1946-1947," Art Collins, Jeff Irvin, and Tracy Harrington, also concluded:

> ...The conspiracy has been centered in and perpetuated by members of Football Corps Squad, until the Fall of 1950 at which time cheating began to spread.

...There was a feeling among some of the cadets who cheated that their actions were known to and condoned by the authorities.

...The impression prevails in the Corps of Cadets that football players are given added advantages in academics, due to academic coaching systems, admission of players to the hospital for the purpose of study, and the unfounded belief that they get advance information from the football office.

...Had this conspiracy continued it would have destroyed the Honor System at West Point.

...The change of regulations pertaining to the marking of the absence card during daytime call to quarters has contributed in some degree to the easy passage of information within the group.

....There are many academic and personnel matters beyond the scope of this Board which provide background for this distasteful state of affairs.

The Collins Board's first recommendation was the ninety-four guilty cadets be "offered an opportunity to resign and be discharged" and, if they refused, "be investigated in greater detail, with the view of sustaining a conviction by court-martial." The three officers also recommended studying regulations governing visiting in barracks to maintain "a closer check on absence from rooms during call to quarters," and more significantly, "a Board of Senior Officers be appointed to study the background which contributed to this unfortunate state of affairs."

This latter recommendation, read by the superintendent and senior officers in the Army staff, set the stage for the Academy's searching self-examination which began August 13, but not before the Collins Board and the July 23-25 Hand Board had been reviewed by General Irving and the Army's senior officers. The Army Chief of Staff attached unmistakably clear direction to the Collins Board's recommendation for a "study" and returned it to General Irving in the form of a letter from General Maxwell Taylor, the former superintendent, and General Irving's immediate boss in the Army chain of command. The letter was dated August 3, the day of the public announcement of the honor violations.

1. The Chief of Staff directs that the Superintendent, United States Military Academy, review all aspects of the recent incident involving violations of the Honor Code at the Military Academy with the view to making such changes of procedure and practice as may be advisable to prevent reoccurrence of such an incident in the future. Specifically, he desires that the practice of giving identical written recitations and examinations to successive groups of cadets be discontinued.

2. With regard to the Honor System itself, he would like this review to verify that the system is simple, clearly understandable and is not a device to affect the enforcement of regulations.

3. It is the impression of the Chief of Staff that a contributing factor to the recent incident has been the separation of the athletic groups, particularly the football squad, from the environmental influences of the Corps of Cadets. He desires the Superintendent assure that in the future the athletes are not so insulated from the rest of the Corps as to be deprived of the sustaining ethical influences of normal cadet life.

On Tuesday, August 7, while the Jones Board was in progress – the first of the two "screening" boards convened to review the cases of the ninety-four cadets – General Irving called a meeting of the Academic Board. The Academic Board, the most powerful and influential board within a system of permanent boards and committees affecting the performance of the Academy mission, and its curriculum, was continuing its heavy involvement in the events of 1951. When the meeting opened at 11:30 in the morning, General Irving summarized circumstances surrounding the impending dismissals of the cadets. After some discussion, he then read General Taylor's August 3 letter.

Considerable discussion ensued. What was the best means of carrying out the directive from the Chief of Staff? How many senior officers would be on the board? Who would select them and who would they be? What would be the mechanism put in place to ensure the needed actions were taken to preclude the reoccurrence of a similar incident, and give the Chief of Staff the assurances he asked?

Before the meeting adjourned at twelve twenty, General Irving gave the Academic Board his decision. There would be three members on the board of senior officers. The Dean of Academics, Brigadier General Jones, would name one member from the Academic Board. The Commandant, Colonel Waters, who was also on the Academic Board and the Board of Athletics, would name a member from the Tactical Department, and General Irving, the superintendent, would select a member.

On Monday, August 13, 1951, the Bartlett Board convened. The letter order convening the three officers, signed the same day, was brief and to the point:

> ...A board...is appointed to review present procedures and practices at the U.S. Military Academy with particular emphasis on those having a bearing on conditions brought out in the present investigation of Honor violations...

> ...Report will be made to the Superintendent with the least practicable delay concerning those procedures and practices which should be examined in detail with a view to material change...

The Board's mission evolved out of informal discussions between its president and General Irving, and is in the final report:

> ...Survey procedures and practices at the United States Military Academy, particularly those which may have any conceivable bearing on the recent honor violations.

> ...Form judgments as to the causes of the incident, both proximate and underlying.

> ...Suggest changes in existing practices and procedures which might prevent a recurrence.

> ...Recommend specific areas of practice or procedure in need of detailed study by subsequent boards with a view to material change.

> ...Consider means of continuing or periodic check that would show up an incipient repetition of a similar incident in its early stages.

The Bartlett Board, as it was later called, got its name from the president, Colonel Boyd Wheeler Bartlett, West Point class of 1919, Permanent Professor and Head of the Department of Electricity. He was selected by the Dean of Academics. The two additional members were Colonel Francis Martin Greene, class of '22, selected by the Commandant, Colonel Waters, and Colonel Charles Harlow Miles, Jr., class of '33, General Irving's choice. Francis Greene was the Director of Physical Education, a position which held the title "Master of the Sword." Charles Miles was the Academy's comptroller, its Fiscal Officer, and was on General Irving's staff. As the junior member of the Board, Miles was named its recorder.

Like Earl Blaik, who resigned his commission in 1922, two years after he graduated from West Point, both Bartlett and Greene resigned their Regular Army commissions, in 1922 and 1926 respectively, during the drastic reductions in the officer corps following World War I. They remained on the Army's reserve rolls, pursuing education and business careers respectively, while Blaik gravitated slowly toward coaching college football. Bartlett and Greene returned to active duty reserves in 1942, during World War II, and were assigned to the Academy. Bartlett was forty-five years of age, Greene forty-two when they returned to the Academy.

By this time Boyd Bartlett, whose home was Castine, Maine, was widely known in academic circles as an educator and engineering scholar at Bowdoin College. His ties to Bowdoin went back three generations in his family, to his great-grandfather, who received an honorary Doctor of Divinity from Bowdoin in 1860. "Red" or "Brick," nicknames he earned as a young man because of his flaming red hair, graduated *summa cum laude* from Bowdoin College in 1917, the same year he entered West Point.

He wore the stars of a distinguished cadet, graduated number three in his Academy class, and received a commission in the Engineers. Also like Earl Blaik, he was in one of the World War I classes which received abbreviated Academy educations – he was at West Point seventeen months, and the War was over eleven days after he graduated the first time. He and the rest of his class were recalled to West Point as student officers to complete the prescribed War Emergency Course, and they were graduated again on June 11, 1919. He was a well-rounded athlete, having lettered in football, hockey and tennis, and was the captain of the hockey team at Bowdoin.

At West Point he lettered in varsity football during the war, when football, on a national scale and at West Point, had been de-empha-

Boyd Wheeler Bartlett
as a cadet, class of 1919,
from Castine, Maine.

Colonel Boyd Bartlett,
Board President,
August 13 - September 7, 1951

(Photos courtesy USMA Archives.)

sized because of the war. He also played hockey. In the 1918 football season, he was on the team with Earl Blaik when the cadets played only one scheduled game. He graduated from the Academy by the time Douglas MacArthur arrived in 1919 to rebuild West Point's shattered and outmoded curriculum, and wasn't influenced by MacArthur's drive to bring needed, stern reforms in the physical education of cadets – reforms that included instituting an intensive intramural athletic program and powerful backing of more aggressive, competitive Academy intercollegiate athletics.

Above all, "Brick" Bartlett was a scholar and educator when he came back to West Point in 1942. He arrived with impressive academic credentials, including a Bachelor of Science in Civil Engineering from Massachusetts Institute of Technology, a Masters and Ph.D. in Physics from Columbia University and, while on a sabbatical from Bowdoin, did post doctoral work in atomic physics at Germany's Munich University in 1934-35. By the time he was appointed as president of the Bartlett Board in August of 1951, "Brick" Bartlett's flaming red hair was almost white. He looked the part of a distinguished

302

officer, gentleman, and professor at the nation's oldest national Military Academy.

Francis Martin Greene was born in Brooklyn, New York, and entered the Academy with an appointment from Wappingers Falls in the state's 8th Congressional District. When he was recalled to active duty in 1942, he left the position of District Manager in the Newburgh District of the Central Hudson Gas and Electric Corporation. He had been with Central Gas and Electric since leaving the Army in 1926, progressing to District Manager several years after taking a Public Utility Management course at Harvard University.

"F.M.," as he was known as a cadet, was an outstanding athlete while at West Point, lettering three years in football, and was captain of the team his first class year. He also captained the wrestling team, and played lacrosse. Contrary to the circumstances of Boyd Bartlett, Francis Greene had seen General MacArthur's reforms begin to take hold at West Point before he graduated in 1922. His outstanding work in physical education at the Academy, after he returned to active duty in 1942, earned him an honorary degree as Doctor of Science from

Francis Martin Greene
as a cadet, class of 1922,
from Wappingers Falls, New York.

Colonel Francis Greene
Bartlett Board Member
August 13 - September 7, 1951

(Photos courtesy USMA Archives.)

Boston University and the honorary degree of Master of Physical Education from Springfield (Massachusetts) College.

Charles Harlow Miles, Jr., born in Philadelphia, came to West Point's class of '33 from Wenonah Military Academy in New Jersey, where he obtained a Congressional appointment from the 1st District. Harlow, as his classmates called him, was on the Academy swimming team his plebe year and the track team four years, earning monograms two years in track. He went into the infantry on graduation from West Point, later transferred into the Finance Corps, and was at Hickam Field, adjacent to Pearl Harbor, Hawaii, when the Japanese attack brought the United States into World War II. He served three and a half years in the South Pacific and remained in the Finance Corps throughout the rest of his career. As Fiscal Officer at the Academy, he was West Point's "treasurer," and as such, also served as treasurer for the Department of Athletics. Most of the Department's operating funds were non-appropriated funds, that is, funds not appropriated by Congress. Non-appropriated funds were generated through athletic ticket sales, Army Athletic Association memberships, donations, and additional retail activities associated with intercollegiate sports.

Charles Harlow Miles, Jr.
as a cadet, Class of 1933,
from Wenonah, New Jersey.

Colonel Harlow Miles
Bartlett Board Member
August 13 - September 7, 1951

(Photos courtesy USMA Archives.)

Noticeably absent from the Bartlett Board was anyone from the Department of Athletics, or the Board of Athletics. The Commandant, the Dean and two professors from the faculty, plus Earl Blaik and the Graduate Manager of Athletics, were on the Board of Athletics. The two professors from the faculty were Colonel Gerald A. Counts, Professor and Head of the Physics and Chemistry Department and Colonel T. Dodson Stamps, Professor and Head of the Department of Military Art and Engineering.

There were only two senior, active duty officers within the Department of Athletics. One was Colonel Philip H. Draper, Jr., class of '29, a World War II veteran who arrived at the Academy on July 17. He was replacing the Graduate Manager of Athletics, Colonel Orrin C. Krueger, class of '31, who was still at West Point, preparing to go to his next assignment. The Graduate Manager of Athletics worked under Earl Blaik's supervision, in Blaik's role as Director of Athletics, and served as a non-voting recorder on the Athletic Board. The position, which included executive management of the Army Athletic Association, as well as treasurer for the Association, also gave Draper direct access to the superintendent. But considering all that had transpired since the honor violations had been uncovered, and the organizational arrangement that put him under Blaik's supervision, he obviously wasn't a logical choice to serve on the Bartlett Board – if he was ever considered.

There were two assistant Graduate Managers of Athletics, Colonel Russell P. "Red" Reeder, Jr., class of '26, and Colonel Elliot W. Amick, '38, both retired officers who had been seriously wounded in the European theater of operations during World War II. Both were reassigned to West Point in 1945 after recuperating sufficiently from their wounds. They later retired from active duty, and remained as greatly admired and respected civilian employees in the Department of Athletics.

Earl Blaik, the Director of Athletics, in addition to head football coach, was given the Director's responsibilities in 1946 while General Taylor was superintendent. When he became Athletic Director, Blaik was seriously considering leaving the coaching profession for a more lucrative career in business. He was completing his five-year contract as Army's first civilian head football coach. He was forty-nine years old in 1946, and the Army, as it had following World War I, was forced by budget considerations to sharply reduce its active duty force after World War II. Blaik was returned to inactive reserve status after being called to active duty as a lieutenant colonel during the War.

There were no retirement or pension plans for the head coaching position, and Blaik was painfully aware of the pressures, volatility, and insecurity of college football. His business, coaching, and athletic experience, plus an increasing scarcity of qualified senior, active duty officers to fill the position of Athletic Director, made Blaik a logical choice for General Taylor to engage the Army coach's considerable talents, with increased responsibilities and remuneration.

Though Blaik was Director of Athletics and Chairman of the Athletic Board, as well as head football coach, and clearly was a powerful voice in directing the Academy's intercollegiate athletic programs, he was a civilian employee. Because he was a civilian employee, he didn't have the power of a vote in recommending athletic policy to the superintendent. However, the lack of a vote was more an inconvenient technicality, because Blaik's influence reached far beyond the confines of West Point by the time he completed his next five year contract in the spring of 1951.

His position as Director of Athletics, as well as head coach, gave him direct access to the superintendent. He was independent of the Academic and Tactical Departments, as well as the Chief of Staff of the Academy, a "direct access" organizational arrangement he negotiated with the president at Dartmouth in 1933, and with West Point superintendent, Brigadier General Eichelberger, who brought Blaik back to Army in 1941.

Blaik insisted on reporting directly to the superintendent, or college president, because he firmly believed it was the only way a football program would ever receive the support required to be successful. In practice he was equal in seniority to the Commandant and the Dean of Academics, a circumstance that had apparently become an increasing irritant to a number of senior officers, although they never openly complained to the superintendent.

What's more, Blaik had an old friend and associate in Washington, D.C., assisting him in securing the annual, late spring appointments for West Point's athletic recruits. He was retired Colonel Lawrence McCeney "Biff" Jones, class of August 1917, the highly successful former Army head coach, and later Athletic Director, who brought Blaik to West Point as an assistant varsity coach in 1927.

Finally, Blaik's success, his winning records in a collegiate sport that drew huge crowds each fall to equally huge stadiums throughout the land, garnered him much press, public attention, and considerable public admiration. He was a celebrity in his own right, whether or not he wanted such status.

Thus, by the spring of 1951, Blaik was both well respected and established at the Academy – at least among most at the Academy, and didn't hesitate to exercise his considerable influence at West Point, in the Army, the halls of Congress – and with sports writers. He was sufficiently well established that he signed another contract not long before the existence of organized cheating was divulged to Colonel Harkins the first days of April.

The lack of recent intercollegiate athletic experience, and a steadying Athletic Department voice on the Bartlett Board, while perhaps reasonable and logical on the surface, created additional stresses and strains and complicated efforts to heal the wounds inflicted on the Academy by the honor scandal. Most notably, the natural tensions existing between the Athletic Department on the one hand, and the Academic and Tactical Departments on the other, were made more difficult by the composition and conduct of the Bartlett Board.

While its purposes were to identify contributing causes and find remedies to avoid a reoccurrence of organized cheating, the Board was not formed as an investigative body. Its members didn't consider themselves fact finders. Instead, the Board considered themselves a "study" group, searching for underlying and proximate causes for the scandal. They relied on thoughts, opinions, beliefs, and ideas to reach conclusions, findings, and recommendations. They explained their rationale for their approach to the overall problem in their explanation of "THE NATURE OF INVESTIGATION REQUIRED," their explanation of why and how the investigation would be conducted, which precedes their report – but was written at least in part after the investigation method was chosen and applied.

Analysis of the report of the board which originally investigated the honor violations (Collins Board) suggests the great importance of psychological factors in the incident. If the ideal of the honor system is strong in the mind of the normal cadet, strong motivation is required to break it down. The idea of standing by a friend, of not telling on another, is commonplace in the world at large. Many a boy who would not himself cheat would not inform on another boy who did.

...To obtain information about beliefs and opinions a mere study of matters of record is not sufficient. It is necessary to talk to people, to find what is in their minds. The opinions and attitudes of one segment of the community interact with and influence those of other segments. For that reason you will observe in the procedure we have followed a considered and deliberate attempt to mirror the thinking of the West Point community, the Corps of Cadets and those agencies and individuals who deal with cadets. We have tried to determine whether or not attitudes and opinions which might have bearing upon the incident itself are strongly and widely held. Finally we have made an effort to compare current beliefs and opinions with those of the past to determine if significant changes or trends are evident.

The Board members first "read carefully" the Collins Board report. They then "for convenience in...[their]... survey of practices and procedures... subdivided the whole field into separate areas, each of which appeared initially to have a possible bearing upon the honor violations." They selected eleven areas of investigation, of which six involved intercollegiate athletics. The remainder included the honor system, academic procedures, additional instruction and tutoring of cadets, the cadet duty committee, cadet aptitude system, and privileges and extra-curricular activities.

The proceedings were deliberately informal to avoid the appearance of a courtroom or an inquisition. In explaining the conduct of interviews, the Board's report said its "desire [was]...to obtain frank and uninhibited discussion of the problems at issue." Further, because the Board was to deliver a report with "the least practicable delay," and the mission given them indicated they could identify additional areas other boards could examine in greater detail, the Board elected not to investigate in depth the various matters which Collins Board testimony clearly indicated needed to be considered. Instead, those re-sponsibilities were assigned follow-on boards and committees who would receive the benefit of the Bartlett Board's work, and could pro-ceed in greater detail investigating assigned areas.

Boyd Bartlett and his two Board members worked at a hectic pace, completing approximately seventy-nine interviews totaling three hun-dred seventy-nine officers, cadets and civilian employees by noon on August 29. A substantial number of interviews were in groups. Two more officers were interviewed on September 4 and 6, respectively. One was Colonel Paul Harkins, the former Commandant, who had su-pervised the undercover investigation to obtain evidence of the honor violations, and convened the Collins Board. Two members of the Bartlett Board interviewed him in Washington, D.C., on September 4.

People interviewed by the Board came from every Academy orga-nization involved in educating and training cadets at West Point, and included a large number of cadets. Among the three hundred eighty-one interviewed were: two hundred sixty-seven cadets, nearly all in-terviewed in groups; one hundred four officers, of whom twenty-three were interviewed in groups; and ten civilians, all interviewed individu-ally. Among the ten civilians were Earl Blaik, three of his assistant football coaches, and four Academy athletic coaches from other inter-collegiate sports. Another assistant football coach interviewed was Captain Johnny Green.

Among the cadets interviewed were the entire Cadet Duty Com-mittee, all of whom were first classmen, on August 16; eighty-eight men in the class of '54, in six groups, on August 20, at Camp Buckner; and thirty-six men in the class of '52 who were on the Buckner train-ing detail, at Buckner, on August 24.

The Board also interviewed Colonels H. Crampton Jones, Charles J. Barrett, and James W. Green, Jr. Colonel Jones, class of 1916, and a field artillery man, was the Academy's Inspector General and was presi-dent of the "screening board" which convened August 3.

Colonel Barrett, Professor and Head of the Department of Foreign Languages, was president of the second "screening board" which be-gan its work on August 9. In addition to his marvelous facility for lan-guages, Charles Barrett had had a remarkable career as a soldier before becoming a professor and department head at West Point. He served as an enlisted man, a motor cycle courier, on the battlefields of France during World War I, before he caught the first boat home from Europe after the War to enter West Point on November 26, 1918. He graduated number one in his class of 1922, and was First Captain in the Corps of Cadets. He went into the engineers on graduating, later transferring to the artillery. During World War II he became the chief of staff of the 84th Infantry Division, and later was chief of division artillery. He distinguished himself in combat in the European Theater, winning the silver star for heroism, and later was promoted to brigadier general and given command of division artillery. Before coming to West Point as a professor after the war, he twice previously served on the Academy faculty. Like thousands of other officers who served in World War II, after the war, when massive force reductions occurred, he reverted to the grade of colonel before returning to West Point as a professor.

Colonel Green, class of '27, was a professor in the Department of Electricity, in Boyd Bartlett's department, and had been given duties as "coordinator" between the two screening boards conducted by Colonels Jones and Barrett.

To each individual and group interviewed, the Board explained its purpose, using the word "mission" rather than purpose. The people who appeared were referred to in the report as "witnesses" rather than interviewees, although there was no sworn testimony. The Board described the barest of essentials about the cheating incident to its witnesses – though Jones, Barrett and Green were obviously well versed on Collins Board proceedings and results because of their roles in the screening boards. The Bartlett Board, nevertheless, cautioned each witness any information obtained from its proceedings was "confidential." The caution was consistent with the treatment of the Collins Board report, which was later classified "confidential." The Bartlett Board went on to tell witnesses there was "specifically" no intent to obtain additional names of cadets that might be involved, nor was information derived to be used in any form of disciplinary action.

For the cadets interviewed, who were randomly selected, the Board developed standard questions for uniformity of treatment. The three officers used care to avoid interjecting their opinions in discussions with cadets, and framed questions to learn "the cadets' ideas on various matters pertinent to the mission [of the Academy]."

The Board's days were long and arduous from August 13 until the report was signed shortly before noon Saturday morning, September 7. But they did break from their fast paced routine on two of three weekends, adjourning two of three Saturday afternoons and reconvening the following Monday mornings. Board sessions generally began from 8:00 to 9:00 in the mornings and lasted frequently until 10:30 or 11:00 at night.

The morning of the first day, the Board received the superintendent's approval for its agenda, which Boyd Bartlett had been discussing in advance with General Irving. The three officers then listened to a briefing on the Collins Board proceedings, and interviews began that afternoon, with six of the cadets found guilty by the Collins Board.

Tuesday, August 14, the Board interviewed sixteen people, including all three members of the Collins Board; Colonel Waters, the Commandant; Colonel Krueger, the soon to depart Graduate Manager of Athletics; and eleven members of the Cadet Honor Committee, men in the new first class, the class of '52.

The next day, nine men interviewed with the Board, including the three Collins Board members, back for a second session. From that day forward the pace picked up as Boyd Bartlett pushed through "Transition Week," and on into the beginning of the 1951-52 academic year. Transition Week was traditionally the last week in August, when the Corps returned to West Point to organize into its twenty-four permanent companies for the academic year.

The afternoon of Wednesday, August 29, the three officers began concentrating on completing their report, including findings and recommendations. They worked every day until the report was complete on September 7. Football practice officially began Saturday, September 1 and academics began the following week.

The Bartlett Board's "informal" conduct of its proceedings was useful, in that it allowed people interviewed to express themselves freely. They didn't hold back their feelings and opinions about what had caused the honor incident. As a result, the Board also became a forum which exposed long pent-up frustrations involving the three departments, as well as professional and personal differences between some of the strongest and most influential personalities at the Academy. Unfortunately, the frank, uninhibited, and sometimes acrimonious discussions, while useful in obtaining ideas and opinions about the causes for the cheating incident, did little to foster better understanding and communication between the three major departments charged with educating and training the Corps of Cadets – particularly as to understanding the role of intercollegiate athletics in the education of cadets – future officers.

In spite of the Bartlett Board's best efforts, to the men in the Department of Intercollegiate Athletics and on the Athletic Board, the Bartlett Board's membership and the questions it posed to all who appeared before the three officers gave the appearance of the Athletic Department on trial before the Academic and Tactical Departments and the superintendent's staff. In some respects, while the Bartlett Board's mission was to search for causes contributing to the honor incident, its composition and conduct unfortunately added more misunderstanding where communication breakdowns and barriers already existed. It would be nearly six months before interdepartmental relations would show positive signs of healing.

The added stresses and strains weren't debilitating, however. It became apparent later that General Irving, in reading and responding to the three officers' September 7 report, recognized the Bartlett Board was excessively zealous in finding "over-emphasis on football" the basic cause for the "misalignment of values" which produced the scandal. General Irving acted as buffer by ignoring and withholding their more strident comments in the report, and altering the means of implementing some of the Board's recommendations. Fortunately, in the case of the Bartlett Board, its closed, "restricted" proceedings helped General Irving avoid excessive adverse effects on interdepartmental cooperation as rebuilding progressed.

To their everlasting credit, however, Colonels Bartlett, Greene, and Miles did recognize the powerful emotions and increased tensions loosed at the Academy by the cheating scandal, and the contentiousness voiced in the Board's proceedings. After the three officers' considerable study and deliberations, they identified "Team Work Within the Command" as one of several "Additional Areas Requiring Investigation."

Though no additional investigation was ever done on this recommendation – General Irving decided "normal staff action," and no doubt leadership, would be the way to improve team work. It was a recommendation that was to become the watch word, the flag, the battle streamer, for all to hear in the two and a half years ahead, as the Academy struggled toward recovery from the honor incident.

REPORT – AND MORE TO FOLLOW

When public institutions are wracked by scandal, or perceived scandal, there is need for extensive investigation – sometimes several investigations – which generate numerous records and reports. The honor incident of 1951, the first of its kind in the history of the nation's oldest Military Academy, was a classic example of an incident needing deep, thoughtful examination. The Bartlett Board's responsibility was to conduct such an examination.

The Board convened while the white hot glare of the press was still on the Academy after the public announcement of August 3, though the announcement made no mention the investigation was in the making. The Board in turn spawned several other committees and studies, all intended to make the necessary changes in practices and procedures to preclude recurrence of a similar incident.

The classified letter from General Taylor to General Irving kept the Board's proceedings and records "behind closed doors," a reasonable precaution that would permit people interviewed to candidly offer their observations, opinions, and beliefs without fear of repercussions.

While keeping the Bartlett Board proceedings restricted was reasonable to ensure open discussion from interviewees, and the Board was following an agreed method of study, the Board's composition, in terms of its ability to look at the Academy mission from a broader perspective, was less than circumstances and events required. As a result, the Board enmeshed itself in what it concluded were the underlying and proximate causes of the honor violations – "misalignment of values" caused by "overemphasis on [winning] football" – and based its findings and recommendations largely on those conclusions.

What was going on outside the Bartlett Board surely had a profound effect on the opinions, beliefs and ideas of all who came before the Board to be interviewed. After August 3, when the public announcement concentrated attention on the origins of the scandal, the Academy became a flash point for rumors. Though rumors were numerous and detailed information and facts were scarce, there was no doubt about where attention was centered – the football team. Open discussion and speculation at the Academy centered largely on the football team's role, why and how they had involved themselves in organized cheating, and how they managed to involve so many other cadets. By the time the Bartlett Board convened ten days later, emotions were running high and disillusionment, embarrassment, frustration, and anger were running deep. It is little wonder that by the time the Bartlett Board began its deliberations on August 13 many fingers would be pointed at Earl Blaik, the football team, and the Department of Athletics. They collectively became the lightning rod for all the static electricity which had been rapidly building inside the Academy the preceding four months – and more slowly for two or three years preceding the undercover honor investigation of April and May.

There was no question the honor incident had its origin in the football team and had spread from the team "like spokes from the hub of a wheel," nor that the football team had become somewhat separated from the Corps of Cadets and had, along with its coaching staff and many others, developed a fierce, detached pride in winning. But there were too many other questions which needed to be answered, questions the Board didn't address.

For example, though it was apparent from Collins Board testimony that more than a few cadets who weren't on the football team, or had no direct connection to it, had known that honor violations were in progress and had been repeatedly told during honor training they were "honor bound to report any violations they observed," they stood by and allowed the cheating to continue. They waited until someone else made the difficult choice of divulging the existence of organized cheating. Why did this happen, and what might be done to avoid a similar contagious breakdown of the honor system? The behavior couldn't be explained away entirely by the "overemphasis on football" with all its perceived pressures, team induced loyalties, or "misaligned...values."

And what of all the football players who hadn't cheated, were known "supporters of the honor code," and were deliberately avoided by the men engaged in the practice?

Hidden in the Bartlett Board's interview records were obviously sharp differences between the three major departments about roles, contributions, authorities, and responsibilities in performing the Academy mission. Some of the antipathies and finger pointing expressed may have been more effects rather than contributing causes in the incident, but interviewees were unabashed in their comments about progressively deteriorating interdepartmental relationships they had felt long before the honor violations were finally divulged. Interviews were replete with criticisms of the Athletic Department, voiced by members of the Tactical and Academic Departments. Athletic Department members fired back in kind, and members of all three departments gave numerous accounts of vexing disagreements that from their perspectives hadn't been satisfactorily resolved.

In the text of the report, the Board did hint at the Academy "leadership's" failing to notice and improve deteriorating interdepartmental relationships, apparently an oblique reference to the late General Moore, General Irving's predecessor, and Earl Blaik's and Paul Harkins' unresolved differences, which, according to Blaik, began soon after Harkins arrived at West Point. In another section the Board carefully explained that Academy leaders hadn't been caught up in the "subtle" growth of the overemphasis on winning football.

Yet the Board report didn't highlight those disputes and conclusions or probe deeply into why they occurred, how they got started – or how they may have affected the institutional environment, and hence contributed to the football team's "segregation" from the Corps of Cadets. Only one Board recommendation touched tangentially on the interdepartmental strife that had gradually increased in the years prior to the incident – the recommendation for additional study of interdepartmental team work.

What were the backgrounds and experiences of the young men entering West Point within the two to three year period following World War II? Might the answers to that question bear upon their attitudes toward a "cadet honor code and a cadet honor system"? This in turn might lead to other important questions about what education and training is needed to ensure cadets understand and accept the practical, daily application of the high ideals and standards the honor code and Academy motto express.

And what of the honor code and honor system, and their relationship to leadership and character? Beyond the questions posed in General J.

Lawton Collins' entreaty to consider whether the system was being used to "enforce regulations" or had grown too complex, was there a need to look deeper and more broadly at the honor code and system? Were the code and system acting as deterrents to cadets' reporting transgressions because there was too much fear of consequences? – a point Earl Blaik vigorously argued. What needed to be done to change the attitudes that say "you never tattle, you never tell on a friend, roommate, or teammate – no matter what"? Was the honor system really the sole province of the Corps of Cadets, as had been long held, or was there a wider institutional meaning, purpose, and setting for both the code and the system?

Wasn't integrity crucial to successful leadership and command? And wasn't the Academy, above all else, training leaders of character who would aspire to command, and be able to make difficult decisions that would clearly and consistently reflect integrity and strength of character? What had been missing in the education and training of the eighty-three cadets that allowed them, in their minds, to separate the principles of leadership, command, and integrity – and rationalize cheating to "beat the academic system"?

What of the inferences in Collins Board testimony that some cadet companies were electing representatives to the Honor Committee without serious consideration of the purposes, duties, and responsibilities of Committee members – and who such men ought to be?

There were clear, repeated inferences in Collins Board testimony that there had been something wrong in the manner some cases were being investigated and adjudicated by the Honor Committee, because of actions by one, two, or maybe three members of the Committee. If that were so, what was needed to avoid such problems in the future? A perception in the Corps of Cadets that the Honor Committee acts unjustly, winks at breaches of honor, or dodges its responsibilities can be devastating. On September 4, Colonel Paul Harkins would tell two members of the Board that all hadn't been well with the Honor Committee since the preceding fall. While there might be innuendo about "football sympathizers" or protective influences at work, as was implied in the Collins Board testimony, the implied questions needed carefully investigated answers.

Clearly, there was need of immediate attention to excesses in the Department of Athletics. Equally clear, there were other complex forces and factors at work which set the stage for the disaster. Some of those factors and forces had their origins in changes in priorities and emphases in the Academy curriculum wrought by World War II and the reforms that followed.

But unfortunately, the Bartlett Board made some flawed assumptions in the body of their study and accorded too much weight to opinions and beliefs, and other educators' outspoken and sometimes shrill frustrations over the "corrupting influence of big time college football." In describing the effects of "overemphasis on winning football," the Bartlett Board laid bare some of their assumptions and sweeping generalizations that facts wouldn't have supported if the Board had delved more deeply into how and why young men who played football chose West Point.

The assumptions were rooted in what Boyd Bartlett, "F.M." Greene, Harlow Miles, and others believed about young men talented and skilled enough to complete a demanding academic curriculum and play major college football in the 1940s and '50s, not what thoughtful research and investigation would have shown.

> ...It is difficult to arrive at a complete understanding of all influences upon the football player unless one studies their cumulative effect as a whole. It is doubtful whether responsible individuals at West Point have ever had the complete picture presented to them at a single time and place. To obtain such a picture one must start with the young man who is a high school football player of regional or even national reputation and follow through his experience in detail, bearing in mind a number of factors which operate strongly from the very beginning. Such a young man well may have been over-rated by his teachers and others in his local community from the very start. His academic preparation may be more superficial than his grades indicate because of the time he has devoted to athletic activities. He will usually have received attractive offers to play football for numerous competing colleges. If representatives of the West Point Football Office are successful in inducing him to choose West Point in the face of this competition he will already have been inclined to think he is doing West Point a favor rather than the reverse. On top of this initial motivation he receives

most of the preferential treatments listed in detail in ...[this study]. Under these circumstances it is understandable that he can develop the feeling that the football team 'is the rock on which the Corps was built.' There is here a fertile psychological atmosphere for the feeling to grow that he and his mates are different from other cadets, and that the authorities want him to stay proficient in academics at all costs. Such a cadet could readily fall prey to the delusion that it was all right for him to cheat or to aid a teammate to cheat.

From such generalizations and assumptions, the Board built a carefully constructed logic path from the questions it posed and the opinions and beliefs it gathered to the conclusion that an overemphasis on winning football, which misaligned values in the performance of the Academy mission, was the underlying and proximate cause for all that happened. In rendering a report which sharply criticized the overemphasis on winning football, and pegging its recommended actions largely to those conclusions, the Bartlett Board virtually foreclosed answering other important, relevant questions about contributing causes.

The "Specific Conclusions..." of the Board were "...the following factors may have been operative in deflecting the Military Academy from its mission...": football recruiting practices, special tutoring for Academy candidates who were football recruits, physical and psychological separation of the football squad from the Corps of Cadets, pressures exerted by the football office on other agencies, academic eligibility requirements for cadets participating in intercollegiate athletic competition, special academic tutoring of football players, athletic policy with respect to football competition, and the organization of intercollegiate athletics at West Point.

The Board also concluded "...there are certain other areas, only tangentially connected with our mission, which our study has indicated might well be subject to further review...": the honor system, team work within the command, and discipline and morale within the Corps of Cadets.

In the end the Board made nine recommendations, each one entailing another study, board, committee, or review.

...A Board of Officers be appointed to study in detail those practices connected with the solicitation of appointments for the recruiting of football players with a view to establishing appropriate policy for the future... that this board study as a corollary the associated problem of preparation of football players for the entrance examinations and recommend policy in that area.

...A Board of Officers be appointed to study the entire problem of the physical and psychological segregation of the football team from the rest of the Corps of Cadets...

...The academic eligibility requirements for competition in varsity athletics be reviewed by the Academic Board or an appropriate committee thereof.

...A Board of Officers, including representatives from the academic faculty and the Department of Tactics, be appointed to study the entire system of tutoring and additional instruction of cadets, including football players, with a view to recommending a comprehensive policy in this area.

...The superintendent review the entire policy for and organization of intercollegiate athletics at West Point in light of this report.

...The Corps of Cadets be asked through its appropriate representatives to review its honor system to make sure that in all respects it is simple, clear, and unencumbered with unnecessary technicalities, and that concurrently a committee be appointed by the Commandant of Cadets with academic officer representation, to study those regulations which impinge on the Honor System.

...A Board of Officers, including at least one member of the permanent professor group not a member of the Academic Board, be appointed to study the whole area of inter-agency and inter-departmental unity in presenting Military Academy policies to the Corps of Cadets both officially and unofficially by all members of the staff and faculty.

...The Commandant of Cadets be directed to review the effect of privileges and extra-curricular activities upon the discipline and morale of the Corps of Cadets and report his findings to the superintendent.

...A committee be appointed to study in detail ways and means of detecting the existence of any incipient conspiracy against the honor system in the future.

The Bartlett Board had unwittingly set the stage for insufficient follow-through on the honor incident of 1951. The only department subject to considerable study, albeit somewhat superficial, had been the Department of Athletics. The Board attempted to perform an academic study, complete with a free exchange of ideas, opinions, and beliefs, all in a closed but collegial setting – rather than conduct an intensive, in depth, fact-finding investigation that reached across all departmental and organizational boundaries into the Academy's heart and soul – where the institution had been hard hit by the organized honor violations.

The Board spawned additional studies and reviews internal to other departments; studies and reviews further removed from and unrelated to the important and voluminous information contained in Collins Board testimony. And through their findings, conclusions, and recommendations the Bartlett Board had injected biases, unfounded beliefs, and opinions into work that should have been hard headed and factual – intended to find contributing causes and remedies, wherever they might be.

There was no criticism of the Academic Department, leaving the Department's procedures virtually untouched, and finding they "...are essentially sound and healthy except for...minor points...noted...." They specifically noted "...no evidence of weaknesses in regulations or policies governing admissions, administration of academics and the procedures for eliminating deficient cadets." This must have been puzzling and disconcerting to General Irving because the report highlighted what it believed were faulty athletic recruiting policies and practices, and questionable qualities and academic capabilities in some of the young men recruited. Ironically, the Academy's Admissions Committee was chaired by Boyd Bartlett, with other members from the Academic Board, and each year the Committee reviewed the records of every cadet candidate for the Academy, and had the opportunity to reject proposed appointments if records and entrance examination results were marginal or didn't meet standards.

The Board found no significant areas of weakness in the honor system, concluding it was "basically sound and operates effectively," going on to say "we believe that it would be desirable for it to be reviewed to make sure that in all respects it is simple, clear and unencumbered with unnecessary tie-ins with regulations...." It made no mention of honor education or training, an important element of the honor system, leaving that question to the Commandant, the Cadet Honor Committee, and the two boards or committees it recommended to review the honor system.

The Board concluded the fundamental causes for the scandal lay elsewhere, leaving the review of Academic policies and procedures and the honor system to other committees and boards composed of members of the departments responsible for those systems, thus setting the stage for a less intense internal self-examination of what went wrong and why.

If winning football were de-emphasized as the report seemed to argue, would that likely avoid a recurrence of similar honor incidents – organized cheating – in the future? The answer was "no." The Bartlett Board had presented a faulty dilemma in their solution to a deeply rooted problem of organized cheating. Their solution seemed to say, "Winning football and honorable behavior are incompatible. They cannot exist together at the United States Military Academy. Good, honest men, working together are incapable of preventing the excesses and corrupting influences that might be wrought by winning college football."

Their logic didn't hold water, and others, including General Irving, would recognize the answer lay primarily in strengthening honor training and education for all cadets, including football players, and in bringing the Corps of Cadets and its football team back together.

Given the Bartlett Board's findings and recommendations, concentrated primarily on the overemphasis on winning football, the follow-on boards and committees which looked at the large number of other "procedures and practices" hadn't the incentive to look deeper

321

for other "contributing causes" or to believe their changes in practices and procedures contributed directly to curing the underlying or proximate causes of the incident.

None of the Bartlett Board's omissions, nor the omissions of the several follow-on boards and committees, stopped or permanently slowed the rebuilding of the Academy and its football team. Though the Board wasn't entirely successful in coming to grips with the disaster that befell the Academy, it did highlight important contributing factors needing attention, and set in motion changes that helped the Academy toward refurbished institutional pride, and a brighter future.

Nor did the Board's omissions obscure or lessen the severity of the mistakes made by the eighty-three young officer candidates who, tragically, had to resign and leave West Point that summer and fall. The eighty-three, and others, knew their actions were wrong, yet they persisted in their behavior, some compounding their mistakes with conspiracies of silence and lying under oath.

Colonel Paul Harkins was right when he said in his June 2 memorandum for record, his parting written words about the honor incident as he prepared to leave his only assignment to the Military Academy,

> ...having discovered...[the organized cheating], proved it to be a fact, my duty to honest cadets, to the Academy, and to the country was to expose it, expel it, and let the chips fall where they may. I think when the air clears, the whole thing will have a very salutary effect on West Point and the country.

In his interview before the Bartlett Board in Washington, D.C. on September 4, Paul Harkins was pointed in talking of his disagreements with Earl Blaik – without ever mentioning Blaik by name. He was equally outspoken about the way he believed Blaik looked wrongly at the role of football at the Academy and the team's relationship to the Corps of Cadets and the Academy. His comments indicated he probably would have agreed with the Bartlett Board's description of the underlying and proximate causes for the scandal.

But his interview was also full of information about serious trouble in the honor system, as well as deep division between him, Earl Blaik and his football coaches – the Tactical Department and the Football Office. He believed Earl Blaik and his coaches were asking for "spe-

322

cial treatment or privileges" for the football team that were unwarranted, treatment other cadets didn't receive. He recited in detail several instances of such requests by Blaik or his assistant coaches, and told of football players who had been caught up in the cheating, who had committed what clearly seemed to be honor violations that the honor committee didn't act on as Harkins believed they should.

There obviously had been sharp disagreements, not satisfactorily resolved for either Blaik or Harkins, and they had quite literally shouted some of their disagreements at each other – over the phone. According to Harkins, Earl Blaik once threatened in a phone conversation to "report" Harkins to the superintendent, General Bryant Moore, General Irving's predecessor. They also frequently took their disagreements, separately, to General Moore, and without finger pointing, asking for decisions favoring the one who came to see him. It was a feud that could only spell trouble if not resolved – and it eventually did contribute to serious trouble.

And undoubtedly, Paul Harkins would have agreed with the ringing, emotional, "Summarizing Remarks" written in the Bartlett Board report, on the last two pages just prior to the report's recommendations.

We conclude that West Point can do no other than continue intercollegiate football with all the vigor at its command provided this is done with the same integrity which has been used in our handling of the cheating scandal. Our football teams should have the best coaching it is possible to obtain. They should have the enthusiastic support of every member of the West Point community.

From our student body of twenty-five hundred physically fit young men we can build a team whose spirit and determination, win or lose, will give further example of the courage and character of our alma mater.

West Point has so far been forthright and courageous. We believe we have the admiration of honest people. We shall not, however, retain that admiration if we indulge in policies which cast public doubt upon sincerity or upon our determination to place high principle above expediency. In the cheating scandal we have emphasized repeatedly that we do not condone a guilty individual because others may have engaged in the same practice and escaped punishment.

The Military Academy is going through one of the greatest crises in its history. At a time when intercollegiate athletics are in a period of over-emphasis and corruption, West Point has dismissed the major portion of its football team on a matter of high moral principle. We are convinced that this action has been regarded by the responsible public as a forthright example of the highest type of moral courage and institutional integrity. West Point has an opportunity to turn a potential catastrophe into an example of constructive leadership. The Military is about to enter its sesquicentennial year, with plans already publicly announced for a dignified and adequate observance. Attention will continue to be focused upon West Point throughout the next twelve months. The Military Academy can take a leading part nationally in the reevaluation of educational objectives.

Earl Blaik would not have agreed with all the Bartlett Board's summarizing remarks. His interview before the Bartlett Board was tense, sharp, painful, defensive. Before appearing in front of Boyd Bartlett, Francis Greene, and Harlow Miles on August 17, he fought hard, at first, to retain the young men who had violated the honor code, in hopes they could learn from their mistakes while remaining at West Point – that they could be salvaged and become the outstanding officers he was convinced they could be. Then, once he realized that defense was hopeless, he would fight for honorable discharges for the implicated cadets, that they "would go out with their heads held high" to begin new lives. His last defense, after the screening boards and the Bartlett Board completed their work, was the hope some might be reappointed. It was a desperate fourth down gamble that he also lost.

And suddenly, now that all the young men were leaving, he found himself fighting to avoid de-emphasis of football, fighting to remove the stain brought to Army football by the cheating scandal. "Never give up," he would tell his players, and now he had to remind himself of those words, over and over.

Deep down in Earl Blaik there must have been a glimmer of light in all the darkness. There must have been something, or someone, who

324

told him Earl Blaik must make some changes, that he had been partly responsible for what happened. But a proud man who seldom tastes defeat doesn't make such an admission easily. Recognition of his responsibility in the scandal came slowly, grudgingly, but it came – in small steps, almost imperceptible. But they were there.

Earl Blaik wasn't alone, as he would soon learn. The Academy was beginning to rebuild, led by a quiet, thoughtful superintendent who worked to find and heal the wounds. The Corps of Cadets and those who taught them – and the Army football team – would soon begin pulling together in ways and with strengths few anticipated.

CHAPTER 9

PLEBE YEAR:
HARD LESSONS AND LAUGHTER

"Taps." Bedtime for West Point's newest class, 10:15 Tuesday evening, July 5, 1950. This was the first day at the Academy for the class of 1954. "Beast Barracks" was what they called it, the kind words cadets and tradition had used to name New Cadet Training the first summer at the Academy.

As the recording of the bugle's mournful sound came over loudspeakers signifying "lights out," Bob Mischak lay exhausted on his bunk bed at the end of his first day at West Point. Recruited by Vince Lombardi and Captain Johnny Green to play football at Army, he wondered, What have I gotten myself into?

Bob was experiencing what most of us felt in every new fourth class. And he was asking the same question, probably repeated in similar fashion every year for 149 years by nearly every new cadet to enter the Academy. A year later, on Tuesday, July 3, 1951, when our class entered, the questions were the same. The answers were slow to come in a life immediately accelerated to a furious pace, a complete blur. That first Tuesday was an experience which blocked memories, leaving us with fleeting images we could recall later only with the help of photographs.

THE FIRST DAY OF "BEAST BARRACKS" –
WHAT'S IN A NAME?

The objective for our first day was deceptively simple, as it had been for Bob Mischak a year earlier:

To train new cadets sufficiently in the school of the soldier without arms so that they will be able to march in company formation to and from the Oath of Allegiance Ceremony to be held at 1700 [five in the afternoon]...

What it took to get to Trophy Point for the Oath of Allegiance that day was quite another matter. It wasn't simple for those who had no prior military experience. Undoubtedly, it wasn't simple for the first classmen, the men on the "Beast Detail." It just appeared that way, though everything the first classmen did seemed effortless.

To men fresh out of high school or a year of college, with no prior military experience, the upperclassmen looked like gods. They were exactly what I'd seen in the Academy catalog and on movie screens: impeccably dressed, immaculate in appearance, ramrod erect, "official" sounding, stern, commanding – absolutely perfect to the untrained eye. But their presence made for a long, long day.

Some of my classmates found humor in the first day, but not me. I was far too serious, didn't accept my own mistakes well, and spent more time berating my errors than learning from them. Fortunately, by year's end, my tendency to be so hard on myself for my mistakes would be one of many difficult but necessary lessons I learned. Though I saw little humor in the first day of Beast Barracks, and precious little that whole first summer, I did learn to laugh before the year was out. I had to, or I wouldn't survive.

Earl Blaik labeled the cadets' sometimes comical circumstances "situational humor," a label not originated by the Army coach. Situational humor kept us laughing, mostly at ourselves and at West Point, this place that seemed most serious. And there were men in every class who captured the humor in their story telling, writing, and art. *The Pointer* magazine, the cadet corps' publication – drafts of every issue reviewed by the Tactical Department – was a major outlet for humor needed in tightly regimented cadet lives.

* * *

On the train from Denver, Colorado, to West Point, which arrived on Monday, Dick Auer, John Martling, Wayne Smith, and I, all from Colorado, and others who had boarded the train at various stops along the way, had become acquainted. We told each other what we'd read or heard about Beast Barracks. When we pooled our knowledge, we thought we knew what we were in for. We didn't.

"Tuesday, no later than eleven o'clock in the morning, at the east sally port of central cadet area," said the reporting instructions. With several other soon-to-be classmates, I arrived at the appointed location near ten o'clock. As instructed we brought a minimum of clothing and personal belongings. Most of us carried one small overnight bag. New York's early summer heat made sports shirts and trousers right for the occasion – our "official welcome." The thought didn't enter my mind this would be the last time for nearly a year I would wear civilian clothes – "civvies." I was too unsettled to think about anything but that day.

As a small group of us entered the sally port, an arched passageway beneath the four-story west academic building, we paused briefly and gazed at the quadrangular-shaped central area ahead of us. Cadet barracks and the academic building immediately above us enclosed the area on four sides, except for openings through sally ports beneath barracks on the north and west sides of the area, and openings between cadet barracks and the academic building, into south cadet area and north onto Jefferson Road. The entrance to the first division of cadet barracks, where Douglas MacArthur lived as a cadet, was the first door left of the exit onto Jefferson Road from central area.

There were Army enlisted men, dressed in clean, crisply starched, khaki uniforms, shuttling back and forth between the sally port entrance and central area, escorting small groups of classmates from the sally port, to be met by cadets at prearranged locations in the area. A series of signs marked the locations.

Our attention was drawn to unusual, noisy activity in central area. We were unable to see that the same activity extended into south area, to our left, out of view from inside the sally port. In central area we could see first classmen, members of the senior class, in dress gray over white, the traditional gray wool tunic coat trimmed with black braid, and worn with pressed, white, heavily starched duck trousers. As in the West Point catalog's pictures, they were wearing white gloves. The first classmen's caps were of the same gray wool, trimmed in black. The sun reflected off highly polished black shoes, shining black, patent leather cap bills, and gold-plated metal Academy emblems centered just above the cap bills.

Most first classmen were giving commands to small groups of three to six new cadets, who were still in their civilian clothes. The majority of new cadets were wearing what appeared to be luggage tags tied around one belt loop on the right front of their trousers. They didn't have their overnight bags with them, as we did. They were already learning the rudiments of close order drill. "Ri-i-ght! Face!" "For-ward! March!" "Deta-il! Halt!" "Le-eft! Face!"

Other first classmen were standing at attention, usually a few paces in front of small groups of new cadets who still had their overnight bags with them, but no luggage tags on their belt loops. The first classmen were giving a series of commands. My classmates were responding, some quietly at first, then more loudly with each instance the first classman spoke. These new cadets, too, were standing at attention, but with every command given they seemed to be undergoing a change. It was plain the position of attention for the new cadet was going to be markedly different from that of the first classmen. My classmates who were undergoing this strange metamorphosis stood progressively more rigid, straight, chests puffed up and out, chins and stomachs sucked in. None were smiling.

As the verbal exchanges continued, the new cadets' responses to first classmen's commands became louder and more rapid. The shouted commands, responses, and echoes reverberated noisily throughout the cadet area. Finally, with overnight bag in hand, trying to control the growing number of butterflies in my stomach, I walked the last few steps through the sally port toward the cadet area. At the entrance, enlisted men from the 180 2d Special Regiment, which participated in military training of Academy cadets, were greeting small groups of us to escort to waiting first classmen.

"Follow me, gentlemen," one of them said, after organizing a few of us into a column of two's. We walked behind him, into the world of West Point, inside the cadet area.

An enlisted man from the 1802d Special Regiment escorts six new cadets into south area of cadet barracks to begin Beast Barracks, July 3, 1951, the day the class of 1955 entered the Military Academy. Left to right: unidentified, William W. Harris, Richard A. Fontaine, Fred G. Knieriem, C. J. Miller, Jr., and Jerry M. Gilpin. (Photo courtesy USMA Archives.)

He took us to a designated point in the area, motioned us to a stop, and with a few words, quietly indicated the waiting first classman would tell us what would be next. "Good luck, gentlemen," the soldier said, and turned to walk back to the sally port to retrieve the next group of new cadets.

The first classman wasted no time. He stood smartly at attention, to the left of our column, and in a loud voice commanded, "Le-eft! Face!" Our response was far less than military. We were ragged in turning to face him. "Misters! Drop those bags!" Again, the response was ragged, imprecise, not in unison, lacking "snap." Some of us set our overnight bags down carefully, as though there were items inside that might break if we dropped them. Some took the command more literally and dropped their bags, but not with any enthusiasm.

It didn't take long to realize we hadn't responded the way he wanted. "Pick up the bags!" was the next command. Again, a ragged response. Some casually bent down to grasp the handles of their overnight bags. The remainder moved a bit more rapidly. Then came our first glimmer of what was to be the manner of response to upperclassmen the remainder of plebe year.

"Give me your attention, Misters. You are to learn to obey orders without hesitation, without question, promptly, immediately, instantaneously, together, smartly, as a unit. Your mission in life for the next year is instantaneous and unquestioning obedience. You can't learn how to give orders until you've learned how to receive them, obey them, and know how it feels to receive and obey orders – any order, good ones and bad ones. Now let's see you move with speed."

"Pick up those bags!"

We moved only slightly faster.

"Too slow!" the first classman growled.

"Drop those bags!" – "A little better, but I said 'drop them', not set them down. Now let's try it again."

"Pick 'em up!" – "You're still too slow!"

"Drop 'em!" His tone became more insistent.

"Pick 'em up!" – "Misters, you can do better than that! Faster!"

"Drop those bags!"

"Pick 'em up!"

"Drop 'em!"

"All right, let's learn about the position of attention expected of a new cadet. The 'Brace.' Posture training – and believe me, Misters, you need it! Stand erect! Stand tall! Straighten up! Now, suck your guts in! Take a deep breath! Suck 'em in! You can exhale, but keep those ponderous guts sucked in! Pop your chests out! Get 'em out there! Look proud! Now, pull your chins in! Don't tilt your heads down, Misters. Keep your heads up! Pull your chins straight back! Let me see some wrinkles under those chins – lots of wrinkles! Now, straighten your arms and clamp your elbows in! No – no! Don't lock your elbows! Straighten your arms – clamp those wings in, Misters!..." And on it went.

This was the beginning of the first day. At the instant I heard and responded to his first commands, I wasn't aware my life was beginning to change, rapidly, and would never be the same.

Processing! That's what they called it! Commands shouted. Close order drill. "For-ward! March! Col-umn right! March! Col-umn left! March! To the rear! March! By the right flank! March! By the left flank! March! De-tail! Halt! Hand – salute! Ready – two! Pa-ra-a-de! Rest. De-tail! Atten-tion! Dress ri-ight! – Dress! Ready! – Front! For-ward! March! Get your shoulders back! Pull your chin in! Pop your chest up! Keep your eyes straight ahead, Mister! Come off gazing around! Keep your eyes up! Not on the ground! Tuck your butt under you! Keep your back straight! Keep your arms straight! Don't lock your elbows! Don't lock your knees! But keep them straight! Turn your palms in along the seams of your trousers! No, not like a toy soldier! Cupped normally, with thumb and forefinger lightly touching one another along the seam of your trousers! When you're marching, swing your arms six inches to the front, and three to the rear! The commands came in an endless stream.

We marched everywhere: to the line in front of tables where first classmen checked our names on rosters, gave us company, barracks division, and room number assignments, handed us tags to be tied to our trousers, and a laundry bag apiece, in which we were to carry items of clothing we would be issued in the cadet store that day; up to our rooms where we left our overnight bags; to the barber shop where our hair was shorn to crew cuts within what seemed like seconds – the fastest haircut I'd had in my whole life; to the cadet store where we were measured for shirts, trousers, shoes, caps, dress coats, overcoats

331

– every conceivable item of clothing; the "Blue Book" – cadet regulations, manuals and pamphlets of instruction; more close order drill. Then came the march in small groups to the cadet dining hall, Washington Hall, for our first meal as cadets, dinner.

We were still in our civilian clothes. Our first acquaintance with Washington Hall was a profound shock, just like the whole first day. Not much eating at the first meal as a cadet – or at a lot of other meals plebe year. Nevertheless, at that first meal in Washington Hall an incident occurred that would bring a lifelong memory, and my first taste of humor. I didn't think it was funny at the time, but Bob Meisenheimer did – for a brief moment. He laughed when he shouldn't have, and it cost him dearly, in humor and lost weight. And so did the rest of us at the Fourth New Cadet Company table.

There were eight new cadets and two upperclassmen at the ten-man table. Our entire class was in one wing, the west wing of Washington Hall. The table commandant, the upperclassman seated at the head of the table, spent most of the twenty-five minute meal explaining the intricate rules governing conduct of new cadets at the table, and the responsibilities and duties of the three sitting at the far end of the table.

He then let us begin eating, attempting to use the rules he had just explained. We didn't do well. Remembering all the courtesies and steps we were to go through to get a bite of food in our mouths wasn't easy. And no sooner had we begun than he ordered everyone, beginning with my classmate to his left, to introduce himself by giving his name and home town.

Bob Meisenheimer was sitting diagonally across the table from me in the next to last chair. He was the third man to give his name and home town – "New Cadet Meisenheimer. Peoria, Illinois, Sir!" he said loudly. He then resumed eating.

On around the table introductions came. I knew the name of my home town would cause a stir. It was unusual, right out of the old west, and identified a place less than a village in size. In fact, it was a rural post office stop and train depot en route to the old 1890s mining town of Creede, Colorado.

"Mister, you're next."

"Sir, New Cadet McWilliams. Wagon Wheel Gap, Colorado."

By this time Bob Meisenheimer had a mouth full of food. He burst out laughing, scattering food across the table in front of him. When he laughed, the mirth spread quickly around the table to our classmates, who tittered and giggled, some with food in their mouths. All eight of us were fighting to control the laughter. Fortunately, consistent with the rules of etiquette the table commandant had given us, I swallowed my food before introducing myself and was looking straight at the upperclassman when the commotion started. I turned my head forward and looked down at my plate. I kept my composure, but the two first classmen sitting at the head of the table didn't.

Stony, icy silence. Then the explosion.

"Knock it off, Mister Meisenheimer! You think that's funny? Sit up! Jam your chin in! Sit up, Dumbguard! What's so funny? You think you're some kind of clown? You get hold of your emotions! Do you understand? You don't laugh at your classmate's home town, or your classmate! You think you're something special because you're from Peoria, Illinois? Well, you're not! Sit up! Pop your chest up! Get your neck back. Wipe that smirk off your face! All you men, wipe the smirks off! Sit up! Keep your eyes down on your plates!"

For the balance of that first meal we sat at attention, and were severely dressed down for the impertinent show of emotion. Our first meal at West Point was over, and Bob ate little the rest of the week.

Noon meal, July 3, 1951, the first meal in Beast Barracks and Washington Hall for the class of 1955. (Photo courtesy USMA Archives.)

Every meal the first classmen were all over him. He spent most of the week sitting at attention, reciting plebe knowledge.

For the rest of the week everyone at the table paid for failing to control their emotions. As for Bob Meisenheimer, he never forgot that first meal in Beast Barracks, and neither did I. He insisted with good humor that Wagon Wheel Gap, Colorado, had been responsible for a miserable Beast Barracks diet for him and an inordinate loss of weight, at a time when weight loss would come easily enough as a result of the daily routine.

What we didn't know was the first classmen were probably fighting to control their own emotions – struggling to keep themselves from laughing with everyone else at the table.

New cadets in south area of cadet barracks, learning the rudiments of close order drill from the class of 1952, preparing for the Oath of Allegiance ceremony later the first day of Beast Barracks, July 3, 1951. Note the clock – it's just after the first meal at West Point for the class of 1955. Also note the supervising Tactical Officer, standing on the stoops at the head of the stairs, beyond and to the left of the clock, observing conduct and progress of new cadet processing. Directly behind the officer is the entrance to K Company, 1st Regiment, cadet barracks division, with the K Company orderly room windows to the left of the division entrance. (K Company headquarters during the academic year.) (Photo courtesy USMA Archives.)

Along with all we were trying to digest that first, hectic day – except food – was learning how to respond verbally to directions or questions from the Beast Detail. After the noon meal, close order drill continued, along with many more instructions from the first classmen, while they prepared us for the Oath of Allegiance ceremony.

"What's your name, Mister?" I had been asked that morning before the dinner meal.

"Bill McWilliams," I said, in a somewhat strained, but friendly response. The upperclassman's response was swift and sharp.

"Get that tone of familiarity out of your voice, Mister! I'm not your friend! Your name is 'New Cadet McWilliams, Sir!' Do you understand?"

"I understand."

"Yes, Sir! is the proper response."

"Yes, Sir."

"Louder!"

"Yes, Sir," I answered more forcefully.

"Louder! I can't hear you, Mister!"

"Yes, Sir!" I shouted. I was becoming tense and frustrated.

"That's better. Now, what's your name?"

"Cadet McWilliams."

"What did I say your name was?"

"New Cadet McWilliams!"

"Wrong! Say your name again."

"New Cadet McWilliams!"

"Wrong! How do you address me?"

"New Cadet McWilliams, Sir!"

"Louder! I still can't hear you!"

"New Cadet McWilliams, Sir!" I shouted. By this time sweat beads were forming on my forehead.

"Mister McWilliams, 'Yes, Sir. No, Sir. No excuse, Sir,' are your only answers to questions asked by upperclassmen. You may be asked to explain facts and circumstances when you fail to perform your duties properly. But facts and circumstances are never excuses for failure. There is no excuse. Do you understand?"

"Yes, Sir!"

"The only exceptions are when you are asked to recite plebe knowledge, or provide other information, such as responding to oral quizzes

335

about your training. Plebe knowledge is in *Bugle Notes*, issued to you and your classmates today. Start studying it – tonight! 'Bone up' on – memorize – everything in *Bugle Notes*, beginning with your chain of command."

During the day we received our first issue of cadet uniforms, instructions on what uniform to wear at the Oath of Allegiance ceremony, as well as how the uniform was to be worn. Included was a collar stay inserted underneath the black tie, to keep the shirt collar from having a curled, disheveled appearance when we wore ties, which were

Beast Barracks processing lines, July 3, 1951. New cadets in the class of 1955 wait at the next station in processing, on the steps – and stoops – in the older, central area of cadet barracks. They are, left to right: David L. Pemberton, Louis T. Tebodo, Dale D. Patterson, Donald M. Buchwald, Chester H. Pond, Marshall W. Dickson. (Photo courtesy USMA Archives.)

to always be tied neatly and tucked into our shirts between the third and fourth buttons. We also received written instructions on how every item in our rooms would be displayed, what was authorized in rooms, and what wasn't. In between all the processing activities, close order drill continued.

With the issue of cadet clothing we began the change into uniformed cadets, complete with white T-shirts, white boxer shorts, black socks held up with black, elastic garters which fastened below the knees around the upper calves of our legs. With the change into uniforms, off came our civilian clothes, to be folded and placed in storage, along with our overnight bags and nearly all their contents. On went the gray gym trousers, with wide, black stripes down the outside of the trouser legs, broad white gym belts, and black shoes, in which we continued the day's processing and preparation for the Oath of Allegiance.

And with the little available time to ourselves, as we stormed through the first day of Beast Barracks, we began organizing our rooms and, finally, briefly meeting our roommates. Harvey Garn, from Sugar

Metamorphosis from civilian to Beast. At left, New Cadet Thomas C. West is measured by first classmen for uniform items the first day of Beast Barracks. Later in the day, new cadets began the change into uniformed cadets: John E. Rudzki, Irvin G. Katenbrink and Richard D. McCarthy. July 3, 1951. (Photo courtesy USMA Archives.)

City, Idaho, and Al Bundren, from Knoxville, Tennessee, and I met one another for the first time that afternoon and began settling in for a tough two months.

A few minutes before 5:00 in the afternoon on July 3, the class of 1955 formed up in central area, as a provisional regiment of six companies, to march to Trophy Point for the Oath of Allegiance. To the gray, wool gym trousers with the white gym belt was added charcoal gray, tropical worsted wool shirts. We wore black ties, black socks, and black shoes – not uniforms designed for hot summers in New York. The "Beast Detail" wore dress gray over white, with white gloves.

When Fourth Company assembled for the first time, in front of our barracks, we heard our cadet company commander, Mr. Shea – Richard T. Shea, Jr., of Portsmouth, Virginia – giving commands. From the outset, his voice, his conduct, the manner in which he carried himself, and the way he related to others gave him a remarkable presence – especially to awe-struck new cadets who had no prior military experience. We heard much more of him before plebe year was over.

The march onto Jefferson Road, thence left onto Thayer Road, and past the Plain to Trophy Point, overlooking the Hudson River, was amazing. In one short, rushed day – which was far from over – we actually learned how to march. At least some of us thought we had. There were first classmen telling us we had a long way to go. From first classmen in the rear rank of the companies came constant corrections as the brief march to Trophy Point progressed. Nevertheless, there were feelings of

July 3, 1951, on Trophy Point. Colonel John K. Waters, Commandant of Cadets, briefly addresses new cadets in the class of 1955 at the Oath of Allegiance ceremony, late afternoon the first day of Beast Barracks. In the front ranks, in dress gray over white, is the First Beast Detail, members of the class of 1952. (Photo courtesy USMA Archives.)

pride among us as the Hellcats, twelve musicians playing bugles, fifes, and drums, provided the martial music for the ceremony. This was the first time we marched in company formation outside the cadet area.

The six companies, led by the cadet provisional regimental commander – "The King of the Beasts" – and his staff, assembled in regimental mass on Trophy Point. The front two ranks of the regiment were first classmen. Colonel Waters, the new Commandant, gave a brief talk, and administered the Oath of Allegiance. A retreat ceremony followed, with lowering of the national colors. We then marched back to the cadet area to prepare for supper formation.

The march to our first supper formation was also a harbinger of things to come. The Hellcats played martial music which ensured we continued to learn the rudiments of "dismounted drill," marching in formation without a weapon. The brief march in a column of six companies, from central area, out between the west academic building and the first division of cadet barracks, and left onto Jefferson Road, thence to Washington Hall's front steps, was an eternity. First classmen in the Company's rear rank continued what we had first experienced during the march to Trophy Point, an unending stream of corrections fired at new cadets who detracted from the desired appearance of absolute precision and perfect company unity.

The booming voice of big, athletic-looking Dick Inman, who was a squad leader and drill instructor, "Come off bouncing, Mr. Smith! You're out of phase! Dig your heels in!"

Don Lasher, a rifle and physical education instructor on the Beast Detail, "Get that arm swing right, Misters! Six inches to the front, three to the rear! Cover down! Where are you gazing besides at the back of the head of the man in front of you? Can't you see you're not covered down in this file?"

And before the company leaves central area to turn toward the dining hall, from just beyond the right flank comes the loud voice of the first sergeant. Checking successive ranks for alignment and spacing, he commands, "You man, fifth man in this rank! Check to your right and left! Can't you see you are a half step ahead of the entire rank? Pay attention! Where did you leave your mind and eyes – in your room? Wake up, Mister! You man, in the left flank file! You're falling out! Brace up! Ram that chin in! Slap it in! Get your neck back! Suck your gut up! Pop up your chest!"

Every minute, every step of the way to the dining hall, in excruciating detail, first classmen pounced on every faulty movement, every show of emotion, every inappropriate action, every incorrect statement or response to a question, anything that didn't meet standards.

The supper meal was no better than the dinner meal that first day. Cadets were assigned to numbered tables, one of ten or eleven tables allotted to each company, in a designated area of the east wing of Washington Hall. We double-timed up the front steps of the great Hall and to our assigned seats. There was much milling about as we searched for our table areas and numbers, which were on cards mounted on small stands on the tables. First classmen were everywhere, giving directions to the men in their companies, keeping the traffic of bobbing heads and slow, almost jogging, double-timing feet, moving in an orderly flow. When we finally found our tables we stood behind our chairs, in a brace, until the cadet regimental adjutant ordered, "Take seats."

We had begun to learn at dinner that meals were rituals, organized in infinite detail – as was everything else at West Point. They were unforgettable mixtures of imposed discipline, self-discipline, courtesy and table etiquette training, and to hard pushed, hard working new cadets, slow, deliberate starvation. The entire meal period was twenty-five minutes. New cadets sat stiffly at attention on the front half of their chairs, eyes down and focused on their plates.

We ate each meal slowly, in a carefully prescribed pattern: cut off a piece of food no larger than a sugar cube and convey it to the mouth with a fork lifted from the plate in a curved motion, enabling us to keep our eyes on the fork. Before chewing, as slowly and deliberately as cattle, we returned our forks to the plate by the same route and placed our hands in our laps. After chewing and swallowing the small bite of food, we then repeated the process.

One new cadet at each table served as "gunner." He sat opposite the table commandant, at the far end of the table. He was responsible for waiters keeping the table supplied with food, dishes, napkins, and silverware. Another cadet, designated the "coffee corporal," was responsible for remembering hot beverage preferences of upperclassmen, and keeping their cups full. He sat in the seat to the immediate right or left of the "gunner." Across the table, facing the "coffee corporal" was his cold beverage counterpart, the "water corporal." This carefully thought-out system enabled one waiter, an enlisted man, to serve four tables – forty men – with one four-tiered, wheeled cart, on which he brought everything needed from the huge kitchen. The entire Corps of Cadets – twenty-five hundred men – could be fed at one sitting in twenty-five minutes, a measure of the size, organization, and efficiency of the kitchen.

While we ate or performed duties, the two or three first classmen at each table continued the litany of instructions begun at noon, and for the first time, asked for recitations of plebe knowledge. The responding new cadet immediately "halted eating," "sat up" (braced), and "sounded off" his reply. A halt in eating meant promptly chew and swallow the meager bite, or if the fork was en route with food, return it to the plate, and hands in the lap.

"Mr. Bundren, how is the cow?"

Silence, as he chews more rapidly, swallows, and sits in a brace, eyes looking down at his plate, unable to answer the question.

"Look at me when I'm talking to you, Mr. Bundren."

"You know your plebe knowledge?"

"No, Sir!"

"Why not?"

"Sir, I didn't have time..."

At the end of a day when we had been shocked into numbness, the first classman knew his question couldn't be answered satisfactorily.

"That's not the right response, Mister. Do you understand?"

"Yes, Sir!"

"What's the correct answer?"

"No excuse, Sir."

"All you men, give me your attention. You better start boning up on your plebe knowledge – fast. Mr. Bundren, you drive around to my room on 'calls' tomorrow, before supper formation."

Before the week was out all of us had our noses in *Bugle Notes*, memorizing what we were instructed to memorize, learning to think, recall, and respond under pressure. What's more, we had had the pleasure of our first "call" in our squad leader's or assistant squad leader's room.

Calls were an additional means of disciplining and training new cadets and plebes who failed to meet standards throughout the remainder of the year; a means of enforcing discipline, instructing, and training; and for those few upperclassmen who had other reasons for their behavior or other ideas about leadership, intimidating, agitating, getting even, and harassing – without issuing delinquency reports and awarding demerits. Calls were ten to twenty minutes before assembly for supper formation. Precisely at the appointed time, the new cadet was expected to knock on the upperclassman's closed door and wait outside until invited in. "Drive in, Mister!"

We could count on our uniforms and personal appearance being inspected in minute detail. Then came the grilling on plebe knowledge. We were to be veritable fountains of plebe knowledge, recited without error – or receive the pleasure of additional calls.

Before the week ended, there weren't many in '55 who didn't know the answer to "How is the cow?"

"Sir, she walks, she talks, she's full of chalk, the lacteal fluid extracted from the female of the bovine species is highly prolific to the nth degree."

"Mr. Meisenheimer, how many lights in Cullum Hall?"

"Three hundred and forty lights, Sir."

"How many gallons in Lusk Reservoir?"

"Ninety point two million gallons, Sir, when the water is flowing over the spillway."

Another, far more significant item of plebe knowledge was asked of us later in the week, a quotation packed with meaning for the lives of American soldiers and the officers who must lead them.

"Mr. Garn, give me the definition of discipline."

"Sir, Schofield's definition of discipline:

'The discipline which makes the soldiers of a free country reliable in battle is not to be gained by harsh or tyrannical treatment. On the contrary, such treatment is far more likely to destroy than to make an army. It is possible to impart instruction and to give commands in such manner and such a tone of voice to inspire in the soldier no feeling but an intense desire to obey, while the opposite manner and tone of voice cannot fail to excite strong resentment and a desire to disobey. The one mode or the other of dealing with subordinates springs from a corresponding spirit in the breast of the commander. He who feels the respect which is due to others cannot fail to inspire in them regard for himself, while he who feels, and hence manifests, disrespect toward others, especially his inferiors, cannot fail to inspire hatred against himself.'"

Major General John M. Schofield, a graduate in the class of 1853, who had received the Medal of Honor during the Civil War, gave this classic definition of discipline in an address to the Corps of Cadets on August 11, 1879, while he was serving as the Academy's nineteenth superintendent. The words were powerful, said much about both discipline and leadership, and were not to be forgotten.

We also learned at the meal table that a failure to correctly respond usually resulted in prolonged periods of "sitting up" by the offending new cadet. First classmen at the table verbally jumped on the hungry offender with corrections and additional commands for recitations of plebe knowledge, which also meant less food.

Like everything else for me, meals were to become a dreaded ordeal.

"Call to Quarters," study period, began at 7:15 in the evening, after we returned from supper. We had to be ready for the next day. Put away all the clothing, equipment, and books we were issued. Or-

342

ganize our rooms, consistent with instructions and handouts we received. Practice and memorize the movements of close order drill. Study handout material, plebe knowledge, the "Blue Book" – the cadet book of regulations. Learn what our duties and responsibilities would be in keeping our rooms and ourselves in "inspection order." Fold our new, military issue underwear, socks, handkerchiefs, pajamas, and other uniform items, to exact widths for display in prescribed locations in lockers. Hang up in prescribed order all the uniform items to be displayed in our lockers. Look at the next day's schedule to prepare for training, where we had to be, what time, and wearing what uniform.

Just learning how to read schedules and instructions took time. There seemed a mountain of items to learn and little time to study. As we worked that first night, we began to get acquainted and talked quietly of the day's events and tomorrow's schedule. Getting to know one another. Talk of home and family.

Before Taps sounded at ten thirty that night we had one more formation to attend – shower formation – our introduction to Beast Barracks' military and personal hygiene.

The day had been long and hot. We were tired, wrung out. The constant push, push, push had taken its toll. Several times during the day I realized my arm and leg muscles were so rigid and my mind so intent, my legs quivered. At times I shook so hard my knees knocked, an embarrassing secret I hoped no one noticed. I sweated profusely, partly from the summer heat, but mostly because of swarming, growling, shouting, commanding first classmen and the day's intense, hectic activities. A shower would feel good.

But we received more of the same. Even showers were planned in excruciating detail and supervised under the watchful eyes of first classmen.

We assembled by squads in the hallways outside our rooms, up against the wall, with our squad leader and assistant squad leader in charge. We wore cadet issue bathrobes and slippers. First classmen wore white, short-sleeved shirts with black ties tucked in as we had worn our ties during the Oath of Allegiance ceremony, and the white duck trousers they'd been wearing all day, with brass-buckled black belts, black socks, and black shoes.

We carried white bath towels neatly folded in half and draped over our left forearms, which were extended horizontally, straight forward from our bodies. In the left hand each held a closed, plastic soap dish containing a bar of soap – also issued.

The first classmen led us down the three flights of stairs from our third floor hallway, to the basement of cadet barracks – "the sinks" cadets called them. There we stood in a brace – the new cadet's exaggerated position of attention – as we had done all day, and were, from that day forward, required to do from the moment we stepped out of our cadet rooms into the upperclassmen's world.

We stood in line, eyes fixed on the back of the head of the classmate in front of us, while the first classmen gave us instructions for showering.

"All right, Misters, you each have two minutes in the shower. As your turn comes, step out of your slippers and set them in order next to the wall. Hang your bathrobes and towels on the hooks you see outside the shower door. Time begins when you are under the shower. Be careful walking in and out on the wet floor. Soap yourselves thoroughly in the shower, and rinse thoroughly. Dry off out here, put your bathrobes and slippers on, and post back to your rooms. Remember, you have two minutes! We'll keep time. All right, first man – Go!"

"Move out! Go! Go! Go! Go!" – and the countdown began. "You have one minute – thirty seconds – ten seconds – five seconds – time's up! Next man! Go! Go! Go!" So began shower formations in Beast Barracks 1951.

When Taps finally played over the loud speakers, and we turned out the lights, there was little conversation in the room. We were too tired to talk. Every muscle in my body was sore, reminding me how I felt after the first day of each season's high school football practice. It had been a wild, day, and we knew tomorrow, July 4th, would be more of the same. The nation's Independence Day might be celebrated, but it certainly wouldn't be a day of independence or celebration for us. I didn't stop to ask myself what this was all about. Like hundreds and thousands who had gone before, including Bob Mischak the year before, the first day of Beast Barracks left me exhausted. I simply wondered what I'd gotten myself into.

As tired as I was, I remained tense, and tossed and turned most of the night, sleeping fitfully. I was wide awake at 5:50 the next morning when, for the first time, we heard the loud boom of the reveille gun on Trophy Point, followed immediately by the startling sound of the class bell clanging in the cadet area. And as if that weren't enough to wake us up, the sounds of the Hellcats came fast behind the class bell, greeting us with ten minutes of their version of morning music, with bugles, fifes and drums, which ended with the playing of "Reveille."

Precisely five minutes after the first clang of the class bell, the bell rang a second time, and while the Hellcats were serenading us, the first of my classmates selected to perform the duties of minute caller, standing in the second floor front hallway, began calling – shouting – the minutes. This duty was to ensure punctuality, if there were any doggedly, sound-sleeping cadets who hadn't heard all the other ruckus.

There was a minute caller in each "division" of cadet barracks, nineteen Beast Barracks divisions in all, a number which assured a considerable volume of noise. Each division served as an excellent loud speaker for shouting minute callers. A division was four floors with connecting stairways, each floor with four cadet rooms. Latrines were in the basement in the central area of barracks where we were. Covered stoops – similar to porches – extending around the entire perimeter of cadet barracks, were outside the first floor entrance to the division. There were no connecting hallways between divisions, except through the basement sinks, where there were locked storerooms as well as showers and latrines. To go into the next division, a cadet normally went down stairs, out the division door, on to the stoops, to the division he intended to enter.

The minute caller's responsibility was to crisply, loudly, in a commanding voice, count down the minutes from "first call" to "assembly." First call was five minutes after the class bell sounded the start of reveille formation. The end of the last bugle note, which immediately followed a third ringing of the class bell, was assembly.

All cadets were to be in their designated places in company formation at assembly. New cadets were to be in formation five minutes before assembly – at first call.

"Sir, there are five minutes until assembly for reveille formation! The uniform is 'Drill B'! Five minutes, Sir!"

The minute caller had hit a brace the instant he double-timed out of his room to his duty post. His position of attention was subject to the scrutiny of drowsy, but eagle-eyed first classmen en route from their rooms to inspect fourth classmen during the five-minute interval before assembly. When the minute caller believed the time was approaching to call the next minute he snapped his left forearm smartly to the horizontal, at eye level, to observe the second hand of his watch.

Exactly one minute after the five minute call, "Sir, there are four minutes until assembly for reveille formation! The uniform is 'Drill B'! Four minutes, Sir!" And so the countdown continued.

Because new cadets were to be in formation five minutes before assembly, we had either to be up and dressed before reveille, which was against regulations, or so well organized that we could leap out of bed at the boom of the reveille canon or the clang of the class bell, get dressed, and still make it to our assigned places in five minutes. We double-timed from our rooms into formation, where members of the Beast Detail were waiting, now wide awake, ready to ensure we were on time and to inspect our personal appearance.

Drill B meant fatigues, with helmet liners and combat boots, merely one of what seemed an endless array of uniforms. Whatever the prescribed uniform, the standards for appearance were always high. We had to be impeccable in appearance. In this case, fatigues were to be freshly laundered, starched, and pressed; shirts properly bloused and tucked into trousers to give a trim, tapered appearance; shirt line and trouser fly perfectly aligned; belt buckle polished to a brilliant shine without any scratches, and aligned with the shirt line and trouser fly. A clean white T-shirt evident on the upper chest, at the V neck of the fatigue shirt. Spit-shined combat boots with fatigue trousers properly bloused and tucked into boots. The helmet liner must fit properly, squarely, and couldn't be casually placed on the head. No jaunty tilt. No lint on the uniform.

Woe to the new cadet who didn't meet standards, or was in such a "ratty" -rat race – he failed to remember all the checkoff points in his personal appearance. First classmen seldom missed a new cadet's mistake, and we quickly learned to check each other over before we left our rooms to "fall in" at formations. If the new cadet answered a question facetiously or lacked a serious tone in his voice when a first classman corrected him, a pointed dressing down occurred, with a clear reminder of the unwarranted show of emotion. If the errant new cadet was blatantly rebellious in his response, exhibited anger or surliness, it wasn't unusual to see several highly agitated first classmen gather around him, and verbally gnaw on him from all sides, while intensifying the inspection, as if examining him with a microscope. Calls or delinquency reports – demerit slips – could be meted out along with tough, sometimes embarrassing verbal corrections.

By the last note of assembly the squad leaders checked their squads to see if anyone was absent and began their reports, with a hand salute to their platoon sergeants.

"First Squad, all present and accounted for, Sir."

"Second Squad, all present and accounted for, Sir."

"Third Squad, New Cadet Johnson, emergency sick call, Sir."

Each of three platoon sergeants then reported their roll check to the company first sergeant, who relayed the report to the cadet company commander. "All present or accounted for, Sir."

When Mr. Shea dismissed the company from reveille formation, the day's rat race began in earnest. Double-time back to the room to shave and brush our teeth if we hadn't completed these daily requirements before reveille.

There were three or four men to a room and one sink in each room. Efficient organization and teamwork in all tasks were essential to allow one man at a time at the sink. Make beds, straighten rooms, dust, sweep, empty waste baskets, clean mirrors, open room windows to the proper height, and ensure every item in lockers and hanging on alcove hooks was properly displayed – to be ready for room inspection. Check the uniforms we were wearing to again stand the rigorous standards of inspection at breakfast formation. To ensure everything got done on time during the morning rat race, new cadets would often get up well before reveille to quietly get a head start on all that had to be done.

At 6:25, assembly for breakfast formation sounded, and the six companies, led by the provisional regimental staff of first classmen, marched to Washington Hall for another busy meal with little time to eat.

Welcome to West Point, new cadets!

347

LEARNING TO BE SOLDIERS AND CADETS – FUTURE OFFICERS AND GENTLEMEN

Summer military training for cadets was the responsibility of the Tactical Department. The Commandant of Cadets, Colonel Waters, was the Department's boss when our Beast Barracks began.

Beast Barracks training, like every part of the Academy's education, was planned in infinite detail, and in its planning and execution, factored the war into our training. On Monday, July 2, the day before our class entered Beast Barracks, the *New York Times* listed the names of 496 more battle casualties in Korea. Released by the Department of Defense a day earlier, the list included sixty killed, 387 wounded, twenty-eight missing in action, and twenty-one injured in battle-zone accidents.

Training Memorandum Number 24 had been published in late April, shortly after the undercover investigation into the cheating ring began. Signed by the Commandant's personnel and training officer, the S-1, the memorandum provided the needed sense of mission and urgency to our training, described methods of instruction, and encouraged officers and cadets responsible for the training:

> Every opportunity will be taken to utilize the lessons learned and the experiences gained in World War II and Korea in presenting instruction to cadets. Wherever possible, combat lessons will be utilized to increase cadet attention and assure the assimilation of the subject material. Instructors should keep abreast of the latest trends and developments in tactical doctrine and equipment to ensure cadets receive up-to-date instruction.

Another, more detailed memorandum, published the same day, described mission statements and objectives for training all four classes. The intense first two months for the fourth class were to train the class of '55 as though we were recruits in the Army. They were also to prepare us to be cadets, accepted officially as plebes in the Corps of Cadets. The acceptance ceremony, our first parade as "official" members of the Corps, was at the end of Beast Barracks during reorganization week, when we moved into our permanent companies to begin the academic year.

Officers and cadets in charge of the various blocks of Beast Barracks instruction knew of their responsibilities well in advance. Officers assigned to the Commandant's first regiment staff, and first classmen from both cadet regiments, were responsible for Beast Barracks training. Second regiment officers and first classmen from the two regiments were responsible for training given the new third class, '54, at Camp Buckner, when the class returned from leave. Materials such as lesson plans and training aids, used in similar blocks of instruction in previous years, and filed in vaults, storerooms, and film libraries, were available as references – along with recommendations for improvements from the previous summer, all intended to simplify and speed necessary changes in lesson contents.

From June 25 through July 2, except on Sunday, July 1, officer and cadet instructors went through instructor training, working up completed lesson plans, discussing and conducting dry runs, critiquing, and doing final, full dress rehearsals of their presentations – to include rehearsals for our first days of processing and orientation.

As a result, Beast Barracks ran like clockwork on a schedule which squeezed training into every minute of every weekday and most of Saturday, leaving us little time to think or talk. When we weren't receiving lectures or demonstrations, practicing close order drill, doing mass calisthenics, rifle calisthenics, and bayonet drill, we were marching from one location to another – academic building, gymnasium, athletic field, Plain, field house, or other locations of instruction. We were studying, practicing on our own when time permitted, and performing many additional fourth class duties and responsibilities that would be ours in the coming academic year.

There were 317 scheduled hours in the period July 3 through August 27. Subjects included: Articles of War and the new, 1951 Uniform Code of Military Justice [Military Law] (two hours); Automotive Vehicles, which included drivers testing on one-quarter, one-half, and two-and-one-half-ton trucks (five hours); Defense Against Chemical Attack (one hour); Dismounted Drill – close order drill in squad, platoon, and company formations – (nineteen); Equipment, Clothing and Tent Pitching (six); Field Sanitation and Personal Hygiene (six); First Aid (five); Interior Guard Duty (six); Marches and Bivouacs (fifty-seven); Military Courtesy and Discipline (six); Academy Orientation (eighteen), which included the Honor System (four hours), orders, regulations, history and traditions of the Academy, history of the Army, organization of the Armed Forces, absence cards, cadet budget, awards and decorations; Basic Field and Combat Training (twenty-five); Weapons (ten), on the M-1 rifle, including nomenclature, care and cleaning, mechanical training, and one thousand inch familiarization firing of five rounds per man; Physical Education (thirty-five); Administration (one hundred fifteen), to include preparation for inspections of rooms and in ranks, reviews, parades, and processing – including visits to the cadet store for fitting, altering, and issuing uniforms; and "free time."

Then there were all the hours of study and preparation which occurred during evening "call to quarters," 7:15 until Taps, and on into the nights by unauthorized hallway lights after Taps.

Every day in Beast Barracks we were to listen, study, and respond to questions. Ask questions if we didn't understand. Perform tasks. Meet standards. Learn to take orders. Be expert followers.

Sunday, the day of rest, wasn't a day of rest. "There are no atheists in the foxhole," said the men of wars past. Religious training was a necessary part of a future military leader's education and had become a fixed part of cadet life when Sylvanus Thayer was superintendent. According to the cadet Blue Book, chapel attendance was mandatory: "Attendance at chapel is part of a cadet's training; no cadet will be exempted. Each cadet will receive religious training in one of the three principal faiths: Catholic, Protestant, Jewish."

When we processed into the Academy the first week, we reaffirmed religious preferences stated in applications before entering. Each Sunday we dutifully "fell in" for chapel formation and marched to places of worship. When not at chapel or Sunday meals, we prepared for next week's training, and memorized plebe knowledge.

A few fortunate classmates were able to visit with parents who lived on post or within easy traveling distance. Their time together was on Saturday afternoons or Sundays, spent in Grant Hall or other specifically authorized areas near cadet barracks. Grant Hall was the reception hall for visitors.

Communications with first classmen were carefully circumscribed by military customs and courtesies, and Academy history and traditions. There was a "seniority distance" between new cadets and first classmen akin to the "professional distance" between senior officers and junior officers in the armed forces, or between officers and enlisted men. But the formality existing between new cadets and first classmen at West Point was far more precisely defined, exacting, and remained so with nearly all upperclassmen when we were accepted into the Corps as plebes during reorganization week.

Unless providing plebe knowledge or other requested information, our responses to upperclass inquiries were the usual "Yes, Sir," "No, Sir," or "No excuse, Sir." We were to sound off our responses clearly and forcefully, in an official tone, with not the slightest hint of emotion. That is, unless emotion was ordered as part of the response.

While communications with officers and first classmen were limited, observation – seeing and listening – could never be limited, nor could thoughts, conversations, and the inevitable judgments rendered among roommates in the privacy of their rooms. The principles, ideals, and practices taught us made good sense. They also brought into sharper focus the varying professional capabilities among officers and cadets instructing us, differences between what was taught and the actions of the instructors, words versus deeds, inconsistent behaviors. And it wasn't long before we began to attach nicknames, some humorous, others not so kind, to the men who didn't appear to live up to what was taught, or possessed noticeable, out-of-place idiosyncrasies. A first classman who seemed to have a particularly nasty disposition when talking to or disciplining new cadets might quickly be singled out as a "Hard Ass" or "Hell on Wheels."

Beast Barracks was indeed the great leveler, the harsh, yet democratic training environment in which, no matter our name, birth, background, family history, or accomplishments, we were, in theory, systematically stripped of airs and rebuilt. "Put ons," selfish pride, vanity, boastfulness, arrogance, uncontrolled emotions, intemperate, undisciplined and immoral behaviors were not accepted. Teamwork, personal responsibility and responsibility to our classmates and the organizations to which we belonged, hard work, responsiveness, absolute integrity, consideration for others, courtesy, knowledge, civility, tenacity, determination, moral behavior, and self-discipline were expected, indeed demanded. Beast Barracks was, among other things, intended to wipe away less desirable traits of character and begin our rebuilding, to prepare us to be officers; unselfish, moral, honorable leaders; men of character, and gentle-men, willing to sacrifice in the extreme, if need be.

And Beast Barracks was preparation for the four years to follow before graduation and commissioning. Earl Blaik's confidant, Douglas MacArthur, understood the leveling effect and ideal intended by life and education at the Academy.

> ...the democracy of the Corps assure[s] every individual cadet a standing won by his character and personality irrespective of his social or financial position outside the walls of the institution. Every member of the student body throughout his four-year course wears the same clothes, eats the same food, passes through the same course of study, rises and retires at the same hours, receives the same pay, and starts always without handicap in the same competition.

But young men coming from all walks of life in the late 1940s and early 1950s didn't see the Academy's education and training through the background and experiences of a Douglas MacArthur. High ideals, the great intangibles and virtues, are always a struggle to attain, their practical meanings often difficult to understand and apply in day-to-day living. Young men confronted and shocked by Beast Barracks didn't all respond positively or well. We were, above all else, human, and inevitably fallible. And what was happening at West Point in July, behind closed doors, while we were sweating through Beast Barracks, was living proof we were all capable of serious mistakes – and equally capable of rationalizing and defending our errors.

The Academy's system of education and training, had evolved and matured over a one hundred forty-nine year period. It thrived on young Americans' competitive instincts. Competition, deeply ingrained in the American psyche, was what brought out the best in men. The entire Academy system had been thoughtfully, carefully developed with the best of intentions, and, while intensely competitive, demanded high ideals and standards. Yet, while men and women may compete vigorously, and forever strive for perfection in reaching for the great intangibles, the virtues of good character, there are none who will be perfect. None. And in 1951, neither was the Military Academy perfect, this product of peoples' most excellent, well-intentioned stewardship.

"BUT SIR, I CAN'T SWIM!"

In early July 1951, our second week at West Point, Fourth Company, the company I was assigned to, marched to the gymnasium to take a swim test. As we neared the gym I remembered the words I read while trying for an appointment to West Point. "Cadets must be able to swim 175 yards in five minutes to graduate from the Academy."

I had chosen to ignore the graduation standard for swimming. I had a fear of water. I knew the cause but avoided facing the problem. I had never taken swimming lessons, and I couldn't even dog-paddle. In water I was a rock.

I also remembered, "Every man an athlete," meaning every cadet would compete in intramural or intercollegiate athletics each season during the academic year at West Point. I was well prepared to be an athlete, having come from a family in which athletics had been important to our livelihood. My parents were teachers. Mom taught elementary school. Dad was an athletic coach, and I played football, basketball, and ran track on his teams during three of my four years in high school. I pitched softball in town leagues and played American Legion baseball. I had learned to play tennis, volleyball, and badminton. I could ride horses well, pitch a winning game of horse shoes, and ice skate. No doubt about it. I was an athlete.

I was also an excellent student in school and had been salutatorian in my small town high school class of forty-nine students. Obviously I was pretty smart. Right? Book smart, maybe. Youthfully cocky and overconfident, yes. Experienced and smart? No.

I suppose I believed all the marvelous athletic ability, and my brains, would enable me to pass the 175-yard swim test. Once inside the gym in cadet issue swimming trunks, I realized I might be in trouble.

I had never seen an indoor collegiate swimming pool. This one was long, wide, and at the near end, unnervingly deep. The air was heavy with humidity and the pungent odor of chlorine. My eyes watered. Butterflies fluttered once more in my stomach.

We were in eight lines, twelve men each. The lines were evenly spaced along the deep end of the pool. I was in the line nearest the north edge, which was to my right. A whistle sent one man from each line into the water. Another whistle sounded to end the five minute test period. One first classman supervised each line, counted laps, and acted as a safety observer.

While a classmate just ahead of me was successfully passing the test, the first classman in charge of our line began quizzing me.

"Mister, you know how to swim?"

"No, Sir."

"You don't know how to swim?" He looked at me skeptically.

"No, Sir."

"You must be kidding."

"No, Sir."

Still unsure, he turned and walked away. He paused, looked back at me, and shook his head.

The whistle ended the test period for the classmate just ahead of me. He had been out of the pool for what seemed like a couple of minutes, and had already toweled off. He had easily passed the test.

"Aw, c'mon, Mister, give it a try."

"But Sir, I can't swim."

"What have you got to lose? Give it a try. You can save yourself a lot of trouble if you give it a try, and pass."

"Yes, Sir," I said with a tone of resignation. After almost a week of training, learning to be an obedient follower, I decided it wasn't a good idea to disagree with a first classman. Life could get unpleasant if I chose to argue.

I turned toward the pool and waited for the whistle. When it sounded, I dove in with a noisy, belly-burning splash. I began to thrash wildly, with frantic, ineffective motions that were my best imitations of swimming. I promptly went under, but had the presence of mind to take a big gulp of air before sinking.

My eyes were open. In spite of vigorous flailing, I could see the surface receding above me. I was entering the strange world of watery noises, with distant, muffled sounds of splashes and bubbles generated by my own uncoordinated efforts to swim and the movement of my classmates swimming rapidly away from me. In what seemed an eternity I felt my feet hit bottom. Concrete beneath my feet caused rationality to overcome the fear gripping me. I settled to a squat then sprang off the bottom, turning toward the side of the pool which I could see through the hazy blue green waters. I struggled back to the surface, coughing and spitting what seemed half the pool.

By this time, I had the undivided attention of several first classmen, including the one who encouraged me to try the test. He walked hurriedly down the side of the pool, dropped to both knees, reached out his hand, which I frantically grabbed, and pulled me to safety. As I stood there coughing and shivering, he looked in my eyes, grinned mischievously, and said, "You weren't kidding, were you?"

At the beginning of my second week at West Point my fear of water had been unmasked. For my unwise decision not to solve an old problem and learn to swim before entering the Academy, I missed intramural athletics with my classmates during the rest of Beast Barracks. I couldn't participate in the intercompany field day to occur in late August. No touch football. No softball, volleyball, or speed ball. And absolutely no company swimming team. Instead I was a member of the "special swimming squad," swimming for beginners, three hours each week.

My embarrassment was made more acute when I considered my love of fishing, boats, and ocean-going vessels. I had spent several of my adolescent years near the Gulf coast of south Texas. Many times I fished with my dad in the surf of the Gulf of Mexico, in the intercoastal canal, large freshwater irrigation and drainage canals, and lakes. During World War II, I had been transfixed by sights of tugboats, commercial fishing fleets, and ocean-going freighters gliding slowly out of Port Brownsville through the ship channel, the intercoastal canal, toward Gulf waters where German submarines prowled.

My fascination with fishing, ships, and the calming, distant horizon of the great ocean I knew as the Gulf of Mexico was as important as my love of football, airplanes, and building airplane models. And during the latter part of World War II, both Annapolis – the United States Naval Academy – and West Point had been much in my thoughts for a military academy education. Now I was flunking a beginner's swim test at West Point.

A big, water-soaked pin had punctured my confidence balloon. I couldn't participate in sports activities I knew I could do well and enjoy. And if ever there was a time to enjoy something, anything, Beast Barracks was the time. It was not a good beginning at a place which had been my boyhood dream. So much for my welcome to West Point.

Little more than six months later, at the beginning of our second semester at the Academy, I successfully completed the infamous 175-yard swim. As the years passed I many times laughingly told the story of... walking on the bottom of the pool... while taking the test that would make me a member of the special swimming squad.

There was another side to the story. The embarrassing swimming lessons I received at West Point changed my fear of water to healthy respect.

In October of 1962, during the Cuban Missile Crisis, as a twenty-nine year old captain and fighter pilot in the Air Force, I was flying low level training missions over the Gulf of Mexico, at dawn, out of MacDill Air Force Base, Tampa, Florida. The aircraft was the F-84F, a single engine, single seat, jet aircraft. In predawn hours, in flights of four aircraft, we flew west, twenty-five thousand feet over the Gulf. At planned distances from the coast we made descending turns to easterly headings and an altitude of two hundred feet above the water, increasing speed from three hundred to four hundred twenty knots. We were conducting over water navigation, learning to "coast-in" over Florida, and find our assigned training targets, as we would if we were ordered to bomb targets in Cuba. I thought of that first summer of swimming at West Point as we leveled off a scant two hundred feet above the Gulf, and raced toward the east in the early morning light, light that could cause a cloudy sky to merge decep-tively with a horizon of water.

Two years after the Missile Crisis, while attending sea survival school at Langley Air Force Base, Virginia, I again remembered swim-ming lessons at West Point, as I was hooked into a parachute harness behind a motorized barge in Chesapeake Bay, and dragged face down through water considerably deeper and wider than a swimming pool. I was learning how to escape and survive after a parachute landing at sea in high winds. The fear was gone.

My failure to pass the swim test was one of several self-inflicted disappointments that first summer at West Point. Through my own harsh judgments about my mistakes, I set in motion a downward spiral of performance, almost causing me to resign from the Academy. Not un-til I had a second chance to try out for Army's plebe football team in early October of 1951 did my life at West Point begin to turn around.

There was more than a little humor in what happened to me at the Academy swimming pool in July of 1951, though I couldn't see it at the time. Every year, for many years, the cadet swim test had been a source of laughter in the Corps of Cadets. For the November 10, 1950, issue of *The Pointer* magazine, during Army football's last golden era season, "Mr. Shea, Sir," our Beast Barracks Fourth Company commander, captured the humor of it all when he drew a cartoon to go with the story of Cadet Bill Grugin, Company H-1, who graduated with the class of '51 the month before I suffered my self-inflicted embarrassment. Grugin's story was funnier yet.

LEARN BY DOING – First Classman Bill Grugin... recalls that he spent the afternoons of his first three years at West Point in the Intramural Pool. Bill was one of those non-amphibians who just couldn't get the hang of water travel and had to spend much of his free time working out with the Special Swimming Squad. Bill finally made the grade, though, and this September began his final nine months as a Cadet and not a 'Walrus.'

The perils of a new cadet not knowing how to swim when he enters Beast Barracks at West Point.

The other day the name of Grugin appeared on the roster of the 1st Classmen designated as Gymnasium Instructors. Bill recalled his proficiency in plebe boxing, and hoped to be assigned to that class. The authorities had other ideas as to Bill's qualifications, however, and – you guessed it, – assigned him assistant instructor of Special Swimming class.

In Beast Barracks, while I couldn't laugh at my predicament and was desperately trying to hold on to a hard-earned education opportunity most boys can only dream of, the Academy was descending into adversity. I didn't know about the other, far more serious events in progress at West Point, though we heard rumors. I was entirely too busy that summer to be much concerned with anything but survival at West Point. I had manufactured my own set of adversities. The war in Korea was far away. I had to face my difficulties and grow up.

AN IMPERFECT SYSTEM LEAKS

We, the new freshmen, the new cadets, were the Beasts. The first classmen, the Beast Detail, were our trainers and leaders. They were the officers and non-commissioned officers in our new cadet companies, preparing to be officers in the Army or Air Force less than a year hence. They were also our tormentors, who, it appeared to us, never let us rest or let down. And in our minds throughout every day and in conversations in our rooms, there was no doubt who were the real Beasts. They were the fire breathing first classmen, the members of the Beast Detail.

Though communication between first classmen and new cadets during Beast Barracks was almost totally one way, there were some compensations. We became good listeners – quickly. It was the age-old response of people reacting to perceived mutual misery, hardship, suffering, a loss of freedom or independence. We banded together, listened carefully, and passed information. We found ways to work together to lighten the load. We helped one another. And in subtle ways, some uproariously funny, and others not nearly so funny, we found ways to strike back at the Beasts – or the system. This was the beginning of class spirit, spunk, fight, teamwork, and unit pride – the feeling of accomplishment that comes from conquering shared hardship or challenges – and ourselves.

Some of my classmates developed listening skills to a high art form. They would return from meals and training formations and recite all manner of news, including rumors overheard from first classmen while going through the rigors of a Beast Barracks day.

A number of my classmates knew far more about the burgeoning scandal because of their families. They were the sons of Academy graduates who were serving as faculty or staff members. Some had learned of the Collins Board's investigation – "the something big" – before they entered Beast Barracks with our class. It's never easy for anyone to keep troubles secret, especially if cherished beliefs, friends, or family are affected. The restricted information leaked to roommates and other classmates who were friends.

Word spread from First, Second, and Third New Cadet Companies as well. They were the "flanker" companies, the tall man companies in Beast Barracks, to which most football players and other intercollegiate athletes were assigned.

Consistent with a reorganization of the United States Army, Colonel Sylvanus Thayer, the third superintendent of West Point, and the man considered the Father of the Military Academy, in 1824 reorganized the Corps of Cadets according to height, which gave uniformity in appearance to the Corps when assembled in military formation. It was one of numerous reforms instituted at West Point by Thayer, and one that remained in effect for nearly a century and a half.

The obvious results, to us, were the tallest men in First Company and the shortest men in Sixth Company during Beast Barracks. Assignment to Beast Barracks companies by height affected first classmen on the Beast Detail as well. Men on the Beast Detail, from permanent companies A, B, C, D, E, and F, 1st Regiment, were in First, Second, and Third Companies of Beast Barracks. In the 2d Regiment, the taller permanent companies contributing to the Beast Detail were G, H, I, K, L, and M.

Unintended results of the nineteenth-century Thayer reform included more cadet jargon and firmly held beliefs about the behavior of tall and short men. Flankers, so it was said, possessed *sangfroid*, tended to relax, be more good-humored, more secure in themselves. They would likely be the practical jokers, risk takers, more "non-regulation." "Runts," cadet slang for short men, were opposites. Runts were temperamental, conscious of their small stature, less secure, less willing to take risks, more "regulation." Competition and friendly rivalry between companies, which grew in the years since the Academy's founding, encouraged embellishment and perpetuation of such beliefs in the Cadet Corps.

In the twentieth century another unintended result of Thayer's sizing of the Corps was the majority of intercollegiate athletes assigned in flanker companies. In football, basketball, and several other intercollegiate sports, in which player size was important in competition, the trend was particularly noticeable. The majority of the intercollegiate athletes in those sports could be found in flanker companies. Thus conversations overheard among first classmen in First,

Second, and Third Beast Barracks Companies, whose classmates had been relieved of their summer assignments and were on work details at Camp Buckner, became added sources and grist for our busy rumor mill about the cheating scandal. Word had leaked that an investigation had implicated many cadets, including intercollegiate athletes. The rumor mill continued to flourish as the first month of Beast Barracks passed.

FURTHER DOWN – BUT NOT YET OUT

In late July, about the time the Hand Board was meeting at West Point, my still secret, aimless drift toward resignation accelerated when I tried out for the plebe football team, something I hadn't seriously considered before entering the Academy. My love of football had retreated into the dim recesses of my mind. The expression, "Every man an athlete," had disappeared from my thoughts. The far more serious concern was "Every man a swimmer."

We marched to Howze Field, an athletic field next to Michie Stadium, for what was called Annual Corps Squad Screening. Under the watchful eyes of coaches from the varsity, junior varsity, and plebe teams, we tried out for various intercollegiate athletic squads, including football. The intent was to identify talented athletes the recruiting system had overlooked. I joined classmates trying out for football and told the coach, "I was a quarterback. That's the position I'd like."

A new cadet, worn, intimidated, and unsure of himself, doesn't think clearly when given an unexpected opportunity. Quarterback at West Point was not the position for me. Other positions, perhaps in the defensive backfield, would have been more reasonable, if there was a remote chance I could make the team.

I should have remembered that Army, under Earl Blaik, used the T-formation on offense, with the quarterback as signal caller, ball handler, passer, a team leader, and field general. College players were considerably taller, heavier, and stronger than the average small town high school player. Coach Blaik, like nearly all major college coaches using the T-formation, also looked for T-quarterbacks who excelled throwing both long and short passes, had been seasoned in tough, usually large high school leagues, and had made enough of a mark with their gridiron exploits to be lionized in sports pages of their local press.

My high school experiences hardly measured up. While I'd played quarterback and defensive halfback three years on my dad's high school teams, the teams were from small town, rural schools, usually with two hundred or less students in the top four grades. They seldom received serious notice on any sports pages. If there was such notice, it was seldom in the pages Earl Blaik's assistants would be scanning for major college football talent.

At the time, however, none of that mattered. The old adage, "Nothing ventured, nothing gained," prevailed. My self-confidence in football hadn't been shattered. Why not try? I asked myself.

Passing was the first in a series of skills we demonstrated that afternoon in Corps Squad Screening. My lack of sufficient skills was clearly evident, as was my size. Five feet, eight and one half inches, 142 pounds, and not very fast, rarely describes a major college football player.

I didn't make the plebe football team, a setback I needn't have considered a failure. But I believed I'd failed again, and it added impetus to my downhill slide in Beast Barracks.

I still wasn't aware my late July descent into adversity was paralleling events of the still-hushed cheating scandal, the far more serious reason to search for talent at Corps Squad Screening the summer of 1951. The Army football coaches were already aware West Point's days of football glory had come to an abrupt, crushing end.

ANNOUNCEMENTS CONFIRM "SOMETHING BIG" – MYSTERY ENDS AND FUROR BEGINS

The August 3 public announcements about the "approximately 90 cadets" to be dismissed, were on Friday of the first week of "Second Beast," the second phase of Beast Barracks. There was a complete changeover in the Beast Detail the preceding Monday – a whole new set of faces and personalities among the first classmen when the announcements came. We had to begin anew learning what was expected and how to respond. The change in our cadet leadership came with a changing emphasis in training. We were turning to more soldierly skills, such as field training – tent pitching and bivouacking, map and compass reading, patrolling, tactical formations of the infantry squad, and weapons training.

In spite of the changeover in the Beast Detail, the new first classmen were equally abuzz with rumors. They too had been preparing for one week prior to taking command of Second Beast, overlapping the First Detail, receiving reports and recommendations from them, and unquestionably the latest rumors about what was going on.

Nevertheless, Colonel Waters' August 3 memorandum, with the two announcements attached, ended once and for all the speculation and rumors sweeping the classes of '52 and '55, about what the 'something big' had been. The same memorandum was also sent to cadets at Camp Buckner, the men in '52 on the Buckner detail, and the yearlings in '54 who were receiving the training. But in spite of all that was done and said that day the announcements also opened the floodgates at West Point, and elsewhere, for rumors, more speculation, and endless questions, still unanswered, about who, when, how, where, and most of all, why?

First classmen on the Beast Detail tried hard to avoid contributing to uninformed discussion of what had occurred, at least in front of the new cadets. The *New York Times*, which was our prime, daily source of news, was full of information not found in the public announcements, information provided by news stories from more than ninety reporters who descended on the Academy after the public announcement.

We began to hear and see names of men found guilty who made statements to the press. Colonel Waters and Colonel Leer, the Academy's Public Information Officer, had met with the increasingly angry and restive cadets in the gymnasium August 4, after they had been moved into north area from Camp Buckner.

At the meeting the Commandant explained what was to occur as the screening board progressed and the cadets' departures from the Academy neared. They were given copies of a memorandum explaining the purposes of the screening board and told they would be afforded the assistance of instructors and professors from the Department of Law, all of whom were military lawyers teaching the new Uniform Code of Military Justice and had practiced law under the Articles of War – the Uniform Code's revised and renamed predecessor. They could retain civilian attorneys if they chose and could afford the legal fees. The first of the two screening boards, the Jones Board, was beginning that same day.

Colonel Waters' explanation wasn't well received. The frustrated young men held their own meeting afterward, while Colonel Leer stayed on and listened, and reporters anxiously congregated outside the gym, waiting to talk with any willing cadets. The Academy's refusal to divulge the names of cadets being discharged prompted some reporters to offer money for names to agitated cadets who would speak to them. Amounts offered were rumored to be anywhere from $20 to $200.

Early the morning of August 4, one of the soon-to-be discharged cadets, a widely publicized, admired Army football player, attempted to place a personal, long distance call on one of the government telephones from Camp Buckner, before moving into north area at the Academy. He told the civilian operator, a government employee who worked for the Army's Signal Corps, he wanted to talk with Walter Winchell in New York City. Winchell was a nationally renowned radio journalist, controversial because of a sensationalist style of reporting.

Personal, long distance calls at government expense were prohibited, the operator explained. But the cadet became more insistent, saying circumstances warranted an exception. There were no pay phones remaining at Buckner. (The Signal Corps had removed the pay phones earlier in the week because a severe electrical storm had damaged the phone system, reducing system capacity fifty percent at Buckner.) The operator asked the cadet to please wait on the line while she talked with her company commander, Lieutenant Colonel Winfield Martin, who reaffirmed the cadet couldn't make the call. Later the same afternoon the cadet complained his phone conversations were being monitored. The complaint triggered an investigation.

The investigation proved his complaint groundless, but the incident was one of several, beginning August 4, that shortly led to a heated disagreement between Earl Blaik and Colonel John Waters, the Commandant. The argument took place at the entrance to General Irving's office and ended with Earl Blaik arguing forcefully to the superintendent that something less than restriction to barracks – what Blaik saw as virtual imprisonment – had to be permitted or matters would get much worse in north area.

We in the class of '55 were soon made aware "the 90" were residing in north area barracks. Feelings were high, but our class was shielded from much of the talk and nearly all that was going on in and around north area, with some exceptions.

Occasionally, as Fourth Company marched past the area, on Jefferson Road, our roving eyes caught sight of groups of men inside the area standing on the stoops, leaning over the rails, in civilian clothes, talking and gazing idly at the marching formations. There were times when we overheard whistles, shouts, and catcalls, evidence of frustration and anger aimed our way when we marched toward the gym wearing athletic shorts and T-shirts. One day as we marched past the barracks toward the gymnasium, a few empty beer cans were thrown in our direction from one or two fourth floor windows. The cans fell harmlessly short, and the first classmen leading our company gave no indication they noticed the incident.

The powerful feelings and emotions surrounding the now public scandal were being reflected in our class, though we had been at the Academy little more than a month. By this time we had received training and orientation on the honor code and system. There was already enormous pride in the honor code for those who believed they understood the meaning of honor in cadet life and the life of service they were to lead after graduation.

On August 9, the day the Barrett Board began – the second screening board – and the day Earl Blaik gave his controversial press conference at Mama Leone's Restaurant in New York City, my Fourth Company classmate, Roy Thorsen, from Brooklyn, New York, wrote in his journal:

> There's been quite a bit of talk here for the past week about the ninety cadets who are being discharged because of cheating on exams. To the public this seems like a major catastrophe, which it is, but also that the solution is dubious. The opinion here among the cadets seems to be one-sided. That being that they have contradicted in their actions one of the principles which so sets the Academy apart from other institutions and for this should pay the penalty of which they were well aware. With all the notoriety this case is receiving it makes one feel laid-low that something like that went on among us, but on the other hand one takes great pride in the fact that we are able to admit our faults openly and regardless of the unpleasant predicaments stand by our beliefs and convictions.

Roy Thorsen, as a fourth classman, a freshman, had felt the pulse and heard the heartbeat of the Corps of Cadets. Art Collins, president of the Collins Board, said years later, "The Army can get good officers anywhere, but somewhere there has to be a nucleus of a sense of honor and integrity the Army must have, and if West Point doesn't provide that – well, do you need a West Point?"

WHERE DID ALL THE GIRLS GO?

Girls! Women! They, too, were disappearing from my thoughts in the scramble to survive those first weeks. Cadets often used the terms "Sing-Sing on the Hudson" and "the monastic life" to describe life at West Point. We were learning why, the hard way.

The Corps of Cadets was all male, a fact I hadn't thought much about before coming to West Point. The tightly controlled, regimented environment allowed new cadets little time for thoughts about girls, although it certainly didn't eliminate females from our thoughts and room conversations. The best we could do was talk about the girls at home, and the ones we hoped to meet. It didn't take long to find out who had a girlfriend, a "steady," or fiancée back home, in college, or – lucky devil – going to college or working and living nearby.

We had been told of the cadet hostess, Mrs. Barth, whose office was in Grant Hall, the cadet visitor reception center. She arranged for young women to come to West Point to attend hops – dances – and other social functions permitting mixed company, escorting, and dating. We wondered if we might meet someone as attractive as the first classmen's girls appeared to be.

First classmen were at the top of the cadet privilege ladder. They could take a weekend pass every month, if they were in good standing and had no scheduled weekend duties. When they weren't on weekend leave, and had no duties, many could be seen escorting lovely young women to various locations outside the cadet area. We watched with envy.

To the new cadet, however, life in prison or a monastery couldn't possibly be any worse than the eight weeks of Beast Barracks, feelings heightened by what we daily observed among the first classmen. We weren't permitted the privilege of escorting young women the whole summer. We already knew there would be no going home for Christmas.

There were to be no distractions. None. Not during Beast Barracks. We scarcely caught sight of female face or form. Our crowded training schedule, and the stern, supervisory gazes of ever present first classmen made certain of that. The absence of female companionship, the confusing individual and collective changes we were undergoing, and our lack of understanding of where it was all leading, resulted in exaggerated talk about the reality of our circumstances.

Early that summer, one of our luckless classmates in a flanker company was caught in a rendezvous that eventually resulted in his resignation from the Academy. When confronted by first classmen who observed his indiscretion, he compounded his amorous infraction of cadet regulations by telling them the young lady was his sister. She wasn't, and his lie was uncovered. He left the Academy, confronted with an honor violation.

Hectic days and nights meant time flew, and so did many romances. In less than a month some of my classmates received "Dear John" letters. Some of the "Dear Johns" were hilarious, but they weren't funny at the time. There was truth in the old adage "absence makes the heart grow fonder – of someone else." I, too, found that out the hard way.

I was much enamored of a young lady in Alamosa, Colorado, where I'd attended Adams State College for two quarters before coming to the Academy. I described her admiringly, and in great detail to my roommates, and received a letter or two from her that summer. I eagerly wrote back, but she quit writing in August. No "Dear John," but another blow to pride and confidence, adding to my growing sense of failure and isolation. My feelings were made more acute by what I was observing around me.

It would be a long year in which our contact, and conduct, with the opposite sex were carefully guarded by the cadet "Blue Book." A "public display of affection," such as holding hands or, heaven forbid, putting your arms around a young lady, was worth eight demerits and four punishment tours. Shortly after Beast Barracks I learned about punishment tours.

"SIR, I'M NOT MEANT FOR THE ARMY LIFE..."

As Beast Barracks continued the blur of mind-numbing, exhausting activity, my interest in anything but survival dropped ever lower. Neither the swirling controversies nor torrents of animated upperclass conversations created by the public announcements and cadet mail of August 3 penetrated the feelings I had for my personal predicament at West Point. I was slowly filling up with self-absorption and self-pity.

By the time Taps sounded at 10:30 in the evenings, my roommates and I were weary, but there was more work to do. We turned room lights out at Taps and closed our doors. Hallway lights remained on. Hallway lights were the salvation of new cadets desperately trying to stay with the furious pace of Beast Barracks. After lights out, doors to many fourth class rooms would slowly, quietly open. With ears cocked for the approach of inspecting first classmen or the officer in charge,

the OC as cadets called him, who always seemed to bound up the steps two or three at a time, our preparations for the next day continued. We often studied long into the night. In spite of the risks depicted in Cadet Jack Galvin's cartoon, "After Taps," we spent many of those late hours memorizing plebe knowledge.

The Beast Barracks I so nonchalantly sauntered into little more than a month earlier, became long days of relentless, fast-paced training, and the near perfect insulator against events, inside and outside the world of West Point, except for those events which filled our endless days.

An Army officer, a West Point graduate, visited me in our home on the Rio Grande River in the mountains of southwestern Colorado, not long before I boarded a train in Denver to begin the trek to New York. He came to tell me about life as a plebe and left a copy of *1950 Bugle Notes*.

"I urge you to memorize its contents before entering the Academy. Life will be easier if you do," he said. "The book contains 'plebe knowledge' which upperclassmen will require you to recite frequently in the coming year. Plebe year will be hectic, particularly Beast Barracks."

Unfortunately, youthful confidence triumphed over good judgment. I didn't learn a single line of plebe knowledge before I walked into the cadet area that first day. Not one line. God, I wished I had! Plebe knowledge memorized in advance would undoubtedly have made life easier those eight weeks.

I was losing weight, but I was no different from the rest of my classmates. The mealtime ritual, tough and unyielding, did nothing to help any of us.

Not knowing how to swim sent me to special swimming and shut off my best source for relaxation and accomplishment, intramural athletics – so I thought. I was disappointed and becoming discouraged. I forgot why I came to West Point. I began what would only lead to more trouble, taking first classmen's corrections personally.

My sense of humor was gone. I needed to laugh at myself, others, and the comical circumstances created by the tough training. I began to withdraw, fall silent. Even some of the humorous skits conducted during lectures, designed to hold our attention, make a point, or simply bring some levity into serious subjects, drew blank stares from me. I was taking myself and everything else too seriously.

I felt West Point was isolating me in a sea of hostility, failure, and frustration. On top of everything else, the smart, confident seventeen-year old who was so anxious to leave home and conquer West Point was becoming homesick. Self-pity was the debilitating, unrecognized enemy stalking me, and thoughts of resignation from the Academy slipped more easily to mind.

And I was keeping it all to myself. I told no one. I had quit thinking about anyone else but myself, the exact opposite of what we were being taught, what was demanded of us.

The plebe hike, August 20-25, brought some relief from the self-imposed misery. To leave the formality and the ever present hostility of life in cadet barracks, was a respite from torment. The plebe hike was more relaxed and exemplified the reasons I came to West Point.

In fatigue uniforms, with M-1 rifles, combat boots, helmets, and full field equipment, we went on an extended march for five and a half days, just as soldiers might do deploying into combat. We bivouacked each night, pitching tents, putting into practice what we had received in lectures about field hygiene, care and cleaning of rifles and equipment, first aid, and sentry duty.

Each day of march, we sang as the six companies kept cadence in their columns of platoons. The marching songs were traditional in the infantry, some bawdy, others light and humorous, most learned by classmates who had prior service. There were night field training exercises – map reading and night navigation problems, including simulated night combat patrols against "aggressors" armed with rifles and machine guns firing blank ammunition.

On Saturday, August 25, when we returned to the cadet area, the one week respite from Beast Barracks was over for me. Monday would be an Athletic Field Meet, a day of intramural athletic competition between Beast Barracks companies. But right on the heels of the Athletic Meet came Reorganization Week, when the entire Corps of Cadets returned from summer training and summer leave and moved into their permanent companies to begin academics.

The Second Beast Detail had repeatedly warned us, good naturedly, "if you think Beast Barracks is tough, Misters, wait until the entire Corps comes back for academics. You'll be outnumbered three to one, instead of outnumbering us four to one!" The state of mind and attitudes I'd developed were not receptive to such humor. Instead, I conjured up images of more hostility, isolation, failure, and self-pity.

When Transition Week had come and gone, and I was in Company K, 1st Regiment, the predicted high ratio of upperclassmen, who all seemed even less friendly than the Second Beast Detail, increased my feelings of dread. I'd been so busy I'd quit writing home. Back home in Colorado, my parents were in their annual summer rush, running a small guest ranch they had purchased in 1949 and preparing to shut down for another year of school teaching, which would begin right after Labor Day. Letters came from home less frequently, and I couldn't call them on the phone.

(Courtesy John R. Galvin)

Among the first classmen in K Company there wasn't one familiar face from Beast Barracks. Neither was there a familiar face among the second and third classmen in K-1, nor anyone anywhere in the Cadet Corps, except for classmates I'd met. I was starting all over again, adjusting to new friends who weren't friends, new faces, new acquaintances, different attitudes, different behaviors. I didn't see the humor that Cadet John Galvin in the class of 1954 had seen in Beast Barracks a year earlier. He had it right, but I didn't. For me, entry into our permanent companies signaled the end of the first phase of cadet life, and the beginning of another, even less humorous.

To make matters worse I felt I'd started badly in academics, and walked into a buzz saw in K-1. At the end of my first week in academics, grade reports posted in the north sally port of central area confirmed what I believed to be a terrible start in academics.

Even worse, I was accumulating demerits at a rate I feared would cause me to go over the fourth class monthly demerit allowance. Then bad news came confirming my fears. I had indeed gone over the allowance. I would have to march four punishment tours beginning the following Wednesday afternoon. This was one more sign I was failing, not living up to my capabilities, not the way my parents and friends back home expected I would perform at West Point. None of this was the way I'd envisioned plebe year to be, not for me.

Grades (Galvin)

The morning of September 13, a Thursday, between reveille and breakfast formation, all the tightly-held emotions and feelings – the building anger at myself, frustration, disappointment, shame, shattered confidence, failure, isolation, and self-pity – exploded. I'd kept it all completely bottled up. I hadn't told anyone anything. A fit of crying began. I shook. I couldn't talk coherently to my roommates who kept asking what was wrong. Alarmed, they asked for help from upperclassmen across the hall, who immediately informed the cadet company commander, Robert N. "Bob" Kelly. He came to the room, attempted to calm me, and asked the Cadet-in-Charge-of-Quarters, a third classman, to call the hospital to dispatch an ambulance.

Strangely, the more the company commander and upperclassmen did in my behalf, the more embarrassed and humiliated I felt. This wasn't the way the son of a football coach did his part or should be treated. This wasn't the way a cadet at West Point behaved. This was no way for a future officer to conduct himself. What's wrong with me? I never wanted to see these people again. I just wanted to get out, leave this place, go home, free myself of all the mistakes, embarrassments and failures.

When I was delivered to the post hospital at 6:30 that morning, it was, of course, an escape from the environment I wrongly concluded had caused it all. It was pleasant, nice to be in the hospital. No bracing, plenty of rest, good food, no upperclassmen to torment me with plebe knowledge at meal tables, no marching in formation with upperclassmen telling me all the things I wasn't doing correctly – or not doing at all, no inspections, no responsibility for anything except what the kind nurses, and pleasantly professional doctors asked me to do. I didn't have to go to class, and I wouldn't have to march the four humiliating punishment tours.

In the meantime the doctors began asking me what was wrong. The specifics of what had happened in the barracks had been given them. I added much to their knowledge, little of it having to do with what was really bothering me. I told them of headaches, pains in back and neck muscles, induced by injuries in high school football, and an injury sustained on a trampoline as a result of no training on the device, and some mistaken, unsupervised "horseplay." The physicians conducted tests for trichinosis and other organic diseases which could produce the symptoms I'd described. The tests were all negative, and the doctor informed me of the results the following Wednesday, September 19.

He called me to his office, closed the door, and asked me to sit down. He then asked, "In light of the test results, is there anything else bothering you?" Out it came, some of it, but only what I wanted to tell him.

Sir, I'm not meant for the Army life. For the past two weeks I've been thinking about resigning. For the same period of time, I've been wrestling with the problem of whether my family or friends back home would think me a quitter. I feel I have the ability to finish at the Academy, but I don't feel the goal is the right one for me. This whole thing has been on my mind so much I can't concentrate in academics.

Dr. A. J. C. Doran explained all this in my medical record of September 19, 1951, and went on to say,

> ...Herein no doubt lies the answer to this man's diagnosis. After about an hour interview we both thought that he would be able to go back to duty this morning; however, when the time came for him to leave he felt he could not go out to duty and cried in front of the nurse. I discussed the case with the Chief of Medical Service who felt that the boy had no medical condition warranting a further stay in the hospital and instructed me to send him to duty as planned. As he leaves, he is still confused and undecided on the problem of resigning. The company tactical officer [Major Robert L. Royem] was advised of all this by phone and will see him sometime today.

THE BEST LESSONS OF ALL: FORGET YOURSELF AND DON'T WALLOW IN SELF-PITY

Major Royem asked that I come to his office as soon as I'd been released from the hospital that day, and with his coaxing, the rest of the story came out. The next day he wrote a letter to my parents, who by this time were in Harlingen, Texas, in the Lower Rio Grande Valley, where they had obtained new teaching positions.

> Dear Mr. and Mrs. McWilliams:
>
> I am writing to inform you that William was in the Post Hospital from 13 September to 19 September because of a case of extreme discouragement. His doctor has found no physical difficulty, but rather a feeling of depression.
>
> After discussing William's case with him and his doctor, I feel that his discouragement is due to an over sensitive reaction to certain aspects of training. All Fourth Classmen have difficulty at this time of the year because there is the new element of academics to consider, the time schedule becomes tighter, and there is a greater degree of supervision by the upperclass cadets. In short, it is a period of readjustment to the regular academic sched-

ule. William has been trying very hard, but he feels that he is not doing as well as he should because of demerits that he is receiving and because of detailed corrections by upperclassmen. Actually, his record is average, and should not be cause for alarm. William needs encouragement more than anything, and I think frequent letters from home would do him a world of good.

William was well cared for in the hospital, and is now back on his normal schedule.

If you desire any further information in the future, do not hesitate to write me.

> Sincerely yours,
> R. L. Royem, Jr.
> Major, USAF
> Company Tactical Officer

Major Royem wasn't aware of it, nor was I, but his talk, and the letter to my parents, lifted the weight I had placed on myself. I marched off the last two of my punishment tours on Saturday, September 29, though, in doing so, I missed Army's season opening football game against Villanova in Michie Stadium. Missing the game was one more disappointment, and another memory of a difficult beginning for plebe year at West Point.

But something happened in all that confusion, doubt, and despair about how well I was doing at the Academy that summer. I could make it. I had told Dr. Doran I could make it. I'd never said that to myself, or out loud to anyone else since the first day of Beast Barracks. There were good people all around me. I wasn't alone. I needed to pay attention. I needed to be tougher than I'd been, as tough as I'd been in times past. I needed to quit thinking about me. I needed to think about my classmates, and others, who were trying to pull themselves over the same obstacles, and make major changes in their lives. It was time to grow up – now.

The self-imposed anguish began to disappear.

"'TIS BETTER TO PRACTICE AND LOSE THAN NEVER ATTEMPT PLAYING THE GAME..."

In the month after Beast Barracks ended, I stuffed myself almost daily with a pint of ice cream. In our permanent companies there had been some loosening of restrictions on plebes going to the "Boodlers" – the confectioners, where we could buy cookies, candy, and ice cream. My roommates and I, starved by the rigors of Beast Barracks, decided to supplement our still slim K Company meals. We pooled enough cash each day that one of us could carefully list the purchase orders, take a laundry bag, and stock up on sweets during late afternoon free time – before the supper meal.

My weight ballooned twenty-three pounds to 165 by the first week in October. Now that I was a heftier 165 pounds, I asked to try again for the plebe football team. Given Army's shortage of talent, I might have a crack at the defensive platoon. I might be able to "walk on," as I'd heard many other cadets had done. The request was granted. There were vacancies on the team. Some of my classmates had been moved to "B" squad, or varsity, because of new rules allowing freshmen to play varsity.

So it was the first week of October, after Army's season opening loss to Villanova, when I donned the practice uniform of the plebe Black Knights of the Hudson, "C" squad. As I jogged across the Plain toward the practice field to the north of the tennis courts, I talked with team members – classmates – about the workout routine I should expect.

I glanced down at the black jersey with white numerals fitting snugly over the shoulder and rib pads, and the well-worn but freshly-laundered, dull gold canvas pants which hugged the hip pads strapped around my waist. My head felt heavier than normal because of the old-style, leather practice helmet I wore. I could feel the thigh pads in their pockets inside the pants and the softer knee pads, which were sewed into the pants when they were manufactured.

The black leather, high-topped, cleated shoes dug into the grass turf of the parade ground as I jogged somewhat breathlessly toward the practice field. It all seemed so familiar. The feel and the smell of the pads and other equipment were just as I remembered from high school football.

But this day was far different. I was full of anticipation and excitement, and there were those pesky butterflies again.

Practice began with warm-up calisthenics, about forty-five men in a large circle at more than double arms length between each man. Calisthenics included shouted cadences to boost team unity and conditioning.

I recognized early I wasn't in good physical condition. Breathing was labored, and my lungs began to burn when we ran in place. The team was well conditioned, having begun practice the first day of September, after conducting another prior month of hard conditioning. They were lean and toughened by a West Point coaching staff which had a reputation for fielding one of the best conditioned teams in the nation.

My physical conditioning had suffered since Beast Barracks. I was flabby. The late afternoon pints of ice cream, supplemented by candy bars, had taken their toll. At 165 pounds, the heaviest weight I had ever carried, I wasn't ready to keep pace with the splendid athletes who were my classmates and future varsity members of the Army football team.

I was, however, completely undeterred by the small matter of physical conditioning. This was the best day I had experienced at West Point. Not only was I trying out for the plebe team, but on the field were men whose names I had heard from sports broadcasters on the radio and in movie theaters on Saturday afternoons in the mid-1940s. Among the plebe coaches was Arnold Tucker, Army's great All-American quarterback on the famous Blanchard-Davis team of '46.

On the adjoining practice field was Felix "Doc" Blanchard, three-time All-American fullback on the '44, '45, and '46 teams, Heisman Trophy winner, and co-captain of the great 1946 team, which had played the "Game of the Century" 0-0 tie with Notre Dame's Fighting Irish. He was an assistant coach of the "B" team, the "Golden Mullets," as they ran the plays of the varsity's opponents for the following Saturday. Then there was Earl "Red" Blaik, the head coach, the mastermind, and a dean of college coaching, with his assistants – among them Doug Kenna, All-American quarterback on Army's 1944 national champion team; Captain Johnny Green, who played guard on the '44 team, and was an All-American guard and captain of the 1945 national champion team; and Vince Lombardi, who had been one of the famed "Seven Blocks of Granite" at Fordham University in the mid-1930s – putting the varsity through its paces. I was among some of the greats of the gridiron, men I'd admired, all of whom were football heroes in the eyes of a small town boy.

But I had work to do that afternoon, no time to be too enamored of the football greats around me. Coach Tucker ensured I worked hard.

He had been told my skills, such as they were, were on defense, in the defensive backfield. For the remainder of the practice I worked in the defensive backfield, first at linebacker, then at halfback.

As linebacker I was to fend off blockers and tackle ball carriers as they ran quick opening plays through the right side of the offensive line. A weighty 165 pound linebacker was no match for the speed, weight, and strength I contended with in ends, tackles, fullbacks, and halfbacks which aimed to ensure I wasn't successful in bringing down the ball carriers. On more than one occasion I heard ringing in my ears and saw flickering, small stars dancing in my eyes after hurling myself head-on against blockers or a fast moving ball carrier.

After awhile Coach Tucker moved me to defensive halfback. The offensive drill was to have the quarterback fake a quick opening play on the right side of the line, and pitch the ball to the fullback or left halfback going around the right offensive end. My job was to, once again, come up and tackle the ball carrier, this time as he attempted to circle end and break into the open field, usually at a fully accelerated sprinter's speed. Again I heard ringing in my ears and saw stars dancing in my eyes on several plays as I attempted to bring down ball carriers much bigger, faster, and stronger than I.

At the end of an hour and a half practice we ran wind sprints until I thought I would drop with exhaustion. I was a tired eighteen-year-old, a good tired, with a feeling of accomplishment when the coaches said, "That's all men. Go take a shower." The sweat was rolling down my face. As I walked back across the Plain toward the gymnasium to shower and dress for supper, I realized I was sore everywhere – all over, in every muscle of my body. But I didn't mind. I was happy for the first time in weeks.

I was absorbed in my thoughts when I heard steps of someone coming up behind, to my left side. I felt an arm brush over my shoulder pads and a hand come to rest on my right shoulder pad. I turned my head and there was Coach Tucker, Arnold Tucker, the great Army quarterback of years past, looking straight into my eyes. As his right hand patted my shoulder he said, "Good job, 'Mac'. We'll call you if we need you." He then turned and walked on hurriedly to talk with another plebe.

I replied with a weary, "Thanks, Coach," and a smile, and watched him walk away.

For some reason I wasn't disappointed in Coach Arnold Tucker's response to my one day of glory with The Black Knights of the Hudson. I guess it was in the trying, the being there, that I somehow succeeded. If for only one brief afternoon, I had been a part of a great football tradition. Smaller players than I had succeeded in college football, and maybe had I been more determined, persistent, I would have eventually made the team. Whatever might have been, I didn't choose to pursue the goal further. I felt no regrets.

Instead I mark that day in October of 1951 as a turning point in my attitude about remaining at the United States Military Academy, and indeed, a watershed event in my life. My entire performance at the Academy began to improve, and kept improving throughout my four years at West Point. I began to enjoy learning, and to understand I could succeed.

The following Saturday, October 6, Army lost its second football game of the season, to the Northwestern Wildcats, 20-14, at Dyche Stadium in Evanston, Illinois. Far away, in Korea, Lieutenant David R. Hughes, K Company, 3d Battalion, 7th Cavalry Regiment, was making plans for the next day's assault on Hill 347. *The Pointer* magazine's second issue of the academic year was published the day before the Northwestern game.

On page nine was another cartoon drawn by third classman Jack Galvin. He captured perfectly my feelings, and the feelings of many more cadets, about Beast Barracks, plebe year, and first classmen.

With his many hilarious cartoons, "Cadet Jack Galvin, Sir," would have all of '55 laughing before the year was out, and indeed the entire Corps of Cadets, as we continued the long road back.

"Mister, you and I are going to go 'round and 'round." (Galvin)

378

CHAPTER 10

WHAT'S IT ALL ABOUT?

When Colonel John K. Waters, the Commandant, addressed the new fourth class cadets at our Oath of Allegiance ceremony late the afternoon of July 3, he had since the first days of April been deeply immersed in the devastating, still closely held revelations of organized cheating at West Point. Now that the Collins Board report was complete, with its seventy-three pages of damning evidence, he was struggling, along with other senior officers at the Academy and in the Army staff, with the decisions that had to be made.

Waters, like the great majority of Academy graduates, felt strongly about honor, the cadet honor code and honor system. He had shown his emotions in writing to General Irving on June 11 when he quoted the "Cadet Prayer" in the memo accompanying the Collins Board's report to the superintendent. Although Waters must have been deeply troubled about the effects of the stunning revelations, the afternoon he spoke to us he didn't mention the investigations. The Academy and the Army weren't yet prepared to make the public announcement senior officers knew would be necessary.

If there was passion in John Waters' voice that afternoon, it's doubtful many in our class heard or recognized his emotions. We were little more than halfway through our first day at West Point, a day of shock and disbelief. Nevertheless, as commandants before him had done at the ceremony, as part of his talk he spoke of honor to the Academy's newest class, 1955.

...Cherish the Honor System. By that I mean die a thousand deaths before you lie, steal, or cheat. There is nothing about West Point in which the cadets and graduates alike take greater pride than the Honor System. It is part of the Corps, and remains at all times the individual and collective trust and responsibility of the Corps. You will hear of it frequently; it is a continuing part of your daily life, and willful violation thereof is a basis for separation. In its written or spoken form it is simply 'cadets do not lie, steal or cheat.' It is not a lengthy code, an

379

order, or a directive from higher headquarters, but a way of life of the Corps of Cadets – each cadet with an admiration for truthfulness. You as individuals are a living part of it. Learn about it, absorb it, live it, guard it, and practice it. The plain statement of a cadet or an officer 'I do so-and-so' is as good as his 'I certify on honor.' Measure yourself by this standard, and your relations with your fellows cannot go far wrong....

There were good reasons for Colonel John Waters' strong feelings about the honor code, and its meaning to the American soldier. Born in Baltimore, Maryland, in 1906, the second of three sons, he grew up on a farm in the Greenspring Valley, and attended schools in Baltimore. In 1925 he entered Johns Hopkins University as a premed student; however, after two years he decided he wanted to be a soldier.

He wanted an appointment to the Academy, but was unable to obtain one from his home state. Measures of his determination and will were evident when he learned he could fill an appointment vacancy in Illinois if he met the resident requirement for that state. He met the requirement by driving to Champaign, Illinois, and sleeping on the floor of the state capitol – then entered West Point with the class of 1931.

His four years as a cadet included participation in intercollegiate soccer, lacrosse, and hockey, and his selection as captain of the hockey team his first class year. More significantly, for his first class year he was also selected as First Captain in the Corps of Cadets, a recognition of his splendid leadership abilities.

His first assignment after graduation was to the 3d Cavalry Regiment in Fort Myer, Virginia, where he met Beatrice Ayer Patton, the daughter of Colonel and Mrs. George S. Patton, Jr. She would later become his wife. In 1934 he transferred to the Cavalry School in Fort Riley, Kansas, and after completing the course, was assigned to the 13th Cavalry Regiment at Riley. It was at Fort Riley that he married Beatrice Patton.

Graduating from the advanced equitation course in 1937, he went on to Washington, D.C., to serve as a military aide to President Franklin D. Roosevelt for one year. Then it was off to West Point to serve as the A Company tactical officer. There the Waters' two sons were born.

In 1941, as war clouds gathered over the United States, he and his family left the Academy for Fort Benning, Georgia, where he joined the 2d Armored Division, "Hell On Wheels," and served in the Division's 68th and 67th Regiments respectively. As America's entry into the war neared, the 1st Armored Division, "Old Ironsides," was marked for deployment, and just prior to its overseas movement, John Waters was assigned to the Division. He was about to begin a difficult period in his life, where he would experience betrayal, defeat, imprisonment, and great pain.

After a shakedown period in Northern Ireland, the 1st Armored Division went ashore in the 1942 invasion of North Africa, near Oran, Tunisia. In North Africa, at Kasserine Pass, where American armor was on the receiving end of a disastrous defeat, John Waters' task force, on a hill known as Djebel Lessouda, was overrun by Field Marshall Irwin Rommel's 10th Panzer Division. The task force's position and status had been betrayed by an Arab herdsman.

Waters was interned in Oflag 64 in Szubin, Poland, where he remained until early 1945. As the Soviet Army drove toward the heart of Germany, the Nazis moved the prisoners to a camp at Hammelburg. There, John, as senior American officer, became the prisoners' camp commander. He was responsible for organizing a chain of command, as well as several committees intended to serve the welfare of the prisoners – not the least of which was the escape committee.

Meanwhile, Lieutenant General George S. Patton, Jr., his Third Army driving east to link up with the Soviets, learned of the prison camp holding the Americans. He sent Task Force Baum ahead to attempt a rescue. The Germans, aware of the Task Force's movements and apparent intentions, ordered Waters to march his prisoners to the local train station to move to another location further east.

As John Waters led the column through the compound's gate with a white flag in his arms, a German deserter took them under fire, seriously wounding him, and forcing the column back into the camp. George Patton's Task Force wasn't successful in the rescue attempt. A Serbian surgeon, interned in a nearby camp operated on John Waters and saved him, but the severe wound caused recurring bouts of illness the rest of his life.

On recuperating from his wartime ordeal, Waters returned to duty as aide to Secretary of War Robert P. Patterson in early 1946. He then attended the Army's Command and General Staff School at Fort Leavenworth, Kansas, and the Armed Forces Staff College in Norfolk, Virginia, graduating in 1949. When he came to the Academy to be Assistant Commandant of Cadets that year, he had powerful feelings about life, service to country, and honor.

* * *

We heard nothing more about honor during our summer training until Saturday morning, July 21. General Irving spoke to our class. He recalled and expanded on Colonel Waters' remarks about honor, given us at the Oath of Allegiance Ceremony nearly three weeks earlier. The superintendent also gave no hint of the struggle still going on behind closed doors.

When General Irving spoke to us, the Hand Board had already convened in New York City, and would arrive at West Point to begin interviews the following Monday:

> ...you have to work and work hard to learn not to lie, cheat, or steal. You must test yourself constantly by asking, 'Am I attempting to deceive?' 'Am I attempting to take unfair advantage of my fellow man?' It is difficult not to lie. One is inclined to make a story just a bit better. One is often inclined to say 'No' although the answer should be 'Yes.' In short, honor is not easy to achieve, and it will not be achieved unless you are constantly on the alert against violations of honor. As you get on in the world and proceed with your career and you rub shoulders with people of many lands and many races, you realize more and more that there is something you get at West Point that is really unique – Honor. When General Eisenhower was Chief of Staff, he wrote a letter to the superintendent [then Major General Maxwell Taylor] which refers directly to the matter I am discussing and I would like to read that now because it states what I, too, feel about the Honor System at West Point.

He then read from General Eisenhower's letter dated 2 January, 1946:

I think that everyone familiar with West Point would instantly agree that the one thing that has set West Point aside from every other school in the world is the fact that for a great number of years it has not only had an 'Honor' system, but that the system has actually worked. This achievement is due to a number of reasons, but two of the important ones are: first, that the authorities of West Point have consistently refused to take advantage of the honor system to detect or discover minor violations of regulations; and second, that due to the continuity of the Corps and of the instructional staff, we have succeeded, early in the cadet's career, in instilling in him a respect amounting to veneration for the honor system. The honor system, as a feature of West Point, seems to grow in importance with the graduate as the years recede until finally it becomes something that he is almost reluctant to talk about – it occupies a position in his mind akin to the virtue of his mother or his sister.

...it seems to me... important that individuals now at the Academy, both officers and cadets, clearly and definitely understand that the honor system is something that is in the hands of the cadets themselves, that it is the most treasured possession of the Point, and that under no circumstances should it ever be used at the expense of the cadets in the detection of violations of regulations.

General Irving continued, "I can add nothing to the most eloquent words of General Eisenhower, except to say that you are the instrument by which the honor system must be and can be made most effective."

General Eisenhower's letter to Maxwell Taylor had been written when Taylor was four months into his assignment as superintendent, an assignment spanning a period in which the Academy's curriculum was reconstituted following World War II. Eisenhower's letter followed by a few days Taylor's visit with him while Eisenhower was the Army Chief of Staff. During the visit, the two men discussed the Academy's role and future. They talked of ideas, changes, reforms, and shifts in emphasis needed as West Point emerged from the war and resumed its traditional role in providing officers who study war and are educated and trained to be leaders in the Army.

Included in his comments on the honor code and system was an experience Eisenhower remembered from his cadet days. He related the incident in his letter to Taylor.

> I remember as my most unfortunate experience while I was myself a cadet, an incident where some light bulbs had been thrown into the [cadet] area. The culprits were found by the lining up of the Corps and the querying of each individual as to whether or not he was guilty of this particular misdemeanor. Any such procedure or anything related to it would of course be instantly repudiated by any responsible officer who had the good judgment to visualize its eventual effect on the honor system; but I do think it important that a policy along this line be clearly explained to all concerned at least once a year, certainly by an authority no lower than the Commandant himself.

Using honor to find violations of regulations, by interrogating all cadets without due regard for whether evidence existed as to individual involvement in the offense – this was the thrust of General Eisenhower's "unfortunate experience." His letter didn't say "No regulations should include considerations of honor." The issue was the manner in which honor was used, which preyed on the young men's obligation to tell the truth when asked a question.

The throwing of light bulbs into the cadet area was not a well-considered prank. Although serious consequences might occur were anyone walking in the vicinity of exploding bulbs, there had been none. Lining up the entire Corps to question each member and find the offenders using the cadets' obligation to tell the truth was what rankled Eisenhower and burned the incident into his memory. The officers acted wrongly in using honor to solve a minor problem.

Five and a half years after Eisenhower's letter to Taylor, in testimony before the Bartlett and Greene Boards, cadets had once again raised the question of honor being used to enforce regulations. What's more, they pointed out that the honor system had become increasingly complex and the code was being "trivialized" by truncated and detailed interpretations of honor that resulted in a maze of honor "rules."

In his August 3, 1951 letter to General Irving, Maxwell Taylor, who was now Irving's boss, told the superintendent that General J. Lawton

384

Collins, the Army Chief of Staff, wanted him "to verify the [honor] system is simple, clearly understandable, and is not a device to effect the enforcement of regulations." The results of this direction, reinforced by a recommendation from the Bartlett Board, were additional boards convened in October, after the Bartlett Board completed its work. Two were the Greene Board, convened to examine "Regulations Impinging On Honor," and a committee of cadets in the class of 1952, asked to reexamine the honor system for possible improvements, a system developed by the Corps of Cadets over West Point's nearly 150 years of history.

HONOR CODE AND SYSTEM: EVOLUTIONARY HISTORIES

The honor code that evolved in the years following the Academy's founding in 1802 began long before that year and had its roots in the British Army's officer corps.

"A gentleman's word is his bond," were the words of the British aristocracy – and members of the royalty, the ruling classes in England, the men who also received commissions as officers in the Army and Royal Navy. An officer is a gentleman and "an officer's word is his bond" were natural outgrowths of the relationship between the British ruling classes and the officers who led their armed forces. In short, "An officer's word is his bond" became the Officer's Code of Honor.

The British Army transplanted the Officer's Code of Honor in the Empire's new world colonies through the officers who served in the colonies and supported the British Crown in ruling the upstart, independence minded Americans. When the Revolutionary War occurred, most of the officers leading the Continentals' Army were men who had served in various capacities in the mother country's army, and they instilled the Officer's Code of Honor in the rebel army, in which it became prevalent before the Academy's founding in 1802.

George Washington was the most notable example of such men and later was the most powerful advocate among several American officers who pushed vigorously for establishing a military academy – to ensure a standing corps of professional officers who would study war, lead the young nation's Army in times of national peril, and thus provide greater security for the fledgling democracy.

Congress enacted the 1802 legislation establishing the Academy during Thomas Jefferson's administration, and in the early years after the Academy's founding, the cadet honor code was an extension of the Officer's Code of Honor.

By 1926 the Academy's honor code was a brief, simple statement of three prohibitions, "Cadets will not lie, cheat, or steal." The code, with respect to its first two tenets, lying and cheating, evolved throughout the first one hundred years of Academy history, and was fairly well defined by the end of 1907. Until that year, lying remained the single, central prohibition in the code, although cheating, for a time during the 19th century, was treated as an honor violation.

The first attempt to expand the early code beyond the prohibition against lying came when Colonel Sylvanus Thayer was superintendent. Thayer, honored as the "Father of the Military Academy," primarily for his development of the West Point education system and cadet training programs, considered cheating to be a violation of the Honor Code and announced that violators would be expelled. Cadets apparently subscribed to his prohibition against cheating until some time after Thayer retired from his post in 1833, but, for a variety of reasons, cheating didn't become a fixed tenet of the code until 1907.

Evidence that cadets and the Academy were ambivalent with respect to cheating as a tenet of the code was found in writings which indicated there were several known instances of cheating in the latter half of the 19th century, for which the offenders weren't dismissed. On May 9, 1905, in a letter written in response to a questionnaire about the honor system, sent by the University of Chicago, the Academy adjutant wrote:

> It is not a point of honor with cadets not to obtain information unauthorizedly. By this I mean that if a cadet is ever caught cheating, his punishment, while very severe, does not include necessarily dismissal from the Military Academy." In the same letter the officer went on to explain, "The honor [code] which we have involves this and only this: that the *word* of a cadet is never questioned.

Just two years later, the superintendent issued a written directive that cheating would in fact fall under the honor code. The issuing of the directive reflected the developing "official" nature of the honor code as part of Academy policy.

Though cadets and officers didn't tolerate stealing throughout the Academy's early history, theft was treated as a violation of regulations. Cadets caught stealing or committing other serious offenses received court-martials, and if found guilty, separated as a minimum. In the mid-1920s stealing became part of the code.

The honor system was yet another matter. The system and code were not one and the same. The honor system supported the honor code, and gave the code more specific meaning and application, practical substance, enabling cadets to apply the ideals on a daily basis.

In the first 121 years of Academy history, until 1923, the honor system was *ad hoc* and was no system at all. The cadets enforced the code primarily by tradition, that is, accepted practice or precedent, except in the more severe cases brought to the attention of Academy authorities.

Unfortunately, because there wasn't a formalized cadet honor system until 1923, records of what was done in honor cases were almost non-existent. Little was known about decisions regarding guilt or innocence and how sanctions would be carried out against cadets believed to have violated the code. While Academy officials were, in most instances, aware of the Corps' actions in matters of honor, unless a case stirred public interest, officers remained aloof from disciplinary sanctions meted out by the cadets and wrote or said little about enforcement.

Based on sketchy historical evidence, it appears minor violations of the code often resulted in the offending cadet being directly confronted by the offended party. The issue would then be settled in a duel of some type, the most common being fisticuffs. If Academy officials were made aware of the dishonorable act, the offending cadet would also be punished under the cadet disciplinary system. In very serious cases, the punishment might result in dismissal. Dismissals could be directed by the superintendent without a formal hearing or investigation, although in most cases, it appears a thorough investigation was conducted.

In the late 1800s the cadets began forming grievance committees to study various aspects of cadet life. A Vigilance Committee was created to deal with honor matters. A forerunner of the Honor Committee, the Vigilance Committee investigated possible honor violations and reported decisions to the cadet chain of command. If a cadet was found guilty, the cadet chain of command would often ask the offending cadet to resign. A cadet who elected not to leave the Academy could be reported to the Commandant for an independent investigation. Although the Vigilance Committee had no official recognition by the Academy, its existence was tacitly approved and its decisions unofficially sanctioned.

At some point during the same period, the "cut," or "silence," came into effect in the Corps of Cadets. The Corps, in exercising its assumed investigative and punitive powers, occasionally encountered honor violators who, though asked to resign, refused, and instead requested that officers investigate their cases. Apparently, in a few instances, officers found the cadets "not guilty," and they were reinstated in the Corps. Men found guilty by the Corps' cadet chain of command and reinstated were met with the "cut" or "silence." The silence was still in effect in 1951, when organized cheating was exposed.

In the spring of 1922, during Brigadier General MacArthur's final months as superintendent, the honor system was officially sanctioned by the Academy and the Army. The recognition of the system resulted in the Vigilance Committee becoming the Honor Committee. The 1923 *Bugle Notes* and *Howitzer*, the cadet yearbook, reflected the change in status which was in MacArthur's final, annual superintendent's report to the Army. The two books also reflected the cadets' understanding of the serious responsibilities they were undertaking.

When the system was sanctioned, the Academy and the Army became the punitive authorities for those cases in which cadets were found guilty. At the same time the two-tier system of investigation and judgment became part of the honor system.

The Honor Committee was responsible for investigating honor violations. When the Committee investigated a cadet and found him guilty of violating the code, the case was given to the Commandant. He reviewed it, confronted the cadet with the Honor Committee's evidence and decision, and offered him resignation with an honorable discharge. However, the cadet retained the right to request a hearing before a board of officers. The right to request such a hearing remained in effect into World War II, until the change in discharge policy which occurred in the 1940s.

By the mid-20th century, the honor system was guided by the "General Principles" contained in *Bugle Notes*, the underlying rationale of the "Principles," and their accompanying logic – the way of reasoning about honor and its implications. The Honor Committee and the means of its members' selection were well established, as were its responsibilities and how it operated to meet those responsibilities. A set of interpretations had evolved from questions raised during or after honor cases and changes that had occurred in the regulation of cadet life and behavior.

Among cadets in the years following official recognition of the honor system, Honor Committees' interpretations of the code and its system were sources of frustration and fluctuating levels of confidence in the system. There were periodic complaints the Committees were turning the code into very detailed rules, often seeming like cadet regulations. Such rule-making obscured the true meaning and higher purposes of the code. The chairmen of the 1934 and 1947 Committees both commented on the importance of maintaining the "spirit of the code," and as the 1947 chairman put it, doing away with the "many poop sheets and interpretations that have come down through the years."

In 1950, the duties of the Honor Committee and its honor system interpretations were described in a September 1, nine-page paper entitled HONOR COMMITTEE NOTES 1950, a paper approved by Colonel Paul Harkins. The paper summarized code and system interpretations in thirty-seven subject areas such as absentees, absence cards, academic policy, borrowing, inspections, late lights, leaves, limits, liquor, narcotics, sick call, and visitors. There was also a description of the "silence" as it had evolved since the latter 1800s.

The honor system included training administered by the first class to the Corps of Cadets, to ensure the Corps and its members were the collective and individual custodians of the code and its system, and that both were passed from class to class, generation to generation. Cadets' custodial role, and their responsibilities in maintaining the code and its system, and passing them to succeeding classes, were jealously guarded.

There was, however, another important dimension in the training which was part of the Academy's honor code and system: Honor Committee members' training, their preparation to serve as officers in the Army or Air Force.

The Honor Committee was the sole cadet authority routinely given the considerable responsibilities of investigating alleged honor violations and rendering judgments as to guilt or innocence of accused cadets, judgments that could result in severe penalties. Though the Committee hadn't the authority to discharge cadets found guilty, the Committee in fact carried heavier responsibilities than any other cadet authority sanctioned at West Point by the Army or the Academy.

When the Honor Committee was performing in its investigative role, or formed as an honor hearing panel, it was shouldering responsibilities and authorities similar to those established in the armed services' Articles of War, administrative disciplinary regulations, and regulations governing various fact finding investigations. Thus, in such roles, Honor Committee members, in addition to their role in maintaining the high ideals and standards of the Corps of Cadets, the Academy, and the Army, received excellent training to prepare for similar responsibilities as commissioned officers.

THE COLLINS BOARD REPORT – THE HARD EVIDENCE OF WHAT WENT WRONG

Buried in the testimony of the Collins Board was evidence pointing to what went wrong, and why there was a cheating scandal at West Point in 1951. The questions asked by the Board members, though not contained in the Board record, were questions routinely asked in any thorough investigation, "What, when, where, how, why, and who?" The answers, from young men caught up in activities they knew were not in keeping with the honor code, also led to significant omissions in later investigations intended to find the causes, and take actions to avoid a repeat of the incident.

"How did you become involved in the ring?"

The Company K-1 yearling, *George Hendricks*, the first to testify before the Board, and the son of an Academy graduate, replied:

"I first heard about this ring during Fourth Class year [1949-50 academic year] but initially I had nothing to do with it.... Until that time [shortly after Christmas, the end of the first semester,] I had no trouble with academics but I then began to have difficulty with English. [My roommate, a football player] said that maybe he could help me out and wanted to know if I could help get the Writ [test] questions. I agreed to go along with him and on several occasions I got the answers to Writ questions and gave them to [the roommate who offered to help me with my English difficulties]...."

"Can you tell us how the ring got started?"

"...The background on this thing as I know it is that it was started by football players for football players. Their classes differ so much from the rest of the Corps that they frequently take the Writs after others take them. When they enter as Fourth Classmen they generally work very hard but in case they are having difficulty with academics this system of procuring information is passed on to them and they continue what they consider to be a standard practice...."

"How does the ring work?"

"...The way the ring works is that they have a representative in each company and the man taking the Writ at the early period has to leave the Writ, whether he finishes or not, to bring the answers to the people who will be going to the next class. They meet in the [cadet] area, in the library, or the hallways of the academic building. If they have a little time and there is an hour between classes, they have someone go to the man's room, get the problems, work them out, and then pass them on to the other members of the ring. There are two rings operating, one in the First Regiment, and one in the Second Regiment...."

"If you knew this was not in keeping with the honor code, why did you continue your participation?"

"...It is hard for me to explain this. I know it is wrong. I know we should not have done it but the group that was in it just felt that there was safety in numbers and there were so many finally involved that no action could be taken without disgracing the Corps of Cadets...."

Another third classman, *Ted Strong, Michael Arrison's* roommate in Company H-2 responded to the Board's questions.

"How did you become involved in the ring?"

"I have been the academic coach for [a football player]...in G-2. As his coach I have also helped [his] roommate, [who is also a football player]. I hope to be a football academic coach next year. I noticed that some of the problems that [his roommate] gave me to work out were exactly the same as those given on the next Writ. I worked out the problems for [his roommate] several times and gave the answers back to him. I realized that it was unauthorized information as soon as I took the Writ...."

"Did you know you were violating the honor code?"

"...I realize that I should have turned [his roommate] in for an honor violation and I know that I am guilty of an honor violation in consequence...."

A third classman, a football player in Company D-1, who had lied under oath in his first appearance before the Board, but at Earl Blaik's urging, returned voluntarily to recant his previous testimony:

"I want to retract all that I said on May 29. It was wrong. I was so stunned when I was called in I didn't know what to say. There were so many close friends involved that I couldn't bring myself to tell; however, I realize I was testifying under oath and I want to straighten things out. I got involved in the exchange of unauthorized information during the Christmas Writs plebe year [1949].... I did it to stay in the Academy; I could not have passed without it. I do not know how the practice started. Knowledge of it has been passed on from year to year. Some of the highest ranking cadets have been members of the group.... I know this was a violation of the honor system, but when I knew all those other people had done it and when I wanted to stay here so much it was either get help or go home.... I have discussed unauthorized information at the [training] table in the dining hall and have also passed out papers which contained unauthorized information...."

Another D-1 football player who also hadn't told the truth in a prior appearance before the Board:

> "...although I was aware of the honor system and its requirements, my loyalty to my teammates seemed bigger. I was given the impression that the practice was passed on from upperclassmen.... At first I didn't know how big the thing was. Now I know it is in all classes.... The football players tried to keep it among themselves. Someone said that there was one man on the Honor Committee that would warn us in time. Last night... a number of us had a meeting with Colonel Blaik. He was shocked. He told us to tell the truth and that is why I am back here...."

The eighty-three men who resigned following the events of that spring and summer expressed a variety of motivations and circumstances for their initial involvement and continued participation in organized cheating at West Point. There were men who entered the ring voluntarily, at the invitation of others already involved. There were men who did so out of loyalty to teammates or friends, such as the D-1 third classman whose "loyalty to teammates seemed bigger." They wanted to "help" their teammates or friends succeed in academics.

There were others like *Ted Strong*, the H-2 third classman who simply hoped to become a football academic coach and discovered himself enmeshed in working out answers to test questions and passing them to the men he was tutoring. Some, such as the football player from D-1, explained their actions by saying they desperately wanted to stay at West Point for an education and be commissioned officers, but believed they were unable to meet the academic standards and thus couldn't graduate without cheating. Among them were men who could have succeeded without cheating, but became involved, and rather than exert themselves, relied on cheating to decrease the time devoted to study.

Some learned of the ring's existence, they said, because a few men involved in the ring openly told them of the availability of "help" if they ever got into academic trouble, that there was a way to "beat the academic system" if help was needed. Testimony from two football players indicated two highly respected graduates, who had been football heroes, had told two separate groups of football recruits at the "cram" schools in the spring of 1949 and 1950 about the "help" available. The two, who purportedly heard the graduates tell of the "help" available, later accepted the assistance, which eventually led to their resignations.

There were men who apparently had little or no regard for the training they were receiving or the standards they were expected to adhere to and spoke disdainfully or sarcastically of having to "play soldier" or do "this cadet stuff."

There were a very few men whose sole interest in being at West Point was playing football on a nationally recognized team. They regarded academics and military training as obstacles they needed to overcome or get around – another "system to beat," similar to the "plebe system" – en route to glory on the gridiron. Among them were men who were later described by their fellow ring members as "recalcitrants, 'bums'." "They would do anything to violate regulations, and didn't belong at the Academy."

Other ring members explained they were drawn by their admiration of men who – to their surprise, and usually to their disappointment – were already involved in organized cheating. This was particularly tragic, because the young third and fourth classmen had considered these men heroes, admirable examples to be imitated and followed. Not only had the young hero worshipers been caught up in activities they knew to be wrong, they had suffered the disillusionment of learning their heroes, team leaders, and cadet leaders were tarnished, fallible, or seemed to have darker sides, in contrast to their apparent upright, shining behavior.

Some said the football team was "the supreme rallying point of the Corps" and "was the rock on which the Corps was built" – words they had heard from Coach Blaik, and which to them apparently implied the team was something apart from the Corps, larger, and more important than the Corps or its honor code.

There were instances when roommates, trying to ignore, or wanting to disbelieve what they were seeing and hearing in the confines of their cadet rooms, eventually became participants in the ring. There were other cases where ring members carefully conducted their activities to avoid involving roommates, fearing either the roommate would report them – or simply wanting to avoid entangling them in activities they knew to be wrong. A few, in Board testimony, made a point of identifying a roommate or classmate as unaware of the ring and not involved in its activities.

Some were caught in a web they didn't know existed, somewhat similar to the one that enmeshed the H-2 academic coach *Ted Strong*. Information would be passed to them, apparently valid and above board, and they would begin to suspect, perhaps in the next day's classroom recitation, that the advance information they received was questionable, if not unequivocally improper. By the time they were convinced they were involved in cheating, they faced serious dilemmas they hadn't the will or courage to confront and correct. Their dilemmas invariably were decisions on whether to report close friends, teammates, or classmates.

Frequently, men who found themselves unwittingly involved in cheating expressed dismay when the discovery was made and wanted to extricate themselves by reporting what had happened. They were reluctant non-volunteers in cheating. Ring members would discourage them by persuasively repeating convenient rationalizations, illusions, questionable assertions, rumors, and as matters turned out, some elements of truth that had grown up in the expanding ring. "This is big. You would be bringing down hundreds of cadets. There are even members of the Honor Committee involved." "The authorities know about this. Why else do you think something of this magnitude goes on?" "How do you think we got this football team?"

Intimidation was another tactic used to keep reluctant, involuntary members, or guilt-plagued recruits from exposing the ring, a tactic used increasingly when the ring began to unravel during the Collins Board's investigation.

There were men caught up in the ring who eventually felt enormous guilt over their involvement in activities they knew were wrong. At least two in the class of 1952 made serious attempts to rectify their predicament by going to respected football teammates who were leaders in the Corps. One was particularly disillusioned when he talked with a well known leader on the undefeated 1949 team and was told to "forget it." In spite of his stunning discovery he pursued the matter further and received an even more unsettling revelation when his cadet company commander in the class of 1951 gave him essentially the same response. Instead of pushing on until finding someone who would take action to break up the growing ring, both men made the mistake of acquiescing to the advice – or the company commander's "direction." Their failure to persist and break open the expanding subterfuge cost them dearly.

There was one bitterly angry participant in the ring who believed "the system isn't fair. It will screw you every time.... So why not cheat?" Others, to explain their actions, pointed to cheating as common in colleges and universities, and they saw West Point as no different from any other college or university.

Also contained in the Collins Board's report, in its discussion of the evidence, was a statement – a conclusion – that was perhaps unwarranted, not carefully considered. "The Honor Orientation that these men received was adequate and complete. They knew they were in the wrong; they were surprised and shocked when they first heard about this practice." This statement by the Collins Board, repeated almost verbatim in the Hand Board Report, was apparently accepted without thoughtful examination, until early in 1952, several months after the Bartlett Board had concluded that overemphasis on winning football had been the fundamental cause of the honor scandal.

Aside from the obvious honor code violations inferred by massive, carefully rationalized cheating in academics in the 1951 incident, there was another serious failure in the cadets' support of the honor system, centered on the most controversial and difficult to understand feature of the system. Yet, this feature is the cornerstone, the underpinning of a system that is unique, sets the Academy apart as a special place, and as General Eisenhower said, provides "a system that actually works."

In an unbroken chain, from year to year since 1923, the year after the honor system was officially sanctioned during MacArthur's time as superintendent, the feature was described in *Bugle Notes*, the small book of plebe knowledge handed every cadet when he entered the Academy. Among the six principles in the "General Principles upon which the honor code was founded," were the words: "Every man is honor bound to report any breach of honor which comes to his attention."

During those twenty-eight years the contents of the "General Principles" had changed somewhat, and the order in which they were presented in *Bugle Notes* changed. In 1951 it was the 5th General Principle. But the words were virtually the same as in 1923, the meaning exactly the same.

THE ROAD TO COLLECTIVE DISHONOR: "KEEP SILENT, DON'T TELL – IGNORE THE 5TH PRINCIPLE"

In the Bartlett Board report was a brief, revealing discussion of the 5th General Principle underlying the honor code, although Boyd Bartlett, Frank Greene, and Harlow Miles didn't identify the subject as a principle, and didn't differentiate between honor code and honor system:

> ...If the ideal of the honor system is strong in the mind of the normal cadet, strong motivation is required to break it down. The idea of standing by a friend, of not telling on another, is commonplace in the world at large. Many a boy who would not himself cheat would not inform on another boy who did.

> Beliefs and attitudes of mind govern action. Action follows only after the idea of action has been conceived. We act in accordance with our real beliefs. The fact that an individual or group belief exists has nothing whatsoever to do with whether or not the belief is justified. If the belief is firmly held the action will result as surely from an erroneous as from a correct one.

The Board's brief comments on the 5th Principle were part of the three officers' explanation of how the investigation would proceed, their description of the "NATURE OF INVESTIGATION REQUIRED" – their method of approaching the questions "What caused the 1951 honor incident?" and "How can a repetition of such an incident be avoided?"

The Bartlett Board had made another fundamental error. They had taken what was a contributing cause of the scandal and set it aside, made it a part of their rationale on how to approach the larger questions, and thereby failed to examine it in their investigation. In the preceding five years, while the seeds for the cheating ring were planted, forty-four other cadets were separated from the Academy in individual honor cases. What clearly distinguished this honor incident from all others in the Academy's 149 years was the massive breakdown in cadet support of the 5th General Principle – "Every man is honor bound to report every breach of honor that comes to his attention." If the breakdown hadn't occurred, if cadets who knew or believed cheating was in progress promptly reported their observations, there wouldn't have been a scandal. Instead, for a variety of reasons, they rationalized inaction, which eventually spawned a "conspiracy of silence" and a web of deceit.

Whether the Board's oversight was unintentional or deliberate will never be known. In all probability, it was both, because the honor code belonged to the Corps – which was under the command of the Commandant. It would be the province of the Corps to examine its honor system. It would be the province of the Commandant to determine, in concert with the cadets, how much and what type of honor education and training cadets should receive.

The Board's concentration on "psychological factors," beliefs, and opinions of the "West Point community" to find the causes for the incident, while of considerable value, was an approach to the problem that caused them to stray from more essential considerations and facts, and bring a more rapid closure on the incident's fundamental causes. Neither of their two recommendations to examine the honor system in greater detail gave the slightest hint they took note of the obvious breakdowns in cadet understanding and support of the 5th Principle.

Further, the Bartlett Board's explanation of the tendencies of "normal cadets" as opposed to the tendencies of the "world at large" seemed to indicate they believed the men admitted to the Academy would become "normal cadets" when given four hours of honor instruction in Beast Barracks, plus additional informal lectures and discussions – except for those cadets recruited to play football.

Honor appears a natural, bright, shining, much desired, much sought after, inspiring ideal, but like any great intangible, requires deep understanding, commitment, moral and professional courage – and, in General Irving's words, "hard work" – to become second nature in any person's life.

Without a doubt, cadets involved in organized cheating, and many others who suspected something was wrong, had failed on a massive scale to successfully confront and stop what, by the spring of 1951, had become a disastrous erosion of the concept of honor and its system, which had been developed in the Corps of Cadets over a period of 149 years.

The questions the Bartlett Board should have included in their investigation were "Why did the breakdown occur?" and "How can the Corps and the Academy counter the 'commonplace' belief in not 'telling on others,' and rationally explain all six principles, especially the 5th principle – and help cadets understand that some of the greatest challenges to truth, morality, and the rule of law come from undisclosed, unbridled corruption discovered among friends, roommates, teammates, professional associates, colleagues, and sometimes family?"

Not in the Bartlett Board, on which Frank Greene served, or in the Greene Board he later chaired – which was asked to examine "Regulations Impinging On Honor" – were these questions addressed. Nor did the cadet committee, which had been recommended in the Bartlett Board report "to review its honor system to make sure that in all respects it is simple, clear, and unencumbered with unnecessary technicalities."

Fortunately, however, the Greene Board, which completed its work 16 January 1952, did dig deeply into the honor system and the plethora of regulations and policies sowing uncertainty and frustration among men who supported the honor code and system, and indifference, anger, and cynicism among men already irritated by what they saw as the Academy's demanding curriculum and imperfections.

Frank Greene's Board found fault lines in the administration of academics. Interviews with professors, instructors, and cadets, and examinations of academic policies and practices clearly exposed weaknesses in the administration of academics, weaknesses that could directly or indirectly affect the day to day application of the honor code and system. Department heads, the senior professors, were issuing uncoordinated and differing policies affecting how cadets studied, the references they could use in studying, to what extent they could seek assistance in study assignments, how to properly do "take home tests" and laboratory experiments, and what procedures would be used when classroom recitations and writs were in progress. The academic policies and procedures included wide variances between departments, some contradictory instructions, different definitions and interpretations of terms – for example, plagiarism.

Over the years since the curriculum was reconstituted during General Maxwell Taylor's time as superintendent, cadet company academic coaches, and academic coaches assisting Corps Squad athletes – including football players – had collected and organized into books, by subject, materials from prior academic years to use for study, including sample tests and test questions. They then passed the books on to men in succeeding classes to use in tutoring. Prior years' materials could be used for study, and instructors often would simply reuse the same lesson plans, recitation, and test materials the next year, at virtually the same point in their schedule of instruction. Thus it could appear that men were passing unauthorized, advance information when in fact they were simply drawing on last year's classroom materials and experiences to be prepared for the next day's classes.

Orientations on the cadet honor code and system for new academic instructors were being given by different people within the departments to which the new instructors were assigned. Not surprisingly, orientations differed in content and emphasis and weren't necessarily accurate.

Similar shortcomings were found in the Tactical Department and the Academy staff, where policies, regulations, orders, and instructions affecting virtually every aspect of cadet life were written, published, issued, and used to define or affect cadet behavior.

Tactical Officers, like newly assigned academic instructors, said their orientations on the cadet honor code and system weren't sufficient, or were inconsistent in either content or emphasis.

The Assistant Commandant's duties included an advisory role to the Cadet Honor Committee, and he was to be the point of contact on the Commandant's staff for honor representatives needing support and assistance. Interviews with cadets resulted in the conclusion that the position of Assistant Commandant didn't lend itself to adequate support, advice, and counsel in matters of honor – that the Assistant Commandant often seemed "too busy" to afford the necessary support.

The Greene Board also dug deeply into the questions of regulations "impinging on honor" and the extent to which honor was used to enforce regulations. The Board identified policies, regulations, and other administrative instructions that clearly had honor implications in their day to day practice, and recommended numerous changes to de-couple honor and regulation enforcement.

But the Board also pointedly concluded that the great majority of regulations, and their attendant enforcement, would, of necessity, always have honor implications in their practice, because of signature, certification, and other reporting requirements, as well as the need for accuracy.

The Greene Board did examine in detail the honor orientation given new cadets that summer. They reviewed the scripts for the cadet lectures and exhibited them in their Board report. Notably, the first lecture was on August 4, one day after the public announcement of the cheating incident, and had been scheduled months in advance by the Tactical Department's Training Officer – without knowledge of events in progress leading to the announcement.

The August 4 lecture touched only briefly on the 5th principle. Cadet Louis V. Tomasetti, class of '52, the Sixth New Cadet Company honor representative and the Company B-2 representative during the academic year, spoke to each of the GENERAL PRINCIPLES during that first, formal lecture:

> ...We have considered honor offenses and the results, but how do we know who committed the violations? Is there a police force here to enforce honor? The answer is, of course, no. You and I, the Corps of Cadets, hold responsibility for apprehending the wrongdoers. Every cadet is honor bound to report an honor violation.' That means any time you see an honor violation, you report it – or suffer the punishment of involving yourselves in an honor violation, and subsequent dismissal. If you fail to report an offense both you and the offender are guilty of an honor violation. This may be difficult to comprehend at first, but when you have learned to respect the honor code as we have then you will see the reason. Nothing is bigger than the honor of the Corps. Of this, Gentlemen, we are extremely proud.

Frank Greene's Board recommended several specific changes related to honor training for cadets and honor code and system orientations for the Academy staff and faculty. There were two sweeping recommendations pertaining to cadets' honor training and character development: First, "The entire subject of training in honor be studied by the Commandant of Cadets with a view toward reorganizing the schedule, method of presentation, timing, and content on a sounder basis." Second, "The overall character and moral training at West Point be reviewed in order to determine whether or not it is adequately guided and coordinated and attains the overall desired coverage and results."

In all, the Greene Board produced fifty-five specific recommended changes affecting honor, including: regulations published by the Tactical and Academic Departments; honor administration by Academy authorities; orientation and training of cadets, Academy staff, and faculty on the honor code and system; assignment of an officer to the Commandant's staff to provide advice and counsel to both cadets and officers regarding the honor code and system; the processing of cadets found guilty of honor violations; the type of discharge given cadets found guilty of honor violations; and overall character and moral training given at West Point. All but a very few of the fifty-five recommendations were put in place over the succeeding two years.

The Board report discussed the various scenarios under which a cadet might be separated from the Academy after a finding of guilty for honor violations. It pointed to an obviously sharp disparity in findings of guilty by the Cadet Honor Committee and Boards of Officers, if cadets found guilty by the Honor Committee refused to resign and instead asked to be judged by a board of officers. The Board's review of recent history made clear that cadets refusing to resign and accept a less than honorable discharge were few in number, but when they did, the Boards almost invariably found them not guilty, and they were reinstated in the Corps of Cadets. This was due to the more onerous rules of evidence required in a hearing before officers.

The Board didn't explore the possible effect of the "silence" on the honor system, or its relationship to the punitive authority retained by the Academy and Army, although the silence could be considered a punitive measure. The silence had been exercised in two cases involving cadets in the class of 1951, during their final semester at the Academy, one case Collins Board testimony suggested could have been related to the cheating scandal. Because the code and system "belonged to the cadets," an examination of the silence was considered the province of the cadets. What the Greene Board decided to recommend with respect to the Academy's and Army's punitive authority was a return to the pre-World War II Army policy of offering an honorable discharge, unless the cadets offense was also "an offense against society."

In spite of the important strides made by the Bartlett and Greene Boards in coming to grips with the basic issues evident in the 1951 honor incident, there were three glaring omissions other than the failure to identify and attempt to rectify the breakdown of cadet support for the 5th principle. They were the apparent severing, in the minds of the cadets who became involved in the cheating ring, of the connection between leadership and command, and honor; the failure to clearly recognize a growing tendency to disassociate duty and honor; and the failure to consider the broader perspective and context in which the cadet honor code and system existed.

LEADERSHIP, COMMAND, AND DISHONOR – IN WAR, "THE FAILURE OF A THOUSAND DEATHS"

Louis Tomasetti also spoke of the future to the class of '55 in his lecture about honor that August 4.

...An officer told me of a story concerning three battalions in the attack of an enemy stronghold. The regimental commander gave all battalion commanders orders to move at a certain hour. The right [flank] battalion was to act as a diversionary force, attacking first. At H-hour [the planned hour for the attack to begin], the Regimental Commander called the commander of the diversionary force to find out if they were moving. The right [flank] battalion was late, still in the process of moving up to the attack position. However, [the battalion commander] reported his men were moving, realizing the regimental commander would assume the diversionary attack would have begun. At H-hour plus thirty [minutes] the main body moved into the attack and was repelled with heavy casualties. The diversionary force was still preparing for the attack and suffered no losses. Because of the lack of exactness on the part of the diversionary force commander, defeat resulted instead of victory. We cannot afford to fight this type of war, losing thousands of American soldiers because one commander lacked the integrity to report he had failed in his orders. Remember this story, Gentlemen, for when you graduate you will be in command of a unit and your word concerning the unit's action will be accepted as the truth – be sure it is.

The story told by Louis Tomasetti was of life, death, leadership, command, and honor – the need for absolute integrity in war. None

403

could dispute the argument for integrity in combat. The terrible price of its absence is self-evident. But how do men who are placed in leadership positions reach the point of failing their integrity, even when the price can be so high, and is so self-evident? The answer to the question doesn't come easy.

Was the battalion commander's failure part of a well developed pattern of behavior too subtle to detect? Was it a one time event, an aberration in which pressure, fear, or ambition moved him to avoid or hedge on the truth? Or was his conduct predictable, and no one recognized it, or if they did, chose to ignore it, or because of uncertainty, friendship, or fear of professional consequences, accepted his behavior? These questions were not unlike those faced at West Point, by young officer candidates who, prior to the scandal of 1951, confronted increasingly curious behavior, and mounting evidence of wrongdoing, by men they trusted.

The answers frequently come too late, when disaster is already the result. Such was the case in the wartime incident described by Louis Tomasetti to the class of '55 the day after the public announcement of August 3, and in each case of the eighty-three men who resigned from the Military Academy beginning later that month.

*　*　*

General Eisenhower's 1946 letter to Maxwell Taylor focused on three subjects: the cadet honor code and system, and Eisenhower's belief the Academy should include in its curriculum "a course in practical or applied psychology." He explained,

> ...I realize that tremendous advances have been made in the matter of leadership and personnel management since I was a cadet. Nevertheless I am sure that it is a subject that should receive the constant and anxious care of the superintendent and his assistants on the Academic Board and these should frequently call in for consultation experts from other schools and from among persons who have made an outstanding success in industrial and economic life. Too frequently we find young officers trying to use empirical and ritualistic methods in handling of individuals — I think that both theoretical and practi-

cal instruction along this line could, at the very least, awaken the majority of cadets to the necessity for handling human problems on a human basis and do much to improve leadership and personnel handling in the Army at large.

I am told since my days as a cadet much has been done, particularly in the first class year, in inculcating a sense of responsibility in the man who is soon to be a 2nd Lieutenant. The more we can do along this line, the better for us. The cadet should graduate with some justifiable self-confidence in his ability to handle small groups of men, to organize any appropriate task, and to see it through in a satisfactory fashion with the men under his command....

The outgrowth of General Eisenhower's wish to strengthen the Academy's leadership training was the Military Psychology and Leadership Course. Colonel Russell P. "Red" Reeder, then serving in the Tactical Department, was selected to develop and integrate the course into the Academy curriculum.

Reeder was the highly respected combat veteran, who, leading his regiment, had landed on Omaha Beach on D-Day and lost a leg as a result of wounds received in Normandy a few days after the Allies' World War II invasion of France. Consistent with wartime policies on manning of the staff and faculty at West Point, Reeder was granted his wish to be reassigned to the Academy after recuperating from his wounds.

He developed a course which included principles of leadership generally accepted in the Army. By the 1950-51 academic year the Academy had expanded the course to include all four classes, and heavily emphasized leadership principles in its military training courses. Among them, one in particular was repeatedly mentioned, "Lead by example," a principle derived from a tenet of an ancient moral code, the "Golden Rule" – "Do unto others as you would have them do unto you." In Colonel Reeder's syllabus of instruction, which described the forty-two-hour course first given to the class of 1947 in the final semester of their first class year, he said:

Commencing with the first day that he enters the Military Academy and ending only with graduation, the cadet comes under the leadership influence of every officer and every senior cadet with whom he comes in contact. The teaching of leadership by example and instruction is therefore, as it always has been, the daily responsibility and privilege of each officer [and senior cadet] on duty at West Point.

Reeder knew well the obligation of military leaders to set high moral, ethical, and performance standards for those they lead, and daily live those standards. He had seen, heard, and understood why soldiers will follow officers they admire, especially officers who, in their daily lives, both demand and reflect morality; strength of character and conviction; knowledge; hard, fair, and just administration of training and discipline; and all the virtues which mark true, successful leaders. "What I do and what I say are the same. Do as I do. I will never ask of you what I wouldn't ask of myself."

For reasons difficult to understand, the cadets who were in the upper three classes, and were involved in or were aware of organized cheating in 1950-51, apparently didn't see their leadership roles and responsibilities in the Corps as inconsistent with incipient compromise of their integrity and behavior. Nor did they see the connection between their behavior and the future roles they would undertake as commissioned officers, leaders, and commanders, roles in which soldiers under their command are quick to discern inconsistent, contradictory, and hypocritical behavior, including disconnects between word and deed. In peace and in war, leadership, command, and dishonor together sow the seeds for disaster, for undermining command authority and effectiveness, destroying unit cohesiveness, good order and discipline, and in war, is "the failure of a thousand deaths."

* * *

The honor code and system at the United States Military Academy belonged to the cadets, a point of pride, and, as General Eisenhower emphasized, a factor essential to its success. But the honor code and system lived within both the Corps of Cadets and an institutional setting, the Academy, whose staff and faculty, as Colonel Reeder pointed out, had enormous influence over the attitudes of the young men learning to be officers and leaders. And the Academy lived within yet another institution, the United States Army, which provided both the larger context in which the honor code and system existed, as well as the life and career of service the Academy's graduates entered. This was the Army where, as Art Collins said, the Academy graduate should provide a "core of integrity...."

Although the Greene Board did recommend well considered improvements to strengthen the Academy's institutional setting for the honor code and system, the Board didn't directly address the broader context of life and career in which honor must survive and thrive, contexts the discharged cadets clearly pointed to as explanations or rationalizations for their behavior.

"Cheating occurs in other institutions, in the Army, and in life. What makes this place so special? You can't use it when you leave the Academy. Things are not like that in the 'real world.' You have to adapt to the 'real world.' You know what happens when you 'rat' on someone else."

Indeed, disillusioning collisions with the "real world" were frequently voiced reasons energized, enthusiastic, junior officers lamented in choosing not to continue careers past their mandatory time of service after graduation. Their experiences invariably involved high ideals and standards, and the jolting realization that the same ideals and standards aren't always sought after as vigorously outside the gray granite walls of West Point. The Army and life outside the Academy weren't perfect. And though the Academy wasn't perfect either, there was ever the knowledge, appearance, or perception that people at West Point strove mightily for the ideals and standards to which people would certainly aspire.

A well-defined system existed at West Point to enable honor's practical application in day-to-day life and in cadet education and training to prepare for service. But what of the mechanisms, the means for support, the systems available in life ahead – the future – to help steel the officer candidate against human fallibilities, the weights and pressures of tempting and potentially disastrous assaults on integrity – and sustain an environment in which honor can thrive? There was a need to put the honor code and system into context, relate it to everyday, routine service in the years ahead, and explain the systems and mechanisms available to officers and soldiers to hold fast to honor in the face of inevitable challenges to its survival and growth.

Obviously the Greene Board had no control over the "real world," and neither did the Academy. Nevertheless, the leadership of the Army and the Academy could exercise powerful influences over the effectiveness of an Officers' Code simply by describing to cadets the means and mechanisms available after graduation and commissioning to sustain and strengthen integrity in the Army and Air Force.

A CADET DOES NOT LIE, CHEAT, OR STEAL –
ONLY THE BEGINNING

The four months between the first revelations of massive cheating at the Academy, and the public announcements of August 3, had been long and painful for everyone caught up in the tragic events of that spring and summer. The fire storm of public controversy which followed the announcement made the events seem all the more unbearable for those directly involved. The Army's and the Academy's senior officers, stunned by the public counterattack by some of the young men about to be discharged, and the parents of others among them, were torn between wanting to expose the unrevealed depth and seriousness of some of the cadets' activities, and remaining silent in hopes the public furor and adverse, and often erroneous publicity, would subside.

On August 6, Senator William Benton of Connecticut, rose from his seat in the Senate chamber and asked the presiding officer, the vice president, to be recognized. He spoke at length to the senators present, recounting his experiences in discontinuing collegiate football at the University of Chicago during a six year period when he served as an officer on the University's Board of Trustees.

He urged the Senate to investigate the cheating incident with the intent of overhauling the Academy curriculum, reorganizing the faculty, examining academic standards – which he deemed didn't measure up to the standards of other, nationally recognized colleges and universities – and discontinuing football at West Point and Annapolis. It was a scathing attack, joined in by Senator Hunt of Wyoming, with similar views publicly voiced outside the chamber by Senator J. William Fullbright of Arkansas. Benton attacked collegiate football with a vengeance.

Big time college football has nothing to do with education, and it certainly has not a thing to do with the education of Army or

Navy officers. Such football misleads the institutions. It misleads and betrays the students. It fools the public as to what a university is all about. If West Point and Annapolis were to get rid of the football rat-race, they would strike a blow not only to serve their own interests at this time, but they would strike a blow toward freeing all American institutions from this incubus which corrupts them. They would give courage to dozens of university presidents eager to follow such an example and eager to destroy the evil which has taken over such a strangle hold on much of college life.

Benton went on to say "the 90" were the "victims of a vicious system and not its perpetrators" He followed up his attack with an August 11 letter to Secretary of the Army Frank Pace, reiterating his remarks on the floor of the Senate.

But the threat of a full scale Senate investigation eventually died, fended off largely by a reply from the Army's Department Counsel, who cautioned against precipitate action, encouraged calm deliberation, and reassured Benton that annual reviews of the Academy's curriculum were already in place, and a matter of constant interest to the Army. Senior members of the Army staff were ready, however, with other information should the Senator and his allies press for an investigation.

When, in mid-August, one of the cadets wrote the Army Chief of Staff, General J. Lawton Collins, admitting all were guilty of cheating and pleading for honorable discharges, several senior officers wanted to publish the letter and General Collins' reply. There was anger in Collins' senior staff, caused by the activities of some of the young men about to be discharged.

The officers were incensed for several reasons. First, among the ninety-four cadets found guilty by the Collins Board, the Board's report identified twenty-two as having lied under oath in their first appearance before the Board. The officers were also aware a number of the cadets, after the Board had completed its report, alleged the three officers hadn't conducted their investigation according to military law and regulations, saying the Board used coercion and intimidation to obtain testimony and admissions of guilt, and hadn't advised the cadets of their rights against self-incrimination before they testified. The allegations of a "raw deal" had been so persistent and so effective, before and after the public announcement, that the Army's Staff Judge Advocate General, at Maxwell Taylor's direction, sent affidavits for Collins Board members to sign, requiring the three officers to swear they conducted the Board according to the Articles of War. Then had come the admission and apologies, conveyed to Board president Art Collins by one of the cadets, that the allegations against the Collins Board were deliberately fabricated in a desperate gamble by the cadets to remain at the Academy.

After the announcement of August 3, a few cadets, and some of their parents, publicly attacked the honor code and system. And perhaps worst of all, while a number of the dismissed cadets publicly stressed they "accepted a [cheating] system of long standing," thus rationalizing their participation by pointing to classes before them, "There was [evidence of] a fairly well defined conspiracy to control the Honor Committee," by electing representatives in the class of 1953 who were known members of the ring. The cadets testified that election of two ring members in two different companies was deliberate, and a third had been attempted, all aimed at controlling the outcome of honor hearings.

In spite of the anger stirred among General Collins' senior staff, and the urge to retaliate by publicly disclosing the additional misdeeds of the men about to be discharged, the Army's chief of public information advised against publicizing the eighty-eight cadets' letter to General Collins. He suggested the press would focus more upon the type of discharges the men would receive than their admission of guilt. He went on to point out that publicity surrounding the incident had begun to decrease, and the letter would merely stir additional controversy.

On August 29, in a memorandum to Maxwell Taylor, the Army's Inspector General, Major General Louis A. Craig suggested information about the cadets' as yet undisclosed misdeeds be withheld for other reasons. Referring to the possibility of a Congressional investigation, he recommended the Department of the Army avoid any "counter publicity," pointing to "facts...given no attention in the press...and of great importance at the proper time." Among the facts were the additional, unrevealed misdeeds of the cadets; the Department of the Army's awarding the cadets discharges neither honorable nor dishonorable was essentially the same as the discharge procedure used in the cases of other Army personnel who possessed traits of character inconsistent with their assignments; and the considerable influence Judge Learned Hand's presence as Chairman on the July 23-25 Review Board would have on a Congressional committee, given the Board had supported General Irving's recommendation to dismiss the cadets.

On August 11, the same day Senator Benton wrote defense officials of his wish to open a Senate investigation, three officers at the Academy, in prearranged sessions, met with members of the press, releasing stories briefly explaining the history and purposes of the honor code and system. Captain John S. D. Eisenhower, Dwight Eisenhower's son, who was an English instructor at the Academy, gave his to the *Associated Press*, which subsequently carried the story. Colonel Waters, the Commandant, and Colonel George A. Lincoln, one of two Professors in the Social Science Department explained the honor code and system to the *International News Service*, which also published their presentations. However, the reports explaining the honor code and system were buried in a still moving avalanche of stories about the scandal, most absent any information about honor and its meaning to the Corps of Cadets, the Academy, and the Army.

Two days later Colonel Billy Leer and Major Joseph Cutrona, in the Academy's Public Information Office, began drafting a "roundup" statement they and the Army's Public Information Department believed the superintendent should make, preferably in a brief fifteen-minute appearance before the press. The statement was to summarize and explain what had occurred and answer the numerous questions raised as a result of the August 3 announcement. On August 17, the Army Chief of Staff, at Maxwell Taylor's urging, disapproved public release of the statement. Instead, it became the basis of a letter to the Association of Graduates, published in its entirety in the Association's magazine, *Assembly*, and a letter to the parents of every member of the Corps of Cadets.

The letter answered numerous questions about the cheating ring, how it came about, where it originated in the Corps of Cadets, how it operated, and why it sustained itself and grew to such disastrous proportions by the spring of 1951. However, the letter had other unintended effects.

Most notably, one paragraph, taken out of context, stirred far more controversy and has reverberated down through the years because of the adverse, often harshly critical publicity the Army football team, and football in general, had received following the August 3 public announcement. General Irving wrote this paragraph under the heading:

Army Football

The cheating was concentrated in and associated with the football players. The remainder were their roommates and other close associates. However, this situation has nothing directly to do with football in itself. Rather, these men apparently came to think of themselves as a group apart, and developed a set of values different from those of the remainder of the Corps of Cadets. Athletics are an integral part of our training. We still subscribe to General MacArthur's word: 'On the fields of friendly strife, are sown the seeds that, upon other fields, on other days, will bear the fruits of victory.' We will have a fighting varsity team, completely supported by the Corps. This is part of our building job and we hope for understanding and assistance from everyone in this effort. Meanwhile, we shall take steps to assure that the athletic squads are not allowed to grow apart from the rest of the Corps....

The harshest, unreasoning critics of collegiate football, including critics resident at West Point, seized upon the paragraph as further proof that football, and especially overemphasis on winning football, were the causes of the scandal. After all, if the cheating started on the football team, was controlled by football players, spread to their roommates and associates, then it was only logical to conclude football was the cause – wasn't it? The answer is no.

At the time General Irving published his letter, the Bartlett Board was still in session, ostensibly searching for the causes for the incident. Yet clearly, when carefully read and placed in context with the rest of the letter, that's not what General Irving was saying.

He was merely stating the facts as known at the time he wrote the letter. There was much yet unknown, and the Bartlett Board was still in session to find the facts and determine the causes, though they later fell short of what might have been accomplished. The Greene Board and other inquiries, General Irving, other members of the staff and faculty, the Corps of Cadets, and Earl Blaik would later get much closer to the truth of what happened. The scandal of 1951 had been a long time coming, slowly, insidiously, and with a complex of causes and contributing factors that needed a deep, comprehensive, factual examination, with no premature conclusions or rushes to judgment.

In October, after the Army Chief of Staff had told the resigning cadets they could attend other officer training programs upon securing waivers and approved applications, two of the cadets sought reappointment to the Academy. West Point's Academic Board informed the Army they wouldn't recommend cadets for reappointment who received discharges following honor violations. The Army's refusal to reappoint the two men, though it never became a public issue, triggered a series of questions from the Department of the Army's Counsel, the Department's senior attorney. The questions were posed to prepare for what the Department of the Army believed would be difficult, probing questions following the decision against reappointment. The questions and answers proved seldom, if ever, necessary, but they contained words few in the American public and those who graduated from the United States Military Academy had heard about the honor code and system.

The essential function of the United States Military Academy is to develop character. The character objective is directly connected with a high ideal of service and is fairly well contained within the meaning of the motto of the United States Military Academy, 'Duty, Honor, Country.' The purposes of that high objective have, through the years, been incorporated into an honor code that is respected by the members of the Corps and revered by its graduates. The past failures of individuals to accept that code have always been handled drastically, and far less by command authority than by the Corps itself.

...It follows that any step taken by the Department of the Army, or by Congress, that would have the effect of mitigating the traditional imperative of the Honor Code of the United States Military Academy, would carry with it the death blow to the usefulness of the Academy itself. The Corps of Cadets would still exist as an impressive appearing military organization, but without soul or spirit that has actuated it for over a century, and with nothing in particular to contribute to the Army or to its country that could not be well met by any other existing institutions.

...It is right that West Point cadets should be held to correspondingly high standards of moral and professional conduct.... The West Point cadet is expected by the nation to be something more than ordinary. He should and is required to live up to this reputation.

...The [honor] system is more important than any individual. It reflects the concept that duty must rise above personal considerations in a loyal and well-disciplined Army. It is a guarantee against a self-protective clique in the service following graduation where officers must often stand in judgment over their closest friends and associates.

The answers to the Army Counsel's questions included words taken from a speech given by Secretary of War Newton D. Baker, who served in Woodrow Wilson's cabinet during World War I. Secretary Baker's comments make clear that the honor code and system have their roots in both ethical considerations and in practical military necessity.

The purpose of West Point, therefore, is not to act as a glorified drill sergeant but to lay the foundation upon which a career in growth of military knowledge can be based and to accompany it by two indispensable additions; first, such a general education as educated men find necessary for intelligent intercourse with one another; and second, the inculcation of a set of virtues, admirable always, but indispensable in a soldier. Men may be inexact or even untruthful in ordinary matters and suffer as consequence only the disesteem of their associates or the inconvenience of unfavorable litigation, but the inexact or untruthful soldier trifles with the lives of his fellow men and with the honor of his government, and it is therefore no matter of pride but rather a stern disciplinary necessity that makes West Point require of her students a character for trustworthiness that knows no evasions.

In 1874, during the period when cadets at West Point formed their Vigilance Committee to examine matters of honor, Thomas Henry Huxley wrote in his *Universities, Actual and Ideal,* "Veracity is the heart of morality." As would be written years later, the Military Academy's honor code, given life and meaning by the Corps of Cadets and sustenance by the institutions which educate and train the Corps, is "...a minimum standard of moral and ethical behavior." It is only the beginning, the foundation for what follows. For graduates, the officers, leaders, and commanders, there is inevitably much more required – and hard work – to reach for and hold the necessarily higher standards and ideals that will be asked of them.

CHAPTER 11

COACH: "THE COLONEL"

When Earl Henry "Red" Blaik came to Army as an assistant football coach in 1927, the "golden age of sports" was in full bloom, and in that gilded age, college football was undeniably the most popular spectator sport in the nation. But the game had some serious matters to consider, unpleasant hand me downs from its humbler beginnings.

When Blaik played football on Miami University's varsity in Oxford, Ohio, before World War I, rules were unsophisticated and equipment was woefully inadequate in all the high schools and colleges of the nation, circumstances that would change slowly in the years ahead. Player injury rates were high, and injuries often serious. It was no different when he played at Army in 1918 and 1919. To make matters worse, because of the game's rules, men played both offense and defense, subjecting them to greater fatigue, and increasing the likelihood of injury. Football was a man's game, the manly sport of big, strong, rugged men – and smaller, tough, feisty men who weren't intimidated by size, physical prowess, or brutishness.

There were few rules restraining eligibility, and star college players could play three years at one school and then move on to another, and continue to play. Blaik lettered three years at Miami and two at Army. It was "Iron Man Football," and men who played the game had to be men of iron to survive the battering they endured. Not until 1938 did the Military Academy go to a three year eligibility rule, and that was the result of President Franklin Roosevelt's insistence, as commander-in-chief, that West Point adhere to the same eligibility rules Navy had bound themselves to years earlier.

The Army Lieutenant Earl Blaik served in for two years after graduating from West Point in 1920 was also an Army different from the one he reacquainted himself with shortly before World War II. The Army of 1920-22, when Blaik was in the cavalry at Fort Bliss, Texas, was still essentially the Old Army, one in which officers were often spoken to in the third person by enlisted men. "Sir, does the lieutenant want his horse saddled?" When an enlisted man didn't know an officer, and desired a particular favor he might begin, "Sir, Lieutenant, sir..." and for a moment the officer felt like a knight, or maybe two knights. A soldier who wanted to talk with an officer on official matters saluted smartly and began, "Sir, Private Carson has the first sergeant's permission to speak to the lieutenant."

When it came to orders, enlisted men responded to orders under the pain of severe disciplinary action under the Articles of War if they didn't obey, and officers obeyed the orders of more senior officers, right up the line, almost without question. "Ours is not to reason why. Ours is to do or die."

During that same era when Earl Blaik was a lieutenant, the response to orders, guidance, or direction in America's civilian population was quite different. Though direction wasn't rigidly enforced by military law, and responses weren't as passive as in the military, it was a time when people tended to be more compliant than would be the case when World War II was about to begin. The Army's rapid expansion into the giant it became during the War was already under way when Earl Blaik accepted Brigadier General Robert Eichelberger's call for him to return to his alma mater. The pre-World War II Army was ingesting a whole generation of depression era men who were far less hesitant to ask, "Why?"

When Blaik returned to the United States Military Academy as head football coach in 1941, he and the game of football were still emerging from those eras. In 1942, not long after the United States entered World War II, the NCAA Rules Committee took the first steps toward what was later called two platoon football, by liberalizing substitution rules. Earl Blaik, partly because of an abundance of talent on his World War II teams, chose to have two or more units playing each game, on both offense and defense, rather than specialize with offensive and defensive platoons. Not until he saw Michigan use specialized teams did he adopt the practice at Army. He promptly became one of two platoon football's most vocal advocates. This was vintage Blaik.

He carried many of the earlier eras' trappings, but he was forward looking, incisive, analytical, adaptable, constructively critical about everything to do with the game, and he never stopped learning. He was a brilliant man with a mind like a sponge, complex, articulate, direct, single minded, driven, and a fierce competitor. He had the "winner's disease." He hated to lose a game, especially a Navy game, and Vince Lombardi would say years later that he was afflicted with the same heady wine of victory and the curse that all great winning coaches endure – the torment of losing.

He was a career soldier who wasn't a career soldier in a literal sense, but had a deep, abiding respect for the military life, and the sacrifices it demands. He tirelessly advocated the service life to his players, telling them that it would not gain them monetary reward, but would gain them lifelong education, travel, associations with the finest of men, and a personal satisfaction that had no equal in any civilian pursuit.

Yet he maintained the veneer of the Old Army officer's mystique, and insisted on keeping himself on a pedestal. It helped him maintain a presence, one that inspired respect, if not awe, in nearly all his players, and many admirers. Above all, Earl Blaik was loyal to his "boys," the cadets, the young men who played football for Army – his players – though tough, demanding, seemingly distant, cold, and aloof to them, he would fight fiercely in their behalf.

And oh how he loved the game of collegiate football, his family, and the United States Military Academy, as well as the young men who were willing to "pay the price" to play on his teams. The price was high, but the reward was great, and the decade of the '40s in Army football was proof of those rewards.

But when the crash came in the spring and summer of 1951, "the catastrophe" as Blaik called it, and his championship caliber team was destroyed, he had to make some major adjustments, rethink, reach down and find something within himself, which, for the first time in his life, he wasn't sure he had. He had to say to himself, as he would say to Bob Mischak two years later, "Don't ever give up." Earl Blaik didn't, and neither did a lot of other people, and that was the beginning of a miracle in the adversity and turbulence of the rebuilding years at West Point – a football miracle that had much wider implications.

"…BORN TO COACH…"

Earl Blaik was enamored by football at an early age. Born in Detroit, Michigan, on February 15, 1897, he moved with his family to Dayton, Ohio in 1901. There he first showed signs of what he would

418

become. But his drift toward college coaching was slow, hesitant, almost aimless, for ten years, resulting in late entry into the life he ultimately chose. When he finally paused and looked back at his twenty-seven years as a head football coach, he was somewhat amazed at how he got into such a demanding, unpredictable profession. He would be the last one to say he was born to coach.

> I was fascinated by this magnificent game, football, almost as far back as I can remember, from Hawthorne grammar-school days in Dayton as a fourth-grader, eight years old, in 1905. I was ten when I formed a neighborhood kids' team and appointed myself coach, captain, and quarterback. We called ourselves the 'Riverdale Rovers' after the Rover Boys books we all read.

The Riverdale Rovers played with a round, black, soccer-type ball, rather than the oval rugby kind, and according to Blaik, concentrated on the running game. The forward pass had been legalized in football in 1906, and Blaik remembered "...if we knew about it at all...we chose to ignore it." He conceded that perhaps his bent toward the running game as a boy stuck with him all the way through his years as a head coach.

The Rovers games were almost exclusively with a team of kids from North Dayton, what Blaik described as a tough section of town. He said the games invariably reached a fever pitch of emotion that usually erupted into a free-for-all, and finally, rock throwing from opposite ends of the Herman Avenue Bridge and the adjacent banks of the Miami River. The rock throwing was the Rovers' "only concession to the forward pass," he said.

When young Earl was an eighth grader in 1909, he moved up a bit in football class. The morning of Thanksgiving Day he was invited to play on the line of a Riverdale team of older boys, which included some boys from Steele High School, the high school he later attended. They were playing a game against Oakwood. He earned a black eye he described as a gorgeous "shiner," which he hesitated showing to his father and mother at the Thanksgiving dinner table. Some of the players brought him a steak to reduce the swelling, but it didn't help much. He held off going home for a couple of hours, but had to finally face the music. His parents didn't say much, but they weren't too happy.

Blaik surmised years later that they didn't lecture him too severely for fear of destroying the holiday spirit.

He was a true redhead, "on the auburn side" he said, just like his mother had been, and he maintained he had the traditional argumentative temperament to go with the color of his hair.

His mother Margaret Jane Purcell was of Irish extraction, and had moved to Detroit from Canada with her father after her mother died. In Detroit, Margaret Jane met William Douglas Blaik, who had come to America from Glasgow, Scotland in 1883, at the age of sixteen. He went to Thedford, Ontario, and lived and worked with his dad's brother three years before moving to Detroit to set up his own blacksmith shop at the age of nineteen. They met in 1889 and married one year later. Earl Blaik's older brother, Douglas Livingston Blaik, was born in Detroit in 1893, his sister Mabel in 1907, after the Blaiks moved to Dayton.

Earl Blaik's father was proud of his Scotch heritage. Though he didn't exhibit his pride by talking about the Scots' clans, or wearing kilts, he did take umbrage when anyone Americanized the Scotch spelling of Blaik by changing it to Blake. According to Earl Blaik, every time such an offense was noted, his father would take pen in hand or personally speak to the offender to correct the affront.

It was in Dayton in 1908 that Earl Blaik's dad became interested in politics, one of life's twists of fate that led to Earl's appointment to West Point ten years later. His dad's friend, James M. Cox, ran for Congress, and William Blaik, whom his son described years later as a "thoroughgoing Democrat," helped in his campaign. It was Governor Cox of Ohio who helped get Earl Blaik an appointment to West Point in 1918.

"Red" Blaik's interest in football grew when he went to Steele High School in Dayton. Although he was only five feet four inches tall and weighed little more than a hundred pounds his sophomore year, he noted with some pride he made quarterback on the "scrubs." The next year he was moved to end, and saw some action. In 1913, his senior year, he was starting left end. He was sixteen years old, five-feet-nine, and weighed 133 pounds. As he said, he wasn't a "blue chip" athlete, having failed to make the Steele baseball or basketball teams.

In fact he did as well on the stage in high school as he did in athletics, performing in a senior three-act comedy that drew notice in the

Dayton newspaper. As to academics, his school teachers expressed some frustration, saying he studied just enough to get by, views Blaik in hindsight remembered as correct. He was to later say his preoccupation with emulating local athletic heroes made sports the center, if not the whole, of his boyhood days. But he noted that was generally true of most boys in his younger days.

Like many a boy, he found a football hero in his home town – Marvin "Monk" Pierce, who was Steele's star back and pitcher, and lived only six houses from the Blaiks. When Earl was in Parker Junior High School's ninth grade, "Monk" was a senior at Steele. He worked a year after graduation before going to Miami University, thus he was only two classes ahead of Blaik at Miami when Earl arrived there in September of 1914, a month after "the war to end all wars" exploded in Europe.

At Miami University, Earl Blaik's rather casual attitude toward academics began to change. He entered the pre-law course. His father had great respect for the law profession, and his son liked the prospect of a career involving tough, serious debate, with the outcome affecting the rights and even the lives of men. It appealed to the streak in him that welcomed argument and high stakes action.

Though he studied more than he had at Hawthorne, Parker, and Steele, at the University he encountered a man who wasted little time in telling him – and writing Blaik's father – that he could do better in academics. In his letter to William Blaik, Dr. Archie Young, Earl Blaik's freshman football coach, who was also the Dean of Men and taught mathematics, wrote, "He should be doing 'A' work. If he doesn't do it, you shouldn't be spending money on him." According to Blaik, the validity of Dr. Young's observations "found a response in his dad's Scots sense of value and Presbyterian sense of right." Dr. Young and Earl Blaik's father caused him to speed up in academics, and in his sophomore year his marks began improving. By his senior year, he had all "A's," was an assistant instructor in economics, and occasionally regretted his slow freshman start because it kept him from achieving Phi Beta Kappa recognition.

Several of Earl Blaik's Army football players during the rebuilding years regarded their head coach as "aloof" and hard to reach, descriptions which puzzled him. But they saw something in him that others had seen many years earlier, when he was a junior at Miami University. That year the Miami yearbook *Recensio,* mentioned he was "aloof."

In looking back at that comment about a younger Earl Blaik, he was equally puzzled, and considered the characterization a "discrepancy." He pointed to his expanded participation in extracurricular activities at the University to argue he wasn't aloof: student government, class functions, language clubs, the debate team, and dramatics. In his senior year he was elected president of the Student Forum, president of his fraternity, president of the letterman's club, baseball captain, and a member of the National Honor Society. And to reinforce his belief that he was anything but aloof, he said, "I was appointed, in the gag section of one of the *Recensios*, to the 'Ancient Order of the Sons of Ursus,' for having thrown the bull with profuseness as well as éclat, for various reasons on all occasions."

During his sophomore year he first became aware of a young woman he said was "a charming freshman coed from Piqua, Ohio, named Merle McDowell." He recalled that the names Earl and Merle "go together and so did [they]." Then in a light moment, which reflected the type of humor he enjoyed – usually a play on words – he told of the years when he was head coach at Dartmouth, when they had a cook named Pearl. "You've never really been confused," he said, "until you've lived in a house with an Earl, a Merle, and a Pearl."

Earl Blaik and his future wife met at a small party in Bishop Hall, a residence for girls, to which his fraternity, *Beta Theta Pi,* and the *Delta Zeta* sorority brought their pledges. Merle had started the evening with someone else, and ended up with Blaik. They began to date regularly, a matter of concern to Jimmy Young, the Miami baseball coach, who had become a mentor of sorts to Earl Blaik. Young feared Blaik's studies, athletics, and extracurricular activities would be adversely affected by the romantic involvement. Years later, when Blaik was head coach at Dartmouth and Army, and expressed concern to Merle about some promising football player's romantic interest, Merle would smilingly remind him of Jimmy Young and his forebodings.

The two didn't become engaged until Earl Blaik was a senior at Miami in the spring of 1918, shortly before he left for West Point. He didn't give her his class ring until 1920, the year he graduated from West Point, and they didn't marry until 1923, the year after he resigned

Bluik as a senior at Miami Univer-
sity in 1918, before he entered West
Point later that year. (Photos cour-
tesy Miami University Archives,
Oxford, Ohio.)

Merle McDowell as a senior at Mi-
ami. They met at the University, were
engaged in 1918, and married in 1923,
after he graduated from West Point
and resigned his Army Commission.

from the Army. In recalling their years together at Miami, he humor-
ously pointed out that his "courtship, if that's what it was, was scarcely
a blitzkrieg."

Despite academic improvement, extra-curriculars, and a budding
romance with his future partner in life, Earl Blaik's priority interest
remained athletics. He tried out for basketball, made the squad, but
didn't letter. He did well at baseball, earning letters three years, as an
outfielder.

Football was what he called his "mainspring." Seventeen years old
when he entered Miami, and twenty-one when he graduated, he was
on average a year younger than most of the University's athletes. In
those four years he grew an inch, to six feet even, and added twenty-
five pounds to the one hundred thirty-three he entered with, to a lean
hundred and fifty-eight. He was maturing, and his strength and speed
developed, too, along with a toughness that became an ingredient of

his football philosophy in the years to come. As he would recall, "I was knocked out in games, but never injured seriously enough to miss any playing time."

His speed and maneuverability enabled him to handle defensive end assignments, get down under punts, block and tackle reasonably well, and catch passes. His freshman year, he played on his fraternity team as well as the Miami freshman team, where he recalled they "proved excellent cannon fodder for the varsity." In his sophomore year, 1915, he won the varsity's right end position, and held it the remainder of his time at the University, earning three letters on strong teams that won the Ohio Conference championship two of those three years.

During the 1915 season Earl Blaik learned another lesson which he carried with him the rest of his coaching days. Miami went 6-2 that year. Their two bitterest rivals were Cincinnati and Denison. They beat Cincinnati 24-12 but lost to Denison 14-0. Blaik remembered that their coach that year, Chester R. Roberts, tried so hard to exhort the team to win over Denison that he "began pressing." At halftime, he said he would buy each player a box of bonbons if they won the game. Blaik said the second half, if they were anything, they were worse than in

Earl Henry "Red" Blaik, right end on the 1915 Miami University football team, when he was a sophomore. (Photo courtesy Miami University Archives, Oxford, Ohio.)

the first half. Blaik wrote, "The experience underlined to me strongly the feeling that in a Spartan game like football the worst possible inducement to a malingering player is, literally or figuratively, any form of sweetening."

The 1916 team was regarded by Miami as the school's greatest to that time. They were undefeated and out-scored their opponents 239 to 12. Blaik's last season at Miami, 1917, was under new head coach George L. Rider, his predecessor George Little having left to become a captain in the Army. The sinking of the *Lusitania* in April had drawn the United States into the war, and though the 1917 team was also undefeated and won the Ohio Conference championship, Blaik remembered that everyone's mind was on the war. He made *All Ohio end* that year, and did his first work as a coach. Rider had no assistants and asked Blaik to help with the ends and tackles.

Although the casualty lists from France indicated a fearful toll in the war, during Earl Blaik's senior year, young men still regarded war as adventurous and romantic. Many of his friends were joining up, and he wanted to enlist, but his parents, like most families in that era, had a powerful influence on his decisions. They persuaded him to stay and graduate. At the time the war in Europe was in a horrible, grinding stalemate, and predictions were it could go on for eight to ten years.

Although football had nothing to do with Earl Blaik's decision to go to West Point, he first got the idea from a Miami University teammate who had entered the Academy in 1912, but had been "found" in mathematics. The man was Jackson T. Butterfield, and he played a couple of games on the plebe team in 1912 before he left West Point to enroll in Miami. His brief time at West Point had helped him get a commission as a captain, and he persuaded Blaik that a commission should be his goal as well.

Earl Blaik's first thought as he neared graduation from Miami was to enter Officer Candidate School, and he submitted his application. But before he could be accepted, the Army published a regulation that men coming right out of college couldn't enter Officer Candidate School. It was then that Earl Blaik began seriously considering West Point. With encouragement from a family friend and his father's past political support of then Governor Cox, he landed an appointment from Democratic Senator Atlee Pomerene, as a first alternate.

"THE BEAUTIFUL AND HISTORIC BASTION OF AMERICA" HE REVERED

When Earl Blaik entered West Point on June 14, 1918, the Academy had already been affected by the pressures of wartime priorities, its curriculum reduced from four to three years. The class that had entered in 1914 had graduated in August of 1917. Two days before he arrived, the class that had entered in 1915 graduated. The class that entered in 1916 had been designated the second class – juniors – and were scheduled to graduate in 1920. The class which entered in 1917 was now the third class, listed to graduate in 1921. Blaik's class, the new fourth class, was to graduate in 1922, signaling an anticipated return to the normal four year curriculum.

That didn't come to pass. There was more turbulence and confusion ahead.

At the noon meal on October 3, 1918, Cadet Adjutant Beverly Saint George Tucker read an order that sent the dining hall into pandemonium. The second and third classes, originally scheduled to graduate in 1920 and '21, were to graduate within a month, on November 1. A new plebe class would enter in November, consolidate with Blaik's class, and both graduate in June 1919.

"It may be that because they were going to get to war faster, the second and third classes cheered more loudly than we plebes, but I doubt it," Blaik remembered. "The way we looked at it, we would be in France ourselves in nine months. Nobody anticipated the collapse of the German Army." Blaik's classmates realized that the October 3 order also meant that within a month, after only five months as plebes, they would be transformed miraculously into the supreme, if the only, class at the Academy.

The upperclasses also realized the impact of the order, right on the spot, and immediately decided to jam the eight months they were about to miss of "crawling" the plebes, into fifteen minutes, and then "recognize" them. Recognition meant shaking hands, calling one another by first names, and "falling out" by the plebes. The *Howitzer* for the class of 1920, in which Earl Blaik graduated, said, "Never did we brace so cheerfully – and forget it so quickly when we shook hands all around with the upperclassmen."

The rest of October was hectic, even by West Point standards. The upperclassmen were under great pressure with intensive drills and lectures. The fourth class worked hard to prepare themselves to assume the responsibility of the Corps' customs, traditions, and administration.

When November 1 arrived, the Academy band played *Dashing White Sergeant* and the graduating classes marched across the Plain and stood in salute as Earl Blaik's class passed in review. Two days later the new fourth class entered. Wartime priorities, and a shortage of cadet gray uniforms resulted in the new class being outfitted in olive drab uniforms with leggings and campaign hats circled with orange bands. The unusual uniform combination earned the new class the nickname of the Orioles, and as Earl Blaik observed, "In an organization that had been clad in gray for one hundred seventeen years, they were indeed strange-looking birds."

Blaik's fledgling plebe-upperclassman class didn't get the normal prerogative of indoctrinating the new class. They were kept hard at academics. Instead, the tactical officers reverted to the Beast Detail role and spent three weeks giving the newest class a belated, abbreviated Beast Barracks. Earl Blaik remembered it made the Orioles the least put upon plebe class in Academy history, a fact he also recalled didn't seem to hurt them. General Maxwell Taylor was one among several distinguished members of classes who had relatively brief careers as cadets, similar to the class known as the Orioles.

Then eight days after the Orioles arrived, with stunning suddenness came "the eleventh hour of the eleventh day, of the eleventh month." World War I was over. The Muese-Argonne offensive had pierced the Germans' Von Hindenberg line. First quiet, then came victory celebrations all over the world. At West Point there was a victory bonfire, but little rejoicing in the Corps. The cadets were proud of the American Army, but as Earl Blaik put it, "we also felt...as if the Plain had been pulled out from under us." The great adventure was gone from the profession chosen by Earl Blaik and his classmates. They weren't going to get a job they had been "promised." They were let down.

The abrupt end to the war brought another major change at the Academy. The two classes that had graduated on November 1, were taken on a tour of battlefields in France, and returned to duty as regular officers. The Army decided, however, to return the second of those two classes to West Point in early winter, to resume studies until June.

With their return to the Academy, West Point took on an even more unusual look: one class in traditional cadet gray uniforms, the Orioles in their olive drab and orange banded campaign hats, and the "Student Officer Class" as they were known, dressed in the uniform of Army second lieutenants. Blaik's class, who had been "Lords of the Plain" for a brief time, had to salute them. The closest his class and the Orioles ever got to the war was the following March when the two classes acted as a "Guard of Honor" in the welcome home parade for the 27th Division in New York City.

When Douglas MacArthur, at the age of thirty-eight came to be superintendent at West Point on June 12, 1919, Earl Blaik found a new, deeply admired hero, and would be set on the path toward a lifelong friendship with one of West Point's most famous graduates, who had become a World War I battlefield hero.

In Blaik's last year as a cadet at West Point, he played varsity basketball, baseball, and football, and won the saber as the best all round athlete in his class. He encountered MacArthur many times, as a cadet athlete, and as first classman and chairman of a class committee which first codified the plebe system, a contribution to the Academy Blaik proudly remembered. After the turmoil of war in 1918, the Army football team only played one game.

By the time the 1919 season rolled around, he had grown to six feet and one and a half inches and weighed 182 pounds. The Army team went 6-3 that year, and Earl Blaik once again secured a starting right end position. He felt the game's old enthusiasm had been revived, spurred by a new superintendent who vigorously supported intramural and intercollegiate athletics, especially football.

When he came to West Point in June 1918, like many before him, and thousands who came after, Earl Blaik took that long, worried walk up the road from the West Shore Railroad train depot by the Hudson River. He was greeted by the Beast Detail with the traditional, "Suck up that gut, Mister!" And he remembered asking at the end of the first day, and for many days afterward, the same question that nearly every Academy graduate asks, "Why did I come?"

The answer came to Earl "Red" Blaik years later, when he said, "I would not be a soldier very long and I would never get to war. But I would command troops in the game that is closest to war, and I would know in this a happiness, as well as a tragedy, that I could not have known in anything else."

THE 1919 FOOTBALL EDUCATION

Blaik's final season as a football player at West Point had a profound effect on him. Years later, after his coaching career ended, he recalled numerous stories of the Army coaches that revived football at West Point in 1919. Every story was filled with what he had learned from them about the game of football, lessons he apparently never forgot. Their teachings became a part of him, added to him, drove him, became the foundation on which he built his success.

The head coach and first assistant were Captains Charles Dudley Daly, and Ernest "Pot" Graves, artillery and engineer officers respectively. Graves was brilliant, number two in his West Point class, and later earned the label, "Father of Engineers." Both were Academy graduates in the class of 1905. They had been teammates on outstanding Army teams just after the turn of the century.

Daly, who had also earned his own brand of fame, as "The Godfather of Army Football," had been backfield coach at Harvard during its golden age of the sport, and brought from its head coach a degree of organization, dedication, and psychology heretofore unseen in the game. Along with Daly, when he returned to Army from Harvard, came tactics, techniques, and strategies, all of which shaped his theories of football and coaching. Blaik also saw other factors in Daly he admired and never forgot. Daly possessed good character, purpose, and he was a leader.

Graves, who had served as Army head coach before World War I, was the line coach who preferred being an engineer more than coaching football. He was far too brutal in Earl Blaik's view, a view that he summed up by saying "Pot's" philosophy of line play might best be described as "Spare the blood and spoil the lineman." Nevertheless, Graves taught well the principles of line play, and according to Blaik was far ahead of his time. Blaik learned Graves' principles of line play well and improved on them, without Pot's "let's see some blood."

Perhaps more significantly for Blaik, he remembered the tough, brutal Graves didn't believe in senseless bellowing. "Don't yell at them," Graves preached. "Teach them something."

The psychology Daly and Graves also brought from Harvard rubbed off on Blaik and his teammates. Earl Blaik could still see Charlie Daly, years later, writing his maxims on the blackboard and reading them off in his Boston-Irish and Oxford tones:

"Carry the fight to the enemy and keep it there all afternoon."

"Play for and make the breaks, and when one comes your way, score."

"The team that makes the fewer mistakes wins."

"Press the kicking game. It is here games are won or lost."

"Break any rule to win the game."

To ensure no one misunderstood the last maxim, Blaik explained it "was not a concession to foul play, but a warning that a desperate situation could justify the unorthodox. It would, however, have to be truly desperate." In his years as head coach at Army, Earl Blaik had his own set of axioms which he kept posted on the West Point football team's locker room wall.

It was also under Daly and Graves that Earl Blaik concluded the most valuable strategy taught him was "the value of defense. Football, or any other game, cannot be played successfully without it." What's more, under Daly and Graves, Blaik began formulating his underlying philosophy about the game. In 1919 was when he first concluded "football was war on a limited terrain." Years later he would articulate his fundamental beliefs and logic underlying the game.

> If ...[the game of football] is the game most like war, it is also the game most like life, for it teaches young men that work, sacrifice, courage, perseverance, and selflessness are the price you have to pay to achieve anything worthwhile. It helps teach that the poet who wrote that life is not 'an empty dream' was far closer to home than the modern lyricist who sang that it is 'just a bowl of cherries.'

Another former Army great, a center on the pre-World War I Army teams, was a 1919 assistant coach. He was John McEwan, who later was the cadets' head coach from 1923-25. Among other things, Blaik remembered that McEwan was an English instructor at West Point, in addition to assistant coach, and had coined what Blaik called "a thundering truth" about football and coaching: "You cannot get anywhere in football by going out and exuding an aroma of good fellowship." Blaik exhibited a sense of humor, but an aroma of good fellowship was certainly never one of his trademarks.

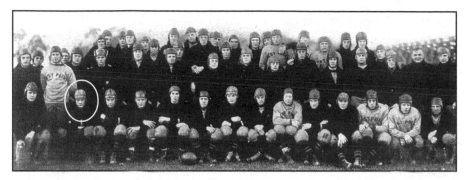

The 1919 Army Football Team. Earl Blaik is circled. (Photo courtesy USMA Archives.)

Lieutenant of Artillery Lawrence McCeney "Biff" Jones was an assistant coach on the 1919 Army team. He had been a standout player at West Point while Blaik was at Miami University. Jones served in Europe during the war, and left his mark on Earl Blaik later, when he brought "Red" to Army as an assistant, beginning the 1927 season. If there were any skills Earl Blaik was well-known for when he reached the peak of his profession, they were his skills for analysis and organization. He became a true perfectionist, but this was especially true when it came to these two abilities for which he gave full credit to "Biff" Jones. And it was "Biff" Jones who – after many years as coach at Louisiana State University and coach and athletic director at Army, Oklahoma, and Nebraska – became Earl Blaik's "man in Washington" – a retired Colonel "Biff" Jones, who helped Blaik shepherd qualified alternate and regular appointments through Congressmen and Senators to help fill Army's athletic recruiting needs in the late '40s and early '50s.

BLAIK AND REBUILDING: THE UNCHANGED FORMULA FOR SUCCESS

By the season of '51, Earl Blaik had a well established set of philosophies about the game of football and coaching. He had a system, a set of methods, a program that had developed a powerful momentum. He knew what he expected of his assistants and routinely hired the best he could find. At the same time he was always studying the game – in all sections of the country – analyzing, learning, looking for innovations that were adaptable to the Academy's demands and the life he knew cadets had to live – and he hired up and coming assistants who were ambitious, dedicated, and who usually brought with them ideas adaptable to the Academy's environment, and its cadet athletes.

After he hired his assistants, he indoctrinated them in his philosophies and methods. He kept a careful watch of their behavior. He corrected them – usually in private – when they didn't follow his coaching and football prescriptions, and welded them into a staff committed to deriving the most and best from available talent – and winning. He ensured that whomever he hired, especially if they weren't Academy graduates, understood the rigors of cadet life, the regimentation, discipline, and academic demands. "Work with the players." "Keep it simple." "Teach them." "Show them how to correct their mistakes." "Don't humiliate them or take their dignity." "Avoid yelling or bellowing," as "Pot" Graves would have said. "They get enough of that in the rigors of cadet life."

He had a well-established pre-practice, pregame, and halftime routine, and practices were planned by the clock – to the minute. Everything was organized to the nth degree, standardized, yet open for persuasive innovation.

There was an intelligence system complete with intelligence analysts – the "scouts" and statisticians – mostly assistant coaches from the junior varsity and plebe teams, and cadets who had exhausted their collegiate eligibility and wanted to help the team. Blaik's whole system of preparation for opponents was highly disciplined and organized, a scouting system which not only scouted opponents in excruciating detail, but took the information gathered and statistically analyzed it with great care. They went one step further, studying each opposing player, looking for strengths, weaknesses, habit patterns that might be exploited, and creating dossiers on them to give Army players who would face them. Navy, particularly, was normally scouted every game of the season. But this wasn't all the intelligence support Earl Blaik and the Black Knights had.

There were hundreds of Academy graduates and former Army players willing to supply information about opponents. These "scouts" included Army officers and Air Force officers, aircrew members, who, on weekend navigation training flights would voluntarily land at Air

Force bases near games Army's future opponents were playing, and attend the games as Earl Blaik's eyes and ears.

Complementing the scouting and intelligence gathering legions supporting Blaik's Black Knights, was his film crew. Wherever they could find film clips of opponents' games – even filming games broadcast on television – they gathered them up, and used them in compiling scouting reports. Occasionally Blaik's assistants showed clips of the next opponents at the Sunday team meeting, which was both a critique of the preceding day's performance, using game films, and the beginning of preparation for next Saturday's opponent.

Additionally, every Army practice session was filmed, and used to analyze players' strengths, weaknesses, and habit patterns. Players generally disliked "Cyclops," but Blaik and his staff used the films to teach, correct, and improve performance. The filming of practices also had another salutary effect. In somewhat good humored irritation, the players often grumbled, "There's no goofing off. Cyclops will get ya'."

An efficient, well organized personnel system was in place, always searching for gifted athletes, providing a never ending stream of information about potential new recruits. Those looking for new prospects were mostly the same men who were interested enough to provide information about Army's opponents, but not always. Family members who knew of or encouraged a son's interest in West Point, or family friends, community leaders, and sports writers who knew of talented football players who were considering the service academies, would write or call the Army coaches and provide names and the schools the boys were attending, often sending press clippings as well. The coaching staff's responsibilities included perusal of sports pages from major newspapers, and some less well-known papers. They searched for names of talented high school players whose game performances were sufficient to be praised in city sports pages. Each year, in this manner, Blaik's staff compiled a list of approximately eight hundred names, which, during the recruiting season established by NCAA rules, could be systematically screened through contacts with high school administrators, teachers, and athletic coaches.

When the list was screened there were usually seventy-five to a hundred who took entrance examinations, of whom a small number would qualify for appointments. Then it would be up to Blaik and "Biff"

Jones, who received some pay – primarily reimbursement for expenses – from the Army Athletic Association, to work with the Academy's Admissions Division and Congress to secure appointments. But there was more to his personnel system.

Because Blaik was also Athletic Director, and nearly all his assistants coached in other intercollegiate sports, the search for new recruits included other sports as well. This was especially important because the Corps of Cadets was small by comparison with student populations at major colleges and universities with which they competed. It was not uncommon that cadets were frequently team members in two and three intercollegiate sports, a tribute to their overall, exceptional athletic skills, as well as their scholarship.

There was a supply and transportation system with cadet managers supporting the team, ensuring the upkeep, repair, and movement of team uniforms, equipment, and supplies, and that orders were placed at the proper time to avoid shortages. The managers were responsible for initiating and following up on requests for team transportation to all games. They were also given numerous supporting details, including roles in management of the team during practices and games.

In preparing for each day's practice Blaik and his staff worked out a tight, detailed time schedule, geared to the intense hour and a half allowed on the practice fields. The team manager's responsibilities included checking the clock against the practice schedule and whistling the next phase of practice.

As Blaik did with his assistant coaches, once he assigned managers their responsibilities, he let them do the job, seldom intervening unless there was a request or if something went wrong. He built a sense of responsibility in the cadet players and team managers at every opportunity. If a manager blew his whistle indicating it was time for the next phase of practice, and Blaik saw a need to continue the activities in progress, he would courteously ask the manager, in front of the coaching staff and the entire team, "May we have another five minutes please?" Of course it was obvious to everyone that the team manager wasn't going to refuse "the Colonel's request." But in making the request, Blaik was reinforcing the manager's authority and responsibility. He was letting his players and coaches know who had the clock and the responsibility to alert all present it was time to move along and get the afternoon's work done.

There was a medical service which included a team physician and a trainer. The team physician was an Army officer detailed to Blaik's staff from the Academy hospital. The trainer was Roland Bevan, a close, loyal friend who, some years after Blaik's 1914 graduation from Steele High School in Dayton, was head coach at Steele. Roland Bevan's high school teams were noted for their conditioning, and as Blaik said of Bevan's attitude toward training and football injuries, "coddle" wasn't in "Rollie's" vocabulary. "Rollie" came to Dartmouth as trainer under Earl Blaik, and Blaik brought him and nearly all his Dartmouth assistants to Army in 1941.

There was preventative medical care and a carefully planned diet, physical therapy and weight training. Injuries always received the best of care during brief periods prior to practice, and during games. Ankles were taped prior to every practice as well as every game. Paul Lasley in the class of '56 became a statistician after the '53 season, when his eligibility ran out, and out of curiosity decided to compute an estimated Army Athletic Association annual expenditure for tape for the Black Knights. The result, $80,000 a year, stunned him. If an injury required a physician's care, Blaik would ensure the player received the best care possible. Civilian specialists were quickly called upon if, in the judgment of "Rollie" Bevan, and sometimes the team physician, the circumstance warranted such care.

The diet for players was carefully considered and maintained with the help of the cadet dining hall staff and the Corps Squad training tables, which had been instituted when Douglas MacArthur was superintendent. Cadet intercollegiate athletes sat on training tables all season long, and thus didn't eat meals with their cadet companies during "on seasons." The maximum allowable number of team members for each sport was spelled out in the "Blue Book," cadet regulations, and Blaik, as Director of Athletics, sent requests for table scheduling and seating arrangements to the Cadet Brigade Commander, the First Captain of the Cadet Corps, prior to the start of each season.

Corps Squad meals were planned to compensate for and support the enormous energy drain normally experienced by intercollegiate athletes, both in practice and competition. High protein diets were staples of sports such as football, basketball, soccer, lacrosse, baseball, hockey and others – because of energy requirements; rapid regeneration of overworked, tired bodies; and healing of injuries. At the same time, "Rollie" Bevan kept a careful watch of players' weight and diet to ensure they weren't getting overweight or filling themselves with pie, cake and other fattening foods.

Football players were the only athletes who were on spring training Corps Squad tables in addition to the regular fall season tables. To some cadets and members of the tactical and academic departments who believed football received unwarranted or excessive attention, spring practice and Corps Squad seating for football players were matters that rankled. Some saw the periods of Corps Squad training tables as evidence of too many football "privileges."

Supporting the Black Knights' coaching staff was a communication system used during games. Assistant coaches positioned in press boxes were connected to the Army bench by telephones. From press boxes they had clear views of play by both teams and could observe player mistakes. With game plans at their fingertips, they could recommend plays to call as games unfolded. In the '51 and '52 seasons, substitution rules were more liberal, the two-platoon system still in effect, and play calling from the bench was easier. That would change in 1953.

The sports information office, headed up by Joe Cahill, was an especially effective and important organization supporting Army football. During the dark days of World War II, the Army team had indeed been, as Earl Blaik often said, the "supreme rallying point" for the Corps of Cadets. This was during an intense wartime education curriculum and military training schedule, which again had been reduced from four to three years.

Like other officer commissioning sources in wartime, the Academy received little press notice. The nation's attention was riveted on the war, and the air, sea, and land battles raging all over the world. What little press notice the Academy received was through occasional references to exploits of its graduates who were already in the spotlight as a result of their leadership positions in the Army and Army Air Force, or in connection with graduates whose battlefield valor had been picked up by the news media.

Beyond that, the Academy's name was kept in the public eye primarily by the exploits of its football team. The public needed victories on the world's battlefields. They also needed sports and entertainment to bring cheer in a time of lengthening casualty lists. A huge effort in entertainment went to the soldiers, sailors, airmen, marines, and coast guardsmen, who needed it more than anyone. But college football, and other intercollegiate sports, though suspended at many college campuses, and affected considerably by security concerns and the war's drain on manpower, couldn't go overseas. Each fall's slate of games remained solaces and sources of pleasure for a nation at war. This was particularly true of college football which continued to be the most popular spectator sport in America.

During World War II, in a very real sense Army football became both the Academy's major source of worthy news, as well as its most important magnet for successful recruiting of cadet candidates – whether they were football players or not. And Army football had also become a center of national attention that seemed to heap even greater glory on a team and its members, whom Earl Blaik already had repeatedly said represented West Point, the Army, and the American people. If not thoughtfully tempered, this was heady stuff for young men who were confident, well prepared, and winning consistently while playing in giant sports arenas before thousands of rabid, thundering, football fans.

The Academy's Public Information Office's emphasis on Army football didn't begin to change until 1949, when Major General Bryant E. Moore became superintendent. A true soldier, an infantryman, and a respected division commander during World War II, Moore had been the Army's Chief of Information before coming to West Point. One of the marching orders he received when he followed Maxwell Taylor as superintendent was to reorient the Academy public information program. The intent was to broaden and shift its emphasis, and throw the public spotlight on the quality of the Academy's curriculum and the performance of cadets on Graduate Record Examinations. The Academy was back in competition for young talent all over the nation for the long haul, and needed to focus the public's attention on the quality of its education as well as the success of its athletic programs.

Moore gave those instructions to Colonel Billy Leer, and he discussed the matter with Joe Cahill, enlisting Cahill's, and no doubt

Blaik's, understanding and cooperation. The shift in emphasis quietly came about, and was increasingly effective as time brought the Academy ever closer to the sesquicentennial year of 1952. That is, until the public announcement of August 3, 1951, when Army football received the most devastating, shattering blow in its long, colorful history.

Undergirding the entire Blaik system, as in any successful intercollegiate athletic program, was the Army Athletic Association. The Association's annual operating budget was large, and football was the economic engine for the entire Army athletic program. Contributions from alumni, especially former athletes who had been financially successful, helped fund numerous activities, including the "cram" school to which Blaik sent recruited athletes each spring to prepare for entrance examinations.

Following the scandal, when Army football came under sharp attack from all directions, Blaik, who was also an astute business man, pointed out that the Army Athletic Association's 10,800 members annually contributed $10,800 toward $500,000 in Association income, and that football accounted for 97 percent of that return. What's more, the Association supported seventeen varsity sports, and over the years had built $2.5 million in Army athletic plants, including Michie Stadium, Smith Ice Hockey Rink, the field house, golf house, Delafield Pond, tennis courts, and a $460,000 addition to the gymnasium.

Although the debacle of 1951 was emotionally draining for Blaik, his coaches, and the Department of Athletics, the system of team support he had methodically built in the ten preceding years remained largely undamaged by the cheating incident. But there had been damage, enormous damage, although of a different character. The Army teams to follow and football at West Point, along with the Academy and the Corps of Cadets, had taken heavy blows that year following the public announcement and the team's destruction. The incident had come about insidiously at first, unseen, creeping, a slow erosion, an infection of an entirely unexpected kind – and Earl Blaik, and many others hadn't seen it coming.

COACHING – HIS ALL-CONSUMING PASSION

Blaik's natural instincts and Army training had molded him into a polished leader. Always cool and calm, the soldierly colonel had schooled himself in self-restraint.... Even with his assistants Blaik was stern and strict, like a general with his aides. He seldom praised his assistants to their face, yet praised them to others. His extreme dedication to winning was also contagious; one could not play for him or coach with him without feeling the same urge to overcome all obstacles on the path to victory....

Surely Earl Blaik knew he was always being observed, by his assistant coaches, his players, the Corps, the men and women who served and lived at West Point, sports writers and commentators, and the thousands of loyal Army football fans throughout the nation. What his players saw and heard during rebuilding was quite naturally far different than the observations of everyone else, including Blaik's assistants.

Almost to a man, his players and coaches accorded him a deep professional respect. He was clearly one of the game's greatest teachers and most successful coaches. To his assistants he was relentless, driven, a master of the game, who could be sharp and blistering with his words, but was always fair and seldom wrong.

He was, as might be expected, much closer to his assistant coaches than his players, and on occasion peeled off his old Army veneer, stepped down off his pedestal, and relaxed, with flashes of humor. But not often. To his coaches, in spite of periods when he let down his guard, he was "Colonel Blaik" or "The Colonel." Even on the practice field, his assistants seldom called him "Coach," and none dared call him "Red," a name reserved only to his closest, personal friends. And to the great majority of his players he was unquestionably an outstanding example as a leader, one who demanded no less of himself than he did of everyone else.

Assistant coach Andy Gustafson, who came with him from Dartmouth and later became head coach at the University of Miami, Florida, recalled Blaik's relentless, hard work and drive for perfection. Every New Year's day, without fail, Army's head coach arrived in his office on the top floor of the South Gymnasium at eight o'clock to begin work for the coming year. It was a habit that made a lasting impression on assistant coaches in a profession already filled with insecurity. Each January 1 when Gustafson was at Miami he picked up the phone and called Blaik's office at eight in the morning, primarily to see if Blaik was at work. Without fail he got his answer, "This is Blaik."

Murray Warmath, who left at the end of the 1951 season to be head coach at Mississippi State, and later, Minnesota, said "He was the most dedicated...hardest-working man I ever knew." Alluding to the Army coach's uncompromising emphasis on football he added, "Blaik's idea of fun in the summertime was to study films."

"Red" Blaik carried his "profession, business, play" obsession with football into the summertime. In Blaik's thirty-two years of coaching he seldom took vacations, and when he did, it was usually after a doctor admonished him he needed to relax. Each summer, however, in late July or early August, he and his coaches, and some of his friends, vacationed for a week in cabins on Bull Pond. Bull Pond is a five-acre lake atop an 1100 foot mountain on the Academy's military reservation, eight miles southwest of West Point. The lake provided good boating, excellent swimming, and only fair fishing. The men swam, fished, joked, ate, and talked football. Blaik recalled, "We needed lightness, silliness, or madness in those days. It was our only respite from an inexorable grind."

A measure of Earl Blaik's never ending drive for perfection is in Vince Lombardi's recollections. Vince's favorite summer recreation was golf. Blaik often played with him, but not enough. A few times Doug Kenna and Vince sneaked off for a late-afternoon round away from the Academy so Blaik wouldn't find out. Once, when he did learn about it, he asked matter-of-factly if they had enjoyed their "vacation."

During the football seasons, Vince remembered each hour and a half of practice as "meticulously organized and intense. We would arrive at that office every morning at eight and by the time we walked out onto the practice field that afternoon we would have worked out every phase and every time schedule for everything." His insistence on precise planning for practice was to obtain the most from available practice time and to relieve some of the enormous pressure already burdening the cadets. Vince went on to say Blaik allowed his assistants no papers on the field. They had to commit every play to memory, including new plays installed from week to week.

Typically, coaches assisting with the offense, working with their group of players – ends, interior linemen, and backfield – had fifteen minutes to give the cadets five new plays. Then the manager blew his whistle to begin the next phase of practice, offensive units assembled, and Blaik, for example, would call the play, "Run number ten," he ordered. Occasionally, an embarrassed Lombardi had to respond, "Colonel, I didn't have time to put that play in," whereupon Blaik simply paused, gave him his bland, Scots stare, and said, "Run number eleven."

Lombardi called it "organization at its highest level." "They took movies of everything," he added, "and after [supper] we would be back in the office and all those pictures would be developed and waiting." Lombardi also remembered Earl Blaik became obsessed with the 1950 upset loss to Navy, which ended a twenty-eight game undefeated streak. "All right," Blaik would say when there was a lull, "let's get out that Navy film." Commenting on Blaik's torment over losing, Vince added he didn't know how many times they looked at that Navy film, but the result was always the same. Navy beat Army again, 14-2, "and that was one of the ways Earl Blaik, the greatest coach I've ever known, paid for what he was."

One summer day Vince and the other coaches just finished studying an upcoming opponent. They intended playing some golf afterwards, but Blaik had other plans for them and ordered the reshowing of the old Navy film. Vince recalled, "You could see the other coaches sneak looks at one another, and although you couldn't hear the groans, you could feel them in the room." Shortly, Blaik was again intently analyzing for his assistant coaches the reasons for the defeat. "Look at that. The fullback missed the block on the end." In their five years together, Vince Lombardi and Earl Blaik watched about thirty-five hundred hours of film, Blaik said, until "our brain lobes felt as worn as the film sprocket holes looked."

Blaik knew he was untiringly intense and gave his assistants little pause. "I always drove my assistants as hard as I drove myself. Our families saw little of us, [which was] part of the price of our mission and the steepest part of all."

The Army coach's austere, patrician exterior belied the churning interior major college football gives all its participants, particularly coaches. Blaik was no exception. To keep the outwardly calm, confident, analytical presence undoubtedly wasn't easy. He was naturally, and of necessity, as Spartan with himself as he was with his coaches and players. It was inherent in his view of leadership – "Set the example." He seldom drank, a restraint caused in part by a "nervous stomach" and the discipline needed to keep a latent ulcer under control, a malady he would be loathe to discuss with anyone.

He disliked public appearances and speeches, tending to stay out of the limelight. But when he spoke publicly, or to the team, he was articulate, an outstanding orator. When he did speak in public and to the press, he was direct and frank. His directness and seeming lack of diplomacy could stir controversy with little effort. He aroused storms of protest when he criticized professional football for "a poor quality of play"; fought vigorously for the two-platoon system; dramatically ended the Army-Notre Dame series when the atmosphere surrounding the games became bitter; and when he publicly clashed with the Academy and Army over handling of the cheating scandal.

When Earl Blaik learned the Bartlett Board had determined an "overemphasis on winning football" to be the underlying cause of the cheating scandal, it must have been a profound shock to him. If there was anything of which he was certain, it was his conviction that winning went hand in hand with football.

> "At the Academy or anyplace else," Blaik consistently argued, "as long as academic standards are not compromised, the aim of the game is to win. You can't tell the youngsters, 'Now, this game isn't as important as you have been led to believe, and winning isn't everything, you know.' When you do that you strip football of its essential ingredients – the importance of the game and the will to win."

TALES FROM THE PRACTICE FIELDS

Except for quarterbacks, team and field captains, and when he saw the need to step in and teach or correct a player or assistant coach, Earl Blaik had little direct interaction with team members. His whole approach to coaching, the philosophies he developed which guided his work, and his personality, kept him distant from his players – just the way he wanted it. He could remain on a pedestal, and that's the way players saw him. As Ralph Chesnauskas, '56, said, "Colonel Blaik was always on a pedestal."

Earl Blaik was the chief executive officer running his corporation, "The Colonel" running his regiment, with his assistant coaches his lieutenants. He was always standing back, gauging the overall performance of the entire organization while keenly aware of how it all fit together, what the organization's goals and objectives were, and every little detail that added to or detracted from overall performance. However, his insistence on indoctrinating his coaches on the Blaik way of doing things, and giving them the responsibility to work with the players, freed him of details that can overwhelm any successful coach.

His detached distance from his players also allowed him to study each one carefully, thoughtfully. And study them he did – constantly. Not just their performance on the field, but their behavior, their willingness to give the most they had to the team. He looked for what he called "soul" in his players, their character – determination, perseverance, motivation to play the game, drive, and loyalty. His analytical mind cataloged the strengths and weaknesses of every team member, and permitted him to capitalize on their strengths to get the most from them for the team while teaching them how to improve or correct their weaknesses.

But none of that kept his players from constantly observing, recording in their minds what they saw and heard, and in nearly every instance forming strong, lasting opinions about this man who so influenced their lives. He was "The Colonel," "Colonel Blaik," their Coach, and seldom did they separate the man from the roles he played. Yet every player saw him through their own eyes, saw something different in this complex, inspiring man. For the cadets, as for Blaik's coaches, he was always setting the example.

With extremely rare exceptions, Earl Blaik wasn't profane, Bob Farris, '56, saying, "He never swore," and Godwin "Ski" Ordway, '55 saying with emphasis, "He *never* used curse words. He didn't even use the word 'hell'." His closest concession to profanity was an infrequent utterance of "Jesus Katie" when he was noticeably upset or frustrated. His use of expletives was virtually non-existent, and he made a point of forcefully correcting team members and coaches who used foul language.

During the pressure-packed rebuilding years, players' tempers often flared, especially among men already inclined to be feisty and temperamental. On one occasion, the team watched in astonishment as their elected captain, Leroy Lunn, a guard noted for his fight and hot temper, got into a scrap with a teammate during a pile up, kicked at him, and let fly with "...you shit!...," whereupon Blaik jerked him from the pile, grabbed the surprised team captain by the front of his jersey, pulled him in close and coldly said, "Don't you ever use words like that on this field. If you do, you'll be off this team."

As to fighting among players, verbally criticizing one another, intemperate outbursts or deliberate acts of "dirty play," he was equally swift in reacting. When fisticuffs got started during practice, he understood the competitive juices were often at work, and he would say with a tinge of sarcasm, "Don't fight with your teammates. You may need them sometime." He had similar but more pointed words for anyone who criticized a teammate for missing an assignment. "Never, never, never criticize a teammate for missing an assignment! That's the mark of a good player, and a good leader on the field."

Blaik was particularly sensitive to his players being labeled "dirty," and immediately reacted whenever he saw one display such inclinations. During the glory years of the '40s, opponents' fans and coaches occasionally criticized his teams for "dirty" play. When there were specifics associated with those criticisms, he investigated carefully and responded, using game films and discussions with the alleged offender and teammates. The reality was that tough conditioning and disciplined, hard hitting by Academy players, who were on average consistently smaller than opponents' teams, were nearly always at the root of such allegations.

In Blaik's mind, he equated profanity and "dirty play" and didn't hesitate to let players know violating either rule would end their days on the Army team. Freddie Attaya, '54, learned firsthand of Blaik's intolerance of competition run amok. Once during a practice session pileup, the Army head coach saw Attaya step in the middle of a teammate's back when he didn't have to. Blaik rushed up to the startled Attaya, jerked him off the pile, glared at him with obvious anger, and said, "If you ever do anything like that again, you're off this team," the same admonition he gave the team captain for using foul language.

When Blaik reacted strongly to his players' conduct on the field, he left an indelible impression, and usually a lifetime memory in the man he confronted. He awed them. His presence was intimidating. One player remarked that "When he looked at you, he seemed to look right through you." Norm Stephen, '54, said, "He had a pair of eyes that could rivet and hold you." Frank Wilkerson, '53, remembered his eyes as piercing, saying "When he looked at you he could read your mind." He added, "He could look in a player's eyes and tell if he was ready to play." Blaik also spoke extremely well, and when he spoke to the team, he was a master. Everyone on the field listened – intently.

Tommy Bell, '55, was particularly awed by Blaik. Tom, the handsome Irish redhead whom Vince Lombardi recruited from New York City's Bronx, recalled, "It was always 'Yes, Sir,' 'No, Sir,' 'No excuse, Sir' when Colonel Blaik spoke to me." His anxiousness to respond to Blaik's calm, restrained drive for perfection, and Vince Lombardi's energetic, bombastic push for explosive running – "Bell, you're loafing!" – often left Tom unsure of just how hard he should throw himself at his teammates in practice. When Blaik or Lombardi called a play in practice, Tommy would ask, "Colonel Blaik, is this full speed?" or "is this for real?" Tom's awed uncertainty was a frequent source of subdued amusement to his teammates – who were equally awed by Blaik, but responded differently than Tom did.

Blaik's tough exterior contributed to some sharply differing views of him among a few team members, but there were characteristics their coach possessed on which agreement was unanimous. Though stern and Spartan, he was a teacher. He would take a player aside and show him what he was doing wrong, and why, and then show him how to do it right. Lowell Sisson, '54, remembered Blaik saying, "Listen, learn, do better next time out." Consistent with his admonition to assistants to "Never yell or bellow," he nevertheless spoke firmly, with authority in his voice when he was teaching.

But he could be tough when he felt the need. He said to tackle Howard "Knobby" Glock, '56, in practice one day, "If I were as big as you and played football the way you do, I'd hang my head in shame." Howard Glock was a terror on the field the rest of the season.

There was always professional respect and admiration of Earl Blaik by team members. They knew he knew what he was doing, and that bred confidence in them. As several players put it, "He prepared us so thoroughly you couldn't help feeling you were ready to play and win."

Al Paulekas, captain of the '52 team, remembered Blaik saying, "If you are prepared, you are going to win. You will lose if you make a mistake." Al Paulekas carried Blaik's words into intercollegiate wrestling with him, where he was a standout as well. Bob Farris, '56, said, "Colonel Blaik made everyone feel so well prepared they couldn't lose. He knew how to make the team mentally prepare."

Though many players felt he was aloof, distant, or unreachable, all clearly agreed Earl Blaik worked through his assistant coaches, and that he had "the best of the best" in his coaching staff. He repeatedly encouraged his assistant coaches to "relax the players" because of the daily pressures of cadet life. Thus team members naturally tended to be much closer to the assistants. Murray Warmath, Doug Kenna, Vince Lombardi, Johnny Green, Paul Amen, Carney Laslie, Paul Dietzel, and Bobby Dodd were the men during the rebuilding years who were in the trenches, one on one, working, toiling with individual team members, to coax the best from them, even more than the players believed themselves capable of giving. Nearly all have warm memories of the assistant coaches who, along with Blaik, though terribly dispirited after the cheating scandal, brushed aside their disappointments and sadness, and worked tirelessly to rebuild Army's football fortunes.

"Knobby" Glock's recollections of Carney Laslie, Blaik's highly respected line coach, were typical. One day "Knobby" was having a particularly good practice day on defense. He broke through three times and tackled ball carriers in the backfield, throwing them for losses. After the third time, Laslie came over to him and said in his Southern drawl, "Howard, you're really havin' yourself a day," whereupon Glock, pointing to a hole in the toe of his shoe, with blood evident around the hole, explained someone had lost a cleat off his shoe and stepped on his foot. Glock told Laslie he was mad and was taking it out on his teammates.

A week later, "Knobby" was having a bad day in practice, not goofing off, but simply not doing well. He heard Carney Laslie call out, "Howard, c'mon over here a minute." He obediently trotted over to Laslie and stood in front of the assistant coach, while Laslie glowered straight into his eyes. Howard was startled when suddenly Laslie stomped on his foot and asked, "Does that tell you somethin'?"

On another occasion, while Army was preparing for a Saturday game, the usual practice ending wind sprints were in progress, a conditioning drill most players, especially linemen, hated. Laslie was blowing the whistle signaling the start of each sprint. Sprints were back and forth between practice field sidelines. After several bursts, when the exhausted linemen turned toward the sideline in the direction of the dressing room, they always hoped to hear, "Take it on in!" On this day though they heard "One more time." There was a chorus of groans.

Laslie responded, "Boys, c'mon over here." When the big, hard breathing linemen trotted over to Carney, he looked them over and said, "I been coachin' a long time. The players on that team we're goin' against Saturday are *big*. You're goin' to be the fastest team on that field or you're goin' to get killed." With that they went back for more wind sprints.

In spite of what some players saw as excessive coldness, lack of humor, and too stern a demeanor, all were keenly aware of Earl Blaik's organizational skills and analytical approach to the game. Team captain Leroy Lunn, '54, had "great, great respect for Blaik" yet described him as "stern, stern, stern" and cold, "lacking in compassion." He also noted Blaik analyzed "the physics of the game" like no other coach he had seen. The Army coach concentrated on the mechanics of the game's fundamentals, and could "break movements, stances, and body positions down like a mathematical equation." He insisted linemen use a stance in which no weight would be on the hand and only the fingertips could actually touch the ground when awaiting the snap of the ball. That would ensure proper weight distribution between the hands and feet, allowing maximum foot movement, hence better mobility. Good, effective blocking and tackling was presented in terms of angles, speed, height, and weight distribution. This wasn't a way of coaching originated by Blaik, but he definitely elevated his teams' use of physics, mathematics, and engineering in the fundamentals of the game.

There were many distinguishing characteristics of Blaik coached teams. One characteristic that stood out among many was individual and team conditioning. Cadets were already, on average, in better physical condition than students at other schools because of the Academy's emphasis on physical aptitude screening for its appointment candidates, and a rigorous curriculum of physical education. Add to those factors intramural and intercollegiate athletics, a controlled diet in the cadet dining hall, a regimented duty and academic schedule, and a demanding academic curriculum. Clearly, the football coaches were working with players who were already in relatively good condition when they encountered Earl Blaik's Spartan, unyielding approach to conditioning.

His tough mindedness consistently stirred grumbling, and some antipathy in his three most notable coaching stints – first as an assistant at Army, then when he went to Dartmouth, and again when he returned to Army as head coach. When "Biff" Jones asked Blaik to come to Army as an assistant for the 1927 season, "Red's" assignment was to be assistant varsity end coach, "coach the passing game and do some scouting." He arrived at West Point from the University of Wisconsin where, "as a favor" he had assisted his former Steele High School coach, George Little. Outside his annual fall interest in the game of football, Blaik remained firmly anchored in his dad's real estate and home building business in Dayton.

In his first season at Army, however, his tough, no nonsense work with the players resulted in a clash between him and Second Lieutenant Garrison H. "Gar" Davidson, a 1927 graduate and Army football standout who came back to the Academy on temporary duty the fall of '27 to help coach the jayvees – the "B" team. Blaik never talked publicly about what happened, and Davidson told of the incident in a never published book written for his grandchildren. Davidson recalled,

In 1927 I was asked literally to head off a revolt among the players. "Red" [Blaik] came to West Point to be assistant varsity end coach and in charge of the passing game. As a disciple of Big Ten football, he had difficulty concealing his poor opinion of Army football and was unduly harsh and unfeeling in his coaching methods. Finally, about mid-season, the situation got so bad that the varsity end squad, men with whom I had played the previous year, came to see me and asked me to intercede for them with the head coach, "Biff" Jones, and to get "Red" to temper his ways, which I did.

Davidson continued, "In 1930 and '31, when I had the 'B' squad which scrimmaged the varsity most every day, it was a problem for me to restrain the scrubs' resentment of his caustic methods, a resentment that bordered on hate."

Davidson made no mention of talking with Blaik before he went to Jones with the players' complaints, and from his recounting of the story, it appears he didn't. If so, Davidson, who was seven years' Blaik's junior, undoubtedly stirred the Scotch-Irish redhead's "argumentative nature" – to say the least. Blaik was well indoctrinated on the Army principles of seniority and chain of command, though he was no longer in uniform, and wouldn't respond kindly to someone who went behind his back to his boss. The 1927 clash between the two men, and their completely different philosophies about football and coaching, opened a rift that later widened.

In 1932 the Department of the Army and the Academy refused to alter their policy requiring an active duty officer to be head football coach, and assigned plebe coach "Gar" Davidson to direct Army football, though Major Ralph Sasse, the outgoing head coach, endorsed Blaik for the position. At the time Davidson was assigned to the Academy as the superintendent's aide.

Blaik was first assistant one season under Davidson before it appeared he may have retaliated in kind following the 1927 incident. When he was selected to be the Dartmouth head coach, he took Army's respected assistant line coach, Harry Ellinger, with him. Davidson heard about Blaik's and Ellinger's impending departure by reading it in the newspaper.

According to Blaik, at the end of the 1932 season, when told of Davidson's selection, he still had no thought of permanently entering the coaching profession, and, to all outward appearances – including his 1933 performance as first assistant – was unaffected by Davidson's assignment over him. In 1933, much to everyone's delight, Army went 9-1 on the season, a tribute to the talents of the entire coaching staff and the team, as well as their ability to work together. Blaik indeed showed neither disappointment nor resentment toward Davidson's elevation to the head coaching position, explaining he had already learned enough about the coaching profession to know it wasn't a secure profession, not nearly as secure as the business he was in with his father.

When Earl Blaik accepted the head coach's position at Dartmouth, he received due notice in the country's sports pages, and a congratulatory letter from Dr. A.H. Upham, president of Miami University. Blaik, commenting on his late-in-life decision to commit to coaching college football, responded,

February 9th, 1934

Dear Dr. Upham:

Thank you very much for your kind letter of congratulations. You are quite right, I had never seriously considered an all-year-round appointment as a football coach. It took three days of debating to assure me that it was a fine thing to do. Although I had the option of making it a part-time position, there was something attractive about Hanover and the college life there that decided us to make the change.

It is surprising the number of Dartmouth graduates I have met who have something in common with Miami. Dr. Hopkins spoke so highly of you, and in conversation with him I learned that he was a very close friend of Jerry Simpson's.

As a graduate of Miami I realize I have not been the best; it has been a long time since I returned to Oxford. I have been fortunate enough to visit many colleges in the last ten years, and I still maintain that Miami is the peer of them all west of the Alleghenies.

With kindest regards to you and Mrs. Upham, I am

Cordially yours,

Earl H. Blaik

When Blaik went to Dartmouth he readily acknowledged his coaching methods didn't endear him to some of the Dartmouth players, as well as a number of vocal university alumni. He recalled his first team meeting in February of 1934. He introduced his assistants as the best in America at their jobs. Next he briefly outlined the system of attack they would use, saying it was the most advanced and useful in football. He then told the prospective team members,

We'll be as successful as you men will allow us to be. If there is anybody in this room who is not ready to do some strong sacrificing, I hope we've seen him for the last time tonight. Because we're going to bring home the bacon.

Later he wrote about what he needed to do to turn the football program around for the Big Green.

> Our major problem at Dartmouth was to replace the spirit of good fellowship, which is antithetical to successful football, with the Spartanism that is indispensable. I believe there is a place for good fellowship. I also believe good fellows are a dime a dozen, but an aggressive leader is priceless.... The play-for-fun approach will lead the player to revolt against the game itself, because play-for-fun never can lead to victory... the only "fun" of the game, if you will, is the soul satisfying awareness that comes not only with victory but also with the concurrent realization that victory more than justifies all the communal work and sacrifice that went into it.

Dartmouth Head Coach Earl Henry "Red" Blaik, 1934-1940. He rebuilt Dartmouth's football fortunes, as he would do beginning in 1941 at Army. With him is Dartmouth Team Captain Gordon F. Bennett, Class of 1937, who led the Ivy League champions the previous fall. As a young 27-year-old physician, Dr. Bennett died after saving the life of his fiancée in Boston's tragic Cocoanut Grove Night Club fire of November 29, 1942. He had been accepted into the American Medical Society that day, and they were celebrating with friends. (Photographer: Safier Studio, Dartmouth College Library.)

He went on to say,

> I suspect my players, at West Point as well as Dartmouth, considered me not only serious but severe. It may be that some who did not understand what I was trying to do thought me a martinet, and a few may have hated me for it. I also believe that most of them, anyhow, after they left school and assumed more and more of the responsibilities of life, appreciated with increasing clarity what I tried to do – what I had to do. I was not heartless, but any revelation of the

slightest sympathy was out of key with the mood I considered urgent if Dartmouth was to regain its lost football respect.

Unquestionably he brought the same philosophy to sagging Army football fortunes in 1941, and kept it firmly in place, right through his entire head coaching career at the Academy. He would use it with re-doubled force and intensity on Army's rebuilding '51-'53 teams. Never ending conditioning, frequent scrimmaging, and hard hitting in practice drills were the rules of the day – along with all the other rules and axioms he enforced.

Don Fuqua, a defensive platoon safety the '51 and '52 seasons, believed Army scrimmaged more than any other team in the nation, noting scrimmages invariably were on Tuesday, Wednesday, and Thursday each week of the season.

Godwin "Ski" Ordway, '55, a "walk on" who started the season on the plebe team in '51, said the scrimmages were brutal. "There was pressure, hard hitting, fatigue, and a lot of fights." "Ski" understood what Blaik was doing. The Army coach was "trying to take nobodies and turn them into somebodies," he said. He also saw Blaik's philosophy as "take people and toughen them." "The toughening wasn't just physical, it was mental," Ordway said.

Players had to accept pain in practices and were expected to play with pain. A willingness to accept and play with pain reflected the toughness Blaik looked for, the "soul" the player possessed, and his motivation to play the game – the price a player was willing to pay. If a plebe showed enough skill, toughness, and determination he would find himself scrimmaging against the varsity, right in the middle of the heaviest hitting.

In the rebuilding seasons, especially '51 and '52, because of a lack of experience and depth on the "A" team, more plebes than in years past were frequently pulled up to the junior varsity, to be what many believed was "cannon fodder" for the varsity. In reality Blaik was always in the hunt for talent, skill, and "soul," and if the plebe "cannon fodder" didn't realize it, Blaik knew he was giving potential varsity players experience and toughening while taking a firsthand look at their performance on the practice field.

Bob Mischak recalled Blaik saying, "Never kneel down. That's a sign of weakness." "Some things were really brutal," Bob said, and there was a particular drill the Army coach used in practice that rankled Mischak. The "bull in the ring" was almost a daily ritual, in which one player would be inside a circle of six or eight teammates who would lunge at him, one at a time, trying to knock the "bull" off his feet. The "bull" was to charge aggressively at his attacker, in counterattack, as soon as his teammate made a move toward him. Contact was frequent, the hitting ferocious and physically demanding, especially for the "bull in the ring." As long as he was inside the ring he absorbed the charges of every man who lunged at him, six or eight times more hits than each of them received. However, there was an unpleasant compensation. Every man in the unit eventually participated in the drill – as the "bull in the ring."

Wind sprints were a dreaded, practice ending drill, common in the game of football. But on Blaik coached teams, wind sprints that caused lung burning exhaustion at the end of a furiously paced, brutal workout, seemed the ironclad rule. From team captain Leroy Lunn's perspective, Blaik's philosophy included physically wearing opponents down. The Army coach's reasoning about what was required of Army players to wear opponents down was simple. Blaik concluded cadets had to be in far better condition, better able to endure – persevere – than their opponents, as well as quicker off the ball.

Don Fuqua remembers assistant coach Doug Kenna intervening with Blaik one day to end what seemed to Kenna an excessively long period of wind sprints at the end of practice. Kenna's relationship with "The Colonel," described by some as akin to a father son relationship, had no effect on Blaik's intensity. He ignored Kenna's intercession and continued until he was satisfied the team was suitably prepared to take to the showers.

Jack "Moose" Krause, '54, remembers wind sprints vividly. "He ran and ran and ran us," Jack said. "Blaik told us 'If you're going to lose, you're going to lose hard.'"

Earl Blaik's Spartan approach to the game included a tough, seemingly cold-hearted attitude toward injuries and how his players, alumni, press, and his team's opponents reacted to what he considered one of the prices of admission to football. At Dartmouth, in addition to his insistence on "no spirit of good fellowship" approach to the game, he also determined to end what he regarded as "softness" in the way Dartmouth dwelled on player injuries.

The custom had grown up at Dartmouth of holding something of a wake over every fallen gladiator in a scrimmage. Boston newspaper stories every week during the season featured who was in the college infirmary. We had to put a stop to that. We believed that pampering and publicizing injuries were not good for winning morale. Games are not won on the rubbing table.

Blaik was no different in his philosophy about injuries to football players at the Academy. Arnold Tucker, Army's great All-American quarterback on the 1946 Army team, assisted in coaching the plebe team during the 1951 season. Since returning to Army he had developed a far better understanding of Blaik's outward appearing attitude toward injuries, versus the reality of his actions. Tucker said the Army coach was deeply concerned for the welfare of injured players and always ensured they had the best of care.

His concern for injuries was, in fact, real, and not simply because he wanted his players to show him they wanted to play, or could play "hurt." His powerful advocacy of the two platoon system of football was based in part on his concern for fatigue and the potential for increasing serious injuries. His experiences in years past, in "iron man" football, told him there were far too many serious injuries and deaths occurring in the game. He considered two platoon football not only a way for more young men to play the game, but an excellent way to ensure against excessive fatigue, and the resulting injuries.

But his players had difficulty in reconciling his expressed philosophy and what many saw as a lack of compassion, an unfeeling coldness he exuded when a player got hurt. More than any other characteristic of the Army coach, this one – his apparent attitude toward player injuries – baffled his team members. A few had encountered in Blaik what they believed was outright irritation, if not insensitivity, with an injured player when the cadet exhibited questionable or unwarranted self-concern following an injury.

During spring practice in 1951, plebe Bob Mischak was invited to scrimmage in the Field House and on Howze Field, against the A and B teams, in preparation for the annual spring scrimmage game against Syracuse University. Bob obviously impressed Blaik, and was put into the game against the Orangemen. While playing defensive end, Mischak received a painful shoulder injury during a pileup. Earl Blaik immediately moved the teams a few yards down the field and continued practice. After assuring that detail was taken care of, Blaik turned, walked back to Mischak and stood, with his hands on his hips, glowering at him. "Well?" he said, as if to say, "Are you going back into practice or are you going to stand here and worry about your shoulder?" The next day Mischak learned he had received a shoulder sprain. If coddle wasn't a word in Rollie Bevan's vocabulary, it certainly wasn't in Earl Blaik's either.

In the Army coach's mind, there was good reason for his outward response to player injuries, but the cadets had difficulty understanding what was in his mind, especially since he chose not to talk about his gladiators' battle scars. As Lowell Sisson said, "Blaik *never* talked about injuries," emphasizing the word never. Not only that, but team members were to take their injuries to the trainer, Rollie Bevan, and the team doctor, rather than the Academy hospital. Bevan and the doctor, usually Bevan, would decide if the player needed to go to the Academy hospital. More than one Army player got a cold reception for violating that policy.

Gerry Lodge saw Blaik's reaction to injuries as a "blind spot" in his work with the cadets, saying the Army coach "didn't accept the fact that injuries slowed you down." He recalled Blaik saying, "If you don't want to go back in, you don't want to play."

Gerry, a big, solid, muscular cadet who was also an intercollegiate wrestling champion, was playing guard on the Army football team during the '51 and '52 seasons. During the week before a home game his "cow" – junior – year he developed a painful case of cellulitis in his right leg. The injury slowed him down and his performance in the game that Saturday was duly noted by Blaik. "Didn't have such a good day today, did you?" the Colonel asked. "No, Sir," Gerry replied.

That evening one of Gerry's roommates was to meet a young lady and take her to a formal hop in the South Gymnasium, but the roommate had been assigned guard duty that would make him late in meeting her. He asked Gerry to please meet and escort her to the dance, and after completing the guard tour he would join her. Gerry agreed to help, put on his full dress gray uniform and dutifully met and escorted his roommate's date to the South Gymnasium. As they were entering the building, who should come walking out but Earl Blaik? Gerry spoke to him. Blaik just glared icily at him as he passed. No answer.

When Gerry returned to his room a short time later, a note was waiting for him saying Coach Blaik had called and wanted to see him first thing next morning. By the following morning, Gerry's leg worsened. The pain was excruciating. On the way to Sunday morning reveille he very nearly passed out, and fell down a flight of stairs. He was taken to the hospital and Blaik soon received word of the junior guard's circumstance. Not long afterward Blaik came to see him in the hospital, and after some preliminary words of greeting and "How are you doing?" said with a faint smile, "If you hadn't been sick you'd have really been in trouble."

As a plebe, after he'd been among eleven promoted to varsity the fall of '51, Howard Glock ran afoul of Blaik's injury policies. He received a painful bruise to his leg during practice one day. While marching in supper formation that evening one of the upperclassmen saw him limping and asked what was wrong. Glock told him of the football injury and the upperclassman instructed him to go on sick call. The doctor found a blood clot in his leg and put him on crutches for two weeks. He realized in obeying the upperclassman's order he'd violated Blaik's and Bevan's rules about going to Bevan or a team physician first, and he found out the hard way about violating team injury rules. Neither Blaik nor Bevan would talk to him for a long time. Although he had felt he was headed for a starting berth before the injury, it wasn't until Navy that he finally started his first varsity game.

Don Fuqua described Blaik's attitude toward injuries, using words the Army coach used. "Spartan, too Spartan," said Don. From his perspective too many players played with injuries when they shouldn't have.

Two players said they had personal knowledge of cortisone and even novocaine shots administered to allow men to play with injuries. If true, this was indeed a questionable practice that put players at increased risk of more serious injury, and was contrary to rules as well as thoughtful judgment. If Blaik knew, encouraged, or condoned such practices by Rollie Bevan, who had been the Army coach's trainer for seventeen years, it was wrong.

There are explanations for his behavior. Blaik did operate with football philosophies which might tempt him to rationalize such action.

It's likely his reasoning was based on his beliefs that "football is the game most like war," and "games are not won on the rubbing table." "Minor" injuries were not to become so important to a man to be an excuse for avoiding the war, the fight, the game. Wounded men didn't leave the field of battle with minor injuries, such as cuts, bruises, sprains, strains, or broken noses. If they did, someone else not so skilled had to step in and fight in his place, and the team, the unit, the platoon would be weakened by his absence. If a player wanted to play, showed the "soul" to play in spite of "minor" injuries like cuts, bruises, sprains, strains, or broken noses, Blaik would help him play, do everything in his power to get the right treatment or alleviate the pain. Ultimately, Blaik's attitude toward the treatment of injuries got him in trouble, and it happened during the time the Bartlett Board was meeting.

Board testimony from senior physicians on the Academy hospital's staff complained of sharp differences between Army physicians at the hospital and the Football Office regarding the treatment of injuries. Boyd Bartlett correctly called the differences and the Football Office's actions to the attention of General Irving in a board letter covering "matters outside the purview" of the Board's charter.

Rollie Bevan was at the center of the disagreement. It was, at the time, representative of differing medical and scientific approaches to the treatment of athletic injuries, as well as a tendency on Blaik's part to spare no Athletic Association expense to rapidly obtain the best specialist treatment possible for an injured player – usually from doctors in New York City, without consulting with physicians at the Academy hospital, and on some occasions without consulting with the team physician assigned from the hospital. The problem was compounded by Rollie Bevan's jealous "stay out of my business" devotion to his role as Blaik's trainer, and an unrelenting, hard nosed, "no coddling" attitude toward football injuries.

The differences in scientific and medical approaches to injuries added to the problem. In Blaik's and Bevan's view, the tendency of hospital physicians to immediately immobilize strains, sprains, and other injuries needing orthopedic care, or too quickly use the surgical knife were both medically wrong, and good reasons to bypass the Academy hospital and go to specialists who worked routinely with athletes. In fact, time has borne out the orthopedics' approach to healing athletic injuries as better for the general population. Carefully considered exercise soon after the injury, coupled with physical therapy to maintain and rapidly restore mobility and strength is the more common practice for such injuries today.

Of course, Blaik's way of solving such injury problems was not entirely acceptable or wise. Irrespective of the source of a player's injury, the Academy and the Army were ultimately responsible for the treatment received, not Earl Blaik or Rollie Bevan. On the other hand, the Army physicians should have worked with Blaik and the specialists that were tending to the athletic injuries. After the cheating scandal, it took General Irving's direction to the Academy Chief of Staff to get the two warring parties to work together in the treatment of football injuries.

BLAIK AND THE SCANDAL: COMING OUT OF THE SHADOW

In spite of Earl Blaik's distance from his players, his stern, driven, Spartan, uncompromising reach for perfection, the ironclad control and discipline he exercised over the team, there was near unanimous professional respect accorded him by the Black Knights and his assistant coaches. To most all he was not only a superb football coach, he was a superb leader of young men.

Joe Lapchick, '54, was outspoken. "I would walk through a brick wall for that man. He was inspirational, knowledgeable, wonderful." Jack Krause considered Blaik "a giant of a man. I've never met one like him before or since. He was a man of his word." Frank Wilkerson described him as "a man's man." Joe Franklin, '55, saw him as "a leader among young men preparing to be leaders. He set an impeccable example. He wasn't perfect, but he rose above his mistakes and didn't fixate on them." Norm Stephen, who became a defensive field captain during the rebuilding years, remembered Blaik's teaching and coaching his assistants and players, then delegating responsibilities, delegating well and successfully. John Krobock, '53, who had been a member of the shattered 1950 team and lettered both '50 and '52, said Blaik "managed, and let the assistant coaches do their jobs." To Freddie Attaya "Blaik was a gentleman on and off the field."

As a leader, he pulled his players as well as driving them. Ralph Chesnauskas, '56, recalled the Army coach saved many players ready to give up on their performance, themselves, or the Academy. Ralph was one of those he saved, pulled back from leaving the Academy, as was Freddie Attaya, and Tommy Bell.

But the cheating scandal did many things to Earl Blaik, the master football teacher and leader of young men. He had to look inside himself, be analytical with Earl Blaik, as he had been with each of his players and teams. Aside from being stung by the personal and professional blows given him, he was forced to examine his own role in the events leading to the tragic affair, and continue, behind the scenes, to assist the men who had resigned from the Academy.

He did everything he could to save the cadets who erred so tragically, and failing that, fought to help them "leave West Point with their heads held high," with honorable discharges. He went one step further, working in the background, trying to gain reappointment to the Academy for some, and seeking the assistance of other colleges and universities to enroll the former cadets, to give them the opportunity to complete their education. Now that his friends, and particularly retired General MacArthur, had persuaded him to "not leave the Academy under fire," he had to turn and face the wreckage, the ashes, of a once proud football team, numerous angry Academy graduates who held him responsible for what occurred, and a devastated, emotionally drained coaching staff.

His football program and the support system he had built, his powerful influence and ironclad control of his assistants and his team, would be among many factors that would help him rebuild Army's shattered football fortunes. But there were other questions needing answers.

What could he do better than he had done in the past? What might he and his teams do that hadn't been done? What might he do to close the divide that had apparently grown between a substantial number of cadets in the Corps, and the team, on the one hand, and between the Football Office and restive, dissatisfied members of the faculty, Tactical Department, and superintendent's staff, on the other?

The public announcement of August 3 had loosed a fury pointing to the football team and an "overemphasis on winning football" as both the origins and causes of the cheating scandal. Those conclusions seemed to have been confirmed by the Bartlett Board, which went into session on August 13 and completed its work on September 9. Although the Board report was never made public, its ringing indictment of an overemphasis on winning football as the underlying cause of the scandal was devastating for the Army coach. Now Blaik wondered aloud "What kind of football do they want us to play?" Every fiber of his being, nearly every reason for his life's work, rested on his philosophy about the game of football: "Football is secondary to the purpose for which the player is in college.... Championship football and good scholarship are entirely compatible.... The purpose of the game of football is to win, and to dilute the will to win is to destroy the purpose of the game."

Earl Blaik, like many at the Academy, had slowly developed some blind spots over the years, blind spots that rested on faulty assumptions or a lack of thoughtful inquiry. The blind spots began imperceptibly, probably during the war years when everyone was riveted to the all-destroying monster that was World War II. The Corps of Cadets had expanded dramatically during the war, almost doubled in size since the days that Earl Blaik and most of the faculty at West Point were cadets.

What apparently hadn't been carefully considered was that an expanding, changing institution must pay close, undivided attention to the important, fundamental principles, the foundation and purposes, on which the institution rests, to ensure the principles are kept in the forefront, the standards not weakened, that everyone performing the institution's mission is working toward the same goals and objectives. A rush to expand, especially when eyes are elsewhere, can result in an increasing number of people admitted who do not necessarily see, comprehend, or even agree with the principles and purposes on which the institution rests. The concepts of honor and duty, or any moral or ethical code had to be worked at especially hard, and couldn't be assumed to be well understood, prized virtues, inculcated in the cadet candidate by family, school, and religious faith.

Young men can be jaded and made more cynical by what they see in the world around them, and what they perceive an institution to be – to the point that they can ignore, discard, or rebel against those principles for reasons of "friendship"; individual, group or team loyalty; or perceived overriding ambition or self-interest. People – good people – apparently given "a good upbringing" – can get caught up in activities neither good nor right, and rationalize their behavior as well as deliberately hide their subterfuge.

Young people coming to the Academy, were often "mere boys," still feeling their way, undecided about where they wanted to go and what they wanted to do, and vulnerable to the appeals of fierce group loyalty – even if the group was engaged in less than wholesome endeavors. What apparently hadn't doubled in Earl Blaik's mind was the understanding that winning can occur dishonorably as well as honorably, with subtle and sometimes boastful arrogance as well as reasoned communal sacrifice, integrity, mutual respect, modesty, and humility.

Earl Blaik recognized he had made some mistakes, but he didn't trumpet them. He began, without fanfare, to learn from his mistakes, and do all the things he had always done as a coach and teacher of football. But he did more, some forced on him by circumstances.

From the first day his inexperienced 1951 team reported to practice, he put the scandal behind them – except to drive home the unrivaled importance of integrity. He told them briefly what happened, without rancor or bitterness. He cast no aspersions upon the Academy or the Army's senior officers; though he felt in his heart there was good reason both institutions needed a serious introspective look at what happened, and were in large measure to blame for the scandal. From that time forward he never talked to the team, as a group, about the scandal.

He went to the Corps, at first out of necessity, to find young men eager to become part of Army football's winning tradition, the walk-ons who came forward as they never had before. The tragedy of the expulsions brought "Red" Blaik closer than ever to the Corps of Cadets. However unskilled they might have been as potential major college players, their energy and motivation drew him closer to them. And the team he began to rebuild increasingly believed themselves more a part of the Corps.

Blaik saw the clear need for better honor orientation and training for his football players, and advocated changing methods of honor training to include more informal small group discussions to supplement this training for all cadets. His views would have a positive effect beginning reorganization week in late August of 1952, when the Army

Army Head Coach Earl Blaik being interviewed by Pointer Magazine *staff writer, Cadet John B. Garver, Jr., Class of 1952, for the September 7, 1951 issue, three weeks prior to the first game of the 1951 season, the first year of rebuilding at Army. The caption under the picture read, "Coach Blaik – '...and we'll need every manjack – even your drags....'" (Photo courtesy USMA Archives.)*

varsity football team began receiving additional periods of honor orientations. Undoubtedly, the "cram school" which would no longer be held at the Academy, would also feature an introduction to the cadet honor code and system, as the cadet candidates – athletic recruits – began preparing for entrance exams.

In the meantime Earl Blaik fought furiously to give the resigning cadets a fresh start in life, although the men he now wanted to help had, in a real sense, deliberately deceived him.

The fact he had been misled, along with everyone else, is not hard to understand. The very pedestal of admiration from which he operated, his seeming aloofness, being distant, "unreachable," as head coach had put him at too great a distance from his players. His tendency to communicate primarily through his assistant coaches, quarterbacks, and team captains, and to talk little else but football, unless there was a player in trouble, kept him from knowing far more serious trouble was brewing. Nobody was going to come to Earl Blaik and tell him they were encouraging or helping cadets in violations of the honor code. Blaik wouldn't react kindly to such activities, and no one wanted to disappoint him or incur his displeasure.

Blaik had mistakenly seen the difficulties of his 1950 team primarily through the lenses of what he perceived as envy or jealousy by men who, he believed, resented the football team's success and its resulting national acclaim, and what he saw as a baffling, frustrating lack of support from the Corps of Cadets.

And at first, following the scandal, he had rationalized the young players' serious mistakes. He listened to their entreaties, and accepted at face value that they were telling him the whole truth, when, in fact many weren't. Instead, many of them saw him as the most powerful and influential man in their lives and their sole means of salvation from a disaster largely of their own making. Some were using him as a last hope in their desire to remain at West Point, while others had attempted to deliberately discredit the Collins Board.

Worst of all, perhaps unwilling to admit his misjudgment of the young men he was responsible for recruiting, Blaik allowed himself to be used. The words Earl Blaik wrote years later, a remark made by a man he admired, Charles Franklin Kettering, and repeated frequently by Blaik to aspiring assistant coaches and players, were ironically, most appropriate to his experiences in the spring and summer of 1951. "It isn't what you don't know that gets you into trouble, but rather what you know – for sure – that isn't so."

Earl Blaik knew these young men, "for sure." He knew they were good men. There had to be something else in the system to cause their serious mistakes. They had proved they would "pay the price." He knew he wasn't mistaken about them, but he was, and he erred in his judgment. He had blurted out his misjudgment, when on graduation day of 1951 he told Genera Irving it would be better to graduate sixty crooks than hold up graduation to try and find out who they were.

None of this, however, deterred him from fighting the good fight, right to the end. He repeatedly took trips to Washington with General Irving prior to the August 3 announcement, arguing to Lieutenant General Maxwell Taylor, General Collins, the Army Chief of Staff, and Secretary of the Army Frank Pace his views and recommendations for less severe penalties.

He appeared before the Hand Board which on July 23-25 reviewed the Collins Board proceedings and the superintendent's recommendations to the Army staff. He gave sworn testimony to the Barrett Board, one of the two "screening boards," convened on August 9 to hear matters of extenuation, mitigation, and explanation, and examine additional evidence that might be presented. To the Barrett Board Blaik submitted a list signed by him, of ninety-four names the resigning football players had provided him. The ninety-four were allegedly men in the class of 1951 who the soon to be discharged cadets told Blaik participated in, knew of, or encouraged the cheating. Undoubtedly, some of those on the list were the same men named in late June by cadets who had already testified before the Collins Board.

Earl Blaik, like many people, seemed to look everywhere to find reasons for what happened, and wasn't looking hard enough at the men who failed themselves. He stubbornly believed that all the team members' pre-entry recommendations from school administrators, coaches, family and community members bore the truth of what these young men were, as well as being accurate predictors of how they would behave, and what they would become. What had happened had to be the fault of the institution in which they came to receive their education, and an honor system that had gone wrong.

Worse, Blaik reasoned the Academy had an errant, arrogant regard for its honor code and system, that Academy officers believed their concepts of honor were somehow "better than honor taught in other schools," or that cadets came to West Point already taught honor by their families, and the Academy couldn't add to what they already learned. He failed to consider the uncompromisingly high standards the Academy demanded of its cadets and graduates, and either lost sight of, ignored, or didn't understand the origins or underlying, rational bases for the honor code and honor system, or simply got caught up in his determined defense of the men about to resign from the Academy.

His response to his own reasoning about the why's of the scandal was a football coach's response. As he normally did with a player who erred in the game of football, he argued for their being the ones to learn from their mistakes, become teachers of honor to the Corps of Cadets – by their exemplary misconduct, and be reinstated as cadets. His conclusions were proposed solutions that went nowhere, because the honor system was unequivocally clear and unyielding in its penalties. There could be no compromise. The principles espoused in the training of an officer didn't include having the serious offender be a teacher, or a "good example" for men to hear. The men of the Corps who proudly, or otherwise, believed in and abided by the honor code and its system would be even more severely disillusioned had this group of offenders been accommodated, given leadership or teaching roles following all they had done.

Earl Blaik lost his battles, all of them, though he had had a major role in persuading the cadets to reveal the truth of what they knew about violations of the honor code. He had done what any good officer would do for his men. They were in training he reasoned, learning, and deserved another chance, a fresh start after receiving severe administrative penalties. What's more, in his view, men in the classes of 1952 and 1953 were being "scapegoated" if none in 1951 and 1954 received the same stiff penalties, when, from his perspective, there was ample evidence men in those two classes were also involved.

He spared no effort in arguing in behalf of the men in '52 and '53, all the way up the chain of command to the Commander-in-Chief, President Truman. He had taken the extreme measure of speaking publicly in an August 9 press conference in New York City, in defense of the men, providing information about their records of achievement before they entered the Academy.

Earl Blaik, the quintessential competitor, the winner, the man who said "Don't ever give up," now swallowed his pride, his defeats, his mistakes, his embarrassments, his humiliation, his months of struggle, and his personal pain over all that happened, and began rebuilding what had been torn down.

He became once again the good soldier. Having lost his private appeals for a more favorable outcome and having gone to the public to defend the "good names" and "basic integrity" of the men about to resign from West Point, he kept publicly silent for now, though he continued privately to argue for what he saw as needed reforms in the academic and tactical departments, and in the honor system.

He concentrated on rebuilding the Army football team. In doing so, he came to the Corps of Cadets and the Academy for help. He worked with the superintendent toward eliminating the deep divisions and flaws uncovered within the Academy, which the scandal had both exposed and exacerbated. He worked with men who had vigorously attacked what he stood for, which he believed was an important, productive part of an officer's education.

That Earl Blaik once again extended himself to save young men he believed in, to restore their badly self-damaged reputations, at risk of being himself vilified, said much about his humanity, and his future success in rebuilding. That he couldn't see beforehand something deeper was wrong in the group of young men he had trusted, tells of both his humanity and his fallibility. He wasn't perfect, and in his drive for perfection, he paid a price no less painful for him than the young men who let him and many others down. That he swallowed his feelings, buried his frustration, anger, and disappointment, and went on to re-build, tells of the steel, determination, and courage in him – and of the future for the young men who were sweating and toiling on the prac-tice and game fields during those three tough years. He was bringing the future back to Army football more rapidly than anyone thought possible, better days and new glories, with the help of many more than he dared dream possible.

CHAPTER 12

THE SEASON OF '51

The United States Military Academy, Earl Blaik, his assistant coaches and Athletic Department, and the Army football team were starting over. The Academy superintendent, General Irving, the new Commandant, Colonel Waters, the Dean of Academics, General Jones, and all the people on duty at the Academy to educate and train the men of the Corps, were watching in dismay beginning in mid-August as some of the Corps' most promising officer candidates resigned under the dark cloud of scandal.

Among the Corps of Cadets, there were strong cross-currents of emotion which began flashing through class grapevines on August 3. Friends, some of them lifelong friends, classmates, teammates, much admired members of the upperclasses and intercollegiate athletic teams, and in some cases, former comrades in arms, were soon to leave West Point, most never to return. All this was to say nothing of the swirl of frustration, anger, disappointment, and embarrassment surging through the hearts and minds of thousands of Academy graduates scattered throughout the world when the press releases of August 3 echoed through the news media.

As for Blaik, he knew little of the men who would constitute Army's varsity in the coming fall. He had to learn more about them – quickly. His varsity assistants, junior varsity and plebe coaches were more important to him than ever, as was General MacArthur.

On September 4, 1951, a few days after preseason practice began, Earl Blaik wrote General MacArthur,

> ...We have started our football – soon I shall write you when I have a better knowledge of the squad. At the moment the defense looks pitifully weak and our every effort will be to bring it along fast...

Then, in a more personal vein he added,

> ...I do hope you will come to our games. Any desire you may wish to eliminate official parties and crowd congestion is eas-

467

ily avoided by coming to our quarters which is adjacent to Lusk [Reservoir] on a dead end road about one block from the stadium...

Sincerely,

EARL H. BLAIK

General Irving also knew of MacArthur's abiding interest in Army football, and the great esteem which most Army men and the Corps of Cadets held for this living legend among Academy graduates. The lofty station of five star general reached by MacArthur in his forty-eight years of service as an officer hadn't set him apart from the Corps of Cadets. He was one of them, but remained on a leadership pedestal in the eyes of young men yearning to be good officers. MacArthur was a shining example for men at West Point, what an officer should be and do – except for crossing swords with the Commander-in-Chief. And in spite of the abrupt and embarrassing end to his career, Irving knew MacArthur could inspire the Corps and the team with his presence. The old soldier had sent messages and letters of encouragement to the Corps, Army football teams, and their coaches from the far reaches of the Pacific and Asia for years, but hadn't attended an Army game since 1934.

Irving was equally aware of the lack of talent, experience, and depth on the Army team. He knew the schedule would have been tough for the now departed holdovers from the powerful 1950 team, but it would be brutal for the former "silhouettes," "the golden mullets," laboring, sweating, and bleeding to become the Army varsity of 1951. Army had never been noted for looking for too many dragons to slay in one season – just a few choice dragons. This year, however, there were precious few Black Knights prepared to slay dragons – or lesser opponents for that matter. Thus, one day after Blaik had written General MacArthur to say he was starting anew, Irving wrote,

Dear General MacArthur:

May I extend an invitation on behalf of the Corps of Cadets for you, Mrs. MacArthur, and young Arthur to be my weekend guests, twenty-ninth and thirtieth of September?

I express the sentiments of the entire command in stating that your visiting West Point on the occasion of our first football game would be a great compliment to all of us, especially the team.

We have long awaited your return to West Point, General, and I had hoped that I could afford a more formal occasion for you to visit us; however, we feel your presence here at this time would provide the team with the inspiration it needs under the present unusual circumstances.

Respectfully,

F. A. IRVING
Major General, USA
Superintendent

WHERE TO BEGIN?

Indeed there was gloom at West Point and on the Army coaching staff late that summer as preparations for preseason workouts began. The announcement of August 3 shattered all manner of dreams and ideals, and loosed a public furor around the football team and the Academy's entire athletic program. That same afternoon the first of the two screening boards, the Jones Board went into session. Letters, teletyped messages, telegrams, phone calls, and press inquiries were pouring in to the Academy.

On August 9, the day assistant coaches returned from Bull Pond and their annual gathering with Earl Blaik, the second screening board, the Barrett Board, was just getting underway behind closed doors. Members of the Collins Board were as unhappy as Blaik and his coaches. Also behind closed doors, Art Collins, Jeff Irvin, and Tracy Harrington were being confronted with an affidavit prepared by the Army staff judge advocate – the Army's equivalent of an attorney general. The affidavit necessitated they swear to having conducted their investigation according to law and Army regulations, and the Jones Board had been asked, at Art Collins' request, to look into the implicated cadets' allegations against the three officers. In the Athletic Department not even the news of peace talks in Korea could lift spirits on the Army coaching staff.

Paul Amen, John Green, Doug Kenna, Vince Lombardi, and Murray Warmath were emotionally exhausted and depressed. There was little doubt that virtually all returning lettermen from the 1950 team would soon be leaving West Point in disgrace. Years of hard work by the Army coaching staff, and people in the Athletic Department, had been swept away, to say nothing of apparently bright futures for the dismissed cadets. In 1950 the team narrowly missed a national collegiate championship. At spring practice, the promise of 1951 seemed far greater, perhaps second only to the great Davis-Blanchard teams of years past. Now, most of the '50 team members, through misplaced loyalties, errors of judgment, and some grievous mistakes, apparently had forfeited individual and team promise, dreams of service, and hoped-for lifetimes of success in the Army or Air Force.

Worse was a deep concern of Blaik and his staff for the ability to recruit "blue chip" athletes to rebuild the team in the years ahead. They feared the dark stain on Army football would drive away promising young players when the Academy was already struggling to compete with major colleges and universities. Blaik believed it would take ten years to rebuild. MacArthur told the Army coach it would take "at least" ten years and Blaik believed MacArthur wasn't overestimating the time required.

The revised draft law of 1948 was already creating a new obstacle for the Academy in the ongoing competition for gifted scholar-athletes. The law permitted college deferments, not granted in World War II, thus encouraging college age men to enter and remain in colleges rather than expose themselves to the draft. By entering a college and maintaining passing grades, they could avoid military service entirely, as well as an increasingly confusing and unpopular war.

Professional football, though not yet a powerful magnet for college players, was gaining in popularity and salary. Eager, confident "blue chip" high school and college players who wanted to continue playing the game, were beginning to consider the professional game as a springboard to fame and fortune. The service commitment after graduation from the Academy tended to discourage exceptionally talented scholar-athletes who might want to opt for professional football. Graduation required three years active duty in the regular Army, four in the Air Force if the graduate attended pilot training.

On Earl Blaik's coaching staff, Vince Lombardi took the sad news of "the 90's" fate especially hard. His meticulously coached backfield was gone, a backfield that developed an admirable nationwide reputation on the gridiron. But more importantly, as an assistant coach, Vince was closest to the Blaik family. He coached young Bob, watched him grow, mature, and develop skills that could make him a great West Point quarterback and a fine Army officer. Blaik asked his mercurial assistant, his backfield coach, to work with Bob, in part to reduce contact between father and son on the practice field. Blaik wanted to avoid, as much as possible, the appearance of nepotism in selecting a field general for the Army football team.

The close relationships among Lombardi, Bob Blaik, and Bob's parents, was more personal than any assistant coach might ordinarily expect, enabling him to clearly see the anguish the Blaik family was going through, particularly after it became clear Bob would have to resign from the Academy. What Vince saw only added to the wrenching emotional impact on the enthusiastic, energetic, always temperamental "hard nose" of Earl Blaik's staff.

INEXPERIENCE AND FIGHT IN THE SHADOW OF DISHONOR

The scandal also brought embarrassment and additional distractions to the inexperienced players left to pick up the pieces and slog through what surely would not be a pleasant season. Questions of their individual integrity, their honor, though unspoken in their presence, were certain to be subjects of frequent discussion outside earshot, and in the press. And the circumstances suggesting members of the just-graduated class of '51 and the new yearling class of '54 were involved in the cheating didn't help. The fact there had been rumors and unmistakable inferences that many who were involved simply lied to the boards and escaped the fate of "the 90" only deepened the sagging outlook for the men competing for starting positions in the opening game.

Compounding the distractions were strong feelings among the remaining team members, only two of whom were returning lettermen. In spite of the always present undercurrent of competition within the team, there was lingering disbelief, shock, and sadness over the fate of teammates many had come to know and admire. Irrespective of all these factors, the opening game was rapidly approaching, against the Villanova Wildcats, September 29, in Michie Stadium.

In the Corps of Cadets, however, there were other emotions less burdened with the accusations being leveled at the football team. Army football, because of its rise to fame under Earl Blaik, always drew men from the Corps who dreamed of being part of the great winning tradition at Army. They were the annual "walk-ons," the young men, most of whom had played high school football and some who hadn't, who believed they might be able to make the Army team. In June and July, before the controversial public announcement, Earl Blaik and his coaching staff were already aware of the coming disaster and had begun an intense search for additional talent within the Corps.

In early summer, members of the Corps, half their number scattered throughout the country in military training programs or on leave, were still uncertain as to the "something big going on," its seriousness, or magnitude. When the public announcement came in early August, among numerous other effects, it galvanized the latent ambitions of young athletes who believed they might, just might, catch the eye of a discerning Army football coach. Probably never in the history of Army football were there so many walk-ons.

What's more, cadets who simply enjoyed the game, were attracted to the nationally popular spectator sport, Army's winning football tradition, or the annual major college chase for the mythical national championship, knew well the Black Knights were in for a long season as underdogs. The competitive juices were always flowing at West Point. Now they were boosted by a deeply held American tradition – root for the underdog. The "Rabble" was now the Corps' team, with absolute certainty. This inexperienced, "no-talent" group of young men needed help. And what the Corps lacked in football talent for the big time gridiron, it would make up for in noise, thunderous noise. No longer would the shouts of Corps support be fueled mostly by the directed, mechanical roars of each plebe class. The seeds had been planted for the rebirth of "the spirit of the 12th man."

From the superintendent to all the people engaged in supporting, educating, and training the Corps, to the plebes – the lowly freshmen – to the privates in the Army units assigned at West Point, it was clear Army football had been devastated by the scandal. But soon after the public announcement, General Irving, responding to sportswriters' questions, emphatically stated Army would play its schedule the fall of

'51. He went on to say, "We will try to make up in fight what we lack in talent." And fight they did. Years later Earl Blaik wrote, "If they lacked talent, they lacked nothing else."

Blaik had his own problems to wrestle with while he began to re-build the shattered Army team. The battle he had begun the evening of May 29 when he met with the twelve men from the class of 1953 had left him weary, disconsolate. He and Merle were emotionally drained by the scandal's devastating effect on their family, Bob's resignation from the Academy. Blaik also knew he was a public figure, looked up to and admired by thousands of Academy graduates and hundreds of former players on his teams. His accomplishments, and those of his teams, had earned him a deep reservoir of respect in the sports world and secured him a place in the pantheon of football heroes. He was a proud man, and his pride was deeply wounded. Time and time again, speaking of the men who resigned, he said, "I don't condone what they did" yet, in the face of increasingly harsh criticism he fought for their honorable discharge, and their reputations. At Bull Pond, he told his coaches, "I know they will have to go, but I want them to leave with their heads held high." To him, it appeared the men who told the truth of their involvement in honor violations were the only ones who suf-fered the consequences, a belief that wasn't well founded. As the res-ignations continued from mid-August through the first days of Sep-tember, it became clear he had been unsuccessful in achieving what he advocated.

After his press conference at Mama Leone's Restaurant on August 9, Earl Blaik began to withdraw. Previously open and outgoing when meeting with the press, sometimes outspoken, tartly frank, and contro-versial, he pulled back. He was frustrated, obviously hurt, dejected, and almost sad. He communicated to sportswriters through taped ques-tion and answer sessions with one of the Academy's public informa-tion officers, who doubled as a sports information officer, Major Jo-seph F. H. Cutrona. Joe Cutrona and Earl Blaik respected one another and had a rapport which remained strong.

Nevertheless, the season of '51 was relentlessly bearing down on Blaik and his staff. His meeting with General MacArthur at the MacArthur apartment in the Waldorf Towers had given the Army coach the impetus he needed to remain as the Academy's head football coach,

Athletic Director, and chairman of the Athletic Board after the storm of criticism, the '50 team's demise, and the dispiriting resignation of his son. Earl Blaik would not resign, no matter the mistakes he might have made or the responsibilities others chose to call his for the scandal that had destroyed the Army team and torn at the heart of the Academy. He was a competitor, a fighter, a winner, and he would not turn his back on the game he loved or the institution he deeply respected and had served for nearly nineteen years. Nor would he turn his back on the men who had "paid the price" on the Academy's "fields of friendly strife."

PRESEASON PRACTICE: WHO WILL START? WHO WILL LEAD?

As he had done in years past, Blaik began the first day of preseason workouts with a team meeting on the practice field near Trophy Point, the team suited up in clean, unsoiled uniforms. But this year was different. When he called them together, he told them about the cheating scandal, briefly describing what occurred.

He talked of integrity, its meaning and importance, and from that day on steadily drummed into his players and the teams he molded the importance of integrity. Integrity was the word he chose. Perhaps, in his mind, honor had become a word difficult to speak because, for him, the spring and summer of '51 tarnished the meaning of honor.

He ended their session saying, "There isn't a man among you with varsity playing time. It won't be easy. There's much work to be done."

For all his grit, determination, and courage, no one could know, but the men who played for him on the 1951 team, just how the earlier events of that year affected "The Colonel." His drive for perfection in the game of football was always relentless. Now the tough, no-nonsense, seemingly cold and tightly-controlled approach to practice was fueled by all the emotions that tormented him in the preceding months.

He remained detached, analytical, solidly in charge, but, by his own admission, he wasn't the same football coach he was in 1950. He was struggling to hold onto his interest and motivation. He seemed unaware that the pace for the young team was furious, the demands unceasing, more intense; the warm-up exercises, drills, conditioning, and scrimmages tougher; the wind sprints endless, more exhausting.

He was impatient, irritable, harder to play for than ever. Compliments, seldom heard from him when he had great teams performing well, were almost nonexistent in preseason practice. There would be few compliments throughout the season ahead.

Earl Blaik and his assistants had two returning lettermen from the 1950 squad, one declared deficient in academics, unable to play; the other, John Krobock, was injured and unable to play. Though his team was absent experience and depth, he decided if he were to rebuild as rapidly as possible, he must continue the two-platoon system he had imported from Michigan in 1948. This meant his players would continue to specialize on the offensive and defensive platoons. More young players would gain experience more rapidly that way. Besides, Earl Blaik's experiences taught him that "iron man football" inevitably meant more injuries. Playing both ways, on offense and defense, brought fatigue and more serious physical and mental mistakes, as the game clock ticked on in the adrenaline driven, high speed collisions of big time college teams. The likelihood of injuries due to fatigue increased as young men became bigger and faster, and the game more complex and demanding over the years.

He had another problem complicating the team's rebuilding. His young players were even smaller than Army players of previous years. The physical size of Military Academy teams was always constrained by Army regulations as well as Air Force regulations when the Air Force became a separate service in 1947. The regulations governing the height and weight of cadet candidates were based on size, configuration, and internal operating space designed into military equipment. The design specifications for military equipment and weapons for crew or operator function, protection, and comfort consistently lagged population samples of data, clearly depicting trends toward decidedly healthier and physically larger youths in America.

Before coming to Army for the 1941 season, Blaik successfully negotiated with the superintendent, General Eichelberger and the Army staff an increase in the size of men admitted to the Academy. The intent was to allow Academy teams to be more competitive in size. Nevertheless, the service academy's teams would consistently be smaller and considerably outweighed in comparison with their opponents – facts that remain true today. In the fall of 1951, the disparity was even more pronounced because of Army players' youth. On average, the Army team was younger and still growing, still "filling out" in bulk and strength, making players even more susceptible to injury against larger opponents.

But the hallmarks of Blaik coached teams had always been conditioning, quickness off the ball – and in thinking, quick hitting, speed, physical toughness. Successful college coaches consistently used the laws of physics in football. Hit a big man low, fast, and hard to knock him down; get the angle and use it; cross block, trap block, or run right past the overaggressive defender while on offense; submarine, stunt, loop, and slant the charges on defense, sow confusion. But Blaik's use of the laws of physics in every phase of the game was developed to a high art form. He would now need all these innovations and tactics – after first drilling them into the younger than usual, still smaller, still growing team of 1951.

As the days slid by and the calendar brought September 29 ever nearer, the frantic push for the best possible starting lineup increased pressures on both the coaches and the team. Blaik and his assistants, always the masters in identifying emerging talent for the eleven positions on each of the offensive and defensive platoons, were moving men from position to position every practice day, observing, experimenting, evaluating, changing. Yesterday's offensive guard or fullback could be today's defensive linebacker or tackle. Yesterday's defensive end, today's linebacker or defensive halfback; yesterday's offensive left halfback, today's fullback, or backup quarterback. What's more, he was moving men back and forth between the varsity and "the golden mullets." He had a system for the moves between A squad and the "mullets," as he had a system for everything.

The practice uniforms for the varsity were black jerseys with white numerals, with faded, well worn yellow canvas pants. The "golden mullets," the junior varsity, wore yellow-gold jerseys with black numerals, and canvas pants. As the players came to their lockers to dress for practice after the day's academic classes, they knew which team they were on by the color of the jersey hanging in their lockers. In the fall of 1951, the frequency of promotions to varsity increased dramatically, followed by returns to the "silhouettes," "the golden mullets." The players seldom knew from one day to the next what color their jerseys would be, and often what positions they would practice.

The furious pace and intensity of practices, coupled with players' emotionally charged competition for starting berths, made for numerous outbursts of frustration, flaring tempers, and occasional fist fights requiring coaches' intervention. In some cases there were sharp exchanges between assistant coaches and individual players. But no one among the assistant coaches or the players would get heated enough to dare take on "The Colonel." When he spoke, everyone listened. He could nail a player or a coach with a cold, icy stare, and that would be the end of it.

Then, there was the plebe team. The class of 1955 was one of the smallest classes to enter West Point in years, and was sparse in recruited "blue chip" players. Tommy Bell, Johnny Wing, and Freddie Meyers had been turned back from the class of 1954 because they didn't pass all their academic courses, and failed the spring turnout examinations. They were readmitted by the Academic Board after passing reentrance exams for the courses they had failed. They were plebes again, a repeat passage in life not welcomed by any cadet.

But Earl Blaik and Army did have one break, because of rule changes, and so did every other team in the nation. Because of the Korean War, the National Collegiate Athletic Association Football Rules Committee changed eligibility rules to permit freshmen to play varsity football. In other words, players now had four years of eligibility for varsity play, rather than three. Now Blaik could reach down to the plebe team, if talent were evident, and promote a promising plebe to the varsity. He promptly took advantage of the break. At least, if the plebe player were talented enough, Blaik could give the youngster some playing experience, looking toward the next three years – if the player had that kind of eligibility remaining.

The promotion of a plebe to varsity didn't make for happiness among upperclass junior varsity players passed over while vying for starting positions. Some plebes, however, had used a year or more of varsity eligibility at other colleges before receiving appointments to the Academy, and thus their eligibilities were limited.

Then there were the yearlings, the sophomores, the class of '54. They were the best talent of what remained from the destroyed 1950 team, when the men of '54 were still plebes. Yet Earl Blaik believed the recruited players from the class of '54, the yearlings on the '51 team, had less promise than any yearling class since he arrived at Army.

As the practice days slipped by, Blaik became aware of another shortcoming in his knowledge of the young '51 team. One of his nine football axioms, permanently displayed in the team locker room, read, "Good fellows are a dime a dozen, but an aggressive leader is priceless." He practiced what his axioms expressed. He placed great emphasis on team leadership. He especially looked for quarterbacks and defensive signal callers who exhibited strong leadership qualities. His prior Army teams were filled with gifted, tough, ambitious young men, many of whom were natural leaders on the field, men who showed their mettle in rugged high school or college competition, and proved themselves among their Black Knight teammates.

Blaik could teach field leadership to his quarterbacks and defensive signal callers. He could also customarily rely on the judgment of team members to select their captains, and each year the captain-elect for the succeeding year was chosen right after the prior season ended. The captain-elect for the season of '51 was gone. He had resigned, found guilty by the Collins Board. Who among these green, inexperienced players might be team captains?

The team of '51 presented another, entirely different field leadership problem to Blaik. His probable starting quarterback was a recruited, academically turned-back plebe, Freddie Meyers. The whole football program, including A, B and C teams, was in a state of flux from the first day of practice through the entire season. The players were eager, equally if not more hard working than their predecessors, smart, tough, and competing furiously to be starters. They were good athletes, perhaps not the best, but they loved the game. Yet they hadn't worked together in the pressure cooker and glare of major college competition. They hadn't developed confidence in one another, nor the timing and execution needed, and certainly not Blaik's confidence in their judgments and performances on the field. The natural leaders who might emerge as team leaders wouldn't necessarily make the starting team. Blaik's solution to the question of team leadership on the field was to name offensive and defensive captains for each game from among the starting lineup. He was laying the foundation for team leadership for the 1952 and '53 seasons.

Before the scandal, the first game of the 1951 season, with Villanova would have been considered a breather for the Black Knights of the Hudson. The Cadet Corps was not certain what to expect, but most sportswriters knew Army's season opener would not be easy.

BRING ON VILLANOVA!

General MacArthur and his wife, Jean, accepted General Irving's invitation to West Point and the Villanova game. The cadet daily bulletin of September 28 told the Corps of MacArthur's visit the next day. Saturday afternoon before the game, as the superintendent's guest, he received the Corps' salute of "eyes right" as the cadets passed proudly in review on the Plain to the stirring refrain of *The Official West Point March*. Before the review, General and Mrs. Irving hosted the MacArthurs and their son at a luncheon in the superintendent's quarters, which faced the Plain. Young Arthur received a gift of a cadet "tar bucket," the traditional formal parade hat.

The reviews before home football games were in the afternoons, right after lunch, followed by "in ranks" inspections. In the mornings, while cadets attended four hours of academic classes, there were Saturday morning inspections of rooms – SAMIs, the cadets called them. Inspecting were company tactical officers and senior cadet officers, such as cadet company commanders.

Friday evening before Villanova, in central area of cadet barracks, I participated, less than enthusiastically, in my first football rally as a cadet. The cadet daily bulletin on Friday morning also notified us that the evening football rally for the Villanova game, scheduled to begin at 8:15 was to be televised from 8:30 to 9:00 on station WNBT, Channel 4. I wasn't happy about tomorrow's schedule. Aside from the prospects of more inspections and more demerits, I couldn't attend the game. I exceeded the plebe's allowed number of demerits of one per duty day in the first month's demerit period in my new permanent Company, K-1. "Going over in demerits" wasn't a good way to gain early attention from the three upperclasses now outnumbering us nearly three to one.

Home football games always brought an air of excitement to West Point, and in spite of the well-known dearth of experience on this year's team, excitement was running high before the 2:00 kickoff at Michie Stadium. Sunday, beginning at 3:00 in the afternoon, there was to be the first of daily, hour long televised programs that fall, "Take Another Look," showing highlights of leading college football games, and featuring famed sports writer and commentator "Red" Barber. Army-Villanova was the first game to be highlighted and could be seen locally on WCBS, Channel 2.

The bulletin's note concerning the one hour highlights of the game held little interest for me. Fourth classmen didn't have TV privileges. Fourth classmen did have movie privileges on weekends if they had no conflicting duties, or weren't serving special punishments, which consisted of scheduled hours of room confinement or marching afternoon punishment tours in central area.

Saturday evening after the game, the movie *An American in Paris*, with Gene Kelly, Leslie Caron, and Oscar Levant, would be showing in the cadet theater. During Sunday afternoon, we could see *The Desert Fox,* starring James Mason, Jessica Tandy, and Sir Cedric Hardwicke. *The Desert Fox* was Hollywood's screen biography of the World War II German General, Irwin Rommel. He repeatedly displayed his tactical genius to America's military leaders and U.S. Allies during the war. The 10th Panzer Division, one of the armored units in Rommel's command, had been victorious at the battle of Kasserine Pass in North Africa in 1942 and had made the Commandant, Colonel Waters, a prisoner of war.

Saturday morning, the usual pregame flood of visitors and football fans began crowding along the sidewalks and pressing against the roped off area of the Plain to watch the Corps' march in review.

Resplendent in full dress gray with the traditional tall, black hats – tar buckets – cadet officers' silver sabers, and the twenty-three hundred M-1 rifles with fixed, chrome plated bayonets shimmering in the sun, the Corps marched in battalion mass, two regiments of three battalions, each battalion nearly four hundred men. As our battalion, the 3d Battalion of the 1st Regiment, passed in review at "eyes right," I could plainly see General MacArthur, standing to the right of the superintendent, in a light colored business suit, and white shirt with a dark necktie, and cream colored wide brim hat. He was standing smartly erect, at attention, acknowledging each "eyes right" salute.

Despite the greatness of the man we were seeing that afternoon, the history he represented, all the solemnity, tradition, color, pageantry, and marching precision the occasion called for, there was a lighter side to the event. Within the ranks, as always, there were waves of what Earl Blaik called "situational humor" sweeping almost invisibly through the Corps. To cadets, it was not quite, but almost, just another "P-rade." From the moment cadet officers gave the commands, "Forwa-a-a-rd, – March!" to mass the six battalions on the Plain, the quiet chatter and hijinks began. Meanwhile, in the South Gymnasium's football team dressing room, there was quiet, restrained excitement and preparation for the first game of the season.

In our battalion, as in the other five, there were wandering eyes, not always fixed straight ahead as they should be. Nearly all the first classmen, the seniors, were in the lead and rear ranks, or on the battalion, regimental, or brigade statts. They were the cadet officers and non-commissioned officers. Also in the second rank and the rear of the battalions marched the cow corporals, junior class members whose military order of merit were sufficient to earn them the grade of cadet corporal.

The happiest times in parades and reviews for our battalion came immediately after we turned right, out of central area, next to the first division of barracks where Douglas MacArthur lived as a cadet nearly a half century ago. Turning right onto Jefferson Road, we marched a short distance before making a left turn onto Thayer Road, and another left turn off Thayer to assemble on the Plain. The fun began as we passed the throngs of visitors lining Jefferson and Thayer Roads.

First, there were always the pretty girls, the high school "bobby soxers," and more maturely dressed women from colleges and universities, as well as the working girls who were not in college. All came to see West Point and the cadets, or date cadets for the weekend – or so many a cadet thought. The massed battalions of nearly four hundred men each had many a roving pair of eyes, "grading" particularly striking young women along the line of march.

"Mr. McWilliams, do you see that one coming up on the left flank in the black dress with the black hat and the white ribbon for a hat band?"

My eyes swivel to the left, head unturned. "Yes, Sir!"

"Not so loud, Dooguard. Keep your voice down. What would you rate 'er?"

"2.9, Sir."

The standard of measure was the academic grade point scale, with 2.0 passing – though unacceptable, and 3.0, perfect.

"What?! Mister, you've been away from home and women too long. Your standards are shot to hell. Better concentrate on your marching. You're bouncing. You obviously can't look at women and march at the same time. Eyes front."

As soon as the battalion turned left onto the Plain, and marched away from the visitors lining the ropes, upperclassmen in the rear ranks began a staccato of corrections in rifle alignment, file cover, and rank dress. The stream of chatter moved, softly spoken, front and back, left and right through the files and ranks as the review progressed – the ranks eighteen men across, files twenty-one or twenty-two men deep, one behind the other. The cadet officers, sergeants and corporals were sprucing up the battalion's appearance from within. The corrections were being passed up and down the lines: rifles properly carried on shoulders with barrels aligned parallel with the line of each file, rifle butts held so the forearm of each rifleman was parallel with the ground, ensuring, as one looked across the ranks of the battalions, the rifles appeared perfectly aligned; cover of one man directly behind the man ahead in the file, so the entire file sought to be perfectly straight from front to rear of the battalion. Each rank needed to be perfectly "dressed," taking their cue from the right flank file, so the ranks were straight when viewed from either flank. The right flank file was to maintain the perfect one arm's length interval between ranks.

But there were other kinds of chatter, some officious and intense, as well as quiet talk that would send soft gales of laughter through the battalions. There were the plebes, always the recipients of "corrections," especially in parades and reviews early in the year.

My first month in Company K-1 increased my unhappiness about life at West Point. I had passed an all time low when I went to the hospital earlier in September. On September 29, I was not much better. The upperclassmen in K-1 knew I was terribly homesick and disappointed in myself. They mercifully seemed to be leaving me to stew in my own misery.

Some of my classmates were less fortunate. One was in K-1. He was David James, a blonde, crew-cut seventeen-year old, with fair and reddish, pimpled complexion. His face was oval shaped, his build slender and athletic. When he responded to an upperclassman's command, his hallow baritone voice boomed, suggesting he was listening intently, eager to follow directions.

For reasons I never understood, he consistently stirred hostility in upperclassmen. He wore a perpetual grin on his face, no matter how serious things were supposed to be. He appeared to be unable to respond satisfactorily to the smallest tasks given him. The more pressure the upperclassmen applied, the more they yelled at him, the more frustrated they grew with his responses. The grin was a permanent fixture on his face. He seemed completely unfazed, even oblivious to the upperclass storms raging around his performance and behavior. And that day, as every other day, he received unending corrections from all three upperclasses.

Dave James wanted to be on the fencing team. Some cadets who had no understanding of fencing's rigors, often derisively called fencers "fairy stickers," but not in the presence of the man targeted for criticism. The label poked sarcasm at the body positions used in the sport, and was an equally unkind reference to men who seemed to possess feminine characteristics. Such sneering, backbiting comments didn't sit well with most cadets, and suggested a number of characteristics – all bad – in the men who voiced the criticisms.

"What are you falling out for, Mr. James? Wipe that silly grin off your face! Jam your chin in! Pop your chest up! Get that neck back! You haven't got enough wrinkles in that chin, 'Dumbguard.' Pull that [rifle] butt in. Suck your gut up. Butt right. Butt left. You have lint on your dress coat, 'Dumbsmack.' I want to see the beads of sweat rolling down your neck, Mr. James. You aren't working hard enough. What's so funny, Mr. James?"

In a lighter vein, there was the command joke telling. While the Corps was en route to reviews and parades, or while standing at "order arms," "Mister, give us a joke" could be heard among the ranks, even while the Academy band played. " On cue the plebe receiving the command attempted to entertain upperclassmen around him with a joke, one they hopefully hadn't heard. If he were successful, the sharp eyes

483

of tactical officers gathered near the reviewing party could see nearly imperceptible shuddering movements as tittering and giggling rippled through several ranks in a battalion. If the hapless plebe told an "oldie," or the story didn't rouse laughter in the upperclassmen, the young freshman's attempts at humor came to an abrupt end.

Then there were the bets on fainting cadets. During warmer parade weather, the heavy, wool full dress coats, tar buckets, and extended periods standing at attention, parade rest, or "present arms" became oppressive. The nine pound M-1 rifle held perpendicular, at forearm's length, centered in front of the body, while at "attention" and "present arms" added to the heat and misery. The cumulative effects were a few woebegone cadets overcome in one or more of the battalions.

Some would crash to the ground face first, or onto their backs, with little warning. Often the ashen faced, passed out body would lie perfectly still except for breathing; then quietly, unobtrusively, as the color returned to his face, he would get slowly to his feet using his rifle as a crutch, in slow motion put his tar bucket back on his head, straighten himself up, and march off with his battalion for the pass in review.

Usually, however, the crash was preceded by hilarious symptoms. In most cases, it was a plebe whose posture was pressed into excessively tensed muscles and "locked" knee joints by upperclass "corrections." The first symptom would be an ever so slight swaying from side to side, front to back, or in a circle, as the plebe sought to loosen up the restricted blood circulation and fight off the darkness beginning to shrink his peripheral vision. Knees began to buckle. Alert cadets on either side, or behind, would quietly talk to him, giving instructions on how to fight the strange numbness overtaking him, or grab his arms to attempt holding him up with their rifle-free hands. Stricken cadets with greater presence of mind would slowly drop to one knee before allowing themselves to collapse, unnoticed by the crowds gazing from the distance at the orderly masses of men assembled on the Plain. Then, as the band struck the first chords for the pass in review, the recovering cadet would slowly rise to his feet and march off with his battalion.

And frequently, when warning symptoms of an impending faint could be seen early and clearly, from somewhere in the rear ranks could be heard, "See that one four ranks up, two files to the right. He looks like he's going down." Then might come another exchange between the two upperclassmen, a friendly wager on whether or not the hapless cadet was going to crash, slowly sink to one knee, stay down to be carried off by the medics, or be able to rise to his feet and march off with his battalion. It was important to laugh in parades, as in many other situations at West Point.

As for me, I saw little humor in the review. Aside from my growing unhappiness over life at West Point, I had another reason to be particularly downcast the day of the Villanova game, the only day in my whole life I saw General Douglas MacArthur in person. In exceeding my monthly demerit allowance by four, I was to have my second two hours of "walking the area." After the noon meal, while the Corps and all the fans at Michie Stadium watched the season opener, I walked off the last two of four hours of punishment tours in the central area of cadet barracks.

The uniform was dress gray, with the traditional gray wool cadet tunic trimmed in black, over gray wool trousers with the vertical black stripes down each side. With it we wore the gray wool garrison hat with a black patent leather brim, crossed white shoulder belts with a polished brass breastplate, a white waist belt with a polished brass buckle, the waist belt worn over the shoulder belts, which held the black cartridge box at the small of the back. A sheathed dress bayonet was at the left hip, and plebes carried the M-1 rifle "at the right shoulder."

An inspection by the Officer of the Day, Cadet Officer of the Day, and the Cadet Sergeant of the Guard preceded the first punishment tour of the afternoon. It was another means of receiving more demerits – if our uniforms were not clean, pressed, and free of lint; shoes, belt buckles, breastplates, and hat brims impeccably shined and free of finger prints; and a dry, oil, lint, and rust free rifle.

For the small town boy from Colorado, who had a longtime love affair with football, particularly Army football, there was no season opener. I remember well the muffled sounds of the Corps' football yells, which I had obediently memorized as part of plebe knowledge, and the echoes of thousands of cheering Army and Villanova fans rolling down the hillsides from Michie Stadium, past the Protestant Chapel into the concrete and granite cadet area. The distant echoes reverberating off the walls of central area barracks deepened my personal disappointment as I marched at the prescribed military cadence of one hundred twenty steps per minute, back and forth across the huge concrete slab covering the entire quadrangle. There was a single arm's interval between me and the cadets to my right and left, as we marched without speaking, eyes straight to the front – except when we could get away with rambling eye movements, gazing back briefly at the visitors who attempted to catch glimpses of the "area birds" while strolling past the east and north sally ports of central area.

I wasn't alone. There were many other cadets, mostly plebes, serving punishment tours that day, but their number didn't bring me happiness. Not only had I failed to make the Army football team, I didn't make it as a member of the Corps, a fan, or admirer of Army football at the opening game of the season, and my first opportunity to see an Army game in person.

THE FIRST OF SEVEN LOSSES

In spite of the perfect autumn day; the roaring enthusiasm, hopes, and encouragement of the Corps of Cadets and all the other Army fans in Michie Stadium; General MacArthur's inspiring presence; Earl Blaik's and his assistant coaches' always thorough team preparation; and the spirited response of the inexperienced Army players, Army lost its first season opener since 1893. The loss to Villanova was also the first defeat at Michie Stadium since Navy won over Army in the war year of 1943. The score was 21 to 7.

It was a day of sweet revenge for Villanova's Wildcats. They hadn't won a game against Army since 1915. From the time Villanova took the opening kickoff and went 84 yards to a touchdown, it was the Wildcats' game. In eight, successive prior meetings, mostly in Army's golden era of the '40s, the cadet football team allowed Villanova not one single point. By halftime it was 14-0, and Villanova added their third and last touchdown in the first two minutes of the second half. They scored almost as many points in that one game as they had in fifteen previous meetings with Army.

Though few could see the bright spots for the cadets that day, there were signs of future success for football trained eyes to see. On receiving the kickoff after Villanova's first score, Army drove 43 yards to the Wildcats' 24-yard line, only to lose the ball on a fumble.

Twice in the second quarter Army had chances to score, penetrating to the Villanova 10-yard line on both drives. The first time they went 48 yards and again lost the ball on a fumble, recovered by Villanova's end, Jack Patrick. Lowell Sisson, Army's yearling left end caught his first pass in an Army varsity game during that drive, a 10-yard throw from Freddie Meyers. Lowell didn't know it at the time, but it would be his first pass reception on the road to a life of wonderful memories of Army football.

The Wildcats' Jack Patrick again frustrated the Army offense on the second penetration, a 44-yard drive to the Villanova 10, when he brought down Freddie Meyers, the Army quarterback, for a fourth down loss on the 20.

With eight seconds remaining in the half, Jim Ryan, an Army guard, recovered a Villanova fumble on the Wildcat's 7-yard line. On the last play of the half Freddie Meyers threw a touchdown pass to Army's left halfback, Tommy Bell, but it was called back. Both teams were off sides prior to the snap from center and the clock ran out before Army could get off another play.

For Tommy Bell, his first crossing of the goal line in a varsity uniform, though the touchdown was called back, was a start toward achieving what Vince Lombardi had said to Tom's little Irish mother in the spring of 1950. "I'm going to make Tommy an All-American." The handsome redhead from New York City's Mt. Saint Michael's High School in the Bronx, was a hard man to bring down, a quick starting, hard-hitting, fearless, powerful ball carrier who earned the nicknames of "Ding Dong," and "Freight Train," because "he loudly huffed and puffed, and rang defensive players' bells, as he ran like a freight train."

Finally, late in the third quarter, a Villanova fumble led to an Army score in the last period. Ray Bergeson, Army's left guard, recovered the fumble on the Wildcats' 19-yard line. Tommy Bell, Freddie Attaya, and Freddie Meyers moved the ball to the two on running plays. Johnny Wing carried it over the middle for Army's lone touchdown. Plebe fullback Dick Reich kicked the extra point.

Late in the quarter, a quarter in which Villanova was heavily penalized, Army again drove to the Wildcats' 24 but faltered with another fumble.

Fumbles, an indicator of a mistake-prone, inexperienced team, had helped undo Army that day, but there was another side to the story of inexperience and fumbles. Army fumbled four times, Villanova eight. This was a glimmer of the rebirth of a quick, hard-hitting, ferocious defense, ever a hallmark of Army teams. Army's hard-hitting defense jarred the football loose from more than one Villanova player.

When the final whistle sounded in the lengthening shadows of Michie Stadium September 29, 1951, deliriously happy Villanova fans stormed the field and tore down Army's north goal post. The Corps, as in years past, accompanied by the Military Academy Band, stood at attention and sang the first verse of the alma mater. The defeat by Villanova was the first of seven losses that fall, four more losses than a Blaik coached team had sustained in the previous seven years. It was a difficult day, beginning a long New York fall, when, for the first year since 1940, Army suffered a losing season, Earl Blaik's only losing season at Army. Freddie Attaya and Johnny Wing were to become two of Earl Blaik's building blocks in the weeks ahead.

On Sunday, in the *New York Times*, accompanying sports writer Allison Danzig's article about the game, there was a photograph of

Army's opening day loss to Villanova, 21 to 7, at Michie Stadium, September 29, 1951. Bob Mischak reaches for thin air as the Wildcats' Ben Addiego crosses the Army goal line. (Photo courtesy New York Times.)

Villanova's speedy young sophomore halfback, Ben Addiego, crossing Army's goal line for a touchdown. Addiego was well beyond the outstretched arms of Bob Mischak, another Army yearling, caught by the photographer in mid-air in a desperate attempt at a flying tackle. No one recalled Bob's leaping grasp at thin air that afternoon. Few would forget his heroics in October, two years later, at the Polo Grounds in New York City, against the Duke University Blue Devils.

MORE PURPLE WILDCATS

Army's football team didn't win the opening game of the 1951 season. Yet Earl Blaik's abilities as a coach and leader of young men, his talented assistants, the feisty desire of his young charges, and the rallying support of the Corps and the men and women serving at the Academy, were already having positive effects.

The second game was against the Wildcats of Northwestern University at Dyche Stadium in Evanston, Illinois. Villanova's Wildcats wore purple. So did this second band of Wildcats. Had the Army team been its old self, Northwestern would have been one of those "few choice dragons" the cadets hoped to slay, but not this year. Army was a three touchdown underdog, and the Corps of Cadets stayed at West Point, their home away from home, many listening to the game on radios.

Game preparations for Northwestern were according to Blaik's usual routine, a routine experienced for only the second time by most of the '51 players. The Sunday afternoon following the Villanova loss, they gathered at a scheduled time in the projection room in the back of the cadet gymnasium. As in any other cadet formation, tardiness was not acceptable for a scheduled appearance in a Blaik run practice, "chalk talk," or "skull session."

First came the Villanova game films. The "newest Rabble" learned how "The Colonel" critiqued and taught his players to learn from their mistakes. These sessions were not unlike an Army combat unit's "after action report," or unit officers' "lessons learned" from combat actions. Except in Blaik's football command, it was more personal and frequently embarrassing. An assistant coach rolled the silent film and narrated the action, already reviewed by the Army coach and his staff. When a mistake by an Army player came on the screen, the narrator would "stop action," "freeze the frame," or reverse and re-roll the film.

The offending player then listened, often with a red face, to a detailed discussion, laced with questions to him or others about the mistake, and why it occurred. Some players, who might have made particularly embarrassing mistakes, and knew their turn was coming for the whole team to see, slid slowly downward in their chairs as the moment of truth drew nearer.

Then came the scouting report for next week's opponent. Typically, the scouting reports, compiled mostly by assistant coaches, were detailed, filled with information about the opponent's strengths and weaknesses. The attention to detail was extraordinary. Not only did the reports highlight patterns of play exhibited by opponents, but included discussions of obvious vulnerabilities and weaknesses of their play, both as a team, and for each player on offense and defense.

From the scouting reports, and the disciplined analyses of the opponent's play, sometimes complemented with available game films from newsreels, Blaik and his staff developed strategies and points of attack for both offense and defense. New offensive plays, or variations developed from previous plays, all aimed at exploiting opponents' weaknesses – and Army's strengths. The scouting report was the first step toward a game plan, the synthesis of all that was known about the next team to be faced, and how to defeat them the following Saturday.

October 6, 1951 was the first football game ever between Army and Northwestern University. Once more the young Army team was well prepared. Before 40,000 fans at Dyche Stadium they very nearly slayed a dragon.

Northwestern scored in the first quarter and took a 7-0 lead on a 42-yard pass play from quarterback Bob Burson to end Joe Collier. In the third period, Army battled back with two quick touchdowns to take a 14-7 lead. The first Army score came when left guard Ray Bergeson broke through Northwestern's offensive line and blocked a behind-the-goal-line punt by the Wildcat's Norm Kragseth. Ron Lincoln, the Black Knights' twenty-year-old yearling left end fell on the ball in the end zone.

The next score came when Freddie Attaya intercepted a Burson pass and raced 42 yards to the Northwestern 44. On the next play, fullback Johnny Wing took a pitch out around Army's left side and scampered the distance for Army's second touchdown. Dick Reich, the plebe second platoon fullback, kicked both the cadets' extra points. Despite the Wildcats' bulging first half statistical edge in yardage gained and first downs, an upset was now in the making.

Northwestern came back furiously in the fourth quarter. They drove 80 yards in 16 plays, with Burson going over from the 1-yard line on a quarterback sneak. It appeared the Wildcats would tie the score, but Norm Kragseth's extra point kick was wide. The ball bounced back harmlessly onto the playing field after hitting the left upright. Army held a 14-13 lead.

With a light drizzle dampening the field and time running out, Burson drove Northwestern 56 yards to the game winning touchdown. The end came for Army with one and a half minutes to go. A 33-yard pass play, from Burson to end Dick Crawford, abruptly shattered hopes for an upset.

So disheartening was this end to a gallant try by players unaware they lacked the necessary skills and abilities, that in the dressing room after the game, Vince Lombardi stood with tears running down his face. Earl Blaik recalled years later, "This was the Vince Lombardi I knew; his professional reputation gave no clue to the emotion that was so strong a part of his makeup."

DARTMOUTH'S FIRST WIN IN '51

In the third game of the season, the Dartmouth Indians took the measure of Army 28-14 before a crowd of 20,242 at Michie Stadium. The two teams had met only twice before, the last, fifty-two years ago, before the turn of the century. Army won both encounters. Like Army this season, Dartmouth had yet to win a game.

Army fumbled five times that Saturday and lost four of them, including the opening kickoff. Twice Dartmouth capitalized on fumbles by scoring touchdowns, the first coming at 2:02 of the opening period, following the "Big Green's" fumble recovery on the opening kickoff.

Peter Joel Vann from Hamburg, New York, the tall, slender, plebe classmate, started at quarterback that day, replacing the still injured Freddie Meyers who left the Northwestern game in the first quarter. This was Pete's first start at Army and a big, exciting step for him.

Dartmouth's opening touchdown came swiftly and caused some confusion among Army's young starting lineup. Although Earl Blaik let his quarterbacks call approximately fifty percent of each game's offensive plays, his pregame routines included his calling the three plays of Army's first offensive series. The calls normally were part of a morning walk around Trophy Point, in which he walked and talked with his players as a means of relaxing them, prior to a mid-morning, protein rich, energy boosting meal that included steak. Before the Dartmouth game, he conjured up a plan to score quickly using Pete Vann's strong, accurate passing arm – a freshman weapon sure to surprise the Indians. The first play was an "out and down" by right halfback Fred Attaya in which he would sprint, angling toward the right sideline, then cut abruptly up field past Dartmouth's defensive left halfback, and get behind him. Pete was to throw a pass into the arms of the speedy, quick accelerating Attaya.

When Dartmouth recovered the fumbled opening kickoff on the Army 24, they attempted almost exactly the same play Blaik had called, an out and down. Although Dartmouth quarterback Jim Miller's pass was good for only six yards on the first play, the Indians marched quickly to score, running three times in succession at Army's defensive left side.

On the sidelines, before Army received the return kickoff from Dartmouth, Pete Vann and Freddie Attaya wondered if there was to be a change in Blaik's pregame call since Dartmouth had run virtually the identical opening series. After quickly explaining their reasoning, they asked, "What are we going to do, Colonel?" Blaik's reply was, "Run just what we planned." The first play of the series was to be Pete Vann's first forward pass in an Army game.

Everything went exactly as planned at the start of Army's first offensive play against Dartmouth. Freddie Attaya outran Dartmouth's pass defenders and broke into the open two or three steps beyond them. Pete Vann, dropping back behind good pass protection, exhibiting near flawless poise in the classic moves of the T-formation drop-back passer, threw a perfect strike into Freddie Attaya's hands at the Dartmouth 47. Freddie was in the clear, at full stride, flying toward the Dartmouth goal line, and dropped the ball.

Army never seemed to recover. Pete Vann remembers that pass to this day, because it was his first forward pass at Army. It was a perfect throw, and disappointingly slipped through Freddie Attaya's hands, a sure touchdown that never came to be.

For entirely different reasons Freddie Attaya never forgot the dropped pass from Pete Vann either. It was a terribly embarrassing mistake from which he nearly didn't recover. He had played on the Louisiana State University freshman team in the fall of 1949 before an unusual chain of circumstances intervened to bring Freddie to West Point. An Academy graduate who was acquainted with Earl Blaik, and brothers Freddie deeply admired, had influenced him toward the Academy the spring of 1950, before the Korean War broke out. He was a confident player, tough, a gracefully fleet-footed ball carrier. When he ran, he leaned forward slightly, like a sprinter, his knees lifting high, his legs churning in an easy fluid motion.

But the humiliating dropped pass in the game against Dartmouth shook him. He turned inward as the weeks rolled by and found himself sitting more and more on the sidelines, both in practice and in games. His confidence had taken a blow and it was showing. His enthusiasm waned. But all was not dark that day, in spite of the loss to Dartmouth. Fred's dropped pass was a miscue that eventually became a lesson, taught him by Vince Lombardi and Earl Blaik, a lesson that some weeks later turned his performance around and set him afire on the gridiron.

Fred Attaya's muff of Pete Vann's first pass wrecked Army's hopes for a quick tally. Later in the period, Army's Bob Guidera blocked a Dartmouth punt, but the cadets were stopped on the Greens' 6-yard line. Before the period ended, Army, through Bob Mischak's return of a punt, two completed passes by Pete Vann, a 23-yarder to Paul Schweikert, an 8-yard toss to Lowell Sisson at the start of the second quarter, brought Army within striking distance of a score. Two plunges by Attaya, and a 1-yard smash over Army's left guard by Dick Reich brought Army's first touchdown. Reich kicked the extra point.

Except for one more exciting play by Army, the rest of the afternoon belonged to Dartmouth. The Indians kept up a one-touchdown-a-quarter pace the rest of the game. Army's fumbles and pass defense were once more the team's Achilles heels.

A more obvious bright spot at Michie Stadium the afternoon of October 13 was Bob Mischak. The heavily-recruited former high school tailback, who had grasped for thin air as Villanova's Ben Addiego crossed Army's goal line in the opening game, brought everyone in Michie Stadium to their feet on the next to last play of the

Dartmouth game. In an electrifying 97-yard kickoff return, he ran straight down the center of the field for a touchdown. He crossed the Dartmouth goal line with two seconds remaining, Dick Reich kicked the extra point, and it was all over. Bob Mischak's better days were still ahead.

FIGURED TO WIN – THEN ANOTHER LOSS

On October 18 Earl Blaik wrote General MacArthur,

This is a crucial week for our cadets as we need a victory over Harvard to rally the players, the Corps, and our Alumni...

There is no doubt the squad is carrying a great psychological load, and the ever present controversy which has stirred the Academy as well as the country undoubtedly has adversely affected the squad. In other words it is so apparent to me that they operate under a cloud that every effort is being directed toward releasing them from under abnormal tension.

Army lost its fourth straight of the season to the Crimson of Harvard, 22-21, before 14,000 fans at Harvard Stadium, after, for once, being favored to win. Fumbles and other mistakes were again the Black Knights' undoing. Six times they coughed up the ball, and six times Harvard recovered, once capitalizing with a touchdown.

Army started well, despite an early bobble, and jumped into a 7-0 lead on the strength of a 50-yard run by Fred Attaya, followed by Dick Reich's extra point. Harvard came right back with a touchdown on a short pass after a sustained drive, tying the score 7-7 in the second quarter, and adding eight more points to take a 15-14 lead at the half. The margin of the Crimson's victory was in a safety scored in the second quarter when Harvard broke through and blocked a punt by Fred Attaya. Fred recovered the ball in his own end zone, avoiding a touchdown, but the safety gave Harvard two points.

Time and again Army drives were halted by fumbles, offsides and illegal motion penalties that nullified long gainers. Once in the fourth quarter, when they desperately needed a first down to keep alive hopes of winning, a substitution error resulted in a twelfth man on the field and cost them the first down.

Dick Boyle and Pete Vann both received tries at quarterbacking as Blaik sought a signal caller to move the Black Knights against Harvard. Freddie Meyers, who had emerged as the Army coach's choice for quarterback, was still hobbled by injuries he received against Northwestern.

In the critical last ditch push to win with four and a half minutes remaining, Army, in a series of running plays, drove from their 36 to the Crimson's 43. The ground game had taken too much time however, and the Cadets had to go to the pass. Pete Vann overshot an open Ray Tensfeldt on one throw, and Harvard's Don Cass batted down the next one just in time to prevent Lowell Sisson from making the catch. Harvard took over on downs and ran the clock out.

In 1951, Earl Blaik considered Pete Vann immature, "a mere boy" he called him. Pete had a long road ahead with Army's football master, but he was learning.

The Harvard game was the first time the entire season that Army gained more total yardage than their opponents, yet nothing seemed to go right. This was the first Harvard win over the Cadets in ten years and their highest score against Army since 1900. It was also the last game in a series of 32 that began in 1895. Harvard now had the edge in their series, with 17 wins, 13 losses, and two ties. It seemed the loss at Northwestern the second game of the season had had a far more devastating effect on the Black Knights than anyone suspected. The team appeared to be regressing, going from bad to worse.

COLUMBIA'S LIGHT BLUE:
A FORECAST OF THINGS TO COME

On the last Saturday in October, the Columbia University Lions, coached by Lou Little, came from New York City to Michie Stadium to play before 20,349 spectators. The sportswriters said Army might as well give up if they couldn't beat Harvard, but Coach Blaik promised in the *New York Times* on Thursday, "Army will not give this game away."

In his letter to MacArthur before the Harvard game, Blaik had said,

We play Columbia on the 27th of October and I hope that you will make plans to be with us that day. I know your presence will be an inspiration to the players and to the coaching staff...

Still struggling to retain the status and high esteem of Army football in the wake of the cheating scandal, he went on to say,

> I shall write you more in detail about our entire football situation quite soon. I am not sure whether the school of de-emphasis [of football] is going to take over or not, but I am certain of this fact – if it does, it will be a colossal error. I suppose in the end the decision will be made by General Collins, who seems to have little judgment in such matters.

MacArthur was unable to attend the Columbia game, a game he would have enjoyed immensely. It was an inspired defense that saved a long awaited victory, and turned the tables on a Columbia team in a manner that was inevitably compared with the Light Blue's stunning 21-18 upset of Army in 1947, an upset that ended a 32-game winning streak. The final score in the rebuilding season of 1951 was Army 14, Columbia 9.

After a scoreless first period, Columbia moved 77 yards in fifteen plays, with quarterback Mitch Price driving across for the score on a quarterback sneak. Al Ward missed the extra point try. Seventeen seconds before the first half ended, Johnny Wing, Army's fullback, crashed over for the Cadets' first touchdown. His score came the second play after Army recovered a Light Blue fumble on Columbia's 14-yard line. Dick Reich added the extra point to give Army a 7-6 lead at the half.

Columbia took the second half kickoff, and drove 78 yards from their own 19-yard line to the Cadets' 3, in seven plays, before being stalled by Army's tenacious defense. The Lions settled for a field goal, three points, to take a 9-7 lead. After the ensuing kickoff, the Cadets pushed toward the Columbia goal line only to lose the ball on a pass interception at the Lions' 1-yard line. The Light Blue was forced to punt out of danger but Army got the ball on the Lions' 29, and in eight plays punched across the winning score. The Cadets' Dick Boyle went the last yard on a leaping quarterback sneak over the center of Columbia's line. Dick Reich again booted the extra point.

The stage was set for the dramatic fourth quarter, and the Corps of Cadets was on its feet thundering its support for their underdog "Rabble."

Dick Boyle leaps over center for the winning touchdown against Columbia, at Michie Stadium – Army's first win of the season, 14 to 9, October 27, 1951. (Photo courtesy USMA Archives.)

In the stands was Ernie Condina's mother. Ernie, a first classman, a senior from Company K-1, and former "Golden Mullet" on the 1950 team, was now one of Earl Blaik's defensive left guards. Ernie knew well what it meant to play teams scheduled against Army's varsity of old. Each week in the fall of 1950 he was on the "Golden Mullets," running plays against that powerful Army team, plays the varsity's opponents would run the next Saturday. He considered the "Golden Mullets'" weekly schedule among the toughest football schedules in the nation, with good reason. The 1950 "Mullets" took their hard earned experience to four B squad games that year, and won them all.

And now Mrs. Condina had come from their home in Johnsonburg, Pennsylvania, to see her son play against Columbia University in his senior year at West Point. Prior to the cheating scandal, Ernie had little hope of making the varsity. Then opportunity came knocking in a most unusual way. He wasn't a big man physically, and his mother was deeply concerned, as only a mother can be, about her son and the battering he was absorbing each week on the Army varsity. Sitting with her was Dick Miller, the K-1 honor representative who in April carried the first disclosures of organized cheating to Colonel Harkins. As the furious battle unfolded below on the green grass of Michie Stadium, Dick Miller watched in quiet, bemused admiration as Ernie Condina's mother pulled out her Rosary beads and softly prayed for Ernie's safety. Indeed, the fourth quarter was especially furious.

Eighteen times in the final period Columbia put the ball in play inside the Army 20-yard line. Ten times the Light Blue tried for a touchdown from inside the Cadets' ten yard line, but each time the Army defense held. The climactic last seconds of the game typified the entire struggle. With the ball in Columbia's possession, within inches of Army's goal, and with no time-outs remaining for the Lions, the clock ran out after a desperately rushed third down try for a score.

The final Columbia lunge for a score was a play Al Paulekas and Ernie Condina would well remember. Al was from Farrell High School in Pennsylvania, where his dad coached the high school football team, after being on the 1936 professional Green Bay Packers' championship team. Sid Gillman, himself a former star professional quarterback for the Chicago Bears, and one of Earl Blaik's assistants, had come to Farrell High School in the spring of 1949 to talk to one of Al's better known teammates. While there, Gillman looked at transcripts of other athletes, and learned Al was an "A" student as well as a two-sport athlete – a football player and a wrestler. Al had been recruited by Purdue University for intercollegiate wrestling, and earned an academic scholarship if he chose to go to the Big Ten school.

College hadn't been in the picture for Paulekas. Like many a young man who eventually chose West Point, his parents were of too modest means to afford a college education for their son. When Sid Gillman asked him, "Would you consider an appointment to the United States Military Academy?" Paulekas considered his question a gift from heaven.

Now, on this most exciting day, with more than twenty thousand in Michie Stadium on their feet roaring encouragement for their respective teams, Al Paulekas, Army's defensive right guard, was one of three men who saved the victory over Columbia. As the Light Blue linemen rushed to the line of scrimmage for a last desperate try for a touchdown, Al took the defensive lineman's four point stance, half his body weight borne by his hands and fingers, in the grass just inches outside the Army goal line. To his left was Ernie Condina. With their backs straight, parallel with the ground, poised for the defensive charge, their football cleats at the toes of their shoes dug into the turf well behind the goal line, both were ready.

When the Lion's ball carrier drove at the goal, Al's charge straightened him up, while Ernie's charge was low. Paulekas couldn't tell whether or not his momentum and the direction of his hit was going to keep the ball carrier out of the end zone. But right behind him, lunging forward with even greater momentum, was an Army linebacker, a yearling, Norm Stephen. Norm was the insurance Al Paulekas and Ernie Condina needed that day, and together they kept Columbia out of the cadets' end zone. When the referees untangled the pileup of bodies to learn Columbia failed to score, Ernie Condina lay at the bottom of the pile.

Neither Al Paulekas, nor Norm Stephen knew what lay ahead in the coming seasons, but Ernie Condina, who was a first classman, was rewarded the following week. Earl Blaik, impressed by Ernie's sweaty, determined goal line work that afternoon, named him defensive captain for next week's game against Southern California. Al's fiery, competitive toughness and leadership caused his teammates to elect him captain of Army's 1952 team. Norm Stephen became Earl Blaik's team-inspiring varsity center, defensive field captain and signal caller during the 1953 season, when Army, along with every other college team in the nation, due to changes in substitution rules, returned to "iron man football" – playing their players both ways, on offense and defense.

As the clock ran down to Army's first victory over Columbia that late October afternoon, the Corps of Cadets, which had remained standing almost the entire afternoon, lifted its voices to a roaring crescendo, and poured from the stadium to swarm and congratulate their team. Forgotten was the traditional singing of the alma mater, at least until the on-field celebration was over. The cadets surrounded their gridiron heroes in a sea of gray, slapping them on the backs of their black, sweat-soaked jerseys; shaking their skinned, bruised, grass- stained hands; jumping and dancing with joy. It had been a great, thrilling victory, sorely needed. Less than two years later, the Army defense made another, even more memorable goal line stand against the Duke University Blue Devils.

A hard-hitting, rock-ribbed defense was the apparent key to Army's first victory after four straight losses following the honor incident of 1951. Their team's jarring line play and hard-hitting tackles provoked costly mistakes by the Lions. Columbia lost the ball three times on four fumbles, and earned three frustrating penalties which thwarted scoring opportunities. Using the T and A formations, the Light Blue moved the ball almost at will between the 20-yard lines, but were repeatedly stymied once inside the Army 20. The sting of the upset was made more painful by the sharp statistical disparity between the two teams. Columbia had 21 first downs to Army's 11, and 367 total yards to Army's 183.

But there were other factors at work in Army's win that day. An assistant coach on Earl Blaik's staff would say two seasons hence he "credited Corps spirit with two touchdowns each game the Corps attended." Perhaps not every game this year. But there is little doubt that against Columbia's Lions at Michie Stadium on October 27, 1951, the long road back for the Corps of Cadets and their "Rabble" had begun. They were beginning to unify, come together, determined, gallant, and the Columbia game portended better seasons far sooner than anyone dared dream.

Douglas MacArthur cabled his congratulations to Earl Blaik, the Corps of Cadets, and the team for their stirring victory.

BIG TROJAN HORSES IN NEW YORK CITY'S POLO GROUNDS

November 4, 1951, wasn't a good day to be in New York City. It was cold and raining. The Polo Grounds, normally dry and dusty in the summer, were a sticky, gooey mess soon after the opening kickoff. The Corps of Cadets had to forego the traditional pregame ceremonial march-on, and instead march along the wet, soggy sidelines with cadet raincoats and rain cap covers worn over their heavy, gray, wool, winter uniforms, then file directly into their seats.

As for me, the trip to New York City was exciting despite the rain and cold. I was a small town boy from the mountains and valleys of Colorado coming to "The Big Apple" for only the second time in my life, the first a brief overnight before I rode the train to West Point the morning of July 3, to enter Beast Barracks. What's more, it was my first trip away from the Academy and plebe year's rigors. I rediscovered a more relaxed life outside West Point's cloistered, regimented order and gray granite walls. As frosting on the cake, here we were in the Polo Grounds, home of baseball's and pro football's New York Giants, witnessing two of collegiate football's big name teams go at each other, Army and the University of Southern California, USC. However, in 1951, the only big names were USC's Trojans, and several of their players vying for All-American honors. Army's team of unknowns had completely vanished from consideration by the nation's sportswriters, football prognosticators, and big time coaches – with some few exceptions.

There were always Earl Blaik's colleagues, many of whom respected and admired his professional capabilities. Mostly, they were men whose teams his Black Knights had played. Their coaches never underestimated him or his teams. For all his discipline and straight laced composure, they knew he was full of surprises. He was always willing to adapt. He easily borrowed others' good ideas if the concepts seemed to fit the Black Knights' style, talent, and circumstance. And he never failed to improve on those ideas when he chose to use them.

On November 1, in another letter to MacArthur, Blaik thanked the General for his telegram and wrote of game plans for his out-manned team.

> ...Our only hope of launching an offense against them is with the forward pass and for this reason we shall play Meyers, now recovered, at halfback where his unusual receiving talent may help. A few trap plays to slow down the charge of the Southern California line should keep them from rushing the passer too much. The quick kick as an offensive weapon is included in our plans....

Against USC that day, this young team rolled out another of those surprises of which Blaik had written to MacArthur, and for the first few minutes of the game, the Corps of Cadets roared their approval, believing another miracle upset might be in the making. Blaik disclosed his plan to the team during their traditional pregame walk when he called Army's first three offensive plays. If they won the coin toss for the opening kickoff, they would elect to kick to USC, rather than receive. If USC won the toss and elected to kick off, the Cadets' fullback, Dick Reich, was to quick-kick on the first play from scrimmage. The idea was to keep the Trojans pinned deep in their own territory, where they might turn over a slippery ball to Army via a miscue.

Army won the coin toss and kicked off. The Trojans didn't bobble the ball on Reich's kick or their opening offensive series. But neither could they make a first down. On fourth down, USC punted. Army went on offense inside their own 40-yard line.

Now what? Here were the Cadets, not far from midfield, normally good field position on an exchange of punts. What about the first play from scrimmage? "Kick it," Blaik told Boyle, the quarterback, and Reich, as the eager offensive platoon slogged onto the field. Kick it they did, to everyone's complete amazement – a 59-yard kick that rolled dead in water on the Trojan 2-yard line.

Southern Cal tried three running plays, handling a slippery ball on what was quickly becoming a slimy field. Gaining only eight yards, they were forced to punt. USC's Desmond Koch, standing in punt formation, fumbled and Army's Stanley Kuick recovered on the 2-yard line. The small, shivering crowd of 16,508 was in an uproar. Noise from the jubilant Corps of Cadets was deafening.

In two successive heart-stopping rushes, Army fumbled both times – and recovered both. On third down, Tommy Bell lost a yard, and it seemed the Cadets wouldn't capitalize on the plan that had gone so well. But on fourth down, Dick Reich shot between left guard and tackle for the score. The Corps of Cadets exploded in jubilant excitement. Even though Reich missed the extra point, the Corps and the Army team could scarcely contain themselves. Amazingly, Army, a team of plebes and yearlings with barely a handful of cows and first classmen, who had lost the first four of five games, was leading one of the most powerful teams in the nation, 6-0. It seemed the upset of the year was in the making.

The Black Knights were fighting like furies as the rain began to turn to snow. They seemed unmindful of their reputation as a team. They again held USC and forced a punt. Reich quick-kicked again, the ball rolling dead on the Trojan 5-yard line. The Corps went wild. But the USC attack began to move this time, and after another exchange of punts, they used the quick kick weapon to their advantage. Finally, early in the second quarter, they drove 36 yards in six plays to their first touchdown, and then 53 yards in ten plays before the half, to take a 14-6 lead.

That was it for the day. From there on, the afternoon belonged to the Trojans of Southern California, the Thundering Herd that had brought down the mighty California Golden Bears, Washington, and Texas Christian University, among others. The final score was USC 28, Army 6.

Frank Gifford, the speedy 194-pound T-formation halfback, and single wing tailback, did everything but score a single one of SC's four touchdowns. He punted, passed, ran, quick-kicked, kicked off, made a touchdown saving tackle of Don Fuqua, the Army receiver who nearly broke Gifford's kickoff for a touchdown, and booted all four extra points.

On the statistical side of the game, it was no contest. USC made 23 first downs and gained 446 yards while holding Army to zero first downs and a net of zero yards. The Cadets' were -10 yards rushing and +10 passing. Statistically, Army's beating was far worse than the score.

Aside from Earl Blaik's prescient pregame call leading to Army's lone touchdown, and Dick Reich's sterling performance, Bob Mischak again showed his athletic ability and a speed few noted before the USC game. His shining moment in the mire and muck came in the first quarter when Frank Gifford, throwing from his tailback position in the single wing formation, hit Dean Schneider, the Trojan's left handed, T-formation quarterback on a 41 yard pass play. Schneider gathered in the pass on his 30-yard line, and would have gone all the way had not Mischak overtaken him from behind. Bob's heroics were to no avail that day. The story would be far different less than two years hence.

Two Coaches, One Player, and a Lesson to Remember

As the weeks rolled by after the Dartmouth game in Michie Stadium, Freddie Attaya's confidence in himself struggled to survive. The dropped pass from Pete Vann ate at him like a nagging toothache. He was dejected, discouraged. He no longer was the starting right halfback on the offensive platoon, and in spite of his 50-yard dash for a touchdown in the first quarter at Harvard, he found himself spending more and more time on the sidelines. He knew he was better than that, but he just couldn't convince himself what was wrong.

Finally, one day during practice, he decided he must find out why he wasn't playing more, though he believed he knew the answer. It was that mistake he made against Dartmouth, the first quarter dropped pass that, if caught, might have changed the outcome of the game. He went to Vince Lombardi, and after a few words of explanation asked,

"What's wrong, Coach?"

Lombardi looked at him and replied, "Come with me. Let's go see Coach Blaik."

When they caught up with Blaik, Lombardi said, "Colonel, I think Attaya just woke up. He wants to know why he hasn't been playing."

Blaik fixed his penetrating gaze on Fred for a moment. "You get up too slow after carrying the ball, Attaya. Unless you've just made a long gainer, or scored after a long run, I expect you to be the next man back into the huddle, after the center." Then, with pointed emphasis, he added, "I expect ball carriers to be leaders."

Earl Blaik lit a fire in Freddie Attaya that day and taught him a lesson he never forgot. Freddie had done it to himself. He had taken himself off the starting lineup, and now he would fight to return. Later, he again became a starter in the Army backfield, and he never looked back.

A WIN AND A LOSS BEFORE NAVY

Back home at Michie Stadium the following week, before a capacity crowd of 28,183, including 16,500 Boy Scouts, Army won its second game of the season, 27 to 6 over the Citadel Bulldogs. Despite ten fumbles, five recovered by the Citadel, Army scored twice in the first and third quarters to win.

Fred Meyers, returning to the Cadet lineup after his knee injury against Northwestern, was at quarterback and sparked the Army attack all afternoon. He completed six of nine passes for 145 yards, one a 45-yard touchdown to Lowell Sisson. He added another touchdown when he plunged one yard for Army's last score after Bob Mischak bolted 45 yards with an intercepted Bulldog pass.

The cadets got moving quickly in the first quarter when they drove 75 yards in eight plays, with Johnny Wing bucking over from the 3-yard line. Later the same period, one of two Citadel fumbles on the day gave Army the ball on the Bulldog 35. The Cadets capitalized on the miscue with another touchdown when Tommy Bell raced the last 18 yards down the far sideline to give Army a 13-0 lead.

For only the second time in the season, the Black Knights outgained their opponents in total yards that day, 379 to 198. With Meyers back at the helm, the Cadets' passing game immediately improved, with seven completed passes in thirteen attempts, and no interceptions.

Before a crowd of 40,000 at Penn Stadium in Philadelphia, the University of Pennsylvania Quakers provided the final test before the season's traditional climactic game against Navy at Municipal Stadium in the City of Brotherly Love. Penn-Army was an old rivalry with hard fought games of trench warfare. On offense, George Munger coached teams still ran the old single wing formation with an unbalanced line. It was football's power formation held over from the days of "iron man football."

The Quakers always played Army tough, and in this first rebuilding season, Penn was a two touchdown favorite. Penn won for the first time since 1942, Army having won six, with two ties, in the intervening years. The final score was Penn 7, Army 6.

Army began moving for its only score, when, late in the first quarter, Bob Mischak stopped a Penn drive with recovery of a Quaker fumble on the Cadet 26-yard line. On first down, Dick Reich used the quick kick weapon for the first of three times that afternoon, and the Quakers' safety, Bob Evans, inadvertently kicked the ball forward. A scramble for the ball ensued and Bill MacPhail of Army recovered on the Penn 27-yard line.

The quarter ended with Freddie Meyers plunging four yards. When play resumed, runs of three and nine yards respectively by Tommy Bell and John Wing gained a first down on Penn's 15. Then it was two passes, one to Mischak, the second throw 12 yards to Lowell Sisson, giving Army a first and goal on the 2-yard line. Three times Army lunged at the Penn goal, and on the third try the Quakers were called offside, giving Army the ball an inch away from the goal line. Freddie Meyers was successful on his third attempt at a quarterback sneak.

Dick Reich had been good on 13 of 15 placement attempts for extra points before his try against Penn. This time he missed wide left. Army still led 6-0 at the half, and another upset appeared in the making. Once more, it was not to be.

Early in the second half Dick Reich's quick-kick magic worked again when he boomed another one 63 yards. Penn couldn't move the ball, and their return punt netted Army an advantage of 18 yards on the exchange of kicks. But a few seconds later, Tommy Bell fumbled on the Army 32-yard line and Penn's Gerry McGinley recovered. The Red and Blue drove toward the tying touchdown aided by a confusing, illegal substitution error by the cadets, which resulted in a disastrous five-yard penalty. The penalty came on a fourth down play just short of Army's 8-yard line, moving the ball to the Army three. From there, Penn punched over the tying touchdown and didn't miss the extra point, giving the Quakers the margin of victory, 7 to 6. Army's record stood at 2 wins and 6 losses.

NAVY MAKES IT TWO IN A ROW

If I was awed by the Polo Grounds in New York City and the Army-Southern California contest, Army-Navy in Philadelphia's Municipal Stadium on December 1, 1951 was beyond my wildest imagination. I marched in my Company, K-1, as the Corps of Cadets massed all twenty-four companies in the Brigade in our traditional pregame ceremonial march-on. It was a balmy, spring-like day with temperatures near the 60s, and our plebe class, '55, was on its second football trip with the entire Corps to see a great national spectacle on the gridiron.

I well remember my amazement and the pride I felt as our company marched into the stadium to the strains of *The Official West Point March*. As I took in the sight of the huge stadium, still only about two-thirds full, the feeling was one of wonder. This was the stadium I had seen in newsreel clips on Saturday afternoons at our small town theaters in San Juan, Texas, and Los Alamos, New Mexico, when West Point was still a boyhood dream for me. And the thunderous roars, the music, and echoes I knew I would hear that afternoon were the sounds I had heard over radios as the Army-Navy game was broadcast each fall during and after World War II. Army football and its heroes had been magnets to me, though I hadn't any serious expectations of playing on any team which wore the glory mantles of Davis, Blanchard, and Tucker.

The sights and sounds of the day swept me up, so much so that I took little notice of the realities of the 1951 game. It was a contest between two teams with records among the season's worst in major college football. Yet, with only a win and a tie behind them, the Navy still outclassed the cadets. Army had spirit. Navy, which was also rebuilding under its former professional football coach, Eddie Erdlatz, had spirit, plus experience from the tough Navy schedule and the leadership of juniors and seniors on its team.

Municipal Stadium, normally filled to its 102,000 capacity for the game, had several thousand empty seats for this year's classic, a reflection of the two teams' season records. President Truman, who attended the 1950 game, was enjoying a hard-earned rest, vacationing in Key West, Florida, playing golf, while senior members of his cabinet and the military establishment took time out from their wartime duties to attend the game. In spite of the thousands of vacant seats, the huge throng of spectators and nationwide television and radio coverage made it one of the major events of the year. For me, a crowd of more than 96,000 in a huge stadium was a sight never seen in or near small, rural towns of the southwest – or in cities of the southwest for that matter.

And for nearly all the Army team that year, the sight was equally awe-inspiring, but for reasons somewhat different than mine. Few had suited up for the 1950 classic, and only one had played in the huge stadium the year before. Team members who were plebes in 1950 had witnessed the spectacle from the stands, gazing in wonder at what they were now seeing from green turf in the center of the expansive arena as they went through pregame warm-ups. The new plebes, like Pete Vann who had made the varsity, were seeing it for the first time with its 102,000 seats nearly filled, although Blaik, as in years past, had brought the team to see and work out in the empty stadium the day before the game.

The giant horseshoe-shaped stadium was imposing enough, but the monstrous crowd added to its intimidating size. The largest crowds the Army players had seen all season were 40,000 at Dyche Stadium against the Northwestern Wildcats in Evanston, Illinois, and a crowd of equal size at Pennsylvania's Franklin Field. Each were less than half the number in the unfilled Municipal Stadium.

To make matters worse, there was but one player on the team that could say from experience, "Don't let the size of the place or the crowd bother you. Concentrate on the game and your assignments, one play at a time." He was Carl Guess who, on this day, specialized in kicking. In 1951, coaches were the only ones who could speak authoritatively of experience, and their words were not sufficient. There was nothing quite like the reassuring words of a tested, hardened group of team-mates, preferably lettermen on the field, to help green players through the jangled nerves and "butterflies in the stomach" that every big time college game brings to newcomers, particularly the Army-Navy game.

This was it!! This game was what Army and Navy teams looked to every year as the climax, the apex of their seasons. Their respective seasons, good or bad, rode on the outcome of the annual classic. A bad season was no longer a failure for the conquerors in the annual service academy match-up.

Despite the endless distractions bedeviling him that season, Earl Blaik worked hard to prepare Army for THE game. Hoping against hope, he cooked up another surprise for the opening kickoff, aiming for a quick touchdown, and momentum that would throw the middies off stride. The plan was a "squib kick," an attempt to "shank" the foot-ball downfield at an angle so that the ball would bounce erratically away from Navy's speedy kickoff return specialists. The idea was a bounding, hard-to-control football, permitting defenders to more rap-idly cover the kickoff return, and possibly result in an Army recovery of a fumbled ball. Elwin "Rox" Shain, '54, prepared for the kick all week.

The game began badly for Army. The plan backfired. The kick was "shanked" too much and dribbled along the ground a scant few yards. Instead of Army recovering the opening kickoff well down field, Navy recovered the ball at midfield. The middies opened their first offensive series running the Wing-T offense, the formation Army scouts had seen all season. Then Eddie Erdlatz hit the Cadets with his surprise, pulling a play book page from the distant past, the old "Notre Dame box" formation made famous by Knute Rockne. The "box" was a power formation similar to the single wing, and Erdlatz wanted to take ad-vantage of the formation's strength against the young, hard-hitting Cadet defense, putting three and four blockers in front of Navy ball carriers on running plays. Navy scored in twelve plays, without throwing a single pass, and during the drive ran four straight plays against the Army right tackle to move the ball to the 1-yard line. The Navy coach sent in sophomore Jack Perkins to replace Fred Franco, and Navy lined up with Perkins at fullback in the "box" formation. Perkins bucked over for the touchdown, and Ned Snyder kicked the extra point, as he did following five more middie touchdowns that day. Navy led 7-0, and was on its way to a 42-7 Army rout.

On the ensuing kickoff, Dick Inman, one of the few first classmen playing for Army, fumbled when he was hit hard on the return, and Navy recovered on the Cadet 27. After being set back by a five yard penalty, the middies promptly drove 32 yards for another touchdown for a 14-0 lead. Army hadn't run a single play from scrimmage and was two touchdowns behind.

Before the first period was over, John Raster, the only plebe on the entire Navy team, made the longest touchdown run in the history of the rivalry, returning an intercepted Army pass 101 yards for a Navy touchdown and a 21-0 lead.

The interception was a crushing blow, because Army was driving. John Rogers had recovered a Navy fumble on the middies 42-yard line. A pass, Freddie Meyers to Lowell Sisson netted 14 to the Navy 28. Johnny Wing broke loose for ten more yards, and the Army offense pushed on to the 6 before the disastrous interception occurred.

Once more, nothing seemed to be going right for the 1951 Black Knights. Navy scored again in the second quarter, driving 36 yards to take a 28-0 lead at the half. The middies nearly scored another touchdown just before the intermission, after recovering Don Fuqua's fumble on the Army 36. A 21-yard pass from middie quarterback Mike Sorrentino to Don Fisher, followed by Sorrentino's run to the Army one, fell just short of a score. The clock ran out before Navy could run another play.

The Navy game of 1951 did have its lighter moments in the lengthening shadows of an otherwise gloomy afternoon. Once, following a frustrated Navy thrust toward the cadets' goal, Army's offensive platoon stormed onto the field deep in its own territory. Blaik called Fred Meyers over and gave him the first play to call in the huddle. Fred ran onto the field and into the huddle. There was some hesitation, and some obvious confusion as the clock ticked on and Freddie glanced back toward the sidelines at the Army coaches. Finally, about to be penalized for delay of game, Fred called for a time-out, and sprinted hastily to the sideline for a quick conference with "The Colonel." To an astonished Earl Blaik, he said, "What was the play, Coach? I forgot." To which the surprised, stammering Earl Blaik replied, by turning to Vince Lombardi, "I don't know. What did we call?"

At afternoon's end, Navy had scored the highest number of points by the winning team in the history of the rivalry, which began in 1890. The middies had surpassed Army's point total of 40 in 1903. The victory was Navy's second in a row after a prolonged drought without a win since 1943. Erdlatz didn't "run up the score" that December afternoon. He played every man on his squad, and increasingly gave his younger players experience as the second half wore on.

The Army dressing room, for good reason, was quiet and subdued afterward. The season had been the worst since 1940, the year before Earl Blaik came to West Point to resurrect Army's football fortunes.

There were always good performances to talk about following a game, and Earl Blaik talked to reporters about Fred Meyers, his quarterback. "I liked his spirit even when the game looked lost in the very first period." Fred, a "turned back" plebe, was a standout in a difficult season, though he was hobbled a good part of the season following the Northwestern game. He was a strong-armed, accurate passer, who needed to work on quick release of the ball to avoid interceptions. He was big, and an effective runner on quarterback "keepers." He had promise for next year.

Not mentioned that day was Dick Reich, also a plebe. Dick injured his knee in the first quarter and left the game. He, too, was a standout throughout the season of '51, a promising player for next year, with his size, strength, and running power at fullback, as well as outstanding kicking abilities.

Ironically, for all their good performances on the gridiron, no one knew at the time that Fred Meyers and Dick Reich, two of Army's apparently most promising players, would both be gone from West Point before the '52 season began.

Earl Blaik, struggling for something good to say after the Navy game and all the other sad events of that year, also didn't mention Bob Mischak who caught a 32-yard pass from Fred Meyers against Navy; Lowell Sisson, also a consistently effective pass receiver; and John Wing, another ball carrier who had done yeomen work all season. And

The 1951 Army Football Team: Corps spirit and the team that didn't know it lacked talent laid the foundation for a football miracle. Front Row: Kuick, Jelinek, Tensfeldt, Rhiddlehoover, Inman, Williams LA, Bergeson, Guess, Gregory, Weaver. 2nd Row: Schweikert, Wing, Rogers, Lunn, Haff, Paulekas, Fuqua, Ziegler, Storck. 3rd Row: Rutte (Mgr.), Manus, MacPhail, Bell, Boyle, Stephen, Lodge, Attaya, Lincoln, Woodward (Mgr.). 4th Row: Dorr, Wilkerson, Krobock, Rose, Meyers, Harris, Ryan, Kramer. 5th Row: McGinn, Chamberlin, Mischak, Vann, Sisson, Reich, Shain, Glenn, Guidera. (Photo courtesy USMA Archives.)

certainly he didn't mention "the mere boy" Pete Vann, who played little in the last few games, or men such as Al Paulekas, Leroy Lunn, Gerry Lodge, Norm Stephen, Fred Attaya, Tommy Bell, Frank Wilkerson, Neil Chamberlin, Bob Guidera, Jerry Hagan, Myron Rose, or Ron Lincoln. Nor could anyone remotely believe that all these men, and more to come, were laying a strong foundation for a totally unexpected turn of events.

The 1951 season had been a season of adversity for the Army football team. Earl Blaik, writing years later, put the '51 Army team's accomplishments into perspective. "I did not dream, nor did they, that they were exploiting adversity, that this was the stumbling, seemingly pointless, almost ludicrous beginning of a story that two seasons later would end as a football miracle."

At West Point that year there was an early "gloom period," the name cadets normally ascribed to the winter months between the end of Christmas leave and the beginning of spring leave in March. For plebes, as the year before, when the powerful 1950 team had been upset by the Midshipmen, the 42-7 defeat by Navy had special meaning.

A tradition had grown up through the years which said that a victory over Navy resulted in the fourth class being allowed to relax until after Christmas leave, to "fall out" and not have to endure the pervasive regimentation of the fourth class system. It meant a period when the freshman didn't have to "brace," that is, stand at an exaggerated position of attention, and meals could be enjoyed because we didn't have to sit at attention on the front half of our chairs while eating. There were other smaller benefits which, in sum, could make a plebe's life seem considerably more pleasant. But they were not to be. The crushing defeat by Navy meant the plebes at West Point would not get to "fall out" – not after the football season of 1951.

CHAPTER 13

150 YEARS OF HISTORY AND TRADITION

With the December 30 airplane crash still fresh in our minds, and the memories of 1951 receding into mists of a post-Christmas gloom period, the Military Academy began its Sesquicentennial celebration. At noon meal on Saturday, January 5, 1952, a dignified, impressive ceremony in Washington Hall inaugurated six months of celebration. A cadet "Band Box Review" in central cadet area followed the inaugural, which included speeches by Secretary of the Army Frank Pace; retired General Lucius D. Clay, former American Military Governor of Germany; and Governor Thomas E. Dewey of New York, who, in a stunning upset, had been defeated by Harry Truman in the 1948 presidential election.

Few in the class of '55 paid rapt attention to the words and history we heard in Washington Hall. Plebes lived from day to day, event to event, one ceremony to the next, one academic class to the next, one meal to the next. Actually, the Academy had been planning the celebration since 1947, when Major General Maxwell D. Taylor was still superintendent.

This was the sesquicentennial year. For 150 years the Academy had grown and matured. Though accumulating rich history and tradition as the years passed, it had adjusted to demands of the future and a dynamic, technologically expansive, rapidly shrinking world. During the same 150 turbulent years, succeeding generations of Americans relentlessly pushed westward to the Pacific and entered the world stage, while always pressing their Republic to be more democratic. They were the same generations from whom elected and appointed national leaders, and Army men, came to be stewards of an institution that served its people well throughout the entire period. West Point, the nation's oldest national military academy, could claim all manner of heroes in its history, and each year, wherever graduates gathered for reunions, could call up powerful emotions.

The Sesquicentennial occurred in an era somewhat similar to the era of the Centennial. In 1902 America was also in a nasty, undeclared war – in the Philippines. Yet there were some major differences in the two periods in Academy history. In 1952, World War II, the most devastating conflagration in human history, remained a fresh memory for graduates. For many, World War I, the first great convulsion of the 20th century, held more distant, equally clear memories. Another war – Korea – was in progress, and yet another, the Cold War, a far different and perhaps more ominous struggle, provided the backdrop for the fighting in Korea. The threat of another world war, between the Soviet Union and its allies and the United States and its Western allies, had become a reality in scarcely five years, in the midst of an accelerating nuclear arms race.

Nevertheless, the Korean War's off-and-on negotiations offered hope for an end to the conflict and a reduction in East-West tensions. Korea set the tone of the Sesquicentennial, but didn't obscure the tightly held loyalties and other emotions expressed during the celebration.

THE CELEBRATION'S BEGINNING

In the early fall of 1947 there was considerable interest among Academy graduates and officials in the coming Sesquicentennial anniversary. General Taylor asked for a number of senior officers at West Point to serve on a Sesquicentennial Steering Group. They were "to consider ways and means, outline policies, and draft the necessary plans" for the observance. Special Order Number 7, Headquarters U.S.M.A., dated 10 January 1948 activated the Steering Group, which included: retired Brigadier General Chauncey L. Fenton, president, Association of Graduates; retired Colonel Hayden W. Wagner; Colonel Herman Beukema, Professor of Social Sciences; Colonel Boyd W. Bartlett, Professor of Electricity – who was later to chair the Bartlett Board; and retired Lieutenant Colonel William J. Morton, Jr., Librarian.

The Steering Group was active for the ensuing five years, formulating policies for the Sesquicentennial observance. The Group held sixty-six formal meetings, discussing each aspect of policy and providing recommendations to the superintendents. Minutes of the meetings went to Academy organizations to inform them of progress in Sesquicentennial planning.

Colonel Wagner resigned after a short time, due to the pressure of other business. Colonel John W. Coffey, Professor of Ordnance, replaced him, but due to Colonel Coffey's untimely death in Germany in March 1951, Colonel Lawrence E. Schick, Professor of Military Topography and Graphics, replaced him. As the celebration drew near and the Steering Group's work increased, Colonel Oscar J. Gatchell, Professor of Mechanics, was added to the Steering Group.

The first order of business was to consider objectives the Sesquicentennial should achieve. The Group decided the occasion should serve as a rededication to the ideals of the Military Academy developed over the 150 years since its founding. The activities planned for the Sesquicentennial period would be dignified and serious, centered on the theme, "Furthering Our National Security."

> Traditionally, through its assigned function of training officers for the armed services, the Military Academy has been an important link in the chain of national security. As it completed a century and a half of service to the nation, West Point faced a future in which its mission and its performance were likely to have even greater meaning for the country than ever before. The Sesquicentennial year, therefore, would be an especially suitable time to examine the relationship between education and national security, to review and appraise the Military Academy's contributions to this field, and to attempt to increase public understanding of West Point's ideals and purposes.

When the celebration began in January 1952, early planning for the occasion appeared prescient. In spite of gathering Cold War clouds, the Steering Group's work began in a period of relative peace, two and a half years following World War II. In the intervening years leading to the inaugural ceremonies, world tensions increased dramatically and were given grave impetus by the outbreak of the Korean War. Then, as a result of the 1951 cheating scandal, some of the Academy's most cherished ideals seemed to come under sharp public attack. Not only did the Steering Group's early decisions and planning appear wise and foreseeing, world events and the nation once more at war added special significance to the Sesquicentennial celebration.

With the basis of the observance established, the Steering Group next drew up a program of events. At its second formal meeting on November 23, 1948, the Group decided on a plan of "Basic Policy and

Occasions," to present to the superintendent. The plan included the following: celebration events would be scheduled between January 1 and graduation on June 3, 1952; there would be an education conference, a military training and education conference; alumni would have a special day set aside for celebration; a scholarly history of the Military Academy would be published; the climactic event would be a Jubilee. The Jubilee would consist of an academic convocation, a review by the Corps of Cadets, and a formal dinner, with speeches by distinguished guests.

General Bryant E. Moore, the new superintendent, approved the program shortly after his arrival at West Point in early 1949 and submitted it to the Army Chief of Staff for official sanction. When General Moore approved the Steering Group's program, basic planning had matured sufficiently to begin the more detailed planning necessary to make the celebration a success. As a result, in June 1949, additional committees formed which gradually increased in number to thirty-five by the time the celebration began. The Army Chief of Staff approved the general plan for the Sesquicentennial on January 11, 1950, less than six months before war came to the Korean peninsula.

Part of the planning and the next stage in organizing was to create the temporary office of Sesquicentennial Director. The duties of managing and directing all the business, planned activities, and correspondence necessitated the assignment of an officer who would devote his full time to successfully completing the celebration. The officer selected was Colonel William E. Crist, who reported for duty on February 20, 1950.

With Colonel Crist's arrival, advance preparations for the Sesquicentennial accelerated. Developing a list of universities, learned societies, military academies, and distinguished public figures to be invited to one or more events was among several items needing earliest attention. Decisions about the final program of Sesquicentennial events, plus the technical and administrative assistance from various Academy-based and other organizations were necessary as well.

Months of discussion resulted in a list of 3,310 names of people to be notified of the impending celebration, including representatives of universities, learned societies, military academies, and government, who would be invited to participate. The means of informing all those institutions and individuals of the impending celebration were formal Sesquicentennial announcements, dispatched on November 1, 1950, over General Moore's signature. The announcements left West Point as the final month of Army's football season began and the ill-fated team of that year drew closer to the end of Army's golden era on the gridiron – while far away, in Korea, David Hughes and the 7th Cavalry Regiment, and all the United Nations' forces, were just beginning to learn of the CCF's powerful, slashing move south from the Yalu River.

The announcement read:

The United States Military Academy at West Point, in anticipation of celebrating its Sesquicentennial Year in nineteen hundred fifty-two, sends greetings to its friends and fellow institutions of learning.

In eighteen hundred and two Thomas Jefferson, following advice of George Washington, John Adams, Alexander Hamilton, and others, established a national military academy on the Hudson River at West Point. From the beginning the academy dedicated itself to the task of training young men to be military leaders imbued with ideals of integrity of character and unswerving devotion to the nation. The newborn republic, at that time, was engaged in a difficult experiment in popular government, which was giving hope to freedom-loving people everywhere. The success of that experiment has vindicated our faith in the ability of men to reason together to attain their common ends. The United States Military Academy, as it approaches its Sesquicentennial Year, is mindful and proud of its share in the nation's long struggle through peace and war to hold to the ideal of Government by the People.

The problem of our times is how best to preserve and develop this ideal, which has become a tradition for democracies in a free world. Surely all our national resources, physical and intellectual and moral, must be turned to account if our own democracy is to live and grow. West Point has therefore chosen as the theme of its Sesquicentennial Celebration: "Furthering Our National Security." We believe that study and discussion of this theme will disclose ways and means for the most effective use of our national resources in the service of democracy. The Military Academy has selected the period from January to June, nineteen hundred and fifty-two, as the time for special observance of its Sesquicentennial Year. We hope you will join us in marking that period of celebration.

> Bryant E. Moore
> Major General, U.S.A.
> Superintendent

General Moore also addressed a letter to all officers, cadets, enlisted, and civilian personnel at West Point. He pointed out the importance of the occasion:

> The Superintendent wishes all organizations, activities, and individuals to plan to participate actively in the Sesquicentennial. The success of the celebration will depend upon the wholehearted cooperation of everybody at West Point.

The following January 16, General Moore received his short notice assignment to Korea, as General Ridgway's I Corps Commander, which led to General Moore's death and subsequent interment at West Point in early April.

As the Sesquicentennial announcement list was being prepared and the volume of work increased through the summer and fall of 1950, the staff of the Sesquicentennial Director reached its full complement of six officers, authorized by the Department of the Army. For reasons of efficiency, Colonel Crist organized and divided responsibilities among the activities: administration, editorial and publications, publicity, and services. Primary responsibilities were assigned each officer in one of the activities, and all became thoroughly knowledgeable of the policies and procedures, to assist wherever needed.

In the summer of 1951 Colonel Crist received a promotion to brigadier general and left the Academy for Korea. When he left, a solid foundation had been laid, and his successor, Colonel Thomas W. Hammond, Jr., put into effect the policies and plans developed preceding his assignment.

Following General Moore's November announcement, a groundswell of interest was evident in personnel assigned to West Point, academic leaders, government and military officials, and the public. Hundreds of congratulatory messages poured in from all over the world. Many included suggestions on commemorative and ceremonial activities for the celebration. The Director and the Steering Group screened the suggestions, keeping in mind the theme, "Furthering Our National Security," the celebration's fundamental purposes, and the dignified and modest tones intended for the observance. The suggestions added several major events to the program.

The first was dedication of a new portrait of Robert E. Lee, who was second in the graduation order of merit, among forty-five men in the class of 1829. A group of friends and graduates of the Military Academy offered to present his portrait to West Point. Since 1952 marked the one hundredth anniversary of General Lee's assignment as Academy superintendent, the Steering Group and the superintendent accepted the offer. General Lee's portrait would be displayed in the main room of the Library, next to General Grant's. The date for the unveiling ceremony was January 19, 1952, Robert E. Lee's birthday.

Next, the Steering Group decided to mark the official opening of the Sesquicentennial with suitable activities. West Point was integral to United States history. Cadets came from the people of the entire country and its territories. The Steering Group agreed the inaugural ceremony should be dedicated to the forty-eight states, and all the territories of the Union. In honor of the occasion the National Guard Association of the United States offered to donate to the Military Academy a magnificent set of state flags, to be hung in Washington Hall, the cadet dining hall. Major General Ellard A. Walsh, president of the Association, would formally present the flags at the inaugural luncheon on January 5, 1952. The evening of the Inauguration was set aside for a joint concert by the Military Academy Band and Cadet Glee Club in New York City's Carnegie Hall, for the benefit of the Army Emergency Relief Fund.

A supplementary goal in the celebration was to inform the public of the Academy's role in national security. A number of planned activities centered on this goal. Inviting two Reserve Officer Training Corps cadets to the Academy from each senior college in the nation was one of the activities. Each invited cadet was to receive an orientation on the Academy's methods of education and traditions, and exchange ideas with West Point cadets. The purpose was to foster closer relationships between future Regular and Reserve officers. There would be 478 students from 239 colleges divided into four groups, each to spend an extended weekend at West Point.

Patriotic and learned organizations expressed desires to hold their annual meetings at West Point in honor of the observance. In the final program of events were meetings of the New York State Historical Association, the American Ordnance Association, the National Security Industrial Association, the Engineering Drawing Division of the American Society for Engineering Education, and a selected group of Boy Scouts.

Shortly before start of the Sesquicentennial period, the Steering Group met for a final review of the entire program. The November 1951 review began as Army's football season was coming to a close, and committees and boards examining every aspect of cadet education and life worked toward eliminating causes of the '51 honor incident.

The Steering Group's review took into account the gravity of the international situation and the ongoing war in Korea, and responding to needs for greater economy, reduced planned expenditures. The Group eliminated relatively costly events, retaining only those considered essential to the observance, or believed to be self-supporting. The Steering Group canceled the conferences on education and military education, and a proposed Sesquicentennial Exhibit at Smith Rink. A series of lectures by distinguished national speakers, on the Sesquicentennial theme "Furthering Our National Security," replaced the conferences, and a permanent, scaled down display at the Visitors' Information Center replaced the more expansive Exhibit originally planned.

With the schedule set, the Sesquicentennial Director and his staff began planning the final details. Plans contained a program outlining activities for each event, recommended guest lists, supply and transportation needs, and information necessary for efficiency and success of the event.

Using the plans as guides, the Sesquicentennial Office prepared invitations to selected guests for the major events: the inaugural, the Lee portrait presentation, Founders' Day, and the Jubilee. With the invitation was a leaflet containing information on the event, and a return questionnaire to assist the Military Academy in planning details of guests' receptions and accommodations.

All thirty-five committees formed to support the celebration, all the Academy's service organizations – including transportation, engineer, supply, housing, food service, communications, and others – hundreds of people, worked together, cooperating to make the Sesquicentennial a success.

Other senior people and agencies in the Department of Defense joined in supporting the Sesquicentennial. Secretary of the Army Frank Pace, Jr. and Chief of Staff of the Army J. Lawton Collins encouraged and enthusiastically supported the celebration, as did organizations under General Collins' command. The Adjutant General's Exhibit Section was notably helpful with technical advice and assistance in designing and manufacturing Sesquicentennial plaques, exhibits, and other decorative material.

Because of the national significance of the occasion, invitations went to General Matthew B. Ridgway, General Alfred M. Gruenther, and Air Force General Lauris Norstad to serve on a Sesquicentennial Executive Committee. They accepted and served in an advisory capacity to the Steering Group. When General Moore and General Crist left for Korea, each received and accepted invitations to be members of the Executive Committee. And serving *ex officio* was Major General Frederick A. Irving, General Moore's replacement as superintendent.

The last organization to become a part of the Academy's celebration was a West Point Sesquicentennial Commission. Composed of prominent government and military leaders, the Commission was in keeping with accepted practices of colleges and universities observing important anniversaries. Secretary of the Army Pace invited their participation, and General Collins signed Army General Orders Number 44, 23 April 1952, establishing the Commission:

UNITED STATES MILITARY ACADEMY
SESQUICENTENNIAL COMMISSION

1. The United States Military Academy at West Point, New York, will commemorate the one hundred and fiftieth anniversary of its establishment during the period beginning 1 January 1952 and ending 3 June 1952. A series of events has been sched-

uled in celebration of the sesquicentennial anniversary which culminates in the Sesquicentennial Jubilee on May 20, 1952.

2. In order to provide a fitting and appropriate observance of the one hundred and fiftieth anniversary of the founding of the United States Military Academy and to facilitate such a commemoration, there is hereby established a commission known as the United States Military Academy Sesquicentennial Commission and to be composed of the following-named persons who have expressed their interest in contributing to the observance of the anniversary:

Honorable Harry S. Truman, President of the United States of America, Honorary Chairman.

Honorable Alben W. Barkley, Vice President of the United States.

Honorable Thomas E. Dewey, Governor of the State of New York.

Honorable Sam Rayburn, Speaker of the House of Representatives.

Honorable Robert A. Lovett, Secretary of Defense, Chairman.

Honorable Frank Pace, Jr., Secretary of the Army.

Honorable Dan A. Kimball, Secretary of the Navy.

Honorable Thomas K. Finletter, Secretary of the Air Force.

General Omar N. Bradley, Chairman, Joint Chiefs of Staff.

General J. Lawton Collins, Chief of Staff, United States Army.

Admiral W. M. Fechteler, Chief of Naval Operations, United States Navy.

General Hoyt S. Vandenberg, Chief of Staff, United States Air Force.

General Lemuel C. Shepherd, Jr., Commandant, United States Marine Corps.

Vice Admiral Merlin O'Neill, Commandant, United States Coast Guard.

Major General F. A. Irving, Superintendent, United States Military Academy.

Rear Admiral Arthur G. Hall, Superintendent, United States Coast Guard Academy.

By Order of the Secretary of the Army:

J. Lawton Collins
Chief of Staff, United States Army

OFFICIAL:
Wm. E. Bergin
Major General, USA
The Adjutant General

INAUGURAL

Cold, wintry weather blanketed West Point the morning of January 5. A brief, heavy snowfall in mid-morning covered the Plain and surrounding countryside. The snow added to the beauty of West Point without seriously affecting plans. By noon, when most of the guests had arrived, the skies were clearing.

Many distinguished guests participated in the inaugural. Included were representatives of most of the states and territories, senior officials of the Department of the Army and National Guard, and many West Point graduates. Officers of the West Point garrison and the entire Corps of Cadets were also present in Washington Hall to witness the event.

For many guests this was their first time at the Academy, and the scene greeting them was reminiscent of a picture post card. As they arrived, cadets from their home states escorted them to Cullum Hall for registration and a welcome cup of hot coffee. From Cullum Hall, overlooking the Hudson River, the cadets escorted them to Washington Hall where General Irving and the speakers for the occasion officially greeted them.

Guests then witnessed the noon meal formation of the Corps of Cadets outside Washington Hall. At twelve o'clock, with the Corps called to attention and "present arms," the air reverberated with a nineteen gun salute in honor of the state governors present. Following the honors, the Corps marched into the huge dining hall.

Washington Hall is one of the most imposing structures at West Point. A large mural on the entire south wall of the building sets off a high ceiling, supported by deep cross beams. The mural depicts in brilliant colors great battles and great military leaders in world history. A beautiful stained glass window on the west wall portrays the life of George Washington.

The hall was especially redecorated for the Sesquicentennial observance, with the walls painted gray, the ceiling beams brown. The ceiling in the south and west wings was a deep maroon, and in the center wing, blue. Heraldic devices decorated the beams. Pendants of the south and west arches carried the dates 1778 and 1802, the years the Continental Army was first stationed at West Point, and the Academy founded. The original design of the hall provided places for mounting heraldic flags. In the inaugural ceremony the National Guard Association of the United States was to present 52 state and territory flags to the Academy, and the flags were in place, ready to be unfurled at the appropriate moment.

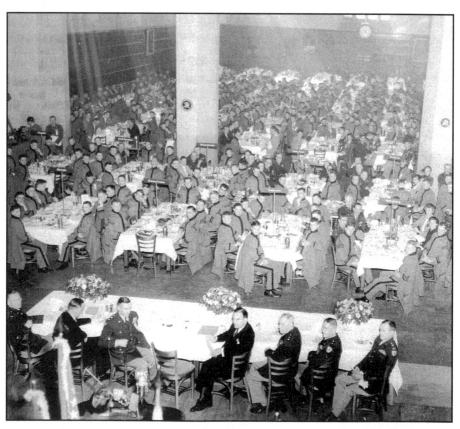

Inaugural Ceremonies Luncheon in Washington Hall, Saturday, January 5, 1952. (Photo courtesy Signal Corps.)

Within the dining hall, on the speaker's dais erected in front of the large center door, sat General Irving; Governor Dewey; Mr. Pace; General Clay; General Walsh; Colonel Herman Beukema, representing the Sesquicentennial Steering Group; and Colonel John K. Waters, the Commandant of Cadets.

Delegates, and cadets from their home states escorting them, sat at tables surrounding the dais, the tables ordered according to each state's admission to the Union. Officers of the garrison and the Corps sat at tables filling the remainder of the vast hall.

At 12:45, immediately following the luncheon, the Cadet First Captain, Gordon D. Carpenter, of Silver City, New Mexico, gave the command, "Battalions, Attention," and the inaugural ceremony began.

Chaplain Frank E. Pulley gave the Invocation:

Almighty God, who hast taught us that righteousness exalteth a nation, we give thee thanks for the guidance Thou hast given the leaders of the Academy for one hundred and fifty years. We acknowledge that it has been Thy grace and strength which has enabled many sons of West Point to serve honorably and with distinction their God and country.

We humbly pray that Thou wilt continue these blessings as we march forward into the coming years. Inspire us, O God, to fulfill our mission in achieving Thy will for mankind, thereby attaining the peace which passeth all understanding. In Christ's Name, we pray, Amen.

Following the invocation the superintendent gave the following, brief address of welcome:

Distinguished Guests, Ladies and Gentlemen:

On behalf of the United States Military Academy, I extend to each of you a most cordial welcome to West Point and I thank you for honoring us with your presence at this, the opening event of our Sesquicentennial year.

One hundred and fifty years ago, our nation was still young. Four million people lived in the sixteen states of the Union. Ten cadets reported to instruction when the Academy was formally opened; two men were graduated in the first class.

The nation and the Academy grew together. Today's Corps of Cadets is drawn from the one hundred and fifty million people of the forty-eight states, the territories and the District of Columbia. The Corps is truly representative of our nation.

It is indeed appropriate, therefore, that this Inaugural Ceremony of the Sesquicentennial emphasize the national character of West Point.

West Point has been training officers for service with the Army since 1802. It is particularly fitting that we have with us today, as the chairman of our Inaugural Ceremonies, the Secretary of the Army. It is my great pleasure to present to you the Honorable Frank Pace, Secretary of the Army. Mr. Pace.

The Secretary of the Army spoke the following:

General Irving, Distinguished Guests, Members of the Corps:

It was a hard decision for me to make as I contemplated what I might say to you of the Corps about West Point, in the short period of time that I have to speak today. I felt there were many things that I might say about West Point to you, but I think uppermost in my mind lies the fact that West Point is truly a democratic institution. West Point draws men from every state in the United States and from our territories. Likewise, it draws men from the enlisted ranks of the Army, from the National Guard, from ORC and by competitive examination. The choice of men who come here is in the best tradition of the United States of America.

When you come here and spend four years at this institution, your ranking in the class is not based upon what you were before you came here, but is based upon the determination of those who evaluate your capacities as to how you qualify in terms of leadership, in terms of character and integrity, as well as in terms of academic accomplishments. And most of you, coming to West Point as others have come before you from humble background, will go out taking with you not only the confidence that comes from your training at this great institution, but the fundamentally important feeling of democracy that

526

is essential in great leadership which is the basis of a great officer.

Next, Mr. Pace introduced retired General Lucius D. Clay, class of June 1918, who delivered the following speech:

Secretary Pace, Governor Dewey, General Irving, Distinguished Guests, and Gentlemen of the Corps of Cadets:

No honor means more to me than the privilege of speaking to you on the 150th Birthday of West Point as a representative of its graduates. I accepted this with deep humility.

The return of a graduate is always a pilgrimage to seek the inspiration which renews faith; the will to rededication of self. We do not return to view again the massive Gothic structures rising from the Plain, as impressive as they may be. We do not return to sit again at the feet of great teachers, even though they created the academic atmosphere which pervades West Point. We do come back because in the presence of the Corps, we feel again its deep, undefinable spirit of service. Only the Corps of Cadets has for 150 years received the heritages of the past, molded them to the present, and carried them forward as the traditions of the future.

Even in this 150th Birthday year it does not become one of its graduates to extol the accomplishments or to evaluate the contributions which they have made to the growth of America. This is the task of others, although rightfully we may be proud of the record.

It is for us to remember what West Point has meant to us. We can render no higher tribute. The dedication to service which is the spirit of the Corps cannot be defined. Only those who have lived it can realize its lasting influence. From this spirit, long before they were ever expressed in words, came the ideals Duty, Honor, Country.

We live today in a world shockingly devoid of ideals. The threat of force dominates. Unscrupulous leaders in many countries enslave their peoples and conspire to expand a false ideological empire. Even in more enlightened countries, government must defend itself from those who would use it to their own advantage. Cynicism has replaced genuine and healthy sentiment. We dare not speak of ideals lest we be regarded as naive and unsophisticated. Yet America was founded by men of high ideals who were not afraid to express them as principles for a new kind of government and a new way of life. Perhaps at no time in history has a return to ideals been so needed.

To be sure, ideals are unattainable goals. If they were easily attainable; if they could be lived by each of us every hour in every day, they would not be ideals.

Duty, Honor, Country – are ideals of consecrated service. In full measure they are unattainable. Yet, proudly accepted and resolutely supported by the Corps of Cadets, they have marked out a pattern of life for thousands of its graduates. Perhaps no one person has ever fully lived up to them, but few who have belonged to the Corps have faltered without resolving anew, to be worthy of the long line of graduates who created and gave these ideals to the Corps.

Gentlemen of the Corps, West Point's unassailable strength of purpose can be felt only in your presence. To be with you is to sense your vitality and your devotion to the soldier's concept of duty. The maintenance of high principles, the determination to live in their concept rest entirely in your hands. The Secretary, the Chief of Staff, the Superintendent, old graduates, can do little to see that each class in turn receives the unsullied inheritance of the past. Only the Corps can set the ideals to which it will adhere; only the Corps can determine those who are unworthy and do not belong.

Let us not be confused by the failure of the few. Let us not lower ideals so that they may be attained by all. Ideals, attainable by all, are not a goal for proud men. Let us remember, too, that for each who fails, the hundreds who strive to achieve these ideals prove their worth.

As a representative of those who were once cadets, I know I speak for the vast majority of the living and the dead. The glory of West Point is in the ideals which it has set for those who have belonged to the Corps. Throughout the 150 years in which it has lived, the Corps, as a whole, has never failed to cherish and to give unsteadied devotion to Duty, Honor, Country. The Corps of today will maintain and strengthen its integrity of purpose for the Corps of tomorrow.

When 150 years have become 200, whoever stands where I stand today will feel then as I do now, that he is in the presence of a group of dedicated young men, willing not only to die for their country, but live and die for it in high integrity and unswerving purpose. This is the living West Point.

Secretary of the Army Pace introduced Major General Ellard A. Walsh, president of the National Guard Association, who said in part:

...No man could feel otherwise than highly honored and thrilled beyond measure in being invited and permitted to participate in the ceremonies incident to so famous an occasion in the life of this Institution, and I feel deeply so. But I am also conscious of a feeling of great humility in finding myself here in this famous and hallowed spot viewing scenes which were once so familiar to many who, when their time came, departed from the cloistered halls of this Institution to achieve undying fame on fields of battle throughout the world and acquire immortality as a result of their contributions to the well-being and security of the United States of America. I also have the feeling, and it is indeed consoling, that as long as this Academy and similar institutions continue to exist and adhere to their mission of developing leadership and leaders when the hour of need arises, all will be well with us as a people and a nation. I voice the prayer that for generations, and even for centuries to come, the several States, the Territories and the District of Columbia will be privileged to send to the Military Academy the finest product of the youth of the Army and Air National Guard thereof as they have been privileged to do for the past three decades, and who on the day of their graduation will become officers of the United States Army and the United States Air Force. I salute those graduates who came from the Guard to this place and who have long since departed; those who are here now, and those who come hereafter, and wish them every success in the profession of arms which they have chosen as their career.

A moment ago I mentioned that it is a privilege and pleasure of our Association to extend to the United States Military Academy, its Faculty and Corps of Cadets, through the medium of a Resolution, its felicitations and congratulations, and which in a moment I shall be only too happy to present formally to the First Captain of the Corps of Cadets. However, when all is said and done, a Resolution, no matter how beautifully it may be drawn and engrossed, is merely something of the moment, to be filed away when it has served its purpose and ultimately forgotten. Thus it was that the Delegates to our Seventy-Third General Conference of 1951 felt rather strongly that the Army and Air National Guard should recognize this great occasion in the life of the Academy in a more fitting and permanent manner, namely, in the form of a memorial which would endure for generations and serve as a permanent token of recognition on the part of the National Guard of the significant contributions which have been made by this Institution and its graduates to the well-being and security of the Country.

Now, therefore, as President of the National Guard Association of the United States, on behalf of the several States, Territories, the District of Columbia, the present Army and Air National Guard, nearly a half million strong, and those members of the National Guard who have graduated or who may hereafter graduate from this Academy, I present to the First Captain of the Corps of Cadets this Resolution which was adopted by the National Guard Association of the United States of good will and appreciation, and I further present to the United States Military Academy and the Corps of Cadets thereof on this Fifth Day of January in the year of our Lord Nineteen Hundred and Fifty-Two, in connection with the inauguration of the Sesquicentennial observance, a Memorial consisting of a Stand of Flags of the several States, the Territories and the District of Columbia, to be displayed within the confines of Washington Hall, and which said Academy is to have and to hold and may the kingdom, the Power and glory remain with it forever more.

The flags of the states and territories were simultaneously unfurled as General Walsh presented the Resolution to the Cadet First Captain, while the USMA Band played the National Anthem. Cadet Carpenter thanked General Walsh, saying:

> On behalf of the United States Corps of Cadets, it is my pleasure to accept these flags and to express the gratitude of the Corps to the National Guard of the United States for this commemoration of our one hundred and fiftieth year. The flags symbolize to us that the Corps of Cadets, like the National Guard, is drawn from all the states and territories. Thank you, sir.

Mr. Pace then rose and introduced Governor Thomas E. Dewey to the audience. Governor Dewey delivered the following address:

> Secretary Pace, General Irving, Distinguished Guests and Men of the Corps of Cadets:

> I am delighted to take part in this historic salute on the 150th Anniversary of the finest Academy in the free world for the finest officers who ever represented free people – the soldiers who have preferred peace but never lost a war. It is an honor to represent the forty-eight States and the Territories whose flags are being presented here today by the National Guard Association. I am equally honored to speak as the Governor of the State in which West Point has carried out its appointed mission in this century and a half and to extend the affectionate greetings of their neighbors to the Corps of Cadets.

> But really I should rather speak to you as just another son of a Cadet. For it was sixty years ago that my father was a member of the Corps. The unflagging devotion of the Cadets to their country was a by-word in my home all my life, as were their unequaled standards of integrity, honor and dedication to the cause of peace through strength. It has seemed to me a little sardonic that tens of thousands of examinations in which thousands of men took part under an honor system should have gone without comment while a recent exception in a century old tra-

dition should have aroused such controversy and comment. It is the exception that proves the rule. Every good American is deeply proud of every one of the Cadets here today, of the traditions you maintain and the administration of the Academy.

You of the Corps of Cadets will be the future officers of the sons of all America. You abhor the tradition of totalitarian nations which give to their officers the brutal power of life and death over men. You are trained in discipline and expect discipline from your men; yet you are also trained to have a tender interest in the welfare of the men in your command as well as the needs of conquered and allied peoples. The graduates of this Academy are today feeding millions of refugees from war and misery; they are helping to rebuild stricken nations all over the world. Trained as leaders in war, you will be, as your predecessors today are, the best examples of free government in action and builders of peace.

Of all the college men in America you of the Corps of Cadets at West Point are almost unique. You are not here for the purpose of training yourselves to improve your earnings or your social, economic or political status. On the contrary, you know that your earnings will never be comparable to those of your friends who go into almost any other professional career. For some reason or other the fiscal policies of the National Government regardless of the party in power have been unfailingly niggardly to the officers of our armed forces. You know that you and your family may live for years in sub-standard jerry-built housing erected in haste in the last preceding war. You know that promotion often will be desperately slow.

One of the speakers here today is one of the great generals of American history, the man who saved Berlin for the free world in the post-war crisis, today one of our greatest business executives. For sixteen years he lived and raised his family on the pay of a First Lieutenant and towards the end of that period saw his salary cut still further in a wave of government economy. Sometimes I wonder how men's patriotism can survive such treatment by their government. Yet whenever the country is in peril it expects the regular Army to produce overnight from its long forgotten, underpaid Officer Corps a Pershing, a Marshall, an Eisenhower, a MacArthur, a Clay, a Patton, a Bradley, a Somervell, an Arnold and all the others of that long list of great men. It was they and their associates in the other armed services who produced the organizational, engineering and diplomatic skill and the qualities of sacrificial leadership of the forces of freedom which saved the world twice in this century.

Men of similar quality and pure-minded patriotism are serving their country all over the world today in ways that too few notice and too few care about.

Alumni of the United States Military Academy are representing our country at its very best in difficult and delicate posts far from home. Unlike many other Americans overseas, they have not commandeered the best houses in the countries where housing is so short. They are often living in the poorest quarters. Boredom and loneliness have not led them to drink or misconduct. I have seen them in Europe and in the Pacific and I have watched the reactions of native peoples in a dozen countries just within the past years. You have a right to be deeply proud of them. Every American should be profoundly grateful to the diplomacy, the gracious conduct and the sympathetic understanding given by our military missions to the many nations whose course is still being shaped by their reaction to individuals who represent the conflicting ideologies of freedom and totalitarianism.

Our military missions have welded the armed forces of Greece to the point where she was able to repel Communist conquest. They have waded in rice paddies in water up to their shoulders as observers in desperate jungle warfare in Indo-China. They have patiently and without credit labored to build the strength of our friends and allies all over the world to the point where Communist aggression will not dare strike.

Under Generals Ridgway and Van Fleet our armed forces may well have saved the world from wholesale conflagration by the brilliance of their leadership under impossible conditions in Korea. As you mourn the loss of your brothers who have died in Korea, you may take deep pride in the knowledge that their sacrifices saved at least for a time a free area of 300 million people in the Pacific.

It is my deepest hope that our civilian leaders will move quickly enough to guarantee the integrity of the crisis area of Indo-China and Southeast Asia to prevent another Korea there. If we wait until the Red Chinese monster again invades, it may again be too late. If the free nations act now there is still time to make it perfectly clear that we know that the free world is indivisible, that aggression must be resisted everywhere, else we shall be unable to resist it anywhere. In this desperate struggle to build strength to prevent a third World War, West Point graduates are adding luster to the name of the United States Military Academy. The amazing progress in the defense of Europe in the past year, the leadership in Korea and the military advisors throughout the Pacific are showing statesmanship in action which will save the world if it is to be saved.

As Cadets you have consciously chosen a life of sacrifice, of discipline and of service. But you are joining a company of great men who have done more to save human freedom in this world and to prevent its destruction in the future than any other group. There can be no finer contribution than a life of service in the cause of human freedom.

While the regular services are the backbone of our defense, they can never carry the whole burden. It seems to me therefore particularly appropriate that at this celebration of the Academy's Sesquicentennial, this impressive display of flags of States and Territories should be presented by the National Guard Association. For the National Guard is our largest ready reserve for defense. When I was in Korea last summer, I was proud to meet National Guard units for the State of New York and to find them rated first class by regular generals.

From a long range standpoint we obviously must have a universal military service and a means of constantly feeding its graduates into all branches of the armed forces including the National Guard. The increased cooperation between the regular services and the National Guard is a good sign. It must be implemented by legislation, by adequate arms and authorization for much more extensive training. I hope that will come more swiftly in the near future than it has in the recent past.

Meanwhile the country can be secure in the knowledge that here at West Point honor, courage, truth, compassion and loyalty are engraved in the hearts of men who will be free government's stoutest defenders. The cause of human liberty will rest secure in the hands of the men of the Corps of Cadets and their successors in the generations to come.

When Governor Dewey's address ended, Mr. Pace stated the benediction would conclude the ceremony. Father Joseph P. Moore, Catholic Chaplain, gave the following prayer:

In the name of the Father, and the Son, and the Holy Ghost, Amen.

O God who of old who didst lead the wise men by the light of a star, look with favor upon us gathered here to inaugurate the Sesquicentennial Year.

May the light of Thy wisdom guide us in observing it, so that the aims and purposes of the Academy may be more widely known and more clearly understood, its dedicated service to God and country may be more efficiently sustained; and by Thy blessing of God – Father, Son and Holy Ghost – descend upon us and remain with us forever, Amen.

The guests left Washington Hall and Cadet Carpenter dismissed the Corps of Cadets. A half hour later the Corps formed in the central area of cadet barracks for a review before assembled guests and state delegates. After rendering "Honors" to the reviewing party, the Cadet Brigade Adjutant read United States Military Academy General Order Number 1 for 1952:

On 16 March 1802, President Thomas Jefferson signed the Act of Congress establishing the United States Military Academy at West Point. In the one hundred and fifty years since the founding of the Military Academy, its graduates have served the nation in every field of endeavor.

To commemorate this important anniversary, the six months from January to June 1952 are hereby designated as the Ses-

Inaugural Review: New York Governor Thomas E. Dewey; Secretary of the Army Frank Pace, Jr.; General (Ret) Lucius D. Clay, class of 1915; Major General Walsh, National Guard Association President; Major General Frederick A. Irving, Academy superintendent; and Colonel John K. Waters, Commandant of Cadets. (Photo courtesy Signal Corps.)

quicentennial Period. Appropriate ceremonies, meetings, lectures, and a jubilee convocation will be held during this period in accordance with instructions, programs, and schedules issued by this headquarters.

The review's high point occurred when General Irving placed the black, gold, and gray Sesquicentennial flag in the custody of the Corps, for the duration of the celebration period.

When the review ended, cadet escorts took their guests on guided

Superintendent General Irving, presents the Sesquicentennial Flag to the First Captain, Cadet Gordon Carpenter of Silver City, New Mexico, at the Inaugural Review. (Photo courtesy Signal Corps.)

tours of the post. They returned for a 5:30 buffet supper in the West Point Army Mess and then departed for New York City for the final event of the day.

The concert in Carnegie Hall completed inaugural activities that evening. The United States Military Academy Band, directed by Captain Francis E. Resta, and the Cadet Glee Club, led by Lieutenant Barry Drews, were the centerpieces of the evening for the capacity audience of over twenty-seven hundred people. Guest soloists for the two hour concert were Miss Marguerite Piazza and Mr. Walter Cassel of the Metropolitan Opera Association. Among highlights of the evening were three compositions written especially for the Sesquicentennial: *100 Days*, an overture composed by Captain Resta, *Israfel*, a tone poem by Chief Warrant Officer H. L. Arison, Jr., of the USMA Band, and *West Point Suite*, by the noted French composer Darius Milhaud.

Presented for the benefit of Army Emergency Relief, the concert brought in $6,000 for the Fund's charitable activities. The West Point Society of New York cooperated with the Academy in publicizing the event and the sale of tickets. The Society and its secretary, retired Colonel George DeGraaf, were largely responsible for the event's financial success.

The Sesquicentennial Inaugural of January 5, calling upon the Academy's history, traditions, fundamental purposes, and future, set the tone for the coming months in our second semester of plebe year. But there was a pall hanging over West Point, despite the uplifting speeches in Washington Hall and the concert in New York's Carnegie Hall, a concert the Corps of Cadets couldn't attend. The following Wednesday at 1:13 in the afternoon, the train pulled into the West Shore Railroad Station down by the Hudson River, carrying the caskets bearing the remains of three of the nineteen cadets killed in the crash of Air Force 266 – and a mournful ceremony began.

PRESENTATION OF THE LEE PORTRAIT

The second major event of the Sesquicentennial was Saturday, January 19, the unveiling of the portrait of General Robert E. Lee in the Main Room of the Military Academy Library.

Over a period of several years, many friends and graduates suggested adding a portrait of Lee in his Confederate uniform to the fine arts collection at the Military Academy. As pointed out in General Maxwell Taylor's dedicatory talk that afternoon, as late as 1930, charges of the Academy's disloyalty to its southern graduates, had once more briefly become an issue, when the United Daughters of the Confederacy sought to hang a painting of Lee, the Academy's ninth superintendent, in his gray uniform, among portraits of superintendents in Washington Hall. At that time the superintendent agreed to accept the portrait of Lee, but in the blue coat he wore as the superintendent in 1852. The decision was of "obvious propriety," as General Taylor explained in his talk, because Lee's portrait would have been in company of superintendents' portraits.

Advocates of the Lee portrait presentation in 1952 believed it would symbolize the end of sectional differences in the United States.

In 1950 Mr. Gordon Gray, former Secretary of the Army, then president of the Consolidated Universities of North Carolina, and retired Army Major General Russell L. Maxwell, Academy class of 1912, formed a committee to make the presentation a reality. Dr. Douglas Southall Freeman, the noted author and Lee biographer from Richmond, Virginia; famed publisher Mr. William Randolph Hearst, Jr.; Mr. William Clayton of Houston, Texas; and Mr. Wharton Weems, also of Houston, served on the committee.

General Bryant E. Moore, while superintendent, enthusiastically supported the proposal. He solicited opinions from distinguished military officers and ex-superintendents of the Military Academy, and most supported the proposal. After due deliberations the decision was to display Lee's portrait beside General Grant's, painted in 1866 by Paul Louvrier, and already in possession of the Military Academy. The joint exhibition of the portraits of these two great Civil War leaders would be "a splendid expression" of what the Academy stands for – national unity.

Ulysses S. Grant and Robert E. Lee are enshrined in the pantheon of Academy heroes. In the minds of most historians and nearly all graduates, both exemplify the high ideals of the Military Academy, devotion to principles and integrity of character. General Grant's portrait depicted him in the uniform he wore at the height of his fame, as commander of the Union Armies. It seemed fitting that General Lee be portrayed in the uniform for which he is remembered, when he was commander of the Confederate Armies.

With the project approved, the West Point Museum Board and the Art Advisory Committee cooperated with the Lee Portrait Committee, and both groups unanimously recommended distinguished American artist, Sidney E. Dickinson, be commissioned to paint the portrait.

Born in Wallingford, Connecticut, on November 28, 1890, Mr. Dickinson's talents were evident early in life. Teachers and masters such as Bridgemen, Volk, and Chase directed his growth and development as an artist. In 1927 he was elected to the National Academy. He was a consistent winner of prizes and awards in the exhibitions of the National Academy of Design, the Pennsylvania Academy of Fine Arts, and others of national importance, and was a member of the Century Club and the Allied Artists of America. His work was on display in a number of art museums and galleries of the United States. One of America's leading portrait painters was adding to the Academy's already substantial art collection.

For weeks Mr. Dickinson researched and studied General Lee's life and characteristics. After completing his preparations, including the examination of many photographs, he began the portrait. When complete, the work spoke eloquently of Mr. Dickinson's talent, and proved to be more than a mere likeness of the Confederacy's renowned military leader. The portrait reflected the character, personality, and great quality of soul of Robert E. Lee.

The Library, at the time of the presentation, was one of the few buildings remaining at the Academy which were in existence when Lee was superintendent from 1852 to 1855. Redecorated for the occasion the Main Room of the Library included new indirect lighting, which replaced antiquated chandeliers. New, dark paneling covered the wall on the south side of the room. The Lee portrait hung on the west side of the large window in the room, and the Grant painting hung on the other side. Similar colonial type, simply molded, gold leaf frames brought both portraits into agreeable relationship.

At 4:30 in the afternoon, with approximately 125 guests present, the ceremony began. In addition to General Taylor, two members of the Lee portrait committee who had a major share in bringing the portrait to West Point, Mr. Gordon Gray, Chairman, and Major General Russell L. Maxwell, Secretary, were among the guests. Mr. Gray was the second principal speaker, giving the presentation address. Mr.

George Hearst represented his brother, committee member William Randolph Hearst, Jr. Other honored guests were superintendent General Irving, the Dean of Academics, Brigadier General Harris C. Jones; the Academy Chief of Staff; the Commandant of Cadets; members of the Sesquicentennial Steering Group; the artist, Mr. Dickinson; Mr. DeWitt M. Lockman of the Art Advisory Committee; and two great-grandchildren of General Lee, Miss Anne Carter Lee Ely and Mr. Hanson Edward Ely III. The two direct descendants of General Lee unveiled the portrait.

Following brief remarks, General Irving introduced General Taylor, whose dedication address, in part, follows:

> ...It is easy to misstate or to overstate the significance of this return of General Lee to West Point. We recall the storms of criticism over the role of our graduates in the Armies of the South, which swept over the Military Academy after the close of the Civil War. Then, as now, the Military Academy had its critics, often more distinguished by the intemperance of their attacks than by the justification of their charges. It was said that West Point had fostered the leaders of the Rebellion; that far from inculcating loyalty and devotion to country, the Academy was a hotbed of treason which should be blotted from the land...
>
> To some, this delay in acknowledging our Confederate graduates may seem excessive prudence, or indeed a concession to the bugbear of sectionalism long since departed from the land.... To most of us it appears high time to lay aside such historical blinders in viewing our graduates, and acclaim what every schoolboy knows – that Robert E. Lee was not only a distinguished graduate of West Point, a Superintendent who contributed notably to the development of the Academy, a brilliant officer in the United States Army worthy of being offered the supreme command, but also the immortal battle leader of the Confederacy, whose deeds will stir men's souls as long as future generations find time to read the history of this country.
>
> Past refusals to acknowledge the Confederate General Lee suggest that West Point has sat in judgment upon General Lee's political opinions. This judgment, I believe, West Point has no

right to make. The historic role of the Academy is to form military leaders, men of character, capable of leading the citizen soldiers of our country to victory in battle. It is by their success or failure to meet this standard that we may fairly judge our graduates, not by their politics or personal opinions. If our southern graduates fell into political error, as some will aver, they did so with nearly half the American Nation of their time, with whom they staked their lives and fortune for a common cause they felt to be right. Had they been cowardly, dishonest, evasive or recreant to their principles, then West Point might judge them derelict to the standards of the Corps. But they were intrepid, steadfast, honorable soldiers who brought world renown to American arms on many a hard fought field of war. It is proper for West Point to take pride, not of necessity in the rightness of its Confederate graduates but in their uprightness.

In bringing General Lee home to West Point, we could have found no place more suitable for his portrait than this panel in the Cadet Library. It is one of the few buildings in the West Point of today which General Lee knew. This Library, the Central Barracks, the houses of the Superintendent and the Commandant across the plain, and the old Chapel now in the Cemetery – these are all that remain of Lee's West Point. Elsewhere than here the General might feel ill at ease. Here he is at home among his books and among his friends.

As he looks about this room, General Lee sees much that is familiar. There on the gallery are the pictures of many friends and associates of former days. In that corner is General Scott, Lee's revered Commander-in-Chief at the time of the fateful decision to leave the Army of the United States and join the South. It was to Scott that he wrote his letter of resignation, saying: 'Since my interview with you on the 18th instant, I have felt that I ought no longer to retain my commission in the Army. I, therefore, tender my resignation and request that you will recommend it for acceptance. I would have presented it at once but for the struggle it has cost me to separate myself from a service in which I have devoted all the best years of my life and all the ability which I possess. During the whole of that

541

time, more than a quarter of a century, I have experienced nothing but kindness from my superiors and a most cordial friendship from my comrades.'

In addition to Scott, there is Thayer the father of the Military Academy who was General Lee's superintendent during his Cadetship. It is not recorded that Cadets (past or present) form abiding friendships with their Superintendent but Lee always expressed the deepest admiration for his, this great educator, Sylvanus Thayer. Probably closer to Lee were the four professors yonder who served at the Academy when he was Superintendent – Professor Weir of Drawing, who painted one of three portraits made of Lee before the Civil War; Mahan, Professor of Engineering, father of Admiral Mahan, the famous Naval historian; Bartlett, who taught Natural Philosophy and was nationally known as an astronomer, and finally, Professor Bailey of the Chemistry Department. Indeed, General Lee might well hold a meeting of his Academic Board in this room and have a quorum to decide the issue of the day....

General Taylor, General Maxwell, Mr. Gray, Mr. Dickinson, and Mr. Ely at Lee Portrait Presentation, in the Main Room of the Military Academy Library, Saturday, January 19, 1952. (Photo courtesy Signal Corps.)

In unveiling this portrait of General Lee today let us rejoice in the symbolism of the unity which the occasion affords. Today there is no North or South, no East or West, one people proud to honor two such leaders as Lee and Grant. It is true that we have other divisions, other troubles which distract and distress. But when our burdens seem heavy, let us draw new strength and inspiration from the fortitude Lee and Grant displayed in the terrible days of the Civil War when the country was in greater danger than at any time in our history. If the schismatic forces seem to rend us and to spread doubt of ourselves and our leaders, let us remember the words of General Lee to a mother who after the Civil War sought his advice in rearing her sons. 'Abandon all these local animosities,' he said, 'and make your sons Americans.' Let us hope that future generations of West Point men gazing upon the noble features of Robert E. Lee will be moved to become better cadets, better officers, better citizens, better Americans. Thus will they serve our country until that day arrives which General Grant bespoke, saying, 'I believe that our great Maker is preparing this world in its good time to become one nation, speaking one language and when Armies and Navies will no longer be required.'

After General Taylor's address, the superintendent introduced Mr. Gordon Gray, who made the presentation address, in which he thanked the various participants in the work to bring the portrait to the Academy. General Irving gave brief remarks of thanks and acceptance, followed by the National Anthem, which concluded the ceremony.

Thus, at the Sesquicentennial Celebration, ninety-two years of separation ended, separation that began when "a band of brothers" took leave of one another beginning in 1860, to go by different paths, and lead the Armies of the North and South in the bloodiest war in the nation's history. Myths had grown up over the years since the Civil War, that Grant and Lee met and knew one another in the Mexican War, or on other occasions formed a friendship before finding themselves antagonists leading opposing armies in the Civil War. Evidently, none were true, for after the war, Lee was asked about his prior encounters with Grant when they were wearing the Union blue. He said he might have met him in Mexico, but didn't remember, and denied any friendship or professional encounters while Lee still wore Union blue, or at any time prior to Lee's surrender at Appomattox April 9, 1865.

The symbolic, invitation only, Lee portrait presentation at West Point, on January 19, 1952, wasn't widely reported in the press, perhaps intentionally, and went unnoticed. Nevertheless, the event reached back in time to heal one more, but not the last, of the great Republic's Civil War wounds.

MARCH 16: FOUNDERS' DAY

David Hughes was on his way back across the Pacific on the Japanese ship *Oturu Maru* in March, writing his letter to Captain John Flynn, as the United States Military Academy prepared for the annual Founders' Day celebration. Earl Blaik and his coaching staff had completed their annual recruiting work, and were preparing to start spring practice. Dick Shea and Dick Inman, both first classmen in the Sesquicentennial class were in Corps squad training for spring's outdoor track season, having completed the winter, indoor season. Dick Shea, the track team captain, was soon to begin setting more records in his last season for Army, and Dick Inman, too, was destined for a record setting performance at the Penn Relays in Philadelphia.

Behind closed doors, Academy officers and cadets continued to work on committees and studies intended to complete actions necessary to prevent a repeat of the prior year's devastating cheating scandal. Progress was slow, and the results of their work wasn't described publicly to the Corps of Cadets. Many small changes were made, quietly, with no fanfare, no notice taken, and no connection explained or evident between the events of 1951 and the corrective measures taken.

The Academy's 150th anniversary fell on a Sunday. To graduates and former cadets of West Point, this Sunday was perhaps the most significant day in the entire Sesquicentennial observance.

Traditionally, graduates celebrate Founders' Day wherever they may be, gathering in groups of four or five, or a hundred and more, to restore friendships and revive memories of cadet days. This Founders' Day was more serious. Many graduates were facing enemy fire, and others were scattered in trouble areas throughout the world, most notably at the ready adjacent to East Europe's "Iron Curtain," the line separating an expanding Soviet empire and the Western democracies. From Korea to Germany, from Alaska to South America, the men of the United States Military Academy paused in their work to participate in this important anniversary.

General Irving, the superintendent, opened the Sesquicentennial Founders' Day observance sending messages of greetings to West Point graduates throughout the world. He spoke for the officers and cadets of the Military Academy, and accompanying his message was a salute from the Corps.

Wherever you may be on Founders' Day of this Sesquicentennial year, the officers and cadets of the Military Academy join hands with you to rededicate themselves to the ideals of Duty, Honor, Country. West Point salutes its sons the world over.

F. A. IRVING
Major General, USA
Superintendent

Message from the Corps of Cadets:

We men of the Corps of Cadets salute the men of the Corps of the past on Founders' Day of the Sesquicentennial year. We treasure the heritage you have given us; and we pledge ourselves to uphold the traditions and high standards of the Corps in fulfillment of the trust you place in us. We join hands with you today to dedicate ourselves to the ideals of West Point: to Duty, Honor, and Country.

Replies poured in from every part of the United States and numerous foreign countries. From graduates came affirmations of loyalty, while non-graduates sent their congratulations.

Three hundred graduates and former cadets in Tokyo, Japan, join you in rededication to the ideals of Duty, Honor, Country, and in commemoration of our alma mater's anniversary.

– RIDGWAY, April '17

Heartfelt thanks for your message of 8 March. All West Point graduates in the Eighth Army in Korea appreciate this opportunity to join with the officers and cadets of the Military Academy – and sons of West Point throughout the world – in commemoration of our alma mater's anniversary. We proudly return the salute as we rededicate ourselves to the ideals of Duty, Honor, and Country.

– VAN FLEET, '15

Three hundred members of the Long Gray Line assembled in Frankfurt, Germany, join you in a toast of the 150th anniversary of the founding of the United States Military Academy. May we serve the cause of Duty, Honor, Country as faithfully and well through the coming years as those who have gone before.

– PERRY, '17 HUGHES, '22
MURPHY, '19 STROTHER, '31

I welcome this opportunity to extend the congratulations of the Second Army on the occasion of the 150th anniversary of the U.S. Military Academy. Throughout its existence, the Academy has imbued the Cadet Corps with ideals of integrity of character and devotion to duty. It is that very inculcation that has been responsible for the high degree of leadership which has inspired countless American citizen soldiers. In this period of international tension, the U.S. Military Academy stands as a symbol of the free world's fight against tyranny. It can mark its Sesquicentennial celebration with the assurance that every loyal American is appreciative of its great contribution to world peace.

– EDWARD H. BROOKS
Lt. Gen., Commanding
Second Army

On behalf of the trustees, the alumni, the faculty, and the students of the Pennsylvania State College, I wish to extend to the staff and the Corps of Cadets of the United States Military Academy our most sincere congratulations on the Sesquicentennial anniversary of the founding of the Academy. I know that Penn Staters everywhere share with all Americans a warm pride in the great tradition of courageous and unswerving devotion to Duty which successive generations of West Pointers have established in the service of our country.

– MILTON S. EISENHOWER, *President*
Pennsylvania State College

On behalf of all members of the British Army Staff Washington I should like to take this opportunity of congratulating the United States Military Academy on the occasion of its one hundred and fiftieth anniversary. Coupled with these birthday greetings may I extend to all West Pointers every good wish and good luck in the future.

– W. H. STRATTON

The Regiment of the Cadet Midshipmen of the United States Merchant Marine Academy sends its heartiest congratulations to the Corps of Cadets on the 150th anniversary of the founding of West Point. May your next 150 years be as illustrious as have the past 150 years.

– WILLIAM G. RANDALL

Messages came to the Academy from alumni organizations and groups of graduates at the following locations:

Aberdeen Proving Ground and the Army Chemical Center, Maryland; Alabama; 4th Army Area; The Army War College and the Central Pennsylvania Area; Camp Breckenridge, Kentucky; Central California; Chicago, Illinois; Columbia, South Carolina; Columbus, Ohio; Connecticut; The Engineer Center, Virginia; The Florida West Coast; Fort Benning, Georgia; Fort Bliss, Texas; Fort Bragg, North Carolina; Fort Hood, Texas; Fort Knox, Kentucky; Fort Leavenworth, Kansas; Fort Meade, Maryland; Fort Monroe, Virginia; Fort Riley, Kansas; Maxwell Air Force Base, Alabama; Miami, Florida; Michigan; Mitchell Air Force Base, New York; The Monterey Peninsula, California; Philadelphia, Pennsylvania; Portland, Oregon; The San

Francisco Bay Area; Washington, D.C. and Wright Patterson Air Force Base, Ohio. The Antilles; Austria; Berlin, Germany; The European Command Communications Zone; Fort Kobbe, Canal Zone; Hawaii; Iran; 7th Infantry Division in Korea; Mexico City, Mexico; Nurnberg, Germany; Okinawa; The Philippines; Pusan, Korea; Sendai, Japan; Stuttgardt, Germany; Thailand, and Trieste.

Many cities and states honored the Academy. New York's Governor Thomas E. Dewey proclaimed the week of March 15-22 as "West Point Week." Governors Douglas McKay of Oregon, Sidney McMath of Arkansas, and Frank J. Lausche of Ohio proclaimed March 16 "West Point Day." The Massachusetts House of Representatives and California's State Legislature passed resolutions extending to the Academy the congratulations and felicitations of the peoples of the two states. Philadelphia Mayor Joseph S. Clark, Jr. proclaimed March 16 "West Point Day."

At the Academy, the Corps of Cadets, officers of the West Point garrison, and hundreds of graduates participated in special ceremonies intended to emphasize the occasion's solemn spirit. Members of various West Point Societies were present. Representatives of the Societies – local chapters of the Association of Graduates – came from as far away as California.

Early in the morning, at the West Point cemetery, a group of graduates and cadets performed a quiet ceremony at the tomb of Sylvanus Thayer. The Cadet First Captain, Gordon Carpenter; Colonel J. A. McComsey, Secretary of the Association of Graduates; Mr. J. L. Grant, representing the New York West Point Society; and Mr. R. D. Reynolds, of the St. Louis West Point Society, placed a wreath at the grave of the "Father of West Point."

At Mt. Vernon, Virginia, a group of cadets escorted by Major R. B. Shea visited the tomb of George Washington, clearly the best known and nationally admired of the Academy's founders. Cadet H. L. Van Trees, class of 1952 and first in the graduation order of merit, placed a wreath on Washington's tomb, with this statement:

On this Sesquicentennial anniversary of the founding of the United States Military Academy, the Corps of Cadets and graduates of West Point honor General George Washington as one of our founding fathers. Ever mindful of our obligations to the Nation, we pledge ourselves anew to the ideals of our alma mater; to Duty, Honor, and Country.

At West Point, there were special services at the Cadet Chapel and the Catholic Chapel. Former Chaplain, Right Reverend Monsignor George C. Murdock, celebrated the Solemn High Mass and preached the Founders' Day sermon. The Right Reverend Arthur B. Kinsolving, Bishop of Arizona and former Academy Chaplain was guest preacher for Protestant services in the Cadet Chapel. Bishop Kinsolving's sermon topic was "The Part of Religion in the History of West Point."

After chapel services, guests and graduates assembled in Washington Hall, where they sat at tables surrounding the speaker's dais directly in front of the center door. The Corps marched in and Gordon Carpenter called cadets and graduates to attention as the superintendent's party entered the Hall. Seated on the dais with General Irving were General J. Lawton Collins, Army Chief of Staff, principal speaker; General Hoyt Vandenberg, Air Force Chief of Staff who presented a plaque to the Military Academy on behalf of the Air Force; Lieutenant General Willis D. Crittenberger, Commanding General, First Army, the master of ceremonies; Major General H. C. Hodges, class of 1881, the oldest graduate present; Brigadier General Chauncey L. Fenton, president of the Association of Graduates; Brigadier General Harris C. Jones, Dean of the Academic Board; and Colonel John K. Waters, Commandant of Cadets.

After the noon meal, Carpenter again called everyone to attention, and the ceremony began. Following the invocation, the Cadet Glee Club sang *The Corps*, and General Irving made welcoming remarks, which included some light humor regarding old grads and present day cadets:

...One old grad suggested that perhaps we should have spread these visitors throughout the Corps. Perhaps this would have been a good idea because we are very proud of our present-day product, and we like to show them off on every occasion. However, on second thought, it is just as well that we did not because I clearly remember the proclivity of old grads upon occasions like this to regale cadets with the measures they found effective during their own cadet days to outwit the Tactical Department. I am sure that the Commandant of Cadets feels that the present-day generation needs no coaching in this respect...

The superintendent then introduced Lieutenant General Crittenberger, the master of ceremonies, who, General Irving pointed out as having the distinction of being "the graduate with the most years of service now on active duty with the line."

General Crittenberger introduced Air Force Chief of Staff Hoyt S. Vandenberg, class of 1923, who made the following remarks, then presented a plaque to the First Captain, Cadet Carpenter:

Distinguished guests, gentlemen:

The Air Force acknowledges its immeasurable debt to the Military Academy and in this memorial recognizes the continuous contribution of West Point men to the United States Air Force since its very beginning.

Forty years ago those founding fathers of American military aviation, Lieutenant Hap Arnold, Lieutenant Thomas Milling, and Lieutenant Frank Lahm, were taught by the men who invented the airplane, the Wright brothers. The first man to lose his life in an airplane was Lieutenant Selfridge of the class of 1903, while flying with Orville Wright in the first airplane demonstrated to the United States Army. What is now known as the United States Air Force began with a handful of men such as these.

Military flying was not even considered a career in those days but only as a special assignment. Moreover, the physical risks were extremely great. But an unbroken succession of men from the Point were willing to accept not only the physical risk that went with the conquest of the air, but also the professional uncertainties that went with the new service. The development of American leadership in the air cost heavily, not only in unremitting work and struggle, but also in the lives of the many high-spirited pioneers of the early days of aviation.

At the present time, the rising cost of air power, and indeed of all armament, gives serious concern to the American people. But on this occasion, we might dwell for a moment upon another aspect of the cost: the sacrifice of the pioneers in both peace and war, many of whose names are inscribed on the Roll of Honor in Cullum Hall. We of the Air Force feel that the United States Military Academy can count its contribution to the building of the Air Force as one of the proudest achievements of its first one hundred and fifty years of service to the Nation.

It is therefore a privilege to present to the United States Military Academy this modest but significant memorial to the sons of West Point who chose the Air.

The audience applauded as First Captain Gordon Carpenter approached the rostrum to accept the plaque and respond:

General Vandenberg, it is a great honor for me to accept this plaque from the Air Force on behalf of the cadets and graduates of the Military Academy. All of us – the cadets of today and the cadets of the past – truly appreciate the generous gift you have given us today. Thank you.

The master of ceremonies, General Crittenberger, then announced that taped messages had been received. He said,

I regret that time will not permit the reading of all of these greetings. However, we do have recorded messages from several individuals. The first is from General Matthew Ridgway, Class of April 1917, now in Tokyo. General Ridgway.

General Ridgway's recorded message:

I contemplate with profound respect the one hundred and fifty years of service to our country by this anniversary of the United States Military Academy. Through this century and a half, the concepts of 'Duty, Honor, Country' so jealously guarded by the Corps of Cadets have shown vitality on every one of the Nation's battlefields, have had an effect on the Nation's highest councils, and have guided the conduct of lives wholly dedicated to our Republic. Hallowed names of the Academy's sons owe their greatness and enduring fame to the ideals and traditions nourished by the Corps. The living and dead of our generation have nobly made their contribution to the honorable part our Nation has taken in assuming world-wide responsibilities. Nothing seems to me more certain than that young gentlemen now members of the Corps will in years to come write proud new achievements into this unfolding record. With reliance upon God and with full faith in the righteousness of principles as expressed in our motto, the sons of West Point will continue to place service before self, to accept more of sacrifice than of emoluments, hardship than of ease and to regard integrity in public affairs as the one indispensable foundation upon which the Nation's future can rest securely. On this foundation and with these principles our fellow citizens yet unborn will continue to have the assurance that the United States Military Academy will remain a faithful and loyal instrument of the American people and a strong bulwark of their government and freedom.

The next recorded message was from Chairman of the Joint Chiefs of Staff, General Omar N. Bradley, Class of 1915.

For a century and a half West Point has trained men whose lives are dedicated to 'Furthering Our National Security,' the theme of the Sesquicentennial celebration. Graduates of the United States Military Academy will continue to learn strategy and tactics of war. But in this democracy of ours we know our professional soldiers and airmen are first devoted to the preservation of peace. The past accomplishments of West Point graduates in their service to the United States provides real inspiration for our Nation on this anniversary. But the future, not the past of our Country is of greatest concern. We want to continue to enjoy freedom. To have it and hold it, we must protect it. Now and for some time to come, the military life is going to be part of all of our lives. Each young man must necessarily spend some time in the service of his Country. All the young Americans who serve, with the exception of those who enter the Navy, are going to depend to a greater or lesser degree upon the graduates of West Point, to the code of performance and the standards of leadership throughout the Army and Air Force. The future generations of American soldiers deserve the best leadership this Nation can offer. With this great demand for our best young leaders, we can rightfully ask, 'What does West Point have to offer these intelligent and energetic young men? Is it worthwhile for the young man himself?' Let me answer my own questions. West Point takes the richest and the poorest, the farmer's son and the factory worker's boy, from all walks of life and gives them an excellent academic education and unparalleled opportunity for service to their Country. The leadership training at West Point is typically democratic, typically American, and all American. For one hundred and fifty years West Point has graduated men in whom America and the Armed Forces can take great pride. On behalf of the men and women of the Armed Forces, I congratulate West Point, my own alma mater, on one hundred and fifty years of service to the Nation and wish her well for the future.

General Crittenberger then introduced the taped message of General of the Army Dwight D. Eisenhower, Class of 1915, the class "the stars fell on." General Eisenhower's recorded words came to the Corps from Paris:

With all the other men of West Point, I proudly join today in salute to our alma mater. Her record of public service over the past century and a half is written in the annals of a thousand of our Nation's battlefields and in the accomplishments of all her sons. But to her own, the true meaning of West Point is a deeper and more personal thing. It is of the heart and of the spirit. From cadets she made soldiers – good soldiers; but more than this, she has made better men of all of us. She has pointed out the true path to a rich fullness of life in the service of America, its people and its cherished ideals. She has kept high the flag before the world and in our hearts.

The next recorded voice heard on the speaker system was from Secretary of the Army Frank Pace, Jr., who sent his message from Washington, D. C.:

As Secretary of the Army, I am privileged to work daily with Army officers who are alumni of the United States Military

Academy. When I speak about West Point therefore, I speak from intimate personal experience with the men produced by the Academy. To my mind there is no fairer way to evaluate any institution of learning. The mission of West Point is to train professional leaders capable of rendering a life-time of service in our Regular Army and Air Force. For one hundred and fifty years the men of West Point have furnished living proof of the success with which the Military Academy has single-mindedly pursued its mission. Its graduates have exerted their influence on every major campaign waged by American soldiers or air-men. From the Mexican War to Korea, West Pointers have filled with distinction many of the high command positions in our Army and later in our Air Force. It is fair to say that their con-tribution played some part in moving Prime Minister Churchill to say, 'It remains to me a mystery, as yet unexplained, how the very small staffs which the United States kept during peace were able not only to build up the Armies and the Air Force units but also to find the leaders and vast staffs capable of han-dling enormous moves and of moving them faster and farther than masses have ever been moved in war before.' In between wars the West Pointer has proved himself the highest type of public servant, living and working according to the code of 'Duty, Honor, Country', in which he has been reared. The Mili-tary Academy has grown with our Nation, rendering peacetime as well as wartime service throughout all the years of its exist-ence. Today the business of every Army officer is to aid in pre-serving peace. This calls for a breadth of understanding of eco-nomic and geopolitical problems which goes beyond pure mili-tary knowledge. West Point has adjusted and must continue to adjust itself to the requirements of its graduates in a changing world. It is clear that the caliber of its graduates provides the ultimate test of an educational institution. That is why on Founders' Day, I pay tribute to West Point – molder of men.

The Chief of Naval Operations, Admiral William Fechteler, United States Naval Academy Class of 1916, next spoke to the Corps with a greeting recorded in Washington:

At a time when the need of leadership among men is receiving so much attention in the free nations in the world, it seems particularly appropriate to me that our attention should be focused upon the United States Military Academy at West Point; and I am honored to have the privilege of extending the Navy's congratulations to an institution which has been so devoted for one hundred and fifty years to the vital task of developing leadership.

The history of West Point can be written in terms of the men of West Point, men whose names are familiar to every student of American history, men like Grant and Lee, Goethals and Pershing, MacArthur and Eisenhower, Ridgway and Van Fleet. From generation to generation, this heritage of leadership has been preserved at West Point despite the rise and fall of the popularity of professional military education. Graduates of West Point have been in the forefront of the defense of democracy throughout the years in response to the demands of this freedom-loving nation for leadership built upon the pillars of the Academy's motto: Duty, Honor, Country.

Whatever the demands of the future may be, the accomplishments of the past are evidence enough that your United States Military Academy stands ready to perpetuate its enviable record of leadership and service to the United States and to the free world.

General Crittenberger then introduced the oldest graduate present, Major General Henry Clay Hodges, Class of 1881. The old soldier received an ovation from everyone assembled in the great hall. Following the introduction, the Cadet Glee Club serenaded General Hodges and the entire assembly with *Army Blue*. The master of ceremonies next introduced the principal speaker, the Army Chief of Staff, General Collins, who addressed the gathering:

Never before in history has it been so important that the soldier have a broad understanding of the relationships between all the factors that make up our national security.

In the early days of this great institution which we honor today, a graduate of the Military Academy needed above all else a pioneering spirit in order to shoulder his responsibilities in the opening of our new land. Because West Point was one of the early seats of learning in this country, and the earliest source of engineering knowledge, its graduates were called upon to play a great part in the development of our young nation.

As our frontiers moved westward many West Pointers led the way guiding and guarding, mapping and surveying as they battled Indians and an unknown terrain. The migrant trickle became a steady flow to the Pacific over the trails they had blazed. At the same time our Engineers were improving our rivers and facilitating the safe berthing of our growing ocean trade in our harbors.

It would be difficult to find a major railroad in the United States which does not owe its original construction, at least in part, to the efforts of graduates of this Academy. The excellence of their labors was so widespread that they were also called upon for such foreign projects as the Panama and Cuban railroads and the Mexican line from Vera Cruz through Mexico City to the Pacific Ocean.

Quite naturally, also, graduates of this earliest engineering school were called upon to pass on their knowledge to schools in other parts of our country. By 1860, forty graduates were professors of mathematics and sixteen were professors of civil engineering in colleges and universities spread over twenty-one states. Equivalent contributions were made in other fields such as theology, astronomy and navigation.

As the nation and its arts and sciences expanded, graduates also needed a greater appreciation of the more intricate implements of war. They had to study the 'new developments' of their day – the revolver and the repeating rifle, the telegraph and the balloon, the new maneuvers of infantry and the radical employment of cavalry. The lessons learned at West Point were now to stand them in good stead in leading men in new battlefield tactics where the spirit of the offensive and the doctrine of mobility were to ensure our national survival on global battlefields of later years.

As the country grew larger, the world grew smaller with every advance in communications and our problems of national security became interlinked, perforce, with world problems of security.

Today, a graduate entering the service of his country needs an even broader training than ever before to understand the complex relationships between all the factors which make up our national security. He needs to comprehend electronics as well as men; economics as well as weapons; diplomacy as well as tactics. He needs to know these things not only because they are part of the science of war, but because they are more closely related than in the past to our national security, and to the survival of the free world.

It is only natural that as things change the Army's tasks change. In some cases we have been required to assume new non-military responsibilities simply because we could not escape them.

Such was the case at the close of World War II when the responsibility for the establishment and the administration of government in Germany, Japan, Korea, Trieste, Italy, Austria, and the Ryukyus was thrust upon the Army. These tasks involved all the complex civil functions which are normal to everyday community life.

New forms of government had to be set up. Laws had to be written, enacted, and administered. Normal police functions had to be reestablished. Millions of people had to be fed, clothed, and sheltered. Their health and sanitation measures were a constant concern. Millions of children had to be put back in school – and some reeducated – under our supervision. In Germany alone, the Army Youth Assistance Program aided in rehabilitating almost a million youngsters. Hundreds of thousands of displaced persons and prisoners of war had to be repatriated. Public utilities had to be rebuilt. Newspapers and radio stations had to be reorganized, and recreational facilities revived.

Of course, the Army could not accomplish all these tasks without the help of many civilians, some commissioned especially for these purposes, but most of the executive directors perforce had come largely from the regular corps of officers.

Lieutenant General Crittenberger, seated, while General Collins, Army Chief of Staff addresses the Founders' Day Luncheon, in Washington Hall, Sunday, March 16, 1952. (Photo courtesy Signal Corps.)

While we would like to have escaped some of these responsibilities, and while we felt that they should be turned over to civil authority as soon as possible, nevertheless they became our problems; and it was several years following World War II before we could concentrate on our military jobs alone in many areas. And through it all, officers of the Army needed a wide knowledge of issues well outside the military field; they had not only to guard and further the interests of our nation but they had to recognize the everyday needs of a conquered people.

As some semblance of stability returned to the postwar world, we were denied the firm basis for peace for which all of us so fervently hoped. We were denied this because a new aggressor, backed by the largest Army in the world, was, and still is, pursuing imperialist expansion, even beyond that of past Emperors, Czars, and Kaisers.

This new menace is using every device known to man to gain his aggressive ends: propaganda, subversion, infiltration, and – where advantageous – the barbarous use of satellite military force. As a result, war-torn nations have been strangled in their efforts to regain political and economic stability, and many free nations have been forced to spend beyond their means to bolster their defenses and guard their freedoms.

It was clear then, in the postwar world, that such an unprecedented challenge required unprecedented effort to meet it. And the scope of the effort requires unprecedented understanding on the part of young men – particularly young officers – entering the service of their country.

The program of the American people is designed to prevent the catastrophe of another world war. However war could be thrust upon us, and we must be prepared to win it with all the resources combined forethought and superior science can muster. All of our preparations are defensive ones to provide mutual security through common effort.

The American program for peace encompasses:
first, active participation in the United Nations;
second, regional security arrangements within the framework of the United Nations – like the North Atlantic Treaty Organization and the Organization of American States;
third, military assistance programs by which we furnish military aid to increase the strength of friendly nations and thus strengthen the defenses of the free world;
fourth, economic assistance programs to aid certain friendly nations to stand on their own feet in areas where human welfare is the first essential in the struggle against communism;
fifth, an information program including the Voice of America, to tell the people of the world the story of freedom;
sixth, the rebuilding of our own armed forces.

This program is so vast that it challenges the imagination and wisdom of all of us, but upon it rests the hopes of the world; and we must make it work! Most of you now in the Corps of Cadets will at some time take part in one way or another. It is imperative, therefore, that you understand all aspects of this undertaking and that you prepare yourselves not only in a technical military way, but on a far broader basis in order to grasp the full meaning of the political, economic, psychological and moral factors involved.

Following World War I, men tried to formulate a method of ridding the world of the curse of war. The League of Nations, in the opinion of many, could have accomplished this. But men could not agree on ways to "put teeth" into the League and it foundered primarily on the reluctance to commit, ahead of time, armed forces to meet armed aggression. World War II was the result.

Out of the chaos of World War II rose the concept of the United Nations. It envisaged the creation of a world instrument capable at least of giving to all its members security against aggression. Of course, it is imperfect, as all human creations are, but still there is no better instrument at hand for preventing another world war. The world has again been rent by a counterfeit philosophy which purports to offer freedom, but which is really based on slavery and the complete subjugation of the individual to the state. The free world, holding to the dignity of the individual to be supreme and seeing the gulf widen, found it necessary to meet the challenge by adopting additional security measures.

The North Atlantic Treaty Organization is one of these measures. In it lies a tangible assurance of collective strength which, we hope, will deter an aggressor from launching war. The security of Western Europe is vital to the security of the United States, and the defense of it is, in effect, a defense of the United States – a factor which is often overlooked. The importance of the productive capacity of Western Europe should not be underestimated. If Western Europe falls under communist domination, the free world will lose the industrial efforts and scientific skills of more than 200 million people. The coal and steel of the Ruhr, the Saar, and the Lille areas would be lost, and the economic scales might be tipped against us. The free world now outproduces the communist nations approximately 3 to 1. But if Western Europe were lost, Soviet productive capacity would exceed that of the free world.

These are important facts to the soldier, as well as to the economist, and he must be the first to recognize that the true foundation of modern armed strength is great industrial capacity. This is a fundamental relationship which the young officer today must learn quickly.

During the past ten years, I have had to make many trips to various parts of the globe. I never return to the United States without a deeper impression of the fact that an empty stomach is more likely to cause conflict than a loaded rifle, and is even more likely to turn despairing people to communism.

Greece is an example of the necessity for a close relationship between our military and economic programs. Here was a country trying to rebuild its economy in the very shadow of the Iron Curtain without military strength to preserve that economy. Here is a country to which the rest of the world can look for an example of the effectiveness of coordinated military and economic aid in enabling free nations to withstand aggression.

Prior to our entering the picture, the Greeks could not cope with the guerrillas. The farmers of Greece could not gather their crops and the Greek people were not getting enough to eat, because just as soon as the grain was ready for harvest the guerrillas would swoop down from their mountain strongholds and seize it. What they couldn't seize they destroyed.

The Greek government did not have the means to strengthen their armed forces so that they could put an end to the guerrilla menace. Yet without adequate military security the people of Greece could not produce either the crops or the goods which would, in turn, produce the revenues necessary to maintain the troops. Thus a vicious circle was established which could result only in those chaotic economic conditions which are so favorable to the spread of communism.

Our combined economic and military missions, under the wise and efficient leadership of Ambassador Grady and General Van Fleet, finally brought an end to these conditions. The Greek people, with our military and economic assistance, rose, as with their ancient might, defeated the guerrillas, and then set about solving their internal problems unhampered by the constant menace of active communist military forces within their very borders.

These results could not have been achieved if our American representatives, military and civilian alike, had not shown good judgment, restraint, patience, and a thorough understanding not only of Greek sensibilities but of the dovetailed relationships between military and economic aid.

Somewhat similar conditions existed in Turkey, although here the threat of Soviet armies was closer at hand.

Again our purpose was to assist in creating efficient, well-balanced modern armed forces with the capability of resisting aggression, without imposing an unbearable drain on the Turkish economy.

Here, too, our American personnel had to develop a broad understanding of the intricacies of Turkish life and customs, and an appreciation of the fine balance between military requirements and economic capabilities. How well this is being done was brought home to me recently when a working newspaperman went out of his way to tell me personally his reactions to the job our military mission is doing in Turkey.

He had been traveling in Turkey and had visited some Turkish outposts where our young officers and enlisted men were assisting the Turks in training. He said that he happened on a small group of American officers in a remote town. These men not only had the confidence of their Turkish comrades-in-arms but they had also earned the trust and faith of the villagers by their high standards of personal conduct and their sympathetic approach to existence in that Turkish town, which had few if any of the conveniences of modern life.

I saw similar evidences of the success of our missions in other countries while on my recent trip around the world.

Those Americans working for our government abroad – both military and civilian – are invaluable in assisting us to gain an insight into the cultures of other peoples, and they to ours; and they are fostering an ever-growing mutual understanding between Americans and all freedom-loving people.

Some of you young graduates and many of you who are about to graduate will be taking their places. Today, almost half of our forces are overseas and it has never been more necessary that the comportment of every man in uniform be above reproach.

I know it may seem trite to mention the importance of every officer and soldier overseas being an "Ambassador of Good Will" for his country, but nevertheless it is true. More damage to the cause of freedom could be done by one mistake of judgment on our part than by a thousand propaganda acts or threatening gestures by those who deliberately plan them.

Taken together, all aspects of our national security are unavoidably intricate but the rebuilding of our own Armed Forces is a vastly complex problem by itself. We are on the threshold of new developments that challenge the imagination and ingenuity of the most optimistic planners. Guided missiles, atomic weapons, jet aircraft, powerful new ammunitions, and many other advances are all creating a new framework of defense in which the technical sciences play a greater role than ever before.

The man in uniform today has need to consult continuously the man of science. And similarly the scientist must keep close touch with the soldier who has to use new weapons. Theory and practice – vision and reality – must be combined in the best interests of the country's security.

The military man must satisfy himself that a new weapon will do the job in the manner necessary without an inordinate expenditure of the nation's resources, including productive manhours, raw materials and sheer dollar cost. On the other hand a new weapon or concept should not be accepted or rejected solely on a theoretical basis without careful examination of its battlefield potential through the eyes of experienced soldiers.

In all of this there is a danger that we may become so enthralled by machines and weapons systems, that we may lose sight of the fact that the man – the individual soldier – is the supreme element in combat. That is the reason why the foundation of our system of discipline is the same as the very foundation of our system of government: the preservation of the dignity of the individual.

Our Officer corps is dedicated to the belief that our high standards of character and integrity must be maintained. We recognize, honor, and preserve the dignity and identity of the humblest soldier. The most fundamental relationship in the Army is that which exists between the officer and the men entrusted to his care. We have perhaps the most democratic Army in the world. But its performance on the battlefields of Korea has clearly shown that it does not lack discipline. The fact that our discipline is a reasonable and not a rigid thing, accounts largely for the magnificent performance of our Army, because it does not dull the resourcefulness and initiative of the individual soldier.

These then are some of the varied relationships which an officer today must know. These complex problems are indeed a new and greater challenge to the young officer. They require of him a keener imagination, a more vigorous mentality, and a broader understanding than in the past.

But, above all else, you must bring to the solution of these problems the highest standards of character and integrity. It is in this field that West Point must continue to make its greatest contribution to the Army and the Nation. While machines and arms may be multiplied and changed, these essential human qualities remain changeless and priceless. However much we may treasure the technical knowledge, tactical skill, or academic foundation we may acquire at this Academy, its high standards of integrity are our most sacred heritage. This country has, in the past hundred and fifty years, come to know and expect such standards as the hallmarks of West Point.

So, as we rededicate ourselves this Founders' Day to 'Duty, Honor, Country,' let us be more resolved than ever to maintain those immutable concepts of character and integrity we learned as cadets at West Point.

In so doing, I am confident that we can do our part in carrying forward our nation's program of security and for peace.

General Crittenberger took the rostrum and thanked the Army Chief of Staff for his inspiring remarks and then played brief, taped, congratulatory messages sent from Key West, Florida, by president Truman, and from Washington by Secretary of Defense Robert A. Lovett.

The Cadet Glee Club sang the alma mater and General Crittenberger made brief, concluding remarks. Bishop Kinsolving gave the Benediction to end the festivities inside Washington Hall. Afterward, cadets and guests left the Hall, assembled at Thayer Monument and participated in memorial exercises honoring Colonel Thayer. Brigadier General Harris Jones, Dean of the Academic Board, gave the principal address and dedicated a bronze wreath permanently affixed to the base of the monument.

In his address, he gave a brief history of Thayer's life and influence on the Academy.

...Few educational institutions have been so profoundly influenced by one man as was West Point by Thayer.

A graduate of the Military Academy in 1808, Thayer served with credit in the War of 1812 and was brevetted Major for

Brigadier General Harris Jones, Dean of Academics, giving his dedication address at the Thayer Monument, Sunday, March 16, 1952. (Photo courtesy Signal Corps.)

distinguished and meritorious service. In 1815, he was sent to Europe by a foresighted War Department to study military education and organization and to observe the tactics of European armies. Upon his arrival in France he learned, to his professional disappointment, that the battle of Waterloo had taken place just two days earlier; – it was possibly the only time in his career that he was ever late to a military formation. He was much impressed by the Ecole Polytechnique, the great French school for military engineers whose memorial stands here as a neighbor to his own. Thayer's thorough study of its organization, methods of instruction and curriculum was to prove fruitful indeed in his next assignment.

In the spring of 1817, when he was not yet thirty-two years old and had but ten years of commissioned service, Captain (Brevet Major) Thayer received orders to return to the United States and to report for duty as Superintendent at West Point. Here he found that the neglect of his official superiors and the inadequacies of his predecessor had combined to produce a sorry state of confusion bordering upon disintegration.

...No doubt some of his policies and methods which served their purpose well in his day have gradually been modified or superseded with the advancing years. Nevertheless, it is quite extraordinary how many of the distinctive practices which make up the 'West Point System' as we know it were initiated by Thayer. To his genius we owe the establishment of the Academic Board, the Office of the Commandant of Cadets, the organization of the Corps of Cadets as a tactical unit. He introduced the 3.0 grading system and the demerit system (for which the present-day cadet may perhaps not thank him), the small classes sectioned according to ability, the regulated schedule for study, recitation, drill, recreation and sleep, and many another procedure which we now take for granted as standard, up-to-date doctrine.

But our admiration for these specific contributions, well though they exemplify his consummate skill as an organizer, administrator and disciplinarian, should not detract our attention from the broader basis of Thayer's claim to the title of 'Father of the Military Academy.' What made him a truly great educator was his devotion to three principles which still constitute the foundation upon which everything worthwhile and distinctive at West Point rests.

These are: first, character, – complete integrity and devotion to duty; second, a curriculum as broad as is consistent with the requirements of professional training; and third, insistence that each cadet exercise his own faculties to the utmost of his ability.

Sixty-nine years ago this June, General George Washington Cullum and a group of graduates unveiled this monument to Colonel Thayer. General Cullum presented the statue with these words:

> 'To Colonel Thayer, who has achieved so much for military science and the glory of his country; who was always true to himself and his trust; and who, with pride could point to the graduates of this Academy as the jewels and adornments of his administration, as did the noble Cornelia to her Gracchi sons...we come this day to offer in memoriam a similitude of his living self...may this monument, reared in loving gratitude to the 'Father of the Military Academy,' on this historic plain of West Point, ever stimulate the men here educated to win a like recognition for merit and patriotism.'

We meet here on this Founders' Day of the Sesquicentennial Year to acknowledge once again our indebtedness to this great leader. On behalf of the Association of Graduates of the United States Military Academy, I present this bronze wreath as an enduring indication of the rededication of graduates the world over to the ideals of the Military Academy: to Duty, to Honor, and to Country.

With the dedication of the wreath at Thayer's Monument, the Sesquicentennial Celebration of Founders' Day ended.

THE SESQUICENTENNIAL JUBILEE

The Sesquicentennial celebration's climax was on Tuesday, May 20, when hundreds of distinguished guests from all over the world gathered at West Point to participate in Jubilee activities. Spring football practice was over, track season was near an end, and the Corps' four classes were beginning final examinations. Plans were nearly complete for summer military training for the three new upperclasses, and entry into Beast Barracks for the class of 1956 was a month in the future. Two weeks hence was graduation for the class of '52, and in Korea the war raged on.

Numerous college presidents and other prominent educators, representatives of the Diplomatic Corps, senior officers of the Armed Forces, and alumni of the Military Academy were among the delegates to the Jubilee. The President of the United States; the Secretaries of the Army, Navy and Air Force; the Chairman of the Joint Chiefs of Staff; the Chief of Staff of the Army; and the Commanding General of Supreme Headquarters, Allied Powers in Europe (SHAPE) were among many government officials attending. Foreign dignitaries included ten ambassadors; fifty-eight military, naval, and air attaches; the commandants of twenty-three foreign military academies; and representatives of many colleges, universities, and cultural institutions.

Guests began arriving Monday afternoon. Officers welcomed them at the registration and information center established in Cullum Hall, and escorted them to the ball room where cadets completed their registration. Charles J. Barrett, head of the Foreign Language Department, ensured foreign delegates were provided interpreters if needed.

Cadets individually briefed guests on the Jubilee program and gave each a kit containing tickets for all events, a Sesquicentennial medallion, and information on the Jubilee and the Military Academy. Guests also received housing assignments for the duration of their visit to West Point, residing in either the Thayer Hotel or at the quarters of officers and their families who volunteered to offer accommodations. Guests arriving Monday afternoon received invitations sent by the superintendent, to attend an informal buffet supper that evening at the West Point Army Mess.

Plans called for an outdoor convocation at the natural amphitheater on Trophy Point. Delegates were to assemble in the north area of cadet barracks, put on their academic robes, and march through the east sallyport, across the Plain between two ranks of first classmen, to Trophy Point. A large speaker's platform had been erected in the amphitheater, and seating arrangements made. Flags of the many nations represented in the audience, and of government officials and military officers participating in the convocation, were to decorate the amphitheater. The schedule included a retreat parade and review in honor of the president. Plans abruptly changed, however, altered by unseasonable weather.

Early Tuesday morning a cold, hard rain began falling on West Point. It continued unabated throughout the morning, increasing in intensity. At noon General Irving announced the inclement weather schedule would be followed.

The inclement weather schedule, though planned as an alternative, changed the convocation from Trophy Point to the Field House, indoors. Because there were preparations that could only be completed on the day of the convocation, redesignating the ceremony's location caused major, last minute changes for the Academy's service organizations supporting the Jubilee.

The heaviest load fell on the Academy's housing and transportation services. The fair weather plan called for delegates to walk from one event to another throughout the day. The rain necessitated delegates and their families be transported from Cullum Hall to Washington Hall for lunch, thence to the Field House for the convocation, and back to Cullum Hall. The weather also caused numerous changes in guests' travel plans, necessitating extended stays, and rearrangements of departure schedules.

Registration of guests at Cullum Hall continued Tuesday morning while final preparations progressed. Guided tours for guests already registered included sites of historical interest at West Point, with stops at a new Visitor's Information Center constructed especially for the Sesquicentennial, the Cadet Chapel, and Trophy Point.

At 10:30 in the morning the presidential party arrived by special train at the West Shore Railroad station. General Irving officially welcomed him to the Military Academy. The president and his party then proceeded to Trophy Point where Mr. Truman received a twenty-one gun salute and inspected the guard of honor. Next came a brief visit of the superintendent's quarters, followed by the president's tour of the post, including a stop at the Cadet Chapel, where Mr. Frederick Mayer gave an organ recital for the president. Mr. Truman, noted nationwide for relaxing by playing the piano, most notably *The Missouri Waltz*, responded by briefly playing the huge organ.

From the chapel the presidential party moved to Washington Hall, where they watched the Corps march in for the Jubilee luncheon. Guests joined the Corps for lunch, and the president sat at a table with nine cadets from Missouri: Cadets Harry L. Van Trees, class of '52, who on Founders' Day had laid a wreath at the tomb of George Washington; Robert F. C. Winger, '52; James H. Elliot, '53; Wallace W. Noll, '53; James A. Kriegh, '54; Jack R. Logan, '54; Louis C. Wagner, Jr., '54; Loomis L. Crandall, '55; and Robert V. Tompson, '55.

Following the luncheon, Mr. Truman spoke briefly, informally, his remarks directed toward his cadet audience:

General Irving, Distinguished Visitors, and the Corps of Cadets in West Point:

I am having a wonderful time today. I was telling my Missouri cadet friends here that the only ones who are suffering and having trouble are the ones who are my hosts. It is always a nuisance to have the President of the United States around. He has to be certain of special treatment which is for the Presidency and not for the individual. You must always bear in mind that the Presidency of this great Republic of ours is the greatest office in the history of the world. It is the most important office in the world and the man in it must do everything he can to cause all the people at home and abroad to respect that office for what it is. It came and was forced upon us because we did not want to assume the responsibilities. We refused to assume that responsibility in 1920, and the Second World War was the result. But beginning in 1938 and when Hitler went into Poland it began to dawn on the people at the head of the Government of the United States that we had a place in the world that had to be filled. We are trying our best to fill that place in the world. You young men who will be the future Generals and the men who will help form the military policy of the United States, have a responsibility which you will have to assume just as soon as you finish your education.

Now this is a great school. I was just telling my young friends from Missouri here that this school has produced some of the greatest men in the history of our country. This school has made a contribution that is one of the greatest in the history of the country. I am proud to be your visitor today on that account. I am just as proud, I must interpolate here, of the Naval Academy and its cadets, and I want to see you become the leaders and the citizens in our military setup that you should. General Bradley this afternoon is going to give you a lecture on what it means to be a graduate of this school and what your responsibilities are and what our military policy really is.

Now I did not intend to give you a lecture on citizenship and the Presidency of the United States, but I thought maybe you would be interested in knowing that the President himself, an individual like everybody else, must keep his eye on the ball in an effort to attain respect for the office that it deserves. It took me a long time to understand why people would come in and see me and be timid or scared and couldn't talk, and then I had to remember back when I was in the United States Senate and when I used to have to go and call on President Roosevelt once or twice a week. I was always scared and embarrassed because I was before the greatest office in the world and one of the finest men also, who ever lived and occupied it, and in order for me to understand how people feel when they come to see me, I have to remember my experience myself. I couldn't appreciate it for a long time, but I do now and I try my level best to make people feel that they do not have to be afraid of the President because he is only interested in the welfare of the whole country. He has nothing else to do but to see that the country runs as it should and to see that we keep our friends in the world so we won't have a Third World War and so you won't have to go and be cannon-fodder. I hope you remember that.

After the luncheon delegates proceeded to the Army Theater, where robing booths for groups of twenty-five delegates had been erected.

When the more than 430 delegates were robed, they boarded buses in a tunnel underneath the gymnasium and were taken to the Field House preparatory to the academic procession. At 2:00 PM, the Grand Marshall, Brigadier General Chauncey Fenton, president of the Association of Graduates, signaled the start of the first academic procession in the history of the United States Military Academy. When the procession began, the Military Academy Band's music echoed through the immense Field House, later accompanied by waves of applause from spectators as the distinguished educators in their colorful academic costumes and the armed forces officers and members of the Diplomatic Corps in brilliant uniforms made their appearance.

The convocation speakers, followed by the superintendent and the president, who would give the principal address, were the last to arrive in the procession. A hush fell over the entire audience when the president took his place on the platform. Then, General Fenton, the Grand Marshall, announced the convocation would begin.

After the national anthem, the Right Reverend John B. Walthour, Bishop of Atlanta and former Academy Chaplain, gave the invocation. General Irving gave a brief welcoming address, and then introduced the first speaker, Dr. Karl T. Compton, educator, scientist, and the president of Massachusetts Institute of Technology from 1930 to 1948. Dr. Compton gave a congratulatory address, centered primarily on the quality of the Academy education, and its relationship to the changing technologies of warfare. He had served on a board of consultants in 1945, when the Academy, emerging from World War II, was preparing to reconstruct and reform its curriculum, and expand from a war shortened, three-year to a four-year program. He praised the Academy education, quoting briefly from the board report:

> ...'The board was very much impressed by the enthusiasm of the teaching which they witnessed and with the evidence which was presented to them of the efforts which have been made by the Academic Board and the instructor staff to take every possible advantage of advances in the art of teaching through their own conferences and through examination of methods used at other educational institutions.'

He went on to say,

...In fact, in an earlier study, former President Ernest M. Hopkins of Dartmouth College and I, who were civilian educators on the Board, were so impressed, (and I may confess surprised), at the enthusiasm of the teachers and their eagerness to take advantage of every suggestion and opportunity for improving their effectiveness, that we wrote a joint letter to the Superintendent expressing our sincere admiration of this aspect of West Point's educational program. I mention this specifically because there has been some impression over the country that the teaching program at West Point is overstereotyped and inflexible. We found the situation to be quite the opposite and, as I say, we were so impressed that we supplemented our participation in the formal report of the board of consultants with this supplementary letter...

He concluded by telling a story to highlight the changing nature of technology, warfare, and the military profession.

...In...illustrating the continuing development of the military profession, let me describe an experience which I had in England in 1943 when I was a member of a radar mission sent over by the Joint Chiefs of Staff. One week-end during our stay I visited my old professor and friend, Sir Owen Richardson, of Nobel Prize fame for his researches on the emission of electricity from hot bodies. He had moved his family and his more valuable scientific equipment from London to the greater security of his country home at Alton, Hants. His farmhouse was beside an old Roman military road. At intervals along this road were movable concrete road blocks for use in case of an invasion. In the fields beside the road were wire structures and entanglements as precautions against airplane or glider landing of enemy troops. Flying overhead was an almost constant stream of airplanes on their missions to and from their objectives on the Continent.

Civil defense orders had gone out to all the property owners to dig bomb shelters, and several workmen were digging an underground shelter beside Richardson's home. In the excavation they uncovered a lot of chunks of iron about as big as one's fist and not much more accurately spherical. These were cannon balls from an ammunition dump of Cromwell's army, which in that neighborhood had fought its last battle.

I could not but be impressed here by the contrast. The old Roman road, built by invading Roman legions, the round cannon balls of Cromwell's army, and every evidence of modern technical weapons and methods of warfare, all are visible in this one spot.

This impressed upon me the increasing complexity of warfare, along with the increasing complexity of so many aspects of modern life. The original military schools trained in formation drill and taught military engineers how to build highways and bridges. The modern military school has a far more complex program, even though its basic objective is essentially the same....

When Dr. Compton completed his address, General Irving announced the Cadet Chapel Choir would sing *The Corps*. When the choir completed its rendition, General Irving introduced the next speaker, Chairman of the Joint Chiefs of Staff, General of the Army Omar N. Bradley, class of 1915. He spoke of the American soldier, leadership, and the leadership American soldiers expected of their officers:

...The principles of military action which are instilled here...[at the Academy]...allow Americans to exercise their ingenuity and their imagination. Beyond the leadership and the heroism is the all-American loyalty and patriotism – the courage, valor, and self-sacrifice that is typical of every generation in this country since the first settlers founded the first colonies.

When the chips are down, it is to the American soldier that we turn for our credo of military leadership. The greatest soldier in the world could never win a campaign unless he had the kind of leadership in his make-up which best suited the men he had to lead. Certainly, the 'Prussian-type' of discipline would not win the cooperation and respect of young Americans. Nor would the political commissar supervision of the Soviets ever encourage an American platoon to fight the kind of battles that the first battalion into Korea had to fight.

Americans need the kind of leadership that is typical of a growing, colorful nation, made up of honest, patriotic, and rugged individualists. It is from our 'G.I. Joe' of World War II, and his long line of valorous ancestors, that we can derive our credo of military leadership.

From my experience with the American soldier, I believe that what he wants in a leader sums up about like this:

First of all, he wants a leader who knows his job. The American soldier is a proud one, and he demands professional competence in his leaders. In battle he wants to know that the job is going to be done right, with no unnecessary casualties. The non-commissioned officer wearing the chevron is supposed to be the best soldier in the platoon, and he is supposed to know how to perform all the duties expected of him. 'G.I. Joe' expects his sergeant to be able to teach him how to do his job. And he expects even more from his officers.

Of course, he wants a leader who is fair. When the battle missions are tough and dangerous, he wants them assigned as evenly as the battle situation permits. Actually, he demands that his company and his platoon get their just share of the tough assignments. He wants to do his job.

If there are privileges, he wants them parceled out as evenly as possible. He basically despises any favoritism on the basis of race, or color, or creed. He recognizes that the responsibilities of leadership are often rewarded with some of the privileges of accomplishment. But he doesn't want those privileges abused.

'G.I. Joe' also demands that his leaders be energetic and forceful. Even though he may be willingly lazy himself, he wants to see his leaders on the job. Then he'll do his part, under the most trying of circumstances.

The American soldier expects to be well-supplied and well cared for. In the city or on the farm, American boys have seen the good things of life provided in what we might consider a routine fashion. When a nation is joined together in a super-human effort for defense, he expects super-human results in the line of weapons, ammunition, food, and other equipment. He expects good leadership to include good planning and supply. At the same time, when difficulties of the situation are apparent, and he has to make sacrifices, he is tolerant of the lack.

I was always conscious of the fact that the higher in command that an officer rises, the more faith and trust his soldiers have to place in him. No soldier dares to ask the division commander whether the attack is well-planned. No private asks the regimental commander if his battalions are properly supported with artillery fire. A great deal of trust is placed in the leadership of an American fighting unit by its men. To meet this trust, an American leader must be well-trained, well-disciplined, and professionally competent.

Finally, and perhaps most important is loyalty. On this factor alone, battles are won or lost. To be really effective, loyalty must go three ways: up, down, and sideways. Loyalty to every man who is in command demands a serious effort to do what he orders, and even to carry out what you believe would be done if he hasn't ordered it.

Loyalty on the part of the leader, to the men whom he leads, is the only adequate repayment for their loyalty to you. This means an understanding ear for their needs and complaints, an honest answer to their questions, and an honest and courageous presentation of their just needs to higher authority.

Loyalty sideways is voluntary cooperation with the man on your right and your left. This is the real fiber of American teamwork. This is the principle of democracy – do for him what you expect him to do for you – that works in battle when all else fails.

Fairness, diligence, sound preparation, professional skill, and loyalty are the marks of American military leadership.

If every graduate of West Point – and every young officer in the service – gives the American soldier the inspired leadership that he deserves, the United States Military Academy will always be worthy and respected.

I could wish nothing greater for the Academy and its graduates on this occasion of its Jubilee anniversary.

At the conclusion of General Bradley's address, General Irving announced the Academy Band, under the direction of Captain Francis E. Resta, would play the *Sesquicentennial Fantasy*. Next, to meet broadcast timing needs for President Truman's address, the superintendent told the audience the program had to be modified slightly and announced the singing of the alma mater.

When the singing ended, General Irving introduced President Truman, who addressed the convocation. The address began slowly, with lightheartedness and good news for the Corps of Cadets. Then came serious, brief, but optimistic reports to the nation on the status and American conditions for armistice negotiations on the Korean peninsula, aims of American economic and military policies abroad, the threat Western nations faced, mutual security policy, and the need to continue strengthening the United States' armed forces. But when he came to the end of his address, there was much more for Americans and West Point cadets.

General Irving, General Bradley, Dr. Compton, Honorable Secretaries of the Army, Navy, and Air Force, and distinguished guests:

I want to make a statement, just two short statements before I start my regular talk. I had luncheon at noon with nine Missouri cadets and I want to say to them – and I am saying it very publicly – that I have not had a more pleasant luncheon in many a day. I appreciate it very much.

I have another statement to make. You know the President has several official positions in which he works. He's President of the United States and Commander-in-Chief of the Armed Forces in the United States, and he is the social head of State, and he is head of his party. I am going to work in two of those capacities right this minute. Under the Constitution, the President has the power to pardon anybody of anything but impeachment, for he could not pardon himself. So exercising my authority, as President of the United States under the Constitution, I direct the Commandant of West Point to relieve all special punishment that is going on on Post today, and as Commander-in-Chief of the Armed Forces of the United States, I direct General Irving to carry out that order.

Now you know I am in a sort of position that Senator Barkley found himself in one time. I want it to be distinctly understood that I enjoyed immensely the two speeches that were made before me. They were wonderful. I hope I can make half the contribution that either of those wonderful men made. Now, Senator Barkley was the last on a program, and Senator Barkley likes to speak. He took his watch out and then he picked it up and put it to his ear and shook it, and some old gentleman out in the audience said, 'Senator, if it's stopped, there's a calendar behind you.' Well, you are not going to need a calendar because this is not that long and I hope that it won't be as boresome as you may anticipate it will be.

It is a real pleasure for me to be here today, and join in celebrating the establishment of the United States Military Academy at West Point a hundred and fifty years ago.

This Academy was started during Thomas Jefferson's first term as President. The United States at that time was relatively small and weak, and surrounded by dangers. We had just fought a limited and undeclared war with France to protect the freedom of our commerce and shipping. We were engaged in fighting another limited and undeclared war with the Barbary Pirates for the same purpose.

Jefferson, like Washington and Hamilton and other leaders of our young Republic, knew very well that a strong military establishment was vital to the preservation of American liberty. And these patriot leaders knew also that you cannot have effective military forces unless you have well-trained, well-prepared officers. They all knew how Washington had to struggle and experiment all through the Revolution to find officers who could take troops into battle and lead them to victory. That was why they wanted a military academy, as an essential part of a strong, permanent national defense organization.

But there was a great deal of opposition to starting a military academy in this country. It took twenty years of argument and persuasion after the Revolution was over before the Academy could be started. Now listen to this. And it was finally started largely because Jefferson took the position that if the Congress didn't authorize a military academy, he would set one up himself.

The argument over establishing a military academy was part and parcel of the argument over whether the United States should have a strong national defense. That argument has continued, of course, right down to the present day, and much of the debate after the Revolution is very, very modern. They are making the same old arguments today as were made about the military academy when Jefferson was trying to start it.

There were a lot of people in 1800 who said that a strong national defense would cost too much; that we couldn't afford it, and we ought to find some magic formula for achieving security without having to pay for it. That point of view is not only echoed today – it is loudly shouted in the newspapers and the halls of Congress.

Fortunately, these arguments did not prevail against hardheaded common sense of men like Jefferson. The Military Academy was set up; and this country has had occasion to be thankful many times since then that our early leaders had so much foresight.

The Military Academy has repaid this country many times over for every cent it has cost. We have learned from experience that, while it may be expensive to maintain a strong national defense, it is much more expensive not to have one. Time and again, we have allowed our armed forces to dwindle down to a fraction of what they should have been, and then we have had to pay enormously – in money and in lives – because of our lack of preparedness. And there are people right now who want us to relax and cut down on our defense program. They are just as wrong as thy can be. We must pay the cost of preventing a world war – or we will surely have to pay the immensely greater cost of fighting one.

The other fear of the early opponents of the military academy has also proved groundless. Our country has never become warlike or aggressive.

This is partly because our Constitution nailed down so firmly the principle of civilian control over the military. The most important means by which this was done by providing in the Constitution that the President, who is the civilian head of the Government elected by the people, shall be commander-in-chief of all military forces. Many Presidents, including the present one, have demonstrated that those words in the Constitution mean just what they say.

But, in addition to this, the spirit of our people has never been warlike. Our people came to this country to find peace and freedom. That is what we have always wanted. That is what we want now, and that is what our national policy is designed to preserve.

But there is a vast difference between being peaceful and being passive. We want to achieve peace. But we know we can't have it unless we are willing to stand up for our rights.

We know we can't have lasting peace unless we work actively and vigorously to bring about conditions of freedom and justice in the world. That is what we are trying to do. And we are having to do it in the face of a concerted campaign of threats and sabotage and outright aggression directed by the Soviet Union.

The policies of the Soviet Union are exactly the opposite of our own. We want to establish equality and justice and the rule of law among all nations. They want to establish domination and dictatorship and rule of force over all countries. The leader wants physical control of the individual and also control of his soul. This makes our situation – the situation of all free nations – difficult and dangerous in the extreme. But I am firmly convinced that it does not necessarily mean a third world war.

The free countries can, by proper and adequate defense measures, make clear to the Kremlin that aggression would be doomed to failure.

And the free nations can, by economic and political means, build up their strength so as to be safe from communist infiltration and subversion.

But strong and active as we may be, we cannot avoid the risks and sacrifices. They are inherent in the situation and we cannot wish them out of existence. The course of events is not completely in our control.

In Korea, we had no choice but to meet armed aggression with military force. If we had not met aggression head on, the United Nations Charter would have been reduced to a scrap of paper. If communist aggression had been allowed to succeed in Korea, the communist conquest of all of Asia would have been simply a matter of time. If the United Nations had failed, and Asia had fallen, we would have been well on the way to the disintegration of freedom in the whole world.

But that did not happen. The valor and sacrifices of United States fighting men – together with the forces of the Republic of Korea and contingents from 15 other countries – has beaten the aggressors back within their own territory. Our Army, led in large part by men trained here at West Point, has done a superb job. From the time our men were first sent into action, in the gallant rear guard defense down to the Pusan perimeter – from then right on up to the present the United States Army in Korea has been magnificent. And the men who have fought with them, from the Air Force, the Navy, and the Marine Corps, and from the armed forces of other free countries, have been just as brave and effective.

Last June, eleven months ago, the badly battered communists offered to confer about a military armistice in Korea. We were willing to conclude such an armistice. We still are. We don't want any more fighting than is necessary. But we were not interested, and we are not interested now, in any armistice that involves selling out the principles for which we are fighting.

Patiently and skillfully, General Ridgway and his negotiating team, headed by Admiral Joy, have worked to bring about an effective armistice. They have done a masterful job in the face of great provocations. They have met the threats, and abuse, and outright lies, all with great self control and an unyielding insistence on the essentials of a just and honorable armistice.

Gradually, the communists have come to realize that we will not sacrifice our principles to obtain an armistice. We do not know whether they will finally agree on an honest and workable armistice. So far, they have agreed to some of the points that must be covered. They have agreed that the armistice line across Korea must be a defensible military line determined by the location of opposing forces. They have agreed that no reinforcements shall be brought into Korea by either side during the armistice. They have agreed that an inspection commission shall observe the carrying out of the armistice terms – and are apparently willing to withdraw their request that the Soviet Union be one of the mutual inspecting nations.

Up to now, however, the communists have not agreed on a fair and proper exchange of prisoners of war. The communists have continued to insist that all the prisoners we have taken must be handed over to them – regardless of whether or not they are willing to be sent back behind the Iron Curtain, and regardless of what their fate would be if they were sent back.

It is perfectly clear that thousands and thousands of the prisoners we hold would violently resist being returned to the communists because they fear the slavery or death that would await them. It would be a betrayal of the ideals of freedom and justice for which we are fighting if we forced these men at bayonet point to return to their ex-masters. We won't do it. We won't buy an armistice by trafficking in human slavery.

We do not know whether the communists will accept that position. We may not know for some time yet. Negotiations are continuing under General Clark's direction. We shall remain ready to reach honorable settlements by peaceful means. But we must also be alert and ready to meet treachery or a renewal of aggression if that should come.

During these months of armistice negotiations in Korea, the communists have increased their military strength. They have more men there than they had a year ago, and many more tanks and planes.

But we have consolidated and increased our strength in Korea also. The morale of our men is high, and our units are well trained, well equipped, and at a peak of combat efficiency. The troops of the Republic of Korea are far better trained and equipped than they were a year ago, and are capable of carrying a much larger share of the defense of their country.

The situation in Korea is still difficult and uncertain. Everybody should understand that. But everyone should also understand that the sacrifices of the United Nations in Korea have brought tremendous gains toward a world of law and order.

The plain fact is that the communists have utterly failed in their objectives in Korea.

The communist aggression has failed to shatter the United Nations. Instead, the communist attack has made the United Nations stronger and more vigorous and has demonstrated that it can and will act to defend freedom in the world.

The communists failed to win a cheap and easy victory in Korea. Instead, they have suffered more than a million casualties, and have used up enormous amounts of war material – and they are back behind the line where they started.

The communists failed to establish tyranny over the Republic of Korea. Instead, the communist aggression has brought devastation to North Korea – a terrible warning to other satellites in the Soviet empire of the cost of aggression.

Furthermore, the communists failed to break the will of free men in other countries. The attack in Korea was supposed to warn other countries that they must yield to the demands of the Kremlin – or else. The communist aggression did show the world that the Kremlin was ready and willing to try to extend its power by military conquest. But the effect of this was not to send the free countries into a panic of fear. Instead, they immediately stepped up their plans for building military forces, and began to get together on concrete and definite defense arrangements.

As a result of the resistance to communist aggression in Korea, the Kremlin knows that free men will stand up and fight against aggression. As a result of the resistance to communist aggression in Korea, free men around the world know that if they stand up for what is right, they will not be deserted by the United Nations. And, as a result of resistance to communist aggression in Korea, the free countries are infinitely better pre-pared to defend themselves than they were two years ago.

Our own defense production has risen sharply. Our produc-tion of military supplies and equipment is more than three times what it was a year ago. For example, in January 1952, six times the dollar value of ammunition was delivered as was delivered in January 1951. In electronics and communication equipment, five times as much was delivered.

The production of one of our most important fighter planes was four times as much this spring as it was last. We now have several thousand tanks of a new model which is very much better than previous models. Our Navy has taken a hundred ships out of mothballs and has a sound shipbuilding program underway.

An atomic artillery piece has been developed and tested and will have to be reckoned with in the future. The Navy is work-ing on its first atomic powered submarine. Our over-all atomic production program is in excellent shape.

In all the vast complicated field of combat vehicles and mili-tary weapons, the research and preparation of the last several years is paying off. The goods are being delivered to the hands of men who are ready to use them in defense of freedom – both in our own forces and among the many trusted friends that we have all over the world.

The improvement in defense production is not the only indi-cation of an improved situation in the world.

In the Far East, Japan has rejoined the family of free and democratic nations. The communist insurrection in the Philip-pines has been brought under control. In Indo-China, the forces of France and the Associated States have succeeded in holding the communists in check. The people of Indo-China are mak-ing progress in the creation of national armies to defend their own independence. Countries like India and Pakistan and In-donesia are making real headway in creating the conditions of economic growth that must underlie solid and stable progress.

In Europe, great steps toward unity are being taken. The Schuman Plan and plans for the European Defense Community are moving forward. We are working to reach final agreement on a new relationship with the Federal Republic of Germany. This will make it possible for Germany to take her place alongside the other independent countries of Europe as a full and equal member of the community of nations.

These are very remarkable developments. Countries like France, Germany, and Italy, Belgium, Holland, and Luxembourg, with centuries of rivalry between them, are now starting to work together. They are developing common economic and political institutions; they are merging their military forces into one great defensive system.

No wonder the Soviets are trying to block this advance. No wonder the current communist propaganda line is trying to persuade the countries of Western Europe they should stay separate and weak, instead of joining together for strength. The Kremlin knows as well as anyone else that in union there is strength – and that a united Europe can frustrate the Kremlin's dearest wish of absorbing the European countries one by one into the Soviet empire.

I don't think people of Europe are going to be fooled by this Soviet propaganda. I believe the firm and concrete steps the Europeans have already taken, over the opposition of the Kremlin, are clear indications that they are not going to be stopped now. I think the Europeans are going to continue to move toward closer union – for they know that is the way of strength and progress for them and for the whole free world.

I have been speaking of the progress that is being made. But I don't want anyone to get the impression that there is any basis for relaxing or letting up. These signs of progress are not evidence that the battle for freedom is won – only that we are on the way to winning it. If we halt or falter now, we could ruin the whole structure of peace and freedom we have been so painfully building.

I have warned the Congress, on several occasions, that the financial support I have requested for our defense effort and for the mutual security program is absolutely necessary. Any cuts in those items would have extremely serious effects. No one enjoys bearing the heavy costs of national security in these dangerous times, but we should never forget how much smaller they are than the costs of another war.

No one should assume that the possibility of world war has become remote. The forces of the Soviet empire are large, well-trained, and equipped with modern weapons. The Kremlin's desire to dominate the world is obviously unchanged.

But I believe we are well on the way to preserving our freedom without paying the frightful cost of world war. We are on the right track. We must go ahead.

If we are to succeed, we must have steady nerves and stout hearts. There is no easy way out, no quick solution. But we have with us the overwhelming support of the free countries, and the powerful moral forces of liberty and justice. We are using the strength God has given us in this great and wonderful Nation to win the struggle for peace and freedom throughout the world.

The young men here at West Point are called on to play a great part in the tremendous effort we are making. You are being trained for a career of which, in these times especially, means service for the great good of your Nation and the welfare of mankind. Your opportunities are great because the task ahead of you is great.

We need – all of us – to draw on the wonderful tradition of resolution and courage which has been cherished for 150 years in the life of the cadets here at West Point.

When applause ceased, Mr. Truman uttered the brief words, "Thank you," and called General Ridgway forward for a surprise presentation of the Distinguished Service Medal. "I would like to speak with General Ridgway a moment please. Would he step up here?"

After General Ridgway came forward, an officer stepped to the microphones and read the following:

The President of the United States of America, authorized by the Act of Congress of July 9th of 1918, has awarded the Distinguished Service Medal, Second Oak Leaf Cluster, to General Matthew B. Ridgway, United States Army, with the following citation:

'General Matthew B. Ridgway, United States Army, has distinguished himself by exceptionally meritorious service to the United States and the free people of the world in positions of great responsibility. At an extremely critical period he assumed command of the United States Eighth Army and of the United Nations Forces in Korea; and through magnificent personal leadership, led these forces in a counter offensive which crushed the Communist advance and drove the enemy north of the 38th Parallel. In April 1951, General Ridgway became Commander-in-Chief of the United States Forces in the Far East and Supreme Commander of the Allied Powers in Japan. In addition to directing United Nations strategy and guiding their Armistice negotiations in Korea with skill and firm forbearance, he supervised on behalf of the Allied Powers the final stages of the rebirth of the Japanese people as an independent nation. In these grave responsibilities he displayed the highest order of physical and moral courage, skillful leadership, and broad understanding. General Ridgway's extraordinary service merits the gratitude, not only of the American people, but of the free peoples everywhere. Signed by the President of the United States.'

"Now General if you will step around front here I will pin this medal on you."

After the pinning, Mr. Truman said, "General Ridgway wants to say a word. I hope you will listen to him."

Mr. President:

Consummate consideration, I believe, was never more evident than in this thoughtful act by you, sir, the Commander-in-Chief,

President Harry S. Truman congratulates General Matthew B. Ridgway upon award of the Second Oak Leaf Cluster to the Distinguished Service Medal. (Photo courtesy Signal Corps.)

in awarding this high decoration in the presence of my wife, my comrades, and this notable assemblage at this great national institution which has served our nation so faithfully for a century and a half. I am grateful beyond words.

General Irving next presented the president with a memento of his visit to West Point.

Thank you very much Mr. President, for those inspiring words. Those of us who have been here and who have been privileged to hear them will remember them for a long time.

Mr. President, it is indeed an honor and a privilege to have you with us today. For many years, the Chief Executives of the United States have visited West Point. George Washington visited here many times while he was Commander-in-Chief. James Monroe visited West Point just before he appointed Sylvanus Thayer as superintendent. Almost all Presidents since Abraham Lincoln have come here at some time during their term of service.

On behalf of the Corps of Cadets and the officers and graduates of the United States Military Academy, I take great pleasure in presenting you, as a memento of your visit, this portrait of our first graduate, Joseph Swift. I hope that in the years to come this portrait will remind you of this Sesquicentennial Jubilee Convocation.

Mr. Truman responded:

Thank you very much General.

This gentleman graduated in 1804. He is no longer with us. He was Superintendent of the Academy, so I understand, for four years, just before Thayer was made Superintendent. He made a contribution to the beginnings of this great institution and I am certainly happy, pleased, and proud to possess his picture because he is one of the famous men in our military history. Thank you very much, General.

General Irving introduced the Army Chief of Chaplains, Major General Roy H. Parker, who gave the benediction. The Convocation recessional followed.

Due to the pressure of government duties, the presidential party returned to Washington. Other official guests were taken to the Hotel Thayer or to officer's quarters for a brief period of relaxation, the first provided them in the Jubilee's full schedule. They assembled once more, at six thirty in the evening, in Cullum Hall, for a reception in honor of the many delegates. After the reception official guests returned to Washington Hall for the final activity in the Jubilee, a banquet.

Many guests had brought members of their families with them to the Jubilee, but Washington Hall was not sufficiently large to accommodate families of official guests at the luncheon or banquet. Thus family members were hosted in the West Point Army Mess at the noon luncheon and in the evening for dinner.

In Washington Hall for the banquet, most officers of the garrison were present, and members of the class of 1952 acted as table hosts for educators, government officials, ambassadors, armed forces attaches, heads of foreign military academies, and Army and Air Force generals.

As for the inaugural in January, there was a large dais just inside the center doors of Washington Hall. Seated on the dais were speakers for the evening, several senior officers of the United States Armed Forces, and members of the Sesquicentennial Commission. Secretary of the Army Frank Pace, Jr., was to give the principal address, and representatives of the various categories of institutions participating in the Jubilee were on hand to deliver greetings to the Military Academy on behalf of their respective groups.

The speakers were: Dr. George D. Stoddard, president of the University of Illinois, of the colleges and universities of the United States; Dr. Robert M. MacIver, noted sociologist, for the colleges and universities of other nations; Vice Admiral W. Hill, superintendent of the United States Naval Academy, for the United States Armed Forces Schools; Brigadier General Nestor Souto de Oliveira, Commandant of the Military Academy of Agulhas Negras, Brazil, for the foreign service academies; Dr. Detlev W. Bronk, president of John Hopkins University, as well as the National Academy of Sciences and the American Society for the Advancement of Sciences, for the learned societies of the United States; and General Marie-Pierre Koenig, distinguished French military leader and scholar, for the foreign learned societies.

General Irving's closing remarks that evening ended the last of the Sesquicentennial's major events. There was much more to the celebration, which added luster to the Academy's one hundred fifty years of history and tradition. But the four events, taken with the Sesquicentennial lecture program, were extraordinary and memorable, particularly for men in the Sesquicentennial class of 1952 who had participated in planning numerous activities to make the entire celebration a success.

Exactly two weeks hence, on June 3, the class of '52 graduated. Some among their number would be the last from within a lengthening list of graduated classes to fight in Korea.

There was a postscript to the celebration. On February 22, 1953, a month after President Truman left office, The Freedoms Foundation honored the United States Military Academy Sesquicentennial with a special award. A jury of distinguished public figures cited the Sesquicentennial "for its significant contribution to the American way of life." The Vice President of the United States, the Honorable Richard M. Nixon, presented the award in an impressive ceremony at the Foundation's headquarters, Valley Forge, Pennsylvania. The accompanying citation read:

> A special Honor Medal has been struck by our distinguished awards jury for the United States Military Academy at West Point, N.Y. Their Sesquicentennial observance program stimulated a national consciousness of the values of duty, honor, and love of country, so today Freedoms Foundation salutes Academy superintendent Major General F. A. Irving, and the West Point Cadets whose integrity and love of country make them the proper heroes of American youth.

PART THREE

TRIUMPH

CHAPTER 14

TO BUILD A TEAM

While the Academy celebrated its renowned graduates, rich traditions, and 150th birthday in the first half of 1952, spring football practice didn't bring much inspiration to Earl Blaik and his coaching staff. All were still reeling from professional and personal loss suffered in the destruction of the Army football team just prior to the 1951 season. Earl Blaik's loss was far more saddening because of the cheating scandal's effects on his family.

In addition to seeing his son resign along with eighty-two other cadets, he had lost every single appeal for leniency and the young men's retention in the Corps of Cadets. He took his appeal public in the August 9, 1951 press conference in Mama Leone's Restaurant in New York City, vigorously defending their individual character, while behind closed doors he battled furiously for disciplinary actions short of dismissal. His public and private stances, and perceived overemphasis on winning football, had brought a storm of Congressional and public criticism, sharp words from many of West Point's faculty and staff members – behind his back, and condemnation among a large number of vocal Academy graduates who saw him as primarily responsible for the scandal.

Though he professed profound admiration for the Academy and its purposes, the criticism and his repeated setbacks in appealing for leniency went against his grain, and frustrated him enormously. "Winning," that's what counts. His attempts to bury his emotions in hard work weren't always successful. His frustration, though muted, was a festering birthplace for bitterness, which couldn't always be contained.

Then had come the disastrous '51 season, when Army football fortunes sank as low as they did in 1940, the year before General Eichelberger brought Blaik back to West Point to revive Army football. Only his endless drive for perfection, and winning, and General MacArthur's and other friends' counsel gave him the will to persist in rebuilding. Football would be his outlet, his means to submerge the criticism, disappointment, frustration, defeat, and sadness he endured. In spite of all he felt, and the periodic setbacks he encountered, his analytical mind always came back to the game he loved.

In spring practice Blaik and his staff surveyed the available talent and recognized there was no team depth. They didn't see prospects for the '52 season to be much better than '51. The returning lettermen had little to their credit but a year of varsity experience. But there were other additions: a bit more maturity, a few more pounds of growth, some added strength, and better skills. Nevertheless, spring practice confirmed what Blaik already knew. It became clear that the best course was to continue the two platoon system, with a small number of more skilled players going both ways – offense and defense. He believed there simply weren't enough talented men to sustain a full two platoon system. He later altered his views.

After spring practice, and as June Week approached, Blaik's mood changed once more. The bitterness he suppressed during spring practice resurfaced. He was reminded again of what might have been.

When he wrote his confidant, General MacArthur, on June 3, 1952, it was graduation day for the class of '52, the day his son Bob would have graduated. Earl Blaik poured out his emotions.

Dear General MacArthur:

This is a sad time for me as graduation cruelly emphasizes the fact that so many good men, including my son Bob, were sacrificed to a dogma of cultism in its narrowest form.

Our pride is great in the young graduates who have brought credit to themselves and West Point in combat duty, even though we know full well that among this group are the known men who originally trespassed on the honor code and indoctrinated others. Yesterday we listened to the platitudes of the speakers, even though we knew full well that the restricted investigation could not cover the fact from either cadets or officers that two classes were clean only because the cursory and shallow inquiry failed to uncover scores of offenders.

Today, unknowingly, they will applaud others among the graduating class only because these young men lacked the courage to be truthful.

And with it all we have some in authority who would direct that all reference to these departed men be expunged from all Academy records – 'to us they don't exist.' How is it possible that in our Army men can reach high authority with so little understanding of human nature and be completely devoid of reasonable sense? Perhaps I am too bitter, but I abhor the hypocrisy of our present Chief of Staff and others who extol the virtues of honor only to give lip service to it on occasions which suit them best...

Please give Mrs. MacArthur my best wishes and remind her that if at any time young Arthur would like to come to West Point for a few days to swim and other fun we shall be happy to have him with us. Bob will be home tomorrow to remain until July when he will be drafted. He is grand with youngsters.

I know that you are very busy but should you at any time want the latest on our football situation I shall be happy to give you the usual briefing.

My very best to you and with devotion

Respectfully,
Earl H. Blaik

On June 9 General MacArthur replied.

My dear Earl:

I have just finished reading your note of June 3d and can appreciate the poignancy of the situation at the Point this Graduation Week. Life deals some strange blows, but the resilience of human character is designed to absorb them without sustaining destruction...

I would be delighted, if not too much of a task upon you, to have the estimate for next season. It will undoubtedly be a rough one for you as the results of last year will not be entirely eradicated until the end of the four year cycle. At that time I am sure you will come into your own again.

My best to you, Earl, as always,

Faithfully,

DOUGLAS MacARTHUR

MacArthur's reply undoubtedly buoyed Earl Blaik. By the last week in June his mood changed again, noticeably. The change was evident in the opening paragraph of his June 24 letter to the Old Soldier, in which he provided his annual football situation report. It was Blaik's typical, thorough, dispassionate analysis of Army's prospects for the 1952 season, but in his first words to the General his enthusiasm for the game of football and the upcoming season was evident, in spite of the fall campaign's bleak outlook.

They are mowing the stadium grass today and the air has that certain scent of autumn which reminds all football addicts that the season is just around the bend. One of the penalties of city life is the absence of nature's reminders, so this letter on the '52 season is a less romantic way of your learning that soon another football season will be with us.

Then followed his analysis of prospects for Army's success.

There are four men who must carry the offense this fall as they represent the only real offensive talent on our squad: They are: Meyers, Vann, Ordway, and Lunn.

By far our best player is Meyers, who has been shifted to LHB [left halfback], a position best suited for his talents.... As LHB Meyers will use the run-pass option play with exceptional skill and this maneuver alone should make our offense 20% stronger....

Peter Vann, a yearling quarterback, is a player with much natural talent as a passer, especially on long throws. He is, however, the most immature youngster I have ever had on an Army squad.... We shall start the season with Boyle and substitute Vann to his benefit at opportune times....

Lunn as an offensive guard has excellent ability...and his presence sets a visual example of what constitutes good performance....

Last fall we had but two centers and they no longer are with us. Fielding Yost often said that the center rush was the number one man to be selected on any squad, but only now have I ever been forced to realize how correct was Yost. During spring practice I selected six men as possible centers and with rare good fortune one of these turned out to be a natural. He is the son of Ordway of '25....

Our defense personnel is far below par. There is only one man of the entire group who normally would make a first West Point team. All priority was given to the defense; therefore this group represents the best available material. I would readily forego the two platoon conception of modern football if in so doing a better defense unit could be assembled. But without material potential our only hope to offset this fundamental weakness is to employ in our scheme of defenses variations, even unorthodox, which will keep our opponents off balance.

We shall miss the graduated class and since our plebes are ineligible, our overall strength is less than last year. The squad is essentially a junior varsity one, though the fact that all the players have experience makes the picture less perplexing than in 1951....,

I need not dwell too much on the teams which will be fielded by our opponents other than to state Southern California, Georgia Tech, and Pennsylvania each has been selected as the #1 team in its respective section. Beyond that our opening game with South Carolina will be a strong challenge as South Carolina has probably the fastest backfield in the southern conference.

In general we shall be equal in material to Columbia and VMI, and progressively outmaterialed by South Carolina, Dartmouth, Pittsburgh, Navy, Penn, Southern California and Georgia Tech.....

He closed his estimate for the '52 season with a slice of his philosophy and a repetition of some Blaik football axioms he kept posted in Army's south gymnasium dressing room.

...Unfortunately, too much experience in losing (gracefully) often lowers the resistance to defeat. Through the years I have found that between equal teams the winning formula is a thin margin above which to remain requires fidelity to fundamental principles and a team faith that abhors mediocrity and moral victories. I have often stated that there never was a champion who to himself was a good loser; there is a vast difference between a good sport and the good loser, but today even at the Military Academy we have a school of thought whose followers believe we should place little emphasis on winning. They have never experienced the pride of accomplishment which only comes from sacrifice and superior performance.

By all rational analysis the season should be a bleak one, but the game of football involves the human soul and this often defies cold analysis. The Navy is wallowing in smugness and football material that would have made an optimist out of Gil Dobie. The cadets will be a hungry team come November 28th. Nothing would please me more than to add to my collection of MacArthur victory telegrams.

With respect and devotion.

Sincerely,

Earl H. Blaik

Because the captain-elect and thirty-six other football players went out the door as a result of the scandal, and Blaik and his coaching staff had had no opportunity to gauge the leadership capabilities of their untried holdovers for the '51 season, there was no election of a team captain that first fall of rebuilding. Instead the coaches named field captains from week to week. With one season's experience behind them and better knowledge of one another, the team was ready to elect its captain for the '52 season.

Guard Al Paulekas, class of '53, won his teammates' confidence. Tough, feisty Paulekas was from Farrell, Pennsylvania. His father played on a world champion Green Bay Packer professional football team before coaching his son in high school. Al, a collegiate wrestler as well as a football player, was undefeated in wrestling his junior year, winning the Eastern Collegiate championship wrestling in the 177-pound weight division. He played on the football B-squad in 1950, earned a starting position on the team's defensive platoon during the fall after the scandal, and in Michie Stadium against Columbia, teamed with Norm Stephen and Ernie Condina on the last-down, game-saving tackle in Army's first win.

When practice began on September 1, Blaik faced even more formidable obstacles in rebuilding the Black Knights' football fortunes. There were few yearlings seasoned with a year's varsity play, and Blaik rated the second and first classmen as no better than good junior varsity material, though most had slogged through the difficult '51 season on the varsity. Contrary to the NCAA rules governing the '51 season, plebes weren't eligible to play varsity in '52, a fact that made the more talented fourth classmen face a harsh first year in Army football – a mixed blessing similar to the ones endured by plebes in the classes of '54 and '55. There were differences, however.

Team Captain Al Paulekas and Earl Blaik on the practice field near Trophy Point, Fall 1952. (Photo courtesy USMA Archives.)

The men on '54's plebe team had had the pleasure of scrimmaging against the powerful varsity Black Knights of the '50 season. The going was rough every scrimmage because there was a great disparity in team skills between varsity and plebes. The next year, a number of plebes from the class of '55 suddenly found themselves promoted to varsity for the '51 season, while the rest scrimmaged against an inexperienced, intensely competitive group of young men, each and every one vying furiously for a starting berth on the decimated varsity. Now, in the '52 season, the plebes in the class of '56 found themselves scrimmaging against still furiously competing varsity players with a frustrating year of defeat behind them. Blaik and his staff continued to drive every man to his limits. Emotions were high and practices hard hitting – and it would remain that way throughout the season.

The lack of depth and experience on the varsity extended to the B-squad, as well. Consequently, Ralph Chesnauskas, Bob Farris, Pat Uebel, Don Holleder, Paul Lasley, Ron Melnick and other plebes in the class of 1956 took a pounding. The pace was grueling, and Ralph Chesnauskas learned early the meaning of tired.

Mathematically gifted, Ralph came from Brockton, Massachusetts. His mother died when he was twelve years old. His father was Lithuanian, and didn't speak English well. Ralph relied heavily on his sisters to talk with his father and to give him guidance and counsel. As a youngster he wasn't interested in football, and didn't participate, not until he reached high school. When he finally did play his sophomore year, he played every game for the three seasons he was in high school, first at end, then at fullback, making all-scholastic the last two years. One fact he was always proud of was that Brockton was also the home town of heavy-weight boxer "Rocky" Marciano. He played high school football with Rocky's brother, and Rocky frequently rode the bus to games.

College scouts from the Southeastern Conference, Ivy League, and independent Catholic schools such as Holy Cross and Boston College expressed interest in giving Ralph a football scholarship. Doug Kenna, Blaik's assistant also responsible for seeing to academic tutoring for the football team, came to visit Ralph twice following the 1951 season. He told Ralph about the cheating scandal, the Academy education, and Army football. Ralph chose the Academy, saying perhaps he could have a role in rebuilding the team. He wanted a good education above all else. He wanted to play football, too, but he didn't want to spend four or five hours a day in practice as he had seen at Boston College and Holy Cross. The one-and-a-half-hour practices at West Point were just fine – he thought.

He didn't count on the demanding cadet life and the resulting fatigue he would feel after a hard-hitting practice against the varsity. He frequently fell asleep at his desk at night while attempting to study. Fatigue in the face of rigorous, unbending academics and military training made life especially hard for many an Army football player, as well as intercollegiate athletes in other sports. At the end of the first semester, Ralph nearly failed English.

The 1952 schedule was one of the most difficult in Army football history. To make matters worse, three of the four starting backfield, right halfback John Wing, quarterback Dick Boyle and left halfback Freddie Meyers, were sidelined with injuries for the home opener against the South Carolina Gamecocks.

South Carolina was rated strong in the Southern Conference, and Blaik opined they had probably the fastest backfield in the conference. As always Blaik was wary and didn't expect what he would see in Army's first game.

A SPECTACULAR BEGINNING

On September 27, in Michie Stadium, the cadets opened their season with a bang. The crowd of 23,479, standing for the opening kickoff, didn't have a chance to sit down before the fireworks began. Yearling Jerry Hagan, a 175-pound defensive halfback returned the opening kickoff 84 yards for a touchdown. Rox Shain came in and booted the extra point and Army was up 7-0 with less than 20 seconds gone in the first quarter.

Two new Army stars were born on that opening play – Hagan and defensive end Neil Chamberlin, who was from Bob Mischak's home town of Union City, New Jersey. Jerry Hagan's run was a hoped-for sign of a great season. Neil threw a key block at the Army 45 to clear the way for Hagan's scamper into the end zone, and his sparkling defensive play harried the Gamecocks all afternoon. If John Wing, Dick Boyle and Freddie Meyers were missed, the cadets' play didn't reflect it. Pete Vann at quarterback, Freddie Attaya at left halfback, and Tommy Bell at right halfback ably replaced them, a forecast of things to come. Paul Schweikert at fullback and Lowell Sisson at left end provided the other sparks for Army's 28-7 victory.

Midway through the first period, Pete Vann did what he loved to do most when he lofted a long pass to Lowell Sisson for a 57-yard touchdown play. Rox Shain again kicked the extra point, and Army was ahead 14-0 when the quarter ended.

The cadets shifted into a more conservative line of attack at the start of the second quarter, emphasizing the ground game. Held to a fourth down at their own 9-yard line, they punted out to the 50. The Gamecocks drove to the Army 26 in five plays. The Southerners scored on the next play when quarterback Johnny Gramling tossed a scoring pass to Clyde Bennet, who got behind defensive halfback Tommy Bell. The score stood at 14-7. The Gamecocks second quarter play picked up, and it appeared the game's momentum might be shifting toward South Carolina. In the final minutes of the half, their fortunes reversed.

Army's John "Moose" Krause intercepted a pass on the Carolina 13-yard line, and in three plays the cadets were in for another score. This time fullback Paul Schweikert plunged in from two yards out, with 25 seconds left before the half. Rox Shain kicked the extra point to boost the cadets' lead to 21-7.

The game's final touchdown came midway through the third quarter when Don Fuqua recovered a South Carolina fumble at the Army 26. The cadets drove 74 yards for the touchdown, the key play coming on a 19-yard quarterback keeper around right end by Pete Vann. Paul Schweikert made his second touchdown of the day when he bucked over to end the long Army march.

The cadets opened the season with a sterling ground game, gaining 237 yards behind crisp blocking. On defense they held the Gamecocks to 128 yards on the ground, while giving up 134 passing yards on an 11 for 19 passing performance by the South Carolina quarterback. But once more it was Army's alert, hard-hitting, opportunistic defense that caused the Gamecocks to make costly mistakes. Army recovered three Carolina fumbles, one leading to a touchdown. The one pass interception by John Krause led to another Army score.

It was an auspicious, spirited beginning for a hungry football team. If Jerry Hagan's long run foretold a much more successful season than sportswriters anticipated, Army's opponent the following Saturday would provide a tougher measure.

"Moose" Krause worked especially hard in practice the week before the trip west to play the University of Southern California Trojans. Army would give away 20 pounds per man on the line, and Moose wasn't big by college football standards. At 185 pounds at the end of the previous season, he had been promoted to the varsity at defensive right tackle for the game against Navy.

The physics and psychology of football said smart play, speed, aggressiveness, fast starts off the ball, and low hard charging by smaller linemen could more than offset size and weight in contests with bigger teams. The idea for the west coast game was to do exactly that – get down low on defense, so low on the charge off the line that you could "submarine" the big USC linemen. The charge included loops, slants and other unorthodox defensive maneuvers which might give Army linemen the edge they needed. Thus Moose worked to perfect his defensive play – and perfect it he did. The sled, the exhausting linemen's training vehicle, was the means Moose used to get himself ready for USC.

The sled is a weighted football training device which builds strength, endurance, and teamwork. There are from two to seven player positions on sleds, each position padded and covered with canvas. The padded, simulated opponent is the width of a single shoulder pad, at the end of a thick, flexible aluminum member bent vertically downward from its length, which is bolted to the floor of the sled, and rises from the floor at an angle to hang beyond the back end of the device. Each padded "opponent" is at the same, short height above the ground, and enables players to take an offensive or defensive stance, as though on a line of scrimmage, and at a whistle or signal count, lunge off the line and hit the padded "player" in front of him. The coach calling signals and observing player performance stands on the sled to give added weight to the device, as well as added friction to the sled's runners, as players sweat and strain to drive it up and down the practice field.

Each practice before the USC game Moose Krause worked tirelessly to charge as low as possible against the sled – the big, imaginary, Trojan lineman in front of him. On every charge he aimed to slam his shoulder pads below the lowest of three buttons which were the heads of brads holding the padding in place. Moose Krause and the Army team were ready.

TROJAN HORSES – AGAIN

Army was off to the Los Angeles Coliseum to play the Southern California Trojans, one of the finest teams in the nation. The temperature on the floor of the Coliseum that day was one hundred five degrees. To make matters worse, the Trojans not only could field a team outweighing the cadets 20 pounds per man, they had far more depth in their squad. Yet, before 48,433 fans, for nearly the entire first half, the cadets frustrated USC's vaunted offense.

Coach Jess Hill of USC substituted several different combinations of players from his deep reservoir of talent, trying to get their powerful offense in gear. Freddie Attaya and Paul Schweikert were able to move the ball in spurts on running plays for Army, but the

Trojans' defense battered the cadets. Neither Army nor USC mounted a sustained drive the first half. Army's defense stubbornly refused to yield, and on offense each team was unable to penetrate beyond their opponent's 35-yard lines in the first two periods. Moose Krause's hard work was paying off. He was making more tackles than any man on the Army team.

Early in the second period, however, Southern Cal's Desmond Koch punted to Army, pinning the cadets deep in their own end of the field. Freddie Attaya then tried a third down quick kick from Army's 8-yard line. Blaik told the team in the hotel conference room that morning he had decided to install the quick kick, which the players hadn't practiced all week. The only practice came in the conference room when "Ski" Ordway, the yearling Blaik moved to center at the start of the season, practiced snapping the ball through quarterback Pete Vann's legs to left halfback Freddie Attaya, while the team was still dressed in their cadet uniforms. The last-minute decision and no practice spelled trouble. To fool the USC defense, Ordway had to keep his head and eyes up, as he normally did to snap the ball directly into quarterback Pete Vann's hands when Pete was immediately behind him in the T-formation.

The snap through Pete Vann's legs, angled slightly to the left was accurate, but a bit slow, and USC's hard charging left end Bill Hattig blocked the kick. "Ski" Ordway still remembers the sickening "thud" of the blocked kick. The ball bounded backward out of the end zone for a Trojan safety. At halftime USC left the field with a 2-0 edge, far less than expected. But Moose Krause and Army were about to be overwhelmed.

In the second half the Trojans' power and depth took its toll. Lindon Crow returned Army's kickoff 37 yards, and SC surged to the cadet 13-yard line, where their defense held once more.

The big USC linemen were finding the secret to Moose Krause's aggressive, hard charging defensive play. Given the weight disparity between individual linemen, they decided to simply sit down on top of the submarining Krause. He was being worn down and tied down on every play. Finally, Blaik pulled him from the lineup, and he sat down on the bench, dog tired. Blaik saw him sit down looking dispirited, and called him over. The Army coach, "the Colonel" of few words, said simply, "Good job, Moose." John Krause's spirits revived. He knew he'd succeeded in doing his best. His admiration for Earl Blaik deepened, and lingers through a lifetime of memories.

Moving from their 13-yard line, Army's yearling quarterback, Pete Vann, bobbled the ball, and Southern Cal's Bob Peviani recovered on the seven. On the second play Jim Sears, the Trojan's tailback, fired a pass to left end Ron Miller at the four, and he went in for the score.

Early in the fourth period, Sears and fullback Harold Han alternated carrying the ball in moving from SC's 45-yard line on the longest scoring drive of the afternoon. Then in the final minute of play, Sears lofted a 37-yard pass to end Jim Hayes, followed by an 8-yard scoring strike in the end zone to halfback Aramis Danboy. The final score was USC 22-0.

The statistical edge for USC came mostly from the Trojan's second half burst, accumulating 299 total yards on offense, including 124 yards passing, against Army's 96 total yards, 42 from 7 completed passes in 13 attempts – with one interception. The Trojans completed 8 of 16 throws, with one interception.

Army had acquitted itself well against a team that had only one regular season loss the remainder of the year, won the Pacific Coast Conference crown, and went on to defeat Wisconsin 7-0 in the Rose Bowl. Their sole loss was to perennial power Notre Dame, 9-0.

BACK HOME TO FACE THE DARTMOUTH GREEN

Sunday, October 12, *New York Times* sportswriter Joseph M. Sheehan wrote ecstatically of Army's Saturday victory over Dartmouth:

> With as devastating an attack as ever was mounted by any of its peerless teams of the Forties, Army ripped Dartmouth to shreds in the first half of their football game at Michie Stadium today.
>
> Scoring on six of seven sallies within the first twenty-five minutes, the astonishing cadets turned what had been expected to be a close contest into a rout of even greater proportions than indicated by the final count of 37-7....

It was a banner day for Pete Vann. Before the home crowd of 18,127 onlookers, he scored one touchdown on a quarterback sneak, connected for two more on passes, and directed the Army offense with "aplomb," as the *Times'* sportswriter effused. He was right. It was a grand day for Peter Joel Vann from Hamburg, New York – as it was for the whole, hungry Army team.

Freddie Attaya scored twice, once on an 18-yard run and the other at the end of a spectacular 67-yard pass play. Other cadet scorers were: Lowell Sisson on a 10-yard pass; Mario DeLucia on a 15-yard run; Paul Schweikert, who tackled Dartmouth's Pete Reich for a safety; and Rox Shain, who booted all five extra points after the Army touchdowns.

Army moved to its first score, a safety, following the opening kickoff. In four plays starting from their 37-yard line, the cadets drove to Dartmouth's 15. Pete Vann and Lowell Sisson hooked up for 42 of those yards as Pete hit him on two passes. After the second grab by Lowell, fullback Paul Schweikert, on a trap up the middle, raced toward the Dartmouth goal line, but was hit hard one step short of the goal line. The ball popped out of Schweikert's grasp into the hands of the Dartmouth captain, Pete Reich, whose momentum carried him into the Big Green's end zone. Schweikert recovered quickly and tackled Reich for the Army safety. The flood gates had been opened.

Before the half was over, Army rolled to a 37-0 lead.

Rules required Dartmouth to kick off from their 20-yard line after suffering the safety. Army returned the ball to the cadet 40, and drove on the ground to a first down at Dartmouth's 12. However, the Big Green's defense, which had held the powerful Penn Quakers to 30 yards rushing the previous week, rose up and stopped the cadets at the Green 4-yard line. The halt was temporary.

Dartmouth was forced to punt and Army regained possession at their opponent's 42. A Vann to Attaya touchdown pass was called back for a man-in-motion penalty. Then Army drove for a touchdown in six plays. Two were passes, Vann to Mischak, totaling 31 yards. Schweikert reeled off 17 yards in two runs, and Pete Vann ended the march with a diving lunge over center for the score.

The next Army score was set up by a 30-yard punt return by Don Fuqua. Taking the punt on his own 30-yard line, Don stormed to the Dartmouth 40. In just four plays Army scored again. After two short gains, Paul Schweikert ran another delayed trap straight up the middle,

Mario DeLucia rams for yardage against Dartmouth, Michie Stadium, October 11, 1952. (Photo courtesy USMA Archives.)

and rushed 27 yards to the Dartmouth 10. Then Pete Vann threw a strike to Lowell Sisson in the Dartmouth end zone. Lowell had beat the Dartmouth defender by a step.

The second quarter brought more misery to the Indians. The period opened with John Krause's recovery of a mishandled Dartmouth pitch-out at the Army 42. From there the cadets went over in eight plays. Pete Vann connected on passes to Schweikert and Attaya, and bull-like fullback Mario DeLucia stormed up the middle twice, bowling over would-be tacklers, the second time into the Dartmouth end zone.

In less than a minute Army scored again, this time on a pitch-out and end sweep by Freddie Attaya on the first play after Ed Weaver's recovery of a Dartmouth fumble at the Green's 18-yard line.

When next Army got the football, they conservatively quick-kicked on first down. The kick by Freddie Attaya wasn't a good one, and Dartmouth began its first serious scoring threat of the game. They fumbled again, however, and Army recovered at their own 22. From there the cadets went into action once more.

After Johnny Wing made a short gain, Pete Vann threw to Bob Mischak for a first down at Army's 33. The next play, Pete faded back and threw a perfect 50-yard pass to fleet-footed Freddie Attaya, who gathered the ball in at the Dartmouth 25 without breaking stride, and outraced all his pursuers for the final Army touchdown.

Though the cadets scored no more points in the second half, Dartmouth found the Army defense almost as formidable as the offense. The Indians reached the Army 23 near the end of the second quarter, but were thrown back past midfield. They were stopped once more in the third period by Neil Chamberlin's goal line pass interception, after penetrating to the Army 7-yard line. Finally, still in the third period, the Big Green scored after recovering a cadet fumble on the Army 28. Quarterback Ross Ellis passed 23 yards, hitting left halfback Jim Donahue in the end zone.

Dartmouth had one more scoring opportunity in the fourth quarter. Another fumble recovery on Army's 16-yard line gave them the ball, but the cadets' Pete Manus intercepted another Ellis pass in the end zone to end the threat.

Dartmouth had come into the game winless in two prior starts, but had given Penn a struggle the week before. The cadets put on a powerful offensive show the first half, gaining 350 total yards. At the finish they had rushed for 246 yards and Pete Vann completed 12 of 21 passes for 204 yards, with one interception. Dartmouth was throttled on the ground, gaining only 69 yards. Forced to a passing attack, they fared better, completing 14 of 26 for 160 yards, with two interceptions. Both interceptions came at critical moments in the game, denying Dartmouth touchdowns.

Army's victory was sparkling, impressive. The cadets had already won two games in three starts, the number of wins equal to all of 1951's wins. Sportswriters were beginning to take notice after the Dartmouth game. One wrote enthusiastically in a column titled "Ahead of Plan at West Point," "Earl H. Blaik said it would take 10 years to recover.... What's Blaik feeding those kids?" However, Army was about to face another dragon. While they were rumbling over Dartmouth at Michie Stadium, Pittsburgh University's Panthers were taking apart Notre Dame's Fighting Irish. Pitt had also taken the measure of Pennsylvania in the opening game of the season.

DRAGONS, PANTHERS, AND ROLLER COASTERS

For contests away from West Point, Earl Blaik arranged stays for the team in quiet surroundings, usually a country club, where the peaceful expanse of fairways and putting greens gave solitude to keyed-up players. As was his normal routine on late Friday afternoons, as part of his pregame preparation, he took the players on strolls around the grounds, and talked with them. Included in the talks were his philosophies about football, Army men who were heroes and admired the game – men like MacArthur – and usually the kickoff and first series of offensive plays which began the Army game plan for the next day. In the evening the team watched an entertaining movie. The intent was to relax players in the interval between the end of the week's practice sessions and game time, and help ensure a good night's sleep.

For home games he normally walked the team around Trophy Point and back across the Plain, the parade ground, toward the South Gymnasium, the site of the Army Football Office and team locker rooms. The stroll around scenic Trophy Point offered a marvelous, panoramic view of the Hudson River Valley and the Academy, and when the fall leaves were turning, the splashes of gold, orange, amber, red, and brown added to the intrinsic, peaceful beauty of the whole area. The walks coursed through grounds rich with military history and its artifacts, monuments to national and Academy heroes, and a quiet setting one would find difficult to match anywhere. Nothing untoward ever happened on those Trophy Point walks, at least not until late one Friday afternoon the fall of 1952.

Typically when Blaik led the walks, the team bunched together with him at the start. However, as the stroll and Blaik's talks progressed, some team members invariably straggled behind. Among the stragglers were a few who deliberately avoided the Army coach's perpetual, analytical evaluation of every facet of player behavior.

On this day as the column wended its way back across the parade route on the Plain, some at the head of the column began stumbling, one or two very nearly falling, as though tripped by an unseen force. Puzzled, the players who seemed to have walked into some sort of trap stopped and looked for the cause. They found nearly invisible piano wires neatly staked into the Plain's turf at exactly the right height and interval to cause a tripping disaster among the ranks of six battalions of cadets, each marching in four-hundred-man mass on parade before thousands of pregame visitors the next morning.

The players reported their surprising find to the Cadet Officer of the Day, and the trip wires were removed. No one ever found out the identify of the culprits who were intending to convert the pageantry of a Saturday morning parade into a comedic mass stumbling and falling by cadets in full dress uniforms. Had the wires gone undetected, the next morning's scene – hundreds of cadets shouldering rifles with glistening, silver-chromed, fixed bayonets, pitching forward to the ground, with rank upon rank following behind, collapsing like a giant, chaotic accordion – would have stirred horrifying embarrassment among the more serious Academy officers, although the pranksters would have undoubtedly found it humorous. Whoever schemed the prank almost found fame, or infamy, in Academy history and folklore. But a football team that had a year earlier been whispered of as being, at heart, an undisciplined, unprincipled group of less than bright athletes, had saved a lot of Saturday morning red faces at West Point.

Undoubtedly, "the Colonel" didn't hesitate in suggesting the trip wires be reported to the Cadet Officer of the Day. But it's also clear the cadets would have done so without his insistence. The incident pointed up something else about the young men struggling to rebuild Army's football fortunes. The team and the Corps were one whole cloth. The pride they felt in one another had no boundaries between them.

* * *

The crowd of 18,850 attending the game against Pittsburgh University was only slightly larger than the previous Saturday. Army faced another dragon on their schedule. The size of the crowd perhaps indicated Army fans weren't ready to believe the cadet football team was on the road to recovery. The cadets were, however, and though the Panthers displayed speed, power, and finesse early, the Black Knights almost convinced Michie Stadium rooters they were in the same class of football teams as the men from Pitt.

On Pitt's first possession, the Panther quarterback, Rudy Mattioli, engineered a 13-play, 80-yard drive for their first touchdown. From that point, throughout the first half, Mattioli dazzled the cadets, guiding a quick-striking T-formation attack. On the opening drive Mattioli handed off to swift, elusive, right halfback Bill Reynolds and hit passes consistently in an Army secondary which seemed unable to cover its territory. In between he pitched out to fullback Paul Chess who swept the cadets' defensive ends. Chess's first score came on this drive, a pitch out and 14-yard run.

Later in the first quarter Army moved 43 yards to the Pitt 19, and seemed headed for a touchdown before the Panther defense smothered Pete Vann's fourth down pass attempt.

In the second quarter Pitt reeled off another long drive, this time going 70 yards in 15 plays, Mattioli passing to Reynolds and Paul Chess, with each of the two ball carriers adding short runs. Chess's second touchdown play was identical to the first but came on fourth down, a pitch out and end sweep behind crisp blocking.

Before the half ended the Panthers added two more points on a safety. Their sophomore punter, Paul Blanda, made a superb kick, out of bounds on the Army 1-yard line. On the next play Freddie Attaya attempted to punt Army out of danger from deep within their own end zone. The center erred on the snap. The ball rolled along the turf, and gave Attaya insufficient time to get the punt off. He avoided a blocked punt, taking a safety instead.

Second classman – junior – Dick Boyle, who had missed Army's first three games due to injury, came in the second half in place of yearling Pete Vann, who had been throttled throughout the first half by a rock solid Pitt defense. There was no more scoring until late in the third period, after Army had gone 48 yards to the Pitt 27, and stalled again. Then the Panthers delivered what seemed to be a crushing knockout blow, when they marched 73 yards to another score. The touchdown came on a 26-yard buck lateral play off the weak side, from a single wing formation, the old football power formation Pitt used effectively several times that afternoon. The score stood 22-0, Panthers.

Not only did Pitt appear to best the cadets in every phase of the game, the breaks seemed to go against Army, with the few penalties the cadets received occurring at inopportune times – one a crucial pass interference penalty that kept alive a Pitt touchdown drive.

But just before the end of the third period Army received its first break, and the game's momentum abruptly shifted. From that time on

Johnny Wing all alone facing three Pitt defenders, ready for a fight,
Michie Stadium, October 18, 1952. (Photo courtesy USMA Archives.)

excitement in Michie Stadium mounted as the cadets kept the pressure on Pitt, and closed the gap.

Linebacker Norm Stephen intercepted a Mattioli pass, and was taken down at the Panther 40-yard line. The quarter ended. The teams swapped ends of the field, and the cadets promptly rolled to their first score.

Their luck changed during the scoring drive, too, when Pitt erred with an offside penalty, giving the cadets an important first down on the way to their touchdown. A minute later, a Panther was called for unnecessary roughness on an incomplete pass from Dick Boyle, and Army had a first and goal on the Pitt 9-yard line.

On the next play hard-running Freddie Attaya stormed to the 1-yard line. From there, big bulldozing, substitute fullback Mario DeLucia rammed over for the score. Rox Shain booted the extra point, and Army was on the scoreboard with seven points.

Then, briefly, it appeared Pitt was going to roar right back and score again. Taking Army's kickoff following its first touchdown, the Panthers moved 50 yards in two plays, both from the single wing formation. First, sophomore left halfback John Jacobs went 14 yards. Then right half Reynolds went 36 yards to a first down on the Army 24-yard line. But this time the Pitt drive was thwarted when Don Fuqua intercepted a Mattioli pass in the Army end zone for a touchback, and the cadets went on the offensive at their own 20.

There were ten minutes left to play and the rest of the game belonged to the cadets. Following a punt, the Army defense stiffened and Pitt couldn't move. Then came another exchange of punts as both teams stalled. After Army punted back to the Panthers, the cadets threw them for three successive losses after Pitt suffered a 15-yard holding penalty. They punted and the cadets took possession at mid-field.

To a thunderous new chant from the Corps, "Unchain Mario! Unchain Mario! Unchain Mario!," Mario DeLucia broke loose for 11 yards. Attaya lunged for 5 more. After quarterback Dick Boyle was thrown for a loss, he hit Freddie Attaya with a pass and a first down on the Pitt 21. Next, on a well executed fake, Freddie Attaya took a pitch out and rambled to the Pitt 8-yard line for a first and goal. The Army stands had come alive the second half, and from the noise, it would appear the Black Knights were about to win the game.

Three plays moved the ball to the 1-yard line. To chants of "Unchain Mario!," fullback Mario DeLucia got his second touchdown of the afternoon, and Michie Stadium was bedlam. Rox Shain kicked his second extra point. There were less than two minutes to play, however, and Pitt held on for a 22-14 win. Their fans stormed the field and tore down one of the two goal posts, while Pitt's magnificent 120-piece marching band put on an exciting, lustily cheered post game show.

The cadets acquitted themselves well against a tough opponent, making a game of it in the last two quarters. Out-gained on the ground 297 yards to 152, they fared better with a strong second half passing attack. Completing only 1 for 9 the first half, Dick Boyle threw 7 completions in 9 attempts the second half, for a total of 93 yards, while holding Pitt to 58 passing yards on 8 completions in 13 attempts. Army's defense snatched two crucial interceptions and clamped down tightly the second half, to make the game exciting. The first half, cadets George Kovacik, Frank Wilkerson, and Ron Lincoln played solid defense when it seemed Pitt was about to run away with the game.

Army stood 2 and 2 in a season shaping up as a roller coaster ride. The following week would be no different, though Pennsylvania's Quakers overpowered Columbia 27-17 the same day Pitt took the measure of Army.

THE LION NEMESIS...

The Columbia Lions were still smarting over being one of only two Army wins in 1951, when Dick Boyle had leaped over the center of the Light Blue's line for the final margin of 14-9 at Michie Stadium. Though Columbia coach Lou Little and Earl Blaik were good friends holding great personal and professional respect for one another, their teams always played one another with unbridled emotion, and the same mutual respect.

Army's most famous alumnus on Saturday, October 25, was Dwight D. Eisenhower. He was president of Columbia University. He also was on a leave of absence from Columbia, running for president of the United States. Taking time out from the final, hectic days of campaigning, he came to the Lions' Baker Field to watch the first half of the game. He sat on the Columbia side of the field and before departing the stadium during halftime, came to the cadets' side to greet the Corps. He then left the stadium while cadets chanted, "We like Ike! We like Ike!" The chant repeated the words seen on thousands of campaign buttons the Republican Party had distributed to their faithful as the presidential election neared.

Unfortunately for Army, the day turned out to be one of the most disappointing of the season. Thirty-one thousand spectators saw the Lions gain a near miraculous reprieve in the final seconds of a game Army completely dominated, but marred with fumbles, pass interceptions, and penalties. The miscues repeatedly frustrated the cadets' ability to put the contest away. Mistakes typically seen in inexperienced teams bedeviled Army all afternoon, and stopped potential scoring drives at Columbia's 8, 15, 25, 26 and finally the 1-yard mark. It was Michie Stadium 1951 all over again, but with the roles reversed.

Fumbles and an interception led to all four touchdowns in the game, the first scored by the Lions within the first five minutes of the first quarter. Freddie Attaya stood deep in Army territory in punt formation, on fourth down. The snap from center was high, causing Attaya to leap for the ball, and he fumbled. He recovered the ball by picking it up and trying to run with it. He got one step before the Lions' right tackle Ernie Gregorowicz brought him down, and Columbia had a first down on the cadets' 17-yard line. Gregorowicz was one of three Lions who stayed on the field the full sixty minutes of play.

Columbia moved quickly to their first touchdown and a 7-0 lead. Right halfback Bob Mercier sped 11 yards around end. Three more rushes put the ball at Army's 1-foot mark, and fullback Bob McCullough, who played an outstanding defensive game all afternoon, threw himself into the Army end zone for the score. Right end Al Ward, who was to become a Lion hero this day, kicked the first of his two extra points.

Army's tying score was set up just prior to the end of the first quarter by Bob Mercier's fumble. The cadets' Ed Weaver fell on the ball at the Lions' 21-yard line. This time the cadets used five plays to tie the game. Freddie Attaya took a pitch out and sped around Army's right end to score from the 7-yard line. Rox Shain booted the extra point. Army took the lead early in the second quarter.

After the cadets kicked off, it appeared Columbia was to be dominated the rest of the game. The Lions were completely stymied, unable to move on the ground. Then, from their 24-yard line, the cadets drove to Columbia's 30, before giving up the ball on downs. But Ed Weaver, one of the few holdovers from Army's 1950 team, intercepted quarterback Mitch Price's pass at the Lions' 34. In one play Army had its second touchdown. Freddie Attaya took a pitch out, and with the aid of an excellent block on Columbia's defensive end, once more sped around Army's right end for the score. Rox Shain kicked his second extra point of the day, and the cadets were up 14-7.

Freddie Attaya breaks into the open against Columbia's Lions, Baker Field, New York City, October 25, 1952. No. 70 is end Al Doremus and No. 11 is quarterback Dick Boyle. (Photo courtesy USMA Archives.)

The Columbia Lions remained bottled up on the ground the rest of the afternoon. Bob Mercier, their clever running, speedy right half-back could get nowhere because Lion blockers weren't able to blunt the charges of Army ends, linebackers, and defensive halfbacks. Hard charging cadet linemen hounded quarterback Mitch Price, as well, causing him to rush his throws and frustrate the Columbia passing attack.

Just before the end of the half, Army started to move again, threatening to up their margin with a 32-yard march to the Lion 10. Dick Boyle threw a screen pass to Mario DeLucia who rumbled 27 of those yards. But a substitute Lion fullback, Jerry Hampton, playing center and linebacker for injured first string center Dave Bueschen, ruined the cadet opportunity by intercepting a Boyle pass.

When play resumed in the second half, the cadets' apparent superiority became more pronounced. The Lions remained pinned in their own territory throughout the third quarter, fighting off Army penetrations threatening to blow the game open.

The cadets took the opening kickoff and rolled to the Lion 8, with Freddie Attaya going 45 yards on a pitch out. A clipping penalty pushed them back, but they made another first down on the Lion 15, only to have the play nullified by an offside.

Army stormed downfield again, going 64 yards to the Columbia 25 on the running of Attaya, Johnny Wing, and Mario DeLucia, who bolted 32 yards with another screen pass. But the cadets were again thwarted.

Ed Weaver gave the cadets another chance when he recovered a fumble by Columbia end Al Ward, after he caught a Mitch Price aerial. DeLucia broke loose for 15 more yards, moving Army to the Lions' 26. Then the Light Blue's Jerry Hampton again recovered a cadet fumble, this time by John Wing.

The only opportunity Columbia had came just prior to the end of the third quarter when Dick Boyle fumbled and hard working right guard Gene Wodeschick recovered on Army's 21-yard line. The Columbia stands went wild, but the Lions were rocked back by the cadet defense and gave up the ball on downs at the 25.

The Black Knights then staged the longest march of the day, surging 74 yards to their opponent's 1-yard line. Key plays were another screen pass to DeLucia, followed with his 25-yard gallop, and end sweeps by Attaya and Wing, behind disciplined downfield blocking which cut down would-be tacklers. The cadets were knocking at the door, and it seemed certain they were about to deliver the knock-out blow. On third down, with a yard to go for a touchdown – another costly bobble. Dick Boyle fumbled, and Columbia's right end, Bob Wallace, recovered to avoid another Army score.

For the first and only time the second half, the Lions began to move the ball, with Bob Mercier breaking loose for substantial gains. But the drive stalled at midfield and Columbia punted. The Lions' hopes soared again when Wodeschick intercepted another Boyle pass at the Army 39, and pushed to the cadet 23 where a rock ribbed defense stopped them for what surely was the last time.

Pete Vann replaced Dick Boyle at quarterback for Army. The starting Army center, Godwin "Ski" Ordway had left the game earlier because of an injury which occurred when "Ski" and Lowell Sisson collided downfield attempting to block the same defensive back. Like starting quarterback Dick Boyle who was accustomed to taking snaps from Ordway, Vann was having trouble with timing on snaps from Ordway's replacement – and fumbled as he attempted to hand off to John Wing. The Lions' left tackle, John Casella, recovered on the Army 33. The stage was set for another thrilling finish in the long history of the Army-Columbia rivalry.

Mitch Price went immediately to work and completed a pass to Al Ward who was brought down on the Army 14. The Columbia stands were bedlam and the cadets on the opposite side of the field were pleading for the Army defense to hold on. From the 14-yard line, Price tried three successive passes, and each fell incomplete.

Time was running out, though the clock stopped with each incomplete pass. Each failure to connect brought groans from the Columbia stands and roaring cheers of approval from Army fans. One last gasp for Lou Little's Lions.

Price took the snap from center and rolled to his right behind a wall of blockers, looking for open receivers. The entire assemblage of fans seemed to hold its breath. He couldn't find one. He turned left, angling toward the line of scrimmage, and spotted both ends, Al Ward and Dale Hopp, uncovered in the end zone. Price fired the ball into the arms of Al Ward, and the Columbia fans went wild. Ward, a Lion hero all day, calmly split the uprights with his extra point kick to tie Army 14-14 with 16 seconds left to play.

It wasn't over yet. The Light Blue not unexpectedly tried an onside kick, but an Army tackle returned the ball from the cadet 45 to the 48, stepping quickly out of bounds to stop the clock. Time for one more play.

Strong armed Pete Vann came in to attempt a long throw toward the Columbia end zone, but the Lions knew what was coming, and with a determined defensive line charge, downed Pete in the backfield. It was all over.

The tie at Columbia was clearly a disappointment for the cadets, who did everything but capitalize on numerous opportunities to put the game out of reach. One of the few times a frustrated Earl Blaik fumed over bad breaks came when, after the game, he complained to the press the injury to Army center "Ski" Ordway was one of the reasons Army didn't win. When "Ski" came limping out of the game with the hip injury, unable to run or drive off his left leg, Blaik had showed his displeasure to Ordway because he couldn't return to the game. Blaik would tell "Ski" in practice the following week, in no uncertain terms, he "*would* be ready to play against VMI next Saturday!"

The Columbia Lions had played gallantly, never giving up – the same determined play that characterized their 1947 win to end the Black Knights' first long winning streak under Earl Blaik, and a form of play typical of Lou Little-coached teams. They had tied Army for the third time in the long, colorful rivalry, the other two ties coming in 1924 and 1939. Mitch Price set a new Ivy League record for passing yardage that day, running his total to 2,764 yards. His record surpassed that of Gene Rossides, who threw for 2,632 yards in his career as a Lion. It was Gene Rossides' passing that upended Army in 1947.

Army gained 266 yards on the ground to Columbia's net of 51. In passing Price completed 12 of 25, with 3 interceptions, for 145 yards. Army connected on 8 of 16, suffering 3 interceptions as well. Fumbles and penalties once again haunted the cadets. Four times they lost the ball to Columbia, and gave away 61 yards on penalties to Columbia's 5.

There was an Academy alumni reunion and brief ceremony before the game, at which approximately 3,000 gathered, including retired

General Dwight D. Eisenhower. Guests included his brother Milton Eisenhower, president of Pennsylvania State College, and Harold Stassen, president of the University of Pennsylvania. At the ceremony Milton L. Comere from the class of 1905 received the Academy's annual award as Alumnus of Distinction.

The roller coaster ride continued at West Point the following Saturday.

VIRGINIA MILITARY INSTITUTE COMES TO MICHIE STADIUM

On November 1, a small crowd of 16,450, including Secretary of the Army Frank Pace, Jr. and Secretary of the Air Force Thomas K. Finletter, looked on as Earl Blaik's Black Knights enjoyed their most productive Saturday afternoon of the last two seasons. Rolling up 532 to 367 total yards on offense, Army was unstoppable by a scrappy VMI football team that held the cadets to a 7-7 tie at the end of the first quarter. The final three quarters belonged to Army, however, and they won going away by a lopsided 42-14 score.

Mario DeLucia's rushes, elusive broken field running by Freddie Attaya, and Pete Vann's pinpoint accurate, long distance passing were key factors in Army's explosive display on offense. DeLucia and Attaya each scored twice while Pete Vann connected on 9 of 18 passes he threw, in totals of 12 for 24 by cadet quarterbacks. The Army defense helped the cause, intercepting 4 passes against a VMI offense which filled the air with 39 aerials, 18 completed.

Ten minutes elapsed in the game when Mario DeLucia, promoted to starting fullback, shot over Army's right tackle 4 yards for the cadets' first score. Rox Shain kicked his first of six extra points. The passing combination of quarterback Bill Brehany and left end Jim Byron accounted for both VMI touchdowns, the first coming on a 26-yard play with 20 seconds remaining in the opening period.

Left halfback Freddie Attaya's first touchdown came in the second quarter, and capped a 51-yard drive by Army. A 27 yard pass from Vann to left end Lowell Sisson set the stage. From the VMI 13, Attaya circled wide around Army's right end, taking advantage of key blocks by Lowell Sisson and "Ski" Ordway, and went into the end zone standing up.

With only 30 seconds remaining in the first half, Army struck again, pushing their lead to 21-7, this time on a 42-yard pass from Pete Vann to right end Bob Mischak, who was standing alone in the VMI end zone.

In the third quarter Pete Vann's long passes hit VMI again. The cadets scored on a spectacular 73-yard play, a perfect Vann to Attaya pass, which Freddie gathered in at the VMI 40-yard line without breaking stride, and outran defenders into the end zone.

Then, in the first minute of the final period, Tommy Bell took a VMI punt and returned it 45 yards to again put Army in scoring position. Mario DeLucia drove across from the seven to push the score to 35-7. The cadets' final touchdown came on a pitch out from Pete Vann to substitute right halfback Bill Purdue who streaked down the sideline 58 yards to score.

Bill Purdue was a high hurdler and sprinter on the Army track team. He hadn't played high school football, though he loved the game. Until the Columbia game, he remained on Army's junior varsity. Earl Blaik, always on the hunt for talent within the Corps of Cadets, persuaded Bill to join the Black Knights. His speed made him the fastest man on the Army team, and his dash against VMI was an exciting introduction to a talent that would bring victory to Army two weeks hence.

If Army had been in the valley against Columbia, they were clearly on the mountain top against VMI – although the team from Virginia couldn't be called a dragon on the Army schedule. Not only were the ground and aerial games efficient, Army fans saw two emerging stars that could bring good things to the Black Knights' future, Mario DeLucia and Bill Purdue. In his first game as a starter, to chants of "Unchain Mario!," DeLucia rushed 110 yards in 17 carries.

The next Saturday, however, could be quite different. Army faced a real dragon, one of the most powerful teams in the nation. When the two teams met, Georgia Tech was on a 21-game streak without a defeat, No. 3 in the nation, and headed to the Sugar Bowl in New Orleans on New Year's Day.

The Tuesday before the Army-Georgia Tech game in Atlanta, Dwight D. Eisenhower, class of 1915, won the presidential election. Harry S. Truman, who had fought as a Missouri National Guard captain of artillery in World War I, and stood firm against the NKPA's assault on South Korea in the summer of 1950, continued toward taking leave of the office in which he had served for nearly eight tumultuous years. The Cold War began in earnest while Harry Truman was president, and American and United Nations men and women fighting in Korea were in our nation's first, bitter "half-war," and the first, bloody shooting war between the world's Western Democracies and the new totalitarians of the 20th century.

ONE WAY TO PAY THE PRICE

Assistant coach Johnny Green was tough as nails and highly respected by Army linemen. The former three-year letterman, two-time All-American guard, and team captain on Blaik's '43-'45 Black Knights, was a line coach. His specialty was defense, but he was small, even by standards of the '40s. Though small he knew how to be effective and, as Blaik's assistant, loved to demonstrate to Army linemen how to be effective. The trouble was, no one liked to be the player Green used to reteach the lesson of cutting large opposing linemen down to size.

Green's technique was to take a low defensive stance, coiled like a spring, a deliberately lower stance than the lineman he faced. At the snap of the ball, he used the power of his legs and speed to lunge forward and upward, bringing his clenched fists into his chest with his tightly folded arms sticking out sideways, level with his shoulders, forearms forward. The upward lunge with that body position was perfectly legal under the rules. The object was to straighten up a slower reacting offensive lineman by slamming a shoulder and forearm into him, and cause him to lose his blocking leverage. The problem was that the defensive maneuver often resulted in a forearm in the opponent's upper chest, or higher.

During one '52 season practice between games, Frank Wilkerson ('53) wasn't performing on defense to Green's satisfaction. Time for a demonstration, one on one, and Green called Frank into position for the demonstration. There were knowing glances among the players watching, because everyone knew what was coming. No one envied Frank Wilkerson.

At the snap of the ball, Green, as always, shot forward and upward slamming hard into a larger, and by no means slow, Frank Wilkerson. This time Green's forearm caught Frank flush on the mouth. Wilkerson straightened up, stopped, stepped back, and looked at Green with a funny, puzzled expression.

Green, "What's the matter?"

Wilkerson shook his head, but didn't answer.

"What's the matter?" Green asked again.

Wilkerson shook his head.

Then Green said, "Open your mouth."

Without saying a word Frank complied. When he did, his lower front four teeth were bent back, laying on his tongue. He quickly closed his mouth again after showing Green the damage.

"Just a minute, Frank. Open your mouth," Green said again.

This time, when Frank opened his mouth, Green reached in, got his fingers underneath the teeth, and pulled them upright. "Now, you're OK. How's that?"

Having completed the temporary repair, Johnny Green promptly sent Frank Wilkerson off to the dentist to get a better, more permanent repair job. Frank was back in business shortly, with one of the first face masks seen at Army – one rubber-coated bar bolted to his helmet, looping strategically across his face.

INTO THE DEEP SOUTH, AGAINST "THE RAMBLIN' WRECK"

Riding their vaunted running game and an improved pass offense, the Yellow Jackets rolled to a 45-6 victory over Army in Atlanta, Georgia. The win came despite Tech's loss of All-American hopeful Leon Hardeman to an injury in the first quarter. After one touchdown in the opening period, the bowl bound Yellow Jackets took complete control of the game the remaining three periods

The Engineers scored first when punter Dave Davis got off an 80-yard kick that rolled dead on Army's 4-inch line. Freddie Attaya punted from the back of his end zone and Jackie Rudolph returned the ball 8 yards to the cadet 25. Two plays later Hardeman swept around Tech's right end and weaved his way 23 yards into the end zone. Pepper Rodgers, the Yellow Jackets' back-up quarterback kicked the first of his six extra points plus a field goal for the afternoon.

Army's opponents then posted 17 more points on the board the second period while the cadets managed one touchdown, their total for the day. The Ramblin' Wreck began moving with slick running and passing and drove to the cadet 6-yard line, where the drive bogged down. On fourth down Pepper Rodgers kicked a field goal from a difficult angle to set a new school record of four field goals in one season. The prior record of three was set in 1940.

Army couldn't move after the kickoff. Following Freddie Attaya's punt, Georgia Tech backs Glenn Turner and Billy Teas churned out most of the yardage as Tech marched 67 yards to another score, with quarterback Bill Brigman going over from the 2-yard line.

Then came Army's lone sustained drive for the afternoon. It was late in the half when the cadets drove 73 yards, mostly on the ground, with Freddie Attaya leading the way. He broke loose for 29 yards and the score. This time, Rox Shain didn't get his extra point. Hard charging Engineer linemen broke through and blocked his attempt.

Georgia Tech scored one more time in the first half to take a commanding 24-6 lead. Roger Frey broke through the Army line to block an Attaya punt, and recovered the ball on the cadets' 3-yard line with ten seconds left. It took one play for Tech quarterback Bill Brigman to throw a touchdown pass to Buck Martin in the end zone.

The Ramblin' Wreck put 21 more points up in the second half, with a touchdown and extra point the third quarter, and two touchdowns and two extra points in the fourth quarter. In the third quarter Brigman hooked up with Martin again on a 13-yard pass into the end zone. George Humphreys capped a 56-yard march in the fourth quarter, ramming over the Army goal from one yard out. Tech's final score came when Pepper Rodgers threw 29 yards to Mike Austin.

It hadn't been a good day for Army, still a depleted team by normal standards, although yearling Pete Vann's passing was sharp and worried the Tech secondary. Freddie Attaya continued to shine with his speed and elusiveness, holding great promise for the balance of the season, and another one ahead.

There was to be some consolation in the defeat handed Army by Georgia Tech. The Yellow Jackets continued their winning record and rambled on to an overall 11-0 season, and defeated Mississippi 24-7 in the Sugar Bowl.

Interestingly, Earl Blaik best remembered the trip to Atlanta for another reason. When the cadets resigned following the cheating incident, Academy graduates and other better known public figures offered help to many, encouraging them to complete their college education. A substantial number went on to some of the finest schools in the nation and quietly enrolled: Notre Dame, Kansas, Kansas State, Villanova, Colorado College, and Mississippi State, to name a few. One entered Georgia Tech. In Atlanta, Earl Blaik received a visit from Dr. Jesse W. Masson, Dean of their School of Engineering.

"I didn't come to see the game," he said, "so much as to have the opportunity to tell you that this boy is one of the finest characters any of us has ever known. He has inspired all of us by his seriousness of purpose and fine mind."

Years later Earl Blaik wrote, "I have long since forgotten the horrendous thumping a fine Georgia Tech team handed our depleted squad that day, but I shall always recall the warmth of those words from Dean Masson."

It was time to go home now, and prepare for another bruising battle with Penn. Two weeks after the Quakers at Philadelphia's Franklin Field would come Navy, at Philadelphia's Municipal Stadium. The day Army got trounced in Atlanta, Navy upset Duke's Blue Devils in Durham, North Carolina, 16-6. A week earlier the Midshipmen fell to Notre Dame in a hard fought game in Cleveland, Ohio, 17-6.

QUAKERS IN THE MUD

The 1952 game against Penn was a classic Black Knight-Quaker battle to the finish, played in the rain on a heavy, muddy field. Penn was favored, and rightly so. They had done well on a typically tough schedule. And they began as though they were going to drive the cadets from the stadium. Army slips, slides, fumbles, and penalties harassed the cadets throughout the first quarter, and the Quakers jumped out to a 13-0 lead in the first thirteen minutes of the game. But in the second half, the legend of Army football would admit a new man to its pantheon of heroes. His name – Bill Purdue.

The mud made the game rugged for both teams, but Penn scored early, at the end of a grinding 55-yard, 10-play drive, nine of them along the slippery, muddy ground. Penn's offensive blocking, led by left guard Stewart Haggerty, was so effective a rout appeared in the offing. Red and Blue fullback Joe Varaitis, tailback Glenn Adams, and wingback Bill Deuber seemed unstoppable, and in spite of the weather and heavy ball, Coach George Munger's team furiously pressed their single wing attack with frequent multiple handoffs and buck laterals. Varaitis plunged the final three yards at 6:40 gone in the quarter. Left end Carl Sempier added the extra point, and Penn was out front 7-0.

Matters got worse for the cadets late in the quarter. In a series of plunges using the power of the single wing formation, Varaitis and Adams moved the Red and Blue to a first down on Army's 18-yard line. Adams missed on a pass, then on a running play reached the 14. The cadets' right end, Fred Bliss, in a fine defensive play, stopped wingback Bill Deuber for no gain on a reverse. It was fourth down and six yards to go for a first down, and the last chance to keep the drive going. Tailback Glenn Adams started around his right end, stopped and threw back across the field, into the end zone to right halfback Bill Deuber for Penn's second touchdown. This time Sempier failed the extra point try because of a low pass from the center. At the time, the miss seemed unimportant. Penn was dominating, Army hardly in the game. More Quaker scores were sure to follow.

The Quakers had another opportunity early in the second period when center Ed Surmiak recovered an Army fumble on the 16, but the threat went nowhere.

Finally, in the second quarter the muddy Black Knights began to move after a long punt by Penn blocking back Ed Binkowski. Starting from their own 20-yard line, the cadets drove for their first touchdown. Freddie Attaya splashed for four yards before Pete Vann threw long to Bob Mischak who made a spectacular

Pennsylvania's Quakers in the mud, Franklin Field, Philadelphia, November 15, 1952. Tommy Bell leaps over downed players, aiming for the Penn ball carrier. (Photo courtesy USMA Archives.)

catch for a 36-yard gain. Attaya added more yardage including a 13 yard romp to bring the cadets deep into Quaker territory. Once more, it was Mario DeLucia who provided the final spark, slogging the last 19 yards on a hand off from Pete Vann. Rox Shain kicked the extra point. The half ended with Penn leading 13-7.

Neither team could score in the third period, with both forced to punt numerous times. There were a total of 26 punts in the game, 14 by Penn. Army did threaten once in the third period, penetrating to the Quaker 9-yard line. DeLucia consistently bulled his way through the muck and the Penn line during the drive, but an offside penalty and a fumbled hand off again thwarted the cadets. Army fumbled nine times in the game, and lost two of them.

In the fourth quarter, Penn threatened again. Adams and Varaitis once more led the Red and Blue attack, and they penetrated to the Army 21-yard line where Army's team captain, Al Paulekas, made a great play, stopping tailback Chet Cornog and blunting Penn's threat.

Two passes by Pete Vann, to Bob Mischak and Lowell Sisson, gained 25 yards, but the Quaker defense tightened, forcing Army's Attaya to punt. Penn, too, couldn't move the ball and about a minute later Ed Binkoski kicked back to Army at the Penn 48-yard line. Four minutes left to play; the winning drive in the rain, mud, and fog was a thriller.

Pete Vann fumbled on the first play but recovered for loss of a yard. Then Bill Purdue, carrying for the first time, ripped off 11 yards. Army had a first down on Penn's 38. Pete Vann then threw a sideline pass to Bill Purdue and Penn's Carl Sempier forced him out of bounds, but not before Purdue had a first down on the Quaker 28. Attaya picked up 16 more yards to the Penn 12. DeLucia rammed through for 6 more yards, to the 6-yard line, then three more off tackle, putting the ball on the three. Needing a first down, Attaya picked up the necessary yard to give the cadets a first and goal at the 2-yard line. Then came the killer for the Penn Quakers.

There was every reason to expect Mario DeLucia to receive a hand off and slam into the middle of Penn's line. Instead Pete Vann executed a near perfect fake to DeLucia heading over tackle, collapsing Penn's defense toward the Army fullback, and leaving Bill Purdue alone to receive a pitch out. Purdue raced into the end zone untouched to tie the score.

Next came the crucial extra point attempt by Rox Shain. With the class of 1953 in the stands shouting "Make that point! Make that point!" and Penn's fans shouting "Block that kick! Block that kick!" it was a moment of high drama in the rain. The holder for Army was yearling Don Smith, a back up quarterback. The snap from center was high, and for a moment it appeared Army's extra point might be ruined by a high snap as Penn's had been by a low snap. But Don Smith reacted, reaching up to save the errant snap, and quickly placed the ball down on the designated spot. Rox Shain put his kick through the uprights for the winning margin of 14-13. The end came for Penn 44 seconds before the final whistle.

There was an emotional story behind Bill Purdue's last minute heroics against Penn that Saturday in 1952. Bill's father was Brigadier General Branner P. Purdue, and the Purdues were stationed in Anchorage, Alaska, when Bill received his appointment to West Point. But this fall his father had entered Walter Reed Army Hospital in Washington, D.C., critically ill with cancer.

The morning of the Penn game, Bill visited his father, sent there in a limousine by Earl Blaik. There was some question as to whether Bill would make the game, but it was his father's wish that he play, and he returned to take the field against the Quakers. Just prior to the game, in the Army dressing room, Bill Purdue learned his father had died. Bill went on to write a glowing page in Army football history in the mud, rain, and fog of Philadelphia's Franklin Field.

Following Army's come-from-behind win over a tough Penn team, Earl Blaik received another of Douglas MacArthur's congratulatory messages. The runaway win over VMI was no tonic for the disappointing tie with Columbia and the defeat the cadets endured at the hands of Georgia Tech's Yellow Jackets. Earl Blaik's November 18 response to the Old Soldier made clear how difficult was the climb out of the ashes of '51.

YOUR MUCH APPRECIATED TELEGRAM WAS MOST EN-
THUSIASTICALLY RECEIVED BY THE SQUAD TODAY.
IT WILL BE READ TO THE CORPS TONIGHT. THE WIN
OVER PENN LIFTED THE SPIRITS OF THE CADETS FROM
A NEAR ALL TIME LOW. THERE IS NO TONIC LIKE A

GOOD WIN. HAVE REGRETTED YOU WERE UNABLE TO
ATTEND GAMES, BUT WANT YOU TO KNOW THE NUM-
BER ONE RESERVATION ON THE BENCH AND IN OUR
HEARTS IS FOR YOU. I SHALL GIVE YOU FULL REPORT
AFTER THE SEASON. MANY THANKS.

EARL

Army-Navy, always the last game of the season, was next – the
great service rivalry renewed. This was the chance for Army to win for
the first time since 1949. Better yet, the cadets could end the year with
a respectable 5-3-1 record if they sank the middies' boat. At West Point
the juices flowed. There were pranks afoot, fun in the making – anything to tease, taunt, and harass the midshipmen of the United States
Naval Academy, to let the whole country know that Army would beat
Navy. Nefarious thoughts and mischievous energies were hard at work
building a team greater than the sum of all its members.

"LET'S PAINT 'GO ARMY...' ON THAT SHIP..."

"Operation Paintbrush" was the brainchild of second classman
Charles Earl Storrs in Company G-2. His classmates and close friends in
'54 called him by his nickname, "The Grin." The nickname was a good
one. He smiled broadly, a winning smile that drew people to him. He
often repeated the phrase from an old song, "Smile, and the whole world
smiles with you." It was his favorite saying, and it reflected his attitude
toward life. Playfulness and a good-humored prankster lived behind the
smile, and he sparked a nighttime sneak attack on a Navy destroyer escort he learned was docked at West Point less than ten days before the
Navy game. Bad judgment on the part of the United States Navy.

Charlie Storrs came up with the idea of painting "GO ARMY BEAT
NAVY" on the side of the DE docked on the Hudson river. He enlisted
three other cadets, all in '54 and all in G-2: quiet, personable, highly
respected William Richard "Bill" Schulz III, who was class president,
a man few would suspect of skullduggery; Richard Hunt Benfer, alias
"The Turk," gifted with the same enviable sense of humor possessed
by Charlie Storrs; and John Alexander Poteat, Jr., also blessed with a
broad smile and pleasing personality, and a man few would suspect of
midnight daring. He was the '54 honor representative in G-2. But when

Six Men from '54: The raiders and photographers of Operation Paintbrush, 24-25 November 1952, South Dock at West Point on the Hudson River. Top, left to right: Commandos Charlie Storrs, Bill Schulz, John Poteat, Jr. Bottom: Richard Benfer, photographers and historians Bob Muns and Willis "Tiny" Tomsen. (Photos courtesy USMA Archives.)

your enemy is the Navy, the midshipmen from Annapolis, surprise, among other factors, is the key to success on the "fields of friendly strife." Good humor and quiet daring are often the hatching grounds for grand schemes.

Storrs, Benfer, and Poteat had something else in common. They all worked on the *Howitzer*, the Corps' class annual. Naturally, the *Howitzer* staff relied heavily on cadet amateur photographers, such as Dave Hughes in the class of '50.

Among the shutterbugs this fall was one Willis Clifton Tomsen, also in the class of '54, known affectionately in his company, M-2, as "Tiny." One look at Willis Tomsen and the contradiction was obvious. M-2 was a flanker company, and its members were the tallest in the 2d Regiment. "Tiny" was anything but tiny. His colossal physique gave him his name.

"Tiny" Tomsen came from a Nebraska farm and loved photography, B-squad football, snacks and to do just enough to get by in academics. Having a good time was the thing. The more serious side of life would

come soon enough. Fun. Laughter. Good friends. That's what's important. When he went to bed Monday night, November 24, 1952, he didn't
know how much fun he was about to have. Neither did Robert Horace
Muns, another '54 man, from Company L-1, whose members resided
far away, among the 1st Regiment runts in South Area.

Bob Muns served in the Air Force two years in Japan before receiving an Arizona senatorial appointment to the Academy. Like "Tiny"
Tomsen, he had many interests including photography and frequently
carried a camera in his sojourns around the Academy. More active on
the *Howitzer* than "Tiny," Bob was serving as an associate photographer for the third year at West Point.

The rendezvous for Storrs, Schulz, Poteat, and Benfer was midnight,
but planning and preparations were necessary for any group of commandos about to launch an assault. Storrs had learned where his painters
could get a rowboat. There was one tied up at the Hudson River dock not
far from their intended target. The needed contacts were made to ensure
the rowboat's availability and obtain the owner's OK for its use.

Charlie Storrs lined up paint brushes and paint, a conspicuity orange, a type of paint used to highlight objects and equipment for reasons of safety. The paint had the added quality of being readily visible
at dusk, twilight, or in darkness at low light levels.

Each man had a role in the scheme. Storrs would be the painter. Two
were to remain on the oars, to keep the rowboat positioned alongside the
ship, which was tied bow upstream, north, against the Hudson River
current. The fourth was the lookout and Charlie's helper when needed.

Great care had to be taken to avoid arousing suspicions among the
ship's watch, the officer of the watch and the guard detail which occasionally prowled the deck to maintain security. A cadet lookout could
keep a wary eye on the ship's watch and any indication the ship's company had detected the intruders. Though there was a war in Korea,
Charlie Storrs and his raiders were counting on the ship's watch to be
less than fully alert. After all, the ship was forty miles or so up the
Hudson River, north of "The Big Apple." Why should they expect an
attack from North Korean or Chinese "frogmen"?

The plan was to row very quietly far enough out in the Hudson to
remain well hidden by sound and darkness, then come into their target
from center stream, on the ship's starboard side, which faced toward

the river. After easing up to her side, near the bow, they would drift slowly alongside, toward the vessel's stern, while Charlie Storrs stood in the rowboat and painted as rapidly as possible – a delicate but awkward task in any rowboat on any body of water – especially while avoiding unwanted bumping noises between rowboat, oars, and steel hull. That problem could be avoided, too, largely by tying cushions or life preservers along the gun'ales of the rowboat to ensure against the sounds of wood thumping on steel. The task was made more delicate because the oarsmen must maneuver constantly, steadily, to keep Charlie positioned near the ship for painting, while avoiding dumping him in the cold waters of the Hudson.

The four darkened their faces with grease paint or charcoal and wore a cadet uniform item, the hooded, black, wool parka, over other warm clothes. Camouflage for night operations.

Midnight rendezvous came and went, as did the entire undertaking. Mission complete. Well done. Pats on the back. Congratulations all around – almost. There remained one problem. A daring deed must be recorded for posterity. Thus it was that right after breakfast the morning of Tuesday, November 25, "Tiny" Tomsen, and then Bob Muns, were told of the previous night's mission. "Tiny" decided to undertake an *ad hoc* damage assessment mission involving a reconnaissance patrol. Clearly, wasn't it to the advantage of the entire Corps of Cadets and the Army football team, and especially the Brigade of Midshipmen, that the event receive proper exposure in the press – and wherever else the story might be told?

Never known to shirk good, clean fun, the two intrepid, amateur photographers promptly departed for the Hudson River's South Dock. But all didn't turn out as hoped. Nighttime under carefully planned conditions designed to hide from the enemy while doing him damage, is one thing. Early daylight reconnaissance patrol is quite another. Suicide.

Someone on the DE saw their launch and its two-man crew of erstwhile photographers and alerted the Officer of the Watch. When "Tiny" and Bob Muns returned to the dock, a ship's officer and accompanying sailors greeted and escorted them to the executive officer. The rest of the story is in official records.

On returning to the Academy, "Tiny" received a cadet delinquency report, a demerit slip, a product of the disciplinary system first instituted in the early 19th century by Colonel Sylvanus Thayer, "The Father of the Academy." In terse official language, the slip informed "Tiny" of his transgression. The offense was "sluggable," normally carrying an award of at least "8 and 4," eight demerits and four punishment tours – four hours marching on the cadet area on Saturday and Wednesday afternoons, when cadets were normally given one of their most treasured privileges, "free time." If they were fortunate enough to be eligible to serve confinements in their rooms in lieu of punishment tours, confinements normally wiped out Sunday afternoons as well.

But "Tiny" had other gifts in his bag of talents, not the least of which was his sense of humor. He decided to confront the report head on. When a "sluggable" offense was written up on a delinquency report, true to Sylvanus Thayer's policies, regulations required cadets to submit an explanation of report. Cadet jargon had long since given the explanation of report the label, "B-ache," short for "Belly Ache," a label expressing the pleasure cadets received in explaining the facts and circumstances for violating regulations. The cadet system of discipline made no allowances for excuses. "No excuse, Sir..." had been the required answer for any mistake, right from the beginning of Beast Barracks, where cadet life began. Facts and reasons, nothing more.

"Tiny" properly reasoned there was a way out of his predicament, that his mission had been in behalf of the Corps, Army football, the United States Military Academy, the United States Army, and against the midshipmen from Annapolis, at the most auspicious time of the fall semester, Army-Navy Week. Typed on Friday, one day before the game, his "B-ache" was a masterpiece, a classic.

UNITED STATES CORPS OF CADETS
WEST POINT, NEW YORK

28 November 1952

SUBJECT: Explanation of Report

TO: The Commandant of Cadets
U S Corps of Cadets
West Point, New York

1. The report, "Absent at 0755 Mechanics of Fluids class, 25 November 1952," is correct.

2. Shortly after breakfast I was informed by an ingenious classmate, Cadet Schulz, Co G-2, that the destroyer escort docked at South Dock had been painstakingly decorated during the previous night. Since this ship was leaving West Point during mid-morning, I believed it my duty as a Howitzer photographer to get to the scene immediately and record this incident for posterity. At once I obtained the services of my colleague photographer, Cadet Muns, Co L-1. Because the inscription was on the starboard side, it became necessary to procure a launch to attain a vantage point from which a picture could be taken. Upon returning to the dock, after taking several pictures, we had sufficient time to return to class on time. However, we were accosted by a member of the watch from the ship and informed that the executive officer demanded our presence in the wardroom immediately. Said officer asked for our equipment, whereupon he destroyed our film and made no offer for reimbursement of same. Giving us a pair of dungarees each to wear, he called the Boatswain's Mate, who informed him that the paint and scaffolds were being prepared. Shortly we were over the side and painting over in battleship gray the inscription "GO ARMY BEAT NAVY." This duty took approximately 70 minutes, at the completion of which we were released. The time was approximately 0930 when we returned to barracks.

3. The offense was unintentional.

> WILLIS C. TOMSEN
> Cadet Pvt, Co M,
> 2d Regt, 2d Cl

There were some happy endings to "Operation Paintbrush." Willis Tomsen, who wasn't particularly fond of academics, missed an academic class which wasn't exactly a favorite of most cadets – an engi-

Covering the evidence – temporarily. "Tiny" Tomsen, left, and Bob Muns paint over GO ARMY BEAT NAVY on a Navy destroyer escort. South Dock at West Point, on the Hudson River, November 28, 1952. (Photograph courtesy of Willis C. Tomsen, class of 1954.)

neering class, Mechanics of Fluids. He didn't receive a "slug" and walk the area, nor did anyone else among Charlie Storrs' midnight raiders. "Tiny" and Bob Muns managed to hide some film from their Navy pursuers, exposed film, and there exists one faded print of a Navy destroyer escort docked at West Point, Tuesday morning, 25 November 1952, proudly wearing the masterpiece painted by Charles Earl Storrs and three G-2 classmates from '54.

There was another piece of good news. Though "Tiny" and Bob Muns had the unpleasant duty of obliterating "The Grin's" masterpiece with battleship gray paint, the Navy blundered again. Virtually the entire ship's company stood, leaning over the starboard railing, looking down and smiling broadly as other photographers – not Navy photographers – caught "Tiny" and Bob on film completing the task of painting over "GO ARMY BEAT NAVY."

There was one other irony which helped the cadets' cause. The photographer who caught "Tiny" and Bob painting over the cadet masterpiece apparently retouched one of his photos, "repainting" in black

"Go Army Beat Navy," and sent it to New York City newspapers. The picture appeared the next day in the *New York Daily News*. Mission accomplished in a roundabout, unplanned, equally ingenious way.

The hijinks of Charles "The Grin" Storrs and his three midnight raiders, and shutter bugs "Tiny" Tomsen and Bob Muns, was one more step in the long road to pulling Army football from the ashes of the '51 scandal. They were men of '54 and they were still stoking a new fire smoldering inside West Point. Their audacious raid did inspire. On Tuesday, at noon meal in Washington Hall, the Corps' first captain took note of the painting expedition and complimented the raiders. All week more pregame hijinks continued, but it wasn't quite enough in the ups and downs of Army football 1952.

NOT THIS TIME EITHER...

New York Times sports writer Allison Danzig began his November 29 dispatch from Philadelphia, "President Truman was on the side of a winning Eisenhauer today as Navy defeated Army at football for the third successive year." He was referring to Navy All-American defensive left guard Steve Eisenhauer who battered Army's offense all day, while on the same day Dwight D. Eisenhower landed in South Korea, keeping his campaign promise to go to the battleground and personally assess the war. In Philadelphia, another Army and Navy fought another kind of war.

It was the seventh and final appearance of Mr. Truman at an Army-Navy game as Commander-in-Chief of the Armed Forces, and he sat among a huge throng of 102,000. With him were most of his cabinet members, scores of flag officers, senators, and congressmen, in the year's most distinguished gathering for a sports event.

The day was near perfect for football players, but slightly cold for fans. Weather clear and brisk, field dry and firm. But fans appeared ready for the falling temperatures as the afternoon shadows lengthened. They had dressed warmly.

There was a memorable coin toss ceremony. President Truman arrived two hours before game time, accompanied by his daughter Margaret. Mrs. Truman canceled her plans to attend.

A Yale football hero of by-gone years, Referee Albie Booth, escorted the two team captains, Navy's left end, John Gurski, and Army right guard, Al Paulekas, to the presidential box for the coin toss. Joining them were coaches Earl Blaik and Eddie Erdlatz, and all the game officials. Each was presented to the president, and the team captains were then presented freshly minted coins.

Navy was the host team, and President Truman sat on their side of the field. He missed catching the coin he tossed, and his bobble was considered a bad omen for Navy. Army won the toss, considered another omen. Contrary to Earl Blaik's normal bent, he had instructed Paulekas to have Army receive the opening kickoff if the cadets won the toss.

The Navy's kickoff to the cadets was the start of one of the most unusual games in the history of the rivalry which began in 1890. On the first play from scrimmage, Army's fullback Mario DeLucia fumbled and linebacker Jack Wilner recovered at the cadets' 27. The midshipmen in the stands went into a frenzy, but didn't have time to sit down before, on the first Navy play from scrimmage, fullback Fred Franco fumbled and Army right guard Joe Lapchick fell on the ball. The two teams lost the ball eight times on fumbles, and five times on pass interceptions. Except for a first quarter score, Navy reached Army's 2, 6, 14, 16, 27, 32, and 35-yard lines without crossing the cadets' goal.

In the fourth quarter Navy drove 59 yards to within inches of another score but couldn't get the ball into the end zone. Time ran out before they could get off a fourth down play. Army's stubborn courage on defense was the only factor keeping Navy from registering a point total much higher than the final score of 7-0.

No sooner did Army get the ball back on Franco's fumble, than the cadets were right back in a hole. Freddie Attaya punted short and the ball rolled along the field to the cadet 35, with Army players hovering around the ball. Navy linebacker Joe Gattuso waited for the ball as it rolled along the ground, then abruptly snatched it up near the sidelines and began a weaving run that carried him all the way to Army's 2-yard line before he was brought down.

It appeared Navy was going to score in the opening minutes of the game, and the Navy stands were bedlam. Earl Blaik had predicted that should Navy score early, the game might become a route – part of his reasoning for putting the Black Knights on offense at the kickoff. Not this game. In this early, deep penetration, the cadet defense made per-

Army protects that goal line stripe against one of Navy's many penetrations into the cadet end of the field. Municipal Stadium, Philadelphia, November 29, 1952. No. 81 is Neil Chamberlin, No. 46 Tommy Bell, No. 22 Don Fuqua, No. 74 Ron Lincoln. (Photo courtesy USMA Archives.)

haps its finest stand of a long afternoon. Four times Fred Franco and halfback Phil Monahan stormed at the Army goal line, and four times the cadets held fast, yielding not one foot. The Black Knights took over the ball on downs right where the midshipmen began, on the 2-yard line.

The Navy came back threatening again. Halfback Frank Brady took Freddie Attaya's punt on the Army 48 and ran it to the 31. Five times in succession Fred Franco carried, moving the middies to the cadet 14. Army stiffened once more. On fourth down Navy faked a field goal and attempted a pass. John Weaver, kneeling to take the snap from center, stood up and fired a short pass, which was deflected into the hands of Navy guard Bob Lowell, an ineligible receiver on the 15-yard line. Army took over on downs.

There was no stopping Navy on the next march, which came late in the first quarter. They moved 65 yards to their lone score, Fred Franco reeling off 45 yards on one run, when defensive halfback Tommy Bell was blocked out of the play. Army's left defensive end John Krobock saved the touchdown with a leaping, diving tackle that barely tripped up Franco. Right halfback Phil Monahan got 18 more to the 3 on a quick hand-off. Two running plays gained a yard, then Monahan followed big, bruising, fullback Fred Franco through right tackle, and shot into the end zone. Navy's Ned Snyder kicked the extra point.

The middies appeared irresistible on their touchdown drive. Army's offense had been completely smothered. The cadets must be in for another crushing defeat. Then came the second period and some of the wildest, most bewildering play ever seen in the long, historic rivalry.

First, Navy's left halfback Fischer fumbled and Ed Weaver recovered for the cadets at the middies 29. The Corps of Cadets roared in excitement, and took up a chant, imploring a touchdown. Army could get no further than the 20, and Navy took over on downs.

The middies drove 48 yards to the Army 32. Navy quarterback Bob Cameron passed to right end Jack Anderson for a 10-yard gain, but Anderson fumbled when he was hit and the ball bounded back to the Army 36 before Navy's captain John Gurski recovered. The play resulted in bruised ribs for Gurski; he left the game, unable to return.

Two plays moved Navy to Army's 32, then Bob Cameron fumbled attempting to pass and Neil Chamberlin fell on the ball for the cadets. On the next play Pete Vann fired a pass to Bob Mischak, who caught it, only to fumble. Jack Raster recovered for the middies. Again Navy could go nowhere, and punted back to Army. Pete Vann faded back and threw a long pass, and Navy's John Weaver intercepted. Two plays later the middies' Fred Franco fumbled a hand off from split-T quarterback Steve Schoderbeck and Army captain Al Paulekas recovered. The very next play, Freddie Attaya let the ball get away from him and Joe Gattuso fell on it at the Navy 40-yard line.

Finally, Navy began to move without giving up the ball on fumbles. With Franco and Monahan carrying and Cameron passing, they made three successive first downs, reaching the Army 16. A spark in the drive was Anderson's spectacular diving catch of a Cameron pass, for a 9-yard gain. Then, at the 16, Cameron passed again, only to be intercepted by Army's Ed Weaver at the 2-yard line.

Times reporter Allison Danzig wrote tongue-in-cheek, "The action slowed down a bit as the second half got under way." Action didn't slow down.

Navy promptly intercepted another pass from Pete Vann. On the next play middie right halfback Frank Adorney fumbled and the cadets' Ron Lincoln recovered at Army's 48. Pete Vann passed again and Navy linebacker Tony Correnti intercepted – then fumbled, and the ball rolled back nearly 20 yards to the middie 15, where his teammate John Weaver recovered.

Navy's attack again stalled, and Bob Cameron punted. Dick Boyle came in, replacing Pete Vann at quarterback, and passed. His luck was no better than Pete's. Navy's Charley Sieber intercepted and raced to Army's 38. Cameron then fired a 30-yard pass to left end Jim Byrom, standing all alone, and Byrom got all the way to the Army 9-yard line before being brought down. Three runs netted three yards to the six, and Navy lined up for a fourth down field goal try. With the ball placed down at the 13, the kick sailed wide of the goal posts, and Navy came away empty handed one more time.

In the fourth quarter Brady returned an Army punt to midfield, and pushed on to the cadet 35. The cadets stopped them again and Bob Cameron punted past the Army goal line giving the cadets the ball on their 20. Then came the Black Knights' longest advance of the afternoon.

Pete Vann passed to Lowell Sisson for 14 yards, and 24 more to Bob Mischak. The cadet stands seethed in anticipation of a reviving offense and a possible tie after all Navy's missed opportunities. The cadets couldn't go beyond their rivals' 34-yard line, however, and gave up the ball on downs. With that final frustration, hope faded, and Navy began moving one more time, all the way to fourth down and inches from another touchdown when the game ended.

The statistics reflected Navy's dominance on both offense and defense. They had the ball for 83 offensive plays, Army 49. The middies made 17 first downs to Army's 4, and rolled up 258 yards rushing to the cadets' 55. Freddie Attaya, who had developed a reputation as one of the most dangerous runners in the east, was held to 38 yards. Mario DeLucia could gain only 26, and speedy Bill Purdue failed to make a yard in three carries.

Navy's pass defense was equally dominating against an Army aerial attack led primarily by Pete Vann, who had shown increasing poise and accuracy as the season progressed. Army threw 21 times, completing only 5, with four interceptions. The hard-charging Navy line repeatedly harassed and pressured both Vann and Dick Boyle, and the middie secondary tightly covered cadet pass receivers the entire afternoon. Navy used the pass sparingly, but connected on 6 of 9, with only one interception, for a total of 65 yards – surpassing Army's 51 yards through the air.

The pageantry, as always, was exciting, and the competition of pranks and fun between the Brigade of Midshipmen and the Corps of Cadets once again made the game a spectacle of memories. Televised across the nation, and beamed around the world by radio to American armed forces wherever they were stationed, the game brought cheers as the war in Korea dragged on.

There was another disappointment at the stadium that day. Johnny Green, Earl Blaik's much admired defensive line coach was on his way to a new job which would lead him, eventually, to the head coaching position at Vanderbilt University. Army players had learned to respect and admire the tough former captain of Blaik's 1945 national championship team. He had been with Blaik as an assistant coach through the glory years after the national championship, and now had struggled with the team during the first two years of rebuilding. When the team buses pulled away from Municipal Stadium that day, Johnny wasn't on board and instead stood outside waving good-bye to the Black Knights of 1952. The tough, grizzled Army assistant who had experienced the sharp reversal of Army fortunes in 1951 and helped start the rebuilding had tears in his eyes as the buses pulled away.

In spite of Army's disappointment against Navy and the cadets' roller coaster season, something special was happening at West Point. The cadet football team had acquitted itself well against nationally ranked teams, far better than anyone expected.

In the final poll of the year, Navy closed its most successful season since 1945 with a 6-2-1 record and a No. 17 national ranking by the United Press board of coaches. Georgia Tech climbed to No. 2, and Southern California No. 5. Pittsburgh, with its three losses, played a tough schedule and tied with Ohio State for No. 15.

Better yet, men like Pete Vann, Bob Mischak, Lowell Sisson, Tommy Bell, Freddie Attaya, Gerry Lodge, Jerry Hagan, John Krause, Leroy Lunn, Norm Stephen, Dick Ziegler, Godwin "Ski" Ordway, Bob Guidera and others had gained another year of varsity experience in one of the toughest schedules in Army history. There would be unexpected losses before the start of the 1953 season. Men like Mario DeLucia, Ron Lincoln, and Neil Chamberlin – tragically injured in a jeep accident the next summer – would be missed, as would the graduating seniors in the class of 1953, such as John Krobock, Al Paulekas, Frank Wilkerson, Myron Rose, and Don Fuqua, who had lived and ambivalently suffered through the storms and adversities of the '51 scandal.

There was to be one other loss which would be felt by the '53 Black Knights, Robert J. "Bob" Guidera. Bob was a promising tackle, fast and hard hitting, who lettered in '51 and '52, and like Neil Chamberlin, was much admired by his teammates. Bob's knees had been subject to repeated injuries throughout his years of football, twice requiring surgery. In his first class year, he would be virtually unable to play the entire season, his injuries so serious he later couldn't pass the graduation physical examination and receive his commission in the Army. Earl Blaik personally intervened with the superintendent to permit Bob Guidera to graduate, though he couldn't receive a commission.

Balancing the departures and losses after the '52 season would be some surprising yearlings coming up from the plebe team which had suffered the continuing frustrations of a still too-small, inexperienced, maturing, mistake-prone varsity, trying to pull itself up by its bootstraps.

More NCAA rule changes were in the offing, changes that would dramatically affect college football for years to come, and give college coaches and their teams all over the country a new set of problems – particularly Earl Blaik, his staff, and the Black Knights.

But there was something else stirring at the United States Military Academy the fall of 1952, and it had once again appeared in the cow class, the men of '54. Growing among those who would lead the Corps of Cadets next season was fierce determination and an optimistic belief that Army football, with the Corps' unyielding, spirited, thunderous assistance, would come all the way back next year. Their members, already busy on the *Howitzer* and *The Pointer* staffs, made their beliefs known. In the class of '53 edition of the *Howitzer*, at the end of their description of the 7-0 Navy victory, they wrote confidently, "Our football rebound was ready to reach the peak in 1953."

The 1952 Army Football Team. Front Row: Meglen, Fuqua, Boyle, Wilkerson, Paulekas, Harris, Weaver, Rose, DeLucia. 2nd Row: Bell, Manus, Hagan, Haff, Lunn, Arnet, Zeigler, Attaya, Schweikert, Wing. 3rd Row: Hammond (Mgr.), Sullivan, Meyers, Stephen, Krause, Walton, Mischak, Dorr, Hicks, Chamberlin, Chambers (Mgr.). 4th Row: Guidera, Ordway, Chance, Kovacik, Doremus, Lodge, Cockrell, Lincoln. 5th Row: Rogers, Ryan, Vann, Lapchick, Spence, Sisson, Kramer, Shain. (Photo courtesy USMA Archives.)

ECHOES OF THE BEGINNING

In the spring of 1952, while spring football practice was in progress, the Army cheerleaders' squad was issued a vehicle popularly known as the "jeep" – of World War II fame. At the North Dock facility on the Hudson River, class of 1954 members of the squad repainted the four-wheel drive vehicle from its wartime color of dull, olive drab to black, gold, and gray – Army's colors. The Academy administration didn't contest the unusual paint job, nor did they contest the restless cheerleading crew's extensive jeep-borne explorations of trails behind Fort Putnam that spring.

And something else was afoot, an idea brought to life for the football season of 1952 by Jay W. Gould III, Benjamin F. Schemmer, and Robert H. Downey, Jr., all from the class of 1954. The three cadets prevailed upon the Academy Museum to lend them an antique cannon. The idea was to fire the cannon at Michie Stadium each time Army made a touchdown, an extra point, a field goal, or won the game. It was a grand scheme and a different twist to Ed Moses' attempted relocation of Camp Buckner's pesky reveille cannon the summer of 1951.

There was, however, a complicating problem for the newest cadet cannoneers. No blank shells were available to use in firing the cannon, so the enterprising cadets obtained permission to store and use large, fuse-lit firecrackers commonly known as "cherry bombs," to be the exploding cannon shells. Thus, for the first time in Academy history, in the fall of 1952, at Michie Stadium on Saturday afternoons, could be heard the sharp blasts of large firecrackers exploding noisily from the muzzle of an antique cannon, into the autumn air. Timing of the blasts wasn't always the best, because the cannon crew had to light the fuses, load the cherry bombs, and wait until the fuses burned their course before the loud roars and puffs of smoke ensued. Sometimes the scoring plays would be called back for penalties when the fuses were already lit, and to avoid embarrassingly inappropriate cannon firing, the quick-thinking crew simply rotated the cannon muzzle downward into the turf, where its sound and fury would be muffled in the soft, green grass of the Army football field.

Another problem occurred. Repeated use of oversized firecrackers in the breech and muzzle of an antique cannon invariably exact their toll, and they did. At the end of the season the three somewhat embarrassed cadets returned the cannon to the Academy Museum – in three pieces. There was no retribution from Academy officials. The three cadets of the cow class completed their precedent-setting work, no matter the lack of technological innovation. From the jeep and cannon of 1952, a tradition was stirring, which would reappear, much improved, one year later.

$$* \quad * \quad *$$

During the 1952 football season, Douglas MacArthur kept close tabs on Army football and continued to encourage a still struggling Earl Blaik, his coaching staff, and the Army team. After each victory he cabled or wrote Blaik, knowing his messages would be read to the team.

When Jerry Hagan opened the season with his spectacular 84-yard return of South Carolina's kickoff, and Army went on to stun the Gamecocks 28-7, Douglas MacArthur cabled Earl Blaik:

A GOOD START EARL. THE BEGINNING IS ALWAYS THE HARDEST PART OF THE ROAD BACK. GO ON AND UP TO THE PLACE WHERE THE BLACK, GOLD AND GRAY BELONGS.

MacARTHUR

On September 29, Earl Blaik replied:

I SHALL READ YOUR GRAND TELEGRAM TO SQUAD. THE ROAD BACK WOULD BE MOST LONESOME WITH-OUT YOUR INTEREST AND SUPPORT.

EARL H. BLAIK

Referring to MacArthur's message years later, Blaik wrote, "That wire may have implied a more immediate climb back than was possible. But it bore an inspiration applicable to the season of 1953 – and, for that matter, to football at West Point as long as it is played."

Army's football miracle was a season away.

CHAPTER 15

BELLS OF LOVE, SOUNDS OF WAR

Above the fireplace at the north end of Grant Hall was a large oil portrait of a faintly smiling General of the Army Dwight D. Eisenhower. His pose was striking. He was standing, right hand on his hip, left hand resting lightly on the top of a wood chair back, bald head almost shining. He was wearing his famous waist length, dark, olive green "Ike" jacket with a small circle of five silver stars on each epaulet. Painted into the background of the portrait, behind him, was a picture hanging on a wall, the famed panoramic view of Omaha Beach in Normandy, France, in 1944, a few days after June 6, D-Day. The background picture, too, was striking.

There were hundreds of ships visible on the English Channel; landing ships unloading streams of men, vehicles and supplies across the beach; and numerous smaller landing craft ferrying fresh troops, wounded men, and supplies back and forth between cargo ships, troop ships, hospital ships, and the beach; and overhead, a few hundred feet above the ships, attached to cables, were dozens of gas-filled barrage balloons intended to keep low flying German aircraft from attacking the giant, anchored armada. A light mounted above the Eisenhower painting's frame brightened and added cheerfulness to the large, somber Grant Hall reception room.

Sworn in as the thirty-fourth president of the United States in January, Eisenhower's likeness, since November's election, gained more frequent attention from admiring visitors who entered the building for the first time. While the painting added brightness to Grant Hall's interior, it brought more formality to a formal reception room filled with trappings of tradition. Underneath the Eisenhower portrait wasn't the most relaxed setting to meet a young lady for the first time, especially on a blind date.

Girls. I hardly remembered what they were – except this was spring, April 25, the last Saturday of the month, and Ronald Earl Button from the class of '54 had asked me early in the week if I'd like a blind date for the weekend. "She's my fiancée's sister," he said, "Helen Collier's sister. Her name is Ronnie, for Veronica." I thought, Veronica, I've never met a girl named Veronica. A lovely name. Helen and Ronnie were two of five surviving sisters from a family of six girls, and their home was southwest Philadelphia, not far from the University of Pennsylvania. Her mother was Irish, and immigrated from Ireland's County Mayo in 1920. Her father was a Scotsman. He came to the United States from Edinburgh in 1923.

Ron Button introduced me to Ronnie Collier in Grant Hall shortly after the Saturday noon meal. She was standing beside Helen at the north end of the large reception room, her back to the fireplace and the soldierly gaze of Dwight Eisenhower. She wore a pale white and dark blue checkered, two-piece dress. The top had broad lapels, adding to the appearance of a fashionable suit. Bending forward slightly, she extended her right hand. I grasped it carefully, lightly. Her hand was small and delicate.

She smiled as she spoke. "Hello, Willy," she said, as we gently shook hands. To the mind of a small town boy, she possessed a young lady's charming sophistication. She was beautiful. Her eyes were green, her shoulder-length hair dark brown, almost black, her skin fair, with traces of roses in her cheeks. I caught the soft scent of a lovely perfume. Her voice was equally soft, yet clear, and carried the sound of a gentle, tinkling bell. Her voice swept me up, and I was smitten. That was that. The beginning for me. The beginning for us. A moment frozen in time, a memory always warm and fresh in the years ahead.

Such was the beginning of love for us, but the story was never the same for any cadet and his future wife. While Ronnie and I met for the first time and talked as we strolled toward Trophy Point on the west side of Thayer Road, we knew nothing of three men, all Academy graduates, and their wives, who were struggling to suppress doubt, and the future's uncertainty. They were Lieutenants William J. "Pat" Ryan, Dick Inman, and Dick Shea, and their wives, Joanne Ryan, Barbara Inman, and Joyce Shea. The three lieutenants had left their wives a few days earlier that April. They were going off to war.

Ronnie and I had never heard the name Pat Ryan, the '51 Cadet First Captain and Brigade Commander, and at the time we knew nothing of his, or anyone else's role in attempting to discover what was wrong in the honor system the spring semester before I entered West

646

Point. Never, in four years at the Academy, did I learn that a few men in my own cadet company, K-1, men whom I admired as upperclassmen, had participated in shattering the organized cheating ring in 1951. What went on behind closed doors before the August 3, 1951, public announcement during our Beast Barracks stayed behind closed doors, as did all the investigations, committee, and board proceedings that followed the announcement.

Nor did we know of Pat's wife Joanne, their first baby, named Patrick, and their loving, worried, daily exchanges of letters after his April departure for an assignment to the 93d Bomb Squadron, 19th Bomb Group at Kadena Air Base in Okinawa. Neither Ronnie nor I had ever talked with a family who had lived their lives in the darker shadows of military service. When you're young, you're uninterested in those less pleasant, less optimistic sides of the life you plan to enter.

Dick and Barbara Inman, Dick and Joyce Shea weren't in our thoughts or words either that late April Saturday afternoon. We were oblivious to the feelings and emotions the Inmans and Sheas had in common with Pat and Joanne Ryan. The Inmans and the Sheas began exchanging the same kinds of loving, worried correspondence as the Ryans the day after the two infantrymen parted with their wives from separate locations, leaving for the west coast and the ports of embarkation in Seattle and San Francisco respectively.

Both Dick Shea and Dick Inman had left for Korea little more than a week before Ronnie and I met for the first time in Grant Hall. Their wives remained at home, Barbara Inman living in an apartment and teaching school in Indianapolis, Indiana, and Joyce Shea at home with her parents in New Milford, New Jersey, a small town near Hackensack. Joyce was almost five months pregnant when Dick left for Korea shortly after mid-April 1953.

As a plebe I'd watched Dick Shea from afar in admiration, first when he was the cadet company commander of Fourth New Cadet Company during Beast Barracks, and later as G-1 Company Commander. I watched with equally boyish admiration as he ran and repeatedly won the mile and two mile events at track meets held at the Academy in the winter and spring of 1952, the last semester of my plebe year. He was captain of Army's track team and had set Academy cross-country, indoor and outdoor records at meets in the northeastern

Dick Shea, one of Army's all-time track and cross country champions, sprints for the finish line in a track meet, Spring, 1952. (Photo courtesy USMA Archives.)

United States. He was considered among the top collegiate fifteen hundred and three thousand meter prospects for the American Olympic team that spring. Though there was talk of Dick's qualifying for the Olympics, the reality was during the preceding Christmas holidays he and Joyce had made their plans.

He also had learned in one of the track meets in which he competed that he was no match for America's strongest prospects for the Olympic team. Lou Davis, a classmate and track teammate who ran the same distance events as Dick, and was his roommate during the Army track team's travels, had become a close friend. The meet was in New York City's Madison Square Garden, and Lou saw Dick Shea run with everything he possessed.

New York was the unofficial world capital of winter indoor track, and the meet featured the equivalent of professional runners, against whom Dick, as a top collegiate runner, had been invited to compete. Most collegiate runners would have been intimidated by the competition, but not Dick Shea. For him there was no alternative to victory, no thought of losing. His only approach to competition in track was to immediately set about determining how he could win.

Lou Davis recalled that members of Army's track team and its followers observed a rare sight that evening – a finely conditioned athlete ran at a pace impossible to sustain for a mile. When there was nothing left in him to give, he literally passed out on his feet.

There were several internationally known fifteen hundred and three thousand meter runners who were consistently competing at three to four full seconds under Dick Shea's best record-setting times at Army. Nevertheless, there was little doubt he could successfully qualify for the Olympics. But he decided to forego pursuit of a place on America's Olympic track team. While he was fiercely competitive, he was probably near his performance peak as a runner. Besides, he was ready for the life for which he had prepared. He had Joyce to consider. There would be no attempt at Olympic glory for Dick Shea. He and Joyce would marry, and he would finally do what he had sought for nearly six years beginning in 1947, when he first saw a circular explaining how a soldier could obtain an appointment to West Point. He was ready for service as an officer in the Army.

They married in the Catholic Chapel at West Point on graduation day in 1952, at 4:30 in the afternoon. As for me, I hadn't the slightest interest in what any graduating first classman did in the spring of 1952. I was too busy thinking about a new life as a cadet without the burdensome presence of the first classmen who had been largely responsible for my plebe and military education that first year at the Academy.

As to any thought of Dick Inman, the same was true. He faded from memory even more rapidly than Dick Shea. After Dick Inman's stint as a squad leader in Fourth New Cadet Company he returned to his permanent company, E-2, and I moved into Company K-1 during his final academic year at West Point. Company E-2 was another world, far removed from 3d Battalion, 1st Regiment. He was a cadet sergeant, a squad leader in his permanent company his first class year, was on the Army football and track teams, but was by no means a star performer.

Long forgotten was his fumble of Navy's first kickoff to Army in the terrible 42-7 pasting the Black Knights received in Philadelphia the last football game of the discouraging 1951 season. The midshipmen had promptly driven for one touchdown after recovering an attempted squib kick by Army at the opening whistle. Matters went from bad to worse when Navy kicked off following the touchdown. Dick Inman received the ball and fumbled when he was hit. Navy recovered the loose ball, this time at Army's 27-yard line, and quickly drove for another touchdown and a 14-0 lead. That was the beginning of a game our entire class would sooner forget, because we didn't get to "fall out" for the three weeks before the upperclasses went home on Christmas leave.

Nor had I the slightest notion how embarrassed and humiliated Dick Inman felt that late fall afternoon. I didn't know how much he loved the game of football, that he had written poetry about it. He came to West Point with high hopes for football glory, and worked doggedly for three tough years to make the starting lineup, a goal he finally achieved his first class year – only after thirty-seven members of the Army team resigned following guilty verdicts for their participation in organized cheating. It was plain Dick Inman would probably never have become a varsity football player at West Point had not the cheating scandal occurred. His hard-won, yet seemingly dubious achievement, was the reward earned by several football players in the classes of 1952 and 1953.

Dick Inman's father, the head football coach at Vincennes High School in Indiana, had taught him the game and encouraged him throughout high school. Dick became a star performer in his home state. When he finally reached his goal on the Army varsity in the fall of 1951, the season was brutal, the most painful season any Army team had endured since 1940. And it was made worse by the cloud of controversy which enveloped Army football after the scandal.

A newspaper reporter wrote a short piece about Dick's returning up the stadium stairs to the Army dressing room after the Navy game. He described the glistening tears rolling down his cheeks as his father waited to hug him.

Even Dick Inman's one moment of glory on Army's track team had gone unnoticed when I was a plebe. In the spring of '52, he was on the cadets' first place, four-hundred-meter high hurdle relay team at the Penn Relays in Philadelphia, a notable achievement in a track meet which each year drew some of the finest intercollegiate track teams in the nation. He and his victorious relay teammates received gold, engraved wristwatches as awards. He wore the watch proudly.

Dick Inman had no prospects of romance when he graduated from West Point. He was unattached, but his family came in force to attend June Week activities and his graduation. Unlike the class of 1950, many of whom were recalled from leave when the Korean War broke out, members of '51 and '52 weren't called back from summer leave – not unless they volunteered in advance of graduation. The war's first summer of crisis was over. What's more, because of the heavy casualties experienced in Korea by the class of 1950, the Army had reinstituted its policy of sending Academy graduates to branch schools after they completed their Academy education. Thus, after graduation in June of 1952, and leave in Indiana, Dick Inman went to the infantry course at Fort Benning. Love and marriage came later, shortly before he left for Korea.

THE STORIES OF THREE LOVES

In the spring of 1948 Pat Ryan was nearing the end of his second plebe year. The former 8th Air Force GI, who, while in England during World War II, had put his name on the supernumerary B-17 gunner's list in hopes of flying into the heart of Nazi Germany, had hit a few bumps at West Point. His life at the Academy had been less pleasant than hoped.

His first plebe year, beginning the summer of 1946, he went out for football. This was the last year of Army's great Davis-Blanchard duo, and the Yankee Stadium "game of the century," 0-0 tie between Army and Notre Dame. Pat apparently was doing well in football. As was the case for many plebe prospects, in practice he was selected to play defense against the varsity. He learned the hard way that greater responsibility often means hard knocks. And he had more than his share of hard knocks the instant he tackled big, fast, hard running, two, soon to be three-time All-American fullback Felix "Doc" Blanchard.

"Doc" Blanchard habitually ran hard in both games and practices. When he took a hand-off from Arnold Tucker, Army's first team quarterback, he ran like a fast moving freight train. When Pat made the mistake of attempting to tackle Blanchard in a practice session, day instantaneously became night. Pat was knocked cold and suffered a concussion. His football career ended abruptly. He wanted to enter pilot training after graduation, and medical records reflecting prolonged periods of unconsciousness were usually disqualifying for pilot training candidates.

But Pat's travails with unconsciousness didn't end when his football career ended. Next came plebe boxing. Apparently not fully recovered from the first head injury, he received another concussion, this time resulting in amnesia. The boxing injury ended his first plebe year, and he returned to try plebe year again, not a particularly joyful repetition under the best of circumstances. However, in the spring of 1948, the second semester in his second plebe year, life at West Point took a happier turn.

Ty Tandler, a yearling in the class of 1950, had two sisters at Briar Cliffe College in Tarrytown, New York. He asked Pat if he'd like a blind date. Thus he met Joanne Ellen Fox, a Briar Cliffe student from Silver Mine, Connecticut, near Norwalk. She was the only child of a well-known portrait painter, and his wife Ruth Bodell Fox, who was a concert pianist. Silver Mine was an arts and entertainment community, and Joanne was a gracious, cultured young woman. Still a plebe, Pat couldn't leave the Academy. She met him at West Point on a Saturday afternoon, and it was love at first sight for both of them.

For the next three years they were inseparable. She came to West Point on weekends when Pat couldn't leave the Academy. As with nearly all young people in love, Pat gave Joanne an endearing nickname, "Foxie," a playful derivative of her family name. When Pat became an upperclassman, eligible for weekend leaves away from West Point, he escaped to Joanne's family home in Silver Mine to spend time with her and her family.

There wasn't any talk of marriage, not where her mother and father could hear, and Pat never proposed. Her parents breathed a great sigh of relief, however, during Pat's 1950 Christmas leave stay at the Fox residence. He was then a first classman, the Cadet Brigade Commander, the Corps' First Captain, a responsibility which placed great demands on his already heavy schedule. One morning Joanne came upstairs to wake him for breakfast. He handed her an engagement ring.

Three days after graduation, on June 8, 1951, they married in the Holy Trinity Catholic Church at West Point. He was a new second lieutenant in the United States Air Force, with an assignment to primary pilot training at Columbus Air Base, Mississippi. His hard won dream to be a pilot in the United States Air Force was coming true at last, and he and Joanne Ryan were newlyweds deeply in love. A life of joy, family, service, and airborne adventure lay ahead.

Friends of the Fox family came from all over the country to attend the wedding, as did members of Pat's family – including his older sister, and his mother and father. After Pat's and Joanne's honeymoon, they began the journey to Columbus, Mississippi. En route they visited his family in West Virginia and Kentucky.

Six months of successfully completed primary pilot training sent them on to Vance Air Base in Enid, Oklahoma. As graduation from basic pilot training neared in August of 1952, Pat was excited at the prospect of finally receiving his coveted pilot's wings – and an assignment to transition into the B-26, a fast, twin engined medium bomber of World War II fame, and a work horse in the United Nations' air to ground campaign in Korea. But other factors intervened.

At Randolph Air Force Base in San Antonio, Texas, where B-29 transition training was in progress, a group of reserve officers scheduled into training refused to transition into the big World War II bomber, until "every single regular officer, active duty pilot had transitioned into combat aircraft." At Randolph six of ten Air Force courts-martial cases were pending "for failure to participate in aerial flights." In essence, the officers were on "sit-down strikes" which resulted from a complex of factors brought by the rapid wartime expansion of the Air Force and the heavy demands on combat crews in Korea.

General Hoyt S. Vandenberg, class of 1923 and Air Force Chief of Staff, identified the factors as "incentive pay...less than it used to be, the death rate going up in combat, and with involuntary recalls there came much pressure from families to keep the persons from flying." The "sit-down strikes" affected Pat Ryan's family and future.

Pat's entire pilot training class at Vance Air Base was held over two weeks while assignments were changed, and men and families rerouted. The entire Vance class was sent to Randolph Air Force Base to transition into B-29s, duties the objecting reserve officers were refusing to perform. Pat left for Randolph, but not before Joanne left for home in Connecticut. Moves were becoming more frequent, training schools shorter in duration. She and Pat agreed it was best that she return home, although they didn't want to be separated, especially at this time of their lives. Joanne was pregnant, and three weeks after she left for home, Patrick E. Ryan was born at the West Point hospital. He was the first of five children born to Pat and Joanne. Three weeks later Joanne and Patrick flew to San Antonio to rejoin Pat.

When Pat completed transition training to become a B-29 copilot, his next, and final assignment before joining an operational bomb squadron somewhere in the world, was combat crew training. Forbes Air Force Base near Topeka, Kansas was the next stop. By this time he was certain he would be sent to the Far East where B-29s were engaged in night operations over Korea.

The introduction of Soviet-made MIG-15 jet fighters into the air war, and increasingly deadly anti-aircraft defenses had forced a change in tactics for the big American Superforts. Prior to the MIGs entering the war, the B-29s were doing daylight bombing, a mission they performed with devastating effectiveness in World War II air campaigns against the Japanese homeland.

At Forbes, Pat trained to become a combat crew member, operationally ready to fly wartime missions. The men joining the crew were to remain together for the duration of their next assignment. Ideally they would fly every mission together, as a crew. As combat crew training neared completion, and they passed all their qualification tests, they were declared "operationally ready," adequately trained to fly combat missions. The next stop was Kadena Air Base, Okinawa, but not before Pat took a hurried leave to Connecticut to return Joanne and their newborn son to be near her parents while he was overseas.

When the day arrived for Pat to leave for the Far East, the parting from his new family wasn't easy. He'd been in England as a young, single GI in 1944, at a B-17 base when 8th Air Force bombers were relentlessly pounding the German heartland, and 9th Air Force tactical fighters and fighter-bombers flying out of other bases in Great Britain were systematically destroying the Luftwaffe. Consequently, the threat to the American bomber base where Pat was stationed was nowhere near as serious as it had been early in the war.

The trip to Okinawa and missions over Korea in 1953 would be quite different, however. There was no doubt Pat was going off to war, a war far more hazardous to him personally than the one he was in as a young airman in England. Although news and rumors continued to flourish about the on and off Korean Armistice negotiations, and a possible end to hostilities, the war ground on.

A loaded bomber is always on the attack – the offensive. Whenever his airplane and crew launched on a mission, every member of the crew knew full well he would be penetrating into aggressively defended target areas. Any aircraft loaded with thousands of pounds of explosive ordnance, destined to attack an enemy's war making potential, will inevitably draw concentrated, uncomfortably accurate anti-aircraft fire, and fighters – if the enemy can commit fighters to their homeland's defense. Pat met combat seasoned B-29 instructors at Forbes, who told every crew member coming through the school of the hazards in flying a huge bomber into the teeth of a ferocious air defense system.

Night air operations are even more unsettling. The world of night combat flying generates its own eerie and disorienting sensations, and a distinctly different set of hazards. The inky darkness, flying an aircraft "blacked out," with no exterior running lights, and interior, red, crew station night lights dimmed to an absolute minimum, shrinks the world outside and inside the aircraft. While darkness provided cover to a penetrating bomber, and usually delayed visual discovery by enemy defenders, radar could guide enemy guns, searchlights could help visual acquisition of the big aircraft, and there were night missions on which there might be no friendly fighter escort.

Thus the parting with Joanne that April day from New York City's Idlewilde Airport was filled with mixed emotions: devoted and passionate young love, sadness at the prospect of a prolonged separation, and the disquieting knowledge that something might go wrong — the nagging possibility that Pat Ryan might not come home.

They went to a wedding on Long Island earlier in the day before Pat boarded a commercial airliner to fly west toward the vast reaches of the Pacific Ocean. The wedding ceremony added emotional contrast to a day already filled with emotion. They were reminded of their wedding less than two years earlier, when war seemed far away, unrelated to their future.

As they kissed, they assured one another they would write every day. Both shed tears. Pat did his best to calm her fears, telling Joanne in the jargon of a well-trained, confident young pilot, "Don't sweat it. Everything will be OK." Now war was upon them, weeks, maybe days away. Joanne wouldn't be fighting. Pat would. But Joanne would be with him on every mission, and both would be in each other's thoughts and prayers.

To help Pat get through the difficult "I'll write every day" was the comfort of knowing another classmate and B-29 co-pilot was saying good-bye to his wife and riding the same airplane with him to Okinawa. He was Frank Fischl, who had been a steady performer on Earl Blaik's 1950 Army football team. Pat and Frank were heading toward a common danger, both, for a time, carrying the same heavy hearts felt down through the ages by men and women who went off to war leaving their loved ones behind.

Korea was the third time in thirty-three years that America's sons and daughters had gone off to war, two of them the most devastating wars in human history. Some of the nation's professional military found themselves fighting in their third major war. For Pat Ryan, Korea was his second time around, but there would be a third time in his future.

*   *   *

In the spring of 1948, the same spring that Joanne Fox met Pat Ryan, Joyce Elaine Riemann met Dick Shea. With approximately one hundred men who were to become his classmates in West Point's graduating class of 1952, he was finishing his one year at the United States Military Academy Preparatory School at Stewart Field in Newburgh, New York, twenty-five miles north of the Academy.

Joyce signed up for a Red Cross dance at the prep school that spring. Another of Dick's classmates was escorting her. She was dancing with her escort when she first noticed Dick sitting in a chair to the side of the floor. He had a cast on his leg, and with a pen and sketch pad was busily sketching the scene before him – the dancers enjoying themselves.

Running, sketching, and writing were three of his passions, until he met another. Joyce and Dick were introduced by the fellow with whom she was dancing. The introduction was first names only. Dick couldn't dance because of the cast. Though a superbly self-disciplined middle distance runner, he injured his heel while throwing the javelin in track practice.

Throwing himself into any athletic interest with complete abandon was typical of Dick Shea. In his prep school barracks building it wasn't uncommon to see him run and jump over rifle racks, which were in the center of the long aisle between lines of bunks. This night, however, Joyce captured his interest, and later in the evening he asked if he could call her. Joyce said yes, but her charm so distracted him he forgot to ask for her phone number. He believed he remembered where she lived.

The next morning Dick told Gray Parks, a classmate and the prep school yearbook editor, "I met a really wonderful young lady last night. I know her first name. She lives in Hackensack, New Jersey – I think." With a little sleuthing through the Red Cross and a New Jersey phone book, he obtained her telephone number. Always the perfectionist and gentleman, he decided the right way to contact Joyce was through the Red Cross chaperone who was with the girls at the dance.

Joyce received a phone call from the chaperone. "There is a Dick Shea, a West Point cadet candidate from the Stewart Field Prep School, who wants your address and phone number. Is it OK to tell him?" She gave her consent. Shortly thereafter he called to say when prep school was complete he would be going home to Portsmouth, Virginia. "I'll be getting the cast off my leg, and want to stop by and take you to dinner before entering West Point."

One week before Joyce graduated from high school, Dick arrived at the front door with his cast off, still walking with a slight limp. Their first date was fun, a delightful evening. Love was in bloom before Dick entered the Academy in July of 1948.

Dick Shea was intense, serious, deeply religious, iron-willed, and a driven competitor. Every morning before reveille he was up, getting ready to attend early Catholic Mass before breakfast. But there were lighter, sensitive sides to him, visible primarily to Joyce, his track teammates, roommates, and cadets who saw his cartoon drawings. His sketches – art – and cartoons in the cadet monthly magazine, *The Pointer*, the *Weekend Pointer*, and the monthly cadet calendar, reflected his ability to see and record the humor in cadet life. His talent was rewarding, satisfying to him and those closest to him. His cartoons complemented his more extensive artistic abilities. He was the primary designer of the class crest, which adorned '52's class rings, and he loved to sketch scenes, which in-

MAIL DRAGGER–Posey Paine, Col. Blaik's niece, recently told of the following embarrassing incident that took place on the occasion of her first date with a cadet: Being quite excited about the whole affair, Posey was ready and waiting in her Uncle's quarters about an hour ahead of the time for her afternoon date. Finally she glimpsed a form in gray walking toward the front door of the quarters, so she ran to the door and flung it open. The mail man was a little startled by her rapid advance, but managed to get out a polite, "Here's your mail ma'am" before Posey retired in embarrassment. Ramsey, B-2

A collection of Dick Shea's Cartoons
from **Pointer** *magazines 1950-52.*
(Drawings courtesy Joyce Shea Himka.)

PLAY IT COOL–While the Army-Navy game is always one of great tension, and both teams play their hearts out, there is always one player who can remain relaxed and still turn in a superior performance. Such was the case in 1929 when Army's great Light Horse Harry Wilson was playing. Harry could be depended on for a cool game of ball, and was casual at the most trying times. In this particular game the score was 0-0 until nearly half-time when Navy scored. When the two teams lined up for the extra point Harry leaned one elbow on the goal post and watched the Navy's kick float between the uprights. Naturally, in the dressing room during the half-time Harry was asked about his unusual behavior. "There's nothing to worry about," drawled Harry, "we're gonna beat these boys." It was a short pep talk, but it worked. With Light Horse scoring one touchdown, and sparking another, the Rabble came through to win 14-9. (Writer unknown.)

SPONTANEOUS RALLY–The Navy game of 1908 was such an upset and thriller that Army's spirit exploded and another custom was born. When the game was over, the cadets swarmed down from the stands and mobbed the team. In those days, the team was taken in horse drawn wagons to a hotel where they showered and changed. The exuberant cadets unhitched the horses and pulled the wagons to the hotel themselves. Upon delivering the victorious team to the showers, the group held a short pep rally in front of the hotel, and after checking a few clocks started off on the run for the train station. 'Way back then, the Corps was allowed very little free time–a few hours at the most. Upon arriving at the station, the rally boys found that the other half of the Corps had gone already, leaving them stranded. The WP training had been instilled deeply however, so the ranking cadet present formed his unit, took a DP count, and arranged for a special train with the Pennsylvania Railroad. In mad chase the pursuing cadets caught the Corps at Weehawken. Believe it or not, not a man was slugged. But there were some consequences, to be sure. The next time the two schools met in friendly combat, cadet guards were posted around the stadium to prevent escape. The Corps was formed up, marched to the station, and entrained–without being allowed to spend any of their trip allowance–fifty cents. (Writer unknown.)

'WAY BACK–In 1908 Army came to the Navy game with a poor record, while Navy, on the other hand, had a fine team which had lost only one game. Hence, the Goat was heavily favored to win. Among the Navy players was a huge tackle, All American and team captain, who was practically the whole Navy line. The game worked back and forth without any advantages being gained until Army recovered a Middy fumble on the Navy four yard line. Army's quarterback, Hyatt, didn't even bother with a huddle, but lined the team up and then called the play through the All American tackle. This play gained two yards, and again without huddling the team lined up. Hyatt glanced over at the Navy tackle and yelled "OK Tubby, through you again!" When the ball was snapped the whole Navy team was ready, but Army scored through the tackle. (Writer unknown.)

ON A PAST SATURDAY NIGHT, Al Mathiasen and some of the boys in D-2 were talking (from the front windows of the 24th Division) to some girls when someone suggested getting some food for poor Al, who was in confinement. The girls, mounted in their chariot, agreed to go after the "boodle."

When they returned, the Officer in Charge appeared on the scene as they prepared to pass the food through an open window.

"Are you trying to have that package delivered to some one?" questioned the

THE LOWEST FORM–At the Naval Academy one is not "quilled" or "gigged" for a breach of regulations, he is "fried"–that being Sailor slang for a Delinquency Report. On a recent Second Class Exchange weekend we met a cute little Navy Femme who had a good working knowledge of the above and most other middie lingo. She told us about her Blue-suit steady whose name is Fish. And we quote: "Those mean old Company Officers are always frying my Fish." (Writer unknown.)

wearer of the yellow band.

"Yes!" answered the girls, and then one added, "Would you deliver it for us?"

What he did deliver later was not food, but something that read, "Contriving to have food brought into barracks." (Writer unknown.)

DOUBLE TROUBLE–In the days when daring other cadets seemed to be the local sport, some pretty wild deeds resulted. Captain Price, of the Department of Electricity, told some stories on his fellow faculty members recently that emphasized this. One particular cadet, known for his daring, was challenged to go to Garrison after taps and bring back a handful of beer bottle caps as proof of his visit. Donning his civvies after taps the bold adventurer took off across the frozen Hudson and made his destination quite easily. Because of his success thus far, the cadet relaxed while he gathered his evidence, and time flew rapidly. With dawn approaching the cadet slipped down to the river to find that the ice breaker had passed and stranded him. Hoping to beat the Hell Cats back to the Point for reveille, said cadet started along the road for Bear Mountain Bridge, and was picked up by an officer he recognized as the "K" Company tactical officer. Realizing the game was over, the cadet relaxed and tried to carry on pleasant conversation on the trip back. Life was quite sad while he waited for his report to come through, but after a month had gone by the worry left, and relief followed. Finally graduation came and the new Second Lieutenant approached the First Lieutenant and said, "Jack, whatever became of the quill for that trip of mine?"

"Well you see, it would have been rather hard to explain to the Comm," Jack said, "just what the Officer in Charge was doing drinking beer in Garrison at three o'clock in the morning." (Writer unknown.)

LOGISTICAL PROBLEM–In order to expedite a hasty departure Second Classmen going on Annapolis exchange trips customarily leave their baggage on the Stoops of Barracks near the Academic Building. This practice precludes a return to rooms before embussing formation. Last Thursday the Stoops in front of A-1 had its usual burden of Cow B-4 bags, but this time there was a new addition. Derrick Samuelson, K-2 Senior, had a long weekend trip and he, too, decided to get a little ahead of the game and leave his bag in a strategic location. Along about 2 o'clock a 2 1/2 ton truck rolled up to take on its load of Maryland-bound luggage, and the stoops were cleared. Upon the Cow contingent's arrival at Annapolis, there followed the usual confusion of baggage claiming. Soon, however, all the Cadets found their bags and found middie escorts to carry them! When the dust cleared the Officer in Charge of the trip section was very much alarmed to find one unclaimed B-4 bag–belonging to a First Classman.

How was the weekend in Dress Gray, Derrick? (Writer unknown.)

LAST SPRING when E-1's Vann Brewster came under the spell of the joys of the warm spring sun, he began to play the evening study hours away on his uke. His wives, Sullenberger and Woodruff, eyed Vann's pleasant reveries with very thinly veiled displeasure. One day Brewster returned to his room in a musical mood and headed for the faithful uke. Groping on the gun rack (Vann's eyes were none too good as a result of cow academics) for his weapon, Vann was somewhat puzzled to find it gone. Subsequent inquiries disclosed that his wives had mailed Vann's uke home to Georgia. Bob Porter, E-1.

THE CUSTOM of throwing yearlings into Lake Popolopen on their respective birthdays is a time honored tradition of Camp Buckner. However, last summer, the custom was not strictly adhered to, due to the close presence of the Tactical Department and the resulting desire on the part of the class of '54 to dodge quill.

Near the end of the summer, however, a refreshing and amazing departure from the normal occurred. On this night, several of the tables were rendering "Happy Birthday" to one of the boys when the Officer in Charge, Colonel Rose, strode up to the Area Commander and inquired with a smirk, "Can that man swim?"

REVEILLE FORMATIONS often turn out to be a session of the blind leading the blind, but some things open even the sleepiest eyes. An experienced late-sleeper can rouse himself with seconds to go and, by dressing all the way to the division doorway, can step into ranks with no pain in time for the report. Howie Hunter of B-2's yearling menagerie will testify that a few extra seconds pay off though. What for? To check that no one has pinned your dress coat sleeves together. (Writer unknown.)

Like wildfire, the statement of the officer was spread through the mess hall. After supper a crowd accumulated at the dock to watch the proceedings. Most would not believe that the tradition would be reinstated without the usual yellow forms being passed out. They were convinced that they were wrong after watching a "double header" take place without a pencil being put to a pad of Form 1s. (Writer unknown.)

661

BLUE BOOK BLUES –Attendant to the post of Room Orderly are a defined number of duties. The insertion of interpolations and counter interpolations in the new tome-size Regulations USCC is not an infrequent task. Co-author Karl Waltersdorf found himself the holder of this title when a plebe, poopsheet in hand, rushed into the room after Tattoo one night. Karl, with a diligence born of a realization of the importance of his office, and with elan engendered by a high sense of duty, applied himself to the rapid completion of the task. He remained unswerving in his application as the mail carrier brought in a second sheet dealing with the chartering of helicopters and safety precautions of the model railroad club. But his tension began to mount as the succeeding leaves came in by pairs and trios. With sweat pouring down his brow, he feverishly clipped and glued, lined out entire paragraphs, ripped out previous addends and replaced them with superseding rules. At the next to the last note of taps he triumphantly completed his work and informed the cadet sub-division inspector that the required interpolations had been posted. The subdiver counter informed Karl that such was not required. He was strapped to the bed until his raging murderous intent gave way to an incoherent sobbing. (Writer unknown.)

DRAGGING PAINS–Plebe Christmas– the best deal since the big Hike, when all heretofore subdued lady-killers come out from behind their collars to drag in the pleasant atmosphere of an upperclass-free West Point. One anonymous (for obvious reasons) Fourth Classman from I-2 decided to ensure himself of female companionship for the holidays and asked three different girls up. One day, having just returned to Barracks from dinner at an Officer's home, he received word that one of the girls had arrived. He promptly hurried to the Hotel. Since she hadn't had dinner, he graciously took her into the Dining Room. After finishing, the couple returned to the lobby where our man spotted Femme No. 2 standing in front of the desk. Hastily telling No. 1 that he had to return to the company for several hours, he shoved her into the elevator and went over to meet No. 2. Again since No. 2 was "simply famished," he offered to take her to dinner. Feeling anything but sharp after this third meal within two hours, he excused himself and returned to barracks.

The rest of the story is a bit hazy, but there was talk of an attempted suicide in I-2 when the CQ notified our Dragoid that No. 3 was waiting for him at the Hotel.

Next chapter: How a cadet managed to distribute three "A" pins during his Plebe Xmas. –Tant I-2.

cluded trees, buildings, homes, and thoroughfares filled with busy people.

An incident which occurred at the end of his plebe year, during June Week of 1949, typified Dick Shea's tongue in cheek good humor, and his iron-willed self-discipline.

Every plebe in '52 knew recognition was near, an event which at West Point from year to year added to the air of anticipation and June Week excitement. Freedom was near for lowly plebes. Adulthood. No more time on the lowest rung of the Academy's four class ladder. Upperclassmen, and particularly first classmen about to graduate, delighted in pouring on the heat, putting on the pressure, heaping additional tasks on plebes as recognition – and graduation – drew near. During plebe year, for example, "calls" were normally a form of discipline administered short of issuing demerit slips, and almost always were ten to twenty minutes prior to first call for meal formations. But during June Week, calls were demanded of plebes when there were no infractions. For the most part, it was all in good fun, although there were always those who seemed to see the end of the first class-plebe relationship as the last time to make a plebe miserable – a time to unleash a last burst of fury in an education system for which some first classmen had no patience or understanding.

During June Week Dick Shea and Ron Obach, Dick's roommate for three and a half years, were told to report to a room of first classmen, "on calls" in full field gear: fatigues, combat boots, helmets, and full field packs. There followed the customary demands for plebe knowledge, and at the slightest hint of an error of any kind, a period of "correction," usually a nose-to-nose dressing down for the lack of knowledge, lack of study, or carelessness, which the error represented. After a few minutes of being vigorously worked over by the first classmen, Dick Shea decided to put a little humor into the situation.

Without warning, he began exhibiting symptoms of an oncoming faint, leaning forward more and more, ever so slowly, his body still ramrod straight, until he fell forward, apparently headed for a face smashing fall to the floor. Down he went, to all outward appearances, a plebe who would soon need medical attention, embarrassingly, in the room of first classmen about to be graduated and commissioned. What the soon to be lieutenants didn't see for a few moments was that a superbly conditioned, sinewy strong Dick Shea had caught his fall with his hands, which were hidden underneath his upper body. There was a short period of astonished silence, as Dick "lay" rigid, about two inches above the floor, apparently passed out cold.

What next? Dick continued to feign his condition for a few seconds while the first classmen's surprised silence and gazes continued. Then the game was up. Dick slowly turned his head, and with a sheepish smile, got to his feet. When the first classmen realized what happened, they failed to see the humor in the situation. Not only did he momentarily fool them, he made the mistake of being unable to suppress the embarrassed smile, which was interpreted as smiling at his own success. Emotion of any kind, especially that which appeared derisive, rebellious, or at their expense, wasn't well received by the majority of upperclassmen.

Then came the acid retort. "Well, Mr. Shea, so you think you're very clever? You'll double time from this room to the Stoney Lonesome Gate and back, in full field gear – now! Do you understand, Mister?!"

"Yes, Sir!"

"Post, Mr. Shea! Move out! On the double!"

"Yes, Sir!" And off Dick Shea dutifully went, all the way to Stoney Lonesome Gate and back, in full field gear, in New York's early June heat. Though his injured heel had kept him from participating in cross country and track all plebe year, he didn't hesitate. He had played intramural football in the fall of plebe year, and joined the Army gymnastics team for the winter, where he won his numerals. He was fit, healthy, and tough when the first classmen sent him on his June Week run.

Dick Shea could exhibit flashes of anger, but he carefully controlled the emotion, turning it to positive, constructive action in his relationships with others. And he couldn't stay angry for long. Ron Obach recalled that one evening after Taps, lights out, when quiet was settling over the company area, Dick suddenly said, "Ron, I'm sorry."

Ron was puzzled. "Why are you sorry?"

"I have to make peace before I go to sleep," Dick answered. There had been an argument earlier in the day, but Ron Obach had considered the matter so unimportant he'd forgotten what the disagreement was about.

Ron was nearly four years younger than Dick, and Dick was always looking out for him, as though Ron were a younger brother. Ron noted his roommate had to work hard in academics, with the same intensity, seriousness, and concentration with which he approached everything he did. And there was always Joyce.

When he wasn't with her on weekends they would write each other at least every other day. Their letters were full of love, family, descriptions of people they met and enjoyed, and their experiences. His letters often contained sketches to describe his observations and feelings. Lou Davis saw in Dick's facial expressions his devoted, passionate love for Joyce, when the two men roomed together on trips with the Army track team. When the rest of the team went out to have a good time, Dick stayed behind and sat quietly at the room's desk, deeply absorbed in writing letters to her.

In 1949, when Rogers' and Hammerstein's musical *South Pacific* began its lengthy run on New York City's Broadway, Ezio Penza, the male lead, sang a dreamy, romantic ballad later recorded by Perry Como. The song went straight to the top of the ten best selling popular records that year. *Some Enchanted Evening* became Dick Shea's and Joyce Riemann's song, the song that wherever they were, whatever they were doing, turned their thoughts to one another and the deep love they shared.

The lovely ballad's lyrics reminded both of their first meeting at the prep school dance.

> *Some enchanted evening,*
> *You may see a stranger.*
> *You may see a stranger,*
> *Across a crowded room.*
> *And somehow you'll know,*
> *You'll know even then,*
> *That some way you'll see her again and again...*
> *Fly to her side and make her your own,*
> *Or all through your life you may dream all alone...*

In March 1950, when he was a yearling, Dick proposed to Joyce. He was in New Milford on weekend leave, and they had seen a movie. They rode the bus to the stop nearest her home, but the walk was long.

The day was cold and windy. Once in the house out of the cold, Joyce made hot chocolate. They talked quietly. It was a romantic day etched indelibly in their memories.

They announced their engagement Sunday, June 4, Joyce's birthday and two days before the class of 1950 graduated. He gave her an Academy miniature, the traditional engagement ring given by cadets. The miniature is a lady's size class ring. For Joyce, it was the 1952 class ring, which bore the class crest Dick designed. The setting was a blue sapphire surrounded by small diamond chips.

A few days after their engagement, Dick, a brand new second classman in the class of 1952, began the summer training schedule, taking leave and then going with one of two groups in his class for training at Air Force bases. In August he sailed down the Hudson River on a Navy troop ship his classmates laughingly referred to as the "luxurious ocean liner, tourist class," the USS *Okanogan*. They were soon to begin CAMID, the annual, joint cadet-midshipman amphibious training exercise near Virginia Beach, Virginia. True to form, as the ship moved slowly down the Hudson from West Point, Dick Shea stood next to the port railing, facing the east shore of the river, busily sketching all the scenic beauty and grand estate homes he could.

Three weeks after Dick and Joyce announced their engagement, while he was still in summer training, the *In Min Gun*, the NKPA, launched their surprise attack across the 38th parallel into the Republic of Korea. War had begun, yet it seemed far removed from their lives and the placid academic environment of beautiful West Point. But the war raged on, and two more graduating classes were to send young lieutenants into the fight.

Their wedding reception after the graduation day marriage two years later was at "A Little Bit of Sweden" on Highway 9W, south of West Point, on the road to New York City. They drove to Vermont for their honeymoon, staying at a cottage on Lake Champlain. There was a canoe for their use and they explored the many islands on the lake. The weather was warm and sunny.

On June 14, after their honeymoon, they started for Joyce's parents' home in New Jersey. In the late afternoon of graduation day's frantic activity – packing, last minute good-byes, loading cars, and breaking out new Army uniforms with second lieutenant's gold bars, Ron Obach had

Dick and Joyce Shea at Ron Obach's wedding reception, June 14, 1952, in Larchmont, N.Y. (Photo courtesy Joyce Shea Himka.)

been in Dick's and Joyce's wedding party. Now, on June 14, after eleven days attending or participating in classmates' weddings, Ron was being married in Larchmont, New York. Dick had promised that he and Joyce would attend his former roommate's wedding.

The travel time from Lake Champlain was more than anticipated. They hurried to the wedding, but arrived uncomfortably close to the starting time. Dick, in a great rush, hoping to make the ceremony on time, stopped his car in front of the church, and ran to the front door. "Is this the Obach wedding?" he asked. "Yes," he was told. "Well, hold everything until we park our car." The wedding began ten minutes late, after Dick and Joyce sat down among the groom's friends and family.

Then it was off to New Milford, New Jersey, and Portsmouth, Virginia, to visit the Riemann and Shea families the remainder of graduation leave. It was a marvelous time, and they both looked forward to their new life.

Next came the Infantry School at Fort Benning, Georgia. They rented an apartment off post. Following Infantry School, they remained at Benning where Dick took airborne training. He wanted to attend Ranger School, but received orders for Korea. When they left Benning near the end of March 1953 for another round of visits with families, Joyce was four months pregnant. They both would rather he stayed with her until the baby was born, but accepted the assignment without question. This was the duty of a professional soldier, his work, and they both understood he would take the assignment and do the best he could.

They stopped by Dick's home in Portsmouth, and went from there to New Milford, where Joyce would stay with her mother and father while Dick was in Korea. While they were in New Milford, they saw Ron Obach and his wife, who were visiting Ron's mother living in nearby Ridgefield Park, New Jersey. Ron, who was in the Signal Corps, had orders for Germany and was on leave preparing to head east across the Atlantic. When the two former roommates and close friends realized it was time to wish each other well, they hugged one another. "Don't go do anything foolish," Ron said to Dick. "Come back." Dick assured him he wouldn't be foolish. He planned on returning.

The day of parting for Dick and Joyce Shea was difficult, too. They had had some serious talks that most young married couples don't have. If anything happened, if anything did go wrong, what would he like Joyce to do? Would she begin a new life? Would she remarry? Where would he like to be buried? The subjects of their conversations, however unpleasant they might be, were necessary, but certainly weren't Dick's way of thinking. He was the eternal optimist who loved life and lived it to the fullest. However, these serious conversations were motivated by the same nagging fears and possibilities which tormented Pat and Joanne Ryan when Pat boarded his flight for Okinawa – and by Ron Obach's remarks to Dick when Ron left for Europe.

Joyce and Dick Shea were at New York City's La Guardia Airport for Dick's night departure less than a week before Ronnie and I met for the first time at Grant Hall on April 25. The boarding area was brightly lit. He was the last passenger to climb the boarding steps to the airplane. As the file of passengers slowly made its way up the steps, there were momentary pauses as passengers pulled tickets from coat pockets or purses, so that flight attendants could check them. With each pause Dick turned to gaze at Joyce, smile and wave, finally turning to wave one more time before entering the aircraft door. He was destined for California, Camp Stoneman, where he stayed a few days before boarding the Navy ship *General W.H. Gordon* in San Francisco, bound for Japan.

He called her several times from Camp Stoneman and continued to write at least every other day. He called once more from Japan. He couldn't tell her on the phone he was assigned to the 7th Infantry Division, "The Bayonet Division."

When he arrived at 7th Division Headquarters, he went through an orientation, as did Dick Inman. The 7th was a proud Division, and had traveled a hard, bloody road in Korea. It had come ashore with the 1st Marine Division at Inchon in September 1950. Included in its ranks were eight thousand Republic of Korea soldiers, nearly half the division strength. Flushed with victory, it had driven into the far northeastern Korean peninsula by the end of November of that year. One of its Regiments, the 17th Infantry, laid claim to the first American unit to see the Yalu River, which marked the border between China and North Korea. When the Chinese stormed south from the Yalu in November of 1950, the Division's 32d Regiment, "The Queens Own" from Hawaii, had been terribly mauled, along with the 1st Marine Division at the Choshin Reservoir the first two days of December.

Dick Shea in South Korea, June 1953, after his arrival in the 7th Infantry Division. (Photo courtesy Joyce Shea Himka.)

Now the 7th manned a sector in the main line of resistance – MLR – for the United Nations' forces and 8th U.S. Army, as well as outposts, trenches, bunkers, and unit command posts in front of the main line of resistance. These positions in front of the main line of resistance constituted the outpost line of resistance, adding to the defense in depth which had been established as the front stabilized nearly two years earlier. The sector included some of the same hills the 1st Cavalry Division had held when the 7th Cavalry's K Company and David Hughes learned their hard lessons about war in the summer and early fall of 1951.

The 7th Division S-1 (personnel officer) assigned Dick to the 17th Infantry, "The Buffaloes," another historic Army regiment with a colorful history. One little known piece of unit history was connected to

Army football. First Lieutenant Dennis Mahan Michie, class of 1892, and the cadet who coached the Army team in the first Army-Navy game in 1890, was serving in the 17th when he was killed in action during the battle for San Juan, Puerto Rico, on July 1, 1898. West Point's Michie Stadium bears his name.

From the 17th Infantry Regiment orientation, Dick Shea went to A Company, 1st Battalion. His battalion commander was Lieutenant Colonel Beverly M. Read, a fellow Virginian and a Virginia Military Institute graduate from the class of 1941. First Lieutenant William S. "Bill" Roberts, Jr. was his company commander. Assigned to B Company, 1st Battalion of the 17th, was Dick Inman, who had sailed from Seattle, Washington, in April.

* * *

Dick Inman was a hurdler and high jumper in high school and at Army, but above all the sports he enjoyed, he loved football. Football was in his blood. During the fall of 1948 he walked on to the plebe team, and made the squad. The first semester of plebe year, English writing assignments were frequent, and in one theme he told of his love for the game. The Army varsity was nearing the end of an undefeated season when, on November 9, he turned in his paper.

...If a player worries about his efficiency in the game, he reveals his desire to win – and if he is to win, he must first *want* to win. As Grantland Rice has said:

'When the Great Scorer comes
To mark against your name,
He marks not if you won or lost,
But how you played the game.'

If a player wants to win, he will play with his heart, as well as with his body.

The football player is more than a machine; he is also a human being. He could not, even if he wished, break away from his feelings. He must take the pounding, sweat, sprains, mud, grime, dust, blood, and fatigue without letting them affect his outlook toward the game. His attitude in receipt of those punishments

670

reflects his real tendency toward becoming a great gridiron performer. He must be able to take what the weather, coaches, opponents, and spectators toss his way with a smile that shows that he loves the game for the game's sake. A player does best what he likes the best, and therefore, to the team the player's love of the game of football is of primary importance. He must like the body contact, the impact of a running man to his shoulders, the little nicks and bruises he gets today and are gone tomorrow, the bloody lip he receives and the chance he gets to 'dish one out,' the dirt that comes off in a hot shower, the thrill he gets by running through a scattered field of men with a ball tucked underneath his arm, the satisfaction he receives from making a good block or tackle, the admiration he demands, the reprimands he deserved, the shouts, the yelling, the moments of uncertainty and the moments of victory, – all must be in his soul, for in football what is inside a man makes the big difference between the 'bench-warmer' and the champion player.

A penned footnote at the end of Dick's typewritten paper reminded his Academy English instructor that Grantland Rice wrote his famous words in the Notre Dame locker room, in South Bend, Indiana. The great Army-Notre Dame rivalry had been played one more time in Notre Dame stadium the year before Dick Inman entered West Point, and was the final game before a mutually agreed end to the series, a series interruption that lasted ten years.

His November 1948 theme paper wasn't the last time Dick wrote of his love for football. English was his best subject. He particularly loved to write, and poetry was the form of writing he seemed to enjoy the most. He later wrote a poem about the game.

The Ifs of the Gridiron

If you can keep your head low when you drive through,
And yet know where you're going – and get there;
If you can call your signals loud and snappy, too,
Yet make your barking and your fight a pair;
If you can play and not tire too soon by playing,
If you can work, yet make your work your play,

If you can start and turn fast, never staying
'Til the other fella takes you out o' the way;

If you can win a ball game like you take defeat,
Letting neither turn your head nor damp your fire;
If you can realize that victory is sweet
And must be bought by actions *and* desire;
If you think, and while you're thinking, get things done,
And, when you do them, do the *best* you can;
And not be satisfied by only winning one,
But win them all, and win them like a man;

If when the chips are down, you give out all you've got,
And get tougher when your progress is the worst;
And know that you're the toughest man on all the lot,
And never *ever* let *that* bubble burst!
If you can keep the confidence of your teammates,
And earn respect from those on the other team;
And learn to do exactly what your coach relates,
However hard or bad the whole thing seems;
If, when two yards are needed, you can make it,
If you give as many bruises as you get,
If you dish out *hard*, as well as take it,
And the dirt and dust, and mud and grime, and sweat;
If you can block 'em high and hard, and tackle low,
And mean it when you get a hold the ball;
If you can keep determination in your soul,
And throw a thrill and scare into them all;

If you can stop bad habits from growing on you,
Yet learn to keep your virtues just the same,
And remain a man, tho' praises are heaped upon you,
And, most of all, if you love your football game, my friend,
And what you win in your later years will tell,
'Cause you will be among the highest in the end,
For you will win the Game of Life as well!

Largely hidden beneath his tough, athletic exterior, and seen in his poetry was a complex, gentle, and sensitive man, a Dick Inman normally seen only by those closest to him – his family and friends. His poetry reflected a deep, serious thinker who spoke not only of a love for football, but also told of the beauty of his home state, the "Hoosier" state of Indiana, his growing up, and revealed his innermost thoughts about war, its ugly senselessness and what he saw as the only way the scourge of war would ever be banished from the earth.

The loving, sensitive side of him was seen briefly by his classmates Christmas of plebe year, when the entire class remained at West Point while the upper three classes went on leave. His family came to visit him during the nine-day break. He had one brother and two sisters, and the youngest, Bonnie Ruth, was three years old. He wrote and spoke of her often, and was lavish in telling of his big-brother love for her. Dick's mother brought Bonnie Ruth to a hop in curls and a pink satin, child's formal dress made especially for the Christmas at West Point. Dick danced proudly with his little sister, and his classmates teased him about "dragging pro," cadet jargon for dating a pretty girl.

A poem, written while Dick was a cadet at West Point, said far more about his love of the littlest lady in the Inman family.

To My Sister Bonnie

The world abounds with hopes of all men
For goodly, beautiful wives
And ladies who would be right for them,
Who'll give them happy lives.
And the thoughts of a man are centered 'round
The celebrities, it seems,
The princesses, queens, and those who are found
Mostly in their dreams.
But my heart goes out to a little girl
With a young and smiling face,
Who fills my heart with a happy love
In the warmth of her embrace,
For the light in her eyes will always be
More dear than treasures of old,

And the lasting joy she has given me
Worth more than a kingdom of gold.

To most of his classmates, however, Dick Inman was outgoing, bubbly, and at times, bombastic. Tough, disciplined, hard working, strong-willed, almost stubborn, he was the picture of a perpetually training athlete. He had a booming voice and a loud, hardy laugh. During the winter, he would storm into his cadet room after a vigorous workout and hot shower at the gymnasium, jerk the windows open while pulling off his cadet gray short coat and scarf, and enthusiastically insist his roommates join him outside to practice wrestling in the snow, or have a snowball fight.

He was, above all, an individual, fully formed, fiercely independent, content in his beliefs. West Point seemed to have little effect on him. He adapted to its rigors easily. Throughout his four years at the Academy, he retained many of the same traits as the young plebe that came to West Point in 1948. To him, although he took pride in the honor code, the code's tenets were insufficient. He possessed strong, unwavering beliefs about morality, entering marriage as a virgin, and the sanctity of the marriage bed. He abhorred profanity, and didn't use it.

Yet, periodically he became quiet, introspective, discouraged with himself. Usually such quiet came from his disappointment at not being as successful as he wished in Army football. There were frequent conversations on the phone with his dad, his high school football coach, from whom he sought counsel and encouragement regarding his self-perceived failures.

His love of football, and his devout but frustrated wish for gridiron success was made even more troubling by the cheating scandal, with its tragic, crushing impact on the Army team. He remained on the junior varsity throughout his yearling and cow seasons, much to his chagrin. But what he didn't know was that he also remained outside the tight circle of varsity and junior varsity team members who were involved in organized cheating.

They had been thoughtful and disciplined in observing their less favored teammates, and talked among themselves about who would be good risks to join in the group's activities. Because of Dick Inman's strong, outwardly expressed beliefs in the ideals and practices of the honor code and system, he had been one of numerous teammates they carefully avoided. And of course he never saw or knew of the contents of the Collins, Hand, Jones, or Barrett Board reports. His name was nowhere to be found in the several board proceedings or among the cadets named as possible participants in cheating. Nor was he called before the Collins Board as a witness.

When the August 3 public announcement hit the national press and triggered the storm of questions and controversy over Earl Blaik and Army football, Dick Inman, like many other cadets and Army players, was home on leave. He told his parents he wasn't involved in the ring, that everything was all right. Indiana newspaper reporters contacted Dick's mother, learned of his conversation with her, and widely publicized his words. He had told his mother what the whole family instinctively knew to be true. He said simply, "I have a clean bill of health."

His experiences on a scandal shattered team and the emotions stirred in him during a period both troubling and traumatic were typical of those suffered by Army football players intensely proud of the honor code and system, and what they were trying to accomplish on the gridiron. For Dick Inman, a young man who so loved the game of football, this period must have been especially difficult.

In spite of the scandal's devastating impact on Army football, there were some pluses for men on the junior varsity, the "Golden Mullets," men like Dick Inman. They would now be afforded an opportunity they otherwise would have been denied had the holdover Army varsity not been swept away. Although Dick was a speedy halfback in high school, and a good high hurdler and high jumper on the Army track team as well, he

Dick Inman wearing his Army varsity football uniform, and the number of a guard, in October 1951 of his first class year. (Photo courtesy Mary Jo Vermillion.)

had spent most of his time playing guard on the Army junior varsity. He wore the number 62 on the Jayvee's, and when 1951 preseason photos were taken of the inexperienced varsity Earl Blaik was going to field, Dick continued wearing the number of a guard – 62. Because of his speed, Blaik later moved him to defensive halfback, and he also received kickoffs. He played little early in the season, but playing time increased enough to earn him a varsity letter, the much sought after "major A."

When the team went to Evanston, Illinois, for the second game of the season, and almost pulled off an upset of the Big Ten's Northwestern Wildcats, Dick Inman suffered the first of two glaring football embarrassments that year. The Northwestern game stung him because his family, always his loyal boosters, had traveled from Vincennes, hoping to see him play. Instead, they saw him keep his parka on, and as the game progressed, slowly lean back and slide down on the bench with his legs stretched out in front of him, heels on the turf. After watching Dick's evident frustration at sitting on the bench far too long on a damp, rainy afternoon in Dyche Stadium, they began to chant. "Send in Inman! Send in Inman! Send in Inman!"

Their hero could plainly hear the chants of his family in the stands behind him, and at one point turned and smiled lamely, acknowledging their enthusiasm. The chant continued, to no avail, either not heard, or ignored, by Earl Blaik and his intent coaching staff. Finally, mounting frustration and sympathy for Dick's bench-bound plight, brought more pointed, indignant shouts. "Send in Inman! It can't get any worse!"

In the last two minutes Army lost a 20-14 heartbreaker to the Wildcats that afternoon. This was the day Vince Lombardi shed tears in the Army dressing room because a gallant, inexperienced team nearly pulled off an enormous upset. Dick Inman never got in the game.

The worst humiliation came in the Army-Navy game, the last game of his collegiate career, when he fumbled Navy's kickoff, and the midshipmen promptly marched to their second of six touchdowns. A player's mistake in an Army-Navy contest can be an unpleasant, lifelong memory. A hundred thousand witnesses seated in the huge Municipal Stadium, the rivalry's grand history, normally the last game in an Army player's career, and the game film's replays seem to keep the mistake before the entire world. Dick's sister, Mary Jo, didn't go to

Philadelphia, though their father did. She saw Dick's fumble in a movie theater in Vincennes and told her brother about it. He remarked ruefully, "My one chance at fame and look what happened."

After graduation leave Dick went to Fort Benning to the Infantry School, one of 174 men in his class to attend. In the fall of 1952, after completing the school, he went to Camp Atterbury near Indianapolis and awaited an overseas assignment. Unlike Dick Shea, he wasn't interested in airborne training. He was anxious to be stationed near home and was actively pursuing an interest in Dr. Wernher von Braun's research into missiles and space at Redstone Arsenal in

Richard George Inman of Vincennes, Indiana, in his first class year at West Point, and a pensive moment at his desk – known as the 1952 poet laureate and their number one apostle of good cheer. (Photo courtesy USMA Archives.)

Alabama. He had been in touch with von Braun, and hoped to be an early participant in the exciting and futuristic field.

He played football on the Camp Atterbury team for a while, as did many newly commissioned former Army players at posts and camps throughout the country. He was happy to be near home and visited his family as often as he could. He was especially happy to be near them for the Christmas holidays. He drove to Indiana University and picked up his sister, Mary Jo, who was in nurse's training, then went to Wabash College where his brother, Bob, was in school, and drove them both home for the holidays.

In January, Dick's close friend and high school classmate, Lieutenant Jack Waters, also stationed at Camp Atterbury, told Dick he'd like to arrange a date with a young lady whom he was certain Dick would be pleased to meet. A benefit dance at Camp Atterbury would be the setting. Dick said yes and met Barbara Kipp from Fort Madison, Iowa.

She graduated from Indiana University in 1952 with a degree in education and was teaching elementary school in Indianapolis. Her work included evening classes, teaching reading and writing to young soldiers at Camp Atterbury – most of whom were draftees – to improve their communication skills to the fifth grade level.

Love at first sight struck again, followed by an unexpected whirlwind romance. Their blind date was followed by a series of frequent dates and meetings. With each encounter, lengthy personal talks drew them closer. Their feelings for one another were rapidly becoming serious, much more than casual friendship. In a quiet, tender moment he said to Barbara, "The jagged edge of our souls fit together." This was the romantic, the poet in him.

Barbara and Dick had much in common. To her delight, she found in Dick a wonderful sense of humor, yet a noble, deep, romantic thinker. She taught English, the subject Dick enjoyed the most. His love of poetry was evident, and he was already working on a book manuscript about world government as the best hope for universal peace. He told her he started the work while he was a cadet, when he wrote a monograph his first class year. The paper was so well done it drew exceptional praise from the Academy's Social Sciences Department, and spurred his already strong interest in the subject.

They discussed marriage, and Barbara said she would marry him if he wished. But he knew he was going to Korea, so he at first hesitated. He was uncertain whether it was right to marry and go off to war, leaving a new bride – an uncertainty repeated millions of times down through the ages of warfare.

During their talks about marriage and the risks of war, he told her he believed "A soldier should lie where he falls." He also insisted that she promise to remarry if anything happened to him. Both unpleasant subjects were foreign and disquieting to Barbara, but she accepted the realistic necessity of such conversations.

He had declined reassignment to Japan to play football and thus possibly avoid the war. Football at Camp Atterbury dampened his enthusiasm for the game. The team included former professional and collegiate players, not well coached or conditioned. The result was rough, coarse football, not Dick's ideal of the game, and a letdown after the excitement of national collegiate competition played before huge crowds.

As the day of Dick's departure for Korea drew nearer, his uncertainty about marriage vanished. He concluded they should marry. The date would be February 28, 1953. Dick, in his usual, take-charge, energetic manner arranged the wedding while Barbara continued her demanding day and evening teaching schedules. They had known each other for six weeks when they said their vows at the Broadway Methodist Church in Indianapolis.

Dick's mother and father attended, as did Barbara's parents. Bob Inman, Dick's brother, was best man, and Jack Waters, who had introduced Barbara and Dick, and Barbara's brother Bill, were ushers at the wedding. Barbara's close friend, high school and college classmate and roommate Joanne Hardy was bridesmaid. Dick and Barbara took a brief honeymoon. Until they found their own apartment, they called Barbara's apartment at 407 North Pennsylvania Avenue in Indianapolis their home. Six weeks after the wedding Dick left for Korea.

Their parting was no less difficult than that of the Ryans and Sheas. He didn't want Barbara to go with him to the airport the day he left Indianapolis for Seattle. After he left, the growing distance between them deepened the love they shared. An almost daily exchange of letters ensued. He received two letters from Barbara in the few days he was at Fort Lewis, Washington, just outside Seattle.

And Dick, who was no less devoted to his family, frequently wrote home – to his parents, Mary Jo, seven-year-old Bonnie, and the entire Inman family. From Fort Lewis he wrote,

23 April, 1953

Dearest family –

I have time this evening to jot down a note or two. Nothing has broken since this morning. Tomorrow we who are on the water shipment I spoke of get our 'water orientation'. Then we will get all the poop on traveling overseas. Today we had a three-hour pre-combat instruction session....tonite I intend to get off a letter to West Point to find out about the Army censorship business as regards my monograph, and a line to Wernher von Braun to see if anything has ever come of my request to get into the guided missile field....

Let's see, Bonnie's birthday will be next Sunday, the 26th, won't it!? Happy Birthday, Bonnie Ruth. I'm sorry I can't send you anything this year, because I'm on the way overseas and we're not allowed to carry very much. I don't have much money with me, anyway, so I'll make up for it on the next Christmas I'm home, OK? Now, you be a good girl and don't hurt yourself riding your bicycle – and write me a letter sometime! I'll be waiting to hear from you.

Well, folks, there's really not much to say.... I'm supposed to have a pretty easy job aboard ship, and that we land in Yokohama, Japan, and go to Camp Drake in the islands before we hop over to Korea.

Will write more tomorrow – Take care of yourselves – I miss you all.

Love and kisses,

Dick

Shortly after arriving in Korea he wrote, "My darling wife keeps flooding the mail with letters. I got eight from her [the day I arrived] at Camp Drake, and *five more* the next day...!"

From Camp Drake he went to the Korean port of Pusan, at the southeastern end of the peninsula, then north, northwest by train, an agonizingly slow 250 miles in eighteen hours, to arrive near 7th Division Headquarters where he would begin his unit orientation. The war torn hills surrounding the village of Sokkogae drew nearer. The tinkling bells of love were soon to mix with the distant echoes of mortar and artillery barrages, and the staccato of machine gun fire and bomb explosions from American aircraft attacking enemy targets along the line.

WELCOME TO THE WAR

Terrain and weather are crucial factors in any ground war of assault and maneuver, and no less crucial to defense. High ground, the highest hills and mountains, road and railroad networks, rivers, bridges, lakes, dams, beaches, coastlines, high mountain ranges, narrow passes and gorges, heavily forested areas, the openness of fields to cross, the ability to take cover in the face of incoming mortar, artillery, direct fire from armored vehicles or automatic weapons – these are all important considerations in any ground war.

The area to the rear of an intractably stabilized front, across which two contending armies have alternately glared, torn, and raged at one another, is strange. The war to which Dick Inman and Dick Shea were introduced was not, in many ways, the war experienced by David Hughes and the men of K Company, 7th Cavalry. The two classmates were entering the eerie life of World War I trench warfare, a trip back in time, whereas the 7th Cavalry's and 17th Infantry's earlier baptisms of fire in Korea had been in a war of maneuver – almost constant movement.

In the early days of the war there had been little time to dig in deeply and provide overhead cover and protection. Every unit and every man had to be mobile, ready to pack up and move with little or no warning. Digging in for cover and a night defensive perimeter was hurried, usually shallow, because there was never enough time. The UN Army hugged the Korean road network as it moved, sent infantrymen ahead, up the hills on either side of the roads to protect flanks and avoid ambushes. While moving in columns on roads, the Army relied heavily on patrols sent ahead, and on the eyes of artillery forward observers and tactical air controllers – pilots assigned to ground forces or who flew in "Mosquitoes," AT-6 spotter aircraft or helicopters – to look ahead and behind advancing or retreating columns.

Artillery and armor units, supply points, staging areas, command centers, tactical communications and intelligence nets, medical units, and administrative headquarters – all the trappings of a field army – had been constantly on the move with the advancing then retreating infantry. Even supporting tactical airfields had suffered the effects of the early days of the "yo-yo" war, setting up operations near the fast moving front as United Nations forces stormed northward out of the Pusan Perimeter following the Inchon landings, then packing up and leaving as the waves of Chinese infantry swarmed southward in November and December of 1950.

Now, when Dick Shea and Dick Inman arrived in May of 1953, the main line of resistance and the outpost lines of resistance had been essentially stable for nearly two years while truce negotiations ground on. The lines were laced with deeply dug, heavily braced and strengthened, interconnecting trenches, bunkers, underground command posts, and caves. There were trails and roads linking defensive positions, one to the other, and to rear supporting areas where units were in ready reserve, and ammunition, equipment, vehicles, and supply marshaling areas were prepared to react to enemy attacks.

There were constant efforts to improve and strengthen the lines, including endless digging and reinforcement with anything that could reduce the devastating casualties and destruction brought by mortar and artillery barrages.

There were backup plans, contingency plans, and redundancies at key points of vulnerability or avenues of enemy approach. Alternative and redundant communications systems and methods were put in place. There was firing in, "registering" – or "zeroing in" – of artillery and mortars for new, pre-planned barrages on likely avenues of enemy assault. Training, orientation of enlisted and officer replacements, rehearsing for raids and patrols, and participation in reconnaissance and ambush patrols were the routine. Regularly scheduled noncommissioned officer leadership schools prepared corporals who could at any moment, in a deadly fire fight, find themselves assuming leadership roles, life or death responsibilities, on the battlefield.

On Tuesday, May 19, Dick Inman wrote,

Dear Folks –

...I am now in Division Forward Headquarters. I am assigned to the 17th Infantry Regiment (the "Buffalo" Regt.) of the 7th Division. We are on line in the western Central Front. We are north of Yonch'on (find on map!) and west northwest of Chorwon – and both the 17th Regt. and the 7th Division are known in Korea as among the best! I'm happy to be in such a good bunch! We finish our orientation on Division level tomorrow, and move on down to Regt 1 HQ. Today we began our orientation.

We went up on the battle line today, and had a good look at the MLR (Main Line of Resistance). We have much better positions than the Chinese along our front, I think. Maybe you've heard of "Old Baldy," "Pork Chop," "Arsenal," and "Snook," and also "The T-Bone." They are all on our front – the first and the last held by the Chinks and the other three by our own forces. The more I learn about this war, the more I like the American Army – I guess the U.S. is the only country in the world who would fight a war like we do – warm showers three miles from the lines, a P.X. and Officers' Club 5 miles back! American air has the sky so controlled that no one worries about or even looks at the sky! All in all, I think things are so favorable for efficient and spirited action as they could be – and I'm eager, sort of, to feel a *part* of it all. I guess in two or three days I'll be right up there, sweating and watching like the rest! We have Ethiopians, Colombians, Koreans and Americans in the Seventh Div. Just a typical U.N. Fighting Force! I feel like a sub on an All-American team, just waiting my turn to carry the ball! If you'll excuse me, tho, I'm not a bit nervous about it, and *in fact*, I'm a little impatient to get up there and get in the game! Maybe that's a dangerous attitude, but I can't help it! I hope you won't worry about me back there!

He told of being responsible for fifty men riding on the train from Pusan. He met

...a boy from Lawrenceville [Indiana] named Perkins – knows Carrie Van Wey and about everybody there in Vincennes! Small World! I told you about meeting Gene Liter, my high school classmate, in Yokohama didn't I? He was a sailor at the port, and when he came walking up to me, I about dropped my teeth! We had a nice talk. It seemed even natural to meet someone from home! My classmates, both high school and academy, are strewn all over the world, and altho I don't like being away from home, I *am* sort of glad to be 'in with the trend', so to speak, and be out serving like the rest! It's sort of the mark of the times! I haven't seen anyone else yet, but I know I shall!...

...The situation here is all right, and I feel very good about everything except missing good old Indiana and all the loving folks there. But I'll be coming back before you know it! I'm just gonna sit back with my arms around my folks and my wife and take a good, clean gasp of that sweet-smelling, non-dusty, clear-skied Hoosier atmosphere, and thank God I'm so lucky! And really, I am!

The *monsoons* will hit Korea in a month or so – and then it'll be nice in the foxholes – whee!! Haw! Rain and puddles everywhere! Good thing we defend the *high* ground, huh!?

Well – must get some sleep, and whip off a quick note to my angel – Take care of yourselves now – and write soon. Bonnie, you take care of our Dad and Mom, now won't you?

Love to all –

Dick

P.S. Has the copyright come in yet? When it does, tell me and I'll 'pull the chain' on things!

In a short time Dick Inman would learn he was assigned to B Company, 1st Battalion of the 17th, as Platoon Leader, 2d Platoon.

* * *

At Kadena Air Base on the island of Okinawa, south of Japan, where Pat Ryan and the men of the 93d Bomb Squadron, 19th Bomb Group, were stationed, was yet another life. To the visitor there was a deceptive calm, an apparent detachment from war, except for the periodic sounds of huge four-engined B-29s as they shuddered to life and thundered into the air northward in the late afternoon and night, heavily loaded with sixty-five-hundred gallons of gasoline, twenty thousand pounds of bombs and fifty-caliber ammunition. While there is an appearance of calm at an air base involved in war, to the people stationed there, such places are always beehives of activity.

The 19th was operating entirely at night over Korea. Their airplanes were painted black. The entry of Soviet-built, often Soviet-flown, MIG-15 jet fighters into the war had taught American bomber crews and the Far East Air Forces a costly lesson. The MIGs weren't equipped with radar for night or all weather attack, but the heavily-armed, propeller-driven bomber aircraft were still no match for the much faster, highly-maneuverable MIGs during daylight bombing operations.

The decision had been made to offset the MIGs' advantages by sending the superforts over targets at night. Though enemy Ground Control Intercept sites could vector – direct turns and compass headings by radio to the MIGs – to possibly intercept the bombers during favorable night flying conditions, visually acquiring the bombers for attack depended on the enemy fighter pilots' abilities to maneuver close enough to see the flames in the big airplane's engine exhaust stacks.

Strangely, on occasion, when MIG pilots acquired the bombers visually, instead of attacking, they would join with and fly off the wing of ingressing bombers. The reasons for such behavior were never entirely clear. It was suspected that the enemy fighter pilots were calling bomber altitudes and speeds to air defense controllers – or perhaps the pilots simply didn't want to attack.

The accepted procedure for these instances was for B-29 gunners not to fire on the "escorting" enemy fighters, for fear there were others in the darkness above that would promptly attack the bomber. Under any circumstance, for heavily defended targets the Americans provided night fighter escorts for the B-29s. The escorts were usually radar equipped, two-seat, single-engined F-94 air defense fighters, which, though not as fast and maneuverable as the MIGs, could engage and keep them at bay with missiles.

Planning and preparations for B-29 missions, the maintaining, repairing, servicing, loading, and launching of aircraft on a pre-planned schedule, and the return of aircraft from missions to begin the cycle anew, are around the clock activities, seven days a week.

The night schedule imposes its inverted, swing shift life on thousands of airmen, officers, and civilian technicians. Sleep during the day, duty at night for most of the Bomb Wing. Flight planning, weather briefings, intelligence briefings, operations briefings, target study, tactics planning, and crew briefings are the order of the day for a "fragged" mission a mission laid on by a secret, teletyped fragmentary order from Far East Air Forces in Japan, to 20th Air Force and its units.

After Lieutenant Paul R. Trudeau's crew, which included co-pilot Pat Ryan, had received their local and operational area checkouts in early May, they were ready to fly their first combat mission over Korea. Paul Trudeau was an experienced B-29 pilot, having flown both P-38s and B-29s in the Pacific during World War II. He had flown numerous bombing missions, all as a co-pilot, and had over three thousand hours in the co-pilot's seat of the B-29, before he became an aircraft commander. Because this was his crew's first combat mission in the Korean war, the 93d's squadron operations officer, Major Ronald Smith, West Point class of 1950, was on board to observe the crew's coordination and actions, and to complete the proscribed in theater orientation. Smith, too, had been an officer and pilot during World War II before obtaining an appointment to West Point.

The night of May 25, 1953, would become the first of two night-time B-29 missions Pat Ryan would never forget. Oil refineries and storage tanks in the east coast port of Wonsan, North Korea, were the targets. Wonsan was one of two ports from which major elements of retreating U.N. Forces had been evacuated in the grim days of December 1950, after the CCF came south across the Yalu River. While they were on their mission to Wonsan on the east coast, near the west coast of North Korea, an escorted mission of B-29s would bomb Sinanju, on the Chongchon River. Sinanju was a heavily defended industrial and transportation center north of Pyongyang, the North Korean capital, and adjacent to the Chinese border.

The Wonsan mission was solo, a single, unescorted aircraft. MIGs weren't deemed a threat as far south as Wonsan, while the mission over Sinanju was expected to encounter MIGs, which launched from air bases north and east of Sinanju, just across the Yalu, in China.

The weapons load for Wonsan was thirty-nine bombs weighing five hundred pounds each, plus six photo flash bombs and 1250 rounds of fifty-caliber ammunition for the airplane's defense. The North Korean air defense system relied heavily on German eighty-eight flak guns which had a fearsome reputation during World War II, but now were made more deadly with retrofitted radar guidance and control. The Soviets had captured thousands of eighty-eights during that war and were supplying them to the North Koreans. The eighty-eight's flak was uncomfortably effective to thirty thousand feet, and there were several batteries in the Wonsan area. The mission plan called for bomb release at altitudes between 29,500 and 35,000 feet, hopefully above the effective range of the Wonsan batteries.

The tactics for the attack on Wonsan were simple, but not exactly low risk, even for night operations. They were to fly a race track pattern over the target, and make six bomb runs, releasing five bombs on each of the first five passes, plus the six photo flash bombs and the last nine five-hundred pounders on the last run, while taking pictures for bomb damage assessment. Six bombing runs afforded even the most poorly trained gunners a reasonable opportunity to score a fatal hit on a large, relatively slow moving B-29, especially if the airplane flew a consistent pattern of ingress to the target. Pat Ryan's crew was in for a shock on their first bomb run on their first bombing mission – which they thought would be "routine," a "milk run."

On their first run at the target, the radar of an eighty-eight battery "locked on." The flashes from the deadly flak pattern immediately spelled trouble, and shock waves from the explosions jolted the big bomber. Seconds after the barrage began, a shell exploded startlingly close to the aircraft, and shrapnel knocked out the number three engine, the right inboard engine, setting it afire. The aircraft commander reacted instantaneously on hearing fire confirmed by the right gunner.

"Fire feather number three!" he commanded on the aircraft interphone. Pat Ryan and the flight engineer promptly initiated shutdown procedures, which simultaneously activated CO2 – carbon dioxide – fire suppressant bottles mounted around the circumference of the engine, inside the engine nacelle and casing. As the huge, four bladed propeller slowed to a stop the blades automatically rotated to streamline into the relative wind to minimize increased drag caused by the shutdown engine, and avoid excessive loss of airspeed. Fortunately, the CO2, sprayed under high pressure directly onto the engine and its accessories, was able to do its job. The fire quickly went out.

Trudeau called over the interphone to the entire crew, "Everybody OK? Check in." One by one each crew member checked in with the aircraft commander. No wounded and no other apparent, crippling damage.

The first pass over Wonsan was an inauspicious beginning, but one of the advantages of having four engines is the mission can be safely continued if there is no other serious damage or injury. A rapid assessment of the situation and Lieutenant Trudeau decided to continue with the remaining five bomb runs. Though flak continued unabated over Wonsan, they received no more hits. Time to return to base.

Another routine for night operations over Korea was the launching of support aircraft such as SB-29s, equipped to drop rescue equipment, including life rafts, in case aircraft had to ditch or crews bail out over water. This night there were two SB-29s flying mission support.

On leaving the Wonsan target area, and consistent with mission procedures, Pat Ryan radioed "Chicago," an American Ground Control Intercept radar site and informed them their aircraft had battle damage and would be returning to Kadena. "Chicago" picked them up on radar and gave heading and distance information for rendezvous with the two SB-29s. With the aid of on board radar, the lone B-29 successfully joined the pair of rescue aircraft.

Paul Trudeau and his crew soon learned that four or five B-29s on the Sinanju raid had also sustained battle damage, some more serious, and that his aircraft would have to be the last in the bomber stream to recover at Kadena. What's more, they learned weather conditions at Kadena weren't favorable.

A decision was necessary. Should they go to an alternate airfield in Korea or Japan, or return to Kadena? An assessment of fuel consumption rates, distance, weather, and available flying time was necessary. The aircraft commander, with his boss, the squadron operations officer on board, wanted to return to Kadena, and believed their margin of fuel was adequate. The decision was to return to Okinawa.

In flying, even more than in other potentially hazardous pursuits, it's the unexpected that can rapidly turn what seems a safe decision into a frightening cliffhanger, or a disaster.

The recovery of the preceding B-29s in the string returning to Kadena was much slower than anticipated. There were thunderstorms, low cloud ceilings, and intermittent rain storms which adversely affected landing visibility. They were in clouds, and the air was bumpy, at times rough, in the vicinity of thunderstorms. The recovering bombers ahead had no choice but to use Ground Controlled Approaches – GCA's, radar – and nearly every aircraft had to make two approaches to get safely on the runway.

Meanwhile, fuel gage readings were becoming worrisome on board Trudeau's plane. He was flying a racetrack-shaped holding pattern, in weather, at decreasing altitudes. Fuel reserve crept below that needed to divert to an emergency alternate airfield, a circumstance compounded by the deteriorated weather. A bona fide emergency fuel condition was looming, with bad weather, and decreasing altitude to ensure a safe bailout – over water.

Finally, when it seemed their turn for landing would never come, they were next in the GCA pattern, and started their approach, which involved a rectangular pattern with two ninety-degree right turns to final approach. As the big four-engined airplane made its first right turn from downwind, a leg and heading opposite the direction of land-ing, to base leg, perpendicular to the final approach approximately four miles from landing, the GCA controller informed Lieutenant Trudeau he had lost them on his radar scope due to heavy rain.

The radar controller's next instruction was normal procedure un-der any other circumstance, but not this night. "Execute a missed ap-proach, I say again, missed approach! Climb to three thousand five hundred feet and return to the Kadena [Radio] Beacon to begin an-other approach." Pat Ryan, who as co-pilot was responsible for radio transmissions, promptly replied, "Negative, Kadena GCA, negative. Our fuel state doesn't permit return to the Beacon for another approach. We'll have to turn immediately to crosswind and begin another ap-proach." Power was increased to remain at pattern altitude and the first ninety-degree right turn initiated for another approach.

Now the situation was critical. Fuel gages indicated tanks were nearly empty. They had to land on this approach. There wouldn't be enough fuel for another attempt. If they missed this time, emergency procedures required a climb to thirty-five hundred feet, fly for five minutes and bail out – if fuel lasted that long.

The navigator called for the radar observer to get on his radar scope in the aft section of the aircraft. At their respective crew stations, both had access to identical scopes, and altimeters to crosscheck altitudes. They could talk Lieutenant Trudeau down, if necessary, using an ARA – Airborne Radar Approach – a less accurate backup emergency re-covery procedure. On their radar they were trained to identify distinc-tive ground-reflected patterns enabling them to "see" and "break out" the runway. If GCA lost them again, they could compare and talk about what they were seeing on their radar screens, provide Lieutenant Trudeau headings to the runway, and call off required altitudes in the descent, altitudes required at specific distances from the runway to assure safe flight on the glide slope to landing.

Nerves were taut as the aircraft began turning right onto final, ap-proximately four miles from the end of the runway, a thousand feet above the water. Suddenly, number four engine, the right outboard en-gine, went full feather and began wind milling down – a potentially dangerous circumstance in a turn. Number four was out of fuel. Pat Ryan saw and felt the engine failure. He didn't tell Trudeau, as the aircraft commander already had his hands full. But he and Trudeau reacted instinctively, jamming hard left rudder to counteract the abrupt, unexpected yaw to the right caused by the two dead engines and increased drag on the right side. It took both pilots to hold the left rudder pedal depressed, with their left legs fully extended, while Pat frantically rolled in full left rudder trim, and the aircraft commander fought to maintain control while increasing power on the two remaining engines, to hold speed and altitude. Throughout the battle to maintain aircraft control they were rapidly cross-checking flight condition instruments, engine instruments, and fuel gages.

Tension, already high in the cockpit, rose to a heart pounding pitch as the aircraft steadied on final approach and Trudeau, Ryan and the flight engineer made rapid adjustments to compensate for the loss of power on number four. While constantly cross-checking attitude, airspeed, heading, altimeter, engine instruments and fuel gages, they glanced anxiously into the inky darkness ahead, hoping to see the glow of runway lights or break out of the overcast.

The radar observer called "Glide slope entry." Three miles, nose lowered gently onto the glide slope, two miles, altitude decreasing, staying close or on the glide slope, on speed, radar observer checking and cross checking on board radar for ground track, and calling out required altitudes. No visual contact with lights on the ground. One mile, still no visual contact with runway lights. At three-fourths mile from the runway, evidence of light ahead of the aircraft, diffused through the clouds from the ground, slightly left of the center windshield. Suddenly they broke into the clear, underneath the clouds, about seventy-five feet above ground and fifty feet to the right of the runway. More power and a rapid, shallow turn, left then right, to safely realign with the runway, using care not to lower the long slender wings so far that they either would catch the runway or ground and cartwheel the airplane in a fiery breakup and explosion.

Muscling the huge bomber quickly wasn't easy. The flight controls were mechanically linked. Yoke slowly pulled back to ease, then arrest, aircraft descent. Pull the power rapidly to idle as soon as the wheels contact the wet runway, use rudder, power, and differential braking to keep the big lumbering aircraft traveling straight down the runway, while braking to taxi speed. That was normal procedure on landing roll out, especially on a wet runway. But not on Pat Ryan's return from his first combat mission over Korea.

As the onrushing juggernaut's wheels made contact with the runway, engines one and two began accelerating, as though "running away." Throttles were rapidly jerked to idle to keep the aircraft from veering to the right. The two remaining engines suddenly went quiet. The beat of the huge propellers slowly decreased as they went to full feather and unwound toward a complete stop. The huge bomber was empty of usable fuel. There was no more.

Lieutenant Trudeau activated the emergency braking system to bring the stricken aircraft safely to a full stop. A momentary silence followed, then a collective sigh of relief as thirteen sweaty, shaken crew members gathered up their gear and prepared to climb down ladders to the hard, blessed surface of the runway.

Welcome to the war. More missions to come.

CHAPTER 16

THE SUMMER OF '53: NEW HEROES

With the graduation of the class of 1953 on June 2, we became cows, second classmen. News from Korea indicated the war might be nearing an end. Truce negotiations, though terribly frustrating, were progressing toward a cease-fire. Maybe it would all be over before the new lieutenants completed branch schools in the fall, or flight training a year and a half hence.

The men of '53, though uncertain about the war's duration, were steeling themselves to follow '52 and earlier classes into the fight. As had been the case in nearly every graduating class during wars in the Academy's hundred fifty years of history, many in '53 earnestly hoped they wouldn't miss this one. There were also men who hoped duty wouldn't require their presence in Northeast Asia. As '53 was graduating, the never ending stream of letters home continued arriving from the theater of the "half-war."

Two days after Pat Ryan's harrowing May 25 mission over Wonsan, Dick Inman turned his thoughts once more to home and family.

> ...a lot has happened since I last wrote you... I'll tell you all I can get down before I retire. I have to rise at 3:00 o'clock...[tomorrow] morning to work my platoon on outpost digging detail out front, so I'll have to make this short...

> Tomorrow night I'll 'tag along' on a patrol to gain some experience. Should be a fair chance to learn something. I've been taking my platoon out each day to dig on the two major Battalion-sector outposts, 'Arsenal' (farther) and 'Erie' (nearer). That has become almost routine in the last few days. Nothing happens very often, although a few mortar rounds came in on us a couple of days ago. Old Joe Chink is funny – he may see a company of men, marching or riding somewhere and just sit and watch them. On the other hand, one man may be walking on the road and the gooks'll throw a whole barrage in on him! Like I say, it's a phony, crazy, out-of-date war and I'm glad the

world doesn't know what it's like. When the Chinese hit, they hit hard, but they don't hit very often...

I wish I could be home to see everyone and play with Bonnie. Sweet little gal, – Barbara loves her as much as I do. And I miss talking and living the good, solid Hoosier life of my former days! I guess I'll be able to pick that up quick enough when I return, tho! I wish so badly I were home with my wife and folks, among the dear old souls of our Vincennes!...

I leave my love and thanks to you for writing as often as you have. I'll write, myself, whenever I can, but time is extremely precious here, and I have loads to do!

But – I'll do my best!

'Til then – don't work too hard, and I'll be thinking of you!

Love to all,

Dick

In mid-April, as both Dick Inman and Dick Shea were preparing to leave for Korea, a fierce fight erupted on Arsenal, then defended by E Company, 2d Battalion of the 32d Infantry, "The Queen's Own." The battle cost eight dead and seventeen wounded. But there had been no major Chinese assault on Arsenal or Erie during the approximate one week rotations of 17th units into and off the two outpost positions – not while Dick Inman and Dick Shea were there.

On June 4 Dick Inman wrote home again.

I'm extremely sorry that I have not been able to write as often as I would like to; I hope there will not be too many occasions in the future which will see similar lapses in my letters to you. Right now...I'm on outpost duty on 'Erie,' the outpost hill out in front of the M.L.R. in the 17th Infantry sector. Erie lies right behind the outpost, 'Arsenal,' which together with Erie, forms a twin combination that the Chinese have been unable to break since 1951.... Presently, Arsenal-Erie is the only place the Chinese have failed to attack along the entire 155-mile front in their recent effort to obtain 'bargaining points' for the current 'peace-talk-truce-armistice-prisoner-exchange' squabbles. I guess they have despaired of gaining anything in our sector, since they've tried so often and failed. They've given it all up as a lost cause, I believe – the capture of the outposts, I mean. I hope they continue that train of thought, since I'm largely in charge of the defense of Erie now. However, Arsenal, the farther of the two hills from the MLR, is in greater danger than Erie is...

Right now we're concerned with digging our positions better – hard! [We're] making trenches, bays for vehicles, tunnels, bunkers, and fighting positions all over the place! Also, we are beginning to erect dummy positions and fake firing bunkers, etc. It's all part of a plan to confuse the gooks, who are in the habit of learning the outlay of our positions down to the last trench and bunker before they attack....

I don't want to scare you...but I've been under mortar fire twice now. Yesterday, the last time, I was sitting on a couple of ammo cans, and a piece of shrapnel about the size of the end of my middle finger missed my old wooden head about two inches and slapped against the ground below me. I was going to send the piece to Barbara, but reconsidered, thinking it would make her worry all the more! So don't tell her! I'll tell her when I get home....

Must close now and inspect my positions. Tell Bonnie to be a good girl and write a letter to me when she can. I miss her and all of you so much. Got the pictures from Barbara that Admiral took – Love, Dick.

In the beginning of June, 7th Division realigned regimental boundaries within the Division's sector. The 17th Infantry Regiment's boundary shifted further left, and the 32d Infantry again took up positions that included Arsenal and Erie, to the right of the 17th, where Dick Inman's 2d Platoon, B Company had been defending and improving defenses on Erie. A few days later, Dick Shea told of the shift to the left in a letter to Joyce. The 1st Battalion was given new responsibilities in a sector that included Pork Chop Hill.

9 June 53

Darling Joyce,

Our battalion has shifted, and we are in a new area now, and I have a new task, much similar to some of the other work

I was doing before, but with more responsibility. Also I have my own log underground bunker, having with me our medics and part of our supply detail. I am fortifications and labor officer, and have two sergeants who oversee about 60 Korean Service Corps laborers. Now we are engaged in digging in deeper, putting sturdy log and earth covers on our trenches, and building new fighting and sleeping bunkers. It is a lot of work, but it is really a good project, for the completed works afford excellent cover from enemy artillery. The KSCs do most of the digging, and our engineer platoon which is attached to the company places in timbers and log walls already notched to fit. So the hole is dug to specifications, the KSCs lug the lumber up to the position, the engineers put the pieces together, then the unit is covered up and the ground leveled. That way the position is difficult to detect, as well as being formidable enough to stand even the heaviest bombardment. It is interesting to move hundreds of yards along our positions completely underground, through tunnels from one underground room to another. The sides are shored like a mine, in fact so much so that one boy who had worked in coal mines in Pennsylvania has been very helpful to the engineer platoon. Jim Deitz, you will remember I saw him the other day, is here with the engineer platoon. We see Dick Inman ever so often. He is strong in his belief that the Reds mean business on peace this time, but I cannot entertain any such hopes as yet. They appear to be playing the old game and only when it happens will I believe it, and then I shall be greatly pleased of course. We have heard, but not read, that the PW [prisoner of war] pact was signed, and that today they recessed after we were all expectantly awaiting further agreement, possible signing of the armistice....

The Chinese had tried three times to take Pork Chop: December of 1952, and the following March and April. April's bloody fight for Pork Chop, Arsenal, and Erie was part of a massive Chinese assault on the entire 7th Division front. The attack and its prelude included now predictable patterns of Chinese behavior the 7th learned through painful experiences, as had nearly every other U.N. unit on the front.

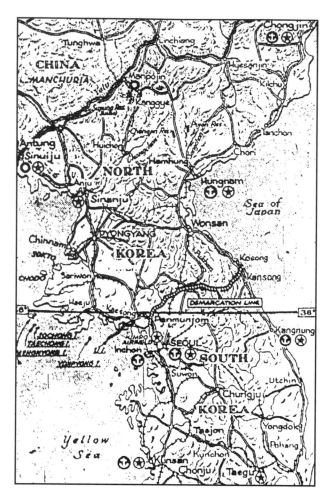

Approximate location of front line in Korea from November 1951 until July 1953 armistice signing. Map below marks location of Pork Chop Hill and surrounding area.

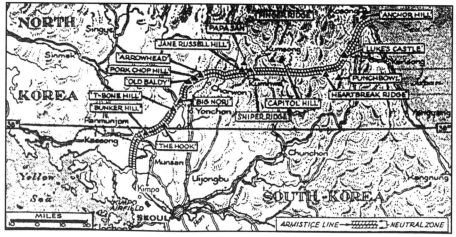

Chinese artillerymen shelled Pork Chop less frequently during the day, nevertheless they made life difficult and dangerous for the hill's defenders. They fired mostly at night, to reduce effectiveness of counter battery fire and avoid the destructive attacks of U.N. aircraft, which prowled the skies primarily during daylight hours. Occasionally, as Dick Inman learned, the stabilized front rattled with brief, but deadly mortar barrages. Sustained Chinese artillery and mortar barrages usually preceded infantry assaults, but not always.

Infantry assaults by the Chinese came nearly always at night, as well, to offset the advantages of American artillery and tactical air support. As American soldiers learned when they neared the Yalu River in early November of 1950, when the CCF entered the war, onrushing masses of Chinese infantry in nighttime assaults as seas of screaming, shouting voices, blew whistles and horns to instill fear in defenders. Dave Hughes and the men of K Company heard those same unnerving sounds as the 1st Cavalry Division reeled southward that fall, and heard them again and again when the front stabilized near the 38th parallel in the late summer of 1951, as the Chinese repeatedly flung themselves against American and United Nations' positions in the vicinity of Hill 347.

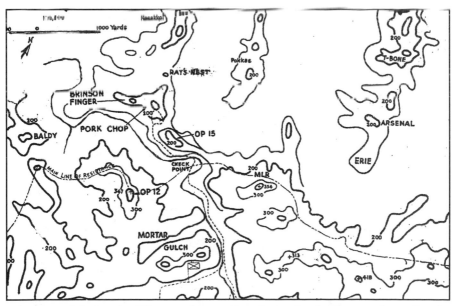

The area of the battle. The supply road leading to Pork Chop is shown by the dotted line.

Pork Chop Hill lay just beyond the toe of Hill 200, and the supply road which served both outposts was under direct observation by the enemy. (7th Infantry Division command report for July 1953.)

Aside from stirring fear among defenders, there was another purpose associated with the waves of noise – cause defenders to open fire, and with muzzle flashes expose their positions. Defenders would then receive a rain of hand grenades flung from out of the darkness, or the eerie, flickering, half-light of flares over the battlefield.

Patrolling and raids by both armies were routine, often resulting in short, sharp fire fights like the one Dave Hughes wrote about in his August 1951 letter to *The Pointer*. Chinese patrols, too, were nearly always at night, while the Americans patrolled both night and day, and usually launched infantry assaults in the early morning hours, to take advantage of artillery and tactical air firepower.

The Chinese patrolled and probed, carefully mapping the 17th Infantry's trench and bunker networks they would encounter during attacks. In the aftermath of April's battle on Arsenal, a detailed, accurate map of the outpost's entire defense complex, to include false firing positions, was found on the body of a Chinese soldier killed in the assault.

When it came to major offensive operations along the front, the Chinese had a predilection to unintentionally disclosing when and where attacks would come. The telegraphing of their punches resulted from a strange contradiction in battlefield behavior, a contradiction the U.N. forces repeatedly exploited. Though CCF tactics – the movement of their forces on the battlefield – reflected the extreme rigidity of an autocratic state, their manner of handling vital information, including battle plans, was democratic to the point of recklessness.

When a battle plan was decided by the Chinese high command, its essentials were passed down the line until finally even rifle squads were permitted to see or read the plan. The theory was the troops got a "good feeling" from knowing the secret. It made them think the plan was the best possible. At least that was the American explanation of Chinese reasoning.

Consequently, when any agent, prisoner, or other line-crosser brought word a Communist attack was imminent, it didn't lessen his credibility that he specified the hour and place. Enough advance information of that type enabled American G-2 – intelligence – to pass the word as to when and generally where to expect an attack. In April the men of E Company, 32d Infantry, defending Arsenal and Erie, were told to expect the Chinese assault to hit the ROK's 2d Division sector on White Horse Hill, a rifle shot's distance to the right of Arsenal. The Americans on Arsenal and Erie knew the attack would engulf the two outposts, and were ready when it came – and it came slightly ahead of schedule.

While there was predictability in Chinese behavior on the battlefield, surprise remained a certainty. Attacks didn't always come with preparatory artillery and mortar barrages. Large formations of enemy infantry units would sometimes crawl silently forward under cover of darkness to positions within a few yards of trench lines, then on signal stand to a crouching position and move rapidly forward to assault and penetrate defenses. Unless soldiers in outguard or listening posts saw the intruders and passed the word – or simply opened fire, the Chinese infantry could be inside the American positions before the alarm was passed. In April, Outpost Dale, in front of the 7th Division's MLR, was overrun in that manner.

Dick Inman and Dick Shea were learning about Chinese and North Korean battlefield tactics, the lay of the terrain to their front, the real meaning of "outpost" and "defense in depth."

An outpost, in military terms, is set up "to delay, deceive and disorganize," and is in front of the MLR. At night, the Americans set up ambush patrols, outguards – "listening posts." The ambush patrols were usually two squads, half a platoon, in front of another platoon's position, which was responsible for a sector in a company's main defensive positions on an outpost. Should the enemy launch an attack, soldiers manning listening posts or lying in ambush, are immediately called back into the platoon's main defensive position to keep the patrols or listening posts from being overrun, strengthen the main defenses, and ensure the soldiers pulled back can resupply themselves with munitions stored within the defensive perimeter.

The entire defense network was tied together with field telephones and radios so commanders, platoon leaders, and squad leaders could control small units deployed in front of the outposts – and connect the network, in turn, to company, battalion, regiment, division, corps and Army headquarters. And there were emergency backup means of communication, if the field telephone and radio systems were knocked out.

Runners, flashlights, and signal flares were alternatives, but weren't nearly as effective, and were slower than radio or telephone. If primary communications failed or were knocked out, prearranged plans and signals could bring supporting fires from heavy weapons platoons and artillery. If isolated groups or individual soldiers ran short of ammunition or were about to be overwhelmed, backup communications and the ability to call for supporting fire, reinforcements, or medics, were made far more difficult, if not impossible. In such circumstances it fell upon individual soldiers to use every ingenuity in their possession, most of it found in their training, to continue the fight and survive.

Thus it was in the stalemated trench warfare of Korea that in the dark of night, fire fights that began with infantry assaults often became a furious, swirling, confusing, terror-filled series of engagements between individuals and small groups of soldiers – bereft of officer leadership – shorn of any connection to the larger, in depth, defensive system, surprising and killing one another with showers of hand grenades, fire from semiautomatic rifles, submachine guns, other automatic weapons, and slashing trench knives or bayonets if ammunition was exhausted.

Kill or be killed. When the enemy penetrates the line, and leaps into defenders' trenches, men on either side of the penetration have to eject the enemy from the trenches, fight their way through toward the safety of larger numbers, more firepower, better defensive positions and better communications, with the likelihood of being able to call for help. Close, face to face encounters occur with terrifying suddenness. In such circumstances quick-reaction, short bursts of semi-automatic rifle, submachine gun fire, or grenades hurled at often unseen enemies, can mean the difference between life and instantaneous death.

At night, when trenches are overrun, the battle becomes a deadly game of cat and mouse, brute force, and shattering surprise, because soldiers can never be certain of the enemy's location in the zig zagging trench and bunker network. They fling grenades or fire weapons in the direction of voices or onrushing knots of enemy soldiers, when they can be seen or heard, then attempt to move on to the safety of larger numbers of Americans, wherever they are believed to be. On this type of battlefield, knowing one's own defensive positions well, and being able to navigate them in the black of night, are as essential as knowing the enemy's whereabouts. Tragic accidents caused by "friendly fire" occur with increasing frequency, because soldiers are unnerved, sometimes trapped, frightened, and fire or hurl grenades at the enemy before identifying him.

Dick Shea's sketching abilities were to be particularly useful to A Company's defense of Pork Chop Hill. On June 14 he wrote Joyce,

> ...my job as I told you, was concerned with improvement and engineering of fortifications. The sketch that I have included is the original one I made by covering every foot of the hill, going into every position in the course of two days. Thereafter I made six tracings, mounted them, covered them with acetate, and made them sector property of the hill, meaning they remain on position when units change. Lt. Roberts, [the company commander], desires that when we take over our next hill, a sort of semi-reserve position when we have completed this mission, I have prepared sketches of those fortifications. I wel-

701

come the opportunities as it keeps me busy, and I like it much, giving my fingers plenty of training.

A day earlier, when A Company returned from outpost duty, Dick was assigned as platoon leader of fourth platoon, the company's heavy weapons platoon. He explained in his June 14 letter to Joyce that the platoon consisted of three fifty-seven millimeter recoilless rifle crews and a section of three sixty-millimeter mortars, and the platoon supported the operations of the company by fire. His platoon numbered approximately forty men. He added,

...We are in reserve at this time, resting contentedly by day behind the MLR, moving up at night as a counterattack force in case the line needs reinforcing. It is good experience and well for me that I take over the fourth platoon at this time rather than when we are occupying an outpost.

Dick Shea was settling comfortably into A Company.

I am well pleased with this company, and with the officers and men. It is difficult to say whether a company is best or not because you do not know the others as well as your own. But I do know we have the much desired spirit necessary and the feeling of warm cooperation needed for a line unit. I would not trade for anything, though I suppose one finds much merit in any group of men this far forward. Lt. Stewart and Lt. Barr are first and third platoon leaders, Lt. Willcox second, Lt. Greenwell Executive Officer and [Lt.] Roberts CO. Sergeant Young is administrative first sergeant and Sgt. Hovey field first sergeant. I will write you a little more about each one in my letters to come....

Again, on June 22 he wrote,

Darling Joy,
    I have a few moments, then I must clean my weapon, and perhaps tomorrow, rather tonight I mean, I will have some more time to write. At present I'm in my sandbag command post (similar to one of those in the water color) at the base of a cliff....

He carefully explained to Joyce that his platoon was still in "semi-reserve," and their sixty-millimeter mortars were indirect fire support for A Company. The mortars were positioned on a ridge to the rear of Hill 200, which was part of the MLR, and with their high, arching rounds could reach the "main battalion hill" – Pork Chop – a thousand yards to their front. A forward observer was on Hill 200, with the company. "He phones back, or radios, where mortar rounds are landing, and adjusts fire from there to get us on the enemy."

With respect to the fifty-seven-millimeter recoilless rifles, he explained they were usually attached to rifle platoons and were direct fire weapons. That is, the high velocity of fired rounds resulted in a nearly flat trajectory, making the fifty-seven essentially a "line of sight" weapon.

Like Dick Inman, Dick Shea took note of the combined forces of the U.N. while he continued to map the company positions and learn the intricacies of the company's defensive positions.

> Yesterday morning I set out at nine o'clock to the right end of our line, the ridge of which 200 is the left end. I went to the Colombian unit on our right, Company 'B', Colombian Battalion, and mapped the positions of the platoon which joins first platoon of our company. I understand why so many people take Spanish in school. I expected great difficulty talking with them, but the military terms are so similar that it was quite easy. Platoon: palatoona; Squad: escuadra; grenadier: granadero; Sleeping bunker: dormitoro (dormitory); Company Commander (captain): Commander (el Capitan). And when I wished to record weapons and fields of fire, if the weapon was not actually there for me to see, I only had to ask for 'machine gun' or BAR, as they understand since they are Americano weapons they use, and to determine the caliber only had to write down 30 or 50 and have them point to the correct one.
>
> In the course of the day I went from their sector down the ridge through 2000 yards of trenches and fighting positions to Hill 200, mapping every position.... By dark I turned [the five foot long map] in to the company commander.

His thoughts then flowed from the more artistic side of him, dreams of longing for home and family, then his wakening to the sounds of gun fire, which seemed at first a part of his dream.

...It is so pretty here. One regrets the caution that must be maintained, but does not forget it. Neither I nor anyone I know of forgets why they are here, even in moments of appreciation of the beauty of this country. The thought always runs: Such a beautiful land, why must this devastation go on; the hills are so bright in the sun, but this is no place to go walking....

Last night I dreamed about [my brothers] Bill and Bob and me, and Pop was holding a big party in the large field in front of James Farm house, a squash field – and we were running all around setting up things, the three of us, so we could give a shooting demonstration he had commanded to back up a boast he had made. I woke up, and there really was shooting going on all right....

Now I say good afternoon...I love you immeasurably, and I am miserably lonely, xoxo Dick

On Thursday, June 25, Dick Inman wrote:

Dearest Family,

I have a little time to write this morning, so I thought I'd drop you all a bit of 'un-newsy' news.

Today marks the beginning of the fourth year of the war in Korea. I am just as tired of the war as I would be had I spent all three years over here. However, things over here are rapidly 'coming to a head', so to speak. You can just about feel it in the air – at least I imagine that I can! Either there'll be a truce, a withdrawal of U.N. troops, a U.N. offensive (very unlikely, I think), a Chinese offensive, or a movement of U.N. troops or U.S. troops into reserve support! 'Somethin's gonna come o' that!', anyway. We'll just have to wait and see – but probably not too long. Again I'll have to say, 'By the time you receive this, you'll know more than I now know!' Things are happening so quickly, y'know.

....I got official notice that my Purple Heart has been awarded to me. I was top of a list of about 60 men on the orders for award! Holy cow! I *am* all right, if you think I may be hiding something. After looking in the mirror, tho', I believe I'll be carrying a scar above and to the left of my left eyebrow. Not too big, probably,

but noticeable. I consider myself blessed that that mortar frag-
ment hit where it did, instead of an inch lower. I'm lucky that I
still have two good eyes! Ah, well – 'the breaks of war...'

The company is busy rehearsing and re-rehearsing an imminent raid on a trouble spot near Pork Chop – don't know
whether or not we'll be permitted to go thru with it or not – it
takes an approval of the – yes, that's right – *Department of
the Army* to put the plan into action! It'll be quite an affair –
we'll have such support that we feel almost flattered – all the
Army has will be turned toward our area – if our company is
chosen for the attack, we will have been given a real honor,
more or less a nod over the other companies in the battalion –
and, as a result, our morale is pretty high!! I feel personally
good because my platoon has been appointed as the platoon
to do the real job – demolitions, and destruction of the positions, plus screening for the entire raiding party. I hope everything turns out all right.

Yesterday and the day before I had time to work on a little
personal affair. About half a month ago I received a letter from a
Major Paul O. Siebeneichen, who is in charge of the guided missiles research and development work at Redstone Arsenal. He
told me to transfer from Infantry to Ordnance Corps, and then
request an assignment to Redstone. Now, I've never had a definite affinity for the branch of Ordnance, but if that's the branch
which handles the Army's missile program, I guess I'll accept it!

So, yesterday and the day before I set the wheels in motion for
a branch transfer to Ordnance. This morning the papers...went to
regiment...[They] have to go to division, Eighth Army, and then
to theatre, and finally to Washington. So, maybe (in a year or
so!!) I'll be in guided missiles finally! I sure hope so. I think that
field offers so much future. I'd kinda like to get into it. Perhaps a
branch transfer would remove me from this Korean mess. On
the other hand, maybe the war will prevent my getting the transfer. If it does come, the Lord knows *when* it will.

Things here don't change. Rumors are running rife all through
the G.I. Armies. Intelligence reports conflict: reported heavy
Chinese movement *south* in some parts – and reports of them

pouring *concrete* bunkers in others; one indicative of offense, the other of defense – who knows?

I know *one* thing! I'd sure like to be home! I would swim in Rainbow Beach with everyone, play with Bonnie, talk with my folks, make love to my wife, and, generally, *live* again! Maybe this Korean business will end before too long. Let's all hope so. Sometimes the future is so uncertain! But I've so far refused to be bitter, although I have pretty good attacks of fierce nostalgia now and then! But that's all included in our privilege to live and love. I miss you all very much.

Write soon. Love,

Dick

P.S. Sorry I dropped the first two pages in the mud.

Dick Inman's second experience with a Chinese mortar barrage earned him his first Purple Heart. When he wrote of the mortar attacks on June 4, he had not told Barbara or his family of the wound or the near miss to his left eye.

His "P.S." referring to the mud on the first two pages of the letter verified Korea's monsoon was in progress. Dick Shea wrote Joyce on Monday, June 29, describing the effects of the monsoon.

...We had been behind Hill 200 all night, the artillery thundered and flashed, and flares lit up the sky and valley, but the incoming was elsewhere and we slept sound, though chilly, under a sky that held no rain for us this night at least. But the creeks and streams are swollen and swift. The monsoon season is upon us, and everywhere the roads and trails turn to mud, then knee deep soup. Yesterday our vehicles began using snow chains to combat the mud. Going to the bivouac area where our company is quartered, Scotty, the driver, drove his jeep along the stream bed, which was smoother, better road than the oozing earth on the bank above. Last night as we were to move at seven, I went ahead with a small detail of men, two jeeps and a trailer, to build a foot bridge across the now raging torrent of the bubbling little stream I have described to you, behind 200. As the men pulled back yesterday at dawn, they had to wade knee deep in water, and we did not want them to get wet clothing and boots and have to sleep out in them all night....

Though rumors and optimism about a truce permeated the U.N. lines, the monsoon increased the likelihood of strong offensive activity by the Chinese. The rain storms that swept the peninsula almost daily rendered American air power less effective, and the enemy would surely take advantage. Dick's June 29 letter to Joyce told of A Company's mission as the monsoon season continued.

...This reserve company has the mission of blocking the valley road leading to our outpost position, the hill [Pork Chop] beyond 200, so each night we must move up in case we are needed. But it has been quite some time since the Chinese bothered our sector. Therefore we only keep our communications, radio and telephone, 'hot' (someone listening on the loop from all CPs) all the time and the bulk of the men sleep. They pull the telephone shifts like guard....

He had begun his letter,

Dearest Joy,

I love you sweet girl – I'm tired and sleepy but I could love you so snuggled in your arms....

Yesterday I went to mass and communion at Regimental headquarters, and I served mass. And afterwards Father Ruesnock invited me to eat with him at the mess there, and we had an enjoyable meal. He is about 32 and very interesting to talk to, and the time went by quickly. I left the company at 10:30 but did not return before 2:00....

...Just this morning I learned that Harold Scharmer is in this company, the first platoon. I saw him lay down a letter addressed to a family in New Milford, and I asked if he lived there. The mail clerk told him I lived there, but I said no but you did. He said he lived there 11 years but now they were in Paramus. I asked if he knew Don and Joyce Riemann and he said that he did, that he played soccer with Don. He's a very fine boy, this morning with his rifle slung across his shoulder, his helmet slightly askance, looking healthy enough, not too tall, slight of build, about 5'8," with a two days growth of blond beard. I will see him more often and will keep an eye on him because it helps when someone here watches out for you. I know I have people taking care of me, and they don't say so, but things indicate that....

Then he turned to the welfare of the forty soldiers in his platoon.

...one comes back from the front...tired, and so we have a few hours each morning of rest. The reason I was up and met Hal Scharmer was because I was trying to obtain cots for the men in the company here I had found sleeping on the ground. They accept most anything, and had I not decided to look into our 12 tents before turning in myself they would have had to pile brush on the muddy floor, about 12 men together. They have such a hard time getting anything they want, if it is not there they do not expect it. I have told them to complain to me about anything they need....

Then he wrote of his duty toward his men, and his responsibilities as an officer,

...The duty of the officer has always been toward the men first, but the trend is toward self-comfort. There are things that seem impossible which are only made so by rank stupidity and laziness – but I have found that a good psychological kick in the right place gets things moving. There is no reason why this war should not be conducted on a 24 hour basis. I am here and my only interest in the thing is that I must do it, and as I must, I will, until as quickly as I can I can come home. What I am doing will make the time go swiftly for I am always busy – not doing what I am told to do, for then I would lay back and ride with many others, but what I know should and must be done, which my fair one, does not include close contact with one individual known affectionately by those in the ranks as Joe Chink. To wait to do only what you are told to, to me appears to be a miserable existence....

Then came Dick Shea's unit spirit, humor, and love of art.

...Yesterday afternoon I spent making three signs to add to those already along our roads. All other units have something:

A BRIDGE TO BUILD, A ROAD TO DOZE
CALL ENGINEERS, YOU BUFFALOS

IT ONLY TAKES A LITTLE MORE TO GO FIRST CLASS. GO BY BUFFALO.

FROM THIS LOFTY VANTAGE POINT YOU ARE PRIVILEGED TO VIEW THE BEAUTIFUL PASTURES NOW ROAMED BY THE WHITE BUFFALOS.

But even our battalion hasn't been mentioned yet. So I made these up with ammunition box tops painted olive drab, with red and gold lettering and gold crossed rifles which I carved out of slats.

A YOUNG MAN WHO WANTS   TO GET AHEAD FIGHTS WITH   ABLE OF BUFFALO ONE

    Last night on the way to the bridge we put them up right at the cross roads near battalion forward headquarters. They look sharp.

    ...Now I will sign off because I must catch a few hours of sleep....

    Good morning sweet, I love you – I answer letters tomorrow. Now I sleep, now I love you, tomorrow I be with you again, soon such a tomorrow, I pray.

All my love, forever and ever,

Dick

    In the three-way correspondence of family, Barbara Inman, still the new bride, wrote Dick's parents frequently. On June 30 she wrote of her fears for Dick's safety.

Dear Folks,
    ...The last time I wrote I told you I hear from Dick about every other day. Then, suddenly his letters stopped coming –

and I *finally* heard from him yesterday. 2 letters. (I imagine you did too.) So, my heart left its crowded place in my throat and stomach flipped back into place. He has been on outpost duty for three weeks – a while back he seemed so pleased that his platoon was commended by two commanding officers and given the critical position on Pork Chop – but he was so thankful in his last letter to be off of that 'hot spot'.

Seven of his men were pretty seriously wounded...he hadn't had a bath since the first of June, ate C-rations for two weeks, and was bone tired. (Said he'd lost weight.) Then, in his second letter, his spirits were high – he was rested, had a chicken dinner, showered, shaved, and even had a new pair of fatigues – then he had to finish his letter abruptly because he had been ordered to take his platoon out to clear some Chinese near the hill. Half the time I don't know what to think – whether he's fighting, or pulled back or what. The latest news broadcasts concerning the fighting on the Central Western Front have pretty much set me on my ear – he drew a diagram once of his position and the ROK 1st Division that has been doing so much heavy fighting were to the left of Dick's Regiment. I can't help worrying and hating this whole mess – but I just *know* in my heart that Dick will be all right. He has been gifted with so much – such a good mind and such capabilities, that he has a much bigger mission in life. Still – I'm so disgusted with 'truce' meetings. Maybe (and it *is* possible) I don't understand many things behind the headlines, but I think the U.S. should pull their troops out of Korea. Well, I won't go into my views on the subject, but the strongest emotion I've felt all day is *Anger*. (Oh, I 'emote' all the time anymore.)...

...To keep me from getting melancholy and bored, Dick also gave me quite a task concerning his book...

...I still write to Dick every day, but I'm sure running low on news – one of these days I'll be listing what I've had to eat for each meal.

Bonnie, Mary Jo tells me that you've become quite an expert swimmer and diver. You put me to shame.

I hope you have some success with your garden.

Write when you have time.

> All my love,
> Barbara

On July 2 Barbara again wrote Dick Inman's mother and father.

> Just me again. I wanted to send these pictures. Doesn't it make you angry and helpless to look at those barren, dismal hills and know that is what we're (Dick) is fighting for?
>
> I haven't heard from Dick since I last wrote you. I certainly hope that by this time they have put him on R&R (rest and rehabilitation).
>
> He's never mentioned to me that he was due for it, however Dick Finch...told me that the soldiers are kept on the front for six months – then sent back for 5 days.
>
> By the way, if you get a magnifying glass and look at the pictures, you can see black dots in the sky – which are either shells exploding or airplanes.
>
> Hot weather is still prevailing here. (That's my news item from Iowa.)
>
> > Lots of Love,
> > Barbara

The same day, Thursday, July 2, Dick Inman wrote home once more.

> Dear Family,
>
> Sorry I haven't written lately, but of course there's always plenty to keep a fella busy around here. I always enjoy getting your letters. Thank you for continuing your writing even tho' I'm not able to answer as often as I'd like!
>
> Things haven't changed much here. We're still up on the line, and tonight I go on combat patrol to the east side of Pork Chop. The 17th Infantry must be indispensable to the frontline duties of the Army! I don't think we'll ever be pulled into the rear! I suppose my grandchildren, and your great-grandchildren, will still be slugging out this slow, 'unreal' war – sometimes things look rather bleak as far as the future is concerned, but sometimes things look fairly (relatively) bright. Of course, whenever I think of coming home, I get a glimmer in these

bleary eyes again! My future is in the hands of God and the ebb and flow of universal fortune! I don't think a man could be in combat long without realizing or learning to believe in a sort of predestination. And I suppose it's true that there's no such thing as an atheist in a foxhole. War has a settling influence on some people, I think. I think a lot of things over here, but mostly I just think about you folks and Barbara and home in general. Sometimes the homesickness in my heart wells up so that I feel I'll burst with it. But I've resigned myself to this type of existence for some time to come. I'll either be so happy to get home that I cry all over myself, or I'll be so numb I'll just sit and stare at the old familiar surroundings in Vincennes and Indianapolis. When the time comes, I'll certainly be glad to get back in the good old U.S.A.!

I've been overseas for some 2 1/2 months now, but it seems like 2 1/2 years! I can just remember what it feels like to walk down a city street and look at the carefree people, the signs, and the busy buildings and streets – traffic! I say I can remember – but I have almost forgotten what it feels like! By the time I return to home and loved ones I shall have forgotten! But 'that's the breaks of war', so they say over here.

You know what I'd like, if you could send it to me somehow – airmail, if possible – a copy of the good old Vincennes *Sun-Commercial*! I'd read it through with misty eyes and trembling fingers, from front to rear! That would be a good parcel – a piece of home, more or less! Could you arrange it, please?!

Tomorrow afternoon I'm going to have my picture taken! It will be taken by a Public Information Office man for an article in *Sports Magazine*. I guess they want pictures of men who were 'prominent' (ahem!) in Army (West Point) athletics and who are now in Korea...I guess I'll give them a real heart-breaking picture and story!

Barbara has the names of possible publishers to whom she'll send letters and copies of the 'book' we have fitted together. I figured she hasn't anything to do and you folks have so much to do that it wouldn't be fair to have you handle all of our business. Now, she may need a lot of help so if you have a good idea of any publisher or 'pusher' for the book, write her a letter and advise her further. It's sorta hard to conduct business from 9,000 miles away! Anything you think up may come in handy so be sure to tell her of any outlet you think of. You have a couple copies of the thesis yourself, so do with them – send them – what and where you will!

Well, I must go to briefing now. Rounds are coming in fairly often right now! See you all soon – Love to all –

Dick

In the steady flow of correspondence from Dick Inman were copies of poems he hurriedly scribbled during free moments. Among them was "My Toys," which told of a boy's toy soldiers and armies, and the reality of war as he now saw it.

When I was quite a little boy,
I used to get a thrill
At playing with my soldier toys
Back in the old sand-hill.
I would wage imaginary battles
While my toy soldiers would run
Thru the make-believe roar and rattle
Of a make-believe machine gun.

And all the day long I used to sit
With my warriors 'round the room,
And use my Red Cross First-Aid kit
To bandage up their wounds,
Or play at being grenadier,
Or armored tank, or plane.
When I was small I had no fear.
In play there was no pain.

But, now that several years have passed,
I have a new outlook.
No longer is this game of war
Come from a picture-book.
For I learned soon after my childhood days,
And it's not been long since I was small,

That war was different from my play.
War wasn't that way at all!

Now the smooth sand-hill is jungle swamp,
And the toy soldier is me.
But where is that pillow-chested pomp
And the royal artillery?

What has become of my clean little toys,
And the beautiful sky blue,
The newly pressed clothes, the juvenile joys,
And the light hearted life I knew?

The roar of the play machine-gun
Is so loud that it hurts my ears
And the fun I had with the First-Aid kit
Has now changed into fears
The fear of grown men living still
With a leg half-missing, or a hand that's gone,
Being carried from the smooth sand-hill
To a first aid station at the break of dawn.

And the shiny soldiers that cluttered the porch
Now splash in the sticky mire
With the tanks and the trucks...And the hot sun's scorch,
And the bombs, and the smoke, and the fire...
And when the little toy soldier runs,
It's not in children's play,
But it's for cover from enemy guns,
'cause wars are played that way!

So the sickness that tugs within my chest
Is not the fever of tropic parts,
But the black disdain of war, and the next,
That's forever in my heart.
For now I know what it's all about
Why good men have to fall –
It's because most people haven't found out
War isn't that way at all.

July 4, American Independence Day, was at hand, and the 8th Army and U.N. Forces had plans of their own.

One night in May the Chinese conducted a reconnaissance in force of the 17th's defenses. They hit every outguard position before firing hundreds of rounds of artillery into the area. They then probed the eastern finger of Pork Chop. An American follow-up patrol cleaned out the area of enemy in daylight, under cover of smoke, and accidentally discovered a system of Chinese fortifications on the north side of a steep-sloped, sharp-pointed hill half way between Pork Chop and Hasakkol. This was the tip-off the Chinese planned to attack Pork Chop in force, but no one knew when. The fortifications could hide a substantial number of soldiers, and a prisoner confirmed the CCF kept a company in reserve in caves within the complex. The Americans dubbed the network the "Rat's Nest."

In mid-June, plans were set in motion to destroy the Rat's Nest. Throughout the second half of the month, increased enemy activity was reported all along the 7th Division front, and was most evident in the area of Hasakkol. Specifically, the enemy was engaged in an intensive fortification program, providing positions from which an attack could be launched against the Division. Observers noted revetments from which Chinese artillery pieces were capable of firing into the Division rear areas, and on June 26 armored activity was first noted in the area.

Both 1st and 3d Battalions of the 17th began rehearsing to raid the Rats Nest, but aggressive ground action was prohibited by higher headquarters in the period from 18-29 June, due to the ongoing truce negotiations. Dick Inman alluded to a raid in his June 25 letter home, saying B Company was rehearsing for an important mission, and his platoon had a central role in the operation. However, hope remained the Rat's Nest could be destroyed without committing ground troops in an assault.

The Americans fired numerous artillery and mortar barrages at the Rat's Nest, and called in air strikes. The artillery barrages included the heaviest, most destructive weapons in the field, one hundred fifty-five millimeter and eight-inch guns. The steep hill shielding the complex gave good protection against artillery and mortar rounds. In military terms the network of caves and defensive positions were in defilade, reasonably well masked from most incoming rounds because of the hill's height and steep slope. Three air strikes failed to hit the complex.

Among the air strikes was a night drop on July 3, a full load of five- hundred pound bombs by Lieutenant Trudeau's B-29, on which Pat Ryan was co-pilot. Diverted from their primary target because of weather conditions and the priority assigned the Rat's Nest, they released the bomb load at an altitude of three thousand feet on a single pass over the area northeast of "Old Baldy." The bomb run paralleled the MLR to ensure a better chance of hitting assigned targets and avoid stringing the bomb load into American positions. An aircraft flying above and parallel to an MLR guarantees sharp enemy reaction. The diverted B-29 mission proved no exception. The big bomber came under heavy fire from CCF defenders and due to battle damage, Lieutenant Trudeau made a decision to recover at Itazuke Air Base, Japan. When they landed there were 168 holes in the aircraft.

Time was running out. If the Rat's Nest wasn't destroyed, the threat to Pork Chop increased. The raid on the complex became a necessity, and Charlie Company of the 1st Battalion was assigned the mission. Able Company was on outpost duty on Pork Chop.

The maneuver plan called for one weapons squad and one rifle squad to create a diversion by approaching the Rat's Nest from the east. Another squad was to block on the east finger to keep the Chinese from reinforcing from Hasakkol. When the diversion was executed the diversionary units were to pull back and the assault leader was to signal for his platoon to begin its sweep from west to east. A third platoon was to follow the assault platoon, in support, placing demolition charges to destroy the complex. The sweep began on signal at three o'clock in the morning on July 4, preceded by heavy American artillery fire, to neutralize the area and cover the movements of the raiding party.

There were few enemy encountered as the sweep began, but the Rat's Nest proved far more extensive than air and ground reconnaissance revealed. Enemy mortar and artillery rounds began finding their mark, temporarily pinning down the raiders. The sweep slowed due to the size of the complex and enemy shelling. The heavy Chinese mortar and artillery fire intensified as the raid progressed. Caught in the open, C Company began taking unacceptable casualties, causing a decision to rush the engagement and pull back as rapidly as possible.

Returning raiders, some of whom withdrew through A Company bunkers on Pork Chop, reported counting three enemy killed in action, and fourteen wounded, with another three estimated killed in action and three wounded. The raiding party carried back twenty wounded and left behind two missing in action.

Lieutenant Colonel Beverly M. "Rocky" Read's 1st Battalion included Able, Baker, and Charlie rifle Companies, and Dog heavy weapons Company. When he was first given the Rat's Nest mission, from that moment on he was deeply involved in the raid's planning and preparation. When Read, a highly-respected, compassionate battalion commander and Virginia Military Institute graduate from the class of 1941, received word of Charlie Company's casualties, he pleaded with Major General Arthur G. Trudeau to permit him to send a small patrol to retrieve the two missing soldiers.

Trudeau, the battle-hardened, yet sympathetic commander of the 7th Division, denied Read's request. The frustrated Battalion Commander, lifted his eye glasses from the bridge of his nose and brushed tears from his cheeks as he listened to Trudeau's denial of the request. Read suffered as nearly all combat commanders do, wondering if he'd made the right decisions and adequately prepared his soldiers for the mission, though he well understood a rescue attempt would likely mean more casualties and a questionable probability of success.

The Rat's Nest was still intact, though damage inflicted by Charlie Company's demolition charges was sufficient to delay an attack by the Chinese. Nevertheless, serious trouble loomed for the men manning Pork Chop. The enemy had made plans, too. A bitter struggle was in the making.

THE LAST BATTLE FOR PORK CHOP HILL

In the nearly two years of on and off truce talks, outposts like Pork Chop had achieved a significance far outweighing their military value. In the high stakes game of negotiations at Panmunjom, Pork Chop undoubtedly was a source of great frustration to the Chinese. Three times in the preceding six months they failed to wrest the hill from the 7th Division. An outpost's primary purposes of "delay, deceive and disorganize" enemy assaults on the MLR were now quite different in the scheme of Korea's static, trench warfare.

Names like Elko, Vegas, Harry, Christmas, and Pork Chop marked the bloodiest spots along the Korean battle line. Each was an outpost never intended to be held at all costs, yet each had become a symbol of dogged tenacity and unbreakable will, great victory or bitter defeat. The men of the 7th Division were proud they hadn't lost Pork Chop in three fiercely contested assaults by the Chinese.

After the April battle "The Chop" was a shambles. The 17th Infantry Regiment reported 402 American and fifty-seven South Korean casualties in that month. Nearly all were on Pork Chop. Trenches and bunkers were unusable, almost nonexistent. General Trudeau, the 7th Division commander, decided to rebuild Pork Chop, make it impregnable if that were possible. The task of rebuilding fell to the 17th Buffaloes. But the April task assigned the Regiment had been handicapped in March when the Chinese took "Old Baldy," rendering "The Chop" far more vulnerable, if not impossible to defend without great cost.

"Old Baldy" was west, southwest of Pork Chop, at the left end of an arc of four, powerful enemy strong points ringing the outpost on three sides. Proceeding from west to east around the arc, the other enemy strong points were Hasakkol, directly north of "The Chop," Pokkae to the northeast, and "T-Bone," further to the east, northeast. All four were higher in elevation than Pork Chop's sharp peaked 234-meter summit. From the tops of all four the Chinese could look down on a hill ill-formed for all-around defense and too loosely tied in to the supporting American positions. And from behind the ring of four strong points the Chinese could pound Pork Chop with mortars. Longer ranging artillery multiplied the fury of the shelling and the unpleasant consequences for defenders. As a result of the hill's position and Chinese frustration at being unable to permanently eject its defenders, day after day more enemy artillery and mortar rounds fell on its few acres than on the rest of the 7th Division.

"Old Baldy's" seizure in March also dictated adjustments and expansion in Pork Chop's defense network. The left flank – the western sector – needed to be strengthened, with greater depth and more firing positions to counter infantry assaults that could be launched from the areas behind Baldy. Approaches from the vicinity of the Chinese strong point, led to the left rear of the American outpost.

At the beginning of July the 17th's 2d Battalion anchored the left sector of the Regiment's front, occupying Hill 347, which Dave Hughes' K Company of the 7th Cavalry had taken in October of 1951. The 1st Battalion of the 17th manned the left center sector, the 1st's responsibility since June 10. The sector included Pork Chop and Hill 200. The attached Colombia Infantry Battalion occupied the right center sector, and 3d Battalion manned the Erie-Arsenal outposts in the Regiment's right sector. On July 2 the Colombian Battalion was pulled off the line for training of its replacements, and 3d Battalion assumed responsibility for defending the Regiment's right sector. E and I Companies, less one platoon each, moved on line to replace the Colombians.

Arrayed against the 17th were elements of the 199th, 200th, and 201st Regiments of the 67th Division, 23d CCF Army. Intelligence reports indicated two companies occupied the Hasakkol complex, with one company on Pokkae, and a battalion on T-Bone.

The 17th Buffaloes, to whom the task of rebuilding Pork Chop had fallen, found the going rough. The pace of rebuilding was slow and tedious, requiring a lot of manpower. Nevertheless, in June the 17th deepened the trenches to six and one half feet, and widened them to allow litter bearers to pass defenders. Deepening the trenches gave far better cover for movement from bunker to bunker without being observed or exposed to direct fire, but had the disadvantage of not permitting defenders to fire out of the trenches except at enemy soldiers immediately above the lips of the trenches – unless there was something to stand on inside the trench. All the while construction progressed, from their advantageous positions, the Chinese could observe every shovel of dirt thrown from the open trenches, every beam nailed into place to build overhead cover.

One company defended the hill day to day. Sleep during daylight hours was essential for men required to be on alert at night, when assaults could be expected. During the nights, in each bunker or firing position, occupants rotated "guard duty," two to three hour periods, watching for approaching enemy, listening to communications nets, prepared to react and warn of enemy activity or assaults.

Sleep didn't come easy during the day. The noise of incoming rounds and rebuilding didn't encourage rest. Soldiers necessarily spent most of their time staying alive, taking cover from incoming mortar and artillery, and keeping their weapons in condition to fight. Lumber, ammunition, rations, and water rode out to "The Chop" on armored personnel carriers, to the evacuation landing, and the wounded and dead rode back. Because of these and similar factors, outpost duty rapidly brought fatigue to defenders.

Prior to Charlie Company's raid on the Rat's Nest, Corporal Dale W. Cain returned to Able Company from Battalion Communications, where he was a switchboard operator. He had been in A Company during the April battle for Pork Chop, when it reinforced the hill's defenders during the bitter two day fight. He was wounded during a short, furious fire fight the morning of April 18, insufficient to warrant evacuation from the peninsula, but sufficient to finally warrant his being sent to "Batt-Comm" until his wounds healed. Lieutenant Colonel Read told Dale the A Company commander, Captain Roberts, and the executive officer, Lieutenant Shea, requested his return to the company, which was on outpost duty on the hill.

He was to report to Lieutenant Shea, be a radio operator for the company, and assist Shea in reinforcing and strengthening Pork Chop's defenses. Shea and Roberts wanted to bring in more radio communications, lay more wire for land line communications, bring in more am-

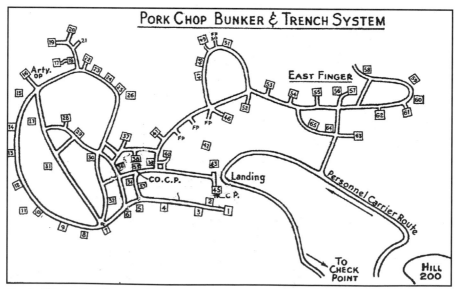

Pork Chop Bunker and Trench System, July 1953. (Courtesy March 1955 The Army Combat Forces Journal.)

munition, C-rations, water, and medical supplies. The company commander had been warned of aggressive offensive operations by the Chinese and, with his executive officer and platoon leaders, was preparing to defend against a strong attack. Dick Shea was supervising the work. For the first time, when Dale Cain came back on "The Chop," he met and began working closely with Dick Shea.

Dick took him to a bunker near the company CP, from where they could see across the valleys north of Pork Chop. He explained to Dale what he wanted him to do, and for the next three days they worked together, preparing to better defend the outpost.

In those three days, Dick Shea made an indelible impression on a young Corporal Cain, who had first served in Korea in the early, frantic days of the war as an underage sixteen-year-old who had successfully lied about his age to enlist in the Army and "go to Korea." He came ashore with the 7th Division at Inchon in September, and again at Iwon, North Korea, on October 29, 1950, before his worried parents tracked him down with the aid of their Congressman, and he was returned to the States and discharged from the Army.

Dale Cain was not to be stopped, however. He wanted to help relieve the suffering of the South Korean people, and when he turned eighteen he promptly reenlisted, legally, with the proviso he return to his old unit. Now, here he was, in the early summer of 1953, once again preparing to do battle.

*   *   *

By the time C Company returned from its July 4 raid on the Rat's Nest, Baker Company received another mission. The night of July 6, Dick Inman was to lead an ambush patrol of twenty men from Baker Company on the front slope of Pork Chop in support of A Company.

It was routine to move ambush patrols into position under cover of darkness, early in the evenings. Their purposes were to lay in wait and surprise enemy patrols attempting to slip into the defensive perimeter, and take a prisoner if possible.

The night of July 5, Dick Inman rehearsed his men for the patrol. B Company had previously taken its normal rotation onto Pork Chop, and except for new replacements, its men were familiar with the map and layout of the hill's defensive network. Ambush patrols normally were composed of men with experience on the line. An ambush patrol wasn't a mission for green replacements unacquainted with defense of the outpost.

The patrol included two elements, an assault element of ten men led by Dick, and a support element of ten men, led by Lieutenant Brubaker. For operational control they were to be attached to A Company's 2d Platoon, led by Lieutenant David R. Willcox, whose platoon anchored the right sector of Pork Chop defenses. Dick Inman's assault element was to move downhill out of the 2d Platoon area, angling to the west into a position in front of A Company, below Finger 21, approximately ten yards below the outguard position, beyond a footpath, known to GIs as a Choggie Trail. The support element was to remain to the rear of the assault element, in position to provide covering fire if the assault element needed to withdraw under attack. Finger 21, a bunker in the defense network, was west of 2d Platoon's left boundary, which was Bunker 49.

Although radio silence was the norm, Dick Inman and his patrol were tied into A Company's communication net by field telephone and radio, and procedures between him and David Willcox included prearranged, brief calls, each signaling specific information or actions required of Dick and his patrol should fighting erupt.

Typically, the rehearsal included various scenarios, or problems, involving enemy contact, and how the patrol would react to the contacts. Reentry and alternate rendezvous locations within A Company's positions were identified, in the event they were called back into the perimeter while under attack.

The evening Dick Inman rehearsed his patrol, Chinese psychological warfare officers blared one of their frequent loudspeaker broadcasts at A Company. From the hillsides to the north, Pork Chop's defenders first heard soft music, which faded in and out with shifts in the summer evening breezes. Then voice broadcasts began, and defenders heard demands they surrender. If they didn't the loudspeakers proclaimed, "You will all die." No prisoners were to be taken. "We will take Pork Chop even if we have to wade through blood."

The threat wasn't new. It was merely another broadcast in what had become standard nightly news over the nearly two years of trench warfare. The Chinese had read off the names on the company roster over their loudspeakers the night A Company rotated onto outpost duty. Nevertheless, the words didn't comfort green replacements just arrived on line, nor was it popular with the more battle hardened GIs.

Dale Cain, now back on outpost duty, had heard it all before, but remained impressed by the lieutenant who'd introduced himself as Dick, Dick Shea. Cain was frequently at Dick Shea's side as he made the rounds to bunkers and firing positions the evenings of July 3, 4, and 5, checking on work to strengthen fortifications, and the status of ammunition stores, supplies, and communications. Dale particularly noted Dick Shea's response when the Chinese loudspeakers were blaring their messages. When he saw or heard the younger GIs expressions of fear, he would reassuringly put his arm on their shoulders, giving them words of encouragement. "Don't let it get to you. Everything's going to be OK. We're in good shape."

* * *

At 8:30 in the evening of July 6, Dick Inman's ambush patrol left A Company's 2nd Platoon defenses, and took up their positions out front of Finger 21. A heavy rainstorm began, soaking the ground and trench networks already wet and muddy in the full blown monsoon season. The storm meant more trouble for weapons daily cleaned of grime splashed into their moving parts. In the confusion and dashes through mud in a night fight, a dropped weapon, stumble, or fall into the muck could quickly render a weapon useless, jammed by wet, soupy grit. This night the heavy rain portended far more trouble than jammed weapons.

Under cover of the storm two battalions of Chinese infantry began moving toward the left and right flanks, and rear, of Pork Chop. At ten twenty-five, Private Robert E. Miller, 1st Platoon, was asleep in a bunker near Able Company's CP, Bunker 35. He was preparing for a turn on guard duty later that night. Corporal Cain and Lieutenant Shea were in their bunker adjacent to the company CP, where Cain, Shea, and one other soldier slept, worked, and operated an additional company radio. They weren't asleep. The radio was on, and Cain, as always, was attentive to traffic on the company and battalion nets.

At the battalion check point behind Hill 200, Private First Class Emmett "Johnny" Gladwell, a medic in the 17th Medical Company, was also in his bunk sleeping, while one or two other medics were sitting at a table writing letters by candlelight. The rain made sleep easier. Johnny, like all medics, was usually called "Doc" by the GIs, a term spoken in tones of admiration and respect. He was in a large, well reinforced medical bunker at the check point. The check point served as a collection point, the first stop for casualties coming off the outpost or MLR. From there the wounded were taken, usually by litter jeep or truck, to the battalion aid station, a mile and a half further to the rear.

Across the road from the medics' bunker was a tankers' bunker, where armored personnel slept – the men who maintained and operated the armored personnel carriers which plied in and out of Pork Chop's evacuation landing, and tank crews for division assigned tanks emplaced or parked temporarily at the check point. A communications bunker was also at the check point, an important part of the battalion's communications net linking the outposts and MLR.

At the battalion aid station, Private First Class Lee Johnson, a West Virginian, and close friend of Johnny Gladwell's, also a West Virginian, was in the large mess bunker sitting at a table writing a letter home by candlelight. Since shortly after ten o'clock, Lee had had an urge to go to the nearby latrine, dug a safe, sanitary thirty yards distance from the aid station, just across a small stream. Because of the rain, he resisted the urge, in spite of growing discomfort. At ten twenty-five he was about ready to relent to nature's insistence, when he heard a sharp, nearby explosion which shook the bunker. He recognized the sound immediately as an incoming artillery round.

Then in quick succession two more explosions, nearer. This was more than occasional harassing fire. Someone ran to the door of the bunker and yelled, "Hope no one in here has to go to the crapper. The place has just been blown all to hell!" Stunned, Lee Johnson was slow to react at first. Then he remembered he'd restrained himself for the better part of a half hour. His rain-imposed restraint probably saved his life.

At the check point behind Hill 200, Lee's friend, Johnny Gladwell felt the effects of incoming at almost the same instant, except the Chinese artillery was far more accurate. The first round to hit the check point was a direct hit on the medics' bunker. The force of the explosion threw Johnny from his bunk onto the floor, blew out the candle, knocked over the soldiers sitting at the tables, and momentarily filled the air with dust and gritty sand. Miraculously, no one was hurt. Their hard, spare-time work strengthening the bunker's overhead cover with extra bracing, and layers of crushed rock and sandbags had done well for the men inside.

At 10:25 the CCF had begun thunderous mortar and artillery barrages all across the 7th Division front. On Pork Chop, in the bunker near the company CP, Private Bob Miller awoke with a start, jolted to sleepy confusion by the sound of artillery slamming into the surrounding hills and valleys. All hell's broken loose, he thought. He left the bunker and entered the trench wondering what was going on. Someone yelled at him to go get his rifle.

Almost immediately Dale Cain's radio came alive with traffic. There were calls indicating the enemy was coming in on the left flank. Trouble was brewing – quickly. Dick Shea hurriedly left the bunker, heading toward the threatened area. It was the last time Dale Cain saw Dick Shea.

Within moments flares fired by American artillery and mortars began shedding their flickering light on the outpost. In Private Bob Miller's bunker was his squad leader, Corporal Charlie Brooks. As soon as the enemy artillery started, Brooks left the bunker to check the readiness of squad members at each of their firing positions, then returned. He got his Browning automatic rifle ready and peered through the firing aperture, rain and flare lit sky, down into the valley below. In the valley, at the foot of the hill he saw what appeared to be hundreds of Chinese infantrymen surging toward the outposts' slopes. Charlie Brooks and the men in his squad, which included seven American and two ROK soldiers, watched and waited.

At 10:38, A Company reported receiving small arms fire. The hill was about to be engulfed with swarms of enemy infantry. Another A Company soldier in Brooks' squad, Private Angelo Palermo, said the attack looked like a "moving carpet of yelling, howling men – whistles and bugles blowing, their officers screaming like women driving their men up the hill." And as usual, among the lead elements of Chinese infantry, were young boys twelve to sixteen years of age, carrying no rifles or submachine guns, but loaded down with grenades and trained to pick up weapons others dropped – American or Chinese – and use them.

As soon as the artillery began, Lieutenant David Willcox called his outguards and Dick Inman's patrol back into A Company's 2d Platoon sector, and started rounds to check his platoon positions. When he reached the area of Bunker 51, where the patrol was to reenter the platoon position, some of Dick's assault element had already made their way into the perimeter, rejoining the support element. Dick was outside the position waiting for the last man to come into the trench. Private First Class Clarence R. Mouser, a Forward Observer from the 17th Infantry Heavy Mortar Company, and attached to the support element, saw Dick reenter the trench.

* * *

When the artillery and mortar barrages began, MacPherson Conner, Dick Inman's and Dick Shea's West Point classmate, was with an ambush patrol in front of the MLR, in the 32d Infantry's sector. Mac had been in Korea three weeks, and was platoon leader of the 1st Platoon, E Company, 2d Battalion of the 32d. He was an observer on the patrol from another company, a step toward completing his battlefield orientation.

A few days earlier he reconnoitered Pork Chop, to become familiar with the terrain in front of the MLR, and the outpost's defense network. En route he encountered Dick Inman, whose company was having a hot meal. Since Mac's company had been eating K-rations for several days he joined them for the hot meal and talked briefly with Dick.

Now, the night of July 6, when enemy artillery opened fire, the patrol Mac Conner was with was also promptly called back into an outpost defensive perimeter, but Chinese infantry didn't attack their position. Mac eventually rejoined his company, in reserve, approximately three thirty in the morning of July 7. By this time, E Company had already been attached to the 17th Regiment, due to a rapidly changing, confusing, deteriorating situation on Outpost Pork Chop. Mac decided to get some sleep, if he could. It's well he did. The nights and days soon became long and miserable.

* * *

The Chinese had moved in under cover of the rain storm with well prepared assault teams, each aiming for specific objectives in Pork Chop's defenses. The move against Pork Chop was one of several coordinated jabs which simultaneously struck at outposts Snook, Arsenal and Arrowhead. When Dick Inman's forward patrol element returned to the A Company perimeter, apparently with no one wounded or missing, incoming mortar and artillery had become intense. In Charlie Brooks' squad, suddenly, without warning, the entire area outside, above, and around their trenches and bunkers seemed alive with enemy infantry. Everyone in the 1st Platoon was in a fight for his life.

On the right flank of the hill, contact was soon lost with the 2d Platoon's leader, David Willcox. After making his rounds to check platoon positions and see that Dick Inman's patrol was pulling back as enemy artillery continued, he returned to his platoon CP. Almost immediately a soldier brought him a report the machine gun in Bunker 53 was knocked out. He and the man who brought the report started for Bunker 53. In the area of the bunker they encountered a flood of enemy soldiers pouring "like a waterfall" into the defenses. The two men engaged in furious fighting to eject the Chinese from trenches near the left boundary of his platoon's sector, as well as the observation post at position number 54, which the enemy had taken.

In the melee David Willcox found himself in two violent, close-quarter fights. He emptied his forty-five and thirty-eight-caliber pistols at close range, killing five or six enemy soldiers in each encounter, before he was attacked by a Chinese soldier who came at him when David emptied his pistols and hadn't time to reload.

David Willcox routinely carried three knives: his carbine bayonet, a hunting knife he brought from the States, and a knife made from the steel of a Samurai sword in Japan. In the ensuing hand-to-hand struggle, David killed the soldier with a knife. To his horror he realized he'd killed a boy thirteen to fifteen years of age, a fact that haunted him for years.

Lieutenant David Willcox, the American soldier with him, and five others eventually took cover in Bunker 53, then were pinned in the bunker, isolated without communications to the rest of the platoon, or the company. They remained in the position for approximately sixty hours, fighting off repeated Chinese assaults.

Dick Inman and his B Company patrol didn't know the enemy also entered the trenches in the right sector, between them and the 2d Platoon's supply bunker. The Chinese had already overwhelmed several bunkers and trench segments in both sectors. Vicious, bloody hand-to-hand and close-quarter fighting was in progress as small groups of soldiers confronted one another in the alternating dark and flare-lit night.

Dick gathered his men and told them 2d Platoon needed their help. He also knew B Company was in battalion reserve, and would be the first company committed to reinforce Pork Chop if the battalion commander responded to the radio request. But first the reassembled patrol must attempt to reach the 2d Platoon supply bunker, where ammunition and grenades were stored.

As they moved toward the 2d Platoon CP, they could hear whistles, horns, and shouting Chinese infantrymen outside the trench and above them. In areas with no overhead cover, enemy soldiers were just outside the lip of the trench throwing grenades into their path. To drive the Chinese back, they threw grenades from the dwindling supply they carried, and fired their rifles and automatic weapons at fleeting individuals and groups of enemy soldiers moving in the flickering, eerie glow outside the trenches. When they approached the observation post at position 54 they met a rain of grenades thrown from inside the position, which had been seized by the enemy. Dick's men responded with grenades, trying to kill or dislodge its occupants.

A soldier emerged carrying a flashlight, yelling "Prisoner! Prisoner!" There was momentary uncertainty as to whether he was Chinese or American, until he was recognized as an enemy soldier. One of the Americans shouted "Kill him," but Dick said, "No," they would take him prisoner. The Chinese soldier came forward with his hands up, and one of Dick's men started to search him. Suddenly, Corporal Harm Tipton, a member of Lieutenant Brubaker's support element, caught sight of a grenade thrown from the darkness. He yelled "Look out!" Every man scrambled and dived for cover as the Chinese soldier bolted and ran around a turn in the trench.

The grenade's explosion stunned but apparently didn't wound anyone seriously. Recovery was quick, and they fired down trench, but were uncertain whether they hit the fleeing enemy. They pulled back a short distance from bunker 54, and Private First Class Mouser noted Dick Inman was limping. He had either turned his ankle or been wounded in the leg, perhaps by the grenade.

Dick Inman decided again to fight their way through to the 2d Platoon supply bunker, but they were again driven back by Chinese gre-

728

nades. Most of the patrol's weapons had quit firing. Ammunition was low, and his men had no more grenades.

He and Brubaker decided to gather the men who had functioning weapons and send the others down the back side of the hill to the evacuation landing. They divided the remaining patrol members into two groups. The plan was to run parallel to the trench, making for the supply bunker and A Company's 2d Platoon, with Dick's group on the lower side, and Brubaker's group on the upper side of the trench. Brubaker and a soldier with a Browning automatic rifle would provide covering fire as his group ran past bunker 54. Dick's carbine and Private First Class Irwin Greenberg's rifle were all that remained operable for the men moving along the lower side. Dick would provide covering fire with his carbine. He radioed the patrol's situation, and probably their plan, to Lieutenant Willcox. He told Willcox he had a broken leg.

They moved out, undoubtedly in a low, running crouch, with Dick first out of the trench, stopping to provide covering fire so his men could rush past him toward the supply point.

Corporal Tipton and the three or four men who had no operable weapons stayed behind, waiting to make a run for the evacuation landing. They watched as both groups left the relative safety of the trench and disappeared into the blinding muzzle flashes, grenade, mortar and artillery explosions, the buzz of bullets, and the shouts and screams of Chinese infantrymen seemingly everywhere on the hill. Brubaker and the man carrying the automatic rifle returned shortly, telling Tipton they couldn't get through because there were too many Chinese. Both had been wounded and headed for the evacuation landing.

Tipton decided to find Dick's group to warn him and went around the lower side of the trench, moving east toward the 2d Platoon area, only to discover the group pinned down by burp-gun fire and grenades. He came up behind Irwin Greenberg, and Greenberg told Tipton Dick Inman had been hit and was lying on the ground about fifteen yards away. Greenberg told Tipton he would provide covering fire with his rifle if Tipton and a soldier from A Company, Sergeant Perra, would make a run to pull Dick to safety.

Irwin Greenberg squeezed off only one round before his rifle jammed, as Corporal Tipton and Sergeant Perra dashed to retrieve Dick.

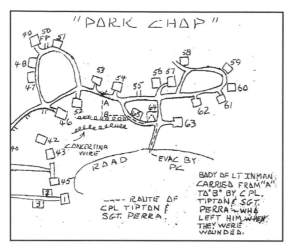

Map depicting the attempt to recover Dick Inman's body the night of 6 July 1953.

He was lying on his back. Harm Tipton got his hands underneath Dick's shoulders and Sergeant Perra lifted his feet. Tipton noted Dick was unconscious and saw what he knew to be a fatal wound on the left side of his head.

They started downhill with him, toward the concertina wire, which they would follow east to the trench leading to the evacuation landing. Chinese soldiers caught sight of them and began hurling grenades. Tipton and Perra had gone only five to six feet, desperately trying to carry him to safety, when a grenade exploded in their midst, wounding both men. Dazed and bleeding, they could no longer carry Dick Inman and staggered downhill to the east, toward the 2d Platoon supply point where medics later treated them and they were evacuated to a field hospital.

The grenades kept coming, more of them. Irwin Greenberg and the men with him made their way past the second row of concertina wire, toward the west, to shelters near the evacuation landing and the tunnel leading to the Company CP. Their rifles had quit firing. They could neither attack nor defend save for their sheathed bayonets, their trench knives. Their circumstance was the same for tens, perhaps hundreds of Chinese and American soldiers isolated from their units, alone or in small groups. Men were moving about inside and outside the trenches of Pork Chop, many with inoperative weapons, encountering one another in the confusing swirl of fighting, often hand to hand, trying, if they were wounded, to leave the field or help other wounded to safety, fighting to take and retake bunkers, firing positions, and trench lines – and survive.

The hillsides, trenches, and bunkers were rapidly becoming littered with dead and wounded. Small clumps of defenders were isolated in bunkers and trench segments, fighting often against overwhelming numbers, running out of ammunition, unable to resupply themselves or join with others for mutual support. Communication between isolated pockets was impossible unless someone had a hand held radio, or was in a bunker which still had an operative field telephone.

From somewhere on the outpost a call came over the Division communication net, "Flash Pork Chop! Flash Pork Chop!" Greenberg heard an American soldier yelling "Take cover! VT's going to come in!"

Within moments American artillery responded – massively. At ten forty-one, sixteen minutes after the first Chinese artillery rounds began crashing into the 7th Division sector, the Division artillery reported to the Division Command Post, "Flash Pork Chop in progress."

'Flash Pork Chop' closed around the hill.

Differing little from the curtain barrage of World War I days, the 'flash fire' of Korean operations was an on-call, tightly sown artillery (plus 4.2 inch mortar) barrage, usually horseshoe shaped and so dropped that it would close around the front and sides of an outpost ridge. The main idea of flash fire was to freeze enemy infantry movement, blocking out the enemy force on the low ground while locking in such skirmishers as had gained the heights. In effect, one battery fired on each concentration, 120 rounds per minute, two shells breaking into the ground every second. High explosive and proximity fuse shells [variable time or 'VT' fuses] were both used in the blast, the balance varying according to terrain conditions. While a flash fire lasted, infantrymen stayed in their fighting positions.

In the usual procedure, a flash fire was delivered with maximum firepower for three minutes, the howitzers then cutting back from twelve to six rounds per tube per minute while maintaining the fire for six minutes....

The VT tore into the Chinese infantry caught in the open, decimated their ranks, and those that survived the killing air bursts took cover or fled. But this wouldn't break the back of a determined enemy. There was only a pause. The battle was just beginning.

To compound what was rapidly becoming a serious situation, at ten fifty-six an A Company call on the radio net reported enemy automatic weapons fire hitting the evacuation landing, which was adjacent to the main trench and engineers' tunnel leading into the Company CP and nearby main supply point. The enemy was in the rear of the defense network, on the south side of the hill. Pork Chop was temporarily cut off.

At two minutes past eleven another call reported enemy coming into the trenches in the left sector, the threatened sector toward which Dick Shea had rushed when the first calls came over the radio Dale Cain was monitoring. Captain William S. Roberts, Jr., the A Company commander, called for immediate reinforcements. The Chinese were around and on top of the Company CP. Before the night was over the defenders counted nine waves of attacking Chinese infantry.

In the meantime at division, regimental, and battalion command posts information and calls for reinforcements were pouring in and decisions came rapidly. The Chinese had launched a major assault on the 7th Division's front, and it was becoming clear Pork Chop was the objective of their main effort. The power of their onslaught had carried into the outpost's perimeter and temporarily cut off its defenders.

Throughout the month preceding the attack General Trudeau mandated the 17th hold a company in reserve, in defilade, immediately behind Hill 200. The 1st Battalion's B Company, Dick Inman's company, was in reserve the night of July 6. Baker Company was ready, and when the call came for reinforcements at two minutes after eleven, they began moving out of their assembly area six minutes later. The short distance to the outpost from behind Hill 200 ensured timely reinforcement. B Company made contact with A Company at two twenty-four in the morning of July 7, and swept over the eastern sector of the hill.

A Company, then B Company, launched successive counterattacks toward the center and left, to retake bunkers, firing positions, and heights overrun in the initial onslaught and the surges of repeated Chinese assaults. In that first night Dick Shea organized two or three such attacks, leading men from both companies.

[B] company retook at least nine bunkers in the left sector and cleared out part of the central sector near A Company's CP. Some time later A Company reported the Chinese had overrun

732

Pork Chop, pulled back slightly, and attacked again, but the right sector, A Company's 2d Platoon, was holding.

Bugles sounded intermittently, heralding successive attacks, and at two thirty-five in the morning of the 7th, A Company reported enemy in the trenches at the CP, the trench blocked, and enemy soldiers placing small arms fire down trench. At two fifty-one another request for reinforcements came, and yet another at three forty-seven, for reinforcements to man bunkers that had been cleared of the enemy by counterattacks.

In the initial assault on the bunkers, trenches and the engineers' tunnel near the Company CP, Dick Shea and others were driven from their positions by overwhelming numbers. A Company casualties rapidly mounted, including the company first sergeant, Master Sergeant Hovey, and other key noncommissioned officers, officers, and men – killed, wounded, or missing.

At 11:00 o'clock the evening of July 6, thirty-five minutes after the attack began, General Trudeau had ordered the 2d

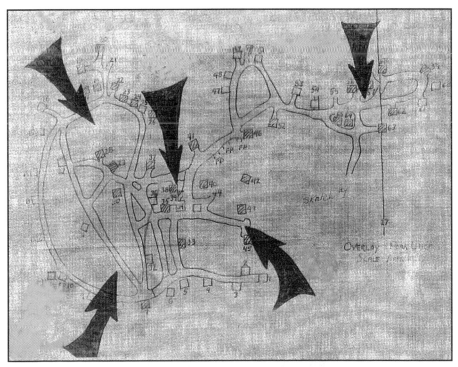

The Chinese assault on Pork Chop Hill.

Battalion, 32d Infantry attached to the 17th Regiment, and a company made ready to move, with the provision the battalion wouldn't be committed to the fight without his approval. The picture on Pork Chop wasn't clear, and there was hesitancy to commit a third company to a defensive perimeter intended to accommodate one company. Nevertheless, evidence mounted that the attack wasn't hit-and-run. Incoming artillery continued heavy, and for the first time in months was hitting the rear area as well as the MLR.

At two forty the morning of July 7, a few minutes after B Company closed on Pork Chop, E Company from the 2d Battalion, 32d Infantry was attached to the 17th Regiment. There was little doubt the company would soon be entering the fight.

* * *

When the enemy assault began, and the waves of Chinese infantry moved against the heights of Pork Chop, Dick Shea, in addition to organizing and leading counterattacks, circulated among soldiers' positions, checking, steadying, directing, and redistributing ammunition. He knew Pork Chop well, having been heavily involved in mapping and sketching the entire layout, and strengthening A Company's defenses. In the dark he was able to move quickly, confidently from one position to another. "Don't worry. Help will be here in the morning. We've just got to hang on till then. We can make it."

As he moved about, he learned which bunkers were in friendly or enemy hands, where ammunition, medical aid, and other supplies and support were needed – information later valuable to Lieutenant Colonel Read and the hill's defenders. During his movements he carried ammunition to positions he could reach, where defenders reported running low in ammo.

Shortly after dawn, a group of Chinese soldiers rushed down a trench line, attempting to expand their hold on the outpost. They overran several of the soldiers Dick Shea rallied, before they encountered him. He used his 45-caliber pistol to cut down four or five of the enemy before he ran out of ammunition. He then ran forward with his trench knife slashing and kicking, killing two more of the enemy.

His raw display of courage caused the remaining nearby Chinese soldiers to retreat.

He took advantage of their momentary confusion and led a group of soldiers in another counterattack down the trench line, chasing the enemy from a number of positions they had overrun. As a result of the attack, Pork Chop's defenders in that area regained better observation of enemy-held slopes and movements, as well as better fields of fire.

JULY 7

The 3:47 call from A Company requesting additional reinforcements to man bunkers retaken from the enemy, was answered twenty-one minutes later when E Company received orders to close on Pork Chop. Lieutenant Colonel Rocky Read, 1st Battalion Commander, with the E Company Commander and a platoon from the company, were the first to arrive about five thirty the morning of July 7.

Lieutenant Colonel Read remained on the hill as the senior commander until the afternoon of July 8, when A and B Companies were finally relieved of their mission to defend Pork Chop. Because of the situation he found in the engineers' tunnel and near the A Company CP, he set up his command in Bunker 45, what had been the large mess bunker, then a supply bunker, near the evacuation landing and the entrance to the engineers' tunnel. Easy Company's commander set up his command post in Bunker 45 where Read was situated, and stayed there until the Company was withdrawn. Wounded soldiers were already congregating in the bunker and the entrance to the tunnel when Read set up his command operation shortly after arriving on the outpost.

* * *

MacPherson Conner was awakened from his sleep. His company commander was already on Pork Chop and wanted the remainder of the company to move forward. Easy Company was to take up counterattack positions inside the tunnel and trenches of Pork Chop. Within the hour, Mac's platoon was transported by armored personnel carriers to the landing – the evacuation landing – and he went into Bunker 45 as his platoon moved inside the tunnel entrance to await his orders.

It was daylight by the time Mac and his platoon arrived on Pork Chop. His commander told him his platoon's mission was to hold the engineers' tunnel. During the furious fighting the night before, the Chinese had several times tried to fight their way into the tunnel. American commanders knew the tunnel was essential to successful defense of the hill, and as circumstances evolved, the engineers tunnel would be the prime location for safely collecting and holding wounded until they could be evacuated by armored personnel carriers.

After receiving his mission, Mac made his way into the tunnel, into what was already a crowded, confusing collection of American soldiers fighting to hold or retake bunkers and firing positions in the defense perimeter. The tunnel was crowded, two men abreast. Mac worked his way forward through the tunnel, reconnoitering the position his platoon would defend, and re-familiarizing himself with the defenses' layout.

There were lateral entrances along the tunnel, through which defenders could enter the web of trenches linking Pork Chop's bunkers, firing positions and CPs. There were openings leading vertically from the ceiling of the tunnel to the surface of the hill above. Through the vertical shafts Mac and his men could hear Chinese soldiers talking. At times the Americans could hear calls in English for defenders to surrender. All too frequently the enemy threw grenades down the shafts or through the lateral trench entrances into the tunnel.

During the lull in fighting the morning of July 7, Mac, his platoon, and A Company men who remained in the tunnel set about improving their position. They stacked sandbags three to four feet high across entrances from the tunnel into the trenches. The sandbags would keep the enemy from throwing grenades through the entrances, and force the Chinese to crawl over the stack, one at a time, enabling defenders to concentrate their fire and pick off attackers. In addition the Americans dug defensive positions into the sides of the tunnel. The recessed positions allowed defenders to take cover while opening up a center lane in the tunnel should the enemy succeed in bringing automatic weapons to bear down trench, as they had the night before. To avoid grenades thrown from above, down the shafts into the tunnel, soldiers were told to stay back from shaft openings.

Along toward noon, the Chinese began throwing grenades at the far, northwest end of the tunnel, apparently intending to fight their way in. Mac Conner immediately responded, rushed forward up the tunnel's slope toward the sound of explosions, and in the darkness, slipped and fell on an ammunition can, cutting his right index finger to the bone. He held the finger, bent, to stem the flow of blood, until the action quieted. When he finally released pressure on the cut he bled profusely. He searched out a medic, and found one in the darkened tunnel, near the A Company CP.

While the medic was taping up the cut, a distinctive, familiar voice heard behind him caught Mac's attention. He hadn't heard that voice in a long while.

"Is that Dick Shea?" he asked.

"Yeah, who is that?" came the reply.

"Mac Conner."

"Oh – Mac Conner!"

Dick, who had been talking with his company commander, Captain Roberts, told Mac the Chinese were coming into the tunnel. They had placed small arms fire down trench, and a burp gun earlier killed the A Company first sergeant. He told him, "A platoon is cut off, and I can't get through to them." He also told Mac, "Dick Inman called in and said he had a broken leg."

Dick Shea couldn't know what happened to Dick Inman in the few minutes following the initial Chinese assault, and the radio call telling of a broken leg. The ferocious, seesaw fighting and disrupted communications net on Pork Chop made it difficult to know what the situation was from moment to moment. There was no way to stop and count soldiers present for duty. Wounded and unarmed soldiers were individually making their way to bunkers and the tunnel near the evacuation landing, in hopes of receiving medical aid or being rearmed to defend themselves.

Now E Company of the 32d gathered in the tunnel to counterattack in hopes of occupying and holding positions that had been retaken after B Company swept over the hill in the dark early morning hours.

Mac Conner's platoon counterattacked to drive the Chinese from the tunnel near the CP, while Dick Shea made another attempt to fight his way to rejoin the isolated A Company platoon on the left flank. He was beginning the second in a series of close range and hand to hand encounters with the enemy.

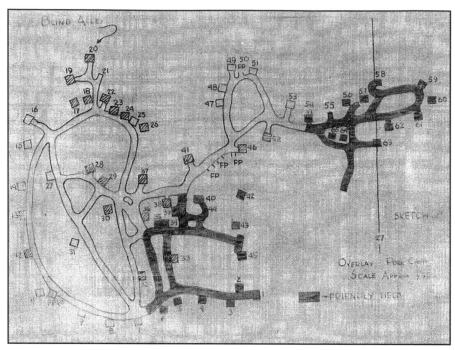

American and Chinese-held positions in Pork Chop Bunker and trench net-work, the morning of July 7, 1953, after the initial Chinese assaults. Shaded areas are Chinese-held positions.

The situation remained chaotic, however. The Chinese firmly held the western sector and the Americans held tightly to the eastern sector and part of the center sector, including the engineers' tunnel. But nei-ther side could advance and hold their gains. Friendly and enemy were interspersed in the trenches and bunkers dividing the two tightly held sectors. Though some bunkers had been retaken by defenders, there weren't enough men available to keep advancing, as infantrymen had to hold bunkers they had retaken.

"Hand-to-hand fighting for the trenches and bunkers of the outpost was intermittent throughout the day. Friendly and enemy forces con-tinued to reinforce." The remainder of E Company, 32d Infantry, came onto Pork Chop via armored personnel carrier, and off loaded into the tunnel near the CP. By mid-afternoon the entire company had joined in the defense of Pork Chop. The Chinese reinforced with groups of five to ten men at a time, slowly building up their casualty-thinned ranks, and their strength.

Sporadic but well-placed enemy small arms, mortar, and artillery fire fell on Pork Chop. The weather continued wet and rainy, with a low overcast which denied the defenders close air support. The Americans used armored personnel carriers to evacuate and resupply the outpost. Without the armored vehicles it would have been impossible to reinforce, since the Chinese systematically interdicted the road to the evacuation landing.

* * *

With the morning light on July 7 came a lull in the fighting. In Charlie Brooks' bunker in 1st Platoon the bitter all night swirl of fighting had taken its toll. No one was dead, but grenades hurled through the bunker aperture wounded nearly everyone inside. One tore off Bob Miller's right leg, just below the knee and shattered his left leg. Another squad member, Paul Sanchez, put a tourniquet on the right leg and bandaged the other, saving Miller's life – but they remained hemmed in the bunker.

Chinese held nearby bunkers and were blocking the tunnel leading past the company CP toward the evacuation landing. Worse, they were getting low in ammunition, with no end in sight, nor any sign of more ammunition, reinforcement, or relief. A second grenade thrown through the aperture exploded between Brooks' legs without doing the serious damage done to Miller.

During the night and on into the day more men came into the bunker, until seven were inside, including Brooks, Miller, Sanchez, one of the squad's ROK soldiers, and a lieutenant. The officer had no helmet on, and was carrying a forty-five-caliber pistol. His conversation was at times confused and incoherent. The men soon realized he was dazed, in shock, out of his head, probably suffering from "battle fatigue."

By late afternoon, Brooks had a half clip of ammunition remaining in his Browning automatic rifle and everyone else was nearly out of ammunition. What are we going to do? was the question being asked earlier in the day. At first Brooks said, "We'll hold on. Reinforcements and ammo will come." Neither came. Now the questions and discussion became more intense. In the middle of it all, the lieutenant put his pistol to the head of the ROK soldier and demanded his helmet. The ROK soldier was angry and frightened. Brooks, to defuse the situation, persuaded the young Korean to give the lieutenant his helmet, telling him, "I'll give you mine."

Finally, late in the afternoon, after discussion of their predicament, they agreed they would have to leave the bunker, somehow, and get to the aid station at the evacuation landing. The only way out, Brooks convinced them, was through the aperture, out onto the loose shale slope on the front side of the hill, and quickly around the hill to the rear. Miller would have to stay. They would get help to him. Though his rifle wasn't operating, there was a half case of grenades they could leave next to him. Thus Charlie Brooks and five other men from the bunker left through the firing aperture and made their way to the aid station and told the medics about Bob Miller, the nature of his wounds, and where he was located.

About dusk on July 7, Charlie Brooks and the other five soldiers left Pork Chop Hill on armored personnel carriers. There would be many years of buried remorse and wondering before Charlie Brooks learned that Bob Miller survived his ordeal.

* * *

Between noon and one o'clock on July 7, Private First Class Jim McKenzie, a medic from the 17th Medical Company, came on Pork Chop via an armored personnel carrier, riding with men from E Company of the 32d Regiment, who were completing reinforcement of the 17th's A and B Companies. Jim had been ordered forward in support of A Company.

When he off loaded from the carrier he went into Bunker 45, Lieutenant Colonel Read's command bunker, and a temporary collection point for evacuating wounded from the hill. In Korea, nearly all medics carried weapons, usually the lighter semiautomatic rifle, the thirty-caliber carbine, and a pistol.

The U.N. forces learned early in the war that, contrary to practices during World War II, medics shouldn't wear white arm bands with the traditional red cross symbol, or the same white circular field with a red cross painted on the center front of the medics' steel helmets. The North Koreans and Chinese disregarded the Geneva Convention and used such insignia as aim points, making medics targets. Consequently, medics weren't distinguishable from GIs, except by the canvas, olive drab bags they carried – which the GIs quickly learned to recognize.

As soon as he arrived in Bunker 45, Jim McKenzie began doing what he'd been sent to do. From the entrance to the bunker a trench led into the engineers' tunnel approximately thirty feet to the west. Propped against the wall of the tunnel entrance was the first of several wounded soldiers waiting to be evacuated. They needed to be moved out of the tunnel into the bunker to clear the tunnel entrance, permit better care, and get them closer to the access road to load onto personnel carriers. Jim began bringing them into the bunker. Like all tasks on Pork Chop, the job wasn't easy, and was hazardous at best.

In the thirty feet of distance to the tunnel entrance, the trench wall height shallowed, making it impossible to stand up and move a wounded man through the area, not without exposing both the medic and the wounded to enemy snipers or skirmishers. To move the first wounded soldier into the bunker, Jim strapped his carbine over his back and crawled on his hands and knees through the shallow area of the trench. He then laid the wounded GI on the ground on his back, with his head pointed toward the bunker. McKenzie straddled him on hands and knees, with the soldier holding onto him with his arms around McKenzie's neck, and then, with much effort, slowly, painstakingly crawled toward the bunker entrance, dragging the wounded soldier underneath him. When he finally reached Bunker 45 and got the man situated, he realized he'd worked hard to get the man to safety. He also discovered he'd lost his carbine. When he went to the bunker entrance and looked up the trench, the weapon was gone. It was needed elsewhere. This was the beginning of more than two days of nonstop toil to save as many lives as possible.

* * *

Approximately noon on July 7, just before Private First Class Jim McKenzie arrived in Bunker 45, General Trudeau, the division commander, and Colonel Benjamin T. Harris, the 17th's commander, arrived on Pork Chop to see the situation first hand. "As a result of an on-the-ground appraisal, General Trudeau was convinced the outpost could be retaken." Plans were made for another night counterattack. Considering the situation and Chinese abilities to observe and call in mortar and artillery concentrations, a daylight counterattack appeared fraught with high cost.

ATTACK BY COMPANY "F"

The Chinese occupied bunkers on the highest ground of Pork Chop. Should the counterattack mission be to sweep the enemy off the hill, or seize and hold the bunkers on the high ground? A night attack held out the prospect of surprise, and could be supported by units already on the out-post. The decision was a night sweep of the hill, with F Company, 2d Bat-

Fox Company, 2d Battalion, 32d Infantry Regiment Counterattack, Night of 7-8 July.

talion of the 32d Infantry given the mission. They had several hours to prepare and coordinate with units defending the outpost.

The F Company plan was to march down the front slope of Hill 347, cross the stream in the valley, then climb and sweep Pork Chop from west to east.

> A night-time single-file march through barbed wire in totally unfamiliar terrain, across a swampy valley and a stream swol-len by heavy rains, was a difficult assignment. But in addition, the company had to climb the precipitous slope of Brinson Fin-ger, make a right-angle turn to the right, and regroup for attack.

While preparations for the counterattack were in progress, the Chi-nese continued to reinforce in small groups in spite of return American artillery, mortar and small arms fire. At dusk Americans on Hill 200 observed a company size group of Chinese soldiers working their way toward the right rear of Pork Chop's defenses. Forward observers called in artillery and eighty-one millimeter mortar fire to break up the for-mation and foil an apparent attempt to seize the evacuation landing.

At 8:23 in the evening of July 7, Company F departed Hill 347 for enemy positions on Pork Chop and began a nearly five-hour move to launch the sweep. At nine fifteen, as they moved toward the outpost, Lieutenant Colonel Read, the 1st Battalion commander, reported a fire fight in progress in friendly trenches and on the personnel-carrier land-ing. Defenders also reported heavy incoming mortar and artillery fire.

JULY 8

Meanwhile F Company of the 32d continued its tortuous one to two mile night trek, and at 1:35 the morning of July 8, was organized for the attack – and suddenly found itself in the western trenches of Pork Chop, almost unopposed. They had surprised the Chinese, who reacted promptly with increased mortar and artillery fire and called for reinforcements.

When F Company reported their situation, they received a change in their mission. They were to stop the sweep and seize the crest of Pork Chop. The Americans moved up the slopes and gained some trenches; however, they eventually ran out of ammunition and grenades, and ground to a halt. At 4:25, F Company reported the enemy had surrounded them. When daylight came, enemy artillery and mortars, which hadn't been located during the night, began saturating F Company's positions, causing heavy casualties. At one point in the relentless shelling Lieutenant Colonel Read counted five rounds a second falling in the area atop Pork Chop, on the slopes, and around the evacuation landing. The earth was pulsating with the roar of explosions, the bunker shook and shuddered, and dust permeated the air inside. Jim McKenzie put his head down and prayed, thinking it would never end.

The 32d Regiment's F company was ordered to withdraw. Remaining elements of the company were able to withdraw down the south side of the outpost, where the wounded were evacuated by armored personnel carrier. The remainder of the company withdrew through the valley between Hill 347 and Pork Chop.

Their losses were heavy, but the timing of their assault spoiled an enemy attack aimed at the evacuation landing and its connecting road. The rest of the outpost's defenders were running low on ammunition of all types. Had the landing fallen into Chinese hands, the entire outpost would have been isolated, starved of resupply, and lost.

* * *

Early the morning of July 8, shortly after artillery fire stopped, Chinese soldiers launched a localized attack down a trench line on a finger of the hill which was across a deep gully from an opening in the engineers' tunnel. Mac Conner and two of his riflemen, hidden inside

743

the tunnel on the enemy's flank, apparently unobserved, could plainly see the attack in progress. Mac and his two riflemen kept a steady fire on the right flank of the assaulting Chinese, wiping out the right half of the attacking element. It was the last offensive action by the Chinese in the vicinity of the tunnel for the duration of Mac's stay on Pork Chop.

A short time later Mac again encountered Dick Shea in the tunnel leading to the A Company CP. Dick's cheek showed a small gash and his neck evidenced a cut or bullet crease, superficial wounds incurred in the hand-to-hand and close-quarter fighting the night before. Mac sported the bandaged finger from his fall onto the ammunition can during his platoon's counterattack to drive the enemy from the tunnel near the A Company CP the morning of the 7th. Dick told Mac that Able and Baker Companies were being relieved of responsibility for defense of outpost Pork Chop, and again expressed frustration at being unable to reach an isolated platoon.

Lieutenant Colonel Read had been alerted to move A and B Companies off the hill. First Battalion was being replaced by the 2d of the 17th. A Company's 2d Platoon had held firm on the east flank, and Lieutenant Colonel Read had told Dick to establish contact with the platoon and come off the hill with them. In the brief second encounter with Mac Conner, Dick Shea was explaining he had to reach the platoon, but had been unable to get through. He was going to try another foray, but didn't know how he was going to reach them.

"Are you OK?" Mac asked.

"Yeah," Dick replied.

Mac Conner knew Dick Shea wouldn't leave Pork Chop without the men he was assigned to bring off the hill. He put his hand on Dick's shoulder. "Don't get yourself killed."

Dick Shea replied earnestly, "Don't worry. I won't, Mac."

The relief of A and B Companies from the outpost was to begin mid-morning, July 8.

* * *

When it became evident the night attack by F Company had failed and troops on The Chop were unable to make progress toward clearing the enemy off the hill, other measures were considered. At midnight, Colonel Harris, the 17th commander, alerted 2d Battalion of the 17th to prepare a plan to counterattack the outpost. The plan, which was later altered, was an attack by two companies, line abreast.

Shortly after daybreak, the 2d Battalion, 17th Infantry, was given responsibility for the sector formerly assigned to the 1st Battalion. A and B Companies had been in a virtual nonstop, all out, exhausting fight for the outpost for nearly thirty-six hours. They had been decimated by casualties, and reinforced with E Company of the 32d, which took many casualties as well. F Company of the 32d had been withdrawn after their counterattack failed to dislodge the Chinese. By the time the 1st Battalion could be replaced they would have been fighting for nearly forty-eight hours without relief.

At a conference at the 17th Infantry CP at 8:00 am, 8 July, General Maxwell Taylor, Eighth Army commander; Lieutenant General Clarke, I Corps commander; Major General Trudeau, 7th Division commander; Colonel Harris, 17th Regiment commander; and Major John Noble, commander, 2d Battalion, 17th Infantry, whose battalion was to make the next counterattack, completely reviewed the situation. The decision was made to launch a two-company daylight attack in the middle of the afternoon. The attack was to be preceded by twenty minutes of preparatory [artillery and mortar] fire ..., with surrounding enemy observation posts to be smoked and enemy mortar and artillery positions neutralized. Troops were to be moved by truck to the checkpoint [behind Hill 200]. Company G was to attack to the east side of Pork Chop and secure high ground, jumping from a line of departure on the forward slopes of Hill 200. Company E of the 2d Battalion was to follow the stream bed [at the base of] the reverse slope of Hill 200, attack up the western side of Pork Chop, and tie in with Company G on high ground. Special instructions were issued to the

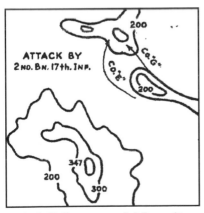

E and G Company, 2d Battalion, 17th Infantry Regiment counterattack, 8 July.

troops to stay out of the trenches until they were on their objective.

* * *

While the Army, Corps, and Division commanders were making plans for the afternoon assault by E and G Companies, more heroics were in progress on Pork Chop. Private Robert E. Miller was being readied for evacuation with the wounded of A and B Companies, and Dick Shea and Jim McKenzie were preparing to assist David Willcox's 2d Platoon. Though the two companies were to be relieved when Major Noble's 2d Battalion took over responsibility for the defense of the hill, A Company's 2d Platoon had to continue to hold the east shoulder of the hill until relieved by G Company in the mid-afternoon counterattack.

Bob Miller admired his company executive officer, Dick Shea. He knew Dick was a "true leader." He was greatly interested in the men's morale, a "do as I am doing" officer. Dick pushed the men hard to be ready at all times and talked to Bob about the subject, as he did others, checking to see that Bob had an adequate supply of ammunition and grenades, and telling him to keep his rifle clean. At five o'clock each day soldiers fired a clip or a short burst of ammunition from rifles and automatic weapons to ensure weapons were clean and in good working order. "Dick Shea," he said, "wouldn't ask anything of soldiers he wouldn't do himself." He would fight beside the GIs, put his life on the line with them. Bob Miller was describing what other men in A Company knew of Dick Shea, and in the first hours of the battle for Pork Chop, Bob Miller would need everything he learned.

* * *

Private Bob Miller's abrupt awakening in the first minutes of the July 6 assault by the Chinese turned into a nightmare. Told to grab his rifle, throughout the night he engaged in a series of close quarter fights with enemy soldiers swarming into the 1st Platoon positions near the company CP. Everywhere he turned, almost without pause, it seemed there were Chinese infantrymen. He repeatedly hurled hand grenades, emptied his rifle and reloaded, firing at fleeting figures and small groups of enemy soldiers. His rifle eventually quit firing, and he hurled grenades and had grenades thrown at him – until the one came through the

746

aperture, exploded at his feet, tore off his right leg below the knee and shattered his left.

When finally, the afternoon of the 7th, Charlie Brooks asked Bob if he could hold on while the rest of the men went outside the trench and bunker network to get assistance for him. They had attempted several times to fight their way through the trenches and tunnel to get aid for him. Miller understood his circumstance and theirs. Miller's rifle was jammed, but he agreed to remain by himself if they would leave him a case of hand grenades. They did, and moved him out of the bunker into a covered section of the trench, away from the entrance, propped against the wall where he could see both directions. Next to him was the case of grenades.

For what seemed like hours he waited. Every time he heard Chinese voices approaching he threw a hand grenade toward the sound. Each time the grenade exploded, the voices stopped. Later, another squad member, Angelo Palermo, who returned to the 1st Platoon after fighting in another sector of the hill, came down the trench, identifying himself to Miller. Palermo was dumbfounded to learn Bob was left behind, alone. Miller explained why and showed him the supply of grenades, telling Palermo how he was defending himself, and noting he'd used up only a fourth of his supply – that he was all right. Palermo said he believed the tunnel was clear of enemy now, and he'd go get a medic. He rechecked Bob's bandages, saying "They'll be OK 'til help arrives," and left.

Again, it seemed hours passed. Day had turned to night after Brooks left. Then Palermo had gone for help. Now it must be near morning. Bob Miller continued to hurl grenades at approaching sounds of Chinese soldiers. Suddenly he heard the roar of a flame thrower behind him, around the corner, coming from the direction of the company CP. He yelled, "What the hell are you doing? Cut that damn thing off." He breathed a sigh of relief when he heard an American voice say, "He's one of us."

Soon he was carried on a litter down the tunnel past the CP. As his litter bearers approached the tunnel entrance he could see reinforcing soldiers lining the walls. He saw the expressions on their faces as he passed them. He knew some, on seeing him, were receiving their first dose of war's reality.

He'd had no food or water since the afternoon of July 6, and asked if anyone had a drink of water. A soldier said, "Here's my canteen. Take it. You need it more than I do." After taking a drink, Bob Miller offered the canteen back to him. "You keep it," he said.

About ten o'clock the morning of July 8, Bob Miller was evacuated from Pork Chop Hill on an armored personnel carrier. He didn't know the soldier who gave him a drink of water and his canteen, didn't ask his name. But over the years he's yearned to know, while inside he's said his thanks to the soldier many times.

*   *   *

The morning Bob Miller was evacuated, Dick Shea and Jim McKenzie both were to make their way to the 2d Platoon area on the right flank of Pork Chop. Lieutenant Colonel Read earlier had told Dick to contact the platoon and prepare them to leave the hill following the G and E Company counterattack that afternoon. The two companies would be reinforcing as well as counterattacking, and would stay on the hill after the attack. Men from G Company would relieve A Company's 2d Platoon. Dick was to come off the hill with David Willcox's platoon.

The Chinese had continued to hold bunkers and trenches between the company CP and 2d Platoon's perimeter, isolating David Willcox and his twenty-five man unit from the rest of the company. The most severely threatened area of Pork Chop was in the left and central sectors. The central, 1st Platoon sector was in the vicinity of the company CP, which was almost directly beneath the crest of the hill. Thus reinforcements coming onto the hill were consistently off loaded from personnel carriers near the entrance to the engineers' tunnel, to shore up the more severely threatened areas, and retake areas held by the enemy.

There were other reasons 2d Platoon hadn't been reinforced sufficiently to be relieved with the rest of A Company. Because the center and left of Pork Chop were under constant pressure from the Chinese, Read couldn't pull men from the most threatened areas to reinforce 2d Platoon, which was holding its own. Additionally, repeated enemy shelling and assaults on the rear of Pork Chop made reinforcement, replacement, resupply, and evacuation of wounded and dead only slightly less hazardous than fighting in the open or from the trenches and bunkers on the front slope of the hill. Persistently troublesome interdicting enemy fire along a one thousand yard section of access road leading to the south entrance to the platoon's defense network hampered efforts to reinforce. Further, the trench leading to the platoon CP didn't offer as much cover from enemy fire as the engineers' tunnel leading to the company CP.

To gather the force necessary to move out of the trenches and bunkers near the company CP and attempt reinforcing the right sector was next to impossible. There were seldom enough men to hold positions retaken with counterattacks. Going outside the defense network onto the northern slopes in localized counterattacks, as when Dick Inman's patrol attempted fighting its way to 2d Platoon positions that first night, would be costly to already beleaguered, exhausted troops. Worse, given the confusing intermixing of American and Chinese held positions, there was an added hazard of "friendly fire" unless Americans chose to identify themselves during counterattacks.

Day or night, it was virtually impossible to use daily passwords, because communications and contact didn't extend to isolated defenders. The certainty of who was to the right, left, in front, or behind defenders vanished the night of the initial Chinese assault, and was seldom restored anywhere on the hill. Voice identification wasn't reliable. Visual identification risked instant death. Defenders were tense, ready to shoot first and ask questions later. When the situation did improve, it could immediately deteriorate in the next Chinese assault, or the next American counterattack.

The morning of the 7th, "Rocky" Read personally intervened to stop Dick Shea from exposing himself to enemy fire during an attempt to eject Chinese from a trench section near the company CP. Now, the morning of July 8, from Bunker 45, Read personally covered Dick with rifle fire to keep enemy heads down and permit Dick to get to the area where the twenty-six men in A Company's 2d Platoon were fighting to hold the hill's right flank. The battalion commander saw Dick Shea safely enter the platoon's perimeter when Dick dashed the seventy-five yards back down the access road and turned left into the trench leading toward the platoon CP in Bunker 64.

Some time later, Jim McKenzie was covered by machine gun fire as he traversed the same route to 2d platoon to provide medical aid and prepare wounded to be evacuated. He reached Bunker 64, where he stayed, tending wounded. Late in the day he would be swamped by casualties as the G and E Company counterattack swept onto Pork Chop and was decimated by enemy artillery and mortar fire.

* * *

The morning of July 8, after Mac Conner's second brief encounter with Dick Shea, he went outside the tunnel with one of his men. Their purpose was to throw grenades uphill toward enemy soldiers who had been repeatedly hurling grenades at the men in Mac's platoon. The foray resulted in the wounding of the man accompanying Mac. He unsuccessfully attempted to pick up and throw a live Chinese grenade back at the enemy, and it exploded in his hand. Nevertheless, Mac decided a more forceful excursion outside the tunnel could get the Chinese off the hill directly above them.

He laid out the plan for his men, and they were successful in driving the enemy off the area immediately above the tunnel. However, the cost was heavy, and it stuck with Mac the rest of his life. One of the platoon's sergeants was killed by enemy automatic weapons fire from much further up the hill.

While Mac and his men were outside the tunnel, and were successful in driving the Chinese off nearby ground, they witnessed a veritable rain of grenades from further up the hill, most falling among the enemy's own soldiers below. Mac's move outside the tunnel made clear to him that numerous Chinese were holed up all over the top of the hill, and a major attack would be needed to dislodge them. He decided to keep to his orders. He didn't want to lose any more of his men. He would take no more offensive actions, not until he was ordered to launch an assault the next morning. While his platoon took no more offensive actions on July 8, before the day was over, Mac Conner was wounded for the first time.

In mid-afternoon, he finally paused to eat some C-rations. He was sitting near an opening at the lower end of the tunnel, spooning fruit from a can, when a mortar round exploded adjacent to the opening, knocking him backward onto the ground. A piece of shrapnel punctured his right arm and lodged in the bone near his elbow. The wound wasn't incapacitating, and he remained on Pork Chop.

* * *

In preparing for the July 8 counterattack, 2d Battalion of the 17th Regiment had been given the best information available as to which bunkers were American-held and which ones were occupied by the enemy. Based on that information, E and G Companies were assigned objectives in the defense network, with E Company driving from west to east across the hill and G Company pressing over the crest, and turning left across the northeast face of the hill in front of 2d Platoon, converging toward E Company.

The afternoon attack began at three forty. Corporal Bob Northcutt was a squad leader in the heavy weapons platoon of G Company, and his squad was carrying thirty-caliber machine guns and all the ammunition and grenades they could on the assault. The company moved across a one thousand yard area, which from Baldy, to the left rear of Pork Chop, could constantly be observed by enemy artillery spotters. The spotters could readily call for fire on any units crossing through the area of observation.

Bob Northcutt remembers well what the soldiers of G Company were told. "If men go down, wounded, don't stop to help them. Keep going. Medics will come along behind and pick them up."

Also attached to G Company was a demolition squad from A Company, 13th Engineer Battalion, which was assigned to the 7th Division. The squad leader was nineteen-year-old Dan Schoonover of Boise, Idaho. The squad's mission was to use demolition charges against hastily erected enemy fortifications, bunkers, and other covered defenses they had seized, as G Company sought to dislodge the Chinese.

When E and G Companies moved off the line of departure, both encountered murderous artillery and mortar fire. Counter battery fire from the Americans couldn't possibly locate and silence all the Chinese mortar and artillery pieces which had been registering on the outpost for weeks. The weather continued overcast, locking out American air support which might otherwise favorably tip the firepower balance.

Nevertheless, G and E companies pressed their attack through the withering fire. In addition to the holocaust of artillery and mortar rounds, Chinese soldiers holding positions in the 2d Platoon sector

751

raked G Company with intense automatic weapons and rifle fire, and grenades.

G Company's officers and noncommissioned officers were struck down, one by one, wounded or killed. With their leadership shredded, the assault lost momentum and bogged down. In spite of instructions to remain out of the trenches and bunkers until on their objectives, the soldiers of G Company had no choice. The awful toll drove them into the trenches and bunkers for survival, with many pinned down in the vicinity of Bunkers 60 and 61. When the attack began to falter, it became once more a series of isolated struggles between small groups of surviving Chinese and American soldiers.

Under the withering Chinese fire, Corporal Bob Northcutt and his machine gunners scrambled up the steep slope toward the left center area of David Willcox's 2d Platoon defensive positions, taking cover as best they could in badly battered, caved-in trenches. They were in for a long night. Running short of ammunition before darkness, they picked up weapons and ammunition from the dead.

For his squad and all soldiers manning machine guns and other heavy weapons, such as mortars or recoilless rifles, life in battle was always difficult and could be terribly brief, especially if they couldn't fire their weapons from inside well constructed and reinforced bunkers, or move quickly to another position if they were not well hidden by other forms of cover. When a fight was in progress, once enemy forward observers identified machine gun or heavy weapons positions, they immediately called in artillery and mortar fire. The pounding was relentless until incoming rounds found their mark – if the crew didn't quickly move after firing a few rounds. Bob Northcutt lost two of his men killed that terrible day, one decapitated by an incoming round, another who completely lost control of his emotions and stormed forward out of protective cover toward the enemy, while Bob screamed at him in vain to come back.

The hail of enemy fire, heavy fighting, and the confusing intermixing of enemy and friendly-held positions soon convinced Corporal Dan Schoonover his squad from A Company, 13th Engineers, couldn't conduct their assigned demolition mission. Instead, in the midst of the desperate fighting on the barren slope, he voluntarily employed his men as a squad of riflemen. He forged ahead, leading them up the hill in the assault.

752

When an artillery round exploded on the roof of an enemy bunker, he ran forward and leaped into the position, killing one Chinese soldier and taking another prisoner. A short time later, when G Company soldiers were pinned down by fire from another bunker, he dashed through the hail of fire, hurled grenades in the nearest aperture, then ran to the doorway and emptied his pistol, killing the remainder of the enemy. His actions enabled troops to resume their push forward to Pork Chop's east crest.

It was late in the day of July 8, about six o'clock, when the counterattack stalled. G Company lost nearly half its men, killed or wounded. The remainder, bereft of officer and NCO leadership, remained isolated, hugging the safety of battered and collapsed trenches, a few in bunkers.

Success of the counterattack was crucial to successful disengagement and withdrawal of A Company's 2d Platoon. Every man in the platoon was exhausted after forty-four hours of almost non-stop fighting. Nearly eighty percent of those able to fight were wounded. If G Company couldn't keep the Chinese at bay and occupy positions vacated by 2d Platoon, relief of defenders on the right flank could stall completely. If the Chinese massively reinforced in that area, which they had shown they were willing to do, 2d Platoon could be completely overwhelmed, causing collapse of Pork Chop's defenses.

No one knows for certain what Dick Shea saw or was thinking when the G Company counterattack faltered. He undoubtedly recognized the stalled attack meant the 2d Platoon relief couldn't possibly proceed as long as G Company remained pinned down and had lost its leadership, as well as its forward momentum. Most likely he recognized the situation had reached a critical stage in which the entire operation's success was questionable. He was that well trained, that well versed, and undoubtedly concluded the 2d Platoon relief couldn't possibly occur unless the men in G Company somehow aggressively resumed the offensive. Dick Shea made a crucial decision.

He ordered another soldier to take responsibility for leading the 2d Platoon off Pork Chop. He then moved to assist and organize approximately twenty G Company soldiers, and led a series of localized counterattacks, keeping pressure on the Chinese and off the men of A Company's 2d Platoon. They confronted an enemy machine gun position. He maneuvered, while under fire, to within a few yards of the position, threw two hand grenades, then moved in after the grenades, firing his carbine as he closed on the enemy. He killed the three-man crew, silencing the machine gun.

It was still daylight. He sensed a lull in the fighting, a hesitation by the enemy, and determined to launch another counterattack to eject the Chinese from the trenches and bunkers they had taken.

Attacks down the trench line weren't effective, the trenches too narrow and crowded to bring the fire power of more than one or two men at a time. The most effective way to break the enemy hold was to move out of the trenches where more firepower could be brought to bear.

He moved from man to man. "OK. We're ready." Then, "Let's go!"

He led a running assault on Chinese positions in Pork Chop's right flank sector, intent upon overrunning them. He was hit, and went down. He got up and continued moving forward, yelling to his men, "Get going!" For the fifth time in less than two days he refused to leave the field to obtain medical aid.

The Chinese held their positions, shifted forces and reinforced to meet the counterattack. Dick Shea, among a dwindling group of attacking GIs was last seen in the eastern sector of Pork Chop, engaged in furious hand-to-hand fighting. They were apparently overwhelmed by sheer weight of numbers, as the enemy continued to reinforce and launched their own counterattack. There was no further communication from him, no way to get to where he'd last been seen, and when 2d Platoon finally pulled off Pork Chop on July 9, he wasn't among them. No surviving American soldier witnessed Dick Shea's death.

* * *

The failure of the 2d Battalion counterattack the afternoon of July 8 didn't mean calling it quits. At the early morning command conference that day a tentative decision had been made to rotate the 17th Regiment's 3d Battalion from the right flank sector of the Regiment's line, and prepare them for another counterattack on Pork Chop should the E and G Company assault fail. The 3d's sector included the equally vulnerable Erie-Arsenal outpost on the tip of T-Bone. There was excessive risk in replacing the 3d's L Company on the often hit Arsenal while a battle was in progress on the adjoining outpost. Instead the 1st

Provisional Company of the 17th Regiment – a company formed from headquarters and service troops, plus other combat able troops from the 17th – replaced I Company on the line. The last of I Company came off the line at 5:30 in the afternoon of July 8, half an hour before the E and G Company counterattack came to a halt, torn apart by Chinese resistance.

At 6:05, K Company of the 3d Battalion was ordered to move to an assembly area near the Pork Chop check point, and fifteen minutes later Major Costigan, commander of the 3d, reported to Colonel Harris and received a briefing on the situation. At 10:30 that night Major Costigan was ordered to prepare a counterattack with two companies, arriving on position at first light. The plan was to send both companies across the cut to the right of Hill 200, then west along the base of Hill 200 to Pork Chop. The intent was to avoid intense enemy artillery and mortar concentrations known to be in the valley south of Pork Chop, and assault from west to east across the outpost.

JULY 9

Because K Company arrived at the line of departure first, they went ahead of I Company, the two crossing the line at 3:15 in the morning of July 9, with I Company fifteen minutes behind. They were in position at 4:15 to begin their attack in their zones of advance.

As soon as the Chinese detected the assault, intense mortar and artillery concentrations began falling on the attackers. When daylight broke, the two companies, line abreast and moving in a near perfect skirmish line, were caught in the open.

*   *   *

Inside the tunnel early the morning of July 9, Mac Conner's platoon had received orders from their company commander to launch an assault of their own, in coordination with the I and K Company counterattack. Another E Company officer was assigned to Mac's platoon to participate in the assault. The platoon had lost both its non-commissioned officers, one killed the preceding day when Mac and his men had cleared the Chinese from the hill immediately above them, the other pulled from the platoon to be the company first sergeant.

When Mac led his platoon out of the tunnel onto the hillside, they moved into the maelstrom of artillery, mortar, automatic weapons and small arms fire already tearing at the ranks of I and K Companies. The officer assigned to the platoon was promptly hit, wounded in the leg, and went down. Mac led on, moving a short distance up the hillside, shouting encouragement, urging his men forward. All the men were wearing armored vests, "flak jackets." He felt something hit him in the left side, and remembered looking to his left, puzzled, wondering what it was. At that instant an explosion on the ground directly below his groin, lifted him off the ground, and spun him completely around. He landed on his back, feet up slope, and slid backward down the hill into the arms of his men.

He was grievously wounded, the pain agonizing. His men pulled him back inside the tunnel, and medics began working on him. He felt paralyzed from the waist down. His testicles swelled to the size of softballs. Because of heavy bleeding, and apparent extensive injuries, the "docs" were fearful of severe internal injuries. They couldn't give him morphine. To make matters worse, the fierce fighting would keep him and the accumulating wounded in the tunnel for several hours, before armored personnel carriers could begin their evacuation. The pain mercifully overwhelmed Mac Conner. He remembered nothing more until he awakened in a Norwegian Mobile Army Surgical Hospital a day or more later.

He then learned he had been first hit in the left side by fire from a submachine gun. A bullet had entered his left thigh, and lodged against his hip bone.

* * *

By 9:00 the morning of July 9, the shelling had inflicted heavy casualties on both I and K Companies. Company I had one officer left and Company K had none. When the attack's forward progress ended, control of the surviving elements of I and K Companies passed to Major Noble, commander of the 2d Battalion, which retained responsibility for defense of the outpost. Major Noble now had under his command elements of Companies E, F, G, H, I and K of the 17th, and E Company of the 32d, plus A Company's 2d Platoon, still holding the right flank with survivors of the July 8th, G Company counterattack.

At daybreak had come the summer sun and sweltering heat. The odor of decaying bodies and blown up latrines, which had permeated the air since the first day of the battle, was more overpowering with each passing day. Water was a precious commodity. Despite the frustration of four failed, major counterattacks and heavy casualties, American commanders believed air attacks on reinforcement and resupply routes and enemy exhaustion from three days of continuous fighting might cause the Chinese to give up first. In hopes of finally tipping the balance, they prepared another attack, this one by C Company, 1st Battalion of the 17th, the same company that conducted the July 4 sweep of the Rat's Nest.

Charlie Company had been defending Hill 200 since the battle began. They needed to be replaced on their position before they could attack enemy units on Pork Chop. Several reliefs and movements of various units freed them to plan and begin the assault. A reconstituted A Company of the 17th replaced them.

While the shift was in progress, the Chinese stepped up their artillery fire. Casualties sustained during the change, plus exhaustion from three days living under constant barrage and oppressive heat, reduced C Company's combat effectiveness. Nevertheless, Charlie Company began the move from Hill 200 by armored personnel carriers, and in little more than an hour, at four thirty-two in the afternoon, was on Pork Chop.

Sergeant Dan Peters, the platoon sergeant in the Company's heavy weapons platoon, was in one of the ten loads of troops shuttled forward for the assault. His platoon's seventy-five millimeter recoilless rifles and mortars weren't used in support of the company. Instead Dan and men in the platoon moved ammunition and supplies forward to sustain the counterattack. Consequently his platoon suffered fewer casualties compared to the three rifle platoons.

The company was to attack up the slope on the eastern shoulder of the hill, to the left of the route G Company had taken the day before, except Charlie Company had the advantage of moving to the assault via armored personnel carriers rather than on foot, which reduced their exposure to enemy fire.

In spite of heavy enemy fire, which took the life of the company commander, their counterattack was the most successful to date. The commander was leading the first element when he was killed near the crest of the hill, and the counterattack temporarily faltered. Lieutenant Frederick K. Tanaka from Honolulu, Hawaii, the executive officer, was leading the second element of the assault. When the first element was pinned down near the crest, Tanaka came forward to lead in a fierce, aggressive assault that surged to the top of the hill.

Joined by Sergeant First Class Richard G. Beacher of Eufala, Alabama, and Sergeant Donald D. Swope of Woodbine, New Jersey, Tanaka and his men ran from one enemy bunker to the next, throwing grenades and firing into the doorways and firing ports, killing or taking prisoner any enemy soldiers that dared show themselves. After securing the crest on the right shoulder of the hill, Tanaka set up machine guns on the company's flanks in order to hold the position.

When Dan Peters came on the hill in an armored personnel carrier, bringing ammunition, he left the vehicle and entered the seven foot deep trench leading toward the A Company, 2d Platoon CP. Enemy fire was heavy – artillery and mortars. Though the trench was seven feet deep, it wasn't covered where he was, and Peters' instinctive reaction was to keep as low to the ground as he could, and crawl on his belly to bring the ammunition where it was needed. He quickly learned he couldn't carry ammunition and a rifle while crawling on his stomach. He leaned his rifle against a wall of the trench and continued the ammunition delivery. When he came back to retrieve his rifle and another load of ammunition, the rifle was gone.

More startling, as he crawled back and forth in the trench, he noticed wounded men were walking past him, standing fully upright. He wondered, What in the hell are those crazy bastards doing? As day wore on toward night, however, Dan Peters, like the walking wounded, became increasingly desensitized to the dangers surrounding him, and was soon walking upright in the hell that was Pork Chop.

Before dark, Charlie Company made substantial gains and cleared the Chinese from most of the right finger of Pork Chop – but failed to clear the enemy from the entire hill. Perhaps their greatest achievement was the relief of David Willcox's 2d Platoon of A Company.

When relief came, Jim McKenzie was still tending wounded in Bunker 64. The devastating casualties taken in G Company's counterattack a day earlier flooded the already overflowing bunker with

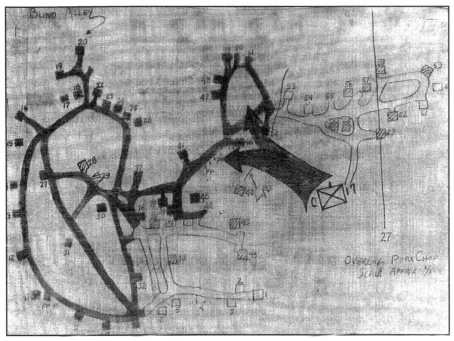

Charlie Company, 1st Battalion, 17th Infantry Regiment counterattack, afternoon of 9 July. Chinese-held positions are blackened in this sketch.

wounded and dying. Among the wounded was another medic who knew Jim. In the dark he called Jim's name, but when the man shined a small light on his own face, McKenzie had difficulty recognizing him because his face was covered with blood.

As relief drew near, Jim McKenzie helped move the wounded and dead from the 2d Platoon CP to the evacuation landing. Jim didn't know it, but the term "Doc," given the medics by the GIs, was more than a mark of respect. Medics were special people, truly angels of mercy, and the soldiers knew it. The medics saw all the horror, all the ugliness, but sheltered a deep well of compassion within them, inside the steel shell which gave them the strength to keep at their difficult task under the most extreme conditions. Among Jim McKenzie's most vivid memories of July 9 was the image of American dead, stacked like cord wood at the edge of the evacuation landing, waiting to be moved on armored personnel carriers. Lying on top was a Latin-American soldier, perhaps Puerto Rican, his face young, so young. His fatigues were clean, freshly laundered, and his combat boots showed little wear and mud. The young soldier's life on Pork Chop Hill had been dreadfully brief.

In Bunker 53, David Willcox and six other men had held firm for nearly sixty-four hours. Most were wounded, and he was in a seriously weakened state when he heard shouts of "We're Americans, Charlie Company!" He had to be helped from the bunker by the other men. He had received thirteen puncture wounds from grenade shrapnel, with one leg being particularly hard hit. He was carried on a litter and placed on the ground just outside the trench leading to the 2d Platoon from the access road. He remembers lying on the litter gazing up at what seemed a beautiful sky he hadn't seen for days. Yet he knew he wasn't entirely safe. He reached for a nearby sandbag and pulled it to cover his groin. A heaven-sent American soldier appeared out of nowhere, and poured some clear, warm water into his dirty helmet. The water tasted marvelous, "like champagne" to David Willcox, in spite of its resulting color.

David Willcox carried an unforgettable, glowing memory with him, among many memories he wished to forget. Near the end of the long ordeal in Bunker 53, there was growing doubt the seven men would get out alive. Water and food were gone. Their ammunition and a case of nearly one hundred and fifty grenades were running low, and they were all exhausted.

Early in the fight they had sealed the bunker door with sandbags to keep the enemy from finding them or throwing hand grenades through the door. Over the bunker's firing port they firmly anchored a wire mesh similar to chicken wire to keep "potato mashers," Chinese grenades, from being hurled through the firing port. There was a Gideon Bible in the bunker. They decided to take turns reading passages out loud.

During a pause in the action, while they were taking turns reading the Bible, suddenly they heard a slight noise at the entrance to the firing port, and looked up to see an American grenade, firing pin pulled, lodged in the wire mesh. Instantaneous, absolute silence followed, and the certainty they were all dead. American grenades were far more efficient killers than Chinese potato mashers.

Then, after what seemed an eternity, just as abruptly as the grenade appeared before their eyes, the wire mesh loosened its grip and the grenade fell from view. The explosion which followed, always startling, was harmless. The men in Bunker 53 knew a miracle had occurred.

JULY 10

In spite of C Company's successful counterattack, the situation on Pork Chop Hill the night of July 9-10 was little different from the night before, and in some respects was worse. Major Noble commanded remnants of seven companies from the 17th Regiment plus E Company of the 32d. He had almost no staff to assist him in controlling the outpost's defense, and communications were virtually nonexistent. American and Chinese-held bunkers and trenches remained intermixed, some having merely changed hands between both sides in the preceding twenty-four hours. The Chinese still held the tail of The Chop, Brinson Finger, most of the hill's rear slopes, and the crest.

American commanders' review of the situation resulted in a plan to evacuate the dead and wounded, and relieve and replace all the men on Pork Chop on July 10. Armored personnel carriers were to be used for the evacuation and relief. The 3d Battalion of the 32d Infantry would replace the departing defenders and prepare The Chop for a battalion counterattack on July 11. The 1st Battalion of the 32d would attack, with a supporting, diversionary battalion attack on Old Baldy – if 8th Army approved.

Circumstances intervened to knock out the plans. As I Company of the 32d arrived behind Hill 200 at midnight on July 9-10 the Chinese launched a series of attacks on Pork Chop which continued until dawn. Simultaneously the enemy shelled the MLR and roads leading to the battalion CP. Commanders thus suspended relief of 17th Regiment units until the situation became clearer. The 17th continued to hold the outpost.

Many forward observers sent to Pork Chop during the preceding days' counterattacks were no longer paired with their company commanders, but with their high-powered radios, were able to pass information to senior commanders. The disconnected pockets of infantrymen were less fortunate. With their battered communications net, they were less able to transmit a clear understanding of the situation. "Piecing together a comprehensible picture received from so many, and of-

761

ten so conflicting reports, was almost impossible." Because there were men from so many different companies on the hill, Major Noble was commanding a large number of individuals, not cohesive units.

Company I, 32d Infantry, assembled just before dawn on July 10 and began moving via armored personnel carriers, with instructions to relieve defenders in the eastern sector of Pork Chop. When they arrived, the threat of a stiff fight didn't materialize, and the relief resumed, K Company of the 32d taking over responsibility for the western sector. Throughout the day, armored personnel carriers, under constant fire, moved men to Pork Chop and returned loaded with weary soldiers, plus the wounded and dead. At six in the evening, relief was complete, with men in I and K Companies of the 32d believed to be the only American units on the outpost.

When I Company of the 32d Infantry relieved the remaining 17th Infantry units on the eastern flank of Pork Chop, Corporal Dan Schoonover from the 13th Engineers voluntarily remained in the area, manning a machine gun for several hours. He subsequently joined another assault on enemy emplacements. When last seen he was operating a Browning automatic rifle with devastating effect until he was mortally wounded by artillery fire.

The night of July 10-11 Lieutenant Colonel Royal R. Taylor, commander of the 32d's 3d Battalion, took command of the outpost's defenders. The Chinese struck again. At 8:50 the night of 10 July an enemy company attacked the K Company sector, and after a forty-one minute fire fight, withdrew.

JULY 11

At 3:30 the next morning a heavy assault by an estimated battalion of Chinese infantry hit the left flank of I Company, and succeeded in overrunning the left and center platoons before a counterattack succeeded in restoring the I Company defenses. Some additional bunkers were lost, but at daylight the 3d Battalion retained control of a substantial part of the defenses.

By Saturday the battle had been raging for five days, virtually nonstop. To regain control of an outpost defended by one company, elements of five American infantry battalions, some with South Koreans in them, were committed to the fight. Intelligence gathered during the action now confirmed at least one entire Chinese division had been committed against the hill's defenders, sometimes two battalions in one assault wave. The enemy seemed willing to accept any cost to hold their gains.

When the daily command conference convened at the 7th Division CP on the morning of July 11, General Maxwell Taylor, the 8th Army commander, reluctantly announced his decision to withdraw from Pork Chop. Belief the armistice was near, and other factors unknown to the men who had put their hearts and souls into the outpost's defense, weighed heavily in the decision. To give up voluntarily, what so many had given their lives to defend, seemed incomprehensible, an almost impossible mission to perform. But planning went ahead for a daylight withdrawal.

Lieutenant Colonel Rocky Read, the battalion commander who provided covering rifle fire for Dick Shea late the morning of July 8, again went forward to Bunker 45 and coordinated the pullback of I and K Companies of the 32d Regiment. Carefully prepared supporting artillery and mortar fire ensured the enemy wouldn't successfully impede the withdrawal. The entire operation was structured to mask the withdrawal, with the relief force feigning another counterattack. By late afternoon most of the troops, including wounded and all the dead who could be found, were off outpost Pork Chop. Among the men in the last armored personnel carrier to leave the hill was Rocky Read.

Also among the last men to leave the hill that day was Able Company's Corporal Dale W. Cain, who had been beside Dick Shea in the three-day period just prior to the battle, had stayed at his post as a radio operator, and fought through five terrible days. When A Company was relieved on July 8, Dale had apparently been separated from the rest of A Company, with other men in the company unaware of his whereabouts and circumstance.

Combat engineers placed demolition charges as the pullout progressed. At sundown The Chop stood abandoned. Exploding demolition charges sounded the start of its systematic destruction.

Casualties in the last battle for Pork Chop Hill were heavy. In the 17th Infantry Regiment three officers were reported killed and eight missing in action, with an additional twenty-one officers evacuated as battle casualties – wounded. Among the officers reported missing were

Lieutenants Richard T. Shea, Jr. and Richard G. Inman. Thirty-four enlisted men were reported killed and 132 missing, with 590 evacuated, wounded in action. The 17th's losses in killed, wounded and missing were roughly equivalent to one entire, fully manned infantry battalion – four companies.

The losses in the 32d Infantry were less severe. A total of eighteen were killed in action, 304 wounded, with forty-one missing – half a battalion. Among those moved to a Norwegian mobile army surgical hospital to the rear of the MLR was a severely wounded Lieutenant MacPherson Conner. On the morning of July 11, Mac received a visitor. General Trudeau, the 7th Infantry Division Commander, told him the 7th had been hit by two Chinese divisions, and the decision had been made to pull off Pork Chop. The outpost would be obliterated by artillery and airpower, and made unusable by the Chinese.

Trudeau had been at the commanders' conference that morning and knew what was coming. The exploding demolition charges left on Pork Chop by the engineers was the beginning. A brief pause followed.

The American withdrawal was skillfully done, albeit painful and emotional for the men given the mission, as well as those who watched. The Chinese recognized the pullback was in progress, and moved to fill the void, intending to firmly occupy their hard won terrain. As they moved to secure the vacated positions, ten battalions of American artillery opened fire at a prearranged time, and the systematic ravaging of the outpost began. Among the hundreds of projectiles falling on the hill were VT fused rounds, which exploded a few feet above the ground and shredded everything beneath and around them with shrapnel. The rest dug more craters and smashed what remained of the defense network, collapsing trenches, splintering the supports for overhead cover, and caving in bunkers. The barrage continued relentlessly, lifting occasionally while American aircraft added more fury with tons of bombs.

On Pork Chop the night of July 11, and most of the next day, the Chinese encountered what David Hughes and the men of K Company, 7th Cavalry had met on Hill 339 nearly two years earlier, when 339 seemed so easy to take – an artillery barrage. But this was far worse, a holocaust. The carnage the remaining Chinese soldiers endured July 11 and 12 ended the last battle for Pork Chop Hill.

The morning of July 12, Dale Cain and six other A Company soldiers assembled in a formation on a hillside in the 1st Battalion area south of Pork Chop. The day before, when Pork Chop was evacuated, Dale had been taken off the hill on an armored personnel carrier and then transported further, via a two-and-a-half-ton truck to the rear area. Captain Roberts was at the morning formation and asked if the seven soldiers would like to talk with a chaplain. Dale Cain didn't learn until years later that A Company had been relieved from Pork Chop on July 8.

* * *

In Indiana, Iowa, New Jersey and Virginia, the wives and families of Dick Inman and Dick Shea anxiously watched news and read of the Chinese Army's powerful offensive in the west central sector of the Korean front. Before news stories appeared on television and in the newspapers confirming the Chinese attack, Barbara Inman had a deeply disturbing dream. She saw Dick in a train station buying a ticket home. When he got on the train, he discovered it wasn't going home. She woke up screaming and crying, which brought her mother rushing to console her.

The comforting flow of letters from both men ceased. Early in the third week of July the parents of both were informed by the Department of the Army that their sons were missing in action. Dick Shea's parents immediately phoned Joyce's home in New Milford, New Jersey. Her dad answered, and heard the news brought in the brief, sad telegram the Shea's received. He then turned to Joyce to deliver what none of them had ever wanted to hear.

Dick Inman's parents assumed Barbara had been notified and expected a phone call. Like Joyce Shea, she hadn't. On the second day, after the Inman's had heard nothing from Barbara, they phoned her. She was at home with her parents for the summer months in Fort Madison, Iowa. The haunting dream had become a real-life nightmare. Now began the heart rending ordeal of two families, and two soldiers' wives, waiting, wondering, hoping, not knowing, praying their sons and husbands would return from the chaos and confusion of a war the entire nation hoped was nearing an end. Less than two weeks later, on July 27, 1953, all along the front of the tortured, divided peninsula in northeast Asia, the guns of Korea fell silent.

Corporal Doug Halbert from the Clearing Company of the 7th Medical Battalion was well to the rear of the MLR. He had heard the distant, ceaseless pounding of artillery from both sides in the two-hour period before the cease-fire. Then it became quiet. He stood gazing north where the fighting had been. The quiet was surreal. Shortly after the cease-fire, he saw flares being dropped all along the MLR, by low flying aircraft, just south and parallel to the truce line.

Robert Hall, a combat Marine, remembers the ending of his ordeal. His recollections speak for all who returned.

The cease-fire was announced for ten o'clock that night. There had been artillery fire all evening from both sides – not a barrage, just a banging away. We were all up for it. The artillery gradually tapered off as the hour approached. Flares were shot up by both sides all along the MLR. Then it became very quiet as even the morons who wanted to be the one who fired the last shot of the war quit cranking off. A few showed flashlights. I turned in about midnight. It had grown very still.

At earliest light the troops came up out of the ground to look. At first we stood in the trenches. Then some climbed up to the forward edges, then to the tops of the bunkers, for a better look. It was unheard of – standing in the open in daylight. An incredible feeling. I think the infantrymen all across the peninsula, on both sides of the line, must have been awed by it. Just the simple, natural act of standing erect in the sunshine. Then to look, and eventually walk through the land ahead of the trenches, a thing that would have meant sure death twenty-four hours before. That's when we began to realize that it was really over.

The killing had mercifully ceased, blessed relief for the infantrymen who, in the war's last months, suffered the brunt of its frustrating, daily grind, and survived. The "half-war," the "yo-yo" war, the brutal trench war, was over. For the American citizenry's half who had not been personally touched by the Korean War, life went on as before.

But the war wasn't over for all Americans. Not for the Inmans and Sheas, not for Joyce and Barbara, and the wives and families of 5,178 missing in action. Not for the families, friends and loved ones of the 54,246 dead, or the 103,248 wounded. The war's torment had just be-

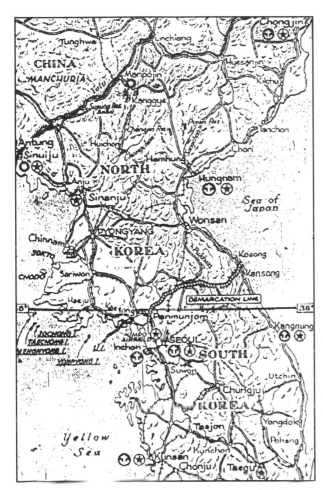

The Demilitarized Zone (DMZ) between the Republic of Korea (South Korea) and North Korea took effect at 10:00 p.m. on 27 July 1953, following prolonged and frequently interrupted armistice negotiations which began July 10, 1951.

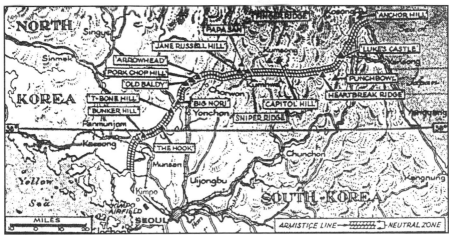

gun. Nor was it over for Lieutenant Pat Ryan and the B-29 crew he served with in the 93d Bomb Squadron, 19th Bomb Group, on Okinawa.

ONE MORE MISSION...

The Korean armistice and the collective sighs of relief following its signing did nothing to bridge the ever-widening divide separating the western democracies and the communist nations of the east, the new totalitarians of the twentieth century. The armistice wasn't a formal end of the war, not a peace agreement. Rather, the document's signers agreed to a set of conditions under which both sides would observe a cease-fire on the Korean peninsula. It was almost *status quo ante bellum*, except a state of war still existed between North and South Korea, and the Republic of South Korea now had the power of the western world, and in particular the United States, firmly rallied to its side. There was massive force in the ROK ready to protect its hard-won, fledgling democracy. Clearly, neither side trusted the other, and both determined they needed to be vigilant to avoid a surprise attack.

The two opposing armies remained "at the ready" as they carefully took the first, halting steps to adjust lines on opposite sides of the DMZ and open up a three-mile-wide "no man's land." The DMZ was a winding line that ran south southwest to north northeast across the 38th parallel – the original demarcation line between South and North Korea. On the ground, United Nations' forward observers kept binoculars and telescopes trained on the Chinese and North Korean armies digging in behind the DMZ. Intelligence listening posts maintained their watchfulness, monitoring electronic communications.

In the air and on the sea, where United Nations and American forces held a considerable advantage, similar measures ensured the agreement held. Navy intelligence gathering vessels plied the North Korean coastline, outside territorial waters, and carrier task forces laden with firepower kept a wary eye on the movements of Chinese and North Korean air, ground, and naval units. Airborne helicopters, Navy carrier- borne patrol and tactical aircraft, and Air Force tactical and strategic reconnaissance and intelligence gathering aircraft flew outside and parallel to demilitarized airspace separating north and south, as well as airspace outside the three miles marking the limit of international waters off either coast of North Korea. While eyes and electronic sensors on board aircraft ensured the truce held, another element of American airpower was at work.

The 19th Bomb Group was among several Air Force combat units flying missions to let would-be attackers know American and United Nations air firepower was still present, waiting and watching to ensure there wouldn't be another surge of communist forces toward the tip of the Korean peninsula. Though not labeled as such, and not generally known to the American

Pat Ryan in the copilot's seat of a B-29, Kadena Air Base, Okinawa, 1953. (Photo courtesy William J. Ryan.)

public, they were peace keeping missions. Most of the aircraft didn't carry bombs, but their would-be attackers could never be certain which ones did and what weapons they might be carrying.

Every twenty minutes, lone B-29s from Kadena and Yakota Air Base, Japan, or B-50s from the island of Guam, flew in a never-ending stream of aircraft up and down the east and west coasts of North Korea over international waters. The turnaround point at the extreme northern end of their eastern flight path was off the coast of the Soviet Union, just northeast of the imaginary extension of the border between North Korea and the USSR – abeam the Russian city of Vladivostok. The voices of Soviet air force "volunteers" and advisors had frequently been heard during the war, in the air and on the ground. Vladivostok was a major transshipment point for Soviet military equipment and people funneled to both the CCF and NKPA.

The B-29 missions were long and arduous, usually about ten and a half to eleven hours, but not nearly as tense as combat missions during the war. In the first days after the armistice, airmen experienced feelings not unlike infantrymen on the ground, mixed disbelief and uncertainty. "Would we be able to fly the length of North Korea's coast, approach Vladivostok, and return without hostile responses?"

* * *

On August 24, nearly a month after the armistice, as transition week was about to begin at West Point, Pat Ryan flew a peace keeping mission. It wasn't his first since the cease-fire.

Late that afternoon, at Kadena Air Base on the Island of Okinawa, Lieutenant Paul R. Trudeau and his crew went through their preflight checklist on B-29, tail number 44-61920. The airplane was a substitute on the mission schedule because "their" airplane was down for maintenance. Pat Ryan noted during examination of 920's maintenance records that number three engine, the right inboard, was a "high time engine." That is, few engine operating hours remained before it must be removed for major overhaul. While high time on an engine wasn't reason to call for another substitute aircraft, it was a factor to be respected, and might, under certain conditions, temper engine operation. The mission proceeded as planned, and Lieutenant Trudeau started the four big engines for the long night's work.

It was another mission off the east coast of North Korea. They carried no bombs, but were armed with the usual load of 1250 rounds of ammunition for the airplane's defense, six sets of twin fifty-caliber machine guns intended to ward off enemy fighters. Radio discipline and other procedures were to be the same as on a combat mission: no outside aircraft running lights turned on, and red night lights inside dimmed to a minimum without making instruments and switch settings unreadable.

The mission profile called for flight across the Republic of Korea, climbing to thirty-five thousand feet, then north along the track paralleling the east coast of North Korea, with turnaround abeam Vladivostok, and return. The airplane weighed more than seventy tons at takeoff and carried sixty-five hundred gallons of high octane aviation gas.

There were thirteen men on board, including the 19th Bomb Group commander, Colonel Lloyd H. Dalton. He was observing the crew's performance, and as group commander, was also familiarizing himself with the mission his command was directed to fly. In addition to the aircraft commander, Lieutenant Trudeau, and co-pilot Pat Ryan, the crew of twelve included: Captain Marvin F. Pelletier, navigator; First Lieutenant Nils P. F. Ahls, visual observer; First Lieutenant Albert W. Junes, bombardier; Staff Sergeant Raymond A. Hagen, flight engineer;

Airman Second Class John R. Thomson, radar operator; Airman Second Class Ronald A. Ersfeld, central fire controller; Airman Second Class Calvin A. Mittelsteadt, right gunner; Airman Second Class Edmond P. Schow, left gunner; Airman Second Class Gary E. McCluskey, tail gunner; and Airman Second Class Norman K. Kelley, electronic countermeasures operator.

The takeoff, climb and cruise northward proceeded without incident. The presence of the "old man," "the boss," the group commander, Colonel Dalton, was palpable, and the crew, aware their performance was being observed, paid careful attention to crew discipline and flight procedures, while remaining alert to the slightest hint of trouble. Trouble came in a most unexpected way.

There were a number of potential hazards associated with late summer flight operations far north, up the east coast of Korea. Abrupt changes in weather occur more frequently in late August, as Siberian cold fronts begin pushing southward with greater speed and clash with the fading, warm, humid summer air typical of the Korean peninsula. Air Force weather forecasters didn't have the means to accurately predict the speed, intensity, and effects of the cold fronts, which brought with them fast-moving, layered, sometimes thick overcasts, and rapid shifts in direction and speed of high velocity, high altitude winds moving further south with the fronts. In the absence of adequate ground or sea-based long-range navigation aids and clear skies for star shots (navigation measurements using a sextant), sudden sharp changes in high altitude winds can wreak havoc with an aircrew's known position during flight. Dead reckoning – when navigators and pilots use airspeed, time, distance, heading, and predicted winds for planning and flying the most basic form of air navigation – often results in rapid drifts off course if actual and predicted winds sharply diverge while en route.

The flight of 920 was routine until they turned and flew north, up the coastline. They flew in and out of clouds at first, then entered a solid overcast. Lieutenant Trudeau and his crew couldn't see the stars above or the lights on the North Korean coast to their left. Occasional turbulence encountered told them they were probably crossing a front, with changing high altitude winds. Unknown to the crew, winds rapidly shifting from forecasters' predicted direction and speed was drastically affecting the airplane's ground track. In thick clouds they didn't have a way to check or verify their position. Altitude and distance from the Korean coastline was excessive, negating the use of radar as a navigation aid. At thirty-five thousand feet the clouds seemed dense, reaching far above them. Climbing to higher altitude to get into the clear above the clouds, didn't appear possible. They dared not descend in hopes of breaking out of the clouds. If they had drifted over land, they could descend into range of anti-aircraft guns. The best course was to fly the planned airspeed, time, distance, altitude, and heading, using forecasted winds.

The clock indicated the big bomber was approaching the planned turnaround point. Pat Ryan had earlier moved from the copilot's seat on the right side of the cockpit, to allow Colonel Dalton to fly the aircraft while Pat grabbed a bite of supper. He was sitting on the door above the nosewheel well, behind the pilots' seats, eating. A few minutes earlier the aircraft had come out of the solid overcast, and was between layers of clouds. Suddenly the layer of clouds beneath them began thinning and within seconds a dense sprinkling of lights appeared directly below, on the earth's surface.

The first reaction was "Must be a fleet of ships." A brief silence. Then, "Damn! That's got to be Vladivostok. Let's get the hell out of here!" Almost simultaneously, without warning, on the left side of the airplane a thump was felt by the entire crew. Pat Ryan had finished eating and sat bolt upright, turning his head and eyes toward the flight engineer. He caught sight of the radar operator, Airman John Thomson. He had been on the radar scope when he, too, suddenly sat up and peered out a window on the left side of the aircraft. Pat Ryan could tell from Thomson's expression something was wrong.

On the intercom, Staff Sergeant Hagen, the engineer: "Manifold pressure on number two engine dropped three inches and decreasing." Almost immediately the left scanner, Airman Second Class Edmond Schow, reported sparks coming from underneath the cowl flaps of the left inboard engine. "Number two's on fire!" John Thomson called.

Someone else shouted "Fighters! I saw tracers!"

Colonel Dalton got out of the right seat, as Pat Ryan moved quickly back to his crew station. Lieutenant Trudeau rolled the big airplane into a hard right turn toward the Sea of Japan, instinctively pushing the throttles to full military power on engines one, three and four.

Airman Schow again called over the intercom, "Sparks coming over and underneath the leading edge of the wing on number two."

Lieutenant Trudeau commanded, "Fire feather number two!" and Sergeant Hagen, the flight engineer responded, repeating on the intercom, "Fire feather number two," as he activated the engine's feathering switches and fire suppression buttons.

The startling chain of events shattered the monotony of the long, droning mission. By the time Trudeau rolled the airplane into its turn, every man on the crew was alert, keyed up. The radar operator John Thomson resumed scanning his radar screen as did the central fire control operator, Airman Ronald Ersfeld, and navigator, Captain Marvin Pelletier. All three anxiously searched for blips on their screens that would warn of additional attackers. The three gunners, the observer, any man who had a crew position from which he could visually search the night sky, nervously watched for a follow-on attack from another Soviet fighter, while Lieutenant Trudeau occasionally glanced at the stricken engine.

While peering into the dark for additional Soviet fighters, Airman Schow also kept a close watch on number two, to see if the CO_2 bottles did their job, and the huge propellers wound down and streamlined into the relative wind. If either event didn't occur, the aircraft was in immediate, serious trouble.

They were back in a solid overcast. Not long afterward came sleet, snow, lightning, and more turbulence. Navigator Marvin Pelletier feverishly recalculated the direction and speed of the winds that had driven them so far off course. When he completed his computations he grumbled, "Jesus, these winds are unbelievable!"

For the entire crew it was May 25 again, their first mission over Wonsan, North Korea. Only worse. Pat Ryan wrote Joanne about that first mission. The letter frightened her, and he didn't write her about any others.

If the fire on number two didn't go out, there would be precious little time. American air and sea search and rescue capability wouldn't be nearby. Unquenched fire in an engine mounted on the leading edge of a wing with electrical wire bundles coursing through it, mechanical control links to the left aileron – crucial to aircraft control, and crammed with fuel tanks even partially full, meant only one thing. Bailout! And soon. The danger of a catastrophic fuel tank explosion or wing structural failure increases with each passing second an engine fire goes unchecked. Once explosion or wing structural failure occurs, swift, usually violent loss of aircraft control follows, and the likelihood of successful crew bailout rapidly diminishes.

There was another problem to consider, though the fire suppressant might do its work. Had the Soviet fighter's attack or ensuing fire damaged the feathering motor on number two? If the huge propeller didn't streamline into the relative wind, and continued rotating, 920 couldn't possibly make it to Japan, and probably not to an emergency divert base in South Korea. One B-29 engine failed but not feathered was akin to bolting a sixteen foot barn door on the propeller spinner.

Because of enormous drag created by four, windmilling, flat-faced prop blades, even with full power on the remaining three engines, a steady descent was necessary to maintain flying speed. Diminished aircraft range, speed, accelerated fuel consumption, and continued high power operation on the remaining engines, made successful emergency recovery a cliff-hanging gamble. It was a long haul from Vladivostok to the nearest South Korean air base. The Sea of Japan was cold any time of year, and every member of the crew wore summer flying suits.

There was relief when scanner Edmond Schow reported the fire out, and the feathered propeller windmilling to a stop. There had been tense moments, however. The initial shot of CO_2 didn't kill the fire. The engineer drew from a central fire suppressant tank for a second shot to snuff it out.

Seconds, then minutes passed, with the crew straining through clouds with radar and eyes, searching for another attacker. None came. Everyone on board 920 could relax a bit more. There were other problems, however.

They knew they had violated Soviet airspace, and had been hit by a Soviet fighter. Their role in what could become a major international incident less than a month following the Korean armistice wasn't a comforting thought. A widely publicized violation of Soviet airspace by an American B-29 bomber was the stuff of diplomatic nightmares and increased international tension in a world filled with all the tension it could stand. What to do?

First, no radio calls telling of the attack. Second, 920 still had three good engines. Continue the mission. Third, when the aircraft recovered, damage would be evident. Colonel Dalton and the crew must report the facts of the incident through the Air Force chain of command as soon as they landed. But events didn't turn out as they planned.

When Lieutenant Trudeau shoved throttles forward on one, three, and four, and rolled into the turn for safe haven, he decided to remain at thirty-five thousand feet for the return flight, to hold as much altitude as possible. The decision required higher than normal power settings on the three good engines, to hold flying speed and altitude. It was impossible to land at Kadena. Higher fuel consumption on the three remaining engines meant 920 would have to recover in South Korea or Japan. The decision was Japan, where longer, safer runways and bomber maintenance facilities were readily available. It would be a two-hour flight to Japan. But there was more trouble brewing.

Pat Ryan thought of the right inboard engine, number three, and its high time, as did Staff Sergeant Hagen, the flight engineer. Continued operation at high-power settings could cause engine failure, and Hagen kept cross-checking engine and accessory instruments to ensure there were no warning signs of failure – a rough or back firing engine, decreasing oil pressure, oil temperature increase, loss of manifold pressure, fluctuating engine rpm or generator voltage. Loss of engines two and three would cause loss of key flight instrument readings, such as airspeed.

The temperature on number three increased. The engine was running hotter than it should, and hotter than one and four. Finally, Sergeant Hagen could no longer ignore the obvious and told Lieutenant Trudeau, "Boss, if we keep this up, with number three overheating, we'll lose it. We have to reduce speed and descend to a lower altitude." The aircraft commander reduced power and started a slow three-hundred-feet-per-minute descent toward twenty-five thousand feet. The temperature on number three retreated along with temperatures on one and four, but remained higher than normal. Finally, 920 leveled at twenty-five thousand feet and flight toward Japan continued. Except for the nagging high temperature on number three, the crew could relax again – somewhat.

Finally, they were close enough to Japan to contact Misawa and check weather at suitable alternate recovery bases. Then came the thunderbolt. Every single recovery base in Japan was below minimum weather conditions for safe recovery. They couldn't return to Korea. Bailout, if not certain, was at least probable. Planning began. The best alternative was to set up for a simulated radar bomb scoring run on Johnson Air Base near Yakota, Japan.

There was a bailout procedure for Johnson, as there was for all bases. It called for navigating across the initial point on the simulated run toward Johnson, then turning to a prescribed bailout heading. Bailout was minutes away. "Chicago" was the ground control intercept radar tracking their progress, and had been notified of their plight.

"Air Force 920, Chicago. You're cleared to Yakota Control."

"Roger, Chicago, Yakota frequency. Out."

They were approaching the initial point for the run toward bailout. The temperature on number three started to rise again, and kept going up. The engine was running hot. Voltage readings on number three generator began fluctuating, a precursor of more serious problems. Failure of a generator, with its rotor turning at an extremely high number of revolutions per minute, could cause engine failure or fire. If the generator "ran away," the rotor's increasing speed would overheat the generator shaft and bearings, causing expansion of the generator's moving parts – until it started to tear itself apart and abruptly cease rotation due to high temperatures and friction. The result would be an explosion of the generator, an accessory mounted on the engine's circumference.

The generator on number three, overheated by the aircraft engine, ran away and disintegrated on 920 that night, and was the last and most catastrophic in a sequence of failures.

Sergeant Hagen gave Lieutenant Trudeau a running account of the deteriorating engine instrument readings on number three on the interphone – not on the intercom – which broadcasted to all crew stations simultaneously. Then he reported, "Number three running rough." Suddenly, Airman Second Class Calvin Mittelsteadt reported orange sparks coming from number three. Another fire. Then the generator let loose. The explosion fired chunks of hot metal deep into the heart of engine number three.

Once again, the aircraft commander ordered, "Fire feather three!" as 920 continued inbound from the IP toward bailout. Procedures called for a turn from their heading onto a prescribed bailout heading, and fly a four minute leg prior to abandoning the aircraft. Hope was the feathering motor and fire suppressant would once more do their jobs, allowing Lieutenant Trudeau to maintain aircraft control. The engine feathered, and for a few seconds, sparks continued to diminish. It appeared the CO2 extinguished the fire. It hadn't.

Bombardier, Lieutenant Albert Jones, was the first to call again to Trudeau, "Number three's on fire!" The engine quickly became a glowing furnace, with a stream of white smoke visible behind the wing's trailing edge. An additional shot of CO2 from the central fire suppressant tank didn't choke the furnace raging in number three. The situation was deteriorating rapidly. Middelsteadt called on the interphone again, "Flames licking two to three feet long, over the top of the wing – orange and blue!"

With two engines feathered, one of them on fire, and the airplane turned onto the bailout heading, Lieutenant Trudeau lowered the nose to hold flying speed and control, yet keep airspeed slow. With the failure of number three there was no airspeed reading. "Gear down!" Trudeau commanded. "Gear down," Pat Ryan responded as he lowered the gear, a necessity for safe escape of crew members from the forward section of the bomber. Bailout was imminent.

Trudeau, fighting to control the airplane and its rate of descent, didn't take his left hand off the controls to ring the bail out warning bell – three rings signaling "Get ready!" Instead, with flames rapidly spreading over and underneath number three engine and the right wing, he pushed the intercom button on the control yoke and warned the crew, "Consider you've received the bailout bell and get ready to bail out!" Molten parts from the disintegrating engine struck the plane's right, horizontal stabilizer – in the tail section – embedding themselves and setting it afire.

Pat Ryan didn't hear the command to get ready. He was in the right seat with his head down, talking to the bombardier, Lieutenant Jones. He hadn't heard Jones' call, telling the aircraft commander number three was on fire. Pat noticed Captain Marvin Pelletier, the navigator, moving much further than normal from his crew station and asked Pelletier, "Where are you going?"

Pelletier replied, "I'm going out!" Ryan turned to Trudeau. "What's going on?" Trudeau said, "They're getting ready to go out. Get ready!"

By this time men in the plane's aft section were abandoning the aircraft through a door on the left side, the side opposite the fire. They didn't wait for the final command to go. Uncontrolled fire on a combat aircraft strikes fear in every crew member because he knows what follows. Men in the aft section of 920 could see flames spreading aft over and under the right wing, near the fuselage, toward fuel tanks – and them.

Pat had no time to transmit a "Mayday" on the radio. It was get out now, before a fuel tank exploded. He moved to the boarding ladder at the aft end of the nosewheel well, ensuring his parachute straps were hooked up and cinched tight. The emergency door above the nosewheel well was already gone, the hatch wide open. Others had dropped through it. Reacting to bailout and survival training, and the certainty of the aircraft's impending destruction, he climbed down without hesitation, his legs extending further into the cold wind of the slip stream. The noise was deafening from two huge engines and the aircraft slipstream, roaring past the open hatch. He pushed away hard, thrusting himself down, free of the ladder. The long, cold, free-fall began. No emergency oxygen bottles for high altitude bailout. No gloves. The temperature was minus sixty degrees centigrade at bailout altitude.

He knew he'd left the airplane below twenty-five thousand feet. Like everyone in the crew, he'd been trained to know pulling his ripcord at too high altitude posed hazards, such as a possible chute failure because of excessive free-fall velocity in the thinner air. If the chute didn't fail, opening shock could be severe, and parachute descent in the high altitude cold and lack of oxygen could render him unconscious. A night parachute landing isn't fun, and to land unconscious is worse.

Like the other men, Pat estimated time to free-fall before pulling the ripcord. Don't panic or react too soon – not easy to do in the dead of night, when fear is already the most powerful emotion at work. Then come the fleeting seconds after pulling the ripcord. It's pitch black and dead silent except for the distant, fading sound of the B-29, about to enter its final plunge.

Will the chute deploy?

Pat Ryan's did, and so did every other parachute that night, except for one which partially opened. Miraculously, every man on board 920 survived his bailout, parachute descent, and night parachute landing. Only one had injuries of any consequence, and by five minutes after ten next morning, the last of the thirteen men was picked up. The crew had been scattered a distance over sixty miles because of the high altitude bailout and parachute drift. For Colonel Dalton, the 19th Bomb Group Commander, it was his third bailout.

Yakota rescue had been alerted at ten fifteen that night, when Pat Ryan initially radioed one engine feathered and bailout pending. At ten forty Yakota rescue was told an airplane had crashed, to begin search, rescue, and recovery operations.

As for Pat Ryan, he landed on a mountain side without serious injury. He saw lights below him. At first he thought they were lights of a truck convoy on a road. He found a steep trail and walked down toward the stream of lights. They were lights along the main street of a small Japanese town. He entered the town and walked up to the window of a school, through which he saw two Japanese men standing at a blackboard, working math problems. Pat knocked on the window, startling the room's occupants. They ran. When they recovered their composure, they came and helped him. One spoke perfect English. He had been educated at the University of California, Los Angeles UCLA.

The two men took him to another village where Pat was picked up an hour and a half later and taken to Yakota Air Base Hospital.

There were some humorous moments after the harrowing odyssey of night bailouts, at least for some of 920's crew members. Training for night parachute landings included emergency procedures in case a crew member landed in a tree at night, hanging the hapless man in his parachute straps an unknown distance above the ground. Procedures required crew members to simply stay put, and remain still, suspended in the chute until daylight came, a precaution to determine how far it is to ground and whether the tree could support a climb down.

Two of the crew crashed through limbs of trees until their parachutes and risers snagged and jerked each to a stop. As they had been trained, each stayed in his chute harness the remainder of the night, not knowing the height above ground and fearful of crippling falls. At morning's light, one found he was suspended barely two feet above the ground, the other six feet.

Another decided he would take a chance and climb down from his night-time perch. Dangling in a tree, he groped for solid branches and cut his parachute harness. He was higher in the tree than he hoped. He tumbled through thirty feet of limbs and branches before thumping to earth, fortunately not severely injured by the fall.

Two members of the crew nearly didn't make it out of the airplane. One of the waist gunners froze at the bailout door aft of the airplane's midsection and refused to go. He held onto the doorway with a viselike grip. When pushed from the airplane he dangled in the cold, violent slip stream, still holding to the door. Speed during the still-controlled descent was 300 knots. John Thomson, the radar operator, kicked at his fingers trying to force him to let go. Consumed with panic-stricken fear and the brute strength which accompanies it, the young airman kipped upward and swung back through the open door into the airplane.

John Thomson, furious and running out of time, shouted, "I'll give you one more chance...!" and kicked him so hard he knocked him through the open door. Thomson's act saved the man's life.

The last man out was the aircraft commander, Paul Trudeau. He went out through the nosewheel well and received a deep gash when buffeted about in the wheel well by turbulence. He was found next morning, face down in the mud, unconscious. A doctor clamped the wound to save him. His hospital stay lasted twenty-four hours.

As soon as possible after he was rescued, Colonel Dalton briefed the 20th Air Force commander, laying out the chain of events leading to the crash of 920. He could speak from firsthand experience. He was on the airplane. Dalton's report, classified, went quickly via encrypted phone and message to the national command center in the Pentagon.

The crash was reported the next day in Associated Press and International News Service dispatches, and hit the *New York Times* as well as other national and international newspapers. The story also appeared in the *Pacific Stars and Stripes*. There was no hint of 920's straying over Vladivostok or being attacked by Russian fighters. The Soviets made no mention of a penetration of their airspace or eastern air defense system.

Early the next morning, Jimmy McCrea, a friend of Pat's, heard a broadcast news story telling of the B-29 accident twenty-five miles southwest of Tokyo, and that the airplane was from the 19th Bomb Group in Kadena. He phoned Air Force headquarters in the Pentagon. They confirmed the story and told him Pat was on board, had been picked up, and was safe at Yakota. McCrea called Joanne's parents to tell her. He learned Joanne was in Bluefield, West Virginia, and after explaining to her parents, called the good news direct to her.

An aircraft accident investigation board convened one week later. Though not classified, accident board reports aren't releasable to the public. Nevertheless, to avoid a possible international incident, extra measures were taken to ensure men and women in the Air Force who had access to the report didn't learn what triggered the crash of 920. The board interviewed each crew member during the closed proceeding, which, under procedures governing accident boards, didn't require sworn testimony. The accident report wended its way up the Air Force chain of command a month later. It contained not the slightest hint of the aircraft's penetration of Soviet airspace, nor did the report tell of the fighter attack which triggered the chain of events leading to the destruction of an American B-29 bomber near Yakota, Japan.

Pat didn't write Joanne about this mission either, or any other while in the Far East this time. She learned about the crash and Pat's safe bailout by nine o'clock the next morning. The Korean War may have been over for Pat Ryan and Lieutenant Trudeau's B-29 crew, but this flight, a peace keeping mission, nearly brought disaster. Peace keeping in areas where powerful forces have once collided in bloody, unresolved conflict poses its own set of dangers.

As the armistice lengthened and held, the Air Force gradually decreased the frequency of such missions near the Korean peninsula. This mission ended safely for everyone on board 920. But it wouldn't be the last mission in Pat Ryan's life of service. There would be more peace keeping missions before Pat went home in November of 1953. And there would be a third war for him – and his family.

MISSING IN ACTION...

As soon as Joyce Shea collected herself following the phone call from Dick's father early the third week in July, she called the office of the Army's Adjutant General, from which the telegram had come to

Dick's home in Portsmouth. She hoped to learn more. Perhaps additional information had come in saying Dick had been found alive. There was none. There were similar reactions by Dick Inman's mother and father in Vincennes, Indiana, and Barbara in Fort Madison, Iowa.

Then, Joyce Shea and Barbara Inman received virtually identical letters within two to three days after the telephone calls. Only the names and dates were different.

17 July 1953

Dear Mrs. Shea:

I regret that I must inform you that your husband, Second Lieutenant Richard T. Shea, Jr., has been reported missing in action in Korea since 8 July 1953. The report states his position was attacked by opposing forces at the time he became missing in action. Telegraphic notification was made to Mr. and Mrs. Richard T. Shea, Sr., parents, designated by your husband as the persons to be notified in case of emergency.

I know that added distress is caused by failure to receive more information or details. Therefore, I wish to assure you that at any time additional information is received it will be transmitted to you without delay.

The term 'missing in action' is used only to indicate that the whereabouts or status of an individual is not immediately known. It is not intended to convey the impression that the case is closed. I wish to emphasize that every effort is exerted continuously to clear up status of our personnel, although under battle conditions this is a difficult task as you must readily realize. Experience has shown that a number of persons reported missing in action are subsequently reported as returned to duty or being hospitalized for injuries...

Permit me to extend to you my heartfelt sympathy during this period of uncertainty.

Sincerely yours,
WM. E. BERGIN
Major General, USA
The Adjutant General of the Army

However Joyce and Barbara felt about the telegrams and letters, both remained hopeful, even optimistic their husbands would be found alive and return home safely. But not knowing was excruciating. Each in their own way began determined efforts to learn more. Joyce wrote the 17th Regiment chaplain Dick had mentioned in his letters home. She wrote Lou Davis, Dick's close friend and track teammate, other friends and classmates, anyone who might remotely have knowledge of his whereabouts and the circumstances surrounding his status as missing in action.

She daily listened to the news and read the names o American prisoners being released by the North Koreans and Chinese. The prospect of learning the worst didn't restrain her. She, like Barbara, was certain her Richard would be back. He, too, had so much promise, so many reasons to live. Their lives together couldn't possibly be so short, and end this way. And there was the child she carried. Dick Shea must see their first born, expected in late August.

* * *

In the absence of any additional information about Dick Inman, George and Becky Inman, his parents, and Barbara pursued answers through other avenues. Telephone calls and letters from friends of his parents went to Congressman William "Bill" G. Bray of Indiana's 7th District, which included Vincennes. On August 15, Mr. A. R. Weathers wrote Congressman Bray asking for help. Weathers had learned the Congressman was leaving on a trip to Korea in September.

> The Inmans are close friends of ours, and their son has been a very close friend of our son for years... It would certainly be appreciated if you could learn any information regarding the circumstances, and if there is a possibility he was captured. I thought perhaps you could go through channels before your departure for Korea next month, and if nothing comes through, to follow up when you are over there....

Barbara formulated her own way of finding out what might have happened to Dick. She would join the Red Cross, complete training, and ask to be sent to Korea. There she would conduct her own search for answers and the whereabouts of the young husband she loved so deeply, and who now seemed to have vanished from the face of the earth. She clung desperately to hope, and rather than return to teaching late that summer, followed her heart into the Red Cross.

* * *

On August 26, 1953, two days after Pat Ryan's nearly disastrous mission over Korea, and while reorganization week was in progress at West Point, Joyce gave birth to Richard Thomas Shea III. There was joy tinged with sadness surrounding the blessed event. A week after she returned home from the hospital, a letter arrived from Korea. It was from Dick's battalion commander, Lieutenant Colonel Read. She opened it, anxious for good news she was certain must come. The hand writing was broad-stroked, heavy, flowing, and filled with emotion.

Korea
August 31, 1953

Dear Mrs. Shea,

I hope that I am not reopening old wounds but I would like to write to you about Dick. I was his Battalion Commander on Pork Chop and can give you some factual information which you may not have. In the early attacks by the Chinese he led to my personal knowledge five counterattacks. In one of these he received a slight wound on the cheek. He gave me much valuable information all during the action. On the 8th of July I personally stopped him from exposing himself in a trench near my CP. At this time I received orders that the remainder of my battalion would be relieved and replaced by a new unit. There were about 25 A Company men in the right sector. Dick's job was to keep these men together and to leave the hill with them. I personally covered his run to this action with rifle fire – he made it safely. Shortly thereafter the new unit attacked – my people were being withdrawn. One of the platoons of the new unit suffered casualties among its officers and non commissioned officers. Dick saw their confusion and carefully turned the remaining men of "A" Company over to another man, reorganized the troops from the other unit and personally led them in an attack in which he was killed.

The battle continued and the ground was taken and retaken many times.

Dick's action was beyond the call of duty in many respects. First, he took command of other troops than his own, he could have evacuated himself, or he could have simply held the disorganized men under cover. As you know any action other than the action he took would not have been characteristic of him. His action saved lives because the people he took were lost without his leadership.

He was clearly the most outstanding officer of his grade I have served with. He was recommended for the Medal of Honor because we his comrades know he earned it and we loved him.

He used to kid me about being too old to keep up with him – my father was track coach at VMI for 30 years and I ran (poorly). I am from Lexington, Virginia and he did say that Virginians should stick together and not let the *wall* Korea builds separate them. At 33, not 23, it was a job keeping up with him.

I have two sons and I hope that someday they will show the same sense of honor and duty that your husband displayed.

He was much more to me than a young rifle platoon leader. His life will be a constant inspiration to me and I shall be a better man for having known him.

My heart is too full to say more and I'm sorry I can't express what I feel.

May God bless you.

BM Read
Lt Colonel
Infantry
17th Inf Regiment

Lieutenant Colonel Beverly M. "Rocky" Read's letter was shattering, devastating. Joyce wept, heartbroken and sick with grief. After she regained her composure, she again called the Adjutant General's office in Washington and told them she had received a letter saying Dick had been killed. They told her he was still missing in action. There had been no confirmation he'd been killed. After her call there was no follow-up letter from the Adjutant General's office.

She wrote Lieutenant Colonel Read, while Barbara and the Inman family continued their search for answers. The days and months would slip by without a letter from Dick Shea's battalion commander. Two men were still missing in action, among many new heroes in the summer of '53. Sadness, grief, hope, questions with no answers, and interminable waiting settled in the homes of two families and two wives, as in thousands more.

Far away, in Korea, in the DMZ, the no man's land between South and North, a deeply scarred place called Pork Chop stood mournful, silent, an outpost without soldiers at the ready in the little that remained of its trenches and bunkers.

The barren shell-shattered crest of an abandoned Pork Chop Hill. (Photo courtesy U. S. Military History Institute.)

At West Point the new academic year was beginning. For all but a few cadets, the war was already a memory. The guns of Korea remained silent. A shaky, but glorious peace had returned in the new era of Cold War. The future was the thing, and it captured young eyes, minds, and hearts. Late summer and the anticipation of an exciting, colorful fall – and football – were in the air.

CHAPTER 17

ACADEMIC YEAR BEGINS

This was no ordinary academic year. Little more than a month earlier the guns fell silent in Korea. The "Yo-Yo War," the American GIs sardonic label given the first ten months of the conflict, was in the distant past. In the spring of 1951 the "Yo-Yo War" passed into history and fighting entered a new phase, the bloody battles of position and attrition which finally became stalemate. Now, the terrible stalemate was ended.

America's first "half-war" was over. "The forgotten war" Dave Hughes described in his August 1951 letter would be forgotten more readily than anyone thought – except for men like Pat Ryan and his B-29 crew, and the hundreds of thousands of soldiers from two recently warring armies which had slowly pulled back from the Demilitarized Zone after the armistice went into effect July 27.

They were still enemies, neither trusting the other, and on both sides elaborate defensive positions were being prepared several miles back from the DMZ, as the two armies carefully repositioned from the no man's land dividing them. For the Americans and South Koreans the new line of defense was Line Kansas, the line they had departed from in August of 1951 in the push to Line Jamestown. The new defensive positions were necessary in the event fighting resumed.

In fact, there was no peace. Only an armistice, a cessation of hostilities, an indefinite, "permanent" cease-fire that had finally emerged out of more than two years of on-again, off-again negotiations. The NKPA, the *In Min Gun,* was returning to *status quo ante bellum*, as was the ROK Army – with some major exceptions. Both foes remained battle ready, armed to the teeth. In the North, the Soviet advisors to the NKPA and their Chinese allies stood behind and alongside the North Korean Army. South of the DMZ, the Americans stood shoulder to shoulder with the ROK Army, while United Nations forces contributed to the war by other allies prepared to withdraw from the peninsula.

In the air and waters on either side of the northern half of the peninsula, United Nations airplanes, ships, and submarines – mostly American – warily patrolled, letting the North Koreans, Chinese, and Soviets know they were being observed, and the United Nations were prepared to reengage, if need be.

The missing status of Dick Shea, Dick Inman, and other Academy graduates in the final weeks of the war wasn't common knowledge in the Corps of Cadets. They had been reported missing in mid-July while most of the Corps was away from West Point, and the class of 1957 was in its second week of Beast Barracks. None of their families or any of us knew of the courage and heroism of either man, and we in the Corps of Cadets wouldn't know until years later. Nor did we know of Pat Ryan's night mission high above the waters East of the Korean peninsula and his harrowing escape from the burning B-29. The experiences of Pat and the other crew members had been sharp reminders to them, that keeping the peace can be as deadly as fighting a war. The reminder had come as everyone's thoughts were turning to peace, and the Academy's four undergraduate classes were ending their summer military training cycles and four weeks of leave.

A tenuous peace had returned, and though some cadets were frustrated, as Earl Blaik and his class had been on November 11, 1918, when they were "cheated of the 'promised' big adventure in Europe," there were few emotional outpourings of celebration or victory, as those which erupted on a massive scale following the two, huge, worldwide conflagrations of the twentieth century.

Korea had been a very different war than World War II, "the good war," when victorious armed forces, millions of men and women, had come home to national acclaim and adulation. The entire nation had been mobilized for World War II. When it ended, "unconditional surrender" meant virtually the complete destruction of the armed forces, political systems, and institutions which had spawned war by the Axis powers, and America's citizens had celebrated like they had been unable to do for years.

In Korea, America had fought a war that for many of its citizens was no war at all. Their lives had been little affected – at least it appeared that way. When the guns finally fell silent in Korea, there were no ticker tape parades, no teeming, excited throngs in Times Square, no nationwide expression of jubilation. Instead there was a collective sigh of relief, intense interest in prisoner repatriation, "brain washing," and "turncoats," and uncertainty about whether the armistice would hold. Attention quickly turned to life as usual, and the Cold War.

PREPARING FOR VICTORY ON ALL FIELDS

The class of '57 began their "plebe hike" early the morning of August 24, the same night Pat Ryan bailed out of the burning B-29 over Japan. A few days later the new class would be "officially" accepted into the Corps of Cadets. They had received six hours of honor lectures, beginning in mid-July, as had the men in '56, two more hours than our class had received in the summer of '51, when the scandal was still behind closed doors. During those two extra hours of lecture the men on the class of '54 Beast Detail expanded on what had been given our class when we were new cadets.

In addition to all we received our first summer, the men in '56 and '57 learned considerably more about the inner workings of the Honor Committee, how the Committee's officers were elected, their responsibilities, and how honor investigations and hearings were conducted. There was a marked change in the way honor investigations were to be conducted, although the change wasn't described as a change, or an outgrowth of the '51 incident.

Prior to the incident, when a company honor representative received information regarding a possible violation in his cadet company, he alone conducted a preliminary investigation to gather facts and determine if evidence warranted further investigation, which might lead to an honor hearing. This had been the procedure used by Dan Myers, Company K-1's honor representative in the spring of 1951, when revelations of organized cheating were first given him by the Commandant, Colonel Paul Harkins, and yearling *Brian C. Nolan*. Dan struggled for two to three weeks in April of that year, trying to find hard evidence to corroborate *Nolan's* stunning disclosures, before the Commandant decided the only recourse in gaining evidence was to ask *Nolan* to join the ring.

Undoubtedly due to information uncovered by the Collins Board, the Cadet Honor Committee established the new investigative procedure. The Collins Board learned that in the spring of 1951 there were successful elections of cheating ring members as yearling class honor representatives in two companies. The new investigative procedure required a subcommittee of three honor representatives to conduct a preliminary investigation, including the representative from the accused cadet's company. Although this wouldn't guarantee a thorough investigation, properly and objectively conducted, it would tend to reduce the likelihood of compromise in an investigation and follow-on decisions.

The new cadets of '57 also were taught more about the relationship between duty and honor, between the Tactical Department and the honor system, and the Academic Department and the honor system. The lecturers additionally told the men of '57 "honor was not used by the Tactical Department to enforce regulations."

During the honor presentations, men on the Beast Detail performed a skit briefly portraying a new cadet asking questions about the cheating incident of '51. The point of the skit was the cadets who resigned had "not lived up to their responsibilities and obligations under the honor code," and that each first class needed to do a better job of imparting honor to succeeding classes. However, none of the lecturers amplified or explained in greater detail the "fifth principle" underlying the honor code: "Cadets are honor bound to report any violation of honor they observe."

As '57 was officially accepted into the Corps during reorganization week, final preparation and editing was underway to publish the September 11 issue of *The Pointer,* the cadet magazine. It was "the ring issue," with the first few pages describing the ideals and craftsmanship associated with Academy class rings. The class of '54 was looking forward to ring weekend and the excitement engendered in every graduating class when their rings, ordered the previous spring, would be delivered.

This year, the new Commandant, Brigadier General John H. Michaelis, just returned from Korea where he had commanded the 27th Infantry Regiment "Wolfhounds," of the 25th Division, presented each first classman's ring to him in a ceremony at Cullum Hall. The receipt of class rings is an annual rite with its origins in the year 1852, when, according to legend, the Academy was the first institution in the country to use a ring as a class symbol. By 1953, the tradition had evolved that rings were presented to first classmen in a Saturday afternoon cer-

John R. Galvin, Class of 1954. Among numerous other talents, John was a cartoonist for The Pointer *magazine and the master designer of '54's Crest, depicted on their class rings. (Photo and cartoon courtesy John Galvin.)*

cmony at Cullum Hall, at the beginning of the new academic year. That same evening a formal ring hop was held — a memorable affair filled with symbolism and romance.

The '54 ring, as in years past, had a distinctive class crest on the side opposite the Academy crest. John Galvin, one of *The Pointer*'s cartoonists who kept the Corps laughing throughout his four years at West Point, was the master designer of the '54 crest. "Strength, character, durability, purity – these are all embodied in the ring which will serve as a constant reminder of the responsibilities which befall the cadet or officer who wears it...," so read the words penned by an unnamed writer and printed in *The Pointer* that September.

With a series of articles, *The Pointer* of September 11 also summarized for its readers the previous summer's military training received by each of the four classes. For '54 it was the Combined Arms Trip which began June 8 with flights via huge, propeller-driven C-124 "Globemaster" aircraft to Wright-Patterson Air Force Base, Ohio, to the research and development center. There the class observed Air Force striking power and received an overview of the extensive research and development capabilities at the center.

From Wright-Patterson, the class went to the Army's Armored Center at Fort Knox, Kentucky, to learn more about the triple principles of armored warfare: mass, shock action, and armor-protected firepower.

Next it was Fort Sill, Oklahoma, with the field artillery where they witnessed firing demonstrations featuring every piece from a seventy-five millimeter pack howitzer to the impressive 280 millimeter atomic cannon – firing conventional rounds of artillery. From there it was off to El Paso, Texas, and the anti-aircraft artillery center, where the men of '54 saw demonstrations of conventional anti-aircraft guns, plus the deadly capabilities of the recently unveiled "Nike" ground to air missile.

The grand finale of the Combined Arms Trip was Fort Benning, Georgia, home of the Infantry. Here was where Captain John Flynn had taught infantry doctrine and tactics after he came home from World War II – his first war, and before he became David Hughes' K Company commander in the 7th Cavalry, in Korea. Though Korea was fast becoming an unpleasant memory, at Benning the class of '54 met the men who did "the meanest and dirtiest job in the Army – seizing and holding a piece of ground." The impression was made lasting when they witnessed a night firing demonstration of an infantry company in the defense, though it was a demonstration absent the return fire by an attacking enemy force. The night firing exercise couldn't possibly replicate the chaos, emotion, and horrors that Dave Hughes' K Company had encountered on Korean hills nearly two years earlier or the bitter fighting on hills such as Pork Chop.

When the Combined Arms Trip was over, some of the class went on furlough, while some prepared to be instructors and trainers for the class of 1956 at Camp Buckner, and others readied themselves for the Beast Detail and the class of 1957. The first of two additional groups from '54 went to assignments on the Recruit Training Detail, at Fort Dix, New Jersey; Fort Jackson, Mississippi; Fort Knox, Kentucky; and Sampson Air Force Base, New York. The men assigned to recruit training served for approximately one month in duties normally given newly commissioned junior officers, duties that energized participants with thoughts of their graduation and commissioning a year hence.

For '55, my class, the summer included CAMID, joint cadet-midshipman summer training, in which the cow class – the juniors – boarded a Navy ship at the Hudson River's South Dock, at West Point, for a cruise ending with amphibious training exercises.

It was June 4, and the USS *Rockwall* lay at anchor, ready for boarding. The Academy Band serenaded the class with *The Tiger Rag*, a tune we were to hear on many joyful occasions. The *Rockwall* was an APA, in Navy parlance – Assault Personnel Auxiliary. Translation: troop transport, small. It certainly wasn't *"The Whiskey"* – the battleship USS *Wisconsin*, on which several members of the yearling class of '54 had taken their brief trip from Camp Buckner, and "mutinied" in the Caribbean. Nor was it the *Queen Mary*, the huge British luxury liner which, during World War II, was outfitted to carry thousands of troops from the United States to the British Isles as the Allies marshaled their forces for the invasion of the European continent.

Stan Harvill, a *Pointer* features writer in our class, noted with good humor, "living on a ship isn't bad – if you don't mind sleeping six deep. A US Navy APA is probably the only ship in the world on which you have to lie down before you can crawl into the sack." He went on to say there were about eight hundred men on the ship, "and the mess hall accommodated them easily – ten at a time."

> Life on the *Rockwall* was routine – we played bridge from reveille until taps with a few minutes out for meals. We missed the usual cartoon with our evening movie, but by that time we could sympathize heartily with Humphrey Bogart as he took the *African Queen* through the rapids.

The first port call was near Fort Lee, Virginia, where we visited the home of the Army Quartermaster Corps, one of the Army's suppliers of combat services and necessities of life, such as uniforms, tents, bedding, and food – including combat rations.

Then it was off to Little Creek, Virginia, where we made amphibious landings. During the amphibious landings we were temporary members of Marine Corps units, some of us assigned as acting platoon leaders.

It was over the side of the *Rockwall,* down the heavy rope nets, into the rising and falling landing craft that were riding the ocean's swells. Then came what seemed like hours of circling in relatively calm water, which still provoked seasickness among an inordinate number of classmates.

Class of 1955 goes over the side of the USS Rockwall *into LCP's, June 1953, in an amphibious training exercise near Little Creek, Virginia. (Photo courtesy USMA Archives.)*

Down the ramps of the LCPs – Landing Craft, Personnel – over the sides if the ramps at the bow of the craft jammed and wouldn't lower – wade through the surf with rifles held high above our heads if in waist deep water, storm the beaches, capture mock fortifications, guided by the steadying advice of Marine lieutenants who were platoon leaders.

Now we had an inkling of what the Army and Marines went through in World War II, when they landed on beaches all over the world, or at Inchon, in Korea – except there was no enemy firing real bullets from rifles and machine guns, or mortar and artillery barrages. For us there were no casualties, just a smattering of seasickness, some scrapes and bruises, and a little embarrassment for a very few who accidentally dropped rifles into the Atlantic while climbing down the nets into the landing craft for LEX I and LEX II – landing exercises one and two.

From Little Creek it was off to Fort Eustis, Virginia, home of the Transportation Corps, the people who operated all manner of Army vehicles, including small, medium, and large trucks, jeeps, and trains to move troops, equipment, and supplies.

Next came Eglin Air Force Base in the Florida panhandle. Classes taking the Air Force trip after the summer of 1950, had split in two groups, one for Eglin, the other to Maxwell Air Force Base, Alabama, the home of Air University Command. But this summer there had been an outbreak of hepatitis in the Montgomery, Alabama area, and the entire class went to Eglin for two and a half weeks.

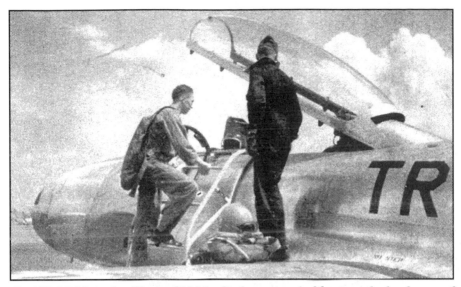

Carl H. McNair, Jr., Class of 1955, climbs a crew ladder into the back seat of a T-33 Trainer for an orientation flight at Eglin Air Force Base, Florida, during 1953 summer military training. (Photo courtesy USMA Archives.)

There we received orientation flights in T-33s or T-28s. The pilots exhibited the cool, detached calm that was legend in the Air Force. "Your aircraft..." they said over the intercom systems, as they shook the control sticks and let excited cadets briefly fly the airplanes.

There were also opportunities for flights in B-50s, a larger version of the B-29; drones – radio controlled, crew manned airplanes – which included a B-17 Flying Fortress of World War II fame; twin engined SA-16s, the amphibians noted for their rescue work at sea; and rescue helicopters. There were even missions in Air Force operated fast boats used in rescue work.

Then came the finale: a massive firepower demonstration at the Eglin range, in which F-86 and F-84 fighter-bombers strafed, dive bombed, fired rockets, and dropped napalm on various types of targets; and a huge, lone B-36 bomber, which dropped a full load of conventional bombs from high altitude in an awesome, thundering string that seemed to go on for miles. Members of the class of '55 so inclined now had their Air Force blood supercharged, ready to fly.

From Eglin the class flew via C-119s, "the flying boxcar," to Fort Bragg, North Carolina, for airborne training. There we received orientations on the basics of being a paratrooper, one of the elites of the Army. We did everything but a parachute jump, including riding in a C-119, sitting in seats on the opposite side of the airplane from a "stick" of troopers, who actually did "Get ready! Stand up! Hook up!" and rush in single file out the open door when the bell rang and the green jump light came on – all to the shouts of the jump master's "Go! Go! Go!" which commenced with the sound of the bell.

The nearest we came to a parachute jump was being sometimes, literally, kicked out of a thirty-four-foot tower, strapped in a parachute harness, its risers connected to casters that rolled on cables slanted downhill from the tower. The fast ride down the cable system ensured we attempted the prescribed parachute landing fall. The leaps – kicks or pushes – from the tower were frequently as hilarious as the parachute landings, which were often embarrassing in demonstrating how we didn't react correctly to the physics of a parachute descent.

After a week at Fort Bragg came Fort Belvoir, Virginia, the Engineer Center, just south of Washington, D. C. There we learned more about the combat engineer, military bridge building, including pontoon bridges for troops and vehicles, and how engineers worked in preparing defenses or infiltrating to breech enemy fortifications in advance of assaults. We also learned of the Army Corps of Engineers' roles in military construction at installations throughout the United States, and overseas, and in dam construction and flood control throughout the nation.

Everywhere we went that summer there was some form of social activity planned to balance the constant living out of B-4 bags and daily training. Washington night life was a welcome relief from the tightly packed schedule. When we next went to the Army's nearby Chemical Center in Maryland to receive a tour and afternoon field demonstration, some of the men who had met young ladies, or had friends and family in the Washington area, attempted return visits afterward.

Unfortunately, as at other places we received training that summer, cadet miscalculation, lack of contingency planning for unresponsive or unreliable transportation services, or just plain too much fun, sometimes resulted in a classmate being retired from his social activities for a time. "Late signing in from privileges," "Late for taps," "Late to formation," "Late to training," "Late to class," or late to anything, weren't offenses smiled upon in military life – for good reason. Depending on the circumstances, cadets late returning from anywhere might spend the balance of summer training, and maybe months during the academic year, confined to their rooms or marching punishment tours in the cadet area.

The final stop for the class of '55, before returning to West Point to begin summer leave, was Fort Monmouth, New Jersey, and the Signal Corps. Some of the subjects taught at Monmouth were teletypes, telephones, radios, television on battlefields of the future, amateur radio, and any other message and signal system we could imagine, and all it took to install and operate them for the Army.

* * *

During reorganization week, before the start of academics, the varsity football team, along with members of their cadet companies, received its second straight year of informal honor conferences, an outgrowth of the scandal two years earlier. The purposes were to review the honor code and system with all cadets, and bring them up to date on any changes in the honor system. As Earl Blaik had advocated, the sessions were informal group discussions rather than lectures, and were moderated by company Honor Committee representatives.

The evening of September 2, Earl Blaik gave a briefing on athletic policy to all cadet captains in the class of 1954, and afterward held an informal discussion with them. They were the new leaders of the Corps, responsible to the Academy and its officers, both as cadet commanders and trainers. They were the men who could "set the example," show the way, exhibit the behaviors of men of good character, set the tone for the coming year. There was no record of the meeting, as it, too, was largely an informal discussion between Blaik, the Army athletic director and football coach, and the men who could help bring the Corps, its athletic teams, including the football team, closer together. In Blaik's mind it was a memorable event, which he described in a letter to General MacArthur the next day.

Last night I spoke to the Cadet Captains, some forty-five in number, about Corps athletics and the regaining of a soul for the Academy. It was a long affair, but I believe much good will come from the open discussion which followed.

Undoubtedly, among other objectives, he sought to explain athletic policies and ask for the cadet officers' help by aggressively supporting a team which had been through two difficult seasons on the gridiron – had been to the wars, sweated, toiled, bled, and wanted badly to win – consistently. Football players were members of the Corps and represented the Corps, Academy, and the Army and Air Force on intercollegiate athletic fields, and it was important that the Corps' leaders knew and clearly understood that.

He likely also sounded a theme he repeated in a *Pointer* magazine interview three weeks later. Without mentioning the 1951 Bartlett Board's damning criticism of football players in general, its bitingly sharp criticism of the "numerous privileges" given Army players, and suggesting ways in which non-playing cadets could help the team, he told the *Pointer* interviewer: "Least of all does the athlete expect his fellow cadets to measure him in terms of a 'privileged' individual... no cadet should or does receive preferential treatment from anyone because he is an athlete, and... idle talk to the contrary is an unpardonable form of gossip." He emphasized all athletic teams must be made to feel by the Corps that they are doing something worthwhile. And, he stressed, this very sense of accomplishment will inevitably produce mutual comradeship which should be present in the Corps.

Blaik's talk with the cadet commanders no doubt also further energized and accelerated a free spirited group of Rabble rousers who were already hard at work. Stuck in their memories was a recurring nightmare: three years without a victory over Navy, an experience no self-respecting United States Military Academy Corps of Cadets should ever endure. Thus, when all the upperclasses returned to begin academics, and the Corps' cadet commanders heard Blaik's talk, plans were well under way in the class of '54 to unify the entire Corps of Cadets and their football team – as an inspired, single unit.

Fast on the heels of the September 11 *Pointer* issue came the issue of September 25, one day prior to the season opener against Furman University's Purple Hurricane. The banner headline on the cover read, "THIS IS ARMY'S YEAR." The cover depicted a main battle tank bursting through the print on a huge newspaper sports page, with team captain Leroy Lunn in his football uniform riding in the tank commander's seat, his head and shoulders protruding through the open turret hatch. The opening line of the lead editorial read, "Football is the central theme of this issue of *The Pointer*."

Meanwhile, on another front, other cadet schemes and plans were percolating, heated to a boil. The "operations group" supporting the football cheerleaders, and the cheerleaders, were hurrying a cannon through production. Ed Moses, now a first classman and Army's head cheerleader, had been a yearling prime mover in attempting to surreptitiously relocate the Camp Buckner reveille cannon during a summer '51 night raid, to "stop its harassing fire." Jay Gould, Ben Schemmer, and Bob Downey, Ed Moses' classmates, in 1952 had originated the idea of firing a cannon at home football games each time Army scored during their games, and when they were victorious over their opponents.

This year Ed Moses, Jay Gould, and Ben Schemmer were vigorously promoting the manufacture of another, improved cannon intended to harass Army's football opponents. But manufacturing cannons, and schemes to generate spirit, fight, morale, emotion, and a general, spontaneous outpouring of football enthusiasm among twenty-four hundred cadets, takes money. And cadets weren't noted for having much cash on hand, due to the tight control the Academy exercised over personal bank accounts and spending habits.

Not to be frustrated by a shortage of capital, the cheerleaders went to Colonel Russell P. Reeder, one of the Academy's most admired battlefield heroes of Normandy, and who, after World War II, established the Military Psychology and Leadership course. Reeder was now, as a retired officer, a Graduate Manager of Athletics. In that position, he also controlled disbursement of the Army Athletic Association's funds. The cheerleaders proposed their organization, including the mule riders, be designated a "Corps Squad," in other words, categorized as an athletic team.

The intent wasn't to gain the "privileges" of training meal tables in Washington Hall, or any other privileges. They didn't need a coach. They didn't need an Officer-in-Charge. Funding, plainly and simply, that's what they needed for their spirit inducing activities. In return, they promised "Red" Reeder an aroused Corps of Cadets like no one had seen in years.

Reeder didn't hesitate. He approved their request, a response Earl Blaik, the Director of Athletics, must have endorsed, since Blaik was ultimately responsible for the Department's budget and expenditures. For the first time in Academy history the football cheerleaders and mule riders were "Corps Squad" members.

Then it was off to the races for Ed Moses, Jay Gould, Ben Schemmer, and the rest of the energetic band of cheerleaders, and their operations group. The goal: Manufacture an Army "victory cannon" and inaugurate the piece with its first public firing outside Michie Stadium, at the Polo Grounds in New York City, where the Black Knights were to play the Duke University Blue Devils on October 17 in the fourth game of the 1953 season.

The first stop after Colonel Reeder gave his approval was an attempt to convince the Infantry to give the cheerleaders a pack howitzer. No luck. Next, they once more tried the Academy Museum, this time asking for any surplus or captured war material they might use, instead of an antique cannon. Pay dirt! – but with a need for more cadet ingenuity.

The Museum offered a German rocket firing tube, large enough to appear as a cannon or field artillery piece when fired. The "rocket cannon" had been captured during the North African campaign early in World War II, when the Allies had defeated Germany's "Desert Fox," Field Marshall Irwin Rommel. The same type of rocket tube had been used in the campaign in which former Commandant of Cadets Colonel, now Brigadier General, John K. Waters was captured at the battle of Kasserine Pass in 1942. Liberation time had come. The cannon would be a symbol for Army's victories on the gridiron.

But how to make a rocket tube into a cannon? That was another question. From the West Point Museum, the cannon conjurers went to the Academy's Ordnance Laboratory, where there was a turning lathe. Their idea was an adapter shell that would hold a sub-caliber twelve-gauge shot gun shell, to be used with blanks. The Laboratory agreed, and became a part of the growing number of Black Knight admirers determined to make this "Army's year."

Then it was down to the Academy transportation motor pool to discuss having the cannon painted black, gold, and gray, Army's colors, the same colors the jeep had been painted in the spring of 1952. After much friendly banter, the motor pool officer, a captain, agreed to their proposal, first explaining how he had cracked his class ring shoving hundred-five-millimeter howitzer shells into the weapon's breach during the Battle of the Bulge in World War II.

During the same discussion someone suggested, and they all agreed, the cannon should be towed into the gridiron wars by the jeep which had been issued to the cheerleaders. Finally, after additional talk, there was a consensus the jeep should be refurbished with a new coat of paint, the black, with gold and gray trim, to match the cannon. Covering the tire on the back of the jeep should be an Academy crest on a wooden plate. Everything had been agreed to – almost. There was one other detail to be wrapped up for the Duke game, other than the manufacture and test of the victory cannon.

The enterprising cheerleader operations group was off to Stewart Air Force Base in Newburgh. The plan? Discuss with the base weather office the possibility of procuring bottles of helium or hydrogen to send a balloon aloft carrying a cadet bathrobe, an offering to mythical gods just prior to the Duke game. There was much to be done and little time remaining until New York City and Duke's Blue Devils in the fourth game of the season. However, getting into the swing of a new academic year didn't stop a spontaneous football rally sparked by the '54 Rabble rousers two weeks before the season opener. A head of steam was building.

The new academic year always brought a few other details to take care of, such as settling back into permanent cadet companies – with the new plebe class of 1957, the new yearling class of '56, the cows of '55, and '54's own responsibilities in the cadet chain of command – plus intramural athletics, intercollegiate athletics, tactics training, late summer parades and ceremonies, all the other extra-curricular activities available to cadets, plus social engagements on weekends – if the men of '54 didn't have military duties or weren't serving some form of punishment.

We in '55 were facing into the infamous cow academic year, reputed to be the bane of all cadets, whether they be "hives" or "goats," cadets' labels for men at the opposite ends of the academic performance spectrum. Cow year, according to cadet lore – for which there was some measure of truth – was to be the most difficult of all four academic years. Cow year, above all others, required concentrated study if we were to survive its rigors.

The supposedly more relaxed yearling academic year was past, when we wrestled with mathematics, physics, chemistry, military topography and graphics, English, and the second year of a foreign language, the only cadet elective course, chosen at the beginning of plebe year from among French, German, Spanish, Portuguese, and Russian. As in each of the four years, there were courses in physical education and tactics, and twice annual ratings in military aptitude.

Our schedules were packed every year, cadet lore to the contrary. We began yearling year with 519 cadets in our class, including men turned back from prior classes. By year's end our number dwindled to 484. Eleven classmates couldn't meet academic standards and were "found" in various subjects, accounting for nearly one third of all the men who left '55 during that year. Several others resigned because of academic difficulties, dispirited or overwhelmed by their inability to keep pace with studies. There was no let-up.

The subjects confronted cow year were heavily weighted toward the sciences and engineering, and for many cadets, not easy: mechanics of solids; mechanics of fluids; electricity – which cadets often labeled "black magic"; social sciences, including macro and micro economics, comparative government, and American, European, and Far Eastern history; military instructor training; and the ever present tactics training, physical education, and military aptitude. And throughout every academic year were "drills and ceremonies" – drill competition, parades, and reviews on the Plain in the late summer, fall, and spring, and "Band Box" reviews in the central area of cadet barracks during winter. The constant stream of visiting dignitaries and award winners ensured there were more than enough parades and reviews, in addition to all the normal reasons for marching the Corps of Cadets before the American public and military.

There were good reasons for parades and ceremonies, which had their origins many years ago. For example, in the Academy's early days, in the summer of 1821, Sylvanus Thayer, fighting to gain public support for the still young institution, sent the then small Corps – a battalion of four companies – on a march all the way from Albany, New York to Boston, Massachusetts. They traveled by boat from West Point to Albany, then marched the 170 miles to Boston. The purpose was to exhibit the Corps to the American populace.

...They marched through Lenox, Springfield, Leicester, Worcester, Framingham, and Roxbury. Some marches were made at night because of the extremely hot weather. The Corps was welcomed wildly by Bostonians and invited to camp on Boston Common. Bostonians presented a stand of colors to the Corps, the first time it had its own flag. The cadet battalion visited Harvard College; another trip took the Corps to Bunker Hill, where they camped for the night.

Perhaps the most memorable part of the Boston trip was a march to Quincy to pay a formal call on former President John Adams. Major Worth [the commandant and officer who led the march] had written Adams to request that the cadets 'might be allowed to pay him their respects', and Adams had replied that the visit would give him great satisfaction. Adams spoke briefly to the Corps. 'It was interesting,' Worth reported to Thayer, 'to see one of the conscript fathers of the land – the wisdom of a century, addressing itself in the affecting language of patriotism, with tremulous tongue and a palsied hand, to the youth of another generation, exhorting them to love of country.' One phase of Adams' short talk was... 'I congratulate you on the great advantages you possess for attaining eminence in letters and science as well as arms. These advantages are a precious deposit, which you ought to consider as a sacred trust, for which you are responsible to your country.'

* * *

Uniformed football practice began Tuesday, September 1, 1953, as the Corps was preparing for the academic year. In spite of an additional year of experience following the painful '51 season, Earl Blaik didn't have a wealth of talent to start practice. In fact he had the smallest Army team in forty-five years.

The scandal had stripped the team of skills and experience, and, with the Korean War, brought immeasurable effects to football recruiting. Blaik, for reasons that turned out to be blessings in disguise, had, of necessity, gone to the Corps of Cadets to find athletes capable and motivated to play major college football. The class of '55, which entered the Academy a month before the scandal hit the headlines, was the smallest plebe class in years. Although fine young players entered the summers of '51, '52, and '53, normal attrition at the Academy always reduced the number of skilled varsity performers.

To compound the team's problems, the National Collegiate Athletic Association dramatically altered substitution rules, effective the 1953 season. The revised rules sharply reduced substitutions allowed during games, turning the clock back twelve years. A 1941 change liberalized substitution, permitting the evolution of specialized two platoon football, teams composed of players specializing on offense and defense. Earl Blaik, after importing the system to Army, became one of its most ardent proponents. In the summer of 1951, after his team was wiped out by the scandal, he reasoned that two platoon football was essential for rebuilding. "Without two platoons in '51 and '52, our situation would have been impossible instead of merely desperate," he said. Now, he had to prepare the team to play both ways, on offense and defense, when they still lacked experience and depth.

What made the rule change more difficult for Army players was two years of rebuilding that had already caused turmoil on the team. The Army coaches began weeding out players less capable of playing both offense and defense. They constantly shifted players from one position to another, and promoted and demoted between the varsity and junior varsity. For the men who completed the 1952 season knowing for certain they would be eligible for this season, believing they would continue in two platoon football, and could become more settled in their specialized positions of offense or defense, this wasn't good news. Now, the furious competition for starting berths began anew.

There was more than one offended first or second classman who believed he had assured himself a starting position on one of the two platoons, who now found himself relegated to a second or third team position, perhaps shunted aside by a talented yearling. To make circumstances more aggravating, revised substitution rules lessened opportunities for playing time and the coveted letterman's major "A" in football – for which he had toiled, sweated, and bled through the first two rebuilding years. The pressure was on again, with no small amount of griping, flaring tempers, and fisticuffs on the practice fields.

Blaik and his assistants continued the never-ending search for the right player in the right position. Bob Farris, '56, became a linebacker on defense, a tackle on offense. Howard Glock, a tackle on defense, moved over to guard on offense. Lowell Sisson's, Gerry Lodge's, and Freddie Attaya's experiences also typified the puzzle the Army coaches worked on constantly, piecing together the best available individual performers to obtain the best possible team performance.

In spring practice Blaik and Lombardi moved Lowell Sisson from end to halfback. His high school senior year in Waterloo, Iowa he had been a standout halfback on a state championship team. He wasn't a recruited athlete at Army, but had "walked on" as a plebe. He was tall and fast, good material for Vince Lombardi's backfield.

Now, during prescason workouts, ten days before the season opener, Lowell received word to report to "The Colonel's" office. Blaik and Lombardi wanted to see him. He was worried. Why do they want to see me? he wondered. Lombardi called him into Blaik's office and said, "Lowell, we want to switch you back to end. We wanted to ask you about it, if you don't object." Sisson was surprised, but relieved, and didn't hesitate in his reply. "I have absolutely no objection. All I want to do is play football. However I can do it is great – at end or anywhere else."

Gerry Lodge, '54, played fullback on offense and linebacker on defense in high school. In his first two years on the Army varsity, he played guard on offense, with some work as linebacker on defense. After the '52 season Blaik called him in and divulged he'd like Gerry to play fullback. "I think you could do both positions," Blaik said. Gerry knew the fullback was an experienced, respected Freddie Attaya. "Sir, why are you doing this?" Gerry asked. Blaik told him, "You win the sprints," meaning the wind sprints the linemen ran as a unit at the end of each practice, "and Attaya could lose out." Gerry said he was glad to have another crack at fullback. Blaik then asked Army's former All-American great, fullback Felix "Doc" Blanchard – a junior varsity assistant – to coach Lodge in spring practice. Gerry Lodge didn't know the reason for Blaik's expressed uncertainty about Freddie Attaya and remained puzzled.

The Army coach was remembering Vince Lombardi's observations of Attaya during the '51 spring practice. Freddie played on defense his plebe season, and in '51 spring practice continued to work on defense, at halfback. During an early spring practice he tackled one of the cadets' star, hard running halfbacks from the 1950 team, and was knocked unconscious, suffering a concussion. A week later Vince Lombardi told Freddie he ought to be a linebacker, and moved him to that position. Lombardi apparently believed collisions between ball carriers and linebackers at or near the line of scrimmage wouldn't be quite so violent. Freddie promptly got his "bell rung" a second time, when he tackled one of his classmates, big, bullish fullback Mario DeLucia.

After the second time Freddie received a head injury, Vince Lombardi made one of his hilarious remarks Freddie never forgot, and was to change his football days at Army. Lombardi told him, "Attaya, you have the brains, you have the moves, the intuition, and the desire to tackle – but your head's too soft." For the next two seasons Freddie Attaya didn't play football on defense. Now, in 1953, Earl Blaik and Vince Lombardi remained skeptical Freddie could avoid additional injuries on defense – though he had well earned their respect for his play on both offense and defense.

The increased two-way play meant the threat of injuries would increase, and as a result, for a team short on experience and depth, injuries to key players could have more exaggerated impacts on overall team performance. The impact of injuries was brought home with a vengeance when Neil Chamberlin, who was Bob Mischak's high school teammate, Academy classmate, and a standout tackle for Army in '52, very nearly died following a jeep accident at Camp Buckner, and was lost for the season – his last year at West Point. Bob Guidera, an outstanding, tough, aggressive, senior prospect and two year letterman at tackle, had suffered a series of debilitating, football-induced, knee injuries and operations, which virtually ended his football career prior to the 1953 season – and nearly ended his cadet career.

These were some of the more difficult obstacles Earl Blaik and his football team had to overcome as academics and football practice were beginning in late summer of '53. But there were other, powerful intangibles at work in the minds of all the young men at West Point as the days of summer slipped toward the fall's season of football.

On the team there was something else. As Earl Blaik was to say, he faced a strange psychological situation as a coach.

Players of the classes of '54 and '55 had been to the wars....
They had lived with the coaching lash, dirt, blood and defeat.
They were afraid of nothing, awed by nothing, eager to do anything asked.... They had only won one really big game and come
gallantly from behind to do it, against Pennsylvania in 1952.

They had suffered eleven losses in two years, as many as all the
Army teams from 1941 through 1950. Yet they had learned from their
mistakes while being seasoned by adversity, embarrassment, and defeat. There was a coalescing team camaraderie not yet visible.

LEARNING ON THE "FIELDS OF FRIENDLY STRIFE"

Earl Blaik, in spite of the changes in substitution rules, and true to
his firm beliefs, redirected his team's practices and intended mode of
play to continue two platoon football – but with a slight modification.
It was back to the way he coached his teams in the glory seasons of
'44, '45, and '46 when he had a wealth of talent. He formed first, second, and third units, which, because of the substitution rules, played
both offense and defense as units.

The revised '53 rules stated that once a player came out of the
game in the first or third quarters, he couldn't return until the next
quarter. The last four minutes of the second and fourth quarters imposed no such restrictions, however, meaning a player who came out
of the game in the first eleven minutes of the second and fourth quarters could return during those final four minutes. The overall effect
was to allow more liberal substitution in the final four minutes of each
half, when play usually was more hectic, with more decisive consequences – including fatigue and increased likelihood of injuries, as
well as game outcome. More significantly, the rules affecting the first
and third quarters, and the first eleven minutes of the second and fourth
quarters clearly destroyed specialized offense and defense platoon football. The final four minutes Blaik could return to his preferred, specialized two platoon football.

The new rules, and two or three units playing both offense and
defense also resulted in different considerations for Army's substitution strategies. None of the cadets' opponents had been through the
complete destruction of its varsity two years past, and would thus be

fielding teams with generally greater depth and more experience than Army's. Substituting a less talented, but rested and fresh Army second unit, against an opponent's experienced, game tested first unit, could be risky at best. Nevertheless, Blaik didn't hesitate. He was intent on continuing to rebuild, using his original methods of playing two platoon football.

The opening game of 1953 was, in one respect, a terribly inauspicious beginning for Army's third season of rebuilding. The disappointing crowd of 8,450 was one of the smallest gatherings at an Army home opener in many years. A former Army football player and renowned graduate of the already famous class of 1915, General of the Army, retired, Omar N. Bradley, and Mrs. Bradley, were in the superintendent's box, guests of Major General and Mrs. Irving. General Bradley was "the soldiers' soldier" during World War II, and he and Mrs. Bradley came to cheer on the Black Knights. They were no doubt surprised by the unusually small crowd. Otherwise, there was a glimmer of something else that day, a promise of better days ahead for the Academy and its football team.

The Furman University Purple Hurricanes were no match for Army, as Earl Blaik used the two platoon system as often as possible, and the Cadets took the Hurricanes' measure 41-0. Furman had won its home opener in Greenville, South Carolina, the week before, but the cadets scored two touchdowns in each of the first three periods, on their way to a shutout.

Gerry Lodge, at guard the season before, was impressive in his new offensive role, as fullback, and scored two of Army's touchdowns. An up and coming yearling halfback, Pat Uebel, scored two more touchdowns, with one each added by another yearling, end Don Holleder, and first classman Paul Schweikert, at halfback.

The game wasn't four minutes old when Army put its first six points on the board. Pat Uebel recovered a fumble by Furman's Ted Yakimowicz, and the cadets paraded 42 yards in six plays. Lodge cracked off tackle for the touchdown at 3:49 of play. Gerry was a first classman, a senior, and this was his first touchdown for Army. Right guard Ralph Chesnauskas, a yearling, kicked his first of five extra points, to put Army up 7-0. But extra points weren't all Chesnauskas did that day. He showed excellent promise and brought smiles to Earl Blaik several times, with his tackling and breaking up the Purple Hurricanes' plays.

At 7:49 in the opening quarter, Army scored again, this time following an opportunity set up by a 15-yard clipping penalty against Furman. This time Gerry Lodge lateraled to Schweikert, who ran 5 yards for the touchdown. Chesnauskas kicked his second placement to make it 14-0.

Late in the first quarter the cadets were off on a 74-yard march that ended 47 seconds into the next stanza. This time, third unit quarterback Bill Cody threw a 26-yard pass to end Don Holleder, who made the catch on the Furman 5 and scampered across for the touchdown.

During the second quarter, Furman managed to get to the Army 30 and 25-yard lines, but hard-hitting cadet defenses provoked fumbles both times. The second turnover by Yakimowicz was followed by a 76-yard march in 11 plays, ending in a 34-yard end sweep for a touchdown by Pat Uebel. Uebel had just scored his first varsity touchdown for Army, a marvelous beginning to what everyone hoped would be great things ahead for him and his teammates.

Pat scored again at 4:11 into the third period. Furman's Paul Stewart fumbled and Bob Mischak recovered to set up the score. This time Uebel dashed around his right end from the 10-yard stripe, into the end zone.

It was Ralph Chesnauskas who intercepted a Furman pass to set up Army's last touchdown, which came with 37 seconds remaining in the third quarter. Gerry Lodge ended the cadets 65-yard march when he went around left end.

Aside from the sparkling play of yearlings Uebel, Chesnauskas, and Holleder, there were other reasons to cheer that day. The cadets had rolled up 352 yards rushing, completed 9 of 15 passes for 115 yards, and made 21 first downs to Furman's 10. For the first time since 1950, on Monday following the win over Furman, Army appeared in the Associated Press poll's top 25 teams, tied at 23 with the University of Pennsylvania.

But the Purple Hurricanes weren't a "Big Ten" team, and next Saturday the Black Knights faced a team from the "Big Ten," traditionally one of the country's most powerful conferences on any athletic field. The opponents were Northwestern University's Wildcats at Evanston,

Illinois, the team the cadets had very nearly upset the second game of the '51 season.

This year's Wildcats had a strong passing attack. During the week prior to the Northwestern encounter, Earl Blaik instituted a new pass defense in hopes of offsetting the Northwestern strength. He later reminisced that the new defense, which failed to stop the Wildcat passing game, had probably placed too much pressure on one of Army's defensive halfbacks. Though the Cadets were the first to score, taking advantage of a 77-yard quick kick by Fred Attaya, and a fumble by Northwestern, the Wildcats took a 33-20 win.

Dick Thomas and Joe Collier, Wildcat quarterback and end respectively, were high school teammates and, together again this afternoon, proved a deadly combination against the cadets. After the opening surprise touchdown by Army quarterback Jerry Hagan, when he recovered a Pat Uebel fumble going into the Northwestern end zone, the Wildcats' passing attack shattered any hopes of a second straight Army victory. Collier scored one touchdown on a pass from Thomas, and set up two more with long catches. Northwestern scored three touchdowns in the second quarter, while the cadets could answer with only one more.

Thomas hit Collier for 38 yards to set up the Wildcats' first score, and fullback Bob Lauter plunged over from the two. The next time Northwestern got the ball, halfback Dick Ranicke raced 41 yards for another touchdown. With 50 seconds remaining in the half, Thomas threw another 7-yard strike to end Dick Peterson. At the half, Northwestern held a 20-14 lead.

The second half got off to a bad start for Army. The Wildcats scored in four plays, the touchdown coming on a Thomas to Collier pass from the 9-yard line. Army struck back in the third period with a touchdown pass from quarterback Pete Vann to end Don Holleder, to close the gap to 27-20. But in the fourth quarter Northwestern put the game away when their second unit drove for another score.

Again, there were some bright spots in the Army loss. The Black Knights played Northwestern slightly better than even in the rushing game, actually out-gaining the Wildcats on the ground, 188 to 173 yards. Army had 15 first downs to Northwestern's 11. In passing Army completed 13 of 24 for 156 yards, with one interception. Few paid attention to Army's last touchdown in a the losing effort, but the Vann to Holleder pass was a harbinger for the remainder of the season.

But it was the Wildcats' passing attack, and explosive long gainers that destroyed the Black Knights that day. Thomas hit on 16 of 23 for 209 yards, and his long throws to Collier were disastrous for the cadets.

Taking advantage of an airplane delay for the flight home Sunday morning, Earl Blaik seized the chance at the hotel to show the team their game films and review mistakes. He emphasized that if they didn't repeat their mistakes they could become "a real fine football team."

During the flight home he reflected on what a great bunch of youngsters they were. They seemed to have believed what he said to them, but he wondered how he could be sure. He had attempted to lift the responsibility for the loss from their shoulders, and placed it on his shoulders, one of the few times he accepted that he and his coaches hadn't adequately prepared their players for a game.

He had other thoughts about the loss to Northwestern, however. He concluded the cadets weren't as well conditioned as they should have been. The absence of any scoring punch in the final quarter against Northwestern, and the fourth quarter touchdown by the Wildcats' second team were evidence. "Their lack of endurance was a contributing factor for the loss," he told sportswriters as the next week's game preparations got under way. That didn't bode well for the cadets' practice sessions. They would be even tougher than usual. Nevertheless, Blaik believed the team was finally beginning to develop the cohesion for which it had been slaving since 1951.

Dartmouth University at Michie Stadium was next, and on Monday Army was nowhere to be found in the Associated Press rankings. Army football was back in the mire of '51, or so it seemed.

* * *

On Saturday, October 10, an injury riddled Dartmouth University team, with two prior losses, played Army before a crowd of 17,525 – double the number of fans that had attended the cadets' home opener against Furman. But Michie Stadium, small compared with most major college and university stadiums, still didn't have anywhere near a full house that afternoon. Among the onlookers were General James

A. Van Fleet, former commander of the 8th Army in Korea, and Lieutenant General Withers A. Burress, 1st Army Commander. The superintendent, Major General Irving, undoubtedly was pleased to have a steady stream of visitors like General Van Fleet, particularly. Men of his stature and reputation were excellent examples and motivators for the team and Corps of Cadets.

But the first half didn't go well for Army. Neither team developed a sustained attack the first quarter. In the second quarter the Black Knights went to Dartmouth's 18, 11, and 27-yard lines but fumbles plagued the cadets, and they couldn't score. Dartmouth was doing an excellent job of stopping Army's offense, though The Big Green couldn't mount a serious threat of its own. The half ended 0-0, and Army didn't score until three minutes remained in the third quarter.

New York Times sports writer William J. Briordy covered the game that day, and wrote enthusiastically of Army's second half performance.

> The Black Knights of the Hudson had more than enough stamina today as they clicked for four touchdowns in the second half to beat Tuss McLaughry's Dartmouth eleven, 27-0....

Briordy didn't report on some noticeable differences in the Army team on the field the first half as compared to the second half, except for Pete Vann's deadly accurate passing. Briordy's headlines for the *Times'* Sunday article included "ARMY TALLIES FOUR TIMES IN SECOND HALF – VANN PASSES FOR THREE TOUCHDOWNS."

Pete Vann did bring sparkle and fire to the Army offense. The fire appeared right at the beginning of the second half, before the first touchdown came twelve minutes into the third quarter. Twice before Army's first score, the cadets drove deep to Dartmouth's 29 and 19-yard lines. When The Green next went on offense, Bob Mischak pounced on a Dartmouth fumble on their 30-yard line. In five plays Army drove for their first touchdown, the score coming on a 10-yard pass, Vann to Mischak. Ralph Chesnauskas kicked the extra point. Army was up 7-0, and the blowtorch was now full open.

Pete Vann led the way again in a 53-yard march that brought Army's second score at 7:14 elapsed in the final quarter. Pat Uebel picked up 16 yards and Mike Zeigler made 17 on a quick opener through Dartmouth's left side, to put Army deep in Dartmouth territory. A 15-

Pete Vann, class of '56, fires a jump pass in second half action against Dartmouth University at Michie Stadium, October 10, 1953. In the foreground is Gerry Lodge, (32), class of '54. (Photo courtesy USMA Archives.)

yard penalty for unnecessary roughness stymied the cadets momentarily, but Vann promptly picked up the pace again. He passed 9 yards to Don Holleder. Then he pitched out to Pat Uebel who sped 22 yards for a first and goal on the 3-yard line. Two tries at The Green's line lost a yard, and then Vann threw to Holleder for the touchdown. Ralph Chesnauskas missed the extra point attempt.

Pete Vann's aerials again drove Army 63 yards to their third touchdown, helped by a 15-yard penalty against Dartmouth for illegal use of the hands. The score came at 11:06 gone in the last quarter, after Pete connected on three passes. The longest throw was a 31-yard hook-up with Bob Mischak, giving Army a first down on Dartmouth's 17. After Gerry Lodge picked up 4 yards, Vann threw another strike, a 14-yard touchdown pass to Mischak, and Howard Glock kicked the point after, to make the count 20-0.

The final Army score came with 1:49 left in the game. Paul Lasley had three prior years of varsity experience at Illinois College before going into the Air Force. He took a long, unusual route to play his last year of eligibility at Army. He "walked-on" at center, on Army's second unit. This day, at linebacker on defense, he intercepted a Dartmouth pass and scampered 42 yards to Army's final touchdown. Ralph Chesnauskas once more kicked the extra point.

It was Paul Lasley's only touchdown in four years of college football, but his score was the exclamation point at the end of a most unusual day for Army football. Army floundered badly the first half, was repeatedly stymied until late in the third quarter, and it appeared once again that the loss to Northwestern, as during the '51 season, might have taken the starch out of the Black Knights. Then came the second half and a seemingly brilliant performance by the whole Army team.

When the smoke cleared from Army's hot second half, the cadets had rolled up eighteen first downs to six for Dartmouth. Two by The Green were in the final 1:43 of the game, when they made their deepest penetration into Army territory. Army garnered 231 yards on the ground to 81 for the Indians. Pete Vann's passing was razor sharp. He connected on 12 of his 15 throws, and no interceptions, for 155 yards, while Dartmouth could manage only 43 yards on 4 for 19. Tall, whip-armed Pete Vann had connected with short jump passes several times, not throwing long as he loved to do. He was uncanny at finding open receivers.

Army intercepted four passes. Though Dartmouth might not have been considered a Northwestern, the cadets' pass defense definitely improved over the previous week's performance.

There were other heroes that day, too. Ends Bob Mischak and Don Holleder were glue fingered catching Vann's sharply thrown passes. Yearlings Ralph Chesnauskas and Bob Farris, at right guard and right tackle, were also stand-outs.

William Briordy, the *Times* reporter, told the Army-Dartmouth story well in Sunday's edition of the New York newspaper, but couldn't know there were other matters affecting the outcome of the game. He knew nothing of the cadets' building enthusiasm for the 1953 season. He had seen and heard the Corps of Cadets respond with roaring excitement to Army's second half play against Dartmouth, but he didn't comment on something else he witnessed on the field, which was stirring the Corps' spirited reaction.

Earl Blaik wasn't a coach who gave emotional fire-and-brimstone speeches at halftime. Rather, he was analytical. He identified mistakes and taught his team how to correct them. Not so against Dartmouth. He was frustrated by a flat, lackluster first-half performance by Army, and it was one of the few times he ever gave his team a severe tongue lashing. He and his coaches had accepted responsibility for the loss at Northwestern, lifted the defeat from their shoulders. But he wasn't going to excuse their first-half work this day. Vince Lombardi and all Blaik's assistants looked on as the Army players were thoroughly raked over the coals by "The Colonel." The man who prepared his team so well, so thoroughly, that "you didn't want to disappoint him," had stung them forcefully, and they responded with a fury.

Blaik's offensive huddle and play routine was meticulous and disciplined in every detail, and required the Army center to be the first man out of the huddle after the quarterback called the play. As the center trotted up over the ball, the rest of the team would pause, then break from the huddle, and jog to the line of scrimmage. Not so the second half against Dartmouth, or any other team the remainder of the 1953 season.

Center Norm Stephen ('54) was so energized by Earl Blaik's blistering halftime remarks, that the first offensive play of the second half he bolted from the huddle, and raced to the ball. The other ten men followed his lead when they broke from the huddle. They stormed to the line of scrimmage, a team of fired up football players, determined to reverse their first-half performance. The Corps responded, and roared their encouraging approval, a pattern that continued the rest of the afternoon.

A new method for calling up "the twelfth man" had been established. Perhaps it was the spark that lit next Saturday's fire.

Now Army was ready to win on any field – with honor.

CHAPTER 18

ARMY VS DUKE: 17 OCTOBER 1953

On Sunday, October 11, the Army coaching staff and team went through their usual post game critique of the preceding day's Michie Stadium performance against Dartmouth. Not a sterling show of strength, particularly the first half, after which Blaik had administered the team a rare, halftime tongue lashing. Next came the scouting report and game plan preparations for Duke University's Blue Devils. As always, Blaik's scouts did thorough, comprehensive work.

This coming Saturday Army needed to slay a dragon, and Earl Blaik had to pull all the stops in team practices for Duke. As he later wrote, "This game linked Army football past, present and future. If ever we were to rise up from the ashes of 1951, this had to be the day of the beginning."

Duke had won four in a row, against Wake Forest, South Carolina, Tennessee, and Purdue. They were steadily climbing in the national rankings, and what the coaching staff and Army players didn't know was on Monday, the Blue Devils would be ranked number seven in the weekly Associated Press poll. By Friday, Army was a ten-point underdog. Duke's winning four in a row came in spite of an injury sustained by their first team quarterback, All-American candidate Worth Lutz, against South Carolina.

Jerry Barger, the reserve quarterback, ably replaced Lutz while he mended, and guided a ferocious Duke ground attack, led by halfbacks Lloyd Caudle and James "Red" Smith, another strong candidate for All-American honors. Barger performed exceptionally well, taking the Devils to wins over South Carolina, Tennessee, and Purdue following Lutz's injury.

"Red" Smith's speed and quickness were major factors in the wins over Tennessee and Purdue. Against Tennessee he ran for one of Duke's three, second quarter touchdowns, led an 80-yard drive on another, and added two extra points. Against Purdue, the two teams were tied 7-7 going into the fourth quarter. One of the two fourth quarter touchdowns by Duke came when Smith outraced and got behind the Purdue safety to catch a long pass from Barger at the Purdue 12-yard line and go in for the score. His quick-starting, flashy running, and speed on defense made him a standout against Purdue, as well as Tennessee, and more than validated the Army scouting reports which said he was dangerous, and had to be stopped.

Duke's defense was equally potent. They were led by All-American left tackle Ed Meadows. In the three games preceding the 20-14 victory over Purdue, the defense allowed only three touchdowns.

If Earl Blaik's scrimmage routine was unusually brutal, the week before Duke was more so. During Tuesday's scrimmage with the "Golden Mullets," Pete Vann, who, with Jerry Hagan, was one of Army's two junior quarterbacks, received a severe ankle sprain. After two seasons under Vince Lombardi's tutelage, Pete was a more confident, accurate, long ball passer – and had fast developed into a ball-handling magician. His injury was a severe blow to pregame preparations. Pete could barely walk. He was unable to practice on Wednesday and Thursday. His play was doubtful.

In spite of Blaik's sixth football axiom, "Games are not won on the training table," physical therapy became vital to Pete Vann as preparations for Duke headed toward a climax. That first evening he was sent to the Academy hospital where fluid was drained from his ankle, and he received a cortisone shot. Wednesday and Thursday he could walk, but he couldn't run.

No practice for Pete, except for play study, upper body and leg exercises that didn't put additional strain on the ankle, and mental practice – while he listened to Vince Lombardi and Earl Blaik coach him and Jerry Hagan on Duke's defense and Army's play-calling strategies. Friday, when the team was to have a light, no-pads workout in New York City, Pete could manage only half-speed. Things weren't looking good for Pete Vann, though his passing against Dartmouth was one of the second half sparks to boost Army to a 27-0 victory over the winless Big Green.

Blaik and his staff worked tirelessly preparing the team for both offense and defense against the Blue Devils. As usual he spoke to his team almost exclusively through his assistants, his quarterbacks, and field captains. There were weaknesses and exploitable patterns of play seen in Duke, and he and his staff steadily pounded home what scouts had observed.

To Norm Stephen, the defensive field captain, Blaik provided signal calling guidelines to counter the Devils' explosive, split-T option running game. The option running game had to be stopped. If it could be stopped, Duke would be forced to pass, not what they had been accustomed to in the last three outings.

In their four prior games, Duke threw a total of 28 passes, only five against Purdue. There was uncertainty whether Worth Lutz, Duke's No. 1 quarterback, would be ready for Army, but the safe way to prepare was assume he would play. Lutz was more disposed to passing, having thrown 17 of the 28 passes, but both quarterbacks were adept at faking and keeping the ball in the option attack, and reeling off yardage in Duke's ground game.

With the option offense and a strong running attack paced by Smith and Caudle, Duke posed an outside running threat, as well, with end sweeps led by big, fast, hard-blocking linemen. They had shown some surprise "razzle dazzle" too, in the form of reverses and double reverses.

Inside the 10-yard line Duke ran 95 percent of their plays between their offensive tackles. Near opponents' goal lines, their quarterbacks tended to run "keepers" off the option, or "quarterback sneaks," reducing the likelihood of fumbles.

The overall assessment of Duke's offensive capabilities and performance set a premium on Army's defensive play, particularly should Duke penetrate inside the 10-yard line, where tackle to tackle the cadets would face their greatest test.

Something else was evident in Duke's encounter with Purdue. The Blue Devils fumbled five times and lost four. This called to mind Blaik's use of the quick kick. If Army could keep Duke backed deep in their own territory with quick kicks, the cadets' hard hitting on defense could cause the Blue Devils to make those same costly errors in ball handling and ball carrying, and ultimately yield Army scores.

The Devils' defense was another matter. They were strong, powerful, and had considerably more depth, enabling more substitutions. Ed Meadows' aggressive charges helped shut down opponents' offenses all season. As Blaik often did against players known for their aggressiveness, he prescribed a counter for Meadows' style of play. Run right past him, or trap him. There was another chink noted in Duke's defensive armor. When an opponent's fullback went in motion prior to the snap of the ball, the Blue Devil defensive secondary failed to adjust adequately and cover him.

For the Army coaching staff, there was a bright spot to offset Pete Vann's ankle sprain. Tommy Bell was returning to action. Tom suffered injuries during the game against Northwestern, and missed the Dartmouth contest. He would be a strong, fast runner at right halfback, alternating with Mike Zeigler.

Meanwhile, to make this "Army's year," plans long ago formulated within the Corps of Cadets, led by the class of '54, were coming to fruition. Concerted urgings toward support of the team were taking effect, fed by Army's 2 and 1 start in the season – though both victories came at West Point's Michie Stadium against opponents considered less than strong: Furman and Dartmouth. The number of cadets, officers, and enlisted men coming to watch practice each week was on the increase, with as many as 200 to 300 now showing up daily to encourage the team. There was an electricity in the air, and the Army cheerleaders, the cheerleaders' operations group, mule riders, and members of *The Pointer* staff, were hard at work, planning a maximum effort to arouse the Corps and bring it closer than ever to the men on the Army team.

As for Blaik, he remembered the encouraging meeting with all the cadet captains in the class of '54 at the start of academics. The meeting was vintage Blaik: "Talk to the Corps' leaders. Enlist their help. Teach them. Coach them." He believed the session, though long, might do some good. He likely hoped it brought a deeper understanding of what the football team meant to the Corps of Cadets, what the Corps meant to the team, and the sacrifices intercollegiate athletes made in representing the Military Academy and the Army.

Whatever he may have felt, the meeting drew him closer to the young men in gray who weren't intercollegiate football players and who, after the 1950 defeat by Navy, Blaik had roundly, if not bitterly, criticized in his letter to General MacArthur. He was now asking for their assistance and support, not demanding. His outlook on the relationship between the Corps and the team was slowly being transformed. He was, for the moment, turning his back on the events of 1951, events he could never change.

Now, this week, before the Duke game, a game Blaik knew was crucial, he spoke to the Corps of Cadets at the noon meal on Thursday. He stirred them as he never had before. Referring to the undercurrent of growing division between the Corps and the football team in the years preceding the cheating incident, Blaik intoned, "The Corps and the football team are one unit. You cannot separate them." The response from the Corps of Cadets was tumultuous.

Cartoonist Milton Caniff, creator of "Steve Canyon," at noon meal in Washington Hall, prepares the Corps of Cadets for Duke. (Photo courtesy USMA Archives.)

At the same meal was cartoonist Milton Caniff, on a platform before the Corps of Cadets doing sketches, helping whip up enthusiasm for the game. Caniff was the nationally-known cartoonist who drew "Steve Canyon," the syndicated newspaper comic strip about a heroic Air Force colonel and pilot who could solve all manner of problems. Steve Canyon was the true, patriotic, handsome, All-American "airman of fortune," always serving the best interest of his country, as he traveled from assignment to assignment all over the world.

The Corps' preparation for the game was more than unusual, including a "silence" imposed by the cheerleaders at the meal following Caniff's sketches and Blaik's talk. Ed Moses, the head cheerleader, the man who masterminded the attempt to rid Camp Buckner and the class of '54 of the pesky reveille cannon in July of 1951, said:

Men, after the team send-off tomorrow morning, we're to remain silent in preparation for the game on Saturday. There'll be no rally in the cadet area this evening. We'll save our voices and our energies. We'll give the team the usual send-off tomorrow morning, but the silence will remain in effect from the

USMA Cheerleaders, left to right, standing: Cadet Ed Moses, 1st Class, of Montivedeo, Uruguay; Cadet John Clayton of Laguna Beach, CA; Cadet Al Worden, 2nd Class, of Jackson, MI; Cadet Billy McVeigh, 1st Class, of Dallas, TX; Cadet Jay Edwards, 1st Class, of Clarksville, TN; Cadet Bill Robinson, 2nd Class, of Brooklyn, NY. USMA Tumblers, left to right front: Cadet Peter Jones, 1st Class, of Vancouver, WA; Cadet Dan Ludwig, 2nd Class, of Berwyn, IL; Cadet Jack Charles, 1st Class, of Ellwood City, PA; Cadet Charles Glenn, 3rd Class, of Oklahoma City, OK. (U.S. Army Signal Corps photo.)

time they depart for New York, through the assembly for our trip to the Polo Grounds, the bus ride, the assembly, and march on – except for the cheers we signal during the pregame ceremony – until the last man in the Corps double time's off the field from the pregame ceremony. When he sets foot on the first step into the stadium, at that instant, we'll explode with a roar that won't stop the entire afternoon. We'll remain on our feet the entire afternoon. Our voices will be heard as they haven't been heard in years.

The cheerleaders put a cork in the bottle of better than two years' pent-up frustration produced by the cheating scandal, with its disillusionment, sadness, and great loss, and football seasons made worse by the all-too-frequent sting of defeat.

Friday morning the Army team left for New York and a final, light, no-pads workout in the Polo Grounds, before the evening's stay at the White Plains Country Club. The practice plan included drop-back passes

– the quarterback's typical three steps back and throw to receivers from the T-formation.

Pete Vann was still questionable at quarterback, able to run no better than half speed. Earl Blaik, ever sensitive to opposing teams' and sportswriters' intelligence gathering, didn't want anyone to know Pete might not be able to play against Duke. Blaik and Vince Lombardi also didn't want Pete to aggravate the injury they hoped he could mend. Each time he took a practice snap from under the tail of center Norm Stephen, Blaik and Lombardi were close beside Pete, one on each side, fading back with him carefully, gingerly, ready to catch him, to ensure he didn't trip or stumble and re-injure his ankle. They were also concealing the injury from sportswriters' prying eyes.

Their caution, and Pete's determination and persistence paid off. The next day Pete Vann played perhaps the game of his life, along with everyone in a small band of cadets who had labored two and a third seasons learning how to consistently win.

ALL QUIET AND READY

On a near-perfect Indian Summer Saturday morning, promptly at 10:40, two long lines of buses, twenty-five buses in each serial, began the trip from West Point bound for New York City carrying the Corps of Cadets, twenty-four hundred strong. The route, after a rest stop at the halfway point, took the Corps down Highway 9W, parallel to the Hudson River, across the River on the George Washington Bridge, to Fort Washington Avenue, West 159th Street, St. Nicholas Avenue, West 145th Street, and left onto Broadhurst Avenue, headed north, on the east side of Colonial Park. We were destined for the Polo Grounds, the home of the New York Giants' professional baseball and football teams.

In K Company, 1st Regiment, the bus ride was quiet. The "silence" requested by the Army cheerleaders was holding. There wasn't the usual exuberant anticipation and excited, noisy talk of a football weekend in New York or Philadelphia. This was different. Restrained conversation was evident, but there was a charged, electric undertone, a churning energy still held firmly in check, ready to explode at the prearranged moment.

The day was golden and couldn't be better for a football game. The afternoon temperature would be in the 70s, comfortably warm.

We brought our long, heavy, wool overcoats on the trip to wear over traditional cadet gray. The officers and first classmen planning the Corps' movement were anticipating New York's normally changeable, sometimes cold afternoon and evening October air. But we left our overcoats in the buses when we arrived in the assembly area for the march-on.

When the long columns of buses came to a stop at 12:55, bumper to bumper, on the west side of Broadhurst Avenue, the two columns had become one, stretching from 155th Street, south to 149th. As drivers in the first three buses shut off engines and opened doors, the Military Academy Band, the Cadet Brigade staff, 1st Regimental, and the Regiment's three Battalion staffs began filing out and assembling on the east half of Broadhurst Avenue. One by one as buses came to a halt, the Corps' twenty-four cadet companies exited for the pregame march-on. After each company assembled, beginning with B-1, they marched forward toward the Brigade staff and A-1, to close up intervals to twelve paces between companies and staffs and await the march into the stadium at five minutes after one.

There were no pregame hijinks, impromptu cheers, or shouted cadence counts during the entire assembly and march-on, as there had been at games in years past. The quiet held fast, and tension continued to build.

The Corps of Cadets, in a column of companies, marches along the field's South sideline for pre-game ceremonies prior to the Army-Duke football game at New York City's Polo Grounds, October 17, 1953. (Photo courtesy USMA Archives.)

When the Academy Band struck up for the march-on, every beat of the drums pounded on the great pent-up mass of emotion which had been slowly but steadily building in the Corps for more than two long years. As the march began, with the band assembled at the intersection of Broadhurst and West 155th, there was an immediate right turn, east on 155th, then after one block, left, north on Eighth Avenue. Two more turns, east to Eleventh Avenue, and north again on Eleventh. By the time the entire column was on the move, the beat of the base drum, one beat per second, marked every other low growl of twenty-four hundred shoe heels contacting the pavement in the one hundred twenty steps per minute march cadence, as the sound of the band echoed down the streets and rolled through the line of companies. From Eleventh Avenue the column turned left into the east entrance of the stadium

When the band entered the stadium, leading the column, it began the strains of the "Official West Point March." The March stirred strong emotions anytime, but today more than usual. We were marching into something special. This was no ordinary trip to a football game.

This would be Army's year. But this couldn't possibly be Army's year unless this was Army's game. The Black Knights had already been set back once, against Northwestern's Wildcats. There could be no more defeats, no more losses. This had to be Army's year.

A MISSION TO REMEMBER

Bill Robinson in Company I-1, class of '55, felt the same emotions we all felt riding on the buses that day. But he and some '55 classmates in I-1 were mulling over a postgame activity, a plan and how it would be carried out to solve a new problem for the Corps of Cadets and their football cheerleaders. The plan would dictate a considerably different evening for Bill Robinson than for the Corps, whose members could simply go downtown and celebrate – celebrate a win over Duke and a night out on the town. Bill's assignment after the game was the result of a well-guarded secret among Army's cheerleaders, and his volunteering to take on the mission.

In the columns of buses and other supporting vehicles en route to New York City, was a refurbished, freshly painted black, gold, and gray jeep, towing a now fully-assembled, brand-new, painted, tested, and hopefully, a successfully baptized "victory cannon." A canvas cover shielded the cannon while being towed.

Saturday, after the game, Bill Robinson would be responsible for moving the victory cannon from the Polo Grounds to a secure hiding place somewhere in New York City, after the cannon's first public appearance. This was the cannon the Army cheerleaders' operations group had painstakingly shepherded through development in the weeks prior to the Duke game. The cannon must be hidden from people who might be sympathetic to Navy's cause in the annual gridiron classic in Philadelphia in late November. There was good reason for caution.

The birth of the victory cannon had come with another bold, cadet plan for the 1953 football season, a plan conceived the previous spring among members of the West Point class of 1954, for the Army-Navy game. From the outset, work on the cannon proceeded cloaked in secrecy, for fear midshipmen would discover the weapon and concoct a scheme of their own to "borrow" it.

Bill Robinson's plan for sequestering the cannon after the Army-Duke football game, would be the final step in the birth of a tradition. The tradition couldn't have picked a better day to begin the now famous, continuous and deafening chant,

GO! — GO! — GO! — GO! — GO! —

To the applause of a small crowd gathering in the Polo Grounds, the Academy Band, leading the Corps, entered the stadium from behind the east goal post, turned left, then right to march down the length of the south sideline of the playing field. The cadet brigade staff, color guard, 1st Regiment and 1st Battalion staffs, followed with the Corps' entire complement of twenty-four cadet companies, with 2d Regiment and Battalion staffs. When the Corps assembled on the field, the 1st Regiment's three battalions were in the north half of the playing field's width, with Company A-1's left flank on the 10-yard line at the west end of the field. The brigade staff was on the 50-yard line and centered on the middle of the regiment's 2d Battalion. The 2d Regiment's staff, and the three battalion staffs with their twelve companies were in the south half of the playing field. As in games past, the Corps was facing toward the stands where our opponents' fans sat. We were about to begin the pregame cheer for Duke.

Standing on a stadium seat, above and behind a growing crowd of Duke fans, was an Army cheerleader, wearing his white trousers and sweater. The big, white letter "A" trimmed in black, was visible on his sweater. Held inverted above his head, in both hands, was a large megaphone. He slowly waved the megaphone from side to side for several seconds after the Academy Band ceased playing, while the Corps stood at attention. It was the prearranged signal, "Give me your attention," before he led the Corps through a hand salute and the "Rocket Yell," the yell traditionally given opponents in pregame ceremonies involving a march-on of the entire Corps. Within the Corps, softly spoken words were being passed through the ranks. "Do you see 'im? He's on the first deck, in the center, right near the top row of seats."

The cheerleaders' commands, signaled "silently," brought an impressive, in-unison response from the Corps, which always delighted football fans unfamiliar with West Point. There was quiet in the stands, and abruptly, twenty-four hundred right hands snapped smartly to the hand salute, a salute to Duke University, and was held for several seconds. Then came the silent command for "two" – and twenty-four hundred right hands snapped back to the position of attention, all in unison.

Next was the "Rocket Yell" with its loud, verbal salute to Duke. Again, the silent commands, well-rehearsed visual signals we had seen cheerleaders give in prior games. In unison, a low whistle, increasing in intensity and pitch:

sssssSSSSS—BOOM!—Ahhh
USMA Rah! Rah!
USMA Rah! Rah!
Hoo Rah Hoo Rah!
AR—MAY! Rah!
Duke! Duke! Duke!

The cheerleader paused on completion of the yell, then signaled "about face," and as one man, the entire Corps executed an about face. Brigade, regimental and battalion staffs quickly double timed past the assembled cadet companies, toward the other side of the field, to positions in front of their commands, and halted. We now faced south, toward clusters of Army fans gathering in the stands around the cadet seating section. Again the hand salute, following silent commands from Ed Moses, Army's head cheerleader, behind and above the Army fans. Then on command, with our right hands, we removed our caps in unison and held them, arms fully extended, at our right sides. Standing bareheaded, we followed Ed Moses' signals, giving the "Long Corps Yell."

Rah! Rah! Ray!
Rah! Rah! Ray!
West Point!
West Point!
AR-MAY
Ray! Ray! Ray!
Rah! Rah! Rah! Rah! Rah! Rah! Rah!
West Point!
Team! Team! Team!

As soon as the signal came to replace our caps, the Academy Band struck up a cheerful "double time" tune, and in columns of two's, cadets from six different companies began streaming from the field into the stands, a company at a time, through five different entry gates. As the cadet seating section filled, from the back of the grandstands forward, from the west end zone to the 45-yard line, the air of anticipation continued to build. Cadets filing into the stands remained silent, calm, giving no hint of what was to come. The last two columns into the stands came from M Company, 1st Regiment, from the northeast corner of the football field. As the last man leaving the field hit the first step into the stands, a thunderous, stadium shaking roar exploded from the Corps of Cadets shattering the relative quiet of the Polo Grounds. It was a roar that continued virtually unabated, as the Corps remained standing the entire afternoon, including time-outs, quarter, and half-time intermissions. No rest today. None. No sitting down, and no respite for Duke's fans or their football team.

The sound was deafening, augmented by reverberating echoes amplified in the stillness of the warm afternoon by the second tier of the stadium. The second deck of stadium seats directly above the Corps had the same effect as the top side of a giant megaphone which aimed the sound toward the playing field and Duke's football fans in the opposite stands. The cheerleaders warmed us up with some of our game time yells, such as the "Whisper, Talk, Shout" and the "Long Fight." Interspersed in the yells were Army fight songs, always begun with the premier and best known of all, *On Brave Old Army Team.*

When the Army football team came back on the field for final warm-ups preparatory to the kickoff, the roar from the cadets became louder. The only semblance of a lull came with the playing of the National Anthem, shortly after the two teams' captains came to the center of the field to join the referees for the coin toss.

Army won the toss, and with the referee's signal came another stadium shaking roar. Army elected to kick off, consistent with Blaik's normal pregame instructions to team captains.

The roar resumed again, fueled by a new sound planned specifically for this game. On signal from the cheerleaders the Band's buglers sounded "The Charge," and the Corps of Cadets punctuated the bugler's last note, as it always had at football games, with a resounding "FIGHT!" Then the band's snare drummers immediately followed with a crescendoing roll of the drums, for three counts, and on the fourth count, all the Army cheer leaders shouted in unison through their megaphones – "GO!" The drum roll immediately followed again, the silent count with the roll, one, two, three – "GO!" Three count drum roll – "GO!" Drum roll — "GO! — GO! — GO!" The Corps picked up the beat. With each three count roll of the drums, voices rested and deep breaths followed, and on the fourth count came deep throated, stadium vibrating thunder. "GO! — GO! — GO!" The sound was relentless, as from a steam driven freight train just beginning its acceleration, with pulsating, explosive exhausts of steam. The beat held steady at a pounding and deliberate speed, and the volume intensified.

The Polo Grounds was a madhouse of never-ending noise, and fans seated anywhere in the stadium couldn't talk in conversational tones and be heard. Army fans seated adjacent to the Corps of Cadets had to shout to one another to be heard. This was the prelude to uncommon, inspired teamwork.

Memories of the 1953 Army-Duke game are vivid for those fortunate enough to be in the disappointingly small crowd of 21,284, a crowd probably reflecting New Yorkers' expectations of a Blue Devil walkover. The walkover was not to be. For everyone at the Polo Grounds that afternoon there was a thrilling, emotion-filled, heart-stopping contest – "a game never to be forgotten," as Earl Blaik wrote years later. Though not attended well or given much advance attention in the sports pages, it was rightly called "The College Game of the Year."

After Army's opening kickoff, the thundering "GO! — GO! — GO!" became "DE-FENSE! DE-FENSE! DE-FENSE!" The fired-up, swarming, gang-tackling Army defense forced a Duke punt after three downs. Once more the bugle, drums, and "GO! — GO! — GO!," but Army couldn't gain a first down and was forced to punt to the Blue Devils.

Duke began to move, and so did James "Red" Smith. He was giving the Black Knights trouble they knew they must avoid. He took an option pitch-out from starting quarterback Jerry Barger, and scampered 30 yards before Pat Uebel dragged him down on the Army 24. But Army stopped the drive on their 25, when Barger was hit hard and brought down for a loss on a fourth down pass attempt.

Now, Army was on the offensive, and again the bugle, slow cadence drum roll, and the stadium-shaking "GO! — GO! — GO!" echoed through the still, Indian Summer air. Army began shredding Duke's defensive line. Jerry Hagan was at quarterback, Uebel and Zeigler at left and right halfbacks, and Attaya at fullback. It was Zeigler on a pitch-out for 12 yards; Attaya off tackle for 17; then Pat Uebel over All-American

Freddie Attaya takes a handoff from Jerry Hagan for a short gain in first half action against Duke's Blue Devils. (Photo courtesy New York Times.)

tackle Ed Meadows for 12 more. "GO! — GO! — GO!" With the ball on Duke's 19-yard line, Freddie Attaya took a pitch-out and rammed for a first down on the nine. "GO! — GO! — GO!"

In came Tommy Bell at right halfback, and on a quick opening hand-off, bolted straight ahead past Duke's Ed Meadows, leaning forward, driving hard, and bouncing off a Duke defender in the secondary, stepped easily into the end zone.

In the stands, the cadets leaped and shouted for joy, with back slapping, arm waving, hand shakes, clenched fists shaking an unyielding, defiant determination.

Army right guard Ralph Chesnauskas, outwardly icy calm, kicked the first of his two extra points to give the cadets a 7-0 lead. The Academy Band struck up *On Brave Old Army Team,* and a chorus of twenty-four hundred men whistled and sang with all the volume they could muster.

After the Army kickoff, the chant once again was, "DE-FENSE! DE-FENSE! DE-FENSE!" Duke was unable to move and gave the ball up on a punt. Freddie Attaya gave it right back to the Blue Devils with the first of several quick kicks. Army, aided with a 15-yard clipping penalty against Duke, pinned the Blue Devils deep at their own 7-yard line, as the first period came to a close. Excitement rose on the Army sidelines, with hopes for a Duke fumble. Hope turned to disappointment as, instead, the men from North Carolina used 13 plays to cover 93 yards, five of them passes by Worth Lutz.

The Duke quarterback was deadly in the face of the fired up, hard charging Army line. He first hit his fullback Jack Kistler, with a screen pass. Then came successive completions to right end Joe Hands, left halfback Nick McKeithan, and yet another left halfback who came into the game, Bob Pascal, who moved the ball to the Army 4-yard line. From the four, Worth Lutz, as the Army scouts predicted, "quarterback-sneaked" into the end zone. Versatile "Red" Smith booted the extra point to tie the game, 7-7.

With Worth Lutz's steady passing performance, shooting the Army defense full of holes, it appeared the game's momentum was shifting irrevocably toward heavily favored Duke. But Army came right back, working its own magic through the air. Pete Vann came into the game, and was throwing. The Army threat fizzled, however, and Duke took over on its own 25 – but was again forced to punt. Mike Zeigler ran the punt back to the Army 37.

Then lightning struck. In three plays Army stormed its way to another touchdown.

Freddie Attaya slammed into the line for two yards. The second play of the drive was astonishing. Pete Vann and Freddie Attaya teamed up for perhaps the most unusual forward pass in Army football history, a pass far more memorable than the one Freddie dropped against Dartmouth in 1951. Pete, a right-handed passer, rolled to his right to throw. Eight yards behind the line of scrimmage, he was confronted by Duke's big, hard charging, left tackle, Ed Meadows. Pete, looking down field, cocked his arm to throw as Meadows closed in, but the Army quarterback had to slow and reverse direction, slanting to his left toward the line of scrimmage to evade Meadows' charge. As he brushed past Meadows, still looking for an open receiver, the Duke player grabbed for Vann's right arm, and Pete instinctively switched the ball to his left hand. Meadows caught hold of his right arm, but Pete, still moving forward and to his left, pulled free from his grasp and threw a short, wobbly, but accurate left-handed strike to Freddie Attaya, good for a 17-yard gain to the Duke 43. The left handed throw by Pete Vann, kept the touchdown drive moving, and sent the Corps of Cadets into a frenzy. "GO! — GO! — GO!"

On the next play, Army struck at Duke's habit of not adequately covering a back in motion. It was another pass. Pete Vann was again hemmed in, but threw a 5-yard flare pass perfectly, wide across the field into the right flat, to an open Pat Uebel who was turning up field, accelerating to near full speed. Pat, the big, handsome, fleet footed yearling left halfback, gathered in the pass from Vann, and aided with key downfield blocks by Mike Zeigler and Ralph Chesnauskas, weaved 38 yards down the sidelines in a beautifully-executed run, untouched, for the second Army score. In the cadet stands there was utter pandemonium as Pat raced into the end zone.

Ralph Chesnauskas again calmly kicked the extra point, and Army moved ahead 14-7. Once more the Academy band struck up the familiar refrain, and there echoed through the stadium the chorus, whistling and singing of *On Brave Old Army Team*. The seven-point lead held as the first half ended.

Except for Duke's second quarter touchdown drive, sparked by Lutz's passing, and "Red" Smith's breakaway 30-yard dash to the Army 24, the vaunted Duke ground attack had been throttled almost completely. Bob Mischak and Lowell Sisson at ends, and Pat Uebel at defensive halfback had repeatedly thwarted Duke's attempts to get outside Army's line on option pitch outs or quarterback keepers.

In their first touchdown drive, entirely on the ground, Army's offensive line repeatedly hit with crisp, fierce blocking and tore gaping holes in the Blue Devil's defense.

Army ball carriers ran with uncommon fury, with quick starts off the ball, driving, lunging, twisting, turning, faking, diving for every yard they could get. The entire team was playing with complete abandon, as Earl Blaik had often told them they must do. As if to emphasize their iron willed determination, they had run right past Duke's big, fast, All-American tackle, Ed Meadows.

Then in the second quarter Pete Vann used the forward pass to stun Duke with a second touchdown. The entire Army line of ends Lowell Sisson and Bob Mischak, tackles Ron Melnick and Bob Farris, guards Ralph Chesnauskas and Dick Ziegler, and center Norm Stephen, with their second unit alternates, were playing inspired football. But it was far from over. A pressure packed second half was in the offing.

*   *   *

Army cheerleaders looked forward to the traditional half time exchanges with opponents' cheerleaders, particularly if their schools were coeducational. The cadets' monastic, highly regimented, fast-paced life at the Military Academy caused their observation of females to be extraordinarily attentive, with speedy judgments.

The usual courtesy extended to "home team" cheerleaders – in this instance, Duke – was for visiting team cheerleaders to first call on the home cheerleaders, who would then reciprocate. However, Bill Robinson realized as soon as the first half ended, this day wouldn't be usual. The Corps' fierce dedication to winning included loud protests as the Army cheerleaders made for the Duke side of the stadium. In spite of cadets' noisy protests, the exchange proceeded.

Bill Robinson, a typical female-starved cadet, couldn't help notice certain of the Blue Devil's cheerleaders, as the Army cheerleaders approached the Duke stands. He was momentarily distracted. They are beautiful! he said to himself. After the two groups introduced themselves, however, the first question asked by a Duke cheerleader wasn't what anyone expected. "How do you do it – come up with so much spirit?"

The question was reasonable and might be asked at any Army game the entire Corps attended. The Corps was always loud and boisterous at football games; however, these Duke cheerleaders probably hadn't attended a game against Army until today. Duke vs Army played their last three contests in 1944, '45, and '46 at these same Polo Grounds, during the halcyon days of "Doc" Blanchard and Glenn Davis. Army won all three handily. They were en route to two national championships, and in separate polls, shared a third national championship ranking with Notre Dame after the Army-Fighting Irish "game of the century" in 1946. Duke fans had no any idea how much spirited scheming, planning, and preparation preceded this game, and what the contest meant to the Academy, the Corps of Cadets, or its football team.

There were other half time thoughts wandering through the minds of Ed Moses, Bill Robinson, and the rest of the Army cheerleaders. The Corps might decide to continue a tradition with one of Duke's charming female cheerleaders when they came to the Army side of the stadium, a tradition reserved for men. Bill shuddered at the thought of an enthusiastic, female-hungry group of cadets beginning the chant "Pass 'er up! Pass 'er up! Pass 'er up!"

The tradition of "passing up" opponents' cheerleaders required the "volunteer" to maintain a rigid body, muscles taut, arms folded across his chest, while the cadets lifted him off his feet, hoisted him high above their heads, held him at arms length, hands on calves, thighs, buttocks, hips, back, neck, head – while cadets crowding below him passed his rigid body, from one row to the next, all the way to the top of the cadet seating section, and then back down. The rite was accompanied by laughter, cheers, "ooh's" and "aah's," as the brave volunteer confidently surrendered his welfare and safety to the Corps of Cadets – whose football team was sworn to soundly defeat his team on the gridiron. The tradition of passing cheerleaders up was always in good jest. Seldom, if ever, was anyone injured, despite those occasions when the trip downhill from the top row of seats became rapid and breathtaking, with the unsuspecting rigid body sometimes rolling over and over as groups of outstretched hands sought to control and protect him against a fall.

Throughout the game, Bill Schulz was standing near the men in Company A-1. He had been among the five cadets who painted "Go Army, Beat Navy" on the starboard side of the Navy destroyer escort the fall of '52. Bill, now a first classman, was on the Cadet Brigade staff, a cadet captain, the Brigade Training Officer, and was the class president of '54. While there was great excitement for the game this afternoon, he overheard a conversation about a plebe in A-1. Word had been received that the freshman's father had just died. The young man had to be told of his father's death and arrangements made to take him directly to the airport on emergency leave immediately after the game. Bill suggested someone needed to go with the fourth classman to the airport and subsequently volunteered to escort him.

Prior to the game, one of Bill's close friends, "Woody" Wilson, a 1953 Kentucky Military Institute graduate, invited Bill on a double date. Bill, short of the necessary cash for a weekend date in New York City, demurred. "Woody" insisted Bill join him and his date, as she would enjoy the company of a third party. "Woody" and Lee were just friends. There was no romantic interest between them. Bill said he would join them and arranged to meet him and Lee at a designated stadium gate after the game. "Woody" and Lee were going to a party at the Roosevelt Hotel after the game. Now, after agreeing to escort the fourth classman to the airport, Bill Schulz didn't expect to be back in time to join the party, due to post game traffic and uncertainty regarding emergency leave arrangements for the fourth classman.

"WE'RE THE HEROES OF THE GRIDIRON GRENADIERS..."

When the two football teams returned to the field to limber up for the second half kickoff, the Corps of Cadets, which hadn't stopped their yells and fight songs throughout the halftime intermission, resumed their thunderous noise, mixed with intermittent chants of "GO! — GO! — GO!" and the bugle call of "Charge," followed by "FIGHT!"

Early in the third quarter, Duke's Blue Devils once more began to move the ball. They were wasting no time. Starting from their 20-yard line after Freddie Attaya quick kicked into the Duke end zone, it was back to the running game. They drove 56 yards entirely on the ground, to the Army 24, sparked by halfback Lloyd Caudle's 25-yard dash. The revived Duke ground attack was relentless, and it appeared a touchdown was imminent. But Caudle fumbled in the next series of downs, and Lowell Sisson recovered for Army on the 25. Two plays later Jerry Hagan fumbled and Bobby Burrows pounced on the ball for Duke, at the Army 28.

From the stands it appeared Duke had been offside on the play, but no penalty was called. The Army stands shouted their disapproval of what seemed the referees' oversight, but to no effect. In five plays Duke was in the end zone with another touchdown. The Blue Devils' Worth Lutz was the whole show, running "keepers" off the option, and scoring his second rushing touchdown. On the 3-yard scoring play, as he fell forward into the end zone, he momentarily fumbled the ball but retrieved it in midair as he fell. 14-13. Duke was about to tie the game.

Duke's "Red" Smith was the place kicker. Up came the cadets' chant: "BLOCK THAT KICK! BLOCK THAT KICK! BLOCK THAT KICK!" The roar from the stands was deafening. The snap from center, the placement, and Smith's wobbly kick sailed wide right of the goal post. Army breathed a collective sigh of relief, and let loose shouts for joy, as the cadets maintained their slim 14-13 lead.

But there was still cause for worry. Following another Attaya quick kick, Duke started moving again, from their 26 yard line. They drove to the Army 37, where the drive was halted as the third quarter came to an end. Army hadn't made a single first down in the third quarter, in part because of Earl Blaik's quick kick strategy. Duke had ground out seven first downs, and it appeared Army's defense was weakening.

The unusually warm Indian Summer afternoon was taking its toll on superbly conditioned young men playing "iron man football." The uniforms of both teams were caked with sweat-soaked mud from the loose dirt and dust of the Polo Grounds' baseball infield, over which the white chalk lines of the east half of the football field were marked. Although both teams were sacrificing heavily in the afternoon heat, Army's white jerseys with black numerals, and gold pants, showed the strains of the furious battle far more than did Duke's royal blue uniforms with white trim. There was another reason Army's uniforms showed more sweat and toil. Duke's team had greater depth and experience, and could substitute more frequently.

There was real cause for alarm in the Army stands early in the fourth quarter, when another Attaya quick kick was called back because the cadets were penalized for illegal use of the hands. His subsequent punt to Duke's Jerry Barger was returned to Army's 28-yard line. The penalty and second punt return gave Duke a 37-yard advance in field position, with respect to Attaya's first kick. But the Army defense rose to the occasion and frustrated Duke's drive at the cadets' 18, when on a fourth-down fake field goal attempt Duke threw a forward pass that fell incomplete.

Now Army's offense began to move, shifting the game's momentum to the cadets just when it seemed Duke was about to subdue them. Once more the men in gray heard the bugle and roll of the drums and began shaking the stadium with vibrating echoes of "GO! — GO! — GO!" Gerry Lodge came into the Army backfield, and he and Tommy Bell reeled off 63 yards between them as the cadets surged to the Duke 19-yard line, all the while eating precious time on the game clock.

Cadets in the stands were ecstatic. On the Army sidelines the coaching staff was excitedly passing instructions and encouragement to the

Led by Pat Uebel (34), hard-running Tommy Bell (46) skirts Army's left end and lowers his head and shoulders to take on a Duke defender. (Photo courtesy USMA Archives.)

team, trying to keep pace with the rush toward Duke's end zone, which lay in the lengthening shadows of the field's west goal post. There was a smell of victory in the air, and feelings of elation began overtaking Army fans and players on the sidelines.

Then the first of two totally unexpected events occurred that very nearly brought disaster to the Black Knights. With fourth down and five yards to go, on Duke's 19, Blaik ordered in a substitute with instructions to kick a field goal. He reasoned that three more points and a 17-13 advantage for Army would at least keep Duke from kicking a field goal to win 16-14, if the Blue Devils were running out of time and Army hadn't scored.

On the bench manning the phones from the press box, where Army assistants were helping with Blaik's coaching chores, was class of '49 graduate assistant Lieutenant Martin "Tiger" Howell. In Blaik's mind, Howell had shown a tendency to take matters into his own hands – "usurp command" in Blaik's words – when things became terse, as, according to Blaik, Howell had done while playing tackle during Army's 1948, 21-21 tie with Navy. Unaware of Blaik's instructions and seeing a substitute running on the field, Howell chased after him and shouted "No! No!"

The player stopped about ten yards from the bench. In the excitement and noise, Army players on the field didn't see the substitute start onto the field and hesitate in response to Howell's shouts. The team huddled, the play was called, and the break from the huddle was under way before there was time to do anything but call the substitute off the field to avoid a penalty. Howell had unknowingly countermanded Blaik's instructions. Army failed to make a first down and Duke went on the offensive.

"Tiger" Howell never forgot what happened immediately after calling the substitute back. Blaik was horrified, but didn't remember the scene. He wrote about it seven years later. Howell recalled Blaik handed him an Army blanket and said, "Take this. You'll need it in Korea." "Tiger" Howell was mortified when he realized what he'd done, and figuratively walked back to the Army bench on his knees. Blaik remembered, "The suffering of the Army side, multiplied by X, could not have equaled the agonies "Tiger" went through [that afternoon]."

Blaik, Howell, and the Army bench turned their attention back to the game, as the chant from the cadets in the stands resumed: "DE-FENSE! DE-FENSE! DE-FENSE!" There were about three and a half minutes to play. The prior 56 1/2 minutes had been enough to wring any crowd dry of emotion. As Earl Blaik said,

> New York could not comprehend what was on the line in this game – but they were being rewarded by one of the classics of Polo Grounds or Army history.

Worth Lutz, Duke's quarterback knew he needed a long gainer. His team didn't have the energy, and there wasn't time left for a long, grinding drive using their split-T offense. The long gainer he called was a double reverse. Rolling opposite the direction of his backfield's movement, Lutz handed the ball off to left halfback Bob Pascal indicating a sweep around Army's left side. Pascal then slipped the ball to "Red" Smith, Duke's speedy All-American candidate, heading the opposite direction. The excellent faking decoyed Army's defensive right side, causing them to over-commit, and Smith broke into the secondary. Lowell Sisson, Army's right end was cut down in a tangle of players as Smith flashed past him, and remembered sitting on the ground, looking up, and saying to himself with a terrible sinking feeling, Lowell, you blew the game... Tommy Bell fell victim to the same tangle of players and Duke blockers, and saw Smith tear past him. After side-stepping a linebacker, the Duke halfback was quickly in the clear, ten yards beyond pursuing Army defenders, sprinting toward the southeast corner of the field.

Earl Blaik years later described reactions on the Army sidelines.

> We on the bench...shot to our feet in sudden silent, stunned consternation. Smith looked home free. I felt our heads were being pushed down once more into the ashes of 1951.

For several seconds a hush fell over the Corps, as the play unfolded below them.

Then from out of the pack of Army pursuers came left end Bob Mischak. He had an angle on Smith, who could neither see nor hear him. Mischak began rapidly closing the gap as Smith crossed the 50-yard line. By the time he crossed the Army 20, Mischak had closed to

3 yards. The Army stands were coming alive, shouting encouragement to him, but it was important that he not commit too soon. Blaik held his breath, muttering to himself, Not yet! Not yet! As Smith crossed the 12-yard line, and Earl Blaik was saying "Now! Now!" Bob Mischak leaped far and high, caught Smith around the shoulders at the 10, and downed him on the 7-yard line.

Earl Blaik remembered Bob Mischak's game-saving tackle the rest of his life, as did the Army team, sportswriters, sports broadcasters, and Army fans present in the Polo Grounds that day.

> In somehow catching and collaring [Smith], Mischak displayed heart and a pursuit that for one single play I have never seen matched. Yet his feat, one of the great defensive plays of football, would have soon been forgotten, had it not been for what followed.

When the two players came crashing into the dirt, both lay there for a moment, exhausted. In the Army stands, bedlam. Duke had a first down and goal on Army's 7-yard line, with the clock running down to three minutes. There was a time out at three minutes, ample time to get off four plays.

In those final three minutes of the game, especially during Duke's eight attempts to cross the Army goal line, it was almost impossible to hear or think because of the constant roar from the crowd. In the radio broadcast booth high in the stands, famed sports broadcaster Ted Husing handed the microphone to a colleague, saying, "These boys have gotten me so excited I can't even finish the broadcast. Take over...." Cadets poured down out of the stands, pressed around the Army bench, and crowded the sidelines at the east end of the field, imploring their defense to dig in. "HOLD THAT LINE! HOLD THAT LINE! HOLD THAT LINE!" rocked the stadium. Duke fans were on their feet, shouting, but their sounds were drowned by the roar coming from the Army side of the field.

Lutz carried the ball on the first play and got only a yard. Smith got two yards, putting the ball on the four. On third down Lutz faked a pitch-out and hit the line. Chesnauskas stopped him a little more than a yard short of the goal line. Fans on both sides of the field were in a frenzy.

839

Would Duke attempt a field goal? Prior to the fourth down play a Duke assistant coach threw a kicking tee onto the field, indicating a field goal attempt. Lutz picked up the tee and threw it back, disdaining the field goal.

Norm Stephen called the defensive signal, preparing for what Army players were certain would be a running play. As the Army linemen dug in, ready to take their defensive stance, they knew their legs and feet would be in the cadet end zone.

Army's always thorough scouting reports once more exploited Duke's habit patterns. Norm Stephen and Gerry Lodge remembered what Blaik had told the team:

> When the Blue Devils get inside opponents' 10-yard line they run the ball between their own tackles 95 percent of plays. And when they get close to the goal line, they run quarterback sneaks.

Gerry Lodge had seen his teammate, left guard Dick Ziegler ('54) playing magnificently all afternoon, often absorbing the energy of three Duke blockers because of his hard charges in the middle of their line. When Norm Stephen called for a five-man line, Gerry was a linebacker. When Norm called a six-man line, Gerry had been on the line, "in the trenches" beside Dick. As the Blue Devils huddled for Lutz to call the fourth down play, Gerry Lodge said to Dick Ziegler, "Remember, he's going to try to sneak."

Lutz did and was met by a converging wall of white, dusty, sweaty, mud-splattered jerseys and a huge stack of bodies. There was once more a lessening of stadium noise as referees, not yet signaling the play's outcome, whistled the ball dead, stopping the clock, and rushed to untangle the pile of bodies. One by one players got up, until the referees could see the ball resting inches short of the Army goal. Worth Lutz had control of the ball, but Lowell Sisson, lying on the bottom of the pile, had reached through the tangle of bodies and was holding his right hand firmly on top of the ball to keep the Duke quarterback from thrusting it forward into the Army end zone.

The referee signaled an Army first down. The cadets had held, in an incomparable, emotion drenched stand, barely two inches from the goal line. The Corps of Cadets leaped and shouted. Pandemonium reigned again. But the game wasn't yet over. Forty seconds remained.

The noise cascaded around Earl Blaik. He called over a substitute but couldn't make him hear what he was saying. He finally got across to him: "We can't risk a running play! It could mean a safety! Tell Vann to have Attaya kick out immediately!"

Suddenly, Earl Blaik felt weak. He had to sit down somewhere, anywhere. He chose the nearest thing handy, a yard marker. He forgot the yard markers were made of foam rubber. When he sat down, he fell unceremoniously backward, on his tail end, with his feet up in the air. Several Army players on the sidelines saw him go down. Startled and fearing he was having some type of seizure, they rushed to him. He quickly assured them he was all right, and they helped him up.

Freddie Attaya backed deep in Army's end zone, in punt formation. Yearling Bob Farris, the cadet right tackle on offense, who earlier in the season was selected to become Army's center on punts and extra points, came up over the ball. An errant snap from Bob Farris could be disastrous, but it was good, and Freddie sent a splendid punt to Army's 41-yard line, to Worth Lutz, who, knowing time was fast slipping away, ran for the sidelines to stop the clock. He went out of bounds at the Army 37. Lutz, a calm field general, husbanded the remaining seconds well. There was still time to get off four passes.

"DE-FENSE! DE-FENSE! DE-FENSE!" the thunder continued. Army set up a deep pass defense, with fewer linemen rushing. Stan Harvill ('55), *The Pointer* features writer, had been down on the field most of the game, assisting the cheerleaders and "victory cannon" crew. He was now among the hundreds of cadets shouting encouragement from the crowded sidelines. But he was caught in the crush and couldn't get near enough to see the action. As everyone knew they would, the forward passes came – three, desperate passes, each batted away by Army defenders. From his position in the rear of the cadet crowd on the sidelines, Stan Harvill couldn't see a thing – just the passes being lofted into the air. The shouting, screaming crowd told him the results of each throw.

Each of the three incomplete passes stopped the clock. Duke had time for one more play. Onto the field came Jerry Barger, the Blue Devils' reserve quarterback, bringing a play called by the Duke coaches.

Worth Lutz remained in the game, and another Duke player left the field. When Duke broke from the huddle, Barger lined up as an eligible pass receiver, with Worth Lutz under center at quarterback. This was it. The last gasp. No more chances.

Lutz threw long, to the end zone where Gerry Barger was waiting. But Pete Vann was there and batted the pass away. Pete Vann, Army's quarterback on offense, had knocked away a pass thrown by one Duke quarterback to another Duke quarterback. The game was over.

From the west end of the field, with a great puff of gray-white smoke, roared the Army victory cannon – "KA-BOOM! – a reverberating explosion which overpowered the noisy celebration of a roiling sea of men in gray. They were storming the field, leaping, shouting,

Deliriously happy Army players and cadets celebrate their stunning 14-13 upset of the Duke University Blue Devils at the Polo Grounds in New York City, October 17, 1953. Identified players are: Lowell Sisson (83), end; Bob Mischak (87), end; Bill Cody (11), quarterback; Ed Zaborowski (58), center; Frank Burd (33), fullback; Peter Vann (10), quarterback. (Photo courtesy Associated Press.)

shaking their fists, rushing to shake the hands of their heroes, slapping them on their backs, lifting them to their shoulders, and carrying them from the field. Not in recent memory had anything like this been seen, not even in the golden days of "Doc" Blanchard and Glenn Davis.

After the game, the Duke University fans sat quiet for a few seconds, shocked, drained, stung by the defeat and what had been so close, yet snatched from their grasp. Worth Lutz, Duke's quarterback said after the game, "The savage cheering of GO! GO! GO! from the West Point stands placed our team in a nervous fright of tension and jitters." The Duke University Band's Drum Major came to the Army side of the field after the game to say, "I had to come over and tell somebody. That's the most beautiful noise I've ever heard. I don't see how you can lose." A Duke cheerleader was later heard to remark, "We can beat the Army team, but we can't beat all twenty-four hundred of them."

"A GAME NEVER TO BE FORGOTTEN."

The scene in the Army dressing room was of jubilant, emotional exhaustion. Players, as usual, had to help one another pull the tight, sweat-soaked jerseys over their heads to get them off. As shoulder pads and rib pads were removed, it became apparent that men who had played nearly the entire game were going to need more help. Sweat from the Indian Summer heat and players' prolonged physical exertion had so saturated the T-shirts worn underneath their pads that the shirts were stuck to their skin. Scissors had to be used to cut and peel them off.

Everyone on the Army team knew they had accomplished something extraordinary. Among the coaches, including the always calm, analytical, almost stoic Earl Blaik, there wasn't a dry eye. Especially for charismatic, mercurial Vince Lombardi, as well as Blaik and Army's entire coaching staff, these were tears of emotional release – expressions of joy rarely seen in tough, hard-bitten college football coaches.

Blaik, his eyes glistening with tears, called the team together and presented Bob Mischak the game ball. As he handed the ball to Mischak, the Army head coach, a man of few but always impressive words, said simply, "Don't ever give up." Bob, a high school tailback recruited by Vince Lombardi and converted to end, had made the incredible game-saving pivotal play of Army's entire season, for Army football, and in Blaik's memory, one of the greatest college defensive plays ever seen.

Though Bob saved touchdowns on at least two other occasions during the '51 and '52 seasons, this one against Duke would be remembered for years by everyone in the small crowd at the Polo Grounds on October 17, 1953. For nearly fifteen years afterward, the story and film clip of Bob Mischak's play against Duke University was part of West Point's Military Psychology and Leadership Course – as an example of the power of motivation.

The win over Duke University was a magnificent team victory, filled with heroics by many players, who, two seasons earlier had been pounded mercilessly each Saturday afternoon, and seldom tasted victory. Nevertheless, there were other factors in this win.

As Earl Blaik often said, "The game of football requires complete concentration." Seldom did his players notice or listen to the crowd, except in Philadelphia's huge Municipal Stadium where the overpowering sounds of fans numbering more than 100,000 were the norm in the annual Army-Navy classic. However, on this football afternoon, many of his players heard the Corps of Cadets as they had never heard them.

Lowell Sisson's experience typified what was different this day. In games past, Army's right end paid little heed to the grandstands and the Corps of Cadets. Earl Blaik had told the team, "You can go out and win this game." Lowell remembered in 1950, as a plebe watching the Army varsity in home games, his and the Corps' cheers seemed mechanical, as though "ordered" by upperclassmen. Today, for the first time, he clearly heard the Corps. While standing on the sidelines during the game, despite its frantic pace and excitement, and the need for total concentration, the Corps' never-ending expressions of unity with the team swept him up in their emotions, caused him to turn, gaze, and admire, and left him unforgettable, lifelong memories. In Lowell's mind, the Corps had won for the team.

Pat Uebel heard the sounds – louder sounds – begin early in the season, against Furman and Dartmouth in Michie Stadium. In his two seasons at Army, one as a plebe, he had never seen the Corps so aroused as on this day.

Bob Farris, who played the entire sixty minutes, heard and felt the power coming from the cadets in the stands. He knew "the 12th man" was there.

Godwin "Ski" Ordway, who alternated at left end with Bob Mischak and Joe Lapchick, was equally taken by what he saw and heard. "Ski," the walk-on who, like Lowell Sisson, persisted doggedly until he made the Army varsity, "felt at one with the Corps, like never before."

Bob Mischak, quiet, calm, reflective, deliberate, rock solid, went through the same intense preparation for Duke as his teammates. But what struck him particularly the week before the Polo Grounds encounter was the excitement and interest exhibited by his roommates, Mark McDermott and Bill Kirby. They talked frequently about the upcoming contest, and to Bob, seemed to be concentrating far more intently on the game than him. Bob Mischak knew the Corps "had the fever," and it was genuine, infectious.

At game's end, deliriously happy men in gray lifted an excited, exhausted Tommy Bell off his feet onto their shoulders and carried him from the field. He didn't notice his white jersey and gold pants were sweat-soaked, streaked with dirt, and caked with splotches of mud produced by his sweat.

His little Irish mother attended the game in the Polo Grounds. She was baffled by the American game of football when Tommy was in high school in the Bronx, but danced around her kitchen with Vince Lombardi the spring of 1950, when Vince came to tell her that Tom had passed his entrance examinations for West Point, and he "would make Tommy an All-American." Now, she confessed, seeing the Corps carry Tommy off the field on their shoulders, she too was "one of his fans."

Freddie Attaya heard Vince Lombardi's shout to him above the roar from the stands, as he ran onto the field with a fullback running play called from the sidelines in a third down and seven situation, as Army mounted a drive. "Run the gauntlet!" yelled Lombardi. Vince was reminding his fullback, as he had in earlier games that season, of a drill he put his backfield through in practices. In the drill, his running backs carried the ball pressed against their gut, and ran hard between two lines of eight players each. The ball carriers' forearms protected the ball, and their hands tightly gripped opposite ends of the pigskin. Every player in the gauntlet aggressively tried to "tackle the ball" as they lunged to knock ball carriers off balance. Urged on with Vince's shouted reminder, Freddie ripped the middle of Duke's line for 12 yards and a first down. When next he came off the field, Lombardi rushed to hug him, telling him he had done a great job.

For Pete Vann, Duke was his "coming of age." In 1951, following the Northwestern game in Evanston, Illinois, Pete had missed the team's flight home, and had to hitch a ride on the airplane with the Academy superintendent, General Irving. Pete was a plebe that year, but the incident was embarrassing. His tardiness and behavior were noted by Blaik, resulting in the Army coach's hard to change belief that Pete was immature, "a mere boy" whom Blaik was concerned wouldn't grow up.

Pete worked tirelessly, often bristling inside with frustration and anger, under tough, sometimes infuriating Vince Lombardi and Earl Blaik, determined to become Army's starting quarterback. He, too, became an Army football hero this day.

> [Earl Blaik] pounded three characteristics into his quarterbacks: 'You must be able to think under pressure. You must be able to transmit common-sense knowledge to your teammates. You must be a leader.' Then he added, as if it were a foregone conclusion, 'Of course you have to know what every one of your teammates is doing on every single play. This is the basis of team confidence in you. When you demonstrate on the field that your knowledge of the game is greater than any teammate, the team will follow your leadership and produce touchdowns.

Although Pete Vann showed flashes of brilliance in passing and ball handling during prior seasons, and in the three preceding games this season, he had shared quarterback duties with Jerry Hagan. Against Duke, he emerged the team leader Blaik sought in his quarterbacks. What's more, on defense, in the role of a throwback to the days of "iron man football," he showed himself dependable in a crisis. In Earl Blaik's mind, Pete was no longer "a mere boy." He was a man, and indispensable to the team.

That evening Pete took the hotel elevator from the floor on which he and his teammates were staying, and headed out for a bite to eat. He was alone when the elevator stopped at the next floor down. The door opened and in stepped Vince Lombardi – alone. As the doors closed,

Lombardi, still elated over the game's outcome, once again complimented Pete on his play that afternoon. Then after a brief exchange about where Pete was going, Vince reached for his billfold and handed Pete a five-dollar bill. "Here," he said, "go buy yourself a beer — and don't tell the Colonel."

For Earl Blaik, "the Colonel," there was another powerful tug on his emotions after the Duke game. Retired Colonel Ralph Sasse visited the Army dressing room after the game, to congratulate Blaik and the cadets for their great victory. Ralph had been the Army head coach from 1930 through 1932, named Earl Blaik his "first assistant," and recommended he be his successor when Sasse requested return to duty with the cavalry. This fall, Ralph Sasse was ill, a man Blaik had been quietly encouraging and helping in his struggle against alcohol. This was the last Army game Sasse saw, and he knew well what Blaik and his team had been through. To Earl Blaik, there would be everlasting, warm satisfaction in knowing Ralph saw one of the great games in West Point football history.

MISSION ACCOMPLISHED

Bill Robinson knew the importance of good planning. He was an Army "brat." His father had retired from the Army. The plan to secure Army's victory cannon in New York City Saturday night after the game wasn't elaborate. Bill had attended high school in Brooklyn, and knew the city well. His family had lived on an avenue near the Staten Island Ferry, not far from Fort Hamilton and the Narrows separating Long Island and Staten Island. He was familiar with the back street short cuts to Fort Hamilton, an Army post, to which there were only two entries, one normally closed. The remaining entry was guarded round the clock by Military Policemen.

After the game, the cannon was hitched to the jeep for towing. Because Bill was a second classman, restricted from driving, he rode in the jeep pulling the cannon. The jeep's driver was an enlisted cannon crew member. Guided by Bill, they wended their way to Fort Hamilton through the back streets of Brooklyn.

He made prior arrangements with the Military Police, explaining the cannon, that it wasn't a real cannon, but also explaining what it was, and its purpose. The Military Police readily agreed to grant entry onto Fort Hamilton, to hide the newly manufactured "weapon." Bill specifically warned them that midshipmen from the Naval Academy might be on the prowl, looking for the weapon, and to keep any middies from gaining entry to Hamilton.

Bill selected a parking place for the cannon between two vehicles near the officers' club. He had asked a friend, a colonel's son who lived in nearby quarters on Fort Hamilton, to assist in securing the cannon. His friend agreed, and once an hour, all through the night, the volunteer sentry checked to ensure it remained in hiding, undisturbed.

On Sunday morning, Bill Robinson drove to Fort Hamilton in the family car, wondering all the way if his important, secreted weapon was still safe. Much to his relief it was, and another driver safely towed Army's victory cannon back to West Point – ready to greet Columbia University's Lions at Michie Stadium the next Saturday, and Navy's midshipmen in November.

For Ed Moses, and the hardy, inventive cheerleaders' operations group, the Army victory cannon became more than they ever dreamed. It had fired its first public blast at "a game never to be forgotten," and though none of them knew it at the time, they started another great tradition that would live among the legends and folklore of Army football.

Ed Moses could smile inwardly with great satisfaction. Maybe he didn't quite succeed in ridding Camp Buckner and his West Point class of that pesky reveille cannon the summer of 1951. But this cannon would certainly leave its mark. It wouldn't wake up twenty-four hundred young men at 5:50 every morning of the academic year, but it would give thunderous wake-ups to thousands of football fans in huge stadiums on fall Saturday afternoons, year after year, every time the Army team crossed their opponents' goal lines – and every time the final score was an Army victory.

* * *

As prearranged, Bill Schulz briefly met his friend "Woody" Wilson, at a Polo Grounds gate after the game. Bill was en route to LaGuardia Airport escorting the Company A-1 fourth classman, and for the first time met and said "Hello" to Lelia McGill, a 1953 graduate of Duke University. He explained to "Woody" he had to take the fourth classman by taxi to the airport and hoped to return in time to join him and Lee at the Roosevelt Hotel.

Bill and the fourth classman left the Polo Grounds for LaGuardia, and Bill waited with the plebe at the flight departure gate until he left on the airplane. Much to Bill's surprise, the trip to the Roosevelt Hotel allowed a brief, pleasant visit with "Woody" and Lee, during which Bill learned Lee worked at *Harper's Magazine* in New York City, not far from the hotel. Now able to pause and converse with Lee McGill for the first time, Bill Schulz couldn't help saying to himself, She's a lovely girl – a doll.

In telling Bill about her work, she referred to the building where the magazine was located as the "Cat House," a label coined by Harper employees. "Woody" had to leave early to catch a plane. Bill volunteered for the second time that day, saying he would walk Lee back to the "Cat House." They sat on the steps of the Harper building, well into early morning, talking, getting to know one another.

All's fair in love, war, and football games. "Love thine opponents' graduates." Bill Schulz and Lelia McGill fell in love the evening of Saturday, October 17, 1953, were engaged the following January 9, and married the next spring, one day after Bill Schulz graduated from West Point – one more wonderful story among many that began in the Polo Grounds that October afternoon.

* * *

The next day, the *New York Times'* Allison Danzig wrote:

...Army stopped Duke short of its goal line in the final minute of play at the Polo Grounds yesterday to save a superbly won 14-13 victory over one of the nation's top-ranking football teams.

...This was Army's shining hour, its greatest victory since the cribbing episode of early 1951 that wiped out the varsity squad. It signaled the return of the Black Knights of the Hudson to their former station among the big powers of the gridiron, and it was a tremendous triumph for Earl Blaik and his coaching staff, as well as for Army's many valiant performers on the playing field.

Gene Ward of New York's *Sunday News* wrote:

> The lean years forgotten and the failures of recent campaigns wiped away, Army's football forces waded into the Blue Devils from Duke yesterday and knocked them from the ranks of the unbeaten with a magnificent display of offensive and defensive team play. And when the final gun popped on this tremendous 14-13 upset, the full Cadet Corps swarmed onto the gridiron and lifted its heroes on a mass of shoulders in a victory jamboree the likes of which had never been seen....

The following Tuesday, Army reappeared in the top twenty-five teams in the Associated Press poll, after tumbling from view following the Northwestern game. Their return to national football prominence was stunningly swift, grand, glorious, almost unbelievable. But there had been far more than a football team, its coaches, or even an overpowering determination to win. From somewhere deep within each of us – in the Corps and on the team – had come an all-consuming fire; a whole whose sum was far greater than its parts; a relentless, unending source of energy we didn't know was in us, a strength that seemed to grow and become more fierce and unstoppable as the afternoon wore on. We would not slow down, wouldn't give up. Never! No matter the circumstance, no matter the field, no matter how hopeless the cause might seem.

Army's victory over Duke in "a game never to be forgotten" was the turning point for the season of 1953 and for Army football. This was Army's year.

CHAPTER 19

THE GREAT AND GOOD TEAM

The fire ignited in the summer of '51 had spread. The army team, the Academy, and the Corps of Cadets were alive with excitement. The conquering of Duke unleashed a latent source of power and energy which continued to build through the fall. Except for Tulane University's Green Wave late in the season, the small band of inspired Black Knights wouldn't hold back the remainder of the season. Neither would the Corps of Cadets, or the men and women who lived and performed duties at the Academy. Football glory, and the glory of victory on the gridiron returned to live at West Point once more, and no one wanted to let go of it.

But this was different. The Corps and the team were what Earl Blaik said they were, "...one unit. You cannot separate them." Beginning the second half against Dartmouth, something electric began rippling through the Corps, the Academy, and their football team. In the Polo Grounds it all came together. Now, Army was 3-1 on the season with four games to go before Army-Navy in Philadelphia, the game the men of '54 had been pointing toward since the stunning 14-2 upset on December 2, 1950. The last Army victory over Navy was a year earlier, in 1949, when the cadets had rolled over the midshipmen 38-0, when Johnny Trent was the team captain, and quarterback Arnold Galiffa an acclaimed All-American.

BRING BACK THE LIONS

The Saturday after Duke's defeat, Army's Black Knights were at home in Michie Stadium. This time, with a win over the nation's No. 7 team behind them, the cadets once more had the college football world's attention.

The stadium was sold out. There were 23,520 fans this October day, a mark of Army's returning football prestige. The small opening day crowd of 8,000 the last Saturday in September, increased another nearly 8,000 each home Saturday. The Army football team was refilling the stadiums it played in.

Among the fans was an American war hero recently returned in the prisoner exchanges following the armistice with the North Koreans and Chinese. He was Major General William F. Dean. His happy, smiling son, William F. "Bill" Dean, Jr., class of '55, sat beside his father in the stands.

General Dean was the 24th Infantry Division Commander in those hectic, disastrous, early days of the Korean War, when the Division was the first American unit committed to stem the tide of the North Korean advance southward, down the Peninsula. Undermanned, underequipped, committed piecemeal, and ill-prepared for what they faced, the 24th was badly mauled by the *In Min Gun*. The

Major General William F. Dean, former 24th Infantry Division Commander in Korea and Congressional Medal of Honor winner, at the Army-Columbia Football Game in Michie Stadium, with his son, Bill, Jr., October 24, 1953. (Photo courtesy New York Times.)

North Koreans captured General Dean in early August of 1950 after overpowering the chewed-up remnants of the Division near Taejon, South Korea.

He and survivors of his shattered headquarters unit attempted to escape southward through the Korean hills, trying to find their way back to friendly lines. He remained in captivity the entire war, suffering torture, virtually unending solitary confinement, and privation as the North Koreans and Chinese relentlessly interrogated him, attempting to gain intelligence concerning American forces, equipment, and methods of operation and training. His determined resistance against his captors nearly cost him his life, but he never broke in spite of the inhumane treatment he and thousands of other American prisoners of war endured. For his actions before and during his captivity he received the Congressional Medal of Honor. He was a genuine hero who survived, and there were many American soldiers who witnessed his valor and lived to tell of his deeds.

The Columbia Lions were 2-2 on the season, and there were vivid memories of the cadets' disappointing 14-14 tie at Baker Field in New York City a year earlier. Worse, the tie had occurred before one of West Point's most famous graduates, Dwight D. Eisenhower – then the president of Columbia University and candidate for president of the United States, now the nation's president.

Lou Little was the Lions' coach. He and Earl Blaik were mutual friends, and the friendship was enriched by a deep personal and professional respect the two men held for one another. Nevertheless, each year their two teams met on the gridiron, they were fierce, determined competitors. Who could forget Columbia's stirring 21-20 upset of Army in 1947, ending the cadets' 32-game undefeated streak? That was the season following the great Davis-Blanchard-Tucker team of '46. And who among those remaining at West Point this year, and had witnessed the difficult season of 1951, could ever forget the stirring last quarter goal line stands when a winless Army team turned the tables on the Lions 14-9 in the cadets' first victory in five outings?

This was another day and another year, however. The stunning 14-13 upset of Duke the previous Saturday was the catalyst for a 40-7 rout of Columbia. The Lions' first two defeats were by close scores. Not this one. It was the worst beating Lou Little's squad had absorbed since 1949, and the score could have been much worse had Earl Blaik not substituted with second and third units most of the fourth quarter.

The Lions' only score came midway in the third quarter when Columbia's Bob Mercier intercepted Pete Vann's pass, intended for Bob Mischak, and went 42 yards in a fine run down the sideline for a touchdown. The cadets already had a 21-0 bulge at that point in the game, and following the interception and runback, roared right back, going 73 yards for another seven points in just three plays.

Once again, sparkling ball handling and passing by Pete Vann, disciplined faking and hard running by Freddie Attaya, Tommy Bell, Pat Uebel, and Mike Zeigler, and devastating blocking and tackling by the cadets' line were the day's markers. Bob Mischak, Lowell Sisson, and Don Holleder at ends added luster with their outstanding play. Army continued to display intensity and determination with their break from the huddle and charge to the line of scrimmage when the team was on offense, the energizing habit begun against Dartmouth after Blaik had blistered the team with his halftime talk. And with every rush from the huddle to the line of scrimmage, the Corps and Army fans responded with crescendos of speech-drowning cheers. Now was mixed, too, the stirring, growling thunder of GO! GO! GO! and the startling roar of the victory cannon, fired after each touchdown and extra point.

The electricity was already flowing at five minutes into the game, when Pete Vann first came on the field, replacing Jerry Hagan at quarterback. On his first play from scrimmage he fired an 8-yard pass to Bob Mischak. Then Bell, Uebel, and Attaya alternated in moving Army on the ground, driving deep into Columbia territory. On the seventh play after Pete's first pass, Pat Uebel bolted 15 yards for a touchdown on the "belly series." The touchdown had been set up one play earlier by a straight-ahead hand-off to Tommy Bell from the Columbia 17-yard line, for a gain of two yards.

The "belly series," or "fifty series," was a thoroughly rehearsed, deceptive set of plays. The series, which could be run to the right or left side of the offensive line, was normally set up by hand-offs from Pete Vann to the fullback, driving low, angling between the offensive guard and tackle. On each "54" play called, which went to the fullback going outside the right guard, the left halfback crossed immediately behind the fullback and between the tackle and end, faking reception of a hand-off from Pete after the ball had been handed to the fullback. Fullback Freddie Attaya's angling, slashing drives and fakes into the line invariably caused the Columbia defense to converge on him, which set up scoring hand-offs to halfbacks Tommy Bell or Pat Uebel.

After one or more successful ground gainers handed off to the fullback, Pete Vann would call "56." The fullback and halfback ran the same pattern, but Pete would shove the ball toward the fullback's belly, in between his forearms, "riding" the fake hand-off forward, then deftly and quickly pull the ball back, turn half to his right, and slip the ball to the left halfback. Both backs ran low, leaning forward, driving hard, giving every appearance each was going to run with the ball, which added to the deception. The defensive tackles, ends, and linebackers, converging on the fullback fake, then learned too late the ball-carrying halfback had flashed past defenders and the line of scrimmage, into the secondary.

Columbia scouts the previous Saturday at the Polo Grounds had seen the stampeding, locomotive-like running of Tom Bell, and Freddie Attaya's flashing speed against Duke, and were ready for both. They weren't ready for the faking wizardry of Pete Vann, Tommy, Freddie, and Pat Uebel together, and they certainly weren't ready for Pat Uebel's hard-driving slants, crossing behind the fullback, and suddenly bursting past a trapped defensive tackle, into the secondary. It was a brilliantly executed play that would taunt, fool, and dazzle opponents' defenses the remainder of the season.

Following Pat Uebel's first touchdown, Ralph Chesnauskas booted the first of his four extra points. Then a third of the way through the second quarter Army struck again, going 84 yards in 18 plays, scoring this time on a relentless, grinding assault on the ground. Moving the ball to the Columbia 7-yard line, Pete Vann again faked, this time to Freddie Attaya lunging toward the Columbia defensive line, and pitched out to right halfback Tommy Bell racing wide around the cadets' left side. Army 14-0 at the half.

The third Army score came at 2:35 into the third quarter, behind Vann's passing, when he floated a perfect 15-yarder over the head of Lions' defender Dick Carr, the Columbia quarterback, into the hands

The Army victory cannon is fired after an extra point against Columbia, October 24, 1953. (Photo courtesy Edward M. Moses.)

Pete Vann cocks his arm to throw deep against Columbia. Michie Stadium, 24 October 1953. In the foreground are Pat Uebel (34) and Tommy Bell (46). (Photo courtesy USMA Archives.)

of Lowell Sisson. Six minutes later Army was on the attack again, when the Lions' Mercier made his 42-yard run with Pete Vann's intercepted pass.

Then right after the Lion kickoff came Army's lightning, three play scoring drive that covered 73 yards. From the Army 27, Pat Uebel gained two to the 29. Then Pete Vann passed to Lowell Sisson at the 44, who shook off defender Dick Carr and continued toward the Columbia goal line. He got past Mercier at the Columbia 25, but was finally caught from behind with a leaping, flying tackle by John Nelson on the Lions' 6-yard line. On the next play Pat Uebel went into the end zone for his second touchdown of the day.

Still in third quarter action, the cadets pushed the score to 34-7 on a fourth down, 8-yard pass to Bob Mischak. Vann again faked handoffs beautifully, twice this time, then hiding the ball on his hip, dropped back calmly and threw to Mischak in the end zone. Freddie Attaya's attempted point after was blocked by Columbia's Max Pirner.

The final Army score came at 11:13 of the fourth quarter when Bill Cody, substituting for Pete Vann, pitched out to Mike Zeigler. The play originated inside the Columbia 10-yard line, and Mike, aided by a key block from Don Holleder, roared around the end for the touchdown. Howard Glock attempted the point after, but the Lions' Mercier blocked the place kick.

Army made nineteen first downs to Columbia's seven. The cadets completely overwhelmed the Lions on the ground and through the air, out-rushing them 270 yards to 46, and passing for 164 yards to 36. Pete Vann threw 18 passes and completed nine for the 164 yards, with the one interception. Lou Little's Lions threw eighteen and completed three, a reflection of the cadets' hard charging defensive line and a pass defense becoming more effective each week.

It had been another grand weekend at Michie Stadium.

THE GREEN WAVE IN NEW ORLEANS

A full stadium and Army's resounding victory over Columbia brought the cadets to comparatively new, giddy heights when contrasted with early season progress in the tough '51 and '52 campaigns. Earl Blaik now had a problem he hadn't seen since the fall of 1950, a team coming off three successive wins, one against a nationally ranked power. He would need to reach into his deep reservoir of coaching skills and call up techniques and strategies he'd almost forgotten.

In New Orleans' Sugar Bowl stadium the next week Army faced the Tulane University Green Wave before 38,000 homecoming fans, who hadn't seen a win yet this season. They had lost five straight and were 14-point underdogs by game time. Their record was deceptive because Blaik knew Tulane's Coach Bear Wolf probably better prepared his team to play using the new limited substitution rules than any team the cadets would face all season.

The Tulane coach had indeed prepared his team well for the rule changes, working them hard on both offense and defense during the entire 1952 season. They were gearing for a fight after five straight losses. Blaik knew his now-confident Black Knights were perfect targets for an upset. To make matters worse it was a game away from home, as was the Northwestern game, with no Cadet Corps to cheer the team on as they had the previous three encounters. If the Corps' support really did equate to two touchdowns as one of Blaik's assistants opined early that season, then the Black Knights were going in to New Orleans with a handicap. Further, Earl Blaik knew how and why Tulane was on Army's schedule. He regarded the Green Wave as a latent football power.

In 1947, when Maxwell Taylor was superintendent and the Academy was again going through curriculum reforms following a world war, Taylor saw the need to develop an athletic policy with respect to scheduling football games. Two years earlier, Army, with their great Davis-Blanchard teams, surged back into prominence as a national champion and a major, crowd drawing independent. What's more, as a result of the team's drawing power, and the people who appointed cadet candidates to the Academy, the superintendent and Earl Blaik were facing increasing pressure to schedule against regional national football powers all across the country.

Powerful, influential people, who were the growing number of congressmen and senators involved in appointing recruited athletes to West Point and Annapolis, under the category II law which had been passed in 1943, were lobbying both Blaik and Taylor to bring Army to their home states to play their favorite teams. Governors who might have recommended recruited service academy players, or had political stakes in their states' economic health – and they all did – would occasionally join the chorus, pressing for Army or Navy to play their favorite teams in their states.

The demands for Army's presence on gridirons far from West Point was a matter of increasing concern to both Blaik and Taylor, senior officers in the Army staff, as well as Academy graduates. Both Blaik and Taylor were thoroughly familiar with the Academy's high standards and the resulting demands on Academy athletes' time and abilities. They also knew of the relatively small cadet population when compared with growing, postwar student populations at nearly all of Army's opponents, and believed Army players couldn't keep up with a grinding week in and week out schedule against major collegiate powers.

Blaik felt his teams could compete successfully against a major collegiate team about once every three weeks. What's more, he was convinced Army should play five games at home on a nine game schedule, preferably alternating weeks at Michie Stadium so the team would better be able to cope with academic demands. Taylor concluded "at least three, preferably four, games will be played at Michie Stadium each year."

Taylor, perhaps more amenable to the pressures being brought to bear, and wanting to draw the best cadet candidates possible to the Academy, also concluded Army should play "six or seven first-class opponents" on a nine or ten game schedule. "About four were to be Ivy League teams and two would be with intersectional opponents." He undoubtedly was also influenced by the fact that during and immediately after the war years, the Army football team's success had been a major factor in capturing the interest and attention of impressionable and talented boys who might be potential cadet candidates. Taylor prevailed in the discussions between him, Blaik, and the Board of Athletics, and signed a scheduling policy that would guide Blaik, the Association Board of Athletics and the Department of Athletics for years. Since scheduling was normally done ten years in advance, the policy formalized what was essentially in effect at the time, and would remain in effect throughout the rebuilding years following the cheating scandal of 1951.

Blaik's pre-1951 problem of preparing Army against dragons one week, and supposedly lesser opponents the next week, was nowhere near as easy as someone unfamiliar with Academy football might suspect. The natural tendency of players to let down after a big win is trouble enough. But the problem is exacerbated when the players know, or think they know, they have "a breather" next week.

Earl Blaik and his staff had to be masters in preparing their teams psychologically for that kind of schedule. The coaches knew top teams in the country, as well as winning teams anywhere, were invariably targets for opponents aching to turn around faltering seasons or earn recognition and respect with huge upsets against the likes of a Notre Dame, a Southern California, or an Army. And while the cadets didn't schedule a dragon to slay every Saturday, the scheduling policy could also set Army up for just exactly the kind of upsets the "easier opponents" would love to claim.

Blaik was worried about Tulane, and he had ample reasons the Green Wave would confirm the coming Saturday. Neither Earl Blaik's disciplined pregame team preparation nor his commanding, penetrating presence could wring a victory from the Black Knights that Halloween afternoon. In a pregame team meeting in the lobby of their hotel, the cadets witnessed his presence at work, and for cadet John Krause, a senior tackle on the team, the incident left an indelible impression.

Blaik was on his feet, talking to the players seated around him, when an obviously intoxicated, noisy, disruptive all-night reveler strolled into the lobby, heading toward the congregation of Army football players. The drunk was smiling the whole time he continued his chatty walk toward Blaik's team meeting. The Army coach, annoyed by the disruption, abruptly stopped talking, fixed his dark, penetrating eyes on the man and glowered as the stranger walked unsteadily toward the gathering. The noisy interrupter finally made eye contact with Blaik and met the coach's riveting stare. The man stopped. Blaik continued glowering at him, plainly displeased by conduct affecting the Army team meeting. The intoxicated smile faded and the slurred chatter ceased. The drunk stood transfixed for a few seconds while Blaik stared at him, unblinking. There was dead silence. Then without a sound from the drunk, and not a word from Earl Blaik, the man simply turned around and walked soberly out of the lobby. The incident, though telling with respect to Earl Blaik's powerful personality, wasn't enough to overcome what awaited the team that afternoon.

From the outset Tulane's defense exhibited an aggressiveness and brutal fury that remained undiminished all afternoon. If Army was inspired against Duke, the Green Wave's homecoming appearance and pent-up frustrations stirred retaliation in kind.

Intensely hard hitting by both teams marked the game's play, particularly on defense. Army fumbled six times and lost five. Tulane fumbled three times and lost all three. Four of the game's fumbles came in the first five plays from scrimmage. Army's fumbles were costly – as were the penalties against the cadets. Though each team received penalties totaling an identical 35 yards, a first quarter cadet offside resulted in the referees' calling back the only touchdown of the entire game.

Army's offense managed only one major threat all afternoon, which came in the first quarter when Pete Vann's passing and Tommy Bell's running paced a drive from the cadet 33-yard line to the Tulane 15. On the next play, Pete threw a perfect strike to Bob Mischak in the end zone, but Army was offside. The cadets started to move again, but two plays later Pete fumbled on the Tulane 10 and the Green Wave's right end, Eddie Bravo, recovered.

The Tulane center and defensive linebacker, Paul Rushing, was all over the field on defense, accounting for half the Green Wave's tackles. Left guard Tony Sardisco's aggressive charges into the Army backfield repeatedly frustrated the cadet ground game, causing a heavier reliance on the passing attack than Blaik and his team liked.

Late in the third quarter, Pete Clement, Tulane's quarterback, sparked a drive from their own 33-yard line, first alternately hitting end Eddie Bravo and then his left halfback with passes, moving to the Army 43 as the period ended. When fourth quarter play resumed, Clement kept the drive moving, pushing all the way to the cadet 8-yard line. Matters looked bleak, but Army's defense stiffened, and Tulane, facing a fourth down, attempted a field goal. Bob Mischak, for the second time in the season, saved an Army game.

The cadet left end and second fastest man on the team was a thorn in Tulane's side all afternoon, and on this game deciding play, came with everything he had into the Green Wave backfield. Completely leaving his feet in a lunging dive, he blocked the attempted field goal, and Army took over on downs. Tulane had a slim chance to score another time during the game, but Army tackle and defensive linebacker Bob Farris intercepted a Clement pass on the cadet 15.

The game ended with Pete Vann attempting to hook up with Army receivers on long passes from the Tulane 40-yard line. The score was a frustrating 0-0 tie.

While the outcome of Army's foray into New Orleans on Halloween was clearly disappointing, there were bright spots in the cadets' performance. Though statistics aren't what decide victory or defeat on the gridiron, and certainly, as far as Blaik and his staff were concerned, were of no importance when compared with the final score, Army had won the statistical battle against Tulane – in spite of the fury of the Green Wave's play.

The cadets gained 170 yards rushing to Tulane's 133. In passing Army connected on 16 of 24 attempts for 135 yards, with one interception. The Green Wave threw four completions in six attempts, with one interception, for a total of 77 yards. The differences in offensive performance gave the cadets a 15-11 edge in first downs.

Perhaps the worst part of the entire afternoon had come when fullback Freddie Attaya, playing in his Louisiana bayou home country, fell victim to the game's hard hitting and suffered a fractured ankle that would end his playing career at Army. Freddie had emerged a bright, shining performer for the Black Knights, a confident, flashing, quick starting ball carrier, a solid blocker and defensive halfback, and a superb punter whose quick kicks had rocked opponents back into their own territory repeatedly in the first half of the season. Blaik had become so confident in Freddie's heady thinking and punting skills, he had privately called him aside and authorized him to quick-kick without it being called from the bench. Freddie Attaya's loss to the cadets was almost as discouraging as the 0-0 tie with Tulane.

Army had escaped with a tie, with a determined, previously unappreciated, inspired opponent, and relearned some old, painful lessons. But this was Army's year. No time for looking back at disappointments or dwelling on player injuries, except for what might be learned from mistakes – to better prepare for the next opponent. Two more games, one at a time, then Navy. Back home to Michie Stadium for the next one.

THE WOLFPACK INVADES MICHIE STADIUM

It was gray, wet, and cold, an altogether miserable day at West Point on November 7, 1953, when the North Carolina State Wolfpack came to Michie Stadium. The 27,000 capacity stadium was sold out a week earlier when a warm, fall New York sun was shining. All was in readiness for the sellout. But 16,000 "sales" were for the annual visit of the Boy Scouts of America to Michie Stadium. The young boys always thrilled to the sight of Army and its collegiate opponents playing in one of the most historic and picturesque settings in the nation.

The victories over Dartmouth, Duke, and Columbia had obviously fired Army football fans with renewed interest. Everyone likes a winner, and Earl Blaik and his coaching staff had already performed a major miracle with the win over Duke and the apparent makings of a winning season for the first time since 1950.

But the weather was a real damper, perhaps augmented by the disappointing tie in New Orleans. Approximately 4,000 Boy Scouts braved the gloomy weather, making the grand total in attendance near 9,500. And on the sidelines sat an injured, disconsolate Freddie Attaya, nevertheless cheering on his teammates.

The field had been covered with a tarpaulin early Friday afternoon to protect against the forecasted cold Hudson Valley rains. This normal precaution against rainy weather was an effective hedge against a damaging quagmire in Michie's grass. A badly damaged turf could occur easily if 22 large men began the game playing on a thoroughly soaked field, and hammered away at one another for sixty minutes. As matters turned out, the tarp didn't keep uniforms from getting wet and muddy quite early in the game, although the field was still dry and fast when play began.

North Carolina State came to Army having lost five games in six starts, not a record to excite sports writers and broadcasters. Earl Blaik, perhaps already looking forward to Navy, and smarting with the loss of Freddie Attaya from a team already lacking depth in key positions, decided to start his second unit, with Pete Vann at quarterback. The decision had an immediate pay-off.

The pregame fight songs, yells, bugles, drums, and thunder of GO! GO! GO!" once again echoed through the air high above the Hudson River, prior to the opening kickoff by North Carolina State. The game's beginning was electric, with the cadets running the kickoff back to their own 41-yard line. Excellent field position. On the first play from scrimmage, Pete Vann faked a hand-off, momentarily freezing the Wolfpack defensive secondary, faded back, and lofted a 39-yard pass to yearling Don Holleder, Lowell Sisson's alternate at right end. Holleder caught the perfectly thrown pass over his shoulder on the dead run at the Wolfpack 20-yard line, and raced into the end zone.

The Army victory cannon boomed mightily and gray smoke belched into the cold, damp air, while a leaping, shouting Corps of Cadets reminded their "Rabble" they were back home in Michie Stadium. Kirk Cockrell kicked the extra point and Army led 7-0. It seemed Blaik made another brilliant move, giving his first unit a rest while providing his second unit much needed experience and confidence, as well as an opportunity to show their first unit teammates they could hold their own. Then at 7:15 of the first quarter, the cadets' second unit fortunes abruptly reversed. The turnaround came following the Army kickoff and an exchange of punts.

Eddie West, the Wolfpack quarterback, who, on defense, also returned their opponents' punts, was back to receive Mike Zeigler's kick from the Army 30. West gathered the ball in at his own 32, slanted through the onrushing cadet punt coverage, shook off two would-be tacklers, and galloped 68 yards for a touchdown. Al D'Angelo, the Wolfpack's right guard, kicked the extra point and the game was all tied up, 7-7.

Earl Blaik promptly put his first unit back in, leaving Pete Vann in to direct the offense. As the *New York Times* reported that day, "...Pete Vann's passing, running, and team direction were superb," and when Pat Uebel, Gerry Lodge, and Tommy Bell entered the contest, their "running was brilliant." The Wolfpack, who came north determined to stop their season's slide, were inspired by the momentary 7-7 tie, and kept up a sturdy defense, trying to hold on. They lasted until just after midway in the second quarter, when Army's patience and persistence paid off.

At 8:46 of the second period Army got its second touchdown, after a long drive on the ground. Gerry Lodge's 33-yard dash around Army's left end, to State's 21 was the spark in the drive. His fullback run, like a hard-driving halfback, was also an indicator of things to come for this big, tough high school fullback and intercollegiate wrestler. Converted to guard, and now back to fullback, he was filling the badly missed shoes of Freddie Attaya. Pat Uebel and Tommy Bell moved the ball to the State 2-yard line behind excellent blocking, and Gerry Lodge rammed over from the two. Ralph Chesnauskas booted the extra point, and it was Army 14-7. But it wasn't over yet.

The third quarter remained scoreless although the cadets continued to roll up the yardage. Twice the Wolfpack had chances to even the score with passes thrown by Eddie West to receivers inside the cadet 5-yard line. Twice the State receivers dropped his passes.

Then, in the fourth quarter, the cadet offense proved unstoppable, both on the ground and in the air. Pete Vann connected twice through the air with Mike Zeigler for long gains, and Gerry Lodge burst over a guard for his second touchdown. The extra point try failed and it was Army, 20-7.

Hard-driving Army fullback Gerry Lodge being brought down by North Caro lina State defenders. Michie Stadium, November 7, 1953. (Photo courtesy USMA Archives.)

The cadets began to move again in the fourth quarter, spurred on by the thunder of "GO!— GO!— GO!," alternately passing and running, paced by Zeigler, Lodge, and Vann. Army moved the ball to State's 2-yard line, and Pete Vann sneaked in for the final touchdown with five minutes left to play. Once more the victory cannon's loud boom. Chesnauskas calmly place kicked the extra point, and that's where the scoring ended, 27-7.

Although the cadets didn't win a noteworthy, high scoring victory against North Carolina State, the Wolfpack defeat had been worse than the score indicated. The cadets made 21 first downs to State's 5, and amassed 346 yards rushing while the cadets' hard charging defense held the Wolfpack to a -20 yards. Pete Vann completed 5 of 11 passes for 112 yards, with one interception, while Eddie West completed 5 of 12 for 88 yards, with one interception.

The Wolfpack handicapped themselves with 91 yards in penalties, while Army, still smarting from their lapses at Tulane, gave up only 15 yards. Nevertheless, Army fumbles again frustrated their offense with four turnovers, while State lost only one fumble. There were other positives for the cadets, however.

In addition to the fire, determination, cohesiveness, and teamwork now evident in the Black Knights, and between the football team and the Corps of Cadets, there were other strong factors emerging to push Army toward one victory after another. Pete Vann had become a magician with his ball handling and faking. More and more as the season wore on, Pete, Gerry Lodge, Tommy Bell, and Pat Uebel were confusing opponents' defenders with masterful faking. This, in combination with quick-off-the-ball, hard charging, aggressive blocking on the line, was punching gaping holes in defensive lines, while Army's ends more frequently broke free of pass defenders to take advantage of Pete Vann's passing. Vince Lombardi's backfield-end combinations were taking Army's offensive game to a high art form, while the cadets' defensive play was spurred by a close-knit camaraderie developing among men playing both offense and defense. The defense, too, had become devastatingly effective.

The cadets' record stood at 5-1-1, already a winning season for the first time since '50. But the next game, the last one before Navy, would not be easy. The University of Pennsylvania, ever a tough, hard playing opponent, stood between Army and the opponent always lurking in the minds of Earl Blaik, his coaching staff, and nearly every Army football team since 1890 – Navy.

QUAKERS – READY AND WAITING

HQ, USCC
WEST POINT, NY
060800 Nov 53

OPERATIONS ORDER 2
ORGANIZATION (Effective 140735 Nov 53)

    a. Commandant: Brig Gen JH Michaelis
CO of Troops: Col JA McChristian
S3: Maj LJ Flanagan
Tac Off A Prov Co: Lt Col GH Mueller
   "   " B "   " : Capt WL Cooper
   "   " C "   " : Maj FB Gervais
   "   " D "   " : Lt Col JS Timothy
Trans Off: Lt Col Herberth, Capt JP Ross
Security Off: Maj W Hausman

    b.   Class of 1954, USCC (Annex #1)

    c.   USMA Band: Maj FE Resta (CO)

1. SITUATION. The Class of 1954, USCC, plus attached troops, will depart WEST POINT, NEW YORK for FRANKLIN FIELD, PHILADELPHIA, PENNSYLVANIA at 140805 Nov 53 to attend the ARMY-UNIVERSITY OF PENNSYLVANIA Football Game.

2. MISSION. To participate in pregame ceremonies and to support the ARMY TEAM in the ARMY-PENNSYLVANIA Football Game....

So read the opening paragraphs of the operations order sending the class of 1954 on a train to Philadelphia the morning of November 14, 1953. The class formed a provisional battalion of four companies, the battalion commanded by the First Captain and Cadet Brigade Commander, John C. Bard. A and B Companies were first classmen from all the permanent 1st Regiment's twelve companies. C and D Provisional Companies were from the Corps' 2d Regiment. The under three classes stayed home. Two weeks hence, the class of '54 would entrain again with the entire Corps and return to Philadelphia for the climactic game of every Army season, against the men from Annapolis.

The class formed up by 7:40 in the morning and began boarding the train for departure five minutes after eight, to arrive at Philadelphia's 30th Street Station, track number one, at 11:40. The Academy Band, with all its instruments in the baggage half of car number one, occupied cars one, two, and three. Guests and Academy officers were in car four, and the class of '54 filled cars five through ten.

When the train rolled into the station right on schedule, the Academy Band and four provisional companies, with their battalion and company staffs, assembled on Arch Street on the station's north side, with the band in the lead at the west end of the column. There was the usual twelve paces between company commanders and the rear rank of the company immediately in front of them. A few moments before 12:15 the band struck up and the column began to move toward the intersection of Arch and 30th Streets, the initial point (IP) for the march to Franklin Field. Turning left onto 30th Street the column marched south, turning right onto Walnut Street, thence west to 33rd. At 33rd the column turned left again resuming its southward move toward Franklin Field.

Precisely at 12:40 in the afternoon, the Academy Band struck up the *Official West Point March* as it turned left and marched into the stadium's southwest entrance. East they came, just outside the south sideline of the playing field, while the gathering crowd that would grow to 47,305 clapped and cheered their approval of the unfolding spectacle. Left, around the east end zone and goal post the column turned, then left again down the north sideline, before the band and all four companies of nearly 150 men each turned left onto the field, one at a time, and aligned themselves flank to flank, ten yards apart. They halted with their front ranks in the middle of the field facing the Pennsylvania fans, and stood motionless, at attention. The band ceased playing.

Then, as the entire Corps had done for Duke fans in the Polo Grounds, led by cheerleaders giving silent, visual commands, the class of '54 rendered a hand salute and gave a "Rocket" yell for University of Pennsylvania fans. After an about face and salute to Army, they removed their caps, and bareheaded, gave the "Long Corps" yell before replacing their caps and double timing into the north stands. Kickoff was at 1:30.

* * *

George Munger, the Pennsylvania Quakers' respected coach of sixteen years had announced prior to their fall campaign that this would be his farewell season at his alma mater. The Quakers, though hamstrung by recent Ivy League rule changes prohibiting spring practice, had had to continue playing a difficult, annual schedule, long ago set by contracts negotiated with their opponents. The prior week, they tangled with perennial power Notre Dame and played the Fighting Irish hard, though losing. The loss increased the possibility Munger would have his only losing season in his last year at Penn. But Earl Blaik and his coaches – and the Army team – knew those factors meant nothing, except the Red and Blue would only fight harder to win. And as if to confirm Blaik's convictions, on the strength of the Quakers' tough schedule and Army's still suspect, premature return to gridiron fame, the cadets came to Philadelphia four point underdogs. There wasn't any doubt that George Munger's team would once more fight the cadets to the bitter end. Army and Penn specialized in thrillers, and they staged another one this day.

Army won the coin toss and elected to kick off. On the opening offensive series Penn began driving and moved the ball to the Army 27. There the drive stalled with three straight incomplete passes. After the Pennsylvanians' initial thrust, neither team could make much headway due to hard nosed defensive play, with the cadets penetrating no further than Penn's 47-yard line before the first big break came.

Right end Lowell Sisson, now doing Army's punting in place of injured Freddie Attaya, booted to the Quakers' Ken Smith, who fumbled. The cadets' center and defensive signal caller, Norm Stephen, pounced on the ball at the Penn 20-yard line. Six running plays moved the ball to the 2-yard line, where Army was faced with a fourth down. On the next play Gerry Lodge bulled his way over left tackle for the score, and Ralph Chesnauskas kicked the first of his two extra points for the day. The cadets were on top 7-0 with 12:52 gone in the first quarter.

Another break, this time for Penn, deadlocked the game at 5:26 into the second quarter. Jim Castle, the Quaker right end retreated off the defensive line and made the lone interception of Pete Vann's passes that afternoon, and headed for the Army goal line. Pete made a heroic tackle, running through a block thrown by Jack Shanafelt to bring down Castle at the Army 8-yard line.

But Penn wasn't to be denied. Joe Varaitis, their big 205-pound workhorse fullback, rammed the Army line for two yards, and left halfback Walt Hynoski cracked into the line for two more. On the next play Penn's routine, well known shift from the T-formation to the old power football, single wing, drew the overanxious cadets offside, resulting in a costly penalty. The ball was at Army's 1. Varaitis slammed over for the touchdown and Penn quarterback Ed Gramigna added the first of his two extra points for the day.

The cadets answered at 11:12 in the second quarter with a 49-yard touchdown drive in six running plays. The spark was right halfback Tommy Bell's dash around Penn's right end, for 32 yards, all the way to Penn's 5. Then came the "belly series" again, twice.

Tommy Bell in action against the University of Pennsylvania Quakers, Franklin Field, Philadelphia, November 14, 1953. The Military Academy Class of 1954 in the stands. (Photo courtesy USMA Archives.)

New York Times sports writer Joseph M. Sheehan called it the "ride 'em" play. As the play had done against North Carolina State, the near-perfect faking usually caused the linebacker, guard, and tackle to commit on the Army fullback, and at the same time either momentarily freeze the defensive end in place, or cause him to also converge down the defensive line toward the fullback. In an instant, the crossing Army halfback would be in the opponents' secondary, often untouched as he leaned forward and rushed low, through the line of scrimmage.

Army ran the play twice in succession, first over the right side, and then to the left. Penn's defense was baffled by Pete Vann's ball handling, and the entire backfield's disciplined faking. The Quakers swarmed on Gerry Lodge to gang-tackle him at the line of scrimmage, while Pat Uebel slanted wide, untouched into Penn's end zone for Army's second touchdown. Chesnauskas converted. Army 14-7.

Penn came right back as though they were going to tie it up again before the half. They took the kickoff and rolled to the cadets' 15 by completing six of eight passes. Then Quaker quarterback Ed Gramigna and left halfback Bob Felver lost the range in the Red and Blue's passing attack, missing on four straight throws.

Army took the ball and couldn't move, nearly resulting in another Penn score. Lowell Sisson, back to punt, received an errant, high snap from center, causing him to be tackled on the Army 10-yard marker rather than getting off the kick. But Penn had time for only one play, and Gerry Lodge stormed through to throw tailback Bob Felver for a huge loss as he was attempting another pass.

In the third quarter Penn picked up where they left off, continuing on the attack, mostly on the passing of Gramigna. They advanced from their 32 to a first down on Army's 20. Hynoski then broke through for what appeared to be a substantial gain, but a hard defensive hit knocked the ball loose right into the hands of Pete Vann, and Army took over again, on their own 12-yard line.

Soon after, the Red and Blue began driving again, and this time they wouldn't be stopped, moving 67 yards in ten plays for the score. The big gainer was a 44-yard pass play from left halfback Walt Hynoski to right halfback Gar Scott, which carried to the Army 18. Penn's progress was slower afterward, but a fourth down pass completion on the last play of the third quarter kept the drive moving. At 1:14 into the final period the Quaker's powerful fullback Varaitis, did the trick, bulling in for the score on his third successive try at the Army line. Gramigna kicked the tying extra point, 14-14.

The cadets roared back to score the final go ahead touchdown following Penn's kickoff. Pat Uebel picked up five yards, and a Quaker offside penalty gave Army a first down on its own 35. Lodge picked up two, and after Pete Vann threw one incomplete pass, he hit Tommy Bell for an 18-yard gain. Tommy picked up three more yards on the ground. Then Pete Vann missed on another pass before he hit Tommy again in the deep right flat, for 27 yards to Penn's 15. Gerry Lodge rammed through to the Quaker 9-yard line. Then, one more time came the "belly series," the "ride 'em," over Penn's left side, and Pat Uebel rocketed into the end zone for his second touchdown of the day. This time Howard Glock kicked the extra point, and Army was ahead 21-14, with 10:32 left to play – still plenty of time for a determined Penn team to come back. It was not to be.

Twice the Red and Blue attempted to mount drives after Army took the lead, and twice the cadets thwarted them with pass interceptions. Norm Stephen intercepted Gramigna to stop the first threat at the cadet 23. When the Quakers got the ball one more time, Pat Uebel stopped the drive before it got started, intercepting another Gramigna aerial.

With another mighty roar of the Army victory cannon, ending the game, the jubilant class of 1954 shouted their approval with voices worn hoarse from shouts, cheers, yells, fight songs, and another thunderous afternoon of "GO! GO! GO!" As for the Army team, it had been another splendid Saturday of victory against a favored, furiously battling opponent.

One of the most astounding statistics to come from the game was fifteen. A grand total of fifteen Army players fought the good fight against George Munger's Quakers that day. Army fielded only four substitutes the entire game: left tackle Joe Lapchick, Jr., center Paul Lasley, guard and team Captain Leroy Lunn, and right end Don Holleder. Every man in the Army backfield and three men on the Army line played sixty minutes of relentless, pounding, "iron man football." In contrast, Coach George Munger kept only right tackle Jack Shanafelt and right guard Jack Cannon in the full sixty minutes, a reflection of Penn's greater depth and ability to more freely substitute.

Both Army's offensive and defensive work was sterling. Penn had consistently put up seven and eight-man lines against the cadet attack, and yet the cadets kept their ground game moving, balancing it with an effective passing attack. Then, at crucial moments they bewildered the Quakers with brilliant faking in the "belly series." On defense, Army lined up in their 5-4-2 alignment against Penn's T-formation, but met the Quakers' "Notre Dame shift," by reshuffling quickly to meet single wing power with a 6-2-2-1.

Statistically, the two teams fought almost even, with Army making 13 first downs to 12 for Penn, and battling for 249 total yards to the Quakers' 242. Earl Blaik's old standby, a hard-hitting ground game, chewed up 172 yards against a strong Penn defense, while the Quakers could manage only 82 rushing yards. A week earlier Penn had managed 131 rushing yards through Notre Dame's vaunted line. Against Army the frustrated Quaker running attack caused Pennsylvania to rely more on the pass, and they threw 27 times, connecting on 11 for 160 yards, with two interceptions. Pete Vann hit on 6 of 14 throws for 77 yards. The Penn defense intercepted one.

Twice Army scored to take the lead and twice the Quakers fought back to tie the game, first at 7-7, then 14-14.

There was another cause for celebration in the Army dressing room that day. Pete Vann.

When Army upended Duke four weeks earlier, Pete showed he was no longer "a mere boy." Not one to lavish compliments, Earl Blaik had faintly praised his play against the Blue Devils. But this day against Penn's Red and Blue, Pete was outstanding on both defense and offense. Blaik noted "He made several clever key tackles, two preventing touchdowns." His field direction and offensive play were equal to his defensive work.

Another little noticed factor in Army's return to "iron man football" was Pete's play selection. Though Blaik developed a game plan for each Saturday, and Pete obviously received masterful coaching from both Lombardi and Blaik, limited substitution also limited play calling from the bench. The Black Knights' small band of fifteen reflected the consistent lack of depth Army's opponents enjoyed, and as a result the Army coach had to rely far more heavily on Pete's field generalship than opponents' coaches needed to with their quarterbacks. Vann's play selection was rapidly improving as the season wore on.

In the fourth quarter against the Quakers he mingled sharp passing to Tommy Bell with alternating smashes at the line by Lodge and Uebel, as Army drove toward the winning touchdown. He then capitalized on the backfield's slick faking to set up Uebel's score with "the 54 play" – one of the play numbers Army used for the "belly series." After the Penn game, Blaik "evoked stentorian cheers [in the Army dressing room] by announcing, 'Today, Peter Vann became a man!'"

The magnificent, unheralded performance by an Army team shrinking in depth with the season's every passing week, set the stage for the game the class of 1954, and the Corps of Cadets had waited for since 1950, the year Army's "golden era of football" had come to its tragic end. The men who came to West Point the summer the Korean War began had never seen a victory over Navy. They had boldly announced in September, "This is Army's year," and laid all kinds of nefarious plans and schemes to bring an end to Navy's string of three victories. Now, from Philadelphia's Franklin Field it was back to West Point for two weeks before the entire Corps would return to "Philly's" Municipal Stadium, and what was shaping up to be another of the year's best football games. Earl Blaik, Vince Lombardi, the entire Army coaching staff – and the Black Knights – couldn't wait to get at Navy.

In the remaining two weeks there were important plans and preparations to make, on the practice fields, in the confines of the cadet area, and in unusual places cadets seldom frequented. There were twenty-four hundred minds always busy, stoking the fires of victory. Among them were some who were always busier than others – and had been for a long time.

A SHAGGY GOAT STORY

The Army-Navy football game, a renewal of one of the great football rivalries in the land was two weeks hence. But this one would be far different because the class of '54 had willed it so. A small group of determined Rabble rousers were at center of the scheming. Several members of '55 were deeply involved in the activities, supporting '54's plans. There had been intelligence gathering since April, contingencies postulated for all manner of situations, and numerous rehearsals by several alternate teams. The idea was train for the future. Instead of scoring a first, the men of '54 wanted to start another tradition, one even better than the victory cannon.

The Pointer magazine enunciated the theme in September – "This is Army's year." The manufacture and rollout of the "victory cannon" was complete. Its announcing blast, preceded by the relentless, deafening chant of "GO! GO! GO!," punctuated the stunning 14-13 upset of Duke, thundered and shook stadiums following each score – and during after-game victory celebrations. Now, intelligence-gathering trips to Annapolis, contingency planning, and rehearsals must cease. Time to deliver on another bold venture.

Earlier in the 1952 football season, the University of Maryland had pulled off a similar caper to the one now being planned, before Bill Schulz and his enterprising band of cadets successfully painted "Go Army! Beat Navy!" on the side of the Navy destroyer escort docked on the Hudson River at West Point. An energetic band of ten Maryland University students tried again to "borrow" Navy's goat on November 10, 1953, at the behest of an unknown cadet who wrote a Maryland accomplice asking that the job be done in time for the Army-Navy game. The attempt was foiled by a Naval Academy security patrol which nabbed two of the Maryland students in the act of breaking the lock on the goat pen.

In 1952, Bill Schulz's artful, nighttime destroyer escort painters returned to cadet barracks without getting caught. The ship's crew nabbed "Tiny" Tomsen and his early morning photography accomplice as they tried to record the artwork for posterity – and gave them the humiliating task of painting over those grand, immortal words of motivation. A light slap on the wrist had been "Tiny's" reward. Seven demerits – but no "slug" and no "bust."

Now, Sunday, November 22, 1953, in the cold, dark, and fog of the wee morning hours in Annapolis, Maryland, payback time was to begin. A heist. The target: Navy's beloved mascot, the long-haired, Angora goat, Billy XII. THE GAME was less than a week away. The plan was to present Billy XII to the Army football team and the Corps of Cadets in time for the final days of football practice. At least that was the plan one small team of cadets had in mind.

A few days prior to that fateful night, Ben Schemmer, one of the heist advocates and leaders, received an unusual summons to the Commandant's office. Inviting Schemmer to stand at ease, Brigadier General John H. Michaelis said he was worried Corps spirit seemed to be peaking too soon before the game. At the time Ben didn't know Ed Moses, the head cheerleader, had received a similar admonishment from a tactical officer, chastising Moses for "not managing the winning psychology correctly." The Commandant similarly wanted to know from Ben how the cheerleaders and their operations group were going to keep spirits up. Ben responded, "Sir, don't worry, we have it all planned." The answer didn't satisfy Michaelis. He wanted specifics. Ben said, "Respectfully, Sir, you don't want to know." The Commandant insisted he did. Schemmer pleaded, "Trust me, Sir, it's best you don't know."

"Mr. Schemmer, I'm ordering you to tell me."

Gulping, Ben told Michaelis they were going to snatch – that is, borrow – the goat over the coming weekend.

Obviously pleased, the Commandant seemed to excuse the larcenous nature of the cadet plan and said in his best, controlled military manner, "Are you sure you can pull it off?" Ben assured him they had no doubt whatsoever. Then Michaelis cautioned, "If you get caught, you know you're going to hang, don't you?" – the same, encouraging words spies and intelligence operatives receive when sent on dangerous or potentially embarrassing missions.

Once Michaelis heard of the plan, he gave Ben an order. When the goat was in the cadets' possession, headed toward West Point, Ben was to telephone Major George Pappas and tell him the goat was en route. "Major Pappas will know what to do.... This is just between you and me," the Commandant added. Ben Schemmer knew Pappas was a major in the Public Information Office, but wasn't well acquainted with him.

Schemmer was puzzled by the Commandant's direction; nevertheless, he assumed Pappas would be told to prepare publicity for the event.

Ben Schemmer and his accomplices had planned in great detail the "borrowing" of Billy XII, including where on the Military Academy reservation the animal would be hidden.

Luck proved the heist team's biggest ally. Jan Le Croy, a mule rider in the class of '54 and L-2 company commander, had just spent an exchange weekend in Annapolis. He decided to go "over the wall" one night – take an unauthorized absence from the midshipmen's dormitory, Bancroft Hall – and misjudged where to reenter the Naval Academy grounds. He found himself underneath the Thompson [football] Stadium bleachers en route to the dormitory, and as good fortune would have it, came upon an occupied goat pen. This was the last-minute intelligence the cadet goatnapping teams needed.

The teams drew straws to see who'd do the honors. Alex Rupp, a cheerleader from Company A-2 and the class of 1955, and Ben Schemmer won the draw. Rupp had been in two rehearsals for the grab. They invited Le Croy, but he demurred. Unknown to Ben and Alex however, another group in '54 had similar designs on Billy XII.

Scotty Wetzel, class of '54 from Company F-2, was on the cadet Public Information Detail with Jan Le Croy. The Public Information Detail was under supervision of the Office of Public Information. The cadet detail's Officer-in-Charge, Major George Pappas, was a public information officer and had been a project officer for the 1952 Sesquicentennial Celebration. Scotty Wetzel was also a member of the cadet Rally Band, a group of seemingly undisciplined, spirited football enthusiasts who performed during halftime intermissions at home games, and at cadet football rallies prior to games. During their happy, "hammed-up" performances, they wore cadet bathrobes and all manner of comical, unmilitary, military uniforms.

Wetzel, it turned out, had a different plan for Billy XII than the one Ben Schemmer and Alex Rupp concocted. He also had a different heist team. Four other cadets were to be part of Scotty's team. One was Al Lieber, Company M-2, who had staked out a "safe house" to possibly accommodate the prisoner. The "safe house" was a relative's farm near Annapolis, where Billy could be kept the entire week before the game. Scotty had worked out the details of his plan after learning of Mary-

876

land University's successful kidnapping of the Navy mascot prior to the 1952 Maryland-Navy game.

Scotty's idea was: once his team made the grab, and the goat's absence was noticed by Naval Academy officials, the finger would immediately be pointed at West Point's cadets. *Give 'im back* the middies would say. Whereupon Scotty and his crew could reply, with their crossed fingers and perfectly straight faces: *What goat? We don't have your goat.* Then with great fanfare Wetzel's heist crew would parade Billy XII into Philadelphia's Municipal Stadium before 100,000 cheering fans, just before the game.

Not so for Ben Schemmer and Alex Rupp. Ben's plan was more elaborate. Bring the goat back to West Point and exhibit him to the Corps of Cadets and their football team – the perfect way to build a fiery head of steam against Navy.

But this was war, and war is an art. The more complex the plan, the more can go wrong, and as usual in the art of war, the biggest obstacles to victory are logistical. In Ben's plan there were other unpredictable factors as well. But first things first.

Ben and Alex needed a professional set of burglary tools. Ben enlisted the help of his brother, Fred, a naval reserve officer who had mixed loyalties. The kit included different-size bolt cutters, handcuffs, scaling hooks, hack saws, lock picks, and two small bottles of chloroform. The chloroform, contrary to myths gathered around this event over the years, was to be used in case the team encountered interlopers during the heist – such as the Navy's Shore Patrol, the equivalent of the Army's Military Police. The reason for using chloroform – if necessary – was simple. The punishment would be much less if chloroform were used instead of "cold cocking" the intruders. Eventually, the burglary kit filled a gunny sack and weighed about seventy pounds. Not an easy load to carry, especially if there's a recalcitrant goat accompanying the sack.

The Army mule riders did some engineering math, and calculated Billy XII's horns spanned fifty-one inches. It's difficult to stuff over four feet of curving goat horn through the back door of an automobile. The solution? A convertible. A courageous enlisted man, a corporal from the Academy Band, volunteered to drive his convertible. With its top down and a constant flow of fresh air, it was just the vehicle needed to transport two energetic cadets and a goat with an overactive thyroid.

So unbeknownst to one another, in the early morning hours of Sunday, November 22, two teams converged on the goat pen at Annapolis, both with the same objective, but two different plans. Ben Schemmer, Alex Rupp, and their getaway driver got there first, while Scotty Wetzel and Joseph U. Weaver, a student from the University of Maryland and the brother of a former member of West Point's class of 1956, were rowing a boat down the Severn. Their route was the same as the University's goatnapping crew had used a season earlier – and another crew from Maryland used twelve days earlier.

Scotty's team had shrunk from five to two, and included the Maryland student because academic deficiencies and assigned duties prohibited weekend leaves for the four cadets who first agreed to assist Wetzel. Scotty's Maryland University accomplice was experienced, however, having peripherally taken part in planning the University's goatnapping plot a year earlier. The early morning air was cold, foggy, and pitch black. There was no moon. Perfect conditions, so it seemed.

About two in the morning, Schemmer, Rupp, and their driver parked the convertible near the seawall along the Severn River. Schemmer and Rupp left the car and cut a gaping hole in the chain link fence which was between the stadium and the road, and ended a considerable distance away, at the seawall along the Severn. They believed the hole was large enough for the goat's wide horn spread.

The two cadets had no difficulty finding the goat. They could smell him from a hundred yards away. To get to the twelve foot high wooden fence, which was the animal's pen underneath the grandstands, they had to cross the Thompson Stadium field, Navy's football field. Fortunately, the dark, foggy night more than adequately concealed the men's movements. When the cadets disabled the lock on the gate, and entered, the animal didn't resist. According to Ben, once leashed and out of the pen, Billy seemed eager to reach the hole in the chain link fence. Schemmer was certain the animal wanted to get away from the midshipmen.

The plan seemed to be going extremely well, but the two were in for several surprises. En route back across the football field they were startled by a loud, totally unexpected whisper. "Halt! Who goes there?" The instantaneous, reflexive thought was Shore Patrol! – the Navy's military police. Instead it was Scotty Wetzel and Joe Weaver, who had debarked from their rowboat, and were making their way toward the goat pen. When the four men finally warily approached and recognized one another, they came to a hurried agreement.

Since Schemmer and Rupp were first on the scene, their plan of action prevailed. There would be no "safe house" for the kidnapped goat. It would be the United States Military Academy for Billy XII, and a showing to the entire Corps of Cadets and the Army football team – with great fanfare. Scotty and Joe would temporarily join forces to help Ben Schemmer and Alex Rupp take the goat to the waiting automobile for the trip back to West Point.

When the four arrived at the fence, they discovered the hole was too small for the goat and its horns. They reluctantly used a flashlight to survey the situation, before struggling to solve the problem – including trying to twist the goat's head to bring him through the hole. The fence was too high to lift the animal to safety. Nothing seemed to work.

There was another problem. Unfortunately, as planned, when Ben and Al succeeded in leashing the goat, they discarded their cumbersome supply of burglary tools, leaving it near the scene of the crime. When they dropped the gunny sack, the bottles of chloroform broke, later convincing Navy officials and midshipmen the goat had been chloroformed for the trip to West Point. But far worse for Schemmer and Rupp, the cutters needed to enlarge the hole in the fence were left behind with the supply of tools.

After more whispered discussion, Scotty convinced Schemmer and Rupp they should return to the tied rowboat, put their "prisoner" in it, row around the end of the chain link fence, and off-load Billy near the parked car. Ben and Al agreed, and led by Scotty and Maryland University's Joe Weaver, they made their way to the rowboat.

The four cooperating heist team members quickly and very quietly loaded Billy in the boat, rowed around a nearby bend in the river bank, past the end of the chain link fence, where they made shore. Then it was out of the boat and into the convertible with the Navy mascot.

Scotty Wetzel and Joe Weaver promptly went their separate way, back up the Severn by rowboat. The getaway driver launched for West Point, sixty miles an hour, headed for the New Jersey Turnpike, with

Ben Schemmer and Al Rupp guarding a leashed, powerfully odorous goat – in the convertible, with its top down.

A top-down convertible sounded like the right idea when plans were made in the spring and summer. Late November not far inland from the cold Atlantic Ocean, was another matter. The cold air at sixty miles an hour was brutal. There was no choice. After bracing against the cold all they could, it was time to put the convertible top *up*.

It was about four o'clock Sunday morning. In addition to stopping the car to put the top up, Ben Schemmer needed to send a prearranged signal to West Point. He and Alex Rupp placed a telephone call each, the first to signal one of their West Point team members the goat was in custody, and to be ready to accept him at the pre-planned hiding place. The second call went to Major George Pappas from Ben Schemmer, as the Academy Commandant had instructed.

Pappas, slowly roused from deep sleep by an incessantly ringing phone, finally picked up the receiver.

"Sir, this is Mr. Schemmer," the caller said. "I'm at a phone booth on the Jersey Turnpike, and we've got the Navy goat." George Pappas couldn't believe what he'd just heard, and asked Schemmer to wait a moment. He went into the bathroom, dashed cold water on his face, and returned to the telephone.

"Say that again, Mr. Schemmer."

Ben repeated he and Rupp had the Navy goat in a car, headed for West Point. Then, remembering his curious conversation with the Commandant, Ben asked, "What do we do now?" The goat-borrowing team knew for sure what *they* planned to do with the goat when they arrived at West Point, but they weren't sure what George Pappas planned to do, and they had no idea what he might have been ordered to do – that might change the cadets' plans. One thing was for certain. The cadets had a complete plan, including how to return the goat to their Navy rivals without an excess of humiliation.

George Pappas, the Officer-in-Charge of the cadet Public Information Detail, was completely surprised. He was the man who often boasted he had the best "G-2" – intelligence gathering position – to know and understand what was going on in the Corps of Cadets. But he had absolutely no warning of the Navy mascot's impending visit to West Point. Recovering his composure, Pappas told Ben to bring Billy to his quarters. Pappas apparently assumed they had no lodging for the goat, that hopefully, by the time they arrived he would have the guest's lodging arranged.

George Pappas' first action after the startling call from Ben Schemmer was to attempt contacting Sergeant Johnson, the Noncommissioned Officer-in-Charge of the Academy's mule mascots. No luck, and still no luck by nine o'clock. Sergeant Johnson wasn't at home and couldn't be located. Time was running out.

General Michaelis had to be notified, and George Pappas headed for the Commandant's quarters. Michaelis answered the door gruffly. Pappas told him he had some information to give him, "completely unofficially and absolutely confidentially." He then described to the Commandant the entourage heading West Point's way, without mentioning any names. The General laughed and asked George what he thought should be done.

Pappas scratched his head for a moment, then recommended Billy be housed in the stable with the mules, that there were cadet plans for an impromptu, completely spontaneous rally at supper that night. Michaelis approved and thought the Academy band should be included. He would arrange that. He added that the superintendent, General Irving, would have to be informed – no ifs, ands or maybes – then looked at his watch and said, "We've just enough time to catch him before he leaves for chapel." Next door to General Irving's quarters they went. The superintendent had little to say. He faced an accomplished fact. "Well, we might as well enjoy this while it lasts." Irving knew far better than anyone that the Navy and most of the Pentagon would erupt when the news spread. As George Pappas left, General Michaelis asked to be kept informed from the time Billy arrived on post until all arrangements were made for the rally.

All the while Major George Pappas was scrambling to prepare for Billy XII's arrival, Ben Schemmer and Al Rupp were wrestling with their own set of problems. Billy enjoyed the car's new found warmth. Lying on the back floor, listening to music on the car radio, feeling the occasional swaying motion of the smooth riding convertible, he rested his chin on the back seat and started snoring. But one problem's solution became the cause of another. An overpowering stench filled the warm interior of the convertible. The nauseating smell was so intolerable that rolling down the car windows seemed an absolute necessity.

Another problem immediately ensued. Every time the windows came down, so did the temperature, and Billy began to stir from his restful sleep. The return trip to West Point was proving to be more than the heist team bargained for. Despite all the obstacles, there was no turning back now. Don't ever give up! There were, however, other important obstacles to be considered in completing the mission. Security.

Technically, there was little doubt laws had been broken. Using the car radio, the fleeing goatnappers scanned early morning news broadcasts for any word Navy's goat was missing. The flash came at six in the morning. State troopers were looking for a getaway car on the New Jersey Turnpike.

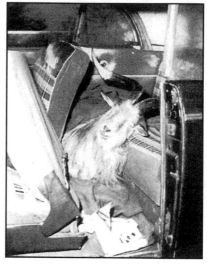

A pause in the saga of the Shaggy Goat. Navy mascot Billy XII in a deep sleep on the back floor of the getaway convertible, early Sunday morning, November 22, 1953. (Photo courtesy Benjamin F. Schemmer.)

Gas was running low when they heard the news, and they pulled into a station to fill up. As they were rolling to a stop, Billy decided to stand up on the back seat. The driver and two cadets heard a loud rip when his horns penetrated the convertible top. Ever curious, the goat turned his head first one way and then another, shredding the canvas fabric. Gaping at the two lengthy horns sticking through the convertible top's frame, the gas station attendant started yelling, "It's them! They've got the goat! Somebody call the police!"

Time for emergency action. Alex Rupp immediately became a station attendant, and pumped gas, while Ben Schemmer ran to the cashier and threw a twenty dollar bill in his direction. They jumped in the car and sped off, turning off to a back road at the first exit. Through an alternate road net they safely entered the back gate to West Point, and went directly to Major Pappas' quarters. Pappas noted the convertible top "was a rag that waved ever so gently in the breeze."

The four men got the now fully awake goat out of the back seat and led him to Pappas' garage. The officer instructed Al Rupp to go to the mule pens and get some hay for Billy to nibble on. While Rupp was gone Pappas told Ben Schemmer of the conversations with the Commandant and Superintendent, assuring Ben he hadn't divulged who was involved. He told Ben the Commandant approved a rally. Schemmer and Rupp told Pappas they and their accomplices had planned for a rally to unveil the goat, that other cadets were already organizing the event. Pappas asked Ben how he planned to get the goat into the cadet dining hall and unveil him before the Corps of Cadets.

He told Pappas the cadets' plan was to bring Billy to Washington Hall from his hiding place and unveil him on one of the cheerleaders' stands just inside the dining hall's main entrance. But the goat had to be hidden from view when the Corps entered the dining hall, or else the surprise would be ruined.

Al Rupp and the corporal returned with hay for Billy, and Pappas told the three men to head back to barracks, shower, don uniforms, and return to his quarters for a late breakfast. While they were gone, George finally tracked down Sergeant Johnson, and he agreed to bring a small truck to the Pappas' quarters around noon.

Meanwhile, Jay Gould, from the cheerleaders' operations group, who had arranged to procure some refrigerator-sized cardboard cartons from the Cadet Store, called cheerleaders and other members of their operations group into action. They moved the cartons to Washington Hall, cut and flattened them, and hammered them in place on three sides of the cheerleaders' platform – to create the appearance of a huge cardboard box – with one side left partially open.

While cadets were assembling the materials for the empty surprise package in Washington Hall, Sergeant Johnson pulled up to the Pappas' quarters, loaded up Billy, and took him to the mule stable. When Johnson led him to a small stall next to the senior Army mascot, Mr. Jackson, the big mule kicked up a fit, braying noisily and knocking his hoofs against his stable wall. Still not showered and in uniform, the weary heist crew came by the stables to check on their guest's welfare and pose for photos.

Word began to get around the cadet area that something was up, but no one knew what. In the meantime one of the cheerleaders came

Al Rupp, the getaway driver and Ben Schemmer pose with Billy XII and Mr. Jackson, the Senior Army Mascot, at the Mule Pen. (Photo courtesy U.S. Army Signal Corps.)

up with a bright idea: four plebes were "volunteered" to be guards for Billy at supper. The plebes reported to the dining hall five minutes before first call for the evening meal. Billy was smuggled through the back door of Washington Hall and lifted onto the platform to join Ben Schemmer, Al Rupp, and their corporal driver through the opening in the cardboard walls. The opening was then closed, nailed in place, and everyone departed to join the supper formation with their companies. The four plebes remained with rifles and fixed bayonets, one at each corner, facing outward, standing guard at parade rest.

General Irving came quietly up to the "poopdeck," overlooking the entire dining hall, one flight of stairs above the main entrance, and awaited the Corps' arrival. With him were two important guests, retired General James Van Fleet, who had commanded the 8th Army and U.N. forces in Korea, and Bernard Baruch, a distinguished senior American statesman and long a special White House advisor and confidant of Dwight D. Eisenhower. What no one knew at the time was that Mr. Bernard Baruch was scheduled to meet President Dwight D. Eisenhower at the White House the next morning.

The minute the mess hall corporals opened the huge doors into the dining hall, the Academy Band began playing the traditional first piece played at all dining hall rallies, *Tiger Rag*. Cadets looked puzzled as they filed through the three large entrances. No rally had been scheduled, no special events, and no announcements made in the cadet area. Especially puzzled were the men filing through the center doors past the huge, oversized cardboard box on the nearby platform. The mystery was heightened by the sight of plebe guards posted at each corner

Al, Ben, and the driver, right, after the Navy Mascot's unveiling in Washington Hall. (Photo courtesy Willis C. Tomsen, Class of 1954.)

of the box. When the last of the Corps filed in, the mess hall corporals, knowing something was up, closed the doors from inside, instead of standing guard outside as they normally did. The band ceased playing.

The Brigade Adjutant Jim Moore, commanded "Taaake Seats!" and from inside the temporary holding pen, Ben Schemmer, Al Rupp and their brave corporal kicked down the makeshift walls. There, in all his glory, was Billy XII, gold painted horns shining, coat carefully brushed, and a rope safely tethering him to a beaming Ben Schemmer. The Academy Band began playing, "Down in Maryland, there's a sailor band..." but few bars were heard that night. Washington Hall exploded in pandemonium, a constant roar of shouts, cheers, yells, and screams. As the full force of what the Corps was seeing swept through Washington Hall, the huge main entrance doors swung open again, and in marched the cadet rally band. They couldn't be heard. The noise rolled through Washington Hall in great waves, growing progressively louder.

Generals Irving, Michaelis, and Van Fleet, Bernard Baruch and astonished tactical officers on the "poopdeck," leaned over the rails, looking down at the bedlam below. Jumping, leaping, dancing, fist shaking, table thumping, whistling, back slapping, hand shaking. Caps went into the air, then dishes, then coats. In

Ben and Al parade Billy XII through the dining hall. (Photo courtesy Willis C. Tomsen, Class of 1954.)

the 2d Regiment area, several tables were literally tossed above the heads of their occupants. In the 1st Regiment area, men in Company A-1 took up the chant, "Take it off! Take it off! Take it off!" – not a demand to take Billy off, but a tradition at rallies wherein anyone giving a speech was being asked to remove his uniform blouse or suit coat before speaking. For fifteen minutes the deafening celebration continued, until the Officer-in-Charge told the First Captain to call the Corps to attention then give the command to take seats *in silence.*

Billy XII's tumultuous welcome to the home of his traditional enemies was probably greater than any he could ever have expected at an Annapolis football rally. When a measure of quiet returned and Ben Schemmer had taken him on a brief, leashed tour of company seating areas, Billy left Washington Hall and was spirited off post to a nearby dairy farm on Storm King Highway, arrangements made by "Tex" McVeigh, another first classman, a cheerleader and mule rider from Company G-2. Everyone knew calls would soon be pouring in from Annapolis and probably Washington, D. C. They were right.

The New York Daily News was hard at work on a story, complete with a picture of Billy XII accompanied by Ben Schemmer and his heist team, proudly displayed atop the cheerleaders' stand in Washington Hall. The paper hit the streets first thing Monday morning.

Army-Navy week was beginning in earnest, but the tale of the shaggy goat wasn't yet ended.

THE GOAT REBELLION

After supper that night, a tactical officer asked where the goat was to be billeted, inferring General Michaelis was interested in the animal's security – and General Irving's two guests might like a personal visit to see the captive. Ben Schemmer and company "took the brass into their confidence, strictly on a 'need to know' basis," he later said. "Confidence was respected – for a while," he added.

Confidence had been respected, probably more by accident than design. The next morning, when the *Daily News* published their front page story, the Army Chief of Staff, now General Matthew B. Ridgway, had already been put into the middle of the goat grab, and laughingly ordered the animal's return to Annapolis. Only a very small number of Academy officers were aware more senior men than the Army Chief of Staff were also involved in recovering the Navy mascot. General Ridgway had received his direction from much higher levels of government.

Reaction at the Naval Academy had been swift when Billy XII's absence was detected. The midshipmen wanted their mascot back promptly, and weren't going to rest until he was safe in his Annapolis home. A substantial number of middies refused to attend classes Monday morning. Orders from their Commandant, superintendent, and even the Chief of Naval Operations in Washington, D.C., went unheeded. A mutiny was brewing at the United States Naval Academy, and authorities there were becoming a bit unsettled. Time to elevate the problem to the Commander-in-Chief, a man well known for his diplomacy and even-handedness in dealing with international allies as well as America's armed services – though he was one of West Point's most renowned graduates.

Bernard Baruch, fresh from his enjoyable evening at West Point, was meeting with President Eisenhower, regaling him about the cadets' Sunday night football rally, when into the office strode the president's Naval aide. A brief recounting of the facts and the Naval Academy's stormy waters would soon have oil poured on them. Down the chain of command, through the Secretary of Defense, Secretary of the Army, to General Ridgway came the directive, thence no doubt to the Army's Deputy Chief of Staff for Operations and Administration, the Academy superintendent, and in turn the Commandant and Assistant Commandant.

Late the same morning, Ed Moses, Army's head cheerleader, received a summons to the Assistant Commandant's office. The usually affable Colonel William J. McCaffrey kept Ed standing at attention and emphatically told him, "Mr. Moses, I am not about to let a goat interfere with the career of the Commandant. The goat's going back. I want to know where you're hiding it." Like all good officers, McCaffrey was carrying out the orders of his seniors without divulging who had given his boss the orders.

An old military axiom was at work. "The time to argue the merits of a directive is before it's given, and if the argument is lost, the subordinate's obligation is to carry out the order as though it were his own. The weak officer's way out of an unpopular order is to blame it on his commander, or 'higher headquarters'." Not Bill McCaffrey. This tough, loyal, no-nonsense officer wouldn't take the easy way out. He would bare his fangs and get the job done, no matter how unpopular the order. Luckily, Ed Moses, for security reasons, wasn't told of the goat's whereabouts. He escaped McCaffrey unscathed.

But persistent inquiry paid off, and cadets who knew of Billy XII's hiding place received orders to return the goat. In the meantime, word spread in the Corps that Billy was being returned to Annapolis. He reappeared briefly, tethered to the clock in the central area of cadet barracks, before Lieutenant Colonel George W. McIntyre, who served as personnel officer and adjutant on the Commandant's staff, whisked him off to Annapolis in a spacious mule trailer, in an armed convoy.

When General Michaelis directed that Billy be returned to Annapolis, he knew the decision wouldn't sit well with the Corps of Cadets. He called in the First Captain, John Bard, told him the goat must be returned, and explained why. He then asked Bard to "get hold of the company commanders and tell them what we've done."

Before he could complete passing the word to company commanders, according to John Bard's recollections, he remembers encountering Ben Schemmer on Thayer Road, between the East and West Academic buildings. John believes the encounter was after he was told the goat was going back to Annapolis, and prior to the uproar which soon followed in central area. He recalls telling Ben of the decision, and in his memory Ben Schemmer responded, "Since I took the goat, let me tell the Corps it was OK to send it back. They'll take it better from me than through the chain of command."

Ben Schemmer remembered the sequence of events quite differently, because of what followed. Ben recalls that he had been in class, and remembers no early afternoon encounter with John Bard. Schemmer had heard rumors the goat was being returned to Annapolis and had been told by other cadets the animal had disappeared from his hiding place. Ben recalls he knew nothing of the order given to return the goat. The facts about the order came later, according to Ben, in the presence of the Assistant Commandant, Colonel McCaffrey.

Whatever the sequence of events were, when Ben Schemmer learned the goat was on the way back to the midshipmen, he felt betrayed, double-crossed. He, Al Rupp, the corporal, and others had taken considerable risk for a worthy cause, and he was most reluctant to accept that the goat had to be returned.

When the mule riders confirmed the goat was gone, and word spread, an increasing number of men in the class of '54 also felt cheated. Especially strong feelings simmered among those who originated and supported the heist. Word also spread to several Rabble rousers who by nature tended to be fiercely independent, and with little provocation, would readily thumb their noses at any command decision they regarded as wrongheaded. Clearly, the authorities didn't understand how close the class of '54 felt to that goat, especially the more rebellious members of the class. It was *"our* goat." They wanted him back.

Early in the afternoon a cadet, to this day name unknown, lit the fire. He was observed on the stoops of south area barracks "rallying the troops," giving a fiery speech to a small but growing crowd of cadets. Obviously frustrated about the goat being returned to the middies, he was reputed to have used inflammatory language, demanding, "Get the Com for returning the goat!" As luck would have it, some of the more impressionable cadets began to respond.

Unfortunately, a witness to the gathering in south area mistakenly identified the fire breathing speaker as Ben Schemmer. Not so. The error was grievous. Ben was in class. Nevertheless, the cadet rumor mill picked up his name, passed it on, and set up the leader of the goat heist for more unexpected trouble. The gathering crowd of Rabble rousers left south area, moving toward central area and the entrance to the Cadet Guard Room, on the first floor, below the Commandant's office.

A short time later, Schemmer was abruptly summoned from Russian class and told to report to the Commandant's office, "on the double." As Ben entered central area he was heartened to see what appeared another spontaneous football rally under way. But this one was unusual. Instead of "Beat Navy!," hundreds of cadets were chanting, "We want the goat back!" The south area Rabble rouser's fire starting speech was burning out of control, not a happy circumstance in a military organization.

Movie cameras from *Pathe' News* were recording the rally. To an outside observer, the gathering appeared more like a mob. Indeed, Ben Schemmer noted there was a bit of jostling at the entrance to South Area Guard Room and the Commandant's office. He could see a classmate bent over on the front stoop, dressed in cadet gray, with red sash and saber. He was a cadet lieutenant, Bill Vipraio, Company B-1, the Cadet Officer of the Guard. He was holding his head, and his nose was bleeding.

What Ben didn't see, high in a fourth floor window above the Commandant's office, was a banner proclaiming "GO NAVY! BEAT ARMY!" It was a banner apparently placed in the window by a secretary who was dating an Annapolis midshipman. The banner had a predictable effect, given all that had transpired that day. The "rioters" had seen the banner and, among other things, were attempting to storm the building to take it down.

Ben also didn't realize that his classmate, Earl Payne, had happened by the gathering "rioters," and seen the traitorous banner. Some of the crowd later concluded that Earl, like any good leader, took the initiative to lead the charge and retrieve the banner. To Stan Harvill, who was standing in the raucous crowd immediately behind Earl, it appeared Earl was pushed into Bill Vipraio. Whatever the facts might have been, the Cadet Officer of the Guard blocked Earl's entrance to the building and was rewarded with some slight physical damage, which accounted for the bloody nose – again, not a good thing to do in a military organization.

As Schemmer bounded up the stairs past the melee, he saw John Bard, the First Captain, and the Cadet Officer of the Day, engaged in animated conversation with Earl Payne and several other cadets. What Ben also didn't know was as soon as the ruckus started, General Michaelis had summoned John Bard to quell it. But things weren't yet under control, and not going well.

Schemmer was surprised to learn General Michaelis didn't want to see him. Instead Ben was directed into Colonel McCaffrey's office. He didn't finish saluting before McCaffrey ordered him to do a U-turn: "Mr. Schemmer, go down and break up that riot." Glancing out the window Ben began to understand why the Assistant Commandant took the rally for a riot. Chants of "We want the Com...shot!" now echoed through central area and into McCaffrey's office.

Ben screwed up his courage and respectfully asked McCaffrey what the ruckus was all about. He didn't believe he'd been in any way responsible for the disagreement brewing in the cadet area. Red faced and obviously distraught, the Assistant Commandant uttered something about insubordination and disobedience of direct orders, but acknowledged reluctantly that the rioters wanted the goat back. Schemmer gingerly asked if McCaffrey would tell him why the goat was gone. Bristling, he replied the goat's disposition was none of the cadet's business. It was "en route back to Annapolis, and that's all there is to it."

McCaffrey's demeanor and state of mind persuaded Ben Schemmer this was no time to press his luck. He dutifully headed back downstairs, uttered what he regarded as "some stupid remarks" to the unruly horde, and the rally band, led by First Captain John Bard and band leader, Company D-1's Perin Mawinney, shepherded the energetic cadet musicians, with a crowd of Rabble rousers, toward football practice.

After things had cooled down that afternoon, General Michaelis once more ordered John Bard to come to see him. He asked John if he had informed the company commanders of the decision, to which he had to answer, "No, Sir." Michaelis wasn't happy, and in light of all the uproar that had occurred, gave him some new direction. The Commandant wanted him to complete what he asked him to do in the first place, and in addition, make sure all the senior cadet officers and company commanders walked around after supper to head off any more trouble. It was well he did.

That evening, while the Corps gathered for supper formation, a sign in the cadet dining hall above the traditional "Beat Navy" banner announced, THE COM IS A PARTY POOP. Another thirty foot banner proclaimed, WE'VE BEEN BETRAYED. Hundreds of cadets wore black socks around the left, upper arms of their gray, wool uniforms, as mourning bands. A pulsating, unceasing chorus of "We want the goat!" began as soon as the Corps entered the dining hall. The uproar continued for nearly ten minutes before the cadets finally responded to the order from the Cadet Brigade Adjutant to "Take Seats!" After supper John Bard apprehended a cadet wearing a black parka, a cadet uniform item intended for cold weather wear for athletics. The cadet, Al Hamblin, Company D-2, was wearing the parka, hood up, making for the superintendent's quarters, intent on burning a cross on General Irving's front lawn.

Ben Schemmer took as a bad omen the banner headlines in Tuesday morning's *New York Daily News*: "Goat Rebellion at West Point." Smaller type below the headline read, "Cadets Protest Navy Mascot Return." Articles also appeared in the *Washington Post, New York Herald Tribune,* and other leading newspapers.

Late that morning, Ben was hastily summoned from another class and ordered to report to the Commandant – for the third time in less than a week. This time he got to see Michaelis himself, an ominous sign. He didn't invite Schemmer to stand at ease but calmly ordered, "Mr. Schemmer, take off those stripes." In his most deliberate military manner he told Ben he was "busted," period. Schemmer was about to protest. He didn't start the riot, at least not in his mind. He helped break it up. Discipline within the Corps was the First Captain's job, wasn't it?

But Michaelis cut him off before he could begin. "Mr. Schemmer," he said, "I can't bust fifteen hundred cadets. I need a symbol. You're it. You should be proud. Take off your stripes." Ben wasn't aware of the rumor he'd been rallying cadets to return the goat to the Corps' keeping. He was once more puzzled by the Commandant's logic, but he couldn't argue with a brigadier general.

As Ben Schemmer left the Commandant's office he wondered. Fifteen hundred cadets? Actually, he said to himself, the riot was only one fourth that size. No doubt Mawinney's sixty-man rally band amplified the magnitude of the goat rebellion in the minds of the tactical officers guarding the Commandant's redoubt. Michaelis' remarks paid tribute to the psychological warfare cadet musicians always inflicted on the enemy.

On Wednesday morning, an Order of the Day appeared on cadet bulletin boards in all twenty-four companies. "The appointment of B.F. Schemmer as a cadet lieutenant is hereby rescinded." No reason given. Others paid a much heavier price.

The heaviest was that of Earl Payne in Company I-2. He got the most memorable slug of anyone in the class of 1954, for "inciting to riot:" eighty-eight demerits, eighty-eight hours, and six months confinement – a form of house arrest. No one ever explained why he got such a big slug. Ben Schemmer also believed Earl didn't incite the riot, if that's what the disturbance was. As far as Ben knew, the riot was already in progress when Earl walked into central area returning from class. According to Ben, Payne was on the tail end of the crowd, the fringes, and got propelled to the front. He didn't muscle his way into the South Guard Room. He was invited there. His only transgression, according to Ben, was to argue with the First Captain and Cadet Officer of the Day – which wasn't entirely accurate.

Earl Payne's punishment was stiff indeed, but was made worse by other circumstances. He, too, hadn't the slightest idea until years later, who really was responsible for ordering Billy XII's return to Annapolis. Nor was he aware of the tasks John Bard and the cadet chain of command had just been given – restore calm, explain the decision, and forestall any further disruptions to good order and discipline. Last, but not least, Earl Payne had just completed serving a stiff punishment for missing taps while on the Recruit Training Detail at Fort Knox, Kentucky, the Armor Center.

Ben Schemmer remembers Colonel McCaffrey worked fast to improve the tactical department's intelligence. A few days after the goat rebellion, another '54 Rabble rouser, Ray "Squash" Cassel, who the following spring resigned and left West Point, planned another protest. The Assistant Commandant got wind of it, called Cassel's L-2 cadet company commander, Jan Le Croy, and asked him if it were true. Le Croy reassured McCaffrey and quickly persuaded Cassel to abandon his protest. McCaffrey evidently was unmoved, and feared a double-cross. The night of the next rally he ordered Le Croy to report to his office with one of his "dissidents" from L-2. Le Croy picked up someone at random, and the two spent an uncomfortable evening trying to make small talk with the Assistant Commandant.

In the meantime, "Tiny" Tomsen, always willing to pitch in and do his part, took up a collection to pay the cost of a new convertible top for the corporal who braved the highways, wintry nighttime cold, a smelly goat, New Jersey State troopers, civic minded gas station attendants, and two cadets on a hilarious, but sometimes not-so-funny mission.

On Monday, November 23, 1953, Lieutenant Colonel George McIntyre became the first Army officer ever to escort a goat down the New Jersey Turnpike under armed guard. When he arrived at Annapolis, his contingent was met by a Shore Patrol vehicle and escorted to the area in front of Bancroft Hall, not far from a large pep rally in progress. After a photograph or two, the goat was whisked off to join the rally. The Naval Academy's Assistant Commandant, Captain (USN) Edwin T. Miller took McIntyre in tow and proceeded to the rally.

The midshipmen were in a frenzy when George McIntyre and Captain Miller arrived about eight in the evening. The Navy's mascot had been safely returned, and was up on a platform with the Navy cheerleaders. While the Army officer was being introduced to Mrs. Miller, and the Naval Academy's Commandant and his wife, Captain and Mrs. Buchanan, a low but insistent cheer started. "We want the Colonel! We want the Colonel! We want the Colonel!" According to George McIntyre, "Mass hysteria took over." The midshipmen began closing around their commandant and his guest as the cheer grew louder and more demanding.

Captain Buchanan turned to McIntyre and said, "I don't know what you're going to do, but I'm going up on the platform – would you please say a word to the midshipmen?" A now surrounded Lieutenant Colonel George McIntyre had no choice. Up on the platform he went, to rousing cheers from the midshipmen.

Captain Buchanan gave a brief talk, then introduced his guest – who hadn't prepared in advance for the occasion. But McIntyre recalled a saying told him by retired Army Lieutenant General Doyle O. Hickey, who had commanded the 3rd Armored Division during World War II, and later was Chief of Staff of the Far East Command during much of the Korean War. The saying came from the "Old Army," and was passed on with wry humor through each generation of the officer corps for nearly forty years. "The only officers who could get superior efficiency reports were aides, aviators, asses, and adjutants."

As George McIntyre moved toward the microphone, the goat moved his head and jabbed him in the leg. George grabbed the goat's horn, and began,

I have learned one thing today, and that is that you have to grab this thing by the horns.

There is a saying in the Army about a certain class of personnel known as aides, aviators, asses, and adjutants. Well, I'm the Adjutant of the Corps of Cadets.

For the past seven hours I've been aide to a goat.

And right now I feel like a bit of an ass speaking at a Navy rally.

I am obliged to wish you good luck in Saturday's game. I can assure you that you will need it because we have a mighty fine team at West Point.

At this point the rally got completely out of hand, and Lieutenant Colonel George McIntyre, West Point class of 1941, was unable to continue, drowned out by the jubilant, loudly shouting swarm of midshipmen.

Navy's mutiny never made headlines. Army's goat rebellion did, and more than one cadet in the class of 1954 did what Earl Blaik said must be done to win. "You have to pay the price." The goatnapping of 1953 wasn't a Pyrrhic victory. The men in the first class had come up with a bold, daring idea, and plans to make the idea work. Men in '55 jumped on board. The tongue in cheek, larcenous act was another in a long string of emotion-charged stirrings that traced their roots to the painful years of 1950 and '51, when adversity roamed the battlefields in Korea, the grounds of a proud institution, and the fields of friendly strife – wherever Army football teams played. The last few, exciting days before Saturday, November 28 were rushing headlong to a climax.

"WHO THE HELL ARE THEY ANYWAY?"

General Maxwell Taylor's 1947 policy resulted in Army's football team taking on dragons roughly every other week. The last and most important dragon, every year, is Navy, and there are always two weeks between the last "other" game of the regular season, and the midshipmen. The extra week's pause allows for special preparations for one of the nation's greatest annual gridiron classics, and a healing of wounds inflicted by the regular season's week in and week out grind. Normally, the week after the last game prior to Navy, Earl Blaik scheduled light workouts following a few days off, then bore down the week preceding the clash in Philadelphia.

Against all dragons, Blaik used his preparation and motivational magic on the Army team. After Sunday's critique of Saturday's performance, the scouts' report on the next opponent, and a summary of what they would work on in practice, the game plan for the next contest unfolded. For home games, on Saturday when the team was suited up and riding the bus from the South Gymnasium to Michie Stadium, he would start firing up his charges. In his deep baritone voice he would snap, "Who the hell are they anyway?" Coupled with his thorough, disciplined game preparation, "Who the hell are they...?" was his way of cutting dragons down to size. It had worked well for thirteen years.

But there were, as always, a couple of wrinkles for this Navy game. He told the players if they beat Navy, he would sing them a song afterwards. He recalled years later, "The small part of me that was *Buttons the Bellboy* had not been subordinated completely, not even in 1951 and those years immediately after."

Earl Blaik didn't make extraordinary tactical plans, except one. He had first classman Elwin Rox Shain, who was on the "B" squad, repeatedly practice a short, diagonal kickoff during the week. Rox had the most powerful kickoff leg on the entire Army team, consistently kicking the ball 10 or 15 yards deeper than anyone else. He had also kicked off in much the same manner at the start of the 1951 thrashing by Navy. If Army won the toss, Blaik, as he had in '51, planned to use the diagonal kickoff, with an erratic, hard to handle bounce of the ball, intending to catch Navy by surprise and cause a fumble. However, for Blaik's own reasons, he didn't tell the first classman he intended to suit him up with the varsity and actually do the practiced kickoff until Rox, a cadet battalion commander, arrived in Philadelphia with the Corps of Cadets. The surprised Rox, called in without notice to suit up for the game, wasn't even on the Army team roster given to sports writers a day earlier.

The long, colorful Army-Navy series stood at 27 Army victories, 22 for Navy, with 4 ties. Navy, still building after Coach Eddie Erdlatz's 1950 debut and the stunning 14-2 upset of No. 2 Army, finished the 1952 campaign with one of its best records in years, 6-2-1, and a third straight win over the cadets, 7-0. The midshipmen weren't seriously depleted by graduation, and possessed both depth and experience when this season began. Erdlatz did confront the same problem all college coaches did in preparing for this season. He had to identify men who could play both offense and defense, which caused major adjustments in starting line ups.

In front of a strong backfield, Erdlatz had a solid line, anchored by a returning, near unanimous All-American guard, Steve Eisenhauer. Steve's quick aggressiveness won him a reputation for being in the middle of nearly every play. In the backfield, the Navy coach had depth and experience as well, strengthened with an up and coming young sophomore named George Welsh, later one of Navy's great, all-time quarterbacks.

The midshipmen started the season with a 6-6 tie against William and Mary, then reeled off impressive wins over Dartmouth 55-7, Cornell 26-7, and Princeton 65-7. They stumbled against tough Pennsylvania, losing 9-6, and were hammered by Notre Dame, 38-7. Then they held powerful Duke to a 0-0 tie and beat Columbia 14-6.

On the strength of Army's recent performance, the cadets were coming into the game slight favorites. The naming of favorites in the Army-Navy classic seldom means anything, but this time it did.

In *The Pointer* magazine, published the day prior to the game, writer Chuck Sterling, stated what many at West Point felt about Army-Navy 1953.

> Tomorrow afternoon the Corps of Cadets marches into Philadelphia's Municipal Stadium to watch a football game. It's been four years since we've won that particular game. During those four years the Corps has taken some of the hardest knocks short of actual combat that could ever hit a military unit. There were those who didn't think the Corps would ever recover from those blows. They thought the Military Academy and everything it stood for was finished. For the first time in history the integrity of the long gray line was in question...

> ...Tomorrow afternoon, radio sets will be tuned on Philadelphia all the way from Berlin to Panmunjom. Graduates will be listening for news of an Army victory. But they'll be listening for something more – something none of them talk about. They'll be listening for evidence that the Corps is on its way back. They want to know that the values which they stand for are still alive in the Corps.

Most of the hundred thousand spectators tomorrow afternoon will be watching a football game and nothing else. They will see a hard fought game and some of the most colorful football pageantry in the United States. Even the Brigade of Midshipmen will be watching a football game and nothing else. Sure, it's a game they want to win rather badly, but if they lose it won't hurt too much. But the Corps will be watching something more than a game. They will be watching eleven men shouldering the task of 2400.

In the same *Pointer* issue was an article titled "The Last For '54," paying tribute to twelve first classmen who, as team members, would be on the sidelines or play in their last Army football game: Halfback Kirk Cockrell; Gerry Lodge, the smart, tough, workhorse fullback, converted from fullback to guard and returned to fullback, also captain of Army's wrestling team; Freddie Attaya who, until his injury, lifted Army's spirits with flashing speed, heady play, and great punting; centers Ken Kramer and Norm Stephen – Norm, the hard-nosed center and team leader who lit Army's offensive fire in the second half of the Dartmouth game, and called defensive signals brilliantly all season long; and halfback Bill Purdue – the Army track star who had never played football until asked by Blaik to come play in 1952, and who, right after his father had died, had been the hero of the '52 Penn game. This season, when Army returned to "iron man football," Bill had requested to be moved to the "B" team to strengthen his defensive play, and had played little on the varsity all year.

Army's two incomparable ends, Bob Mischak, the much sought after high school, single wing tailback recruited by Vince Lombardi, and Lowell Sisson, the walk-on who had struggled his way up Blaik's tough football ladder, both players whom Earl Blaik acknowledged had developed into two of the finest ends on any team in the country; Joe Lapchick, who doubled at tackle and end; small, fast tackle John Krause and guard Dick Ziegler, both solid performers all season – Dick converted from a platoon fullback in 1952, and an excellent basketball and baseball player as well. Then there was team captain Leroy Lunn, who had been affected by injuries most of the season, but never gave up, and more than filled his role as team captain. On this day, Leroy would finally get to do what he'd wanted to do all four years at Army. He started against Navy and played the game of his life.

Army-Navy is always special for first classmen, the seniors. This was their last game, and Earl Blaik each year took note of that fact, the emotion and lifelong memories it would bring, and sought to give seniors playing time even though they might not be regulars.

But this year there was also something special for a yearling from the class of 1956, a walk-on to the plebe team in 1952. He was Paul Lasley, who backed Norm Stephen at center and played linebacker on defense. He had used up all but one year of varsity eligibility at Illinois College before entering the Air Force as an enlisted man and winning an appointment to West Point.

In the fall of '52, he had gone relatively unnoticed in the eyes of Army coaches, and wasn't selected for the following spring's practice roster. Because he wasn't on the roster, he chose to play baseball instead. Then at the start of practice in September, due to a letter from a hometown newspaper sportswriter to the Army coaching staff, Paul was picked up on the corps squad roster for the first season's return to "iron man football." Against Dartmouth he had his day in the sun, intercepting two passes and running one back for a touchdown.

Late Friday afternoon, in Philadelphia, where the team was staying prior to the Navy game, Lasley was told Earl Blaik wanted to see him. He was puzzled to say the least, and walked in to find all the assistant coaches gathered in the room with "The Colonel."

The Army coach said, "I heard you played college ball three years. Is that true?" "Yes, sir," Lasley replied. Blaik paused, gazing at him, and then said, "Hope you enjoy the game tomorrow."

Paul Lasley never forgot that brief encounter. Paul didn't get as much playing time the next day as he wished, and he recalled Earl Blaik spoke to him the grand total of probably ten seconds in the one year he played varsity football at Army. Nevertheless, Blaik's gesture before Army-Navy '53 told him much about the men he played under. To him they were the finest coaching staff in the nation.

* * *

Despite the disappointment of Billy XII's return to Annapolis on Monday, the usual pregame hijinks and preparations for November 28 continued the remainder of the week at West Point. Banners hung beneath upstairs hallway windows in company areas, each proclaiming in its own way how the mule would best the goat on Saturday. There were rallies in the cadet area, spontaneous outpourings of cheers at meals in Washington Hall, and careful preparations to out-hijink the midshipmen in Philadelphia. The team was given a roaring send-off.

General Van Fleet, a former Army halfback, was on hand to fire the Corps of Cadets at Friday evening's big football rally.

"Beat Navy" banner by Company I-2's "Cheech" Favole, a Cuban cadet in the class of 1954. It reads: "El Habana Perfecto. I'd Put Every Tenth I've Got on Army." (Photo courtesy Willis C. Tomsen.)

Retired General James A. Van Fleet speaks to the Corps at a rally in Washington Hall the week prior to Army-Navy 1953. (Photo courtesy USMA Archives.)

That afternoon the traditional Goat-Engineer football game was played on the field where the Army track stadium stood, and the typically large enthusiastic crowd of cadets and officers turned out for the game. There was a legend associated with the game which pitted the top men in the cow class, the Engineers, against the bottom men in the class, the Goats. If the Goats were victorious, Army would defeat Navy the next day. But in recent years the legend had taken a beating, and didn't seem a factor in tomorrow's game.

By the time the Cadet Corps and the Brigade of Midshipmen were in the stands following the pregame ceremonies, the huge stadium was nearly full, with a crowd of 99,616. King Paul and Queen Frederika of Greece were in attendance, as were seven members of President Eisenhower's cabinet, including Secretary of State John Foster Dulles and Secretary of Defense Charles E. Wilson. General Matthew B. Ridgway, the Army Chief of Staff, Admiral Robert B. Carney, Chief of Naval Operations, and General Nathan F. Twining, Air Force Chief of Staff, came to support their favorites, as did General Lemuel C. Shepherd, Jr., Commandant of the Marine Corps. Retired General Van Fleet was there, as was retired Fleet Admiral William F. "Bull" Halsey, a World War II national hero in the Pacific battles against Japan. Army's former football great, Felix "Doc" Blanchard, who had left Blaik's coaching staff to return to regular Air Force duties, was also in the stands. One sportswriter observed, "You can't keep Brooklyn out of any picture. Harold Parrott, formerly traveling secretary of the Dodgers, and now front office 'mahouf,' had seats on the 50-yard line, Army side."

There were other distinguished visitors seated in the stadium. Thirty-four American former prisoners of war were at the game. They had been released by the North Koreans during the prisoner exchanges following the July armistice. The former prisoners had made a vow two years earlier, while still in captivity. The reunion at the Army-Navy game had been the brainchild of Bruce Shawe, a 1946 West Point graduate and tactical reconnaissance pilot, whose airplane went down in enemy-held territory in December 1950. He was in Prison Camp No. 2 when the men made their promise to one another and began formulating their plan. Now the reunion was a reality.

The unofficial chairman of the reunion was Lieutenant Junior Grade J. W. Thornton, a Naval officer who was in the same prison camp. He told the *New York Tribune* writer, Al Laney, the men had originally planned their reunion for the fall of 1951, "but the length of the truce talks pushed it back two years." Friday evening the men and their families had been the guests of the Navy at Willow Grove Naval Air Station, near Philadelphia. Regrettably, Bruce Shawe wasn't well enough to attend the game, remaining hospitalized in San Francisco.

The Brigade of Midshipmen marched on first, nearly thirty-six hundred strong, and the Corps of Cadets second, numbering twenty-four hundred. No sooner had the cadets filled their seating section than the Brigade flashed the first barbed message to the Corps. Near the top of the middie seating section a sign appeared, "Our Ike is here, where's yours?" President Dwight D. Eisenhower didn't attend the game, and the middies were quick to remind their opponents that Steve Eisenhauer, their All-American tackle was present, ready to play.

The president was on a working vacation, playing golf near Augusta, Georgia, while he and Mrs. Eisenhower visited with their son, Major John Eisenhower, and his family. John Eisenhower's father was preparing for a "Big Three" conference in Bermuda the following Friday, with Prime Minister Winston Churchill of Great Britain and Premier Laniel of France.

It didn't take long for the cadets to reply to the midshipmen's barb with their own sign. "Cry baby Navy. The Meanies stole our goat."

The cadets staged a dramatic entrance of their mascot. A cadet sergeant collapsed on the sideline. First aid was administered. In came an ambulance, manned by medical corpsmen. A stretcher was rushed on the field. But instead of the cadet being hauled off when the ambulance doors opened, out pranced the Army mule "Pancho." The collapsed

Corps of Cadets on parade, Municipal Stadium, Philadelphia, Pennsylvania, November 28, 1953. The Brigade of Midshipmen is already in the stands. (Photo courtesy Willis C. Tomsen.)

cadet hopped to his feet and jumped astride Pancho. The "ill" cadet was a mule rider.

During the week, the cadet cheerleaders' operations group persuaded Continental Army Command to furnish a helicopter that would carry a huge banner slung beneath it, with the words "Go Army! Beat Navy!" When the banner was readied for dress rehearsal, the helicopter pilots learned the hard way the rig was too cumbersome and wasn't going to work. There was very nearly an accident. Cadet ambitions were scaled back, and during the pregame hijinks, the helicopter flew low above the stadium, and hovering carefully, descended almost to the field. It bore a smaller sign reading, "Beat Navy!"

Earl Blaik was seldom wrong, and his surprise tactical plan for the opening kickoff once more proved his football genius.

Rox Shain's diagonal kickoff worked to perfection. Navy left end John Reister attempted to advance as he fielded the wobbly, erratically bouncing ball on Navy's 29-yard line. Not in complete control of the pigskin, he was hit hard by Army's center and defensive signal caller, Norm Stephen. Spinning as he fell forward on his back the ball squirted out of Reister's arms toward the Navy goal, and Army's onrushing left tackle Howard Glock recovered on the middies' 31. The Corps of Ca-

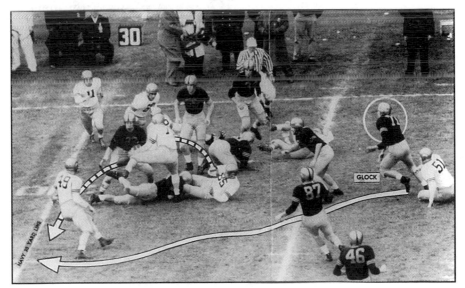

Army's "Knobby" Glock (71) recovers Navy fumble of opening kickoff. (Photo courtesy Urban Archives, Temple University, Philadelphia, PA.)

dets whooped with delight, and the chant immediately started, "GO! GO! GO!" The Army team responded, and at 2:35 into the first quarter, on the seventh play after the Navy fumble, the cadets scored their first touchdown.

Army's Tommy Bell takes a handoff from Pete Vann. (Photo courtesy Urban Archives, Temple University, Philadelphia, PA.)

On the first play of the drive Tommy Bell stormed, snorting and puffing, straight ahead over tackle for eight yards. Pete Vann missed with a pass. Then Pat Uebel took a pitch-out from Pete and skirted Navy's left end for a first down on the 20-yard line. Gerry Lodge bulled his way for eleven yards and another first down on Navy's 9 as the Corps of Cadets went wild with excitement. "GO! GO! GO!"

Navy called time out to make defensive adjustments. But the deafening roar from the Corps of Cadets continued, and the middies couldn't slow the Army attack. On first down Lodge added four more. On second down Pete Vann called Gerry's numbers again, but the middies stopped him for no gain. Then, as against Penn's Quakers, Pete called "54." He faked to Lodge as he moved toward the line of scrimmage with the big Army fullback, pulled the hidden ball back to his body, and eased it to Pat Uebel. Slanting to his right, crossing behind Lodge, Pat rocketed outside Navy's left tackle the last five yards into the end zone, untouched. Ralph Chesnauskas converted, and it was Army, 7-0. The "victory cannon" overpowered the noisy celebration by Army fans. The Academy band struck up the strains of *On Brave Old Army Team*, and the Corps of Cadets loudly sang and whistled its most famous fight song.

When Chesnauskas kicked off after the Army score, Navy's George Welsh caused another flurry of excitement when he fumbled, but recovered on the Navy 25. The midshipmen were unable to move and punted to the cadets. During the middies' first offensive series, starting right halfback Phil Monahan was lost to an injury for the remainder of the game.

Army, likewise was stymied the next offensive series, and Lowell Sisson punted back to Navy, the ball rolling dead on the middies' 35-yard line. Welsh, on a quarterback sneak, went through a large hole to the 44, and followed the gain with another quarterback keeper, snaking three more to the 47 for a first down. Then came what Earl Blaik later described as a "delirious sequence."

Welsh dropped back to pass. The throw bounced off the hands of the intended receiver, into Pat Uebel's midsection and arms at the cadet 33. Pat cut against the flow of Navy's offensive players who had suddenly become defenders. He was doing well until he crossed the Navy 40. There Navy's big left tackle Jack Perkins abruptly stole the ball from the surprised Uebel and headed in the opposite direction, toward Army's goal. He lumbered and fought his way down the sideline to the Army 9-yard line, where once more in his illustrious collegiate career, Bob Mischak saved a touchdown, by catching Perkins and driving him out of bounds.

Navy was knocking at the door. Halfback Bob Craig went off right tackle to the 7. Fullback Joe Gattuso tried a wide sweep and was bumped out of bounds on the 4, but the gain was nullified by a five-yard penalty, placing the ball on the 12. Craig tried a quick-opener but lost a yard. Tommy Bell ended the Navy threat on the next play, intercepting a Welsh to end Don Fullam pass on the Army 3-yard line and bringing it back to the six.

Neither team could mount a sustained drive until midway in the second quarter when Navy's Gattuso kicked to Army and the ball rolled dead at the cadet 33. Once more the thunder of "GO! GO! GO!" reverberated through the huge stadium.

Uebel lost three yards on a pitch-out, and Navy declined an offside penalty against the cadets. Pete Vann dropped back and threw a long pass that went off Pat Uebel's fingertips. Undaunted, Pete next did what he dearly loved to do. He arched another long pass, this one perfectly thrown. Left end Bob Mischak out-raced his Navy defender, Bob Craig, by a step, and gathered Vann's pass in on the Navy 25-yard line. Craig pulled Mischak down from behind on their 18. Uebel ran two straight, the first for six yards, the second for a first down at the Navy eight.

Pete Vann twice called Gerry Lodge's numbers for three and two yards respectively, moving the ball to the Navy three. Then, for the second time in the game it was "the belly series," and Pat Uebel sailed over Navy's left side, untouched, for Army's second score. Ralph Chesnauskas' point after attempt went wide right, and at 10:37 gone in the second quarter it was Army 13-0. The victory cannon boomed, and strains of *On Brave Old Army Team* echoed again in the chilly, gray afternoon air.

Rox Shain kicked off deep to Navy's 2 where Hepworth fielded the ball and ran it out to the 20 before Uebel brought him down. On a quick opener, Bob Hepworth took a hand-off from Welsh and scampered 19 yards. Welsh then threw incomplete to Craig, and followed with a hand-off to Craig good for one yard. Welsh went back to pass again, and unable to find an open receiver, scrambled another 19 yards for a first down on Army's 41.

Navy appeared on the move, but on the next play Gerry Lodge picked off another Welsh pass on the cadet 34. He rumbled down the sideline, and with a key block by Lowell Sisson cut to the inside against Navy defenders closing in on him. Blaik and everyone on the Army bench watched in astonishment as Gerry Lodge, running like a halfback, reversed field and threatened to go all the way for six points, until he was finally hauled down by All-American Steve Eisenhauer on the Navy 7-yard line.

Gerry Lodge tried the middies' line for no gain. Tommy Bell carried inside the five, but on the next play Pete Vann attempted to pass and was thrown for a loss on the Army 15. Chesnauskas tried a 21 yard field goal, which again sailed wide right of the goal posts. Navy took over on their 20, and reeled off 18 yards in two running plays before the half ended.

The midshipmen's Joe Gattuso kicked off to the cadets to start the second half. Neither team could mount a sustained drive in three offensive series. Army, after the kickoff, drove from their own 31 to the Navy 14, but a backfield in motion penalty nullified a 15-yard scramble by Pete Vann and pushed the ball back to the 36. Pete attempted a long

Tommy Bell prepares to receive a pass from Pete Vann. (Photo courtesy Urban Archives, Temple University, Philadelphia, PA.)

pass but Craig intercepted on the Navy 8 and ran it back to their 43. The Navy attack, in turn, stalled and Gattuso punted to Uebel again, and he was run out of bounds on the cadet 33.

Army drove deep this time, all the way to the Navy 14, before Vann lost three on fourth down, and the middies again took over on downs at their 17. They could get no further than their 30 before John Weaver punted to Uebel, who took the ball in at Army's 30. On a 70-yard punt runback, Uebel iced the game.

Bob Farris gave Pat the first key block to spring him loose. Cutting across the field to his left, Uebel angled toward the west sideline, and behind a wall of blockers headed for the Navy goal line. Bob Mischak threw another key block. Almost the entire crowd of nearly 100,000 fans came to their feet, erupting in a roar of noisy shouts as the speeding Uebel turned downfield. Army's Leroy Lunn cut down Navy's Aronis at the middies' 48 as Pat crossed midfield. Lunn's block left three Navy men pursuing close enough to possibly bring down Uebel: right tackle McCool, fullback Dick Padberg, and left guard George Textor. Their eyes fixed on the fleeing Army halfback, none of the

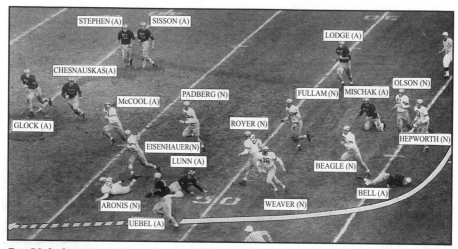

Pat Uebel running 70 yards with a punt return for his third touchdown against Navy, November 28, 1953. (Photo courtesy Urban Archives, Temple University, Philadelphia, PA.)

three Navy players saw Howard "Knobby" Glock, Army's left tackle, angling across the field from their right front. At the 20-yard line Glock delivered a crushing block to McCool, Pat's closest pursuer, and the other two Navy players went down with McCool. Pat Uebel sprinted into the end zone, once more untouched. In the Corps of Cadets there was pandemonium as the "victory cannon" belched smoke and thunder, and the Academy band once more struck up the happy sounds of *On Brave Old Army Team*.

Ralph Chesnauskas booted his second extra point of the day, and it was Army 20-0 with a little over four minutes left in the third quarter.

Army almost followed up with another score when, after the kickoff, aggressive play on defense kept Navy quarterback John Weaver from punting, and the cadets took over on downs at the Navy 28. With Bell, Lodge, and Uebel taking turns on running plays, the cadets drove to the Navy four, where a fourth down lunge by Tommy Bell fell short by inches, and the midshipmen took over on downs.

Army had the ball only six plays on offense the remainder of the game, but it didn't make any difference. Twice the cadets thwarted Navy drives in the fourth quarter, once taking the ball away on Army's 34, and the second time at the cadet 16, after Navy drove to the 7-yard line. There the cadet defense rose up and caused Navy to lose nine yards on three plays, before a George Welsh pass to Bob Craig was batted away. The cadets took over and ran a few plays before Lowell Sisson kicked into Navy territory.

This time, aided by a 15-yard penalty, and then Army pass interference, the middies moved swiftly to the cadet 40. With both teams now substituting freely, and with third string quarterback George Knotts at the helm, the middies finally mounted a scoring drive that ended in success, with 45 seconds to play. Knotts completed four of five passes, his fourth going 11 yards for a first and goal on the Army eight. On the next play, halfback Jack Garrow tore through a large hole in the middle of the Army line, and into the end zone. George Textor kicked the extra point. Army 20-7.

Navy tried a short kickoff in hopes of an Army fumble, but Don Holleder covered the ball at the Army 30. After the kickoff there was barely time for one more offensive series. Gerry Lodge lunged at the Navy line and was stopped. On the game ending play, Pete Vann held onto the ball for no gain. The "victory cannon" thundered a final salute, and the men in gray swarmed the Army team and their coaches in delirious celebration. This was the climax, the magnificent team victory for which everyone had worked so long and hard.

The Army dressing room was bedlam. After a few minutes of noisy happiness, Earl Blaik asked for quiet, that he might speak to the team before the reporters were let in. He paused for a moment, and then after a few brief remarks, said,

...And now, I'm going to keep my promise. I told you if you beat Navy, I

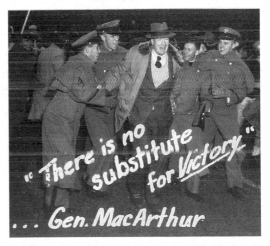

An ecstatic Earl Blaik leaves the football field after Army's 1953 win over Navy. Cadet well wishers in the foreground next to Blaik are, left to right: Clyde W. La Grone, Roman J. Peisinger, Ira Coron, and Leonard L. Griggs. In the background is Rox Shain, the Army player whose game-opening diagonal kickoff preceded Navy's fumble. (Photo courtesy Willis C. Tomsen, Class of 1954.)

909

was going to sing you a song. So here goes: 'Down, Down, Down went the Nayvee—'

That was as far as Earl Blaik got. Shouts from elated team members drowned him out. Years later Blaik wrote, "Somehow, I didn't mind. I never could carry a tune."

The sounds that drowned Earl Blaik's song erupted into back slapping and cheers. Strains of *On Brave Old Army Team,* sung by Army players, echoed through the dressing room after a Navy game for the first time in four years. The lusty singing meant Army had returned from oblivion. The team lifted yearling Pat Uebel on their shoulders while photographers snapped pictures, and one of the Army players held three fingers high in the air, signifying the three touchdowns Pat had scored.

In scoring all three Army touchdowns that day, Pat Uebel set a record. Felix "Doc" Blanchard scored three in the cadets' 32-13 victory over Navy in 1945, the year Army won their second mythical national championship. Another brilliant Army fullback, Gil Stephenson, turned the trick in the 38-0 romp in Army's last win, in 1949, when Johnny Trent was team captain and Arnold Galiffa was an All-American quarterback. But never in the fifty-four year history of the classic had one Army player scored all three of the cadets' winning touchdowns.

But Pat Uebel's day of football glory was a reflection of much more, something that went far deeper than performance on the playing field. Earl Blaik, Eddie Erdlatz, some of Blaik's other professional colleagues, and sportswriters took due note of the team's performance against Navy, and what the cadets achieved throughout the season. Magnanimous in defeat, Eddie Erdlatz told sportswriters, "Army showed us a great team, with everyone in peak form. They're a well-coached team, well-rounded in every department. They lack nothing." To another reporter he said, "We lost and we lost to a good team. We have no alibis."

Earl Blaik again used only fifteen players until the final minutes of the game, when he substituted an additional eight men. Pete Vann played the full sixty minutes, and Erdlatz was effusive in his praise of Pete. "Pete Vann was the best quarterback we have seen all year," rating him above Notre Dame's Ralph Guglielmi. Even through powerful field glasses, he was a magician handling the ball. With their faking during the "belly series" he and Gerry Lodge twice completely fooled spotters, cameras, and sports broadcasters. Pete's wizardry had a similar effect on the Navy line all afternoon.

The statistical game was closer than the score and didn't reflect the efficiency and power in Army's play. Pete Vann connected on 4 of 10 passes for 90 yards, and had one throw intercepted. Three Navy quarterbacks threw 26 times, with 12 completions for 96 yards, and three interceptions.

The cadets out-rushed Navy 171 yards to 143. Tommy Bell, always a hard runner and solid blocker, quietly amassed 75 yards in 13 carries, a better than 5-yard average, and the highest total among all ball carriers that day. Gerry Lodge, the workhorse and decoy who faked so effectively in the "belly series," carried twenty bruising times, most directly into the middle of the Navy line, for 61 yards.

George Munger, Pennsylvania's coach, and Temple University's Al Kawal had more to say. Munger:

[The cadets] handsomely deserved [their] great victory.... Defensively, Army was equally poised and poisonous. It used a 5-man line with four backers-up and two halfbacks deep. Of course, this was varied to include looping lines. And about the only thing Army's defense did do consistently was to rush Navy's passers — a feat Navy couldn't perform against Army either. Yet Army's pass defense produced three interceptions in the first half.

He praised Bob Mischak and Lowell Sisson as "possibly the best pair of ends in the nation." He added, "I salute Blaik for a great rebuilding job."

Al Kawal:

Army operated with electric efficiency yesterday. All winning teams are alert; that's the way you win. But a film of this game would be a great inspiration to any squad in future years. Army was alert on its tackling, its blocking, and its reactions to all situations.

Earl Blaik praised Norm Stephen and Bob Farris for their defensive play as linebackers. "I am very much surprised at what we have accomplished this season," he said. As for stars he couldn't pick one.

They were all great. It was a team performance...a marvelous exhibition of spirit, determination and fine football.... This victory dates back to September 1 when we started practice. Many of these young men who played were the same ones who were trampled here two years ago and to see how far they've come is almost unbelievable.... I did not believe it possible for them to make such a record.

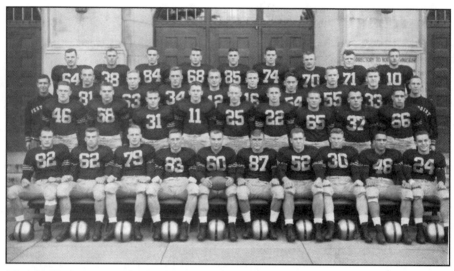

The 1953 Army Football Team: built in adversity, they performed what Earl Blaik termed "a football miracle," returning among the top teams in the nation the third season following the 1951 cheating scandal. Front Row: Lapchick, Ziegler, Krause, Sisson, Lunn (Captain), Mischak, Kramer, Cockrell, Attaya, Schweikert. 2nd Row: Bell, Zaborowski, Wing, Cody, Zeigler, Franklin, Knieriem, Bruno. 3rd Row: Greer (Ass't Mgr.), Doremus, Chesnauskas, Uebel, Mericle, Wynn, Hagan, Lasley, Farris, Burd, Meador (Mgr.). 4th Row: Shannon, Chance, Holleder, Herdman, Ordway, Sullivan, Melnick, Glock, Vann. Not pictured – Stephen. (Photo courtesy USMA Archives.)

"THE TWELFTH MAN LIVES..."

One day shortly after the season ended, Bob Mischak was told to report to the office of the Athletic Director. Quietly, without ceremony, he was handed a certificate saying he had been named to the National Broadcasting Company's All-American team. NBC wasn't one of the nation's prestigious determiners of All-American recognition, consequently Bob didn't, at first, go down in the Academy's record books as an All-American. Neither did anyone else on Army's 1953 team, though Earl Blaik was named "Coach of the Year" by the Downtown Quarterback Club of Washington, D.C., and Vince Lombardi was back on his mountain top, on his way to be offensive coach at professional football's New York Giants.

Yet this small, unheralded team had achieved the "football miracle" of which Blaik wrote glowingly years later. Their miracle unforgettable, yet virtually unnoticed, because the Army team's return to national prominence, back among the great teams of collegiate football, had been so rapid, and so unexpected, that it received little public notice – although the cadets ended the season ranked No. 14 in the nation by the Associated Press poll, and No. 16 by International News Service.

The December 7 issue of *Time* magazine told of Army football's rapid, successful rebuilding, in a brief article under the title "Champions of the East," and included a photograph of Earl "Red" Blaik, his face beaming with a radiant smile seldom seen by his players.

In a letter home on December 15, John Bard, the 1954 Cadet First Captain, wrote his family, "The Lambert Trophy is going to be awarded Sunday in the Dining Hall. This will be the first time [it has been awarded] away from New York [City], and the first time to a football team and its student body."

On December 17 Air Force First Lieutenant John A. Hammack, General Irving's aide, signed a memorandum to "Those Concerned" detailing the "Lambert Trophy Presentation" that would occur on Sunday, December 20, two days prior to the start of Christmas leave. The trophy, first offered in 1936, and symbolic of Eastern football supremacy, had been won by Army in 1944, '45, '46, '48, and '49. But this year was different, as John Bard had written in his letter. The brothers, Victor A. and Henry L. Lambert, of New York City were donors of the award, and decided that this year, they would for the first time, leave the city to present the trophy. What's more they would present it to an entire student body as well as its football team.

As planned, the Lambert brothers and their wives arrived Sunday morning in time for Chapel services and a luncheon at Bear Mountain Inn, a few miles south of the Academy overlooking the Hudson

River. Then the brothers were escorted by cadets on a tour of the Academy and hosted at a social in General Irving's quarters an hour prior to the cadets' evening meal.

That evening in Washington Hall, after supper and a brief five minute speech describing the history of the Lambert Trophy, the brothers presented the trophy to a broadly smiling team captain Leroy Lunn, while newspaper and television cameramen captured the affair on film. Leroy spoke a few words of

Army team captain Leroy Lunn receives the Lambert Trophy, symbolic of eastern football supremacy, in a ceremony in Washington Hall, Sunday evening, December 20, 1953. The two brothers Victor and Henry Lambert first awarded the trophy in 1936. (Photo courtesy USMA Archives.)

acknowledgment and thanks, and in turn, presented the trophy to the Cadet First Captain, John Bard, who after a few words, then presented it to General Irving.

Army cheerleaders led the Corps in a traditional football yell, and the simple ceremony was over.

The next day, buried in the sports pages of the *New York Times* was a brief, five-paragraph article, with the headlines "FOOTBALL TROPHY ACCEPTED BY ARMY." Between Christmas and New Year's, while the three upperclasses were on leave and the plebe class of 1957 was enjoying its first chance to be the upperclass at West Point, General Irving wrote a letter to Victor Lambert.

West Point is indeed honored to be the recipient of the Lambert Trophy and to be the first college to have the presentation ceremony on its campus....

We, of course, share your hope that the Army team will win the trophy again next year; however, we trust that your return visit to the Military Academy will not be dependent on that, and that we shall see you here sometime in the near future....

What brought the 1953 Army football team and the Lambert Tro-phy to Washington Hall the night of December 20 was extraordinary.

Three weeks earlier, on Sunday, November 29, sports pages were filled with stories of the Army-Navy game and were lavish in their praise of Blaik, his coaches, and the small, captivating Army team of thirty-four men who came to Philadelphia's Municipal Stadium a day earlier. But none of the news reports recounted the essence of Army-Navy 1953, or the entire '53 Army season. That was to be found elsewhere – in the brief remarks Blaik made to the team, before he sang his promised song, before reporters were let into the Army dressing room. His words would survive in lifetimes of happy memories. "I never have coached a team," he told them, "that gave me more than you did. I never have coached a team that has given me as much satisfaction. Considering all the condi-tions since 1951, you have done more for football at West Point than any other team in the history of the Academy."

But there was more, much more: A determined band of men in the class of 1954 who wanted to know and remember victories, and led the Corps of Cadets with the thunderous echoes of "GO! GO! GO!" They were the men who fomented a small "Mutiny on the Whiskey," spilled a reveille cannon into a lake, painted and repainted a sign on the side of a Navy ship and photographed it for posterity, conceived the mis-chievous, good-natured goat larceny and rebellion, and the cannon that shook great stadiums with its booming explosions. A Corps of Cadets, whose members in 1951 "walked on" in great numbers to play for their decimated football team, the inexperienced junior varsity and plebe players who became one, inseparable unit of men who refused to re-member they hadn't the skills or experience to play the great game of collegiate football. The Army teams of 1951 and 1952, who felt the sting of adversity, embarrassment, humiliation and loss, sweated, bled, played every game with a fury, to the last second, gave all they had, and more – and laid the foundation for the season of '53. Men in the class of 1955, the smallest West Point class in years, who walked on with all the others, and joined the class of '54's quest for the taste of victory. And '56, with its bevy of talented yearlings who set the grid-iron on fire that fall.

This was a special team, truly a team of "the twelfth man" – many of them, a team of heroes without stars, with different heroes each

915

Saturday, all playing for honor and love of the game. A team about which famed sports writer Grantland Rice said, "They came up the hard way, and there probably has never been a team with a finer spirit."

All, together, became inspired to rise above disillusionment, defeat, and tragedy, and to never give up.

CHAPTER 20

HONOR AND VICTORY REVISITED

A half century has passed since this story began. I periodically browse through the *Howitzers* from those tumultuous years. We were all so young, full of dreams, ideals, bright hopes for the future. Many of us lived our boyhood dreams. Others didn't. They dreamed new dreams, changed their lives, went different directions. Many drifted from the Academy, any association with military service or their classmates.

Then there were the men of Korea, who, like David Hughes, Pat Ryan, MacPherson Conner, and hundreds of thousands of Americans, risked everything and did their duty in the nation's first, prolonged "half-war," the "police action," the GIs' "yo-yo war," "the forgotten war" David Hughes wrote about in his August 1951 letter. They were in a forgotten war some writers and historians said ended in an "uncertain victory." Of the thousands of Academy graduates in service during the war, the great majority came home, many wounded in body and spirit, some maimed for life. There were 157 battle deaths, thirty-four from the class of 1950, nine from '51, and five from the Sesquicentennial class of 1952.

The last to fall from '52 was Thompson Cummings, second lieutenant, Infantry, who was declared missing in action July 24, 1953, three days before the armistice. Thompson had been in cadet Company I-2, was an Army football player on the "Golden Mullets," the B-squad, that first rebuilding season. He didn't win a major "A," didn't become an Army letterman. He never became a nationally known hero, either, though he was posthumously awarded the Silver Star and two Purple Hearts.

But Thompson Cummings was among many magnificent, never to be forgotten heroes of Korea, with Dick Inman and Dick Shea from '52, who gave their "last full measure of devotion" at such a young age. They didn't come home, and many, like Dick Inman, "sleep where soldiers fall." When I put away the *Howitzers* I can still see the faces and smiles of those two men, hear their voices, and the ring of their laughter.

Theirs are among the images and voices of men I knew who fell in Vietnam. Four were classmates. There were more I'd met from other West Point classes. Secretary of the Army Frank Pace spoke of such men at the 1950 graduation, saying that every single class since the Academy's founding had lost graduates in war. Though twenty-one classes lost graduates in Korea, among them men from '50, '51, and '52, the war's end foreclosed participation by graduates in the five succeeding classes. However, those five classes received their baptisms of fire in Vietnam, losing graduates while '50, '51, and '52 added to tolls taken from them in Korea – as did twenty other classes.

Our Company K-1 tactical officer, who, when we were at the Academy, was Captain Herbert O. "Herbie" Brennan from the class of 1946, was one of those men. While I was a cadet at West Point, this handsome young Irishman and then Major Robert L. Royem, our first Company K-1 tactical officer, from the class of 1944, rekindled in me dreams of flying, which eventually led to my seeking a commission in the Air Force.

"Herbie" Brennan had flown combat missions in Korea. Then, in the second war of his life, when he was carrying the responsibilities of an Air Force colonel, his bomb laden F-4 fighter-bomber exploded after a direct hit from anti-aircraft fire during a dive bomb pass near the city of Vinh, in North Vietnam. He was a bright shining beacon in my life, a loving husband, and father of five when he died.

But mostly, the men I knew who didn't return from Vietnam graduated from colleges, universities, officer candidate schools or aviation cadet training. There were no Academy *Howitzers* containing their pictures or the words which described them as young men – although somewhere there are classbooks, pictures in family albums or hanging on walls in homes of people who loved and remember them. But I can still see their faces and hear their voices. Some had been in the Air National Guard or Reserves and were mobilized during the Korean War, the Berlin Crisis of 1961, or the Cuban Missile Crisis of 1962. There were others who lost their lives in military aircraft or vehicle accidents while in training or combat.

Though the losses in Korea and Vietnam were dreadful and left us with heartaches and sadness, these men left us with warm memories, smiles, and gifts to erase the sadness. What gifts they were, and what

918

gifts they passed to us! They honored us with their brief presence, and though they left us, and in our minds, didn't live to see the victory, they did triumph – all of them. The triumph which came to them first, came in another form years later. It was the victory they would wish to see and know – the end of the Cold War.

VICTORY REVISITED

Douglas MacArthur is often quoted at West Point regarding the relationship between victory on "the fields of friendly strife" – the fields of athletic endeavor – and victory in war. His frequent messages to Earl Blaik and the Corps of Cadets during Army's football seasons repeatedly rang with "There is no substitute for victory."

Down through the years these words echo, although, at other institutions of higher learning, and to some extent, at West Point, the word victory is far less frequently used, even on athletic fields. More common is "We will win," "We will achieve our objective," softer phrases, not reminiscent of "V for Victory" and its World War II meaning of unconditional surrender by the Axis Powers – which signified the total destruction of their armies, and the political systems which spawned them.

In 1941, when Earl Blaik's "Uncle Bobby" Eichelberger, the Academy superintendent, asked him to come back to West Point to revive sagging Army football fortunes, Eichelberger's words to the Academy's Athletic Board did have merit. "By the Gods, I believe the cadets deserve a football team which will teach them how to be good winners." Years later, in the heated aftermath of the cheating scandal, some looked askance at those words, implying the superintendent had brought a man to West Point too intent on winning, who would win at all costs. From their perspective, he emphasized winning at the cost of honor and the wasted talents of athletes with great leadership potential, along with many others equally talented, who didn't play football. Blaik's critics and detractors, like Blaik, didn't have all the facts, underestimated, under-investigated, and oversimplified causes which, after years, led to the cheating incident of 1951.

Man-made disasters come slowly, have roots deeply buried, and like their causes, are complex. In any disaster, a clear chain of events leads step by step to the unwelcome end. What, on the surface, might appear valid causes, can be terribly misleading, deceptive. But by careful, determined, objective, factual investigation, the causes for disaster can nearly always be identified.

The game of football does have enormous potential as an educator, and for teaching the real meaning of leadership, winning, losing, and life. In many respects, Earl Blaik's comparison of football to war, as "the game most like war" is valid, and, like many other athletic endeavors, properly used, is invaluable in preparing officer candidates for leadership and decision making on the battlefield, or under conditions of extreme pressure. Physical courage, teamwork, stamina, situational awareness, the need to make instantaneous and correct decisions, effective communication, willingness to learn, loyalty, perseverance against great odds, planning, tenacity, preparation, knowing your opponent, health of individual team members and the team, practice, scrimmage or rehearsal, offense, defense, picking the right people for the mission, relentlessly hard work, willingness to sacrifice for others, and the always powerful drive to win, are characteristics and benefits of playing collegiate football – or many other collegiate sports.

Vince Lombardi said of Earl Blaik, years after the 1950 loss to Navy, that like all other great football coaches, the Army coach was tormented by loss on the gridiron. Blaik geared his whole life and the teams he coached, to victory. He devoted himself to preparations for victory with an unwavering, uncompromising will and determination. Perhaps it was MacArthur's "There's no substitute for victory" that drove Blaik, and later, Vince Lombardi, who often repeated the expression, "Winning isn't everything. It's the only thing." But that's not likely.

All his life, Blaik competed fiercely. He developed a powerful, all-consuming drive for victory – winning, and was a man who would want to excel in any profession he chose. Vince Lombardi was of the same cloth. Both had been raised on winning teams, and passionately believed winning in football not only was symbolic of success and achievement, but was the athletic endeavor that taught the most about life and leadership under the most demanding conditions – in huge stadiums before huge crowds, radio and television fans around the world, in the national spotlight and glare of sportswriters, broadcasters, columnists, Monday morning quarterbacks, outspokenly critical alumni, and fans. If he and his players could prevail under those circumstances and pressures, they could handle anything. Not only that, consistent achievement on the gridiron, winning consistently, was Earl Blaik's vehicle to teach "don't ever give up" – winning in life – very appropriate lessons for the professional military.

Blaik was a master at identifying from among existing talent the best players to fulfill a particular role, to make the Army team the best competitor possible from Saturday to Saturday, under rapidly varying and often difficult constraints. He looked for men with "soul," a word that seemed to accord a religious meaning to men seeking starting positions on his teams. These were the men who would persevere doggedly in spite of apparent limitations such as size, speed, strength, or injuries, who would ignore their limitations, learn, grow, out-think, and out-play their teammates to win a starting position – then on Saturdays perform well beyond what anyone might reasonably expect. These were the men willing to "pay the price," to sacrifice and never give up, achieve over and over again, greatly.

But Earl Blaik learned the hard way in 1951 that a team isn't composed of eleven individuals with souls forged entirely in the competitive fires of football. There has to be a team soul, such as the one that grew from the catastrophe and adversities of that fateful year, and matured in 1953 on that smallest of Army teams. Blaik didn't see what was coming in the seasons preceding '51 because young men who seemed to have the background, desire, and qualities necessary for the Academy and intercollegiate football fooled him. They hid their flawed rationalizations and actions from him and many others, and he was overzealous in protecting them, enabling them to further deceive themselves and rationalize what they were doing. He was too distant, too removed from most of his players – even some of his finest players, to see the whole man.

There is far more to victory than the score or even the victory itself. There has to be team integrity, a moral integrity and force at the core of victory, a sense of victory fairly won – if victory is to be consistently won, have a lasting, positive influence, and be respected.

He also learned that his assumption that the football team was the "supreme rallying point for the Corps," and the Corps was obligated to support the team simply because it represented the Corps and was a winning team, was misplaced. If the team was losing its soul, or the Corps was beginning to lose its soul along with the team, which both were, neither would learn the true meaning of victory. It would all inevitably come crashing down. And it did. Then, and only then, did Earl Blaik realize he had erred in not seeking badly needed direct contact with the Corps of Cadets, at least through their cadet leaders.

That important change began in 1951. The Corps immediately drew closer to the football team because men who hadn't played on the Black Knights, along with recruits in the plebe class, were suddenly the only sources of talent available. Blaik and his coaches came to the Corps for help while many of its members tried "walking on." Additionally, nearly every Saturday afternoon that fall, Army was the underdog, which only fueled the fire and determination to put together a team that would bring victory with honor, and the best kind of pride.

By the fall of 1953, Earl Blaik had become more directly involved with the Corps of Cadets than ever before. The results were electrifying as his small, gritty, inspired band of Black Knights, the class of 1954, and the entire Corps of Cadets were united once more. The emotional victory over Duke University that October 17 was the turning point on the road back, and in the minds of the team and the class of '54, who were the Corps' cadet leaders, made Army unstoppable on the gridiron for the remainder of the season.

Competition. Victory. Someone, or some team wins. Their opponents lose. The competitive drive is deeply embedded in the American psyche and everyday life, and is accepted as the norm in bringing out the best in individual and organization performance. But it can also bring out the worst, if not carefully checked, kept on course, concentrated, kept in perspective.

Competition was endemic to almost everything we did at West Point, as it is in the armed forces, and for good reason. Who are the best soldiers in a squad? Who can be counted on to do the best work, make the mission a success? Which squad is the best in the platoon? Which platoon is the best in the company? Which company is the best in the battalion? Which battalion is the best in the regiment? Which regiment is the best in the division? Which division is the best in the corps? Which is the best corps in an Army? Who is the best squad leader, platoon leader, or company commander? Who from among a battalion's four company commanders can perform the best in leading the assault on a battalion objective?

The word competition wasn't often heard at West Point, except in competitions between companies in drill and ceremonies or intramurals, and, of course, intercollegiate athletics. We knew we were in competition with one another, to some degree, within the framework of the entire evaluation system, because there was an order of merit established using weighted averages. But it was normally accepted as a part of cadet life and not seen as overt, direct competition with one another. Obviously that wasn't true for all cadets, because some saw the system as unfair, emphasizing academics over leadership, physical education, or athletics. Quite often such complaints reflected more what a cadet wasn't particularly interested in, didn't believe important to his future, couldn't understand, or personal frustration, because he couldn't achieve results or knowledge pleasing to him.

"The system isn't fair. It'll screw you," was the bitter lament of one cadet who resigned the summer of 1951 and who had used his view of "the system" to persuade another it was all right to cheat in academics. The order of merit became the graduation order of merit. The order of merit was used in the spring prior to graduation to choose branch and service assignments.

Then there is the ever present competition and tension between some officers within the three major departments – academics, military and administration, athletics – who are responsible, collectively, for educating and training cadets. The cadets, looking at those three departments, often saw one or the other as obstacles on the way to a commission, obstacles which had to be overcome, run around, or slipped through with minimum effort. Or they saw the departments according to their own likes, dislikes, or interests. For some, it was "me against the tacs" – the tactical officers, the disciplinarians. Or it was "me against the academic department," math, English, mechanics, or electricity. It was "the academic wars," "the battle for tenths," the battle to avoid excessive demerits, "slugs," and confinement, to some an endless list of small wars to be won on the road to graduation.

For those struggling to hold on, the competition became serious. "This is my lifelong dream." "I want to be an officer." "I need this education because my parents can't afford one anywhere else, certainly not like this one." For young men with roommates, close friends, and teammates they deeply admired, the competition, the drive to succeed, to win over all obstacles also took on important meaning. "The system is asking too much of him. There's nothing wrong with helping win this battle. After all, isn't this about teamwork, self-sacrifice, loyalty, friendship, lifelong ambitions about to be wrongly discarded by the Academy?"

Earl Blaik was an absolute master at expecting, demanding, preparing, inspiring, and leading young men to greater performance on the gridiron, more than they knew to be possible. During that era, he was undoubtedly one of the best, if not the greatest competitor among the nation's collegiate coaches. On the practice fields and in stadiums on Saturday afternoons he was a marvelous leader who knew and understood the game, both technically and philosophically, and selected and developed the best from talent available, motivated them, inspired them, carefully scouted opponents and prepared his teams to play them, and showed his teams, week after week, and season after season, they could give more than they knew they had, and consistently win. And when matters were at their worst, at the end of the summer of 1951, he swallowed his personal sadness, professional disappointment, and wounded fierce pride, worked through his own self-interest and conflicting emotions, and led his players through their adversities to a football miracle in just three seasons.

While doing all that, he came directly to cadet leaders in the class of 1954, asked for and, whether he knew it or not, had already received their enthusiastic support, which had had its beginning, unknown to either, in the summer of Earl Blaik's greatest personal and professional discontent. Thus in the fall of 1953 was built a stronger, more deeply admired football team than had been at West Point for years. And Blaik, Vince Lombardi, all the Army assistant coaches, the Corps of Cadets, and the Academy, did it with a far smaller pool of talent.

The return to football glory by the Army team of 1953 was an extraordinary achievement. The team was more than worthy of the lavish praises Earl Blaik gave them at the end of the season and the rest of his days. It was one of the great football stories of the century.

For the team members, the Black Knights' victories that season stood among the best of a long, winning, Army football tradition. In winning as they did, each member triumphed over adversity. Though they didn't achieve success entirely on their own, their triumphs were largely over themselves, as well as their opponents, and came from the heart and soul they developed within.

Victories on the gridiron and in wars are far different, however. While football might reasonably be described "the game most like war," it pales in comparison with war's realities.

* * *

Throughout the early fall of 1953, while we in the Corps of Cadets enjoyed the thrill of Army football's return to glory, Barbara Inman and Joyce Shea suffered through the agony of not knowing the fates of their husbands.

Barbara had already taken steps on her own to find Dick Inman. She entered Red Cross training in the summer instead of continuing to teach. She was intent on working her way to Korea to search for him. Meanwhile Dick's parents and friends sought help through other sources. Aside from A. R. Weathers' August 15 letter to Congressman Bray, Howard Greenlee, publisher of the *Vincennes Sun Commercial*, contacted Bray as well, anticipating the Congressman's visit to Korea. Greenlee asked him to seek answers to Dick Inman's status. Might he still be alive?

On Tuesday, October 6, 1953, three days after Army lost to Northwestern in Evanston, Illinois – the site of the Inman family's first joyous trip to see Dick Inman play varsity football, in a game he never entered – Congressman Bray responded to the *Sun Commercial* publisher. The letter Congressman Bray mailed that afternoon brought devastating news, ending dreams for the future, for Barbara and Dick, and the Inman family.

> Hong Kong, B.C.C.
> October 6, 1953

Dear Howard:

I am writing this letter in Hong Kong but will mail it this afternoon in Manila. While in Korea I went to the Headquarters of the 7th Division and talked to one officer and three of the men who were with Lt. Inman. I am also including statements made by these men. My conversations included more than is included in these statements, and in my opinion there is no doubt but that Inman was killed. The Commanding General is already making attempts to recover the body. Lt. Inman was most courageous as was plain from my conversations with those who were with him. As soon as I get any additional information I will notify you...

Will be seeing you in November.

Sincerely,

William G. Bray

Enclosures:
Four sworn statements
One diagram

Now began the unwelcome finality of grievous loss, and a void never filled.

The officer Congressman Bray talked with on his trip to Korea was Lieutenant David R. Willcox, 2d Platoon, A Company, 17th Infantry Regiment. The three soldiers were Private First Class Irwin Greenberg, Corporal Harm Tipton, and Private First Class Clarence Mouser, men who returned from Dick Inman's ambush patrol. Dick's actions led them to safety on Pork Chop Hill, and new leases on life that terrible night of 6-7 July 1953. Each had written sworn statements and signed them September 27.

Dick Inman did his duty leading his patrol. His reactions in desperate circumstances were striking, gallant, courageous, and to those who haven't known war, incomprehensible. He could have ordered the killing of an enemy soldier who, he was to learn almost at the cost of his life and the lives of others, was part of a deadly trap for him and his men. He refused. He could have led his men off the hill instead of attempting to fight to rejoin 2d Platoon of A Company. No one would likely have known the difference if he had done so, and simply returned to B Company, the battalion reserve which soon came forward to Pork Chop to reinforce A Company.

But Dick Inman would know what he had done if he had taken the easier course. Instead, when he saw what had occurred, that A

Company's 2d Platoon defenses and Pork Chop Hill were in danger of being overrun, he didn't hesitate. He chose to lead his patrol to the aid of David Willcox's platoon, to resupply his patrol members with ammunition, and tried again and again to fight his way through.

He could have taken cover in the trenches and bunkers, and waited with his patrol, staying with those who no longer had operating weapons or were out of ammunition, waited for support or relief. He didn't. He was either wounded or injured his leg during the fight in the trenches and bunkers, and had called his status in on the radio or telephone. He could have justifiably evacuated himself to the care of medics and an aid station. He didn't.

Finally, when ammunition was almost gone, and they had no more grenades to throw, he looked to the safety of the men who had no weapons with which to defend themselves. Telling those without weapons to remain in the relatively protected cover of the trenches and bunkers until the right moment to dash to safety on the south side of the hill, he and the rest took their weapons and made one last attempt to fight their way to the 2d Platoon CP and ammunition supply bunker.

His last act was to leave the confines of the trench and bunker network, which had repeatedly been blocked by grenade throwing and burp gun-firing Chinese soldiers, and provide covering fire for the remainder of his patrol who went with him. They went past him toward safety. It cost him his life. He risked his life for his soldiers, without hesitation. He gave them a chance for life before saving his own, and forfeited his own, a selfless act of gallantry and sacrifice, the "last full measure of devotion." His soldiers came first in the final act of his life. "Greater love hath no man...." And in return, Dick Inman's soldiers tried desperately to retrieve him when he was mortally wounded. This is victory from within, triumph from within, honor unbounded, and sacrifice to be remembered.

Dick Inman lived and died the beliefs he cherished. His pen is still, but his poetry lives. He didn't reach his goal of rocketry or missiles, and his manuscript, *Promise for the Future,* was never published, though Barbara tried mightily to fulfill that dream for him. In Korea, he still sleeps where he fell with many others, on Pork Chop Hill in the silent, vacant no man's land called the DMZ. Sleeping where he fell, he had conveyed with lines of poetry to Barbara before he left her, his love for the Vincennes home and the Hoosier state he remembered.

In Korea, Dick Inman was living the early stages of the future's promise of which he wrote in his unpublished manuscript. Though in his mind's eye he saw the promise of world government to end the scourge of war, he was also witness and a soldier in the first war United Nations forces fought to stem the tide of this century's new totalitarians, a tide rapidly expanding to fill the vacuum of misery and chaos left by World War II. Sadly, the winning of freedom and democracy, that form of elected government "of the people, by the people, and for the people" has seldom come without revolution or war.

For the first time in the bloodiest century in human history, eighteen of the United Nations banded together, along with South Korea – and much help from Japan – to stand against the newest totalitarians of the twentieth century, in the first major war within the forty year span of the Cold War. The eighteen nations and the Republic of Korea fought for many reasons, some not so idealistic. No matter the reasons each entered the fight, they saved a fledgling democracy in what many Americans considered a half-war, a forgotten war, and to some, life as usual. Those who fought saved democracy, saved a republic however imperfect it was, made democracy possible in the southern half of an ancient land that had a history of emperors, warlords, and dynasties, and nearly a half century of brutalizing Japanese occupation. There had been no republic and no democracy in its history, until after World War II.

The United Nations and the Korean War, and the sacrifices made there, may one day be looked upon quite differently. All but a few of the eighteen were Western democracies. As in World Wars I and II, nations banded together against dictators, tyrants, and their totalitarian governments and armies, whose lofty, ringing ideals were twisted, distorted, turned inward with terrorizing force, outwardly aggressive, and by their nature worked against peaceful discourse among the nations of the world. While doing so, while fighting the first, major "half-war" or "limited war" they helped all nations avoid World War III.

The call to duty in Korea was abrupt. The war placed ugly demands and impossibly stark choices on American soldiers, sailors, marines, and airmen. It was unbelievably frustrating for professionals in the armed forces who knew through painful experience that once war was joined, the best and quickest way to end it is with overwhelming force, well planned, well applied, and with the will to fight it vigorously and violently if necessary, to keep loss of life to a minimum.

But Korea's circumstances didn't permit that kind of war, and the result was two years of stabilized, bitter trench warfare, and far more bleeding by both sides. For the second time in nine years, our nation wasn't prepared for the war it confronted, was surprised, and paid a terrible price. At the end of the long period of fight-fight, talk-talk, places like Pork Chop Hill became metaphors for the entire war. Utter frustration.

United Nations forces were deliberately held in place, along a line that was defensible, refraining from major assaults designed to break out from the trench warfare, destroy the enemy armies, and tear the enemy political system out by its roots. There were larger strategic issues. The fear of a World War III with China and the Soviet Union allied against the West was real, and increased when China, a nation of 550 million people, entered the Korean War.

Yet, in spite of it all, the sacrifices made have brought the gifts of democracy and freedom to the Republic of Korea, and protected those same hard won gifts for the United States, and all Western democracies – enormous gains by any measure. Democracy has taken root in the Republic of Korea, is growing, maturing, bringing the young Republic more democracy, freeing the enormous creative energies of its people. Northeast Asia remains secure and the world watchful of Kim Il Sung's son, Kim Jong Il, and his totalitarian dictatorship and armies. For in the tradition of ancient emperors, Kim Jong Il, his father's heir, has now become the first communist dictator in history to succeed by family blood. His succession makes him a modern day emperor over half the Korean peninsula.

But the gifts won and protected by thirty-seven months of war on the Korean peninsula are testaments to the courage and devotion to duty of American soldiers, sailors, marines, and airmen, and to one another. Their achievements in holding firm with seventeen other nations, to the fledgling Republic of Korea, and ejecting the aggressors, was the beginning of vigilance and the will to continue standing firm against the dictatorship which still exists to the north. Although on

July 27, 1953, when the killing stopped, there were no celebrations in New York City's Times Square, and few anywhere in America, for the soldiers, sailors, marines, and airmen of Korea, it wasn't "the forgotten war." Korea was "the forgotten victory."

HONOR AND TRIUMPH

In November 1953, as the Army team prepared for its annual clash with Navy and continued the climb back toward recognition and respectability on the collegiate gridiron, Pat Ryan came home from Okinawa and the air war over Korea, for a joyful reunion. He was anxious to see his beloved "Foxie," and their son, Patrick. After some time on leave, they went to his next assignment, the Air-Ground Operations School in Southern Pines, North Carolina.

Pat's experiences were far different from those of David Hughes, Dick Shea and Dick Inman. Early in 1951, as the Corps of Cadets' First Captain, he, like "Ace" Collins, voiced deep misgivings about what he saw and heard in the Corps, and called his observations to the attention of Colonel Harkins. Like Dick Shea, Pat was an experienced former enlisted man when he came to West Point, and later became a member of the class of '51. He had concluded something was wrong in the Corps his first class year, but couldn't pinpoint the causes for his feelings, or find evidence confirming wrongdoing in the Corps. Not until June week did he learn the truth, and the magnitude of the activities that had been so difficult to find, and yet so deeply troubling. And he left West Point after graduation not knowing the full extent of the disaster until the August 3 public announcement.

When he went to war in the skies above Korea, he wasn't a platoon leader. He was a crew member, and second in command of the huge B-29 bomber. Had his aircraft commander been killed or incapacitated in flight, Pat would have assumed control of the aircraft and command of the crew, and been responsible for the mission and safe return of the crew. Fortunately, that never happened. Lieutenant Trudeau, Pat, and the entire crew miraculously survived several harrowing missions to return home. Pat's first tests of perseverance, leadership, and courage as a junior officer did come in Korea. He met them all. More severe tests would come years later, as a squadron commander, in the even more frustrating third war of his life, Vietnam.

* * *

While Pat Ryan was en route home to Joanne and their son Patrick, and on to his new assignment far from the rigors and dangers of combat and peace keeping missions, Joyce Shea was still hoping against hope Dick was alive. She had heard nothing more from the Department of the Army or Lieutenant Colonel Read.

The award of the Lambert Trophy at West Point on December 20, 1953, followed by Christmas leave for the Corps of Cadets, came and went. For the Inmans, Sheas, Riemanns, and Kipps, that Christmas couldn't possibly be the same as in years past. Christmas at the Riemann family home in New Milford, New Jersey, bore an air of melancholy, as did the holidays for the Shea family in Portsmouth, Virginia, the Inmans in Vincennes, Indiana, and the Kipps in Fort Madison, Iowa.

In Indiana and Iowa many questions lingered, though uncertainty vanished about whether Dick Inman was alive. The vague outlines of how and why he died were in the statements included in the letter Congressman Bray wrote in Hong Kong. However, uncertainty remained as to whether Dick Inman's body would ever be recovered and returned home for burial.

In New Jersey and Virginia, there were tenuous shreds of hope while waiting continued for further communications from the Adjutant General about Dick Shea. But no letters arrived for Joyce, not by the first of the year. Then, in January she received a visitor from Able Company.

First Lieutenant David Willcox came to see her at her parents' home in New Milford. Richard Thomas Shea III was not yet five months old. Joyce still hoped Dick would come home despite the shattering news Lieutenant Colonel Read's August letter brought her.

David Willcox had a sad duty to perform. He couldn't provide Joyce new hope, only one more piece of information toward the finality Barbara Inman and Dick Inman's mother and father endured in October. The good things that happened to A Company's 2d Platoon, and the man David Willcox knew Dick Shea to be, fueled in David a powerful obligation and courage to perform the duty of bringing more sad news to Joyce.

He was on the hill with Dick Shea. The two had had frequent contact with one another before the battle. Though not close, fast friends,

931

they had worked together, trained together, planned together, shared the same dangers, fought the same fights when A Company was on the line, and had had a growing respect for one another.

All twenty-five men in David's platoon had survived those first three days of fighting, though all but a few were wounded. Dick Shea hadn't. He wasn't among the 2d Platoon men who came off the hill when Charlie Company relieved them on July 9, nor was he among the G Company soldiers who were relieved on July 10, or men like Dale Cain, who came off the hill on July 11, when the last of the Americans and ROK soldiers were evacuated from Pork Chop.

David Willcox didn't see Dick Shea fall, wasn't near and couldn't see his last personal battle. None who came off the hill and were close to him in the final moments, survived to tell what they had seen. Yet, until the Army verified Dick's death, David couldn't tell Joyce Shea with absolute certainty that Dick hadn't survived, though in David's mind he was convinced. He could tell her, however, that a work party which had braved a heavily mined Pork Chop Hill, in the DMZ, had found a shallow grave on November 12. The grave contained the remains of several men. Facts and circumstances known about her husband's final moments suggested one of those found was Dick. She would be told if comparisons of remains with records verified the remains were his. There really wasn't any hope.

The work party David Willcox described to Joyce was similar to the one in which Sergeant Dan Peters participated on October 28, four months after the last battle for Pork Chop. Dan, a member of Charlie Company fought on the hill during the counterattack that relieved Willcox's 2d Platoon. All twenty men chosen for the work party had been on "The Chop." Early that October morning, Peters and the other nineteen men were issued DMZ passes, transported unarmed to a checkpoint in the northern half of the DMZ, searched by Chinese soldiers, divided into two and three man teams accompanied by an armed Chinese soldier, and rode in trucks with the Chinese to Pork Chop. They recovered the remains of fourteen or fifteen American soldiers in the western sector, the left flank of the hill. Among those found was a soldier Dan Peters had known. The Chinese had found the partly buried bodies and invited the Americans to come retrieve them.

After telling Joyce the shattering news, David Willcox talked with her, telling her about men's reactions in war, their personal responses to the dangers, chaos, and fears that swirl around them. Each man reacts differently, some with towering calm, courage, decisiveness, selflessness, and inspiration under the most extreme conditions. Others are overwhelmed with self-consuming, self-destructive fear, sometimes fleeing, throwing down their weapons, cowering, whimpering, attempting to hide from the grotesque, abject horror they find themselves in – and thus greatly increase the risks to fellow soldiers. Dick Shea was the former, a tower of strength and inspiration.

Joyce was once more numbed by the news David Willcox brought, and remembered few of his words about men's reactions in the extremes of war. The news was simply too stunning for Joyce. It seemed each time she talked with or received letters from men who had known Dick or had information about when last he was seen alive, Dick was killed again. But not in her heart, not in her dearest wish, or her dreams. The desperate hopes and stinging setbacks were almost too much to bear. Yet, she clung to the notion he would be found alive in spite of David Willcox's difficult visit and Lieutenant Colonel Read's August letter, received days after Joyce and Dick's son was born.

* * *

Condolences continued to come in to the Inman's and Barbara, through Christmas and on into January. An unusual and touching letter and package arrived in the Inman home in Vincennes at the end of February. The return address on the envelope was the Director of Athletics at the University of Pennsylvania.

February 23, 1954

Dear Mr. and Mrs. Inman:

Some time ago your good friend, Frank M. Dobson of Carlisle, Indiana, informed me of the tragic death of your son Richard in Korea, and he mentioned certain circumstances of Richard's death that involved the loss of the Penn Relays watch.

We are very proud of our Relay Carnival and very proud your fine son participated in it in 1952 while a cadet at the United States Military Academy.

If you will accept from me on behalf of the University of Pennsylvania a watch identical to that which was awarded your son in 1952, I would consider it a great compliment to the University of Pennsylvania. I send this in the hope that you will place it with other mementos of your son's outstanding military and athletic career.

My sincerest sympathies go out to both of you at your loss of such a fine son.

Sincerely yours,
Jeremiah Ford II

*   *   *

The final answer came to Joyce Shea just prior to Easter of 1954, in the form of another letter from the Adjutant General's office.

21 April 1954

Dear Mrs. Shea:

I am writing you concerning your husband, First Lieutenant Richard T. Shea, Jr., 0 66 428, Infantry, who was reported missing in action in Korea on 8 July 1953.

Information has now been received that your husband was killed in action in Korea on 8 July 1953. The records of this office are being amended accordingly.

The Office of the Quartermaster General, Washington 25, D. C., is responsible for furnishing information on recovery, identification and disposition of remains of our dead. It is customary for that Office to communicate promptly with the next of kin upon receipt of definite information from the overseas command...

I know the sorrow this message brings to you and it is my hope that in time the knowledge of your husband's sacrifice for his country may be of sustaining comfort to you.

My heartfelt sympathy is with you in the great loss you have sustained.

Sincerely yours,
JOHN A. KLEIN
Major General, USA
Acting the Adjutant General

A few days later a second letter came from Colonel John D. Martz, Jr., in the Office of the Quartermaster General. He gave more detailed information about the recovery and identification of Dick's body, telling Joyce he wasn't yet certain when his remains would return to the United States. Martz would notify her by telegram and asked that she inform him where she wished her husband's remains interred.

Then came a second letter from Lieutenant Colonel Read. When he wrote Joyce this second time, he had been reassigned as the executive officer of the 17th Infantry Regiment. Once more, in heavy, broad strokes of the pen, his words spoke of deep personal anguish and regret for all Joyce had been through since Dick had been declared missing in action.

Korea
Easter 1954

Dear Mrs. Shea,

In all my life I have never been more crushed and saddened by the turn of events which caused you so much pain. I hope and pray that Higher Headquarters has given you the information which you dreadfully needed. I have been given permission to write so I am sure that you have been notified.

I know that explanations are difficult if not impossible to understand but I shall give you all of the information I was able to uncover. The hardest thing was not being able to write.

Please don't think ill of the system because in many cases it averts a situation such as yours. First, after receipt of your letter I personally went to every higher headquarters in Korea and begged that information be expedited on Dick. When a casualty occurs, frequently Department of the Army will rightfully hold on the report to relatives until all available witnesses are questioned and sometimes these are in the hospital or unable to speak at that time. Then if witness statements conflict and are not considered conclusive additional checks are made. This is necessary because many times a man may see a friend hit and report him killed. Actually, the man may be in a state of shock and recover or become a prisoner. In some cases witnesses give an oral report to a unit leader and subsequently the witnesses are killed or badly wounded. Their hearsay evidence cannot always be considered conclusive. Both of the above things happened in Dick's case. One of the wounded witnesses is in this status now. Lt. David Willcox. He was not a witness to Dick's actual death but saw him after the action. As you probably know Dick was fighting in a particularly vicious sector where many men were hit, bunkers were collapsed and totally destroyed, and in some cases the whole trench and bunker pattern was wholly unrecognizable to people counterattacking only a matter of hours later. We brought back all our dead and wounded who could be found even though we were under attack at the time our unit was replaced by another battalion. After waiting the prescribed 45 days as regulations prescribe, the regimental Commander Col. Ben Harris and I were informed through personnel channels that Dick's status had been established as killed in action. Obviously, I already knew this from being out there but the letters Col. Harris and I wrote to you were later stopped at higher headquarters for more conclusive evidence. Dick was listed as missing.

I know how you must have felt. I was in despair. Neither I nor my regimental C.O. was notified of the change until your letter arrived which indicated that my [August] letter had been sent on to you but his had been stopped. No criticism implied. People in higher headquarters had thousands of reports pouring in from Eighth Army in the last Chinese pushes. In the vast majority of cases considerate and efficient handling of letters and reports was the rule. Dept. of the Army notified me through channels of your letter and I was told not to write again until the official notification came through. That is why I haven't written. God knows the sleepless and tortured nights have been long for me and I know they must have been dreadful for you. I am still afraid that this is a poor expression of my feelings. As a man who has been in the Pacific three times in ten years and in two wars I can only say that Dick was the finest leader I ever saw in action, and his actions will inspire many men who are left....

I hope that in future years you will allow me to keep in touch with you and your youngster. This is not a stereotyped letter or offer – I am available for any help you may ever need of any type. My heart rides on this pen and I can only say I pray for a full and happy Easter to one whom I know has the same courage and character that her husband so gallantly displayed.

As ever,

BM Read
Lt. Colonel, Infantry
(ex) C.O. 1st Bn, 17th Inf.

Dick Shea was buried in Portsmouth, Virginia, the first week in June 1954, two years after his graduation from the Academy, and their marriage. At Joyce's request, Bob Shea, Dick's brother, escorted his body home for burial when it arrived in the States. Joyce's mother and father, Dick's family, and many friends attended the funeral. She remembers, "Sometimes, when we do all the required moves at a time of sadness, because we have to, it's like walking through a maze. You cannot believe this is reality, but it is."

* * *

One month later, on July 6, 1954, in the auditorium of the First Christian Church in Vincennes, Indiana, Colonel Frederick B. Mann, commanding officer of the Indiana Military District for the U.S. Army, posthumously awarded the Silver Star "for gallantry in action" to Richard George Inman, pinning the award on Barbara's blouse in a brief, moving ceremony. The date marked the first anniversary of the Chinese assault on Pork Chop Hill, and Dick's sacrifice for the men in his patrol. The Inman family, and some friends, were at the ceremony, as were Barbara's parents.

* * *

In May of 1955, in Washington, D.C., Joyce Shea received for her husband Richard Thomas Shea, Jr. the posthumous award of the nation's highest decoration for gallantry in action, the Medal of Honor. Dick Shea had "distinguished himself by conspicuous gallantry and indomitable courage above and beyond the call of duty in action against the enemy near Sokkogae, Korea, from 6 to 8 July 1953...."

Like Dick Inman, Dick Shea could have taken the easier, safer course. Instead he went immediately to the threatened area in the western sector of Pork Chop Hill as soon as reports indicated trouble. He knew the entire defense network perhaps better than anyone in A Company. He had been hard at work strengthening it, preparing for the attack the command knew must come, supervising the improvements, ordering and laying in more ammunition, grenades, communications, food, water, and medical supplies. He had sketched the layout and could move rapidly, confidently from one section of defenses to another, in the dark if necessary.

Not only did he repeatedly engage in fighting, organize counterattacks to retake positions lost, he distributed badly needed ammunition from one sector to another, and at times acted as a runner for Lieutenant Colonel Read and Captain Roberts, the A Company commander. While moving from sector to sector, carrying information, orders, and ammunition to defenders, he brought back reports on the status and locations of defenders, telling the commanders who held which bunkers and fighting positions, information essential to vigorous defense and successful counterattacks in a chaotic, rapidly changing situation.

He fought to eject the enemy from the tunnel and trenches near the company CP. He fought on the left – in the western sector – which was under almost constant attack and threat of collapse the entire time A Company was on Pork Chop. He organized and led counterattacks near the CP, on the left, in the western sector of the hill, and finally, on the right, in the 2d Platoon area, where he organized men from G Company late the afternoon of July 8, and launched yet another series of counterattacks in a bid to complete the relief of A Company's 2d Platoon.

Dick Shea could have taken any number of actions to save himself, along with the men of David Willcox's platoon. As Lieutenant Colonel Read observed, he could have stayed with the men of A Company, and waited in hopes the assaulting G Company force would be reinforced and resume its counterattack. He fought on in spite of wounds.

There were unquestionably other factors urging Dick Shea's last gallant acts of courage and valor. He had been an enlisted man. He knew the enlisted soldier's life. He knew the importance of strong leadership, and in his two and a half months in Korea he had learned the

irrefutable, powerful virtue of good leadership in a crisis. Undoubtedly, when he saw G Company soldiers pinned down, leaderless, and
expected to perform a difficult mission under withering fire, he understood their unspoken need better than anyone else. He could have ignored the responsibility he felt, and asked another to take the initiative
he believed necessary. G Company soldiers weren't his responsibility.
He probably knew none of them. But Dick Shea wouldn't leave them
to their own devices, or void of leadership. There was something else
in his makeup. "If you only do what you are ordered to do, and nothing
more." Dick Shea wasn't that kind of man, not that kind of soldier, not
that kind of officer.

In Korea, as in every war there are men who shine like bright lights
when the situation is desperate, even hopeless. They are rocks, unshakable, calm, deliberate, able to make rational decisions amidst terrifying, irrational, chaotic, life-threatening conditions. They seem to radiate their energies, their souls, and reach outside themselves to help
or lead others, showing absolutely no concern for their own safety, and
every consideration for their men and the mission they are given. They
are golden, marvelous under pressure, and strengthen everyone around
them. Their voices, the fire in their eyes, their presence, though they
might be wounded, lift others to fearless, determined action, urged on
by an unspoken, unbounded confidence in the rightness of their actions. Such men can draw magnificent responses from those around
them, as David Hughes did on Hill 347 in October of 1951, and Dick
Shea did repeatedly on Pork Chop Hill those three ferocious days before he died.

Luck is unpredictable, however. In war, at longer ranges, shells
and bullets are random in their effects. They aren't aimed at specific
individuals. They are barrages or concentrations zeroed in on coordi-
nates marking geographic locations on maps and the earth's surface,
where the enemy is observed or likely passing through; covering or
supporting fire with rifles or automatic weapons aimed at areas or points
where the enemy is known or believed to be; fields of automatic weapons fire pre-planned for machine guns on offense or defense for the
same reasons artillery or mortar barrages occur. Nevertheless, from
long ranges, heavy, concentrated artillery and mortar barrages, and
covering machine gun fire, can decimate infantry caught in the open,

ricochet and rip through openings in bunkers, ruthlessly smash through overhead cover, cave in trenches, and wreak havoc on defenders.

Day or night, at long range, overwhelming numbers of onrushing enemy infantry, are both awe inspiring and fearsome sights. As their distance closes, and if their ranks aren't shredded by artillery, mortars, grenades, and automatic weapons, defenders' confront the harsh reality of close quarter, hand-to-hand fighting with the possibility of being overwhelmed by sheer weight of numbers. This is one of many times fear can destroy defenders' rationality, and increase the killing on the battlefield.

At the end of a final surge in an infantry assault which carries overwhelming numbers into defenders' trenches and bunkers, when showers of hand grenades have failed to stop the advance, matters are far different, and far worse. Pistols, rifles, and submachine guns, fired in a close, seething, vicious fight are aimed point blank at specific threatening targets, the closest enemy soldiers. With ammunition gone or weapons malfunctioning, bayonets, knives, rifle butts, empty weapons, even helmets swung like clubs become brutish necessities for killing and survival.

The temptation to flee, run away, "bug out," take cover, hide – not participate – hovers over soldiers, ready to pounce on and smother the will to stand one's ground and fight. But retreat into a whirlpool of single-minded, all-consuming self-preservation is a disaster for individual soldiers, and potentially far more disastrous for fellow soldiers around them. Mutual support, one for another must be the watchword.

Such actions are intolerable for an officer responsible for leading soldiers who are owed the best possible leadership an officer can give. Officers and noncommissioned officers are expected, and must deliver, a calming influence and correct, rational decisions under the most irrational circumstances. They are expected to radiate a lack of fear, exhibit unbending self-control and consistent steadiness, when inside they may be swimming in fear, terrified.

If in the attack or counterattack, the enemy rushes defenses with overwhelming numbers and firepower, defenders' luck is far more likely to run out. Luck did run out for Dick Shea late in the day of July 8, 1953, in spite of all the good he was, and all the good he had done in his life. And in those final days of his life, virtually everything he did was for others, soldiers from A Company, soldiers whom he had never met from other companies – and whose names he didn't have time to learn. He led and fought for them, beside them, for their safety and survival, for their successful mission, that they would have longer lives.

For Dick Inman, luck ran out early in the battle, about sixty hours earlier. The actions he took were no less gallant, no less courageous, no less giving, than those of Dick Shea's. A bullet found him sooner, in the early minutes of the battle, robbing him of life and actions that might have been equal to Dick Shea's in sustained selflessness, courage, gallantry and devotion. For Dick Inman, there were men who survived and could tell his story, as did the men who fought beside David Willcox in Bunker 53, and told of his gallantry while wounded and growing weaker by the hour.

David Hughes', Pat Ryan's, and David Willcox's luck held, as did MacPherson Conner's and thousands more. They came away and lived, most knowing they were lucky to be alive, blessed, ever thankful. What's more, most who fought and lived have gone through their lives mindful that those who didn't return weren't only less fortunate than them, but their fellow soldiers who fought beside them and fell gave them the gift of longer life, a deeper appreciation for life, and the opportunity to live and breathe the air of democracy and freedom, for which most fervently believed they fought.

As we who survived move toward the twilight of our lives, the memories of lives ended and left behind, are revived, renewed, rededicated, and memorialized. The memories come flooding back in unit reunions, and the pooled recollections of history lived, wars and battles fought, horrors, and the hundreds of magnificent stories of devotion, courage, and valor long buried and seldom discussed. These men come together just as did the men of the Civil War, World Wars I and II, and every war before and since in the history of our nation. The memories of those who didn't return become bittersweet, a mixture of joy, pride, and melancholy. Pride and joy for having served with them, having known them, having been beside them, and knowing what they gave. The tinge of melancholy for knowing they didn't return, were less fortunate, didn't become what they might have been, didn't see their families again, didn't see their children or grandchildren grow up, didn't enjoy the fruits of their sacrifice.

On Pork Chop Hill and hundreds of other hills in Korea, and in its valleys, is hallowed ground, where men from eighteen nations and the Republic of Korea fell. They fell from the air and on the ground. Others rest in the deeps of the waters on either side of the peninsula, in rivers and other inland waters. They honored each of us, as did their wives, sweethearts, and families.

When Dick Shea's and Dick Inman's fates were finally firmly established, long periods of sadness were just beginning for Joyce and Barbara and the families of the two men. For them the Korean War never ended. Like thousands more who lost loved ones in Korea – some forever missing in action – as in wars past, they learned time doesn't heal all wounds. The void is never completely filled, the loss never forgotten. Instead, the wives and families learn to live without lost loves, and children's lost lives of promise, only two of war's terrible outcomes. Happiness returns, if at all, with great difficulty. Solace for losses is never easily found, the sacrifices seldom fully understood or accepted. Carried in their memories and hearts for the rest of their days is the never ending question: "Why?"

To the question, there are answers, none easy, none painless, none complete. The search for answers is spurred by the same courage and devotion their husbands and sons wore so well. Lieutenant Colonel Bev Read wrote of Joyce Shea's courage, character, and devotion all those years ago. She, Barbara, and the Inman and Shea families carried those same qualities. Their strengths came from within, nourished by what others added to their lives.

* * *

Honor. "Cadets will not lie, cheat or steal." As was said in the Department of the Army press release that never became public in 1951, the cadet honor code is a minimum standard, a foundation for the growth of honor far deeper and broader in its implications for life and service. Always at the center in the education we received were the penetrating questions we were to ask ourselves when we contemplated an act with overtones of honor. They were the same questions General Irving gave us in his July 1951 talk to our class. "You must test yourself constantly," he said, "by asking, 'Am I attempting to deceive? Am I attempting to take unfair advantage of my fellow man?'" Selflessness. Sacrifice. The code is only a beginning, is hard won and hard kept.

Invariably, honor's greatest tests come at times least expected, from people believed honorable. Difficult choices and beckoning temptations appear throughout our lives, without warning, in all manner of unexpected situations: When we are alone, or believe we are alone; from among close friends or family; from professional colleagues we know "for sure" are honorable; in large groups when people such as *Brian Nolan* and *Michael Arrison* discover they are recruited, unwittingly enmeshed in, or invited into organized practices we learn too late are clearly less than honorable.

As nineteen-year olds at the United States Military Academy, Cadets *Brian Nolan* in Company K-1 and *Michael Arrison* in H-2 painfully learned honor is easily taught and hard to hold. They were asked to grow up early, quickly, and bear heavy responsibilities in breaking open the cheating scandal the spring of 1951. They did, and remain unsung heroes.

They stood firm for deeply held principles when no one else was courageous enough to look thoughtfully into themselves, question, and act to end activities they knew were inconsistent with the honor code and system. It took enormous courage to confront evidence that friends, roommates, classmates, more senior and less senior class members, and teammates were involved in multiple, organized acts less than honorable, when the nature of the code and system and prevailing attitudes outside the Academy would virtually ensure the two men's actions in reporting the violations would bring their vilification.

Nolan and *Arrison* made their decisions alone, separately, at different times, after due deliberation. Many people later openly castigated them as foolhardy young idealists, "rats," "finks," "moles," "untrustworthy back stabbers," or "tattle tails." They received no rewards except fleeting self-satisfaction in knowing they had remained true to themselves and the code, which had evolved over one hundred forty-nine years. There were no public compliments, and indeed, they received few at all, because their roles in the undercover investigation were exposed before the Collins Board completed its work. In breaking what had become a growing, disciplined cheating ring, and a conspiracy of silence, denial, and lying, a whirlwind had been unleashed upon the two men.

Brian and *Mike* paid dearly, heavy prices in subtle, sometimes crude retaliation and retribution, which kept reappearing throughout their lives, even in their homes or hometown. It is more than irony that these two men, for their courageous acts, suffered far more in their lives than the vast majority of men who resigned from the Academy, left, most quietly, and went somewhere else to begin anew.

Brian Nolan's life was difficult but successful. He was the first of two men who penetrated the ring, and reluctantly accepted the heavy responsibilities given him by Colonel Harkins and Colonel Waters in the pre-Collins Board search for sufficient, hard evidence to open the formal investigation. *Brian* has lived most of his life believing Colonel Harkins hadn't kept a promise, hadn't been truthful.

Brian's friend, classmate, and company mate, *George Hendricks*, from whom *Brian* obtained much of the information Harkins and Waters needed to begin the formal investigation, was the first to testify before the Collins Board. *Hendricks'* testimony was an important factor in the collapse of the ring, and *Brian Nolan* had, prior to the Board, urged *Hendricks* to tell everything he knew. *Brian* was convinced Harkins would let *Hendricks* remain a cadet if he did.

Hendricks wasn't reinstated, instead was among those who resigned. *Brian* was equally convinced Harkins betrayed him, *Brian Nolan*. His conclusions about Harkins were one of many tragic results of the scandal. The reality was far different.

Unknown to *Brian*, Harkins worked hard in the Army staff to persuade senior officers the only cadets who should be discharged from the Academy were the men who compounded their cheating with lying under oath before the Collins Board. Harkins, assigned to the army staff in June of 1951, was serving as Director of Plans for Lieutenant General Maxwell Taylor, Deputy Chief of Staff for Operations and Administration. Harkins was the only man on the staff that fateful summer who had firsthand knowledge of what happened at West Point. He drafted the memorandum Taylor signed and sent to the Chief of Staff, General Collins, advocating all the men who violated the code be retained as cadets and disciplined, except for those who lied under oath. His recommendations were similar to those advocated by Earl Blaik, except Blaik never knew how many lied under oath.

Existing army policy prevailed. The recommendations for decision drafted by Harkins and signed by Taylor, didn't become the decision ultimately made by the Secretary of the Army, Frank Pace. Eighty-three of ninety-four men found guilty by the Collins Board either resigned or faced involuntary separation. All eighty-three resigned and none received honorable discharges when they left the Academy.

Aside from changes in Federal law, there were two important, unspoken issues affecting the decision to involuntarily separate the guilty if they chose not to resign. The first was existing army policy, changed in the mid-1940s to eliminate a double standard with regard to violations of military law involving a court-martial offense for an enlisted soldier, as compared to an Academy cadet. Prior to the change in policy, an honorable discharge after a guilty verdict for a cadet honor violation favored cadets who, like anyone else in the army, were under the Articles of War – the forerunner of the Uniform Code of Military Justice. A similar offense by an enlisted man in the regular army normally resulted in a less than honorable discharge.

The second issue involved a loophole created when the draft law was extended in 1947. Under the amended extension of the law, a draft eligible man could avoid the draft if he was in college, a change intended to avoid drafting World War II veterans who were in college, possibly requiring them to fight again in Korea. An Academy cadet was, in fact, in the Army. Were he separated and given an honorable discharge after an honor violation, he could then transfer to a college or university and avoid conscription; that is, take advantage of a "college deferment" from the draft and also honorably fulfill his service obligation as a cadet at West Point.

From the beginning, and for the remainder of *Brian Nolan's* time at West Point, while he anguished about what he saw as Harkins' betrayal, he received the stout, uncompromising encouragement of his two roommates. To this day, the three former roommates occasionally hold reunions.

Mike Arrison's life at West Point was far less pleasant, although Harkins made no commitments to *Mike* as he did to *Brian Nolan*. On occasion, *Mike Arrison* was openly, deliberately humiliated by cadets and at least one officer, with few willing to come to his defense, for him or honor.

One of his roommates resigned in the scandal. *Mike's* other room-mate didn't participate in cheating, but admitted to the Collins Board he knew about the activity. He had been approached by Dan Myers during the preliminary Honor Committee inquiry in early April and asked to assist in breaking open the ring, but refused to participate, knowing he had friends involved. He appeared before the Collins Board, was found guilty, but was reinstated in the Corps, and graduated.

Art Collins' minority Board report had been successful for *Mike Arrison's* roommate, but brought added grief to *Arrison*. The report argued for the retention of cadets who admitted to knowledge of the cheating, but against whom there was no evidence of cheating. The Army's Judge Advocate General agreed with Art Collins, and over-turned all guilty verdicts involving cadets in similar circumstances. Thus a year later *Mike Arrison* found himself back in the same com-pany with the former roommate, who knew of *Mike's* role in providing evidence in the undercover investigation.

Mike Arrison was also the cadet who had the unpleasant experi-ence of providing evidence about a hometown friend, information which contributed to his friend's eventual resignation in the scandal. The friend quickly became a former friend, returned to their hometown and re-peatedly told the story of *Mike's* role in breaking the scandal and caus-ing his resignation from West Point. *Mike* was figuratively "tarred and feathered" in his hometown.

Word also spread quietly in *Mike's* company, with harsh results. Help finally came for him when he was turned back to the class of 1954, although help was slow in coming. His life at West Point had become miserable, especially after he was returned to cadet Company H-2 with the class of 1954 – the same company in which his role in destroying the cheating ring had been exposed, and from which six cadets, including one of his roommates, resigned.

But his life finally began to turn around, in part because of his new roommate in the class of 1954, a dinner invitation to the Thayer Hotel from Air Force Lieutenant General Laurence C. Craigie, class of 1923, but mostly because of a company tactical officer and part time assis-tant Army football coach, whom *Mike Arrison* admired greatly.

The new roommate was Freddie A. D. Attaya, one of Army's star halfbacks on the teams that, over three seasons, performed "a football miracle." Freddie was in his third season of varsity football, on teams playing for love of the game, honor, and a return to football glory. General Craigie was the father of Jack Craigie, class of '51, the swimming team captain the spring *Mike Arrison* "joined the ring" to assist in the undercover investigation.

General Craigie, on the truce negotiation team when talks began, returned to the United States in 1951 for his last assignment before retiring. When his son told him of *Arrison's* role in exposing the cheating ring, the graduate from the class of 1923 invited *Mike* to the Thayer Hotel for dinner one Saturday and personally thanked him for "saving the cadet honor code."

The tactical officer and part time assistant football coach was tough, feisty Martin D. "Tiger" Howell, class of 1949, who played tackle on Army's undefeated 1948 team. "Tiger" was the young officer who remembered being on the sidelines in the Polo Grounds against Duke in October 1953, when the Korean War was over, and was blistered with a threatened assignment to Korea by a frustrated Earl Blaik. "Tiger" had inadvertently countermanded an order by Blaik for the Black Knights to kick a field goal late in the game.

Dan Myers, class of 1951, and Dick Miller, class of 1952, both from Company K-1, though performing their duties as honor representatives in their respective classes, also reflected courage in upholding the cadet honor code and system. Dick Miller, who was first confronted with the shocking revelations from *Brian Nolan*, was decisive in determining to whom the initial disclosures must go – the Commandant, Colonel Harkins. Dick then assisted in the follow-on efforts, which began in mid-April, to ferret out hard evidence in the ensuing six weeks of undercover investigation guided by Colonel Harkins and Colonel Waters.

Dan Myers, at Colonel Harkins' request, tried for at least two weeks beginning the first two days of April 1951, to do what Honor Committee procedures of that era required: conduct a preliminary inquiry to determine, independent of information *Brian Nolan* provided Harkins through Dick Miller, that there was evidence of a cheating ring sufficient for a full blown Honor Committee investigation.

Dan didn't know that word of his preliminary inquiry had spread. It had. His inquiry quickly became common knowledge among members of the ring. The wall of silence previously agreed to by the disciplined core within the ring was already being erected.

Then, finally, in mid-April, when Colonel Harkins, with General Irving's approval, decided the only way to break the ring and find the needed hard evidence was conduct an undercover investigation, Dan Myers remained actively involved. He met frequently with Harkins and assisted by transporting evidence to the Commandant, evidence gathered by *Brian Nolan* and *Mike Arrison*.

Dan's work in breaking open the cheating ring was determined, crucial to success, and unwavering. As a young man, he carried great responsibilities on the Honor Committee, in the most unprecedented honor case in the history of the Academy.

Both Dan Myers and Dick Miller never fully understood what they were up against. Dan Myers never knew that one member of the 1951 Honor Committee was repeatedly named to the Collins Board by underclassmen in the cheating ring. The man named was serving in a position on the Committee that permitted him, if ring members wished, to select Committee members for hearings. He was able to select members believed sympathetic to ring members who might be brought before an honor hearing. Case histories noted by Pat Ryan imply there were, in all probability, one or two other members on the Honor Committee aware, if not involved in cheating, and sympathetic to those caught in honor violations of any kind.

Dick Miller, who was on the 1952 Honor Committee, never knew that two honor representatives in the soon-to-be new cow class, 1953, were in the cheating ring, and had been elected by design, with the intent to corrupt the Honor Committee in the '52-'53 academic year. The two men were exposed during the Collins Board investigation and resigned. Nor did Dick Miller know that a similar attempt to elect a member of the cheating ring to the 1953 Honor Committee had been attempted in Company K-1, his company. The attempt failed, but the cadet, confronted by the Collins Board with irrefutable evidence of cheating, confessed to cheating and his attempted election as a company honor representative. He subsequently resigned.

As with officers given the responsibility of sitting in judgment of their contemporaries and subordinates, Dan Myers and Dick Miller, as cadets, served extraordinarily well, with great courage, no credit, and no thanks. They didn't shun the difficult responsibilities given them. Both served for integrity and the ideals of truth and selfless behavior, cornerstones of the officers' code and powerful, deeply respected leadership in any setting.

Dwight Eisenhower was purported to have said, "Responsibility is the reverse side of the coin of freedom." Freedom, and the exercise of democracy, demands individual responsibility, and at the Academy, everything we did demanded exactly that – individual as well as unit responsibility. Though the cadet "Blue Book" was the bane and perpetual source for griping and grumbling by cadets, one had only to read the first page and the meaning of laws and regulations were clear. Individual responsibility. Leadership responsibility. Moral responsibility.

Young men and women at West Point must grow up rapidly in four years. The Honor Committee affords the best training and the greatest opportunity available for a cadet to bear responsibilities similar to what he or she will shoulder as a newly commissioned junior officer – and the Honor Committee cadet must bear them well.

Life in the military, beginning immediately after graduation, brings great responsibility to the new second lieutenant. And if the nation is so unfortunate as to be at war when a cadet graduates, the responsibilities can indeed be heavy. Nothing can be more daunting than to realize that, literally, at age twenty-two or twenty-three, you are making life and death decisions for forty-five to sixty men or women under the most extreme, chaotic, and dangerous circumstances imaginable.

If there is no war, military life is only slightly less demanding. It almost immediately insists on individual responsibility and the repeated exercise of good judgment about small units, contemporaries, or subordinates, usually in unpleasant and unpopular circumstances. Examples are participation in administrative disciplinary actions which can affect a soldier's future in or out of the armed forces, conducting or testifying in investigations involving violations of regulations or the Uniform Code of Military Justice; investigating accidents – either to find causes or culpability; and conducting combat training, participating in joint and combined readiness exercises, readiness inspections, or inspector general evaluations – just to name a few. All require searches for truth, no matter how difficult, unpopular, or professionally unpleasant. In short, honesty, integrity, truth telling, and honor.

In military life, the code is a fundamental necessity to guard the nation's security, and guard against those in the military in a democratic republic who might misuse the enormous power and trust for which they are stewards. Honor must be internalized, made second nature, daily practiced, kept, studied, restudied, encouraged, nourished, grown, reinforced, fiercely protected, spread, its roots sunk deep in lives of service.

A young man or woman commissioned an officer in America's armed forces is a second lieutenant or ensign, the most junior grade in a ladder of increasing responsibilities, all the way to the grade of general or admiral. The officer candidate who graduates is graced with the privilege of serving our nation as an officer. To serve as an officer is not a right. There are deliberately high standards for entry. Once a commission is received, the person who fills the office must meet high standards of integrity and performance. Each takes an oath to accept the duties and obligations, swearing to "...support and defend the Constitution of the United States against all enemies, foreign and domestic...." Among the fundamental duties and responsibilities of the officer is to reasonably and justly uphold military law and regulations, and provide soldiers, sailors, marines and airmen the best leadership possible – which all Americans in military service expect and deserve.

Truthfulness, integrity, honor, these are among the most important of the great bonding forces and links between successful, worthy, moral leadership and the greatest responsibilities America's officer corps carry – the lives of soldiers, sailors, marines, and airmen, and the security of the nation.

* * *

During the Korean War a well-known American author, James A. Michener, wrote a novel titled *The Bridges at Toko-Ri*. He was the same author who wrote the Pulitzer Prize winning novel, *Tales of the South Pacific,* a World War II story which eventually led to a Broadway musical, and the popular song, *Some Enchanted Evening* – Joyce and Dick Shea's love song.

The Bridges of Toko-Ri was a story of the incredible deeds of courage and devotion of American naval aviators flying off aircraft carriers against enemy targets in Korea. The central character in the novel was

Harry Brubaker, a young lawyer from Denver, Colorado. Brubaker, in addition to being a lawyer, was in the Naval Reserve, and was called to active duty to serve in the Korean War.

Brubaker was a pilot flying bomb-laden jet fighter bombers – "Banshees" they were called – off the carrier *Savo* in Task Force 77 in the Sea of Japan. Flying missions from carrier decks requires great skill as a pilot, and isn't without considerable hazard, simply in launching and recovering on a pitching deck in the middle of the ocean. Then there is the hazard of air to ground weapons delivery against heavily defended targets. Flying such missions day in and day out, then recovering on the carrier, sometimes at night, with battle damage, can be wearing and emotionally draining. Fear can become copilot instead of God.

After a succession of dangerous missions, Brubaker's fears of not returning to his home and family began taunting him, to the point he seriously considered taking himself off flying status, an intolerable act for a commissioned officer. His fears so consumed him he could no longer hide them, and he talked with senior officers about his dilemma. The carrier task force commander, crusty old Admiral George Tarrant, learned of Brubaker and the torment he was going through, and talked with him. The crusty admiral, referred to by Navy men as "George the Tyrant," persuaded Brubaker to abandon thoughts of taking himself off flying status, and Harry Brubaker continued to fulfill the responsibilities he'd accepted as an officer. Then the admiral, without Brubaker's knowledge, asked the carrier air group commander to take the young officer under his personal supervision, and look after him.

Harry subsequently lost his life, along with a rescue helicopter crew who had volunteered to pluck Brubaker to safety. His airplane was hit during an attack on the heavily defended bridges at Toko-Ri, but he successfully crash landed in enemy territory and got out of the damaged airplane uninjured. The helicopter landed to pick him up but was disabled on the ground by enemy gun fire. Brubaker and his would-be rescuers were all killed by North Korean soldiers while trying to evade capture on the ground, hoping another rescue helicopter would arrive.

The admiral, sitting quietly in the task force commander's chair on the carrier's bridge, was devastated when he first heard the news about the young men who died. There followed an angry confrontation between the admiral and the mission leader, the air group commander who took Brubaker under his wing and did everything he could to protect and rescue Brubaker.

The admiral severely chastised the group commander, who was furious at the "Old Man's" insinuation the mission wasn't well performed, and more could have been done to save Brubaker. The air group commander lashed back at Admiral Tarrant, told him off, and stormed off the flag deck. When the admiral was finally left alone with his thoughts he sat for hours in his chair, all through the night, into the early morning hours.

When he heard the roar of engines preparatory to launching the daily anti-submarine patrol, he asked aloud,

> Why is America lucky enough to have such men? They leave this tiny ship and fly against the enemy. Then they must seek the ship, lost somewhere on the sea. And when they find it, they have to land upon its pitching deck. Where did we get such men?

As for the 157 United States Military Academy graduates who fell in Korea, men like Dick Inman and Dick Shea from the Sesquicentennial class, the same question applies. Where did we get such men? They, like all Academy graduates, belong to the Long Gray Line. They were with 375 more wounded, who, with those who fell, came from twenty-one graduated classes. But those who fell are special, and have a special kinship with one another. They honored each of us in a much larger sense, though they never became what they might have been, never lived to be great captains of American military history, or great national leaders.

The Long Gray Line stretches nearly two hundred years back to the early days of the nation's history. And those who fell in defense of their country throughout those years, also have a special kinship with one another, and with all who sacrificed from every walk of life from every state and territory in the Union. If we who graduated from the Academy, lived through wars' battles, and returned to live full lives, along with men who chose other paths and lived other dreams are members of the Long Gray Line, the men who fell in Korea and elsewhere, and all their fallen comrades, are the heart and soul of the Long Gray Line. Their sacrifices gave us our lives, protected and defended the democracy and freedom we all enjoy. Without them our lives and the world would be quite different as we enter the new millennium.

Their brief lives were songs to freedom, their deaths hymns sung to free peoples, and all those who long to be free. Their songs, their hymns, their triumphs, their honor, came from within, strengthened by the great intangibles of life, passed to them by those gone before.

On January 12, 1951, a poem written by Richard G. Inman appeared in *The Pointer*. He wrote *This Is My Song* in the fall of 1950 while he was a cow, a junior at the Academy, and the class of 1950 was receiving its baptism of fire in Korea. At West Point that fall, Dick was playing football with the "Golden Mullets," the jayvees, while behind the facade of excitement, gloom and evidence of trouble were gathering and Army football's golden era was about to end. When the poem appeared in *The Pointer* in January, Pat Ryan and others had already begun to notice "something was wrong" in the Corps of Cadets, "something big." But Dick Inman's poem began with brightness and optimism, good cheer.

> This is my song.
> Yellow sunlight streaming earthward.
> Illuminating green fields,
> Lending its magic to the glorious petals of young flowers;
>
> A bumble bee monotonously humming
> The droning theme of a lazy summer day;
>
> A delightful, gurgling brook
> And a silver spring that feeds it currents of cool water;
> Tiny, black catfish minnows
> Steadily wriggling their uncertain journeys
> In the slow back-eddies near its bushy, sunlit banks
> Like big flies struggling for life
> In the wetness of the warm stream.
>
> This is my song:
> Meadows of high, thick blue-grass, soft,
> Like cushions of velvet to the laughing children
> Who tumble and roll on it at play,
> Caring nothing;

The family washing flapping slowly
In a limpid breeze that cools no one
During the brilliant morning;

Dirty little torn places and pantlegs patched at the knee;
Small grubby hands from fishing
Or playing marbles with the boys down the block;
A brown little dog trotting faithfully behind;
An orange-chested robin in the cherry tree,
Warbling for all ears the song of his love;

A turtle and a frog sunning themselves on the same log
In the pond beyond the hill;

A deserted tricycle on the sidewalk,
Hot already in the bright, early day.

This is my song:
A lonely, peaceful, windblown hill
Overlooking the city;

A tall, young girl with the apple-blossoms of Spring
Shining from her cheeks,
And a smile spreading her mouth —
The goddess Athene, graceful and stately,
Curly black hair and stone-grey eyes
That sparkle as she looks to me;
Warm arms and soft breast, clean white skin
Undreamed-of sweetness.

This is my song:
High overhead a solitary bird
In ecstatic freedom, dipping and circling
On the powerful wind of the upper sky,
Screeching an unheard song of wildness
That praises the vastness of his home;

Whispering sounds of fervent promise,
The endearing nonsense-talk of youthful lovers
Mingled with grand exchange of things well thought upon —
The important things of living;...

Sleep well brave soldiers.

THE SILENT PENDULUM

Many of the central figures in the events at the United States Military Academy and in Korea during the period 1950-53 are gone, their lives ended, their passing not given wide public notice. Most of the men who were in the Corps of Cadets at the time live on.

None from the classes of 1950-57, either graduates or former cadets, remain on active duty in the nation's Armed Forces. Hundreds from those classes later fought in Vietnam. A large number lived to see the Soviet Union's collapse and the end of the Cold War, and a few were still on active duty to lead the United States Armed Forces and Allied Coalition Forces against Iraq – in the brilliant feat of arms called Desert Storm. The most notable examples were General H. Norman Schwarzkopf, class of 1956, the commanding general of American and Coalition Forces in the Middle East, and General Carl E. Vuono, class of 1957, Chief of Staff of the United States Army.

All who lived the events of this story, like every other man and woman, were imperfect. They were fallible. But they were good men who tried to live the ideals, virtues, and principles taught them in their days as young cadets and officer candidates at West Point. They served their nation well and faithfully.

Major General Frederick A. Irving retired from the Army on August 1, 1954, on completing his assignment as superintendent of the Military Academy. Two months earlier, on June 1, the last '53-'54 academic year issue of the cadet magazine, *The Pointer,* was dedicated to General Irving, the class of '54, and graduates everywhere. Before the magazine went to press an unsigned letter arrived, sent to the editors by a member of the graduating class. The letter was printed on page one of *The Pointer*.

Dear Editors:

Rumor has it that the *POINTER* is going to dedicate an issue to General Irving. Because of the great admiration I have for

General Irving I have written a few lines to him that I hope you will find room for.

To General Irving

We who were watching
Saw honor upheld;
We who were listening
Heard panic dispelled;
We keep in memory
Greatness beheld.

After retiring, General and Mrs. Irving moved to Arlington, Virginia, just outside Washington, D.C. where he was active in volunteer community service. For a year and a half he gave his time to ACTION, a TIME-LIFE foundation which refurbished and renovated communities and towns. From there he took a position at Melpar Corporation, forerunner of a company later known as E-Systems. He remained with E-Systems until reaching the mandatory retirement age of sixty-five.

He was elected chapter president of the Washington West Point Society, an organization of the Academy's Association of Graduates which now has chapters in nearly every state in the union, and several overseas locations.

Like superintendent's before him, General Irving remained deeply committed to the Academy and its mission, principles, and ideals, but studiously avoided involving himself in its affairs and thus complicating the ability of his successors to perform their duties. He offered no advice unless specifically asked.

In April 1976 he received a letter from the Academy's then superintendent, Lieutenant General Sidney B. Berry, class of 1948. The letter gave him General Berry's preliminary impressions of another unfolding, widely publicized academic cheating scandal at West Point. Revelations of this onrushing disaster had surfaced in March when an instructor on the faculty brought the sad news to the superintendent.

9 April 1976

Dear General Irving:

I am writing to provide information about the recently publicized honor incident.

In early March, the Department of Electrical Engineering issued a computer home study problem to approximately 800 cadets enrolled in EE304. The intent of the instructions for this requirement was that each cadet would do his own work. In mid-March, as the papers were being graded, it appeared that some collaboration had taken place.

The Chairmen of the 1976 and 1977 Honor Committees asked that the questionable papers be referred to the Cadet Honor Committee as possible honor violations. Sub-committees of the Cadet Honor Committee interviewed each cadet involved. As a result of these interviews, 101 Second Class cadets will be referred to a full board of the Cadet Honor Committee beginning early next week.

It is important to note that no cadet has yet been formally alleged to have violated the Honor Code. Further, the Cadet Honor Committee has accomplished their work thus far in a calm, professional manner.

As you know, if and when a cadet is found by the Cadet Honor Committee to have violated the Honor Code, he may elect to resign or he may request a *de novo* hearing by a Board of Officers. It is possible that Officer Board proceedings could last through the summer months.

Although this incident appears to be serious, I do not believe that it reflects a widespread lack of concern for the Honor Code on the part of the Corps of Cadets. Cadet opinion continues to be strongly supportive of the Honor Code.

I will keep you informed of developments.

Best wishes from West Point.

> Sincerely,
>
> SIDNEY B. BERRY
> Lieutenant General,
> U.S. Army Superintendent

Below his signature, Berry penned a P.S.:

At present reading, this appears to be fundamentally different from the situation you confronted in that the 101 cadets who will appear before the full Honor Board are spread randomly throughout the Corps. There is no concentration in a particular company, regiment, athletic team or extra-curricular activity. There appears to be no conspiracy involved.

SSB

Because the letter was written soon after first disclosures of the 1976 incident, it indicated a broad, but incomplete grasp of what really happened in 1951, and the Academy would pay dearly for not understanding what occurred or how to approach the calamity about to revisit West Point. As in 1951, the EE304 scandal had been a long time coming, and the warning signs were there. Yet, as in the years preceding 1951, the signs had been vague, uncomfortable, fleeting, largely intangible indicators that people and organizations, in the rush to perform the day to day mission, all too frequently don't see, don't hear, or don't want to believe – until it's too late.

In May of 1985, John A. Hammack, class of 1949, a former Air Force officer who had served as aide to General Irving beginning a year after the cheating scandal, wrote the Irvings a letter. Jack Hammack had been to West Point and had "had the pleasure of spending some time with Col. Blaik " In the letter he told of Blaik's warm, pleasant response to a three page letter Mrs. Irving had written Earl Blaik after the death of the former Army coach's beloved Merle. Vivian Dowe Irving's words deeply touched Blaik. He told Hammack he "appreciated [the letter] and how much he remembered the friendship of the [Irving] family." The friends of many years past were deep into the twilight of their lives and had reached out to one another.

Frederick A. Irving died in Alexandria, Virginia, on September 12, 1995, at the age of one hundred one. From January 22 of that year until his death he held the distinction of being the Academy's oldest living graduate.

Colonel Paul D. Harkins (Commandant), after leaving West Point in June of 1951, served as Chief of Plans in the army staff, under Lieutenant General Maxwell D. Taylor, the Deputy Chief of Staff for Operations and Administration. During that assignment he received a promotion to brigadier general.

When General Taylor was reassigned to Korea as the Commanding General, 8th U.S. Army, near the end of the war, Harkins soon followed, and as a major general, became the 8th's Chief of Staff. After the armistice, he served as division commander in the 45th and 24th Infantry Divisions, and left Korea in 1954.

He rose to the grade of general and served as Commanding General, Military Assistance Command, Vietnam, and Military Assistance Command, Thailand, from 1962 to 1964, in the period prior to America's massive involvement in the Southeast Asian conflict. While in Vietnam he visited the 611th Transportation Company, an aviation maintenance company commanded by then Captain *George L. Hendricks*, who had been the first cadet to testify before the Collins Board at West Point on May 29, 1951.

Captain *Hendricks* gave Harkins a briefing on his company's needs to conduct its mission in Vietnam. The two men engaged in a lengthy conversation. *Hendricks* made no mention of his background or role as a former cadet at the Academy, and Harkins apparently didn't remember *Hendricks'* name, his role in the honor incident, or recognize him. *Hendricks* never saw Harkins again, and the General sent the 611th the additional equipment needed to recover downed helicopters.

Captain *Hendricks*, who was at the time under the supervision of another West Point graduate, then Brigadier General Richard G. Stilwell, class of 1938, was nominated by Stilwell to receive a regular commission. Stilwell knew of *Hendricks'* resignation from the Academy during the scandal of 1951, because *Hendricks* told him. Stilwell's reply was "the Army needs good officers and all my company commanders are regular Army." *Hendricks* received his regular commission and never forgot General Stilwell's compliment or his kindness.

General Paul D. Harkins retired from the Army with thirty-five years of service, in 1964, after completing his assignment in Vietnam. Like many American senior officers who served in Vietnam, he was later harshly criticized for his role in America's longest, most bitterly divisive conflict since the Civil War. He lived to see the United States negotiate an end to its involvement in the Vietnam War, withdraw American forces, bring home its prisoners of war, and terminate support of the South Vietnamese government and armed forces. He subsequently witnessed South Vietnam's defeat and subjugation, and the

United States Army's painful, self-searching examination, and rebuilding after that war. He died in Dallas, Texas on August 21, 1984, at the age of 80.

Earl H. "Red" Blaik remained as Army's head football coach through the 1958 season, when the Black Knights of the Hudson and his "Lonely End Team" went 8-0-1, defeating both Notre Dame and Navy.

The 22-6 win against Navy was a come from behind victory over the last Navy team coached by Eddie Erdlatz, who in 1950, his first year as head coach at the Naval Academy, administered the shocking 14-2 upset of an Army team that was in the running for the mythical national championship – and unknown to all but the men involved, was already deeply immersed in activities that would lead to its tragic undoing.

After the 1958 Navy game, Earl Blaik received this telegram from the Old Soldier, retired General of the Army Douglas MacArthur:

IN THE LONG HISTORY OF WEST POINT ATHLETICS THERE HAS NEVER BEEN A GREATER TRIUMPH. IT HAS BROUGHT PRIDE AND HAPPINESS AND ADMIRATION TO MILLIONS OF ARMY ROOTERS THROUGHOUT THE WORLD. TELL CAPTAIN DAWKINS AND HIS INDOMITABLE TEAM THEY HAVE WRITTEN THEIR NAMES IN GOLDEN LETTERS ON THE TABLETS OF FOOTBALL FAME. FOR YOU MY DEAR OLD FRIEND, IT MARKS ONE OF THE MOST GLORIOUS MOMENTS OF YOUR PEERLESS CAREER. THERE IS NO SUBSTITUTE FOR VICTORY.

The defeat of Navy in 1958, for Earl Blaik, was "a soul-satisfying climax to The Lonely End campaign." It had come at the end of a twenty-five year career as a head coach, eighteen at West Point. He resigned as Army's head coach on January 13, 1959, at the age of sixty-one, after ensuring all his assistant coaches had new contracts.

In the five seasons after the inspired performance of Army's unheralded 1953 team, Blaik compiled a 33-10-2 record with the cadets. While an excellent won-loss total by any standard of college football, those five seasons' results didn't return Army football to the "golden age" of Blaik coached teams in the era of "Doc" Blanchard and Glenn

Davis. Nevertheless, at the twilight of Earl Blaik's coaching career he could look with enormous satisfaction on his final year at Army.

The Black Knights were No. 3 in the nation in the Associated Press Poll at the end of the season, and were once again Eastern Champions. Three players were named All-Americans: team captain Pete Dawkins, halfback Bob Anderson, and guard Bob Novogratz. Pete Dawkins, who was selected a 1959 Rhodes Scholar, was the cadet First Captain and Brigade Commander during the academic year, was also Army's third, and to this date, last Heisman Trophy winner. Bob Novogratz received the Knute Rockne award as "The Outstanding Lineman in the Nation."

When Earl Blaik resigned from his position as Army's head football coach, Lieutenant General Garrison H. Davidson was the Academy's superintendent. He had replaced Lieutenant General Blackshear M. Bryan, who, as a junior officer, was an Army assistant coach with Blaik as the decade of the '30s began. Davidson had been the young officer from the class of 1927, who in 1933 became Army's head coach, selected ahead of Blaik, because Blaik was no longer on active duty, and Army policy wouldn't allow a civilian head coach. Davidson had also sought the head coach's position again in 1940, when Blaik was asked by the superintendent, Brigadier General Eichelberger, to return to Army and reinvigorate the Academy's sagging football fortunes. In 1940, with the change in Army policy, the tables were turned in Blaik's favor, not a circumstance to reduce the rivalry which had grown between Blaik and Davidson.

Davidson's arrival at the Academy on July 15, 1956 brought with it a natural inclination on his part, and probably some guidance from senior Army officers who were Academy graduates, to re-involve himself aggressively in the Academy's athletic policies. In the view of a number of senior officers, if an overemphasis on winning football was the underlying cause for the cheating scandal of 1951, little had been done to solve the problem. Earl Blaik's strong influence, philosophy, and methods were still in place, and he was pushing relentlessly to bring Army football "back up where the Black, Gold, and Gray belong," a phrase coined by the man Blaik admired most, General MacArthur.

When Davidson became his new boss, Blaik had been head coach for fifteen years, director of athletics ten years, and chairman of the Athletic Board seven years. Though Earl Blaik later wrote that he and

Davidson differed little in their approach to the game of football and athletic policies, and that their differences weren't paramount in Blaik's decision to resign, what Blaik probably knew was that Davidson differed philosophically with him in three important ways.

First, Davidson strongly idealized the game of football as played at Army when he was head coach from 1933 to 1938. The three year eligibility rule became effective at Army in 1939, in response to President Franklin D. Roosevelt's direction, and after Davidson left his head coaching assignment. The absence of the three-year rule had permitted Army and many other colleges and universities to recruit star athletes who had previously played three years or more of collegiate football. The rule change caused the Academy to energize a strong recruiting program for intercollegiate athletes, or be noncompetitive "on the fields of friendly strife." Blaik's arrival at Army in 1941 resulted in the annual recruiting drive Army needed to be competitive as an independent, national collegiate power.

Davidson fundamentally disagreed with the practice of recruiting, believing it had corrupted the game – though the practice had been instituted at the Academy under Brigadier General Douglas MacArthur's administration at the beginning of the 1920's "Golden Age of Sports."

Second, when Davidson became superintendent, he took note of the Bartlett Board report which had concluded "overemphasis on winning football" was the "underlying cause" of the '51 cheating scandal, and the scandal had occurred in part due to Blaik's strong personality, "singleness of purpose," and his filling the three positions of head football coach, athletic director, and chairman of the Athletic Board. One of the Bartlett Board's recommendations had been to establish the Athletic Director's position as a military position, as a means of diminishing the influence of the Army coach.

Looked at from another perspective, the Bartlett Board's conclusions were means of ensuring the Athletic Director and head football coach couldn't be the same person, and unless the athletic director's position was awarded tenure, the individual filling the position couldn't have the influence of a permanent professor.

In any case, when Davidson arrived at West Point in 1956, the Bartlett Board's recommendation to separate the head football coach's position from that of Athletic Director had never been carried out, and Davidson was determined to rectify that oversight.

Third, implicit in Davidson's decision to act on the Board's recommendation was his belief that the Board was right in its conclusions, and Earl Blaik was therefore largely responsible for the cheating incident of 1951. "Winning football was overemphasized," although Davidson didn't describe it that way. Instead, he expressed it as "equality for all sports."

This collided head-on with Earl Blaik's beliefs in two ways. First, as at most major colleges and universities, football at Army was the sport which, through its crowd appeal in huge stadiums, provided the financial underpinnings of many other intercollegiate sports. Football, to Earl Blaik's way of thinking, had to be given number one priority in recruiting, in order to sustain the ability to win, and to continue to draw the crowds that would keep Army athletics financially sound. This in turn meant Army needed to allocate a larger number of recruited athletes to football.

Early in 1958, Blaik presented a study to the Athletic Board, explaining that during the '50s the number of recruited football players had varied from eighteen to twenty-four. He argued forcefully that the number of football recruits needed to be increased to thirty-eight, because historically there was a 70 percent attrition rate among "blue chip" recruits. (Once a cadet came to the Academy the decision as to which sport he participated in was his own. Some football recruits chose to participate in other sports – in addition to football, or in a few cases, instead of football. The high loss rate among recruited football players occurred due to: men who, for one reason or another, didn't develop sufficiently to play at the major college level – though many remained on the junior varsity; lost interest and decided not to stay with the game – though they remained at the Academy; or couldn't sustain the effort required to meet the Academy's rigorous academic, military, and athletic standards, and were found or resigned; and football career ending injuries. The latter two categories were a small number of the total losses.)

The Athletic Board, composed of the Dean of Academics, two permanent professors, and the Commandant, recommended thirty-three football recruits. Davidson reduced the number to twenty-eight, a number in Blaik's experience was too low. Blaik was convinced Davidson's decision would place the Army team at a competitive disadvantage due to the likelihood of injuries to key players, and the nearly always much larger schools his teams would have to face.

With General "Gar" Davidson, Earl Blaik would have faced an increasingly uncomfortable professional relationship. He later concluded that Davidson's policy of "equality for all sports" led to a period of drift and lackluster performance by Army football teams.

There was no retirement system in place for Army football coaches in 1959. Earl Blaik made his decision to resign without consulting anyone but his family, and even they, including his wife Merle, didn't know for certain what he had decided, or when he would resign, if he did.

On Sunday, January 11, 1959, Blaik's oldest son, Bill, typed the first draft of the Army coach's resignation letter. The next day Earl Blaik dictated the final letter to his faithful secretary of eighteen years, Miss Harriet Demarest. He could tell she was saddened. He asked her to lock it away, that he might "sleep on it" one more night.

Tuesday morning he phoned Joe Cahill, the Sports Information Director, and asked him to come to his office. He told Cahill of his decision, wanting to alert him so he could prepare the public statement. He walked to the superintendent's office to deliver his letter, but Davidson wasn't there. Blaik simply left it in the superintendent's office. Sometime later, Davidson called him to say he had the letter and was sorry Blaik was leaving. The superintendent then hurriedly called a meeting of the Athletic Board, and the story was released to the press that afternoon. Earl Blaik came home and told Merle of his final decision after he had delivered his letter of resignation to General Davidson's office, and the public announcement had been released.

Earl Blaik had once again given his boss, "Gar" Davidson, short notice of his impending departure. At least this time Davidson didn't read about it in the newspapers as he had after the 1933 season when his first assistant coach, Earl Blaik, had made his decision to accept the head coaching job at Dartmouth.

When word of his resignation at Army hit the press, telephone calls, telegrams, letters, and armed forces messages poured in from across the country, and throughout the world. The first call came from Secretary of the Army Wilbur M. Brucker. One letter of which Blaik was particularly proud read as follows:

The White House
Jan. 14, 1959

Dear Red:

It was with feelings considerably stronger than astonishment that I learned last evening of your resignation. I hasten to say that although I clearly recognize that your leaving will be an irreparable loss to present and future Army football teams, to West Point, to the Army and, yes, to the public, yet I heartily approve of your action.

Few people, I am persuaded, understand the intensity of the nervous strain under which a coach must live who has to train teams that are engaged in highly competitive athletics. It is no wonder that as the years inevitably creep up on all of us, the man occupying such a demanding position sometimes looks toward an occupation where concentration is less rigorous, the pace a bit slower, and remuneration a bit better. What I am trying to say is that it is high time you thought of yourself – and you have already given more of yourself for the sake of our deeply felt loyalties than have most people.

There was some speculation in my morning's paper that your decision might be based on differences of opinion in the Army as to whether or not West Point should play in post-season bowl games. As for this, I have never heard the matter discussed by any other West Pointer except in terms of "whatever Blaik feels is best, I am for it." For my part I never even turned on the television or the radio to keep track of a bowl game. My interest is in the contests of the season, but possibly this reaction is somewhat affected by my knowledge that West Point has not played in bowl games.

In any event, along with the gratitude that I have always felt for the dedicated, selfless and brilliant services you have rendered at the Academy, I send you also my very best wishes for an even better record in the commercial world. I am quite certain that many thousands, particularly including the Cadets who have played under your tutelage, would like to join in an expression of these sentiments.

Please remember me warmly to Mrs. Blaik, and to both of you my personal regards and Godspeed.

As ever,
Dwight D. Eisenhower

Earl Blaik accepted a position with AVCO as a vice president in its management group. AVCO was doing exciting work in the field of missiles and space exploration. As he later explained, his old friend from Dayton, Ohio, Victor Emmanuel, had offered him a position in AVCO ten years earlier, but Blaik decided to stay on at Army. Now was the time to leave. He had brought Army football back to where it should be, given the constraints within which he worked.

In 1959 he wrote a series of articles on current football topics, published by *Look* magazine and the Associated Press. Ninety-six newspapers across the country carried the articles, which ran through September, October, and November. Proceeds from the articles were contributed to postgraduate fellowships, known as the National Football Foundation's scholarship-athlete awards.

Each year there were eight awards given, one in each of eight districts covering the entire United States. Earl Blaik was particularly proud of his ability to contribute to scholarship-athlete awards. The awards were going to young men who were living proof that athletics, scholarship, and leadership are not incompatible, and must always be encouraged as a winning combination.

In 1960 Earl Blaik wrote the first of his two autobiographies, the first co-authored by *Look* magazine sports writer Tim Cohane, the Fordham University alumnus who recommended to Blaik the hiring of Vince Lombardi as an assistant coach in 1949. The work was titled *You Have to Pay the Price*. In 1974, with *You Have to Pay the Price* identified as Part I, the former Army coach wrote a new edition, *The "Red" Blaik Story.*

Both works were filled with marvelous recollections of football history and days gone by in collegiate football, particularly as the game was played at West Point under Blaik. Both works were also sharply critical of the handling of the 1951 cheating scandal, and what he viewed as the Academy's puritanical attitude toward the honor code and system. He strongly implied the need for reform within the institution. He summarized his views about the need for reform when he said, "To restate my views, I believe in the honor code, but I also believe that every reasonable means should be taken to remove inordinate temptation... the 'black-or-white' advocates [of the honor code] place far too much emphasis on the consequences of breaking the code and far too little on an abiding commitment to the moral principle involved."

The 1974 book went one step further and leveled scathing criticism at some of the principal participants in decisions affecting "the 90," whom Blaik referred to as "The 90 Scapegoats." He left no doubt that the underlying issue in his mind was the manner in which the affair was handled by the men responsible for the decisions. In responding to some of the more vocal critics of his role in the handling of the 1951 affair, Blaik, in his books, used a number of sharp-edged labels, none of which endeared him to a considerable number of Academy graduates who, like General Davidson, held Blaik responsible for the '51 scandal – and, who, in some instances, were absolutely convinced Blaik knew of and condoned the cheating while it was in progress.

While such allegations have been whispered, talked, or purported to have been stated in memorandums and letters throughout the years, his actions during those first hectic days of the Collins Board suggest the allegations had no basis in fact. The night of May 29, 1951, he told the twelve yearling football players who came to see him, they must tell the truth about what they had done. That action was a major factor in breaking the "conspiracy of silence" wide open after the first day of the Board's investigation, and bringing down the cheating ring, including his own son. These were hardly the actions of a man who "knew the cheating was going on and condoned it."

Earl Blaik's sadness, pain, and disillusionment over his youngest son's involvement in cheating at West Point healed over the years, and they developed a close, loving relationship. He happily saw Bob go on to be an assistant coach for Murray Warmath at Minnesota, Andy Gustafson of Miami (Florida), and Bud Wilkinson at Oklahoma, with every indication Bob had been bitten by the coaching bug. Yet Bob's father left no doubt he would rather see his son go into business, a far more secure life than the one Earl Blaik chose for himself. And Bob eventually did choose to go into the oil business in Oklahoma, with his older brother Bill, who had graduated from Dartmouth as a petroleum engineer.

Earl Blaik received numerous honors and repeated recognition in the years after his resignation from his head coaching position at Army. In June of 1959, his first alma mater, Miami of Ohio, presented him with an honorary doctorate of laws. In 1966 the National Football Foundation gave him its Gold Medal Award. In 1979 he was enshrined in the National Association of College Directors of Athletics by Citizens Savings Hall of Fame.

In 1982 the Academy's superintendent, Lieutenant General Willard W. Scott, Jr., class of 1948, hired a new head football coach. The superintendent's goal was to turn around a floundering Army football program, and the selection eventually made was Coach Jim Young, who succeeded in rejuvenating Army football in the 1980s. General Scott appointed a committee of well known former Army players to assist him, and asked for their recommendation. He also asked for Earl Blaik's recommendation from among the finalists identified by the player committee.

Blaik agreed, on condition his was a recommendation only, and nothing more. After he made his recommendation he commented, "The coach who comes to Army has to know how to be a winner." The comment was a reference to a major criteria he used in making his recommendation. The remark was also a replay of what he was before he came to Army – a winning head coach, and earlier, an assistant coach who worked under winning head coaches – and as a young man in college and at West Point, he had played on consistently winning teams.

When he hired his assistants at Army, he invariably sought men who had winning records as assistants, or had played on winning teams. For winning was what he always, always prepared his Dartmouth and Army teams to do.

In 1986 President Ronald Reagan awarded Earl Henry "Red" Blaik the Presidential Medal of Freedom. He died in a nursing home in Colorado Springs, Colorado, on May 6, 1989, at the age of ninety-two.

Both he and his beloved wife Merle McDowell Blaik rest in the Academy cemetery near the Old Cadet Chapel, beneath a polished, lightly gold-faced headstone, cut in the shape of a football standing on end, as though set on a kicking tee ready for kickoff.

On September 25, 1999, during halftime ceremonies at the Army-Ball State University football game in Michie Stadium, the Academy named and dedicated the stadium field, Blaik Field, in honor of one of the Academy's most famous and controversial graduates of the twentieth century, and to date Army's most celebrated football coach. Present at the ceremony were many of Earl Blaik's former players, including Heisman Trophy winners Glenn Davis and Pete Dawkins.

General of the Army Douglas MacArthur, a national hero and one of the Academy's most renowned graduates, continued an avid, behind the scenes booster of Army football until his death in 1964. As Earl Blaik had done the two seasons prior to 1953 he often welcomed MacArthur to watch practices, ride on team buses, give the Black Knights of the Hudson motivational talks, and critique the team's performance. As in all his years as Army's head coach, Blaik regularly informed the General of the team's prospects for the season, or preparations for upcoming games – during increasingly frequent visits to MacArthur's apartment in New York City. As Earl Blaik wrote of him,

> ...he... [was] the vicarious coach of football during his long career. In later life, his great joy came from sitting on the players' bench during practice game scrimmage and returning to his New York apartment to write me his thoughts with respect to personnel and the game in general. His handwritten notes are priceless memoranda of detailed observation that would do credit to a professional coach with the benefit of movies. His keen analytical mind, coupled with his memory of play and players, never ceased to amaze me.

To add to General MacArthur's numerous awards and decorations as a heroic young officer in war and one of America's great captains in military history, in 1959 he received the National Football Foundation's Gold Medal Award. In 1963 he was elected to the Helms Foundation Hall of Fame, and in 1964 was the first recipient of the Football Coaches' Association's Tuss McLaughry Award "to the individual who has distinguished himself in the service of others."

While in retirement he wrote his memoirs, *Reminiscences.* Recalling his, the Army's, and America's painful experiences in Korea he wrote, "A great nation which enters war and does not see it through to victory will ultimately suffer the consequences of defeat."

He proffered his military and strategic advice to Dwight D. Eisenhower and John F. Kennedy. Referring to the deepening involvement of America in the war on the Indo Chinese peninsula, he pleaded his Korean War lesson to the nation's young president, urging Kennedy not to involve the United States in a land war on the Asian continent. MacArthur was too late. The nation was already involved, and had been since Harry Truman was president and authorized sending the first seventeen American advisors to Southeast Asia.

Truman's initial commitment to Vietnam was in support of France – two days after he authorized the use of American ground forces in the Korean War. President Eisenhower broadened and deepened America's support of France, as part of his administration's "policy of containment" of an unceasingly restless and expanding communist empire, while the French sought to reestablish their colonial empire in what had been French Indo China prior to World War II.

In May of 1962 General and Mrs. MacArthur returned to West Point where he was presented the Thayer Award, the Academy's highest honor, and gave his unforgettable, stirring speech "Duty, Honor, Country," which has remained an inspiration to Academy graduates throughout the years. At age eighty-two he looked thin and frail, but for this one event he was his old self. He spoke for thirty minutes "without a note, never once groping for words, in rhetoric and phrase never heard before in [Washington Hall]." There wasn't a dry eye in that great hall when he completed his address. This was MacArthur's last and greatest day at the Military Academy he revered and loved.

The week before MacArthur left New York City for Walter Reed Army Hospital in Washington, D.C., early the spring of 1964, Earl Blaik, in spite of the General's obvious frail health, visited with him for an unusually long time. He wasn't anxious for Blaik to leave, though his weakened condition was evident. He walked Earl Blaik to his apartment door and his parting words were, "Earl, I don't have too many days left." Earl Blaik remembered there was no answer to the General's statement, "save a wetting of the eyes."

Before the end of the same week General Courtney Whitney, MacArthur's long time, faithful aide, called Blaik to tell him the General was going to Walter Reed Army Hospital. Would Blaik come to the plane to boost his spirits? The plane was to depart at eleven the next morning. Blaik was there at ten forty-five, too late for the accelerated departure.

When Blaik arrived the plane's doors were already closed and the engines were running. There were two other officers there, on official duty, and Earl Blaik was alone. The former Army coach was thoroughly downcast, and never more dejected, as he searched in vain for a glimpse of General MacArthur.

Suddenly the engines stopped, the door opened, and the General came out to say good-bye to his old friend. Douglas MacArthur and Earl Blaik said their last farewell. The General climbed aboard the aircraft and left. Earl Blaik never saw him again.

General of the Army Douglas MacArthur died in Walter Reed Army Hospital, in Washington, D.C., on April 5, 1964, at age eighty-four.

Vincent T. Lombardi, Fordham University class of 1937, was back on the mountain top of collegiate football after Army's 1953 season. He left the Academy and became an assistant coach for the New York Giants' professional football team. He had spent five of his formative coaching years at West Point. These early years in Lombardi's rise to fame were under the man he later credited as being his greatest football teacher, and the greatest football coach he'd ever known, Earl "Red" Blaik.

In 1987, when author Michael O'Brien completed *Vince – A Personal Biography of Vince Lombardi,* he included in chapter V, "West Point (1949-53)," fascinating stories about the charismatic Army assistant coach. A careful reading of the tumultuous five seasons under Earl Blaik reveals Vince took an emotional roller coaster ride much in keeping with his fiery, mercurial temperament.

Lombardi came to Army in the spring of 1949, after two seasons at his alma mater. One year was as a varsity assistant coach, the second as freshman coach. The Fordham head coach and Vince apparently clashed, causing a strained professional relationship. Vince had had a brilliant record in New Jersey as a high school coach in both football and basketball, and had been a member of Fordham University's famed "Seven Blocks of Granite" during the 1935-36 seasons, when Fordham was ranked among the top twenty teams in college football.

Blaik was curious as to why Fordham didn't better use Vince on their coaching staff, a cause for more than usual caution in Blaik's hiring of an assistant. Blaik interviewed Lombardi three times before signing him in the spring of 1949, though Fordham alumnus Tim Cohane, a close friend of Blaik's and a *Look* magazine sports writer, first recommended Vince to the Army coach.

In the spring of 1949 the Black Knights of the Hudson were still on the mountain top in "the golden era of Army football." In '48 they had gone 8-0-1. Held to a 21-21 tie by an inspired Navy team, they ended the season with a No. 6 national ranking.

In the '49 season, with Vince as the cadets' new offensive line coach, Army went 9-0, and climbed to a No. 4 national ranking. The following year, after Blaik moved him to offensive backfield coach, Army went 8-1, suffering its stunning 14-2 upset at the hands of Navy, ending Army's twenty-eight game undefeated streak. In spite of the defeat at the hands of Navy, the cadets ended the season with a No. 2 national ranking, a measure of the respect sports writers and coaches held for the 1950 team.

Vince Lombardi, like everyone else at West Point, didn't know at the time the loss to Navy marked the end of Army's "golden era of football." In the spring of 1951 the revelations of extensive, organized cheating in the Corps of Cadets shocked and saddened Vince and launched him on a three-year emotional roller coaster ride.

The entire Army coaching staff was drained and devastated by the news. But Vince Lombardi, who was by nature effervescent, excitable, given to elated highs and retreats into himself, was especially hard hit by the destruction of an Army team that had held the promise of the best since the era of "Doc" Blanchard and Glenn Davis. When Vince's gifted backfield was wiped out, along with two years of unending hard work, he had to fight hard to recover.

Lombardi watched uncomfortably as Earl Blaik struggled with his decision to remain at West Point through the storm of criticism directed toward the Army head coach after news of the impending resignations hit the press on August 3, 1951. Like many coaches in the rough and tumble, insecure world of big time collegiate football, he wondered about his own future.

Vince's pain was made more acute by the severe personal anguish the scandal imposed on Blaik and his family. In his role as backfield coach, Lombardi, a man who had deep feelings about family and its meaning, had been drawn closer to the Blaiks than would normally be the experience of an assistant coach.

Bob Blaik, the coach's youngest son, was among those who resigned that summer. In the '51 spring practice he had been preparing for his second and final season as Army's starting quarterback. When Bob's father moved Lombardi to backfield coach after the 1949 season, Vince had taken Bob under his wing and carefully honed the second classman's football skills. He also helped young Blaik relax and work through the inevitable problems of being the head coach's son playing on his dad's team in the intensely competitive sport of major college football.

When the scandal broke, the result for Vince Lombardi was something he hadn't counted on, a sharp professional setback exacerbated by a saddening emotional involvement in the harsh events impacting the football team and the Blaiks' lives.

Vince sought counsel at the West Point Catholic Chapel and through close friends and clergy in New Jersey and New York City. "Should I leave and go back to high school coaching, or back to Fordham?" Eventually, he weathered the crisis, and with the beginning of practice for the tough '51 season, began to come back.

Then in three seasons came "the football miracle." The 1953 Army team, built in the crucible of adversity which the Black Knights had endured in '51 and '52, surged back into the national collegiate spotlight, going 7-1-1, losing only to Northwestern and tying Tulane. The cadets ended the season with the win over Navy, a near perfect game after three years of drought against the middies, achieved a No. 14 national ranking in the Associated Press poll – and won the Lambert Trophy, symbolic of Eastern football supremacy.

In the same three seasons Vince Lombardi built another dazzling backfield – his last one at West Point, and his springboard to professional football glory. He was once more on the mountain top, and after the '53 season at Army, moved on to the New York Giants.

974

Earl Blaik and Vince Lombardi were polar opposites in personality, yet their cautious, tentative beginning as coaches of the same team grew into a lasting, ever-deepening friendship. Blaik was outwardly calm, distant, reserved, austere, and possessed of a commanding presence. A man of few words, he was articulate and analytical, and when he spoke his assistants and his players listened intently. He wasn't a "hail fellow well met" and not one to show affection.

In contrast, Earl Blaik wrote of Vince years later, he "expressed the extremes of enthusiasm much like an electric storm early in the spring." Blaik and Lombardi, in many ways, were perfect complements to one another.

Vince Lombardi's mercurial temperament boiled over frequently with enthusiasm, passion, animated expressions of disappointment or disapproval, intense salesmanship if he had an idea he believed would bring better individual player or team performance, and he didn't hesitate to run up to a player and give him an excited bear hug if the player had executed exceptionally well. He could show affection, warmth, and was as quick to compliment as he was to criticize.

As the years went by Earl Blaik frequently noted that coaching the 1951-53 Army teams brought him the greatest professional satisfaction in his career. Vince Lombardi's step to glory on the professional gridiron began in earnest those three years. Under Earl Blaik he learned what was necessary to turn adversity to triumph, and he applied those lessons to great effect at Green Bay beginning five years after he left West Point.

By observing Blaik, the irrepressible Lombardi improved his ability to organize, discipline, and inspire a team. "My 'football' is your football," he later wrote Blaik. "My approach to a problem is the way I think you would approach it." Recalled Blaik, "He may have learned a few things during our years together, but he didn't learn that magnetism at West Point. It was always in him. You don't put magnetism into people."

Jim Lee Howell, the New York Giants' head coach who hired Lombardi after the 1953 season at Army, said, "If Lombardi can do that kind of job in three years at West Point – he could do a helluva job in the pros where he would have an experienced base to work with."

For the men who played on those Army teams, and particularly in his last backfield, there would be memories of a complex, fiery man, who came to West Point as a "diamond in the rough." They undoubtedly would agree with the words about Vince Lombardi, penned by one of his great players at Green Bay, Jerry Kramer. In his book, *Instant Replay,* Jerry wrote, "Lombardi was a cruel, kind, tough, gentle, miserable, wonderful man whom I often hate and often love and always respect."

In 1959 Vince Lombardi took the helm of the Green Bay Packers, a team that finished 1-10-1 in '58, and had averaged under four victories per year over a ten-year period in which there hadn't been a single winning season. In his first season at the Packers he guided them to a 7-5-0 finish. The rest is professional football history.

Vince Lombardi, too, had returned to glory, and would go on to become a football legend in his own lifetime. He had become the man who would talk and live the refrain, "Winning isn't everything, it's the only thing."

Colonel John K. Waters, Assistant Commandant and Commandant, class of 1931, received a promotion to brigadier general the summer of 1952, and was reassigned to Korea as the I Corps Chief of Staff where he remained until after the armistice was signed. Tragically, his wife, the late General George S. Patton's daughter, died while Waters was serving in Korea. Waters later rose to the grade of general, after having served in Europe as commanding general of the 4th Armored Division, V Corps, and Fifth Army in the period 1960-63. He returned to the United States to command Continental Army Command, and in 1964 became the Commander-in-Chief, United States Army, Pacific, in Hawaii.

His health was affected throughout the remainder of his life following the near-fatal wounds he received in a German prisoner of war camp in 1945. He retired from the Army in 1966, after thirty-five years of service, and lived to the age of eighty-two. He died in Washington, D.C. on January 9, 1989.

Brigadier General John H. Michaelis remained as Commandant of Cadets at the Military Academy until the summer of 1954. On leaving West Point, he served as the Army's Chief of Legislative Liaison in Washington, D.C. Follow-on assignments included Commanding General: U.S. Army, Alaska; V Corps in Germany; Allied Land Forces, Southeastern Europe, in Izmir, Turkey; and Fifth U.S. Army in Europe. He returned to Korea in 1969 to become Commander-in-Chief, United Nations Command; United States Forces, Korea; and 8th U.S. Army. He retired from the Army in 1972, after thirty-five years of service. He lived to the age of seventy-two, and died in Dillard, Georgia, on October 31, 1985.

Lieutenant Colonel Arthur S. Collins, Jr., Collins Board president and class of 1938, received a promotion to colonel the summer of 1951, returning him to the grade he achieved while serving in World War II. He remained 1st Regiment Commander, and Tactical Officer at the Academy until August of 1952. After serving in plans and operations positions in Joint Staffs in Washington, D.C. and Strike Command in Tampa, Florida, he became Chief of Staff, Strike Command. He returned to the Army staff in 1963 in the Office of Personnel Operations, and later was the Assistant Deputy Chief of Staff, Operations.

In 1965, as a major general, he took command of the 4th Infantry Division at Fort Devens, Massachusetts, and deployed with the 4th into the rapidly escalating war in Vietnam. After two more staff assignments in the continental United States and Vietnam, he became the Deputy Commander-in-Chief, U.S. Army in Europe. He retired from the Army as a lieutenant general in 1974, with nearly thirty-six years of service. He died in Washington, D.C. on January 7, 1984, at the age of sixty-eight.

Lieutenant Colonel Jefferson J. Irvin, Collins Board member and class of 1938, completed his assignment in the Tactical Department at West Point in June of 1952. He received a promotion to colonel and was given command of the 179th Infantry Regiment, 45th Division, in Korea. His Regiment was engaged in fighting at various times during the brutal stalemate that ensued in Korea, until the armistice was signed in July 1953. He later served in operations positions in 8th Army and Army Forces, Far East. He was a Deputy Assistant Secretary of Defense, and chief of staffs in VII Corps and I Corps Group. He completed his Army career a brigadier general, and retired in 1971 with thirty-three years of service. His final two assignments were in U.S.

Army, Pacific, in Hawaii, and as Defense Attache in Mexico. He obtained his law degree from St. Mary's University in 1974, and resided in San Antonio, Texas, until his death January 31, 1998.

Lieutenant Colonel Tracy B. Harrington, Collins Board member and class of 1938, received a promotion to colonel while at the Military Academy and replaced Colonel Arthur S. Collins, Jr. as 1st Regiment Commander and Tactical Officer. He was reassigned from the Academy in August of 1953. He served in the Central Intelligence Agency from 1954 to 1957 and again from 1968 until his retirement in 1972. In the intervening period he commanded the 6th Armored Cavalry Regiment, served in the United Nations Command in Korea, and the Army's Office of Personnel Operations and the Joint International Coordinating Staff, both in Washington, D.C. He resided in McLean, Virginia. He died in Fairfax, Virginia, on September 14, 1997.

Colonel Boyd W. Bartlett, Bartlett Board president and class of 1919, remained at the Academy as Professor and Head, Department of Electrical Engineering until his retirement in 1961. Consistent with Army policy at the time, he received a promotion to brigadier general the day he retired. His distinguished career as a scholar and professor spanned thirty-four years at Bowdoin College, Maine and the Military Academy. He was on active duty at West Point for nineteen of those years and retired with twenty-one years of service in the Army. He returned to his home state of Maine, and once again became active in the education and civic affairs of Bowdoin College and the town of Castine. Among other endeavors, he served as reviewing and commissioning officer for the College's Reserve Officer Training Corps. He died in Castine, Maine, on June 24, 1965, at the age of sixty-eight.

Colonel Francis M. Greene, Bartlett Board member and Greene Board president, class of 1922, received an assignment to Headquarters, European Command, and departed West Point and his position as Director of Physical Education – "Master of the Sword" – at the end of May 1952, four months after the Greene Board completed its inquiry into regulations and their effects on honor. While in Europe he also served in Headquarters, Seventh Army. He retired in 1955 in the grade of colonel, and returned to become Director of Personnel Management at Central Hudson Gas and Electric in the state of New York. He later was selected first vice president of that firm. He became active in civic affairs, receiving several awards and citations in recognition of his service. He died in Poughkeepsie, New York, on February 22, 1974, at age seventy-three.

Colonel Charles H. Miles, Jr., Bartlett Board member, class of 1933, remained as the Academy's first comptroller until 1955, when he was reassigned as the Deputy Comptroller, United States Army, Europe. He retired from the Army in 1958 for reasons of health, after twenty-five years of service, and returned to Oceanside, New Jersey, to serve as a summer church superintendent. He later started his own business, a letter shop producing auto-type and mimeograph. Upon retiring a second time, he remained active in civic affairs and volunteer work until his death on May 11, 1984, at seventy-four years of age.

Colonel Charles J. Barrett, second screening board president, class of 1922, remained at the Military Academy as Professor and Head of the Department of Foreign Languages until he retired from the Army on June 11, 1963. As was the circumstance when Boyd Bartlett retired, the day Charles Barrett retired, he was promoted to the grade of brigadier general, the grade he achieved while serving in World War II. He died in Walter Reed Army Hospital in Washington, D.C., at five thirty the afternoon of June 30 of the same year. He was sixty-three years of age.

During his long and distinguished career he had fought in two world wars, the first as a soldier with the 29th Division, a New Jersey National Guard division. In World War II he fought as the 84th Division artillery commander, and earned a Silver Star for heroism. Later, during the Allied advance into Germany, he received the Legion of Merit for his exceptional leadership during the Battle of the Bulge, the German Army's last desperate counterattack before the Wermacht's final collapse. Throughout his career, he served in three assignments at the Academy, his last beginning July 1, 1947.

While he was head of Foreign Languages at West Point, the Department strengthened academic content and teaching methods in French, Spanish, German, Portuguese and Russian. Among the most important contributions were: the introduction of advance courses in French, German, and Spanish; the installation of modernized language laboratories involving the use of sound tapes; and elective courses which gave all cadets the opportunity to study at least one additional semester of foreign language.

He served on several standing faculty and administration committees whose purposes were to maintain and improve the Academy's quality of education. Among the committee assignments was Chairman of the Library Committee, which reviewed plans for the new library. In 1961 he was appointed Chairman of the Athletic Board, a measure of his long interest and participation in all forms of athletic activity in the Corps of Cadets.

Colonel Henry C. Jones, first screening board president and Academy Inspector General, class of 1916, completed his six-year assignment as the Academy's Inspector General when he retired from the Army in 1953. Afterward, he resided in El Paso, Texas, where he was involved in civic activities most of the remainder of his life. He died on August 5, 1988, at age ninety-four.

United States Military Academy classes of 1950-57. Appendix I provides statistical summaries of class histories, but tells little of individual sacrifices and achievements of class members and ex-cadets. Each man's life is, in itself, a story, and each, in some measure, contributed to democracy's defense and the nation's security in the last half of the twentieth century – "the century of the totalitarian."

In Korea, 157 Academy graduates were listed as battle deaths, having been killed in action, died of wounds, declared missing in action and subsequently declared dead, or were executed by the enemy while being held as prisoners of war. The classes of 1950, '51, and '52 accounted for forty-eight of that number. Another nineteen, from the classes of 1944 and earlier, were fighting in their second war when they died, having served during World War II. Lieutenant General Walton H. Walker, Jr., class of 1912, was leading the 8th Army and United Nations forces in Korea, when he was killed in a jeep accident on December 23, 1950. He was fighting in a major war for the third time in his life when he died.

There were 120 Academy graduates who were listed as non-battle deaths in Korea. They died in training accidents, or of starvation, sickness, and maltreatment while in captivity. The classes of 1950, '51, '52 accounted for twenty-four.

Academy graduates suffered 375 wounded in Korea, including many disabled. Among the wounded were 144 from '50, '51, and '52.

Every class from 1950 through 1957 suffered casualties in the longest war in the nation's history – Vietnam. Of the 278 Academy graduates who were killed in action, died of wounds, or lost their lives in prisoner of war camps in Southeast Asia, fifty-three were from the classes of 1950-57.

From those who survived the battles in Korea and Vietnam, or were never on the field in either war, 299 achieved the grade of general officer, including those who achieved the grade serving in the Reserves or National Guard. Among them were ninety-seven brigadier generals, one hundred twenty-four major generals, fifty-five lieutenant generals, and twenty-three generals.

Among the twenty-three generals were three Army Chiefs of Staff, one Air Force Chief of Staff, and one who was the field commander of victorious United States and Coalition Forces in the 1991 Gulf War.

General Edward C. Meyer, class of 1951, was Army Chief of Staff from 1979 to 1983. General John A. Wickham, Jr., class of 1950, succeeded General Meyer, and served until his retirement in 1987. General Carl E. Vuono, class of 1957, succeeded General Wickham, and retired as Army Chief of Staff in 1991, after the Gulf War. These three officers had served as Chiefs of Staff throughout the decade of the 1980s, and led the United States Army as it continued to rebuild and modernize following the painful adversities and lessons of Vietnam.

General Charles A. Gabriel, class of 1950, who fought as a fighter pilot in Korea, and as a tactical reconnaissance wing commander in Vietnam, served as Chief of Staff of the Air Force from 1982 until his retirement in 1986. He, too, served as Air Force Chief of Staff as the Air Force continued to rebuild and modernize after the war in Southeast Asia. He was a backup quarterback, behind All-American quarterback Arnold Galiffa, on Army's undefeated 1949 football team.

General John R. Galvin, class of 1954, who as a cadet kept the entire Corps of Cadets laughing at his delightful *Pointer* cartoons, was

Supreme Allied Commander, Europe, from 1987 until his retirement from the Army in 1992.

A Philippine Islands graduate in the class of 1950, General Fidel V. Ramos, who, as a cadet, played on the Academy's chess club team with Cadet David R. Hughes, became the Chief of Staff of the Army of the Philippines in 1988. He later served as Secretary of National Defense, and in 1992 became president of the Philippines.

From among those same eight classes also came thirteen Rhodes Scholars, seven of the nation's pioneer astronauts, and another who became a posthumous winner of the Congressional Medal of Honor, for his extraordinary heroism in Vietnam during the period 1 to 23 July 1970.

Andre C. Lucas, class of 1954, the second Congressional Medal of Honor recipient among those eight classes, was among the men of '54 who waged a "Mutiny on the Whiskey" – the battleship *Wisconsin* – somewhere in the Caribbean Ocean, early his yearling summer of 1951. With him on the *Wisconsin,* had been Nelson S. Byers, also '54 and Company K-1, who lost his life with eighteen other West Point cadets, in the tragic aircraft accident near Phoenix, Arizona, on December 30, 1951.

A former member of the class of '54, First Lieutenant Numa A. Watson, Jr., son of 1922 Academy graduate, Major General Numa A. Watson, Sr., was killed in action in Korea, on June 22, 1953. Lieutenant Watson's father had fought in Korea in the early, dark days of the War, 1950-51. During that first year of fighting, two years before the Numas lost their son, his father served in the headquarters of the 7th Infantry Division and X Corps, and as Assistant Division Commander, 24th Infantry Division.

The thirteen Rhodes Scholars in the classes 1950-57 were: James M. Thompson, '50; Andrew C. Remson, Jr., '51; Charles R. Wallis, '52; John C. Bard, Ames S. Albro, Jr., and Dale A. Vesser, '54; Lee D. Olvey, John T. Hamilton, Martin C. McGuire, and Harvey A. Garn, '55; B. Conn Anderson, Jr. and Richard D. Sylvester, '56; and James R. Murphy, '57.

Astronauts from those eight classes were: Frank Borman, '50; Edwin E. "Buzz" Aldrin, '51; Edward H. White II and Michael Collins, '52; David R. Scott, '54; Alfred M. Worden and Donald H. Peterson, '55. Among the seven pioneering space explorers, one became the first

American to "walk in space." He was Ed White, who later died with astronauts "Gus" Grissom and Roger Chafee in a deadly fire on the launching pad, while they were practicing launch countdown procedures on board their spacecraft. The accident occurred on January 27, 1967.

Frank Borman, '50, after retiring as a pioneer astronaut and a colonel in the United States Air Force, became president of Eastern Airlines. He later served as the lead member of the Borman Commission, which in 1976 was a commission of inquiry into the Academy's second major cheating scandal in twenty-five years.

Lieutenant William J. "Pat" Ryan, class of 1951, continued his career in the Air Force until, as a colonel, he retired with more than thirty years of service. He served in three wars and saw combat in two. Following his ten B-29 combat missions over Korea, and additional "peace keeping" missions – one of which almost cost him his life   he flew 585 combat missions in Vietnam, nearly all as an airborne Forward Air Controller in O-1 "Birddogs." The O-1 was a light, single engine aircraft, which pilots flew in proximity to enemy targets they could see from their slow, low flying "Birddogs," and from which they guided fighter-bombers to attack the same nearby targets. The airborne Forward Air Controller mission has consistently been one of the most difficult and hazardous aerial warfare missions in this century.

Pat Ryan went to pilot training, and B-29 transition training prior to his missions in Korea. While he and his wife were in a Mississippi restaurant near his pilot training base, he encountered a small group of "the 90." They were former Army football players, and recognized him as soon as they walked into the restaurant. He was uncertain as to their reaction, but they seemed friendly, though he made clear he believed their separations from the Academy were fair and just.

During the conversation, they told him they had received phone calls from Colonel Blaik, asking if they would like reappointments to the Academy, that Blaik "would make them test cases, given their circumstances." Ryan was stunned by the prospect the men might be reappointed to West Point, and subsequently phoned Colonel Harkins in

Washington to tell him of the conversation. Earl Blaik may have pursued the reappointments, but if he did, records revealed his efforts were unsuccessful.

Pat Ryan greatly admired and respected both Colonels Paul Harkins and John Waters, the Commandant and Assistant Commandant while he was First Captain in the class of 1951. He spoke of Harkins as "one of the most honorable, even-minded men...[he]...ever met."

One day, in 1959, eight years after Pat graduated from West Point, on an aircraft parking apron in Germany, he encountered General Paul Harkins. Pat was walking out to his aircraft, preparing to fly a check flight. Harkins, who was on a return trip from Southeast Asia, invited him to lunch. Harkins told him of his recommendations for 500,000 more American troops to fight in Vietnam, and the effects of the Eisenhower administration not accepting his recommendations. He told Pat he had warned, "Every month's delay will mean 500,000 more men needed to do the job."

Before Pat Ryan went to Vietnam in 1967, he served at the United States Air Force Academy from 1963 to 1965. He was in the Commandant's Department, which functioned similar to its counterpart organization at West Point. While Pat was at the Air Force Academy, the youngest of America's national service academies suffered its first massive, violation of its honor code. The incident involved a large number of cadets cheating in academics. For Major Pat Ryan, the Air Force Academy incident was a sad, stinging reminder of young men's tragedies at West Point in 1951.

While serving as a Tactical Air Control Squadron Commander at Bien Hoa Air Base near Saigon, South Vietnam in 1968, the base came under rocket and mortar attack during the violent North Vietnamese Tet Offensive. Pat lost twenty-seven members of his squadron that terrible night. He wrote personal, handwritten letters of condolence to each of the families and loved ones of the twenty-seven, and still gets misty eyed when he remembers their sacrifice.

Colonel William J. Ryan retired from the Air Force's Air Staff in Washington, D.C., in 1981, where he served in the Operations and Readiness Group. Joanne, his beloved "Foxie," and wife for 41 years, died in June of 1992, following a recurrence of cancer. He resides in Alexandria, Virginia, and has five grown children.

Lieutenant David R. Hughes, class of 1950, remained on active duty until his retirement in the grade of colonel, from the Army's 4th Mechanized Infantry Division at Fort Carson, Colorado. After returning from Korea in March of 1952 he served on the staff and faculty of the Infantry School at Fort Benning, Georgia, where he met and married his wife Patsy. He obtained a masters degree at the University of Pennsylvania in 1955, and served on the Military Academy faculty as an English instructor until 1958.

The year prior to his leaving the Academy, he was asked to speak to the Corps of Cadets at a football rally during a meal in Washington Hall. The rally was for the Army-Notre Dame game, the first of a two-game renewal of the great rivalry between the Black Knights of the Hudson and the Fighting Irish – and the first of the final two Army-Notre Dame games Earl Blaik's teams would play.

Captain David Hughes gave a carefully prepared, emotionally stirring talk to the Corps – to no avail. Army lost to the Irish in a thriller the following Saturday in Philadelphia's Municipal stadium. The score was 23-21, decided by a fourth down field goal by Notre Dame's Monty Stickles, who had been turned down for West Point because of his eyesight.

Like many in his graduating class, David Hughes later fought in Vietnam. He commanded the 1st Battalion, 27th Infantry Regiment, 25th Division in that war, winning another Silver Star, the Legion of Merit, fourteen Air Medals, and the Bronze Star Medal for Valor. In Vietnam, as a battalion commander, he once again faced cruel choices war often imposes on those who must fight and lead. He came away with memories of more brave men who didn't live to return after he ordered them into a fight – in a war that couldn't be called a victory.

After his return from Vietnam, he served as Battalion Commander, 2d Battalion, 11th Infantry Regiment, 5th Division, and Commander, 3d Brigade, 4th Mechanized Infantry Division.

He retired from the Army in 1974, and is active promoting "electronic democracy" – the development and spread of computer based technology and communications throughout America's remote regions, and in the far reaches of "Third World" nations. David Hughes' electronic democracy uses spread spectrum technology and is intended as a "grass roots movement" to connect "the little people" – "the salt of the earth" – to one another all over the globe. He works closely with the National Science Foundation to bring these capabilities to Americans and interested overseas nations.

He and his wife Patsy have three sons, one of whom married the daughter of a Chinese physician. Her father was in the Chinese Peoples' Liberation Army, a short distance across the Yalu River, in China, tending wounded Chinese and North Korean soldiers, when Dave Hughes first arrived in K Company of the 7th Cavalry, near Unsan, North Korea in November 1950. David Hughes' Chinese-American daughter-in-law teaches the Chinese language in the Foreign Language Department at the United States Air Force Academy in Colorado Springs.

David R. Hughes and his wife Patsy reside in Colorado Springs, Colorado. He owns a small, but well-known, computer-based business, Old Colorado City Communications.

Lieutenant Richard T. Shea, Jr., class of 1952, posthumous winner of the Congressional Medal of Honor, rests in Olive Branch Cemetery, not far from his boyhood home near Portsmouth, Virginia, half a world away from Pork Chop Hill. On a sunny day in May of 1955, at Fort Myer, Virginia, the Secretary of the Army presented the nation's highest decoration to Dick Shea's widow, Joyce Riemann Shea. In her arms she held their son, Richard T. Shea III, born a month and a half after Dick Shea was killed in action. Young Richard cried during the first part of the thirty-five minute ceremony, until the band began to play and the 3d Infantry Regiment, the Old Guard, passed in review. Joyce, accompanied by Dick Shea's mother and two brothers, Staff Sergeant Robert Shea and William Shea, maintained her composure throughout the ceremony. She remarked she was proud to receive the award and knew someday their son would be equally proud of his father.

On a simple bronze plaque in West Point's Cullum Hall are found the names of Military Academy graduates who have won the Congressional Medal of Honor. Dick Shea is one of two who received the award for their incomparable acts of courage and heroism during the Korean War. The other was Lieutenant Samuel S. Coursen, class of 1949.

On Saturday, May 10, 1958, at a Brigade full dress review on the Plain at West Point, the Army Athletic Association presented a plaque to the Corps of Cadets, renaming the Academy's track stadium in memory of Dick Shea. Shea Stadium overlooks the track where he ran and won nearly every race he entered, and set records that stood for years.

At Virginia Tech, where Dick Shea was a seventeen-year old enlisted man late in World War II, his name can be found engraved on the Cenotaph above the school's Chapel, among six other Virginia Tech recipients of the Medal of Honor – and among the names of 411 Virginia Tech dead on the pylons of its war memorial.

Dick Shea's widow Joyce eventually remarried. Her husband Theodore Himka is a retired Boeing Aircraft Company engineering manager. They are a close, loving family, which includes three sons and their families. The oldest son is Richard T. Shea Himka, and the entire family respectfully and devotedly protects the memory of a man that none ever knew, save Joyce.

In 1973, men from Dick Shea's West Point class of 1952 dedicated a memorial plaque in his name. The plaque is on a wall at what was once Youngsan Army Garrison, in Seoul, Korea, not far from the headquarters of the ROK-US Combined Forces Command and the Republic of Korea Ministry of National Defense.

The 7th Infantry Division, which suffered grievous losses on Pork Chop Hill, has named its outstanding junior officer award in memory of Lieutenant Richard Thomas Shea, Jr., United States Military Academy class of 1952.

The Military Academy class of 1952 will hold its fiftieth reunion at West Point during the Academy's Bicentennial celebration in 2002. Dick Shea's classmates have elected, as their fiftieth reunion gift to West Point, to tear down and completely rebuild a larger, modernized Shea Stadium, using private donations. The stadium will overlook a refurbished, updated track and field where Dick Shea ran to glory a half century earlier.

At the north end of the stadium will be affixed three plaques: the original Shea Stadium dedication plaque; a Medal of Honor plaque as a gift from the class of 1939; and a 2002 Bicentennial dedication plaque containing a bronze relief of Dick Shea and a Medal of Honor medallion. On the field at the north end of the stadium will be a Class of 1952 Memorial Plaza, which will include a flag staff at its center, to fly the national colors. The wall of the Memorial Plaza will be inscribed with the names of twenty-nine men in the class who were killed in action in

Korea and Vietnam, or who died in accidents during military operations. Among the names will be Dick Inman and Dick Shea, who died on Pork Chop Hill during those fateful five days in July of 1953.

Also inscribed on the wall will be the Academy motto, "Duty, Honor, Country"; the cadet honor code, "A cadet will not lie, cheat or steal, nor tolerate those who do"; and General Douglas MacArthur's immortal words, "Upon the fields of friendly strife are sown the seeds that, upon other fields, on other days, will bear the fruits of victory."

Lieutenant Richard G. Inman, class of 1952, who died on Pork Chop Hill late the night of July 6, 1953, less than two days prior to the loss of Dick Shea, still rests where he fell. In his brief but heroic service in Korea he received a battlefield promotion to first lieutenant. First declared missing in action after he was cut down within minutes after the CCF's July 6 onslaught, he was declared dead in October of 1953, following a prolonged search for information by his parents and his wife Barbara.

Barbara, who had joined the Red Cross after Dick was declared missing – in hopes she could be sent to Korea to find him still living – later was transferred to Africa where she continued her work in the Red Cross. She remarried three years after Dick Inman's death, in Tangiers, Africa. She remained happily married to John Colby and was residing in Bridgewater, Massachusetts, when she died on February 1, 1999.

Dick Inman is survived by two sisters and a brother. Mary Jo Inman Vermillion lives with her daughter in Vincennes, Indiana, where Dick grew up. His brother Robert is retired and lives in Sun City Center, Florida. His youngest sister, Bonnie Ruth Wright, about whom Dick penned a loving poem, resides with her family in Keil, Wisconsin.

In February of 1998, a year before Barbara Colby died, she and Mary Jo Vermillion received telephone calls from the Department of the Army, seeking information and DNA tests of Mary Jo, the first of a series of steps toward locating and recovering the remains of a fallen soldier.

For Barbara Colby, the telephone calls from the Department of the Army stirred bittersweet memories of a love long past and a young man who was yet to come home.

The entire Inman family, and their descendants, hold fast to their loving memories of Dick Inman, his zest for life, and his sacrifice.

Lieutenant MacPherson Conner, class of 1952, Dick Inman's and Dick Shea's classmate, is the grandson of the late General Fox Conner, class of 1898. Fox Conner was General John J. "Black Jack" Pershing's operations officer in Europe in World War I, and was considered "the brains of the American Expeditionary Force." MacPherson Conner's grandfather was destined to play a critical role in the life and career of General Dwight D. Eisenhower. Eisenhower met and worked with numerous brilliant, accomplished men throughout his service in the Army and his two terms as president of the United States. Yet, in the twilight of his life Eisenhower would say "Fox Conner was the ablest man I ever knew."

MacPherson Conner was hospitalized for an extended period, recovering from the severe wounds he received on Pork Chop the morning of July 9, 1951. He was awarded the Purple Heart and the Combat Infantryman's Badge. The wounds resulted in the Army's limiting the assignments he could accept, causing Mac's decision to resign his commission in 1955. He went to work in his father's manufacturing company for five years, then went into the electronics business, eventually becoming the sales and marketing manager for North American Phillips, and later president of his own consulting firm. He married, and he and his wife Jane raised a family of five children. They now have six grandchildren. He is retired and resides in Walker Valley, New York.

The Men of Pork Chop Hill. Hundreds of men who fought the last battle for Pork Chop Hill are living, many active in fraternal and professional military associations such as the 17th Infantry Regiment and 7th Infantry Division Associations. Recently, reunion activities have grown in number and attendance. During reunions the men renew their acquaintances and close wartime friendships, remember the battles and their harrowing experiences, recount the stories of courage and devotion they witnessed, and reaffirm their dedication to those who didn't return from the Korean War. Many, late in life, driven by poignant, aching memories, have begun active searches for long lost comrades, unsure if the men who became separated from them during those chaotic five days, survived the war and the passage of years.

Similarly, thousands of searches are being initiated by surviving family members and descendants of those who didn't return from the war. The searches to learn of the lives and fates, the last few moments or hours of loved ones, are spurred in part by the same, poignant motivations of their loved ones' comrades in arms, and the explosive growth in America's telecommunications industry.

Former Corporal Charles Brooks, Bob Miller's squad leader in A Company's 1st Platoon is a retired carpenter in Willow Street, Pennsylvania. Wounded, and his surviving squad members nearly out of ammunition, Charles had had to make the wrenching decision to leave a severely wounded Private Bob Miller behind, and send medical help to him. Charles was evacuated with the wounded from Pork Chop, and received the Silver Star and Purple Heart, never knowing the fate of Bob Miller. For forty-two years he was plagued with the notion he might have left a wounded man to die before help came – until 1995, when he unexpectedly encountered another former squad member, Angelo Palermo, at the dedication of the Korean War Memorial in Washington, D.C. Angelo brought news that Bob Miller had indeed survived. The surprise encounter brought enormous relief and great joy to Charles Brooks. Since that time, Charles and Bob Miller have periodically been in contact with one another.

Dale Cain, who had been beside Dick Shea in the three days before the Chinese assault on Pork Chop, was completely unaware that A Company of the 1st Battalion had been relieved from the fighting on Pork Chop the morning of July 8. Isolated in his bunker with three other men, Dale had been left behind, but survived to leave Pork Chop on July 11. He received no decorations for his five days of heroism on the hill, and not until 1999 did he learn, in a stunning surprise, that throughout the battle, apparently no one from A Company knew of his position or isolation from the company.

Dale left the Army after his enlistment was up and became an Arizona highway patrolman, where he served for twenty-one years before retiring to work in a security company for nine more years. As a highway patrolman he nearly lost his life in a confrontation with a sixteen-year-old escapee from a California Youth Correctional Facility on a lonely stretch of Arizona highway. The youth, who had been carrying a concealed weapon, fired his pistol point blank at Dale, across the top of the patrol car. The bullet glanced off the top of the car and struck the metal highway patrol's emblem mounted on the front of his uniform hat. The ricochet-slowed bullet didn't penetrate Dale's metal highway patrol emblem, but it stunned and confused him. A backup patrolman saved his life by simultaneously felling the young assailant with return fire.

Over a period of years, Dale Cain researched and drafted a manuscript entitled *Korea, the Longest War*, which included his recollections of the war and the recollections of twelve other veterans. Unable to find a publisher in the United States, he wrote a letter to the president of South Korea, which eventually resulted in the book being published in 1997 by the Ministry of Patriots and Veterans Affairs in the Republic of Korea. Copies of the book were distributed to public libraries, middle schools and high schools in the Republic as a not-for-sale edition. The intent of the distribution was to "contribute to enhancing patriotic minds among youth, and to enlightening their recognition on the importance of freedom, and lessons of war." Dale is retired in Buckeye, Arizona, and is active in Korean War veterans' organizations.

Former Private Robert E. Miller, the squad member in Charles Brooks' squad who was left behind and evacuated on July 8, after losing a leg to a Chinese grenade, survived the battle for Pork Chop, recovered from the loss of his leg, and is retired in Morganton, North Carolina. He received the Silver Star for his heroism on Pork Chop, and the Purple Heart for his wounds. He carries with him vivid memories of the battle and those who showed him kindness and compassion, and helped save his life. Periodically he participates in a corporate-sponsored winter ski program for disabled veterans. Bob, too, is active in Korean War veterans' organizations, and remains in contact with A Company survivors he has met at reunions or by mail and phone.

Bob Miller and Paul Sanchez, the man who saved Bob's life with a tourniquet after he lost his leg to a grenade, quite by chance met for the first time in forty-two years when both attended a 7th Division Reunion at Fort Ord, California. It was 1995, the same year Charles Brooks first learned from Angelo Palermo that Bob Miller had survived his lonely ordeal on Pork Chop Hill.

Former medics, James E. McKenzie, Lee H. Johnson and Emmett "Johnny" Gladwell, are among a larger number of men from the 7th Division who have begun holding annual reunions to renew their comradeship and devotion to one another. Each has vivid memories of the endless streams of wounded and dead who came off Pork Chop and the 7th Division front those fateful five days. Among those to whom they rendered life saving care were ROK soldiers and civilian laborers who were supporting the United Nations' forces when the Chinese assault came and wounded Chinese soldiers who were captured during the battle.

Associated with the former medics and other veterans of the 7th Division is David R. Willcox, who as a young lieutenant led A Company's 2d Platoon in the desperate fight for Pork Chop Hill.

After recovering from his wounds and visiting Dick Shea's widow in January of 1954, he left the Army and entered civilian life. He looks back on his three days in the battle, and the memories they bring, as sources of difficult, painful experiences that ultimately deepened and strengthened his faith and brought him into a warm, healing, personal relationship with God. He has since spent much of his life bringing the good news of God and faith into business and corporate life, and the lives of friends and associates. His strong faith and his family sustained him as his beloved wife Janet struggled against cancer and finally succumbed in 1999. David is retired and lives in Glen Ridge, New Jersey.

Robert Northcutt from G Company of the 2d Battalion, 17th Regiment, retired from the Army a master sergeant in 1972, resides in Malta, Ohio, and is vigorously active in Korean War veterans' organizations in and around his hometown. He keeps alive the history of battle streamers earned by the men of the 17th Regiment Buffaloes throughout American wars and campaigns since the regiment's activation during the Civil War.

Former Sergeant Danny L. Peters, C Company of the 1st Battalion, is retired and lives in Manning, Iowa. Dan, whose company successfully relieved David Willcox's 2d Platoon of A Company during the July 9 counterattack on Pork Chop, returned home from Korea to his wife Lois, who was living with her parents and near his parents in Manning. She and Dan had grown up together in Manning and were high school sweethearts. They married three months before he was drafted and went off to war. Upon his return he entered Iowa State University to obtain a degree, with a goal of entering forestry, and he subsequently completed a rewarding career in the U.S. Forest Service. He and Lois have five children and twelve grandchildren. Dan does volunteer work and is also active in Korean War veterans' associations.

Nineteen-year-old Corporal Dan Schoonover, squad leader, A Company, 13th Combat Engineer Battalion, posthumously received the Congressional Medal of Honor for his valor and sacrifice on Pork Chop Hill. Dan was the youngest of four boys, and all of his brothers had entered the service when he went off to war. It was partly to be with his older brother, Pat, who had enlisted in the Army, that Dan enlisted and arrived in Korea in late 1952.

His mother, Mrs. George A. Hess, received a brief telegram from the Department of the Army on July 17, 1953, notifying her that he had been killed in action on July 10. He died, having elected to stay on Pork Chop after the surviving men of G Company were withdrawn a day earlier. In 1982, his remarried mother Peggy Schroeder reminisced about her son who never came home. "He was doing what he wanted to do." She told of talking with one of Dan's comrades. "He said he asked Danny to go down the hill with him [on July 9]. Danny told him, 'Some of these other men are married and have families. I'll stay'."

Cadet Daniel J. Myers, Jr., class of 1951, and Company K-1 honor representative was commissioned a second lieutenant, Infantry, when his West Point class graduated. After attending the Infantry School at Fort Benning, Georgia, he received airborne training and served in the 11th Airborne Division until the fall of 1952, when he was assigned as a rifle platoon leader in Charlie Company of the 1st Battalion, 35th Infantry Regiment, 25th Infantry Division, in Korea. During his service on the battlefields of Korea, the 25th was engaged in fighting north of Seoul. He later was assigned duties on the battalion staff, as an S-2 – an intelligence staff officer, then as a liaison officer to the Turkish Brigade, which fought alongside the 25th Infantry. He received the Bronze Star Medal and the Combat Infantryman's Badge.

After the Korean War he served in the 18th Airborne Corps and the 101st Airborne Division, before obtaining a Masters Degree in Nuclear Physics from Tulane University. He subsequently received assignments related to his nuclear physics education, one at Kirtland Air Force Base in Albuquerque, New Mexico. After an assignment to United States Army, Vietnam, during the war in Southeast Asia, he attended the Army's Command and General Staff College, going from there to an assignment on the staff of the United States Army, Europe. After twenty years of service, he retired from the Army as a lieutenant colonel in 1971 and entered the real estate business. After acquiring and later selling his real estate firm, he became a stock broker for Merrill Lynch. After selling his real estate firm, he eventually relocated to Hawaii, where he lives with his wife Beth and is an author. Dan and Beth, each married and divorced before they were wed, have a total of five children and seven grandchildren.

Richard J. Miller, class of 1952, and Company K-1 honor representative received his commission in the Infantry. He attended the Infantry School and Airborne training, and was assigned to the 508th, then the 187th Airborne Regimental Combat Teams until 1957, when he resigned his commission and left the Army. In 1958 he reapplied for a regular Army commission and returned to active duty. He served as a rifle company commander in the 32d Infantry Regiment in Korea, 1960-61, where he was awarded a Commendation Medal. After obtaining a Masters Degree in Geography, he returned to West Point to teach. After additional professional schooling he served in Vietnam as a battalion executive officer in the 1st Air Cavalry Division, and later was Assistant Chief of Staff in the Division. Following attendance of the Army War College in Carlisle Barracks, Pennsylvania, he served as a battalion commander in the 325th Airborne Infantry Regiment, 82d Airborne Division. After assignments in the 7th Corps headquarters in Europe, and Headquarters, European Command, he returned to Carlisle Barracks for two years before retiring in 1974 in the grade of lieutenant colonel. In 1992-94 he worked as a field assistant, simultaneously, to a Pennsylvania state senator and a state representative.

He and his wife Doris, live in Port Royal, Pennsylvania. They have six children, fifteen grandchildren, and three great-grandchildren

The Republic of Korea is today a free nation, an Asian democracy that has taken root in an ancient culture of kings and emperors, following an oppressive, brutal Japanese occupation from 1905 to 1945, and the devastation of the Korean War. The Republic is a vibrant, thriving nation whose emerging, democratic form of government and free market economy have become one of the great success stories of Asia and the twentieth century.

To the north lies the Peoples' Republic of North Korea, a reclusive, isolated, and still belligerent holdover of Stalinist and Maoist Communism, and the Cold War. It remains a closed culture, perhaps the most secretive nation in the modern world.

For nearly forty years after the Korean War, North Korea was ruled by one man, Kim Il Sung, who rose from the grade of major in the North Korean Army to become that country's president and wartime leader. Before he died in 1994, he designated his son as his successor, in what has become a bizarre, modern-day communist emperor's dynasty.

The tightly controlled state economy is near collapse, staggering under the weight of a Spartan, repressive, militaristic government, that until recently was attempting to develop nuclear weapons. Its people are suffering the consequences of belligerent, armed-to-the-teeth isolation, which has brought increasing malnutrition and slow starvation for many, as resources are diverted to maintain a huge war making capability. Only under threat of total economic collapse has the North Korean government relented, and permitted international aid organizations to render humanitarian assistance to their people.

To the south, along each side of the DMZ, North and South Korea maintain armies totaling nearly one million men each, in defensive positions, always vigilant and wary of one another. Their two armies are backed by air and naval power. They remain in an uneasy peace, frequently punctuated with strident hostility, shooting incidents, and periods of sharply increased tensions.

The United States has maintained ground, air, and naval forces in or near the Republic of Korea since the armistice was signed on July 27, 1953, to help ensure peace, freedom, and democracy for the Republic of Korea, as well as the security of the island nation of Japan, now a strong ally to America and the Western World. The United States Army's 2d Infantry Division, which entered the fight in the Pusan Perimeter during the summer of 1950 is located not far south of the DMZ, ready to fight once more should the North Koreans again invade the Republic of Korea.

In between the two still hostile armies, within the DMZ, lies the hallowed ground of Pork Chop Hill where so many died, and Hill 347, where David Hughes and K Company of the 7th Cavalry Regiment fought so valiantly to scale and hold its heights in October of 1951.

In November of 1997, an agreement was reached to begin talks between South and North Korea, the United States, and the Peoples' Republic of China, in Geneva, Switzerland, aimed at reaching a comprehensive peace agreement finally ending the Korean War. To date, there has been no resolution of their fundamental differences.

In June 2000, the president of the Republic of Korea, Kim Dae-jung and North Korea's leader, Kim Jong Il, met in a three-day summit in Pyongyang, the North's capital. It was the first meeting in history between the two country's leaders and was considered a major breakthrough toward diffusing the hostility between two nations technically still at war. The two leaders signed an agreement pledging to work for reconciliation and eventual reunification. They also agreed to allow reunions of families that have been separated for fifty years by the closed, heavily-armed border, South Korean investment in the North's shattered economy; and a follow-on summit in Seoul, South Korea.

However, the Pyongyang summit did not deal with two major strategic issues: the North's continuing demand that the United States withdraw its 37,000 troops from South Korea and the North's nuclear and long-range missile programs that have led to the American government's labeling it as a dangerous "rogue" state.

The 1953 Army football team became the 1954 team, except for players who graduated or, for various reasons, couldn't play the next season. In '54 Army went 7-2 on the season, losing its opener to South Carolina, 34-20, and the finale to Navy, 27-20. In between those two losses the cadets defeated Michigan, 26-7; Dartmouth, 60-6; Duke, 28-14; Columbia, 67-12; Virginia, 21-20; Yale, 48-7; and Pennsylvania, 35-0.

After the home opener upset at the hands of South Carolina, the cadets caught fire as they had a year earlier. They went to Ann Arbor, Michigan, and thoroughly trounced the Wolverines before 70,000 fans, which included the first class, the then senior class of 1955. When

996

Army played Duke at their homecoming in Durham, North Carolina, the Blue Devils again were undefeated coming into the game, and were anxious to redeem themselves after the stunning 14-13 loss at New York City's Polo Grounds the preceding year. Duke's defeat this time was even worse than the 28-14 score indicated. The cadets faced an undefeated Yale University at New Haven, and crushed them in an awesome display of power. The Eli were on the road to an Ivy League championship.

Army and Navy were ranked No. 5 and No. 6 in the nation when they collided that year, and the cadets were favored. After the dust settled in a sold-out Municipal Stadium in Philadelphia, Navy was ranked No. 5 and Army No. 7 in both the Associated Press and United Press International season-ending polls. The exciting see-saw battle ranked among the great ones of this grand, colorful rivalry. Navy went on to be Sugar Bowl champions the following New Year's Day.

In spite of the loss to Navy and South Carolina, as a football team, Army scaled new statistical heights and earned first team All-American honors in 1954. Their 448.7 average yards total offense was highest in the nation among major college teams, as was their rushing average of 322.0 yards.

Halfback Tommy Bell, '55, guard Ralph Chesnauskas, '56, and end Don Holleder, '56, all stand-outs on the 1953 team, were named first team All-Americans. Pete Vann, '56, the boy who, in Earl Blaik's eyes, became a man in the 1953 Duke and Pennsylvania games, was a second team All-American quarterback in 1954, behind Ralph Guglielmi of Notre Dame. Pete was also No. 9 in Heisman Trophy voting that year, when Wisconsin fullback Alan "The Horse" Ameche won the Heisman. Pete had become a ball handling magician and an acknowledged team leader, his path to football glory mapped out in the years from 1951 through 1953, under the tutelage of Army's charismatic backfield coach, Vince Lombardi.

Although NCAA rule changes played a role in Tom Bell's fortunes as an Army football letterman, he earned yet another distinction Lombardi couldn't have prophesied when he told Tom's little Irish mother in the spring of 1950, "I'm gonna make Tommy an All-American." In the handsome redhead's last season for the Black Knights of the Hudson, Tom became a four year letterman, the only Army player to earn that distinction between Glenn Davis in 1946 and Leamon L. Hall in 1977.

There was another follow-on story for the team of 1953, and it came two seasons later, in 1955, which was the last season the men in the class of 1956 played varsity football for Army. Much of the fire, determination, and inspired play which became the trademark of the unheralded Black Knights of 1953 came from this class, from men like Bob Farris, Ralph Chesnauskas, Pat Uebel, Don Holleder, Paul Lasley, and Howard Glock.

After the 1954 season, Earl Blaik faced a problem he hadn't counted on. Pete Vann had played four years of varsity football, and had been turned back to the class of 1956. Though he lettered three of those four years, he wasn't eligible for a fifth year of varsity play. The constant struggle to find the talent and depth needed at quarterback had turned up a serious hole which needed filling – early. Blaik tried many of his backs at the position, but couldn't come up with Pete's replacement. It was then that he thought of Don Holleder.

Don was a talented pass receiver, not a passer. He threw left handed, and had never played in the backfield. Blaik's idea seemed foolish. He was considering moving the team's most talented end and pass receiver to play the most demanding position in the backfield – with absolutely no prior experience. Nevertheless, he reasoned Holleder was a strong leader, an outstanding all around athlete, a fierce competitor, and a solid runner when he got the ball. Blaik decided to ask him at the start of spring practice to be the team's quarterback.

Earl Blaik's decision would cost Don Holleder a sure second year as an All-American end for Army, and both him and Don a season of controversy and criticism. Holleder was hesitant when confronted with Blaik's proposal, and asked for a night to think before giving his answer. He came back to Blaik's office the next morning, obviously tired and worn. He hadn't slept, Blaik recalled, and looked it, but Don's mind was made up. He would try. He would give up his opportunity to be a two-time All-American for Army. This was the spirit of Army's '53 football team, the team that followed, and the class of '56 at work.

With Pat Uebel as team captain, and Don Holleder at quarterback, Army, riddled with injuries and still lacking the depth it needed, went 6-3 on the season, losing to Michigan, Syracuse, and Yale, but winning in an emotional 14-6 upset over a Navy team that came into the game with a 6-1-1 record. The Don Holleder-led team defeated Navy without gaining a single yard through the air, and stung a team led by All-American quarterback George Welsh, who led the nation in passing and completed 18 of 29 for 179 yards that day. Army gained 283 yards on the ground, 125 coming from team captain Pat Uebel, who scored a record fifth touchdown against Navy teams in his three years on the Army varsity. Don Holleder carried for the other Army score. The cadets ended the season ranked No. 15 by United Press International and No. 20 by the Associated Press.

Throughout the season of 1955, Don Holleder's roommate was Peter Joel Vann, the man who had teamed up with him on so many pass receptions in the '53 and '54 seasons, when Don was scaling the heights to All-American glory as an Army end. Pete helped Don develop his quarterback skills, became an assistant coach to him, encouraged him, and proudly watched him develop into a strong, able replacement for himself.

The two became close friends, and Don Holleder later was best man at Pete's wedding. But there would be something else in Don Holleder's future.

In Vietnam, on October 17, 1967, exactly fourteen years to the day after Army's stirring return to glory at New York City's Polo Grounds against Duke University's Blue Devils, he was killed in action.

Don was in the 1st Air Cavalry Division, the division David Hughes had fought in at Hill 347 in Korea in October of 1951. That day in Vietnam, Don Holleder, as battalion operations officer, was on board a helicopter over an area where a furious firefight was in progress. American soldiers were pinned down, in trouble. There were wounded among them. The wounded needed to be evacuated. A call for assistance came over the radio. Don Holleder didn't hesitate. No time to ponder this one overnight. It was time to act. He responded in the midst of heavy, sustained enemy fire, and directed the aircraft commander to land his helicopter to pick up the wounded soldiers. The decision cost him his life.

When he was laid to rest in Arlington National Cemetery, Earl Blaik and Pete Vann – Don's close friend, former roommate, and teammate – were there, as were Don's wife and four daughters and many of Don's classmates, friends, and teammates. Don posthumously received the Silver Star and Purple Heart.

Today, at the north edge of Howze Field, adjacent to Michie Stadium, stands Holleder Center, a large multipurpose athletic facility for hockey and basketball which is a memorial to Don. In addition to being an All-American football player, he had been a star Army basketball player.

Pete Vann continues carrying the warm, pleasant memories of his devoted friend and teammate through the years. Peter Joel Vann is retired and resides in Dubois, Wyoming.

Army football, like football played at colleges and universities all across the nation, has changed considerably since the era of this story. NCAA rules gradually liberalized substitution after 1953. In 1964 the liberalized changes returned virtually unlimited substitution and the two platoon system, which eventually spawned special teams play. Both are now staples of the game.

Football has grown more complex in style, strategy, tactics, and intricacy of play. Changes have extended into high schools, raising the level of players' skills. Rule making, equipment, player conditioning, and prevention and treatment of injuries have improved, while better diet, disease prevention, and other factors contributing to improved health in the nation's population have also resulted in taller, faster, and heavier players in colleges and universities. It isn't uncommon to see offensive and defensive lines averaging 250-290 pounds in college football, 40-80 pounds more than the decade of the '50s.

College and university student populations have grown dramatically as America's population nearly doubled in the last fifty years. At the nation's three largest service academies, the cadet and midshipman populations – legislated by Congress – peaked at approximately 4400 in the decade of the 80s, and consistent with reductions in the armed forces since that period, have now been reduced to 4000. At West Point, approximately 12 percent of the Cadet Corps are women.

The foregoing, with the dramatic growth and drawing power of professional football, have become factors affecting the ability of service academy teams to keep pace and compete with the mega-university powers of collegiate football. Over the years Congress has increased the active duty commitment of Academy graduates from three to five years. For young men who become powerfully enamored of football and believe themselves capable of going into the professional ranks, the five year service commitment after they graduate becomes a potent inhibitor against their playing at a service academy. Add to these considerations the fluctuating interest in military careers among potential college undergraduates, and the service academies' abilities to recruit scholar-athletes for "big time" collegiate football has become, and will remain, difficult.

While service academies are in fact unique, full scholarship institutions, which have no limits in the number of athletes they can recruit, they have specific and different missions than those of other colleges and universities. What also makes them unique is unyieldingly high standards of performance for "the whole persons" who are to be their graduates. Academy athletes – if they choose of their own free will to remain intercollegiate athletes – must meet all those standards, plus the high standards of performance demanded for successful competition in the sports they choose.

There are, however, constants in collegiate football, as played among America's gridiron powers, and those constants aren't likely to change. The game requires strategy; tactics; intelligence and intelligence gathering; deliberate, high speed collisions between big, strong men; peak physical conditioning and good health; enormous physical courage; concentration; discipline; dedication; hard work; teamwork; endless striving to improve; quick – almost instantaneous – and correct decisions; playing the game by the rules; character; and leadership on and off the field. Earl Blaik was probably right. "It is the game most like war."

Since Earl Blaik left the "fields of friendly strife" at West Point in January 1959, Army has never again appeared among the top ten football teams in the nation, although it did appear among the top 25, at No. 22 in 1984 and 1985 in the Cable News Network Coaches Poll, and No. 24 in the USA Today/Cable News Network Poll and No. 25 in the Associated Press Poll in 1996. Growth in the number of college bowl games, improved won and loss records in the '80s and '90s, and changes in Academy and Army policy since Blaik left Army, have permitted the cadets' football teams to play in four bowl games the last two decades.

Despite all of this, the game goes on at West Point, Annapolis, and the Air Force Academy, played with a vigor, fire, and determination that brings pride to their services, and American collegiate football lovers everywhere.

In 1990, the Academy celebrated one hundred years of Army football. A video was produced and sold, which compiled film clips from Army games of old, with some of its greatest former players and coaches providing their golden reflections on the game, as played by the cadets. Among the men who appeared in the video were two coaches in the twilight of their many years, Earl "Red" Blaik and "Gar" Davidson. The title of the video was *Field of Honor.*

Army's athletic recruits no longer attend a separate "cram school" near the Academy to prepare for entrance examinations after receiving appointments. Rather, they have the option of attending the U.S. Military Academy Preparatory School for one year, along with enlisted personnel and others who have obtained appointments, and are eligible and want to devote an additional year to prepare for the rigors of an Academy education. The "Prep School," as it's called, is in Fort Monmouth, New Jersey, and not only prepares cadet candidates academically and physically for the four years at West Point, but ensures cadets are ready for the demands of an Academy military education, as well as plebe year, cadet life, and the honor code and honor system.

In 1957 the Academy began lightweight intercollegiate football, for men who want to compete in the game, yet aren't big, strong, or skilled enough, to play in Division IA. Army and Navy are in a five team league which plays a double round robin schedule of eight games each fall. Lightweight collegiate football is growing in popularity at West Point.

In 1998 the Army varsity played their first football season as a member of Conference USA, after playing as a major collegiate independent for one hundred eight years. Conference USA includes Southern Mississippi, Cincinnati, Tulane, Houston, East Carolina, Memphis State, Alabama (Birmingham), Texas Christian University, South Florida, and Louisville.

"The 90," who were in reality the eighty-three, continued their lives after they left West Point in 1951. Most went on to colleges and universities to complete their educations. They all suffered some form of pain from their decisions and experiences that year, and for a few the pain was everlasting and debilitating. Yet the great majority served their families, their communities, their chosen professions, and their country, well and faithfully.

When their forthcoming resignations became public knowledge on August 3, 1951, offers of assistance and support, and entreaties for forgiveness, came from numerous sources. The late Joseph P. Kennedy, the father of assassinated President John F. Kennedy, became an anonymous, behind-the-scenes benefactor for twelve men, paying tuition for their completed educations at the University of Notre Dame.

Prior to Kennedy's commitment of support to any former cadets who wanted to attend Notre Dame, Frank Leahy, the university's football coach at the time, received a phone call inquiring of his interest in having former Army players who had resigned come play for the Fighting Irish. After consulting with the school's Athletic Director, and the university president, Leahy declined. In his biography some years later, Leahy said, "They could attend Notre Dame, but wouldn't be permitted to play football for the Irish."

Roman Catholic Cardinal Richard Spellman proffered the support of New York's Archdiocese to any who might wish to avail themselves of Catholic colleges and universities in the area.

A large number of the former cadets scattered to colleges and universities throughout the country, and quietly resumed their educations, most reticent to divulge their West Point experiences. One attended Villanova University, a Catholic university in Philadelphia, Pennsylvania, where he played football for the Wildcats. Others went to Kansas, Kansas State, Colorado College, and Georgia Tech. Two of the former Army football players received All-American recognition before they graduated from college.

At least two former players played professional football, another became an official – a referee in the professional ranks. Another became the head coach of the Los Angeles Rams and took the Rams to the Super Bowl.

Others reentered the military, receiving commissions in the Army or Air Force. At least three retired in the grade of colonel, two in the Army and one in the Air Force. One other rose to the grade of lieutenant general in the Air Force. As a major general, he was on the review-ing stand in Cairo, Egypt, when Anwar Sadat, the Egyptian president, was assassinated in 1982.

The Honor Code is currently described by the Academy as "the *mini-mum standard* of [ethical] behavior required of the Corps of Cadets" and also as "the foundation of the [Corps'] standards and values." It is really many things – a rule of acceptable conduct, a moral and ethical creed, a revered custom of the service, and an important element of the Academy's mission. The honor system, as earlier stated, is the educa-tional and procedural framework supporting the code, and provides the underpinning for the day-to-day practice of the code in cadet life. The practice of the code and its system in cadet life imbues Academy graduates with the precepts of honor which are intended to guide them throughout their lives and careers of service.

Nearly all graduates and cadets continue to admire and believe in the honor code, and the great majority strive to live its precepts – as General Dwight D. Eisenhower wrote in his 1946 letter to then super-intendent Major General Maxwell D. Taylor. The code, still maintained and passed from class to class essentially by the Corps of Cadets, was changed in 1970, and now states "A cadet will not lie, cheat, steal, nor tolerate those who do."

The "silence," instituted by cadets in the nineteenth century, was removed from the honor system by the Corps' vote in 1973. The result was a dramatic increase in the number of cadets who refused to resign following guilty verdicts rendered by the cadet Honor Committee, and instead requested hearings before boards of officers. The elimination of the "silence," though a positive step in improving the administra-tion of the Code, was but one symptom in the beginning of a consider-able period of turmoil that lasted through 1979.

In the twenty-five years following the incident of 1951, there were several outbreaks of multiple, simultaneous honor violations, and one major scandal, in 1976. There was another alleged to have occurred sometime during Earl Blaik's final four seasons at Army, and was de-scribed in the 1974 edition of his autobiography, *The "Red" Blaik Story,* on pages 447-449. There was no evidence that either of the incidents originated on the Army football team or any other intercollegiate athletic activity, as did the 1951 event. If Blaik's account is accurate, he gave the superintendent fundamentally the same advice he gave General Irving the night of May 29, 1951. "You and I know that 174 cadets, highly selected as they are with exceptional background records, are not bad, but the system is." He added, "Don't repeat the 1951 error."

If the incident occurred as Blaik recounted, both it and the 1976 incident were markedly different from the 1951 scandal. Each certainly was different in the way they were resolved, yet both reflected sharply weakened cadet support for the code and system. In both cases they included, ultimately, a widespread breakdown in the underlying principle of the code requiring cadets to report observed, suspected violations of honor by fellow cadets.

The 1976 incident came less than one year after a United States Military Academy Superintendent's Special Study Group on Honor at West Point had identified pervasive, serious weaknesses in cadet support for the honor code and system. In March 1976, the Electrical Engineering Department gave 823 cadets in the Second Class ('77) a take-home computer examination to be returned in two weeks. Known as the Course EE 304 scandal, the incident involved collaboration on the examination, for which specific instructions had been given regarding parts of the exam on which there would be no collaboration. Department review of the papers turned up evidence of extensive collaboration.

On April 4, 1976, the Department forwarded to the Honor Committee the names of one hundred seventeen cadets believed to have collaborated on the assignment. Allegations that the cheating was even more widespread, and that there had been cover ups on some of the honor boards, prompted the superintendent to appoint an Internal Review Panel. As a result, 150 cadets in addition to the fifty already found guilty, were referred to boards of officers. Of these, eighteen elected to resign and 103 were found guilty, including twenty-nine previously found not guilty by cadet boards.

The incident brought unprecedented actions by the Secretary of the Army: the convening of a Special Commission to assess the affair, and the opportunity for readmission, after one year, of all who had been found guilty or resigned for reasons of honor during the 1975-76 year. On September 9, 1976, he appointed the Special Commission "to conduct a comprehensive and independent assessment of the ...EE 304 cheating incident and its underlying causes in the context of the Honor Code and Honor System and their place in the Military Academy."

The Commission, chaired by Academy class of 1950 graduate and former astronaut Frank Borman – who was the cadet manager of Army's 1949 football team – recommended sweeping changes in the honor system. Its report, issued December 15, 1976, contained three general statements of position, the last agreeing to the Secretary of the Army's belief that a second chance to complete their educations was necessary for the cadets involved. The first two statements read:

First – The Commission unanimously endorses the Honor Code as it now exists.

Second – We believe that education concerning the Honor Code has been inadequate and the administration of the Honor Code has been inconsistent and, at times, corrupt. There must be improvement in both education and administration.

Virtually all the Commission's recommendations were implemented, but more turbulence followed in the wake of recommendations for increased "due process" in the investigation and adjudication of alleged honor violations. For two and a half years the pendulum of change swung back and forth in attempts to resolve the question of how more effective, credible, and just "due process" could be put into the system, and how much due process was enough. Effective July 1, 1979, procedures were finally put in place that have since changed little in succeeding years. Among them is leeway for the superintendent, in certain cases, to reinstate a cadet found guilty, thus providing latitude in the "one violation and out" system of days gone by.

Yet, because all young men and women, like those who've gone before, are fallible, imperfect, and have, in some cases, misgivings or have difficulty accepting the honor code, or the system which gives life to the code, a number of resignations and dismissals continue each year. The code's underlying principle requiring cadets to report observed, suspected honor violations by others remains the most difficult, controversial principle to put in practice.

Over the years since 1979, this guiding principle and other aspects of the honor system have continued as subjects of discussion and debate, but remained relatively unchanged until May of 1989, the same month Earl Blaik was laid to rest in the West Point cemetery.

That month a commission on honor, sanctioned by Academy superintendent Lieutenant General Dave R. Palmer, class of 1956, issued its report stating that many of the conditions existing in 1951 were still present, and recommended a total of twenty-five changes to the honor system. The changes were "intended to simplify enforcement of the code; to remove excess and trivial detail; to endorse flexibility of sanctions, enabling rehabilitation of an offender when that is feasible; to reaffirm that enforcement of the code is the responsibility of every participant; to amplify favorable effect of the code in later careers throughout the Army; and to support the code as an exemplar for all public service."

In connection with ethics training, the Military Academy recently began hosting an annual National Conference on Ethics in America, a project initiative begun by the class of 1970. The conference is growing in stature and participation, and provides a national forum for discussion of ethical issues on college and university campuses. Working groups, lectures, and discussions were the venue in the November 1996 Conference, in which fifty-three major colleges and universities throughout the nation were represented.

- In 1957, "Project Equality," begun by the superintendent, General Garrison H. Davidson, resulted in reorganization of the Corps of Cadets. Cadet companies are no longer sized as they had been since 1824, when, among the then four companies in the Corps, taller men were placed in the flank companies, A and D, and the shorter men in the center companies, B and C.

- In 1959 the United States Air Force Academy graduated its first class. Graduates of the Military Academy and Naval Academy were heavily involved in the establishment of the Air Force Academy, and for many years West Point and Naval Academy graduates in the Air Force served on the staff and faculty at America's youngest national service Academy.

- In 1961 West Point was designated as an official national historic landmark.

- In 1964 plebes were allowed Christmas leave for the first time.

- In 1967 New Cadet Counseling Services were established under the supervision of the Office of Military Psychology and Leadership.

- In 1972 a Supreme Court decision ended mandatory chapel attendance.

- In 1973 First Lieutenant Virginia K. Fry became the first full-time female faculty member at the Academy, as an instructor in the Department of Earth, Space and Graphic Sciences.

- In 1974 Cadet Robert E. Johnson, class of '75 became the first Black American to be elected captain of the Army football team; in 1975 he received the American Cancer Society's Courage Award from President Gerald Ford.

- In 1975 pocket-size electronic calculators replaced slide rules in support of math, science and engineering courses.

- In May of 1975 the Academy began plans for one semester exchange programs between the Military Academy, Naval Academy, and Air Force Academy.

- With graduation of the class of 1975 the option of Military Academy graduates to be commissioned in services other than the U.S. Army ended, except on the basis of agreed one for one permanent exchanges between the armed forces.

- In 1976 the class of 1980 entered its largest class to date, with 119 women among 1,479 new cadets.

- Richard Morales, Jr., class of 1976, became the first Hispanic cadet selected the Corps' First Captain and Cadet Brigade Commander.

- The class of 1980 was the first class to graduate women, with 62 of the original 119 receiving diplomas.

- Vincent K. Brooks, class of 1980, became the first Black American selected to be the Corps' First Captain and Cadet Brigade Commander.

- The class of 1985 was the first class to graduate more than 1,000 cadets, with 1,063 receiving diplomas.

- Kristin Baker, class of 1990, was the first woman selected First Captain and Cadet Brigade Commander.

- In the class of 1991, Michael Mayweather was the 1000th Black American to graduate from West Point, and Kimberly J. Ashton was the 1000th woman graduate.

The United States Military Academy continues in its mission of educating and training "the whole person." The cadet's education ranks with the finest in the nation. The balanced, essentially "no choice," 100 percent core academic curriculum that existed in the era of this story, remains balanced, but has been substantially changed to include twenty-six fields of study and nineteen optional majors. The changes began in 1958 with a curriculum study completed and sent to the Department of the Army. The study was started by General Garrison H. "Gar" Davidson, and masterfully steered through a reluctant Academic Board to a successful conclusion. In 1961 a four year transition to a modified curriculum began. Today the thirty-one-course core curriculum represents the essential broad base of knowledge necessary for success as a commissioned officer and also supports the subsequent choice of an elected area of academic specialization. It is, in effect, the "professional major" for every cadet since it prepares each graduate for a career as a commissioned officer in the Army.

The elective fields of study and majors cover virtually all the liberal arts, science, and engineering disciplines one would expect to find in a high quality, selective college or university of comparable size. Cadets may enter any field of study or major without restriction. No special grade point averages are established for entry, and there are no special quotas for particular disciplinary fields.

For most cadets, the path to graduation will be through an elected field of study. Pursuit of the field of study requires each student to complete nine electives defined by the disciplinary field. For those who desire to enrich their academic experience and pursue a discipline in greater depth, a majors program and honors courses are available on a voluntary basis. Cadets electing to major must follow a more structured sequence, usually ten to thirteen electives, and complete a senior thesis or design project.

To graduate, cadets must complete forty academic courses, with an option to voluntarily enrich the experience by majoring with forty-one to forty-four academic courses; successfully complete or validate each course in the core curriculum; complete eight semesters of physical education; complete four military science intersessions; and achieve a cumulative grade point average of at least 2.0. Since the era of this story, the grade point accumulation base has changed from a 3.0 to a 4.0 system.

The **athletic program** is guided by the dictum, "Every cadet an athlete, every athlete challenged." Every cadet at West Point competes in intercollegiate, club, or intramural sports. In addition, each cadet participates in the physical education program.

The value of athletic experience to the potential Army officer continues to be recognized as essential. Still emblazoned in the history, and institutional memory, are the words of General Douglas MacArthur, who as the Academy's superintendent, was largely responsible for establishing the athletic program for the Academy:

> The training on the athletic field, which produces in a superlative degree the attributes of fortitude, self-control, resolution, courage, mental agility and, of course, physical development, is one completely fundamental to an efficient soldiery.

Over the years, the number of intercollegiate varsity sports has fluctuated, leveling off currently at twenty-four, of which men participate in sixteen and women in eight. Nineteen of the twenty-four teams compete in the Patriot League, and as indicated previously the Army football team became a member of Conference USA in 1998. Over one quarter of the Corps of Cadets competes in varsity sports.

A tough, thorough, demanding **military training program** continues at the Academy. Potential officer-leaders must master fundamental military concepts and skills, know the role and employment of elements of the Army, and be thoroughly familiar with the missions of their comrades in arms in America's other armed services. Cadets learn military history, strategy, and tactics, and receive the technical foundation to understand and prepare for the armed services of tomorrow.

Throughout the cadets' four years at West Point, the military training curriculum, in building block fashion, deliberately, thoughtfully, and progressively molds cadets from followers into leaders; from student to trainer and teacher; from the responsibility for one's self to understanding of responsibilities toward classmates, teammates, military units, the soldiers that are the units they will lead when they graduate – and the democratic nation and Constitution they take an oath to defend. They still have the opportunity for "attaining eminence in letters and science as well as arms," as John Adams said so long ago. They can lay the foundation to become both soldiers and statesmen – if their country asks, if they continue to grow, learn, and pursue lives that demand unyielding, high standards of integrity, virtues of character, duty, and a willingness to sacrifice, while receiving deep, life-long, personal satisfaction.

SOURCE NOTES

1 – DISASTER IN A FABLED PLACE

9   Drew Pearson inquiries and General Irving's direction to make the announcement: Joseph F.H. Cutrona and George S. Pappas; author's interviews.

14   William J. Ryan's announcement in Washington Hall, "...a serious violation of the honor code by many cadets...an honor violator is the same as dead": Blaik, *The "Red" Blaik Story*, 464; Proceedings of a Board of Officers convened pursuant to Letter Orders dated 28 May 1951, United States Corps of Cadets, West Point, New York, 46. (Hereinafter referred to as the Collins Board).

15   "...since approximately 1946-1947 a conspiracy within the Corps of Cadets...": Collins Board, 69.

15   "...a small, closely held ring...may have existed as early as the fall of 1944": Proceedings of a Board of Officers appointed by Letter Orders dated 13 August 1951; Headquarters, United States Military Academy, West Point, New York, 20. (Hereinafter referred to as the Bartlett Board).

16   Quotes from newspaper headlines: Special Collections, United States Military Academy Library. (Hereinafter referred to as USMAL).

16-17   Extracts from Congressional Records and national news magazines: *Ibid*.

18   The two "screening boards"... "further inquire into...additional breaches of the cadet honor code.": United States Military Academy, Directives from Colonel R. S. Nourse to Colonel H. C. Jones; West Point, New York, 4 and 9 August 1951. United States Military Academy Archives (Hereinafter referred to as USMAA).

20   On January 16, Major General Bryant E. Moore...early the next morning of January 18 to take command of the IX Corps in Korea: Blaik, *The "Red" Blaik Story*, 277; William Van D. Ochs, Jr., author's interview.

20   General Ridgway...asked for General Moore's release from his Academy assignment.: Blair, *The Forgotten War*, 574.

20   Bill Ochs...the send-off was poignant and subdued: William Van D. Ochs, Jr.; author's interview.

20-21   Description of General Moore's military service: USMA Association of Graduates, *Assembly*, October 1951.

21   In Korea Ridgway expected..., and description of General Moore's death: Blair, *The Forgotten War*, 574, 727.

21-22   "...the Army did all it could...amid the beauty of West Point which Bryant loved above all else...": USMA, *Assembly*, October 1951.

22   The four cadet companies...that day were A, B, C and D Companies, Second Regiment.: Headquarters, United States Corps of Cadets; Orders No. 5, 29 March 1951.

22   Forty-first officer to serve as superintendent: Association of Graduates, USMA; *1996 Register of Graduates and Former Cadets*, 18.

22-23   Description of Major General Frederick A. Irving and his service experiences.: Col. (U.S. Army, Ret.) Frederick F. Irving, author's interview and Public Information Office release, undated; Mrs. Elizabeth Irving Maish and Colonel (U.S. Army, Ret.) William Van D. Ochs, Jr., author's interviews; "The Jap Death Toll is Very Heavy": Norfolk, Virginia *Pilot*, October 7, 1944; USMA, West Point, New York; *1917 Howitzer*, 106; "The Forward Edge": Major General A.S. Newman, *Army*, December 1980.

25   Art Collins...feelings...something's wrong with the honor system: Arthur S. Collins, Jr.; Senior Officers Oral History Program; U.S. Army Military History Institute; Carlisle, PA, 1981, 188-189.

25   Independently, Pat Ryan developed similar concerns: William J. Ryan; author's interview.

25-26   Cadet First Captain...*ex officio* member of the Honor Committee... "open and shut cases" rendered "not guilty": Colonel Paul D. Harkins, Bartlett Board testimony, 41; William J. Ryan; author's interview.

26   One such case occurred: William J. Ryan; author's interview.

27   Harkins...similar actions...in 1946...worked...to strengthen...honor system: Harkin, Bartlett Board testimony, 41.

27-28   There were other cases...no members of the football team on the 1951 Honor Committee... "cadets only" meeting with the entire Corps...to address...honor: William J. Ryan; author's interview.

27   ...injury that ...led to a second plebe year: Interview of William J. Ryan for "1990 Case Study," March 1990, 1.

27   Could there be "plants" or "sympathizers"...deliberately voting against guilty verdicts?: William J. Ryan, author's interview; Interview of William J. Ryan for "1990 Case Study, March 1990, 8.

28   February 19 meeting: United States Corps of Cadets; Daily Bulletin, 19 February 1951.

28   Talked about discipline...honor...I'm convinced we've got members...intentionally acquitting cadets...in a certain group: Interview of William J. Ryan for "1990 Case Study," March 1990, 8. *Ibid*.

28   Convened the Corps... with Pat Ryan... Gordon Danforth... communicated serious concerns...Pat Ryan began receiving comments indicating the meeting had struck a chord: William J. Ryan; author's interview.

29   The glimpse into an unhappy future...Two yearling roommates...Two yearlings at first...Finally, on April 1...: Richard J. Miller, Thomas P. McKenna, *Brian C. Nolan, George L. Hendricks*, author's interviews.

30   Dick Miller's father's World War II experiences: Richard J. Miller, author's interview.

29-30   THERE IS A RING OF CADETS...he asked Miller not to tell...how could evidence be obtained...who would participate in the investigation?: Richard J. Miller, *Brian C. Nolan, George L. Hendricks*, Daniel J. Myers, author's interview; Harkins, Collins Board, Annex C, 1.

34-35   Paul D. Harkins' background, history of service: Association of Graduates, USMA, *Assembly*, March 1987, 160-161.

36   "...True courage of action...appreciation of honor...discrimination between right and wrong...": United States Corps of Cadets, *1929 Howitzer*, 147.

36   "...I AM DETERMINED...EVEN IF I...USE RATHER DEVIOUS MEANS...": Colonel Paul D. Harkins, Collins Board, Annex C, 1.

36   Nearing the end of a successful tour...assignment on the Army staff: Hartle, "1990 Case Study," 9.

36   ...his time as commandant had been fascinating "...brother...father...to 2500 sons,...providing discipline and leadership...": Harkins; U.S. Army Military History Institute, Oral History Papers, 28 April 1974, 38.

37   ...openly expressed concerns for his future in the Army: Arthur S. Collins, Jr.; oral history, 91.

37   Colonel Waters'...involvement in preliminary investigation and all that followed: Harkins, Collins Board, Annex C, 1-4; Daniel J. Myers, author's interview.

38   Almost daily, private meetings with General Irving, in his office: William Van D. Ochs, Jr.; author's interview.

49-40   Description of Honor Committee responsibilities and procedures: Daniel J. Myers, Richard J. Miller, William K. Stockdale, and William J. Ryan; author's interviews.

40   Since 1946...44 individual honor cases resulting in less than honorable discharges: Letter to Commandant of Cadets from Warrant Officer Junior Grade Thomas E. Wertz; "Survey of Honor Violations During Past Five Years," USMA, September 16, 1951; USMAA.

40	Dan Myers' memories of his prolonged involvement in uncovering the cheating ring: Daniel J. Myers, Jr., author's interview.
41-42	Discussion of alternative investigation techniques. Paul D. Harkins, Collins Board, Annex C, 1; Daniel J. Myers, Jr.; author's interview.
42	During the same period... "You don't suppose he's cheating, do you?": Thomas P. McKenna, author's interview.
43	He would "fight fire with fire": Harkins, Collins Board, Annex C, 1.
44	Blaik's thoughts about the team of '51 and "the Davis-Blanchard teams": Blaik, *The "Red" Blaik Story*, 314.
44	Blaik's influence on football players and their perceptions of him: Twenty-four Army football players, author's interviews.
46-48	Description of Blaik's relationship with MacArthur and MacArthur's reforms while he was superintendent: Blaik, *The "Red" Blaik Story*, 488-537.
48	He cabled MacArthur "American public stunned...my affection and devotion to you...": Blaik, *The "Red" Blaik Story*, 493.
49	"Honor is bigger than one man...the Code demands courageous and fearless honesty...General Principles of honor...": USMA, *1951-52 Bugle Notes*, 12-13.
50	"I regret...the rest of my life...": *Brian Nolan*, letter to Earl H. Blaik; *The "Red" Blaik Story*, 453-456.
51	ADVICE AND DECISION: Blaik; *The "Red" Blaik Story*, 453-456; *George L. Hendricks* and *Brian C. Nolan*, author's interviews.
52	Cadet *Nolan* did join the ring...it took a great deal of moral courage...for...he found several of his friends already in it.: Harkins, Collins Board, Annex C, 1.
52-54	A SPREADING UNEASY FEELING OF SOMETHING SERIOUSLY WRONG: Raymond B. Marlin and Lawrence E. Schick, Bartlett Board testimony, 26-33 and 110-112.
55-58	ANOTHER MAN: A SECOND DILEMMA: *Brian C. Nolan, Michael J. Arrison, George L. Hendricks*, John H. Craigie, author's interviews; Recollections of "Sandy" Vandenberg as told to John H. Craigie, October 1996; Harkins, Collins Board, Annex C, 1.
59-64	SPREADING THE NET: Daniel J. Myers, Jr., Richard J. Miller, *Brian C. Nolan*; author's interviews; Harkins, Collins Board, Annex C, 1-6.
65	On May 28...Blaik...wrote MacArthur... "I shall write you about the football prospects...": Blaik, letter to General MacArthur, May 28, 1951; USMAL.
66	"...disaster...a catastrophe for Earl Blaik": Blaik, *The "Red" Blaik Story*, 5.

2 – PROFILE OF A YOUTHFUL CONSPIRACY

68-69   Emery Scott Wetzel's and Pat Ryan's encounter with General Jonathan M. Wainwright: Emery S. Wetzel, Jr., author's interview.

69   ...see and feel the elephant...the class of 1846 in Winfield Scott's invasion of Mexico: Pappas, *To the Point*, 261.

73   The first cadet called to testify was *George L. Hendricks*...K-1: Collins Board, 1; *George L. Hendricks*, author's interview.

73   Names called at reveille...four men from K-1: Daniel J. Myers, Jr.; author's interview.

74   If your young friend... he'll be able to remain a cadet... These were Harkins' words *Brian Nolan* remembered: *Brian C. Nolan*, author's interview.

74   He made no such commitment to *Mike Arrison* or *George Hendricks*: *Michael G. Arrison*, and *George L. Hendricks*, author's interviews.

75   The evidence against *George Hendricks* was overwhelming... *Hendricks* had no idea who had ... he couldn't hold back. He must... tell everything: *George L. Hendricks*, author's interview.

75-78   *Hendricks'* testimony: Collins Board, 2-3; *George L. Hendricks* and John J. Irvin, author's interviews.

78-79   Cadet *Evan A. Corley's* testimony: Collins Board, 3.

80-86   Cadet *Thomas S. Grayson's* testimony: Collins Board, 3-4.

81   Cadet *Theodore K. Strong's* testimony: Collins Board, 4.

82   *Gordon Seabold*, the seventh cadet's testimony: Collins Board, 4-5.

82   Twelve cadets in succession emphatically denied knowledge... of... cheating... or that they... participated: Collins Board, 5-9.

83   Tom Courant is told about the cheating ring and the Collins Board: Thomas E. Courant; author's interview.

83   Guidelines for the proceedings had been... developed... to ensure cadets... not... called if there wasn't firm evidence: Collins Board Memorandum to the Commandant, "Additional List of Cadets Involved in Investigation," June 24, 1951.

84   Board members wondered aloud...was there some strange scheme afoot to discredit the football team?: Comments by Arthur S. Collins, Jr. in Minutes of Athletic Board Meeting, USMA, February 1952; Jefferson J. Irvin, author's interview.

85-86   Lieutenant Colonel Arthur S. Collins' and Board members' backgrounds: Association of Graduates, USMA, *Assembly*, June 1985; Association of Graduates, USMA, *1990 Register of Graduates and Former Cadets*; United States Military Academy, *1938 Howitzer*; 138, 182, 198; Howard H. Danford, author's interview.

85   Art Collins "...sensitive to the improper use of honor to get answers or enforce regulations.": Howard H. Danford, author's interview.

86-87   "In early May...Art Collins...noticed...Colonel Harkins carrying a brown manila envelope... 'How unusual'... Near the end of May...He told Collins...you investigate it and 'let the chips fall where they may.'": Lieutenant General Arthur S. Collins, Jr., Senior Officers Oral History Program, Project 82-4, U. S. Army Military History Institute, 1981, 190-191. (Hereinafter referred to as Collins Oral History).

87   "...as you go through your service...".... he would, for the rest of his life, consider the assignment a compliment to his performance as an officer: Collins Oral History, 172.

88-92   An Emotional Evening: The Story of Two Meetings, and Earl Blaik's first encounter with the honor scandal of 1951: Blaik, *The "Red" Blaik Story*, 288-290; *George L. Hendricks*, author's interview; Arthur S. Collins, Jr., Memorandum for Record, 28 July 1951.

93   "This has been a terrific blow to my father...It has broken his heart.": Collins Board, 15.

94   His catastrophe, which he likened to the devastating March 1913 flood in Dayton, Ohio: Blaik, *The "Red" Blaik Story*, 1-3.

95   The Chaplain didn't want to discuss the matter: *George L. Hendricks*, author's interview.

97   Class of 1951 Aberdeen trip: Headquarters, United States Corps of Cadets, Movement Order Number 4, 23 May 1951; Collins Board, 7-8.

97   First two cadets to testify May 31: Collins Board, 9.

98   Cadet *Evan A. Corley's* testimony: Collins Board, 9.

99-100   Summary of testimony results at end of second day: Collins Board, 9-15.

99   Art Collins' recollections of his former company commander's son: Collins Oral History, 202.

101-103   Change in Collins Board strategy: Collins Board, 16.

102   Description of Corps' organization to attend classes on alternate days: Thomas E. Courant; author's interview.

103-106   A Wall of Silence:...Like those to suffer the consequence... and who knew their secret.: Collins Board, 17-23; Blaik, *The "Red" Blaik Story*, 279-300 and 444-418; William J. Ryan, author's interview.

106-107   The Old Corps' Silence at Work: An Irony: Pappas, *To the Point*, 415; Official Register of Officers and Cadets, 30 June 1951, 71; William K. Stockdale, author's interview.

108 From the class of 1952 forty-three men were called or voluntarily reappeared...the board was convinced thirty-nine were guilty...most admitted involvement...there was sufficient evidence to sustain guilty findings: Collins Board, 39A.

108 Next came testimony from fifty-three men from the class of 1953... The Board found fifty-four guilty... fifteen had falsely testified... in a desperate bid... to remain as cadets: Collins Board, 71.

109 Tom Courant's appearance before the Collins Board: Thomas E. Courant, author's interview.

109-113 TURMOIL BEGINS FOR TWO AND MORE: *Michael J. Arrison, Brian C. Nolan, George L. Hendricks*, Freddie A. D. Attaya, and John H. Craigie; author's interviews.

110 Cadet guard outside *George Hendricks'* door and *Brian Nolan* restricted to his room: *Brian C. Nolan* and *George L. Hendricks*, author's interviews.

111 *Brian Nolan* confronted by *Hendricks* and *DiSantis* and *DiSantis* questioned by FBI: *George Hendricks* and *Brian Nolan*; author's interviews.

119 Ernie Condina offers to be Dan Myers' bodyguard: Daniel J. Myers, Jr., author's interview.

119 There were several incidents during the Collins Board... "air of violence"...Four times Art Collins received phone calls in his quarters...he would place no value in them: Collins, Memorandum for Record, July 28, 1951.

119 Art Collins angered by allegations from outside the Collins Board, including the erroneous information Board members stated the December 1950 Navy game might have been thrown: Arthur S. Collins, Jr., *ibid*.

119 Board members suddenly found themselves on the defensive... about information... that the football team might have thrown the Navy game last December: Collins, *ibid*.

119-120 There was a deliberate attempt to discredit the Collins Board...on August 7... the Board members were required to sign affidavits swearing... one man...personally apologized for the attempt to discredit the Board... Another apologized for the behavior of a small group... who... shouldn't have been at West Point in the first place: USMA Staff Diary entries, August 16, 1951; Collins Oral History, 204.

120 There was an attempt by other men found guilty to implicate additional cadets: USMA Staff Diary entry, August 10, 1951; Memorandum from Colonel Waters to the superintendent, July 7, 1951; Collins Oral History, 204.

121   There were opinions, mostly assumptions, that members of the faculty...were involved: Collins Board, 67.

122   During the Collins Board meetings, several former football players...on the Army coaching staff...were named...it would be virtually impossible to uncover evidence against the officers in question: Collins Board, 62-67.

122   On the Army staff...preparations were made to prosecute officers who had graduated from the Academy for offenses they might have committed as cadets: Memorandum to the Army Chief of Staff from Major General E. M. Brannon, "Court-Martial Jurisdiction to try Officer for Offense Committed as a Cadet," July 20, 1951.

122   On hearing testimony about past years of cheating...the Collins Board called five officers...including John Green...on Blaik's 1944-45 national champion teams: Collins Board, 62-63.

124   Blaik letter to Taylor, 17 June 1951: Army Staff Papers, National Archives.

125   Blaik letter to General J. Lawton Collins, 6 July 1951: *Ibid.*

3 – DECISIONS AND TRAGEDY

128   1 June 1951 briefing to the Academic Board: Minutes of 1 June 1951 Academic Board meeting, USMAA.

128   Colonel Waters' quote from the Cadet Prayer: Waters to Irving letter, USMAA.

128-129   15 June 1951 Academic Board deliberations: Minutes of 15 June 1951 Academic Board meeting, USMAA.

129   Irving's and Blaik's 18 June 1951 trip to Washington and Blaik's letter to Taylor: Academy Staff Diary, Part III, 4 August-10 December 1951; Army staff papers, National Archives.

129   Cadet struck from behind and "...plan afoot to 'stack the list'...": *Ibid.*

129-130   Irving's 27 June return trip to Washington and the conference the next day at West Point: *Ibid.*

130   A court decision...cadets...could be...separated without court-martial: 5 August 1951 letter to the screening boards, *Ibid.*

130   Irving letter to Pace: *Ibid.*

130   Irving conferred with General Collins on July 5, Blaik's July 6 letter to General Collins, and Irving's July 6 letter to Collins: *Ibid.*

130   Plans for public announcement: *Ibid.*

131   Irving's conferrals with Collins and Haislip in Washington on July 9: Army Staff papers, National Archives.

131   General Collins' 18 July letters to members of the Hand Board: Army Staff papers, National Archives.

136-138 The Hand Board and "Parents Committee of 90 Dismissed Cadets": USMAA; Army Staff papers, National Archives.

136   Judge Hand's compliments given to the Collins Board for thoroughness in their work: Jefferson J. Irvin; author's interview.

138   Two Screening Boards: USMAA; Academy Staff Diary, National Archives.

139   Review of cadet discharge cases: *Ibid.*

139   Taylor's recommendations to the Army Chief of Staff: Army Staff papers, National Archives.

140   August 6, 1951 decision and August 9 memorandum for record by General Collins: Army Staff papers, National Archives.

141   Earl Blaik's press conference at Mama Leone's Restaurant, and General Eichelberger's phone call to Blaik: Blaik, *The "Red" Blaik Story,* 295-297.

142-143 Exchange of letters between the cadet writing for the "the eighty-eight cadets" and General Collins: Army Staff papers, National Archives.

144-145 General Collins' and cadet's letters never published and Army Chief of Information comments and recommendations: *Ibid.*

146   Collins Board's analysis of Board Report and extracurricular activities of dismissed cadets: Army Staff papers, National Archives; Jefferson J. Irvin, author's interview.

146-148 Scandal impacts: Collins Board Report; Blaik, *The "Red" Blaik Story*, 279-300.

148   Howard Galloway Brown's father's 6 August 1951 letter to General Irving: superintendent's correspondence, USMAA.

149   Earl Blaik's meeting with President Truman: Blaik, *The "Red" Blaik Story*, 294.

149   President Truman "was a hard-nosed Army rooter": *Ibid.*, 260.

149   Blaik's letter to President Truman: Harry S. Truman Library.

4 – KOREA: ONCE MORE INTO THE FIRE

151-153 David R. Hughes' letter from Korea: "Report...from Korea," *The Pointer*, November 2, 1951, 5.

154   Tuesday morning, June 6, 1950, the Corps of Cadets,...and the graduating class...heard an address by Secretary of the Army Frank Pace...There would be men among those six classes who would eventually be tested on battlefields somewhere in the world: Colo-

nel J. B. Love; "Remember the Price They Paid," *Assembly*, March/April, 1997, 30.

155   The classes of May and June 1861 totaled seventy-nine...twelve were killed in action or died of wounds: *1990 Register of Graduates and Former Cadets*, 283-285.

156-157   In the spring, well before the NKPA invaded the Republic of Korea, a decision had been made by the Army to send the Academy's newest graduates directly to troop duty with Army units,...repetition experienced in the branch schools was becoming a source of frustration and boredom among...young officers anxious to begin...for four years: Collins Oral History, 180-183.

158   The Korean War meant entry into action "as is." No time out for recruiting rallies or to build up and get ready. It was move in and shoot. MacArthur, *Reminiscences*, 335.

158   On September 3, Edmund Jones Lilly III was the first in '50 to be killed in action...before the month of September ended, six more members of the class had died: *1990 Register of Graduates and Former Cadets*, 570-584.

158   By war's end 365 from the graduating class of 670 had served in combat in Korea. Of their number thirty-four died in battle, eighty-four more were wounded, and seven died in accidents while serving in combat units: Love, "Remembering the Price They Paid, *"Assembly*, March/April 1997, 30;*1980 Register of Graduates and Former Cadets*, 822-823.

158   The five preceding West Point classes bloodied for the first time in Korea were only slightly more fortunate than '50...The emergency that had rushed the class of 1950 into the breach in Korea was over. Fewer members of '51 and '52 were sent to the stalemated front...their casualties were less: *1980 Register of Graduates and Former Cadets*, 822-823; *Newsweek*, June 18, 1951.

159   The *In Min Gun* had been taken far too lightly, almost with disdain, by many American commanders: Halberstam, *The Fifties*, 71-72.

159   The soldiers of the NKPA were well trained, tough, disciplined, and well equipped: *Ibid*.

159   By the time David Hughes went to war in Korea, in November 1950, the conflict had gone through several distinct phases...pulled back into...the Pusan Perimeter...mid-September, they went on the offensive...NKPA forced to retreat...retreat became a near route...all the way to the Yalu River, bordering China: United States Military Academy, *Operations in Korea*, 1955, 5-21.

160   David Hughes would see the Yalu River, but only briefly... his introduction to the battlefield was the fighting withdrawal... But he would soon learn he was more fortunate...one of the most illustrious, storied units in all the U.S. Army: David R. Hughes, author's interview.

160-164   Hughes' background, Academy experiences, graduation, leave, assignment to 7th Cavalry Regiment: *Ibid.*

164-165   Hughes' gratitude to Captain John R. Flynn, USMA 1944: *Ibid.*

165   War greeted him with a scene out of Dante's *Inferno: Ibid.*

165-166   Description of the defeat of the 8th Cavalry Regiment at Unsan, North Korea, and the tactics of the Chinese Communist Forces: Blair, *The Forgotten War*, 380-385; United States Military Academy, "Operations in Korea," 1955, Map No. 7.

169-170   Captain John Flynn...called his officers together... to give them the word... what John Flynn had written... where he had been an instructor...he was the consummate teacher and soldier...David Hughes listened, learned, and admired: Hughes, author's interview.

170-172   Description of the UN withdrawal, General Ridgway's and General Van Fleet's assumptions of command: "Operations in Korea," 19-30.

172-173   Description of General Matthew Ridgway's turnaround of the 8th U.S. Army: Blair, *The Forgotten War*, 570-581.

173-174   But he also had to learn much about himself...Ears...developed the habit of jerking his helmet off his head...incoming mortar round: Hughes, author's interview.

174   A hundred twenty millimeter mortar round impacted so close...he couldn't have possibly survived...received scratches when the tail fin broke off: *Ibid.*

175   In an assault on a high hill,...Someone yelled 'Grenade!'...The grenade killed the man lying next to him: David R. Hughes' letter to Mrs. Helen I. Hughes, 1951.

175-176   Enemy tactics and ruses...The carbine had a reputation for jamming... pushing to take high ground... He wheeled to fire his carbine... which missed its target... second attempt to save himself was successful. He killed the enemy soldier: Hughes, author's interview.

176   Get rid of the carbine...found a Thompson submachine gun...rigged the weapon...so he could rapidly retrieve it and fire: Hughes, author's interview.

177   The first independent offensive action K Company participated in... Discomforting thoughts to go with a decoration: Hughes, author's interview; Hughes letter to Flynn, March 1952.

177   Captain John Flynn received serious wounds in June of 1951..He wanted to know what was going on... letters went unfinished... until March of 1952: Hughes, author's interview; Hughes letter to Flynn, March 1952.

178   Once, while his depleted company was preparing to launch an assault... he received approximately 30 replacements... The acquaintance was too brief... no opportunity to see or become acquainted with other men in the company: Hughes letter to Flynn, March 1952.

178   In another long, seven hour fight, when K Company was holding an important outpost... Hughes killed the enemy soldier... the Chinese took three prisoners... the third American soldier was killed by fire from K Company: Hughes to Flynn, March 1952.

179   "Except for several offensives... this was generally the pattern... until the signing of the Armistice...": United States Military Academy, "Operations in Korea," 1955, 43-44.

179   When the truce talks broke off in late August General Van Fleet decided to resume the offensive...launched heavy attacks... west and north of... the 'Punch Bowl'... Further west,... 1st Cavalry Division... British Commonwealth Brigade... advanced on a forty mile wide front: *Ibid.*, 44.

180   "...'get off the road' and 'take the high ground'...": Blair, *The Forgotten War*, 586.

180   The 3rd Battalion of the 7th Cavalry,... This time the fight for 339 was relatively easy... Dave Hughes remembered... Not until the second day...: Hughes' letter to Flynn, March 1952; SLA Marshall, *Pork Chop Hill*, 44 (map).

183-186 Private Edward J. Escalante's last patrol: undated personal paper by Edward J. Escalante. Deposition submitted as part of nomination for decoration for Private Edward J. Escalante; Lieutenant David R. Hughes, 10 December 1951.

188   "...this was merely the prelude...was nicknamed 'Baldy'...'Old Baldy'... 'Chink Baldy'...'T-Bone'...'Alligator Jaw'...": SLA Marshall, *Pork Chop Hill*, 44(map).

188   "...'Bloody Baldy'...": Name of Hill 347, described by Master Sergeant Monroe McKenzie, in a sworn statement accompanying "Recommendation for Decoration for Heroism for David Ralph Hughes," January 11, 1952.

189-195 THE TAKING OF HILL 347...The fight for Hill 347 was over – for October 7, 1951: Hughes' letter to Flynn, March 1952; "Recommendation for Decoration for Heroism for David Ralph Hughes," January 11, 1952.

195-197 GOING HOME...Pork Chop Hill was the name: Hughes' letter to Flynn, March 1952; Hughes, author's interview; "Recommendation for Decoration for Heroism for David Ralph Hughes," January 11, 1952.

5 – BOYHOOD DREAMS

201-202 EXAMPLES, IDEALS AND IMAGINATION – INSPIRATIONS FOR THE FUTURE...That was the place to go. In July 1950 he entered the Military Academy with the class of 1954: Benjamin F. Schemmer, author's interview.

202-206 FOR WEST POINT – THE PLACE, ADMIRATION OF FAMILY, AND LOVE OF THE GAME...Football would become the determiner of his life after graduating from the Academy: Thomas J. Bell, author's interview; Robert M. Mischak, author's interview.

206-207 THEY "UNDERSTOOD" I WANTED TO GO TO WEST POINT...the ability to call up powerful emotions, and "the twelfth man," the spirit of the Corps of Cadets: Edward M. Moses, author's interview.

207-208 I WANT AN EDUCATION, AND TO BE AN OFFICER...hold fast to his dream: John C. Bard, author's interview.

208-209 A SURPRISE AFTER CHURCH, AND THE 'WALK-ON' BEGINS...He received the symbolic sword of excellence as did Earl "Red" Blaik when he graduated from the Academy in 1920: Lowell E. Sisson, author's interview.

210-211 EVER SINCE THE THIRD OR FOURTH GRADE...He knew he was in a tough league, but that didn't turn him aside: Gerald A. Lodge, author's interview.

211-212 A BROTHER'S INFLUENCE...His teammates elected him captain for the 1953 season: Leroy T. Lunn, author's interview.

212-214 BLANCHARD AND DAVIS, BROTHERS HE ADMIRED, AND EARL BLAIK'S ACQUAINTANCE...He had some lessons to learn from both men, some powerful lessons, and he learned them well: Freddie A. D. Attaya, author's interview.

214-216 FOR LOVE OF THE GAME, AND A FATHER'S PERSUASION...with the class of 1955: Peter J. Vann, author's interview.

217-218 HE ADMIRED AN ENGINEER BUT WANTED TO BE A PILOT – TO EUROPE AND BACK IN '44...The road he traveled led him to the role of First Captain...and pilot's wings after graduation: William J. "Pat" Ryan, author's interview.

218-219 "WEST POINT WAS A BOYHOOD DREAM INSPIRED BY MY DAD AND HIS BROTHER"...Howard also became acquainted with Colonel Arthur S. Collins, Jr.....the officer who was Howard Danford's counterpart, his commander-teacher: Howard H. Danford, author's interview.

220-222 A SMALL VIRGINIA FARM – AND A RACE FROM EUROPE TO WEST POINT...Dick was on the First Beast Detail, in Fourth New Cadet Company, our company commander – 'Mr. Shea, Sir...' to us: John K. Swensson, "A Race Well Run...," *Assembly*, Association of Graduates, USMA, Fall 1962, 14-16; "Richard Thomas Shea, Jr.": *Assembly*, Winter 1959, 103-104; Joyce E. Himka, author's interviews.

222-224 IT TAKES MORE THAN A RECORD OF EXCELLENCE...he might never have played varsity, were it not for the tragic honor incident which shook West Point that spring and summer: "Richard George Inman," His parents, *Assembly*, October 1954; 1952 *Howitzer*: United States Corps of Cadets Class of 1952, West Point, New York, 298.

6 – THE END OF ARMY'S GOLDEN ERA OF FOOTBALL

226 From his first season at West Point, in 1941...from 1944 through 1949...there was little doubt it would remain through the 1950 season: Blaik, *The "Red" Blaik Story*, Appendix 1; *NCAA Football, Official 1996 Football Records Book*, 67.

226 After the national championship seasons of '44 and '45, and ...split ranking with Notre Dame in '46...Earl Blaik was named Coach of the Year: Blaik, *The "Red" Blaik Story*, Appendix 1.

227 The greatest collection of college football talent ever assembled...Blaik and Leahy would also join the Hall of Fame: Nelson, *Anatomy of a Game*, 227.

227 Army football records and national rankings: Blaik, *The "Red" Blaik Story*, Appendix 1; *NCAA Football, Official 1996 Football Records Book*, 67.

227-228 During the years preceding the 1950 season eighteen Army players were first team All-Americans...Joe Steffy...won the Outland Trophy: Sports Information Office, *1991 Army Football Guide*, USMA, West Point, New York, 106.

227 "During the war years he assembled probably the best coaching staff the game had ever seen...": Nelson, *Anatomy of a Game*, 227.

228 The University's reputation had its roots in the era Blaik graduated from Miami, before he entered West Point: Blaik, *The "Red" Blaik Story*, 15-16.

228 In Blaik's eighteen years as Army's head coach...nineteen became head coaches at major colleges or universities: Blaik, *The "Red" Blaik Story*, 421-430.

229 By 1914...football emerged...as the most popular of all crowd drawing sports...in the fall of 1914 the Harvard-Yale game drew 68,500 fans in the Yale Bowl...Receipts from football...provided funds to

cover the cost of expanding athletic programs: Nelson, *Anatomy of a Game*, 91, 162.

229   When America's young men returned from war for the 1919 football season, the game became partner in "The Golden Age of Sports"...construction of huge stadiums resumed, spread across the country: *Ibid.*, 162, 172.

231   During the period before the NCAA Football Rules Committee met in early 1942, over 350 colleges and universities had already dropped football...": *Ibid.*, 222.

231   after the country entered the war... major college and university powers continued to dominate the top 20 teams...Among the top 20 at the end of the 1944 season were Randolph Field (3),...Tennessee (12), and Illinois (15)...not one of the armed forces' wartime training centers had a team ranked in the top 20: *NCAA Football, Official 1996 Football Records Book*, 67.

232   In those nine years...two of the four ties were in 1947.": Blaik, *The "Red" Blaik Story*, Appendix 1.

232   The 3rd Division was scrambling to come up to strength in men and equipment, training hard for its overseas move by ship: Blair, *The Forgotten War*, 410-411.

232   "I not only knew where that tackle was going to scratch himself, but when he was going to scratch.": John H. Craigie's recollection of remarks by 'Bunker' Emblad (USMA '51), author's interview.

233   Lowell Sisson, Company A-1,...walked on...it wasn't easy: Lowell E. Sisson, author's interview.

233-234   Ben Schemmer was less fortunate...and the Army team rolled on without him: Benjamin S. Schemmer, author's interview.

234-235   Army was regarded once more as a contender for the mythical national championship...In *Look* magazine's annual preseason review, picked Army and Notre Dame as the top two teams...Army would '...clarify...' Grantland Rice's uncertainty: Ransom E. Barber, *Pointer*, 15 September 1950, 26.

235   "Army may, at some time, hit that ...slump, but, it surely won't be this year...as long as Army teams are Blaik coached.": *Ibid.*, 25.

235   One cadet writer in the September 29 Pointer was critical...bemoaned the air of complacent overconfidence and indifference he heard: *Pointer*, 29 September 1950, 23.

235-41   Two Men in Two Different Roles – Unknowingly Fanning the Fires... the spread of cheating accelerated, until, by the end of May 1951, it had reached epidemic proportions: (Description of differences between Harkins and Blaik) Blaik, *The "Red" Blaik Story*, 279-300,

444-468; Bartlett Board testimony by Earl H. Blaik, Paul D. Harkins, Arthur S. Collins, Jr., Jefferson J. Irvin, Tracy B. Harrington, John K. Waters, and R.B. Marlin; Bartlett Board record, 11-30; Collins Oral History, 188-221; Collins Board testimony from cadets formerly in the classes of 1952 and 1953; Meeting records, Academy Athletic Board, February 15, 1952; A study: "Section 2 of the Act of June 3, 1942": Colonels Counts, Bartlett, and Stamps, 1 August 1951.

241-244 WAR'S SAD REMINDERS...Earl Blaik's feelings were deep and genuine...He felt their loss...and had been a prime motivator toward a career in the armed forces, perhaps more than any other figure in their young lives: *Assembly*, Winter 1959; 1990 *Register of Graduates and Former Cadets*; *Field of Honor, 100 Years of Army Football*; Video, Sports Information Office, 1990; Blair, *The Forgotten War*, 247.

244 THE SEASON MUST GO ON...and a thrashing of lightly regarded...New Mexico...on November 11...The following Tuesday Army was ranked number one: *New York Times*, November 12, 1950; *New York Times*, November 14, 1950.

245 The week after the New Mexico game...Palo Alto...to play Stanford...a 7-0 win. Army slipped to No. 2 in the polls: *New York Times*, November 19, 1950; *New York Times*, November 21, 1950.

245 The team's trip to Palo Alto by plane had been long and tiring...Cold, rainy weather continued, as did indoor practice in the Field House: Blaik, *The "Red" Blaik Story*, 274-275; Blaik letter to MacArthur, 27 December 1950, USMAL.

245-247 Exactly what happened to Johnny Trent that night, no one knows...Johnny, as Coach Blaik called him, was posthumously awarded the Purple Heart, and..., the Football Writers' Association of America named him Football's Man of the Year: Blair, *The Forgotten War*, 412-413; *Assembly*, July 1955; 1991 *Army Football Guide*, 106.

247 When the news of John Trent's loss reached Earl Blaik... "The Colonel,"...must have also remembered the angry shouts ...scream "Slackers"! at Army players...Presumably Army players were at West Point to avoid their obligations to serve their country in time of war: Blaik, *The "Red" Blaik Story*, 237-239.

248 Colonel John K. Waters' thoughts on John Trent's death in Korea and the team captain's talk to the Corps: John K. Waters' testimony before the Bartlett Board, 26-33.

249-250 Description of 1950 Army-Navy game: *New York Times*, December 2, 1950; *New York Times*, December 3, 1950; Blaik, *The "Red" Blaik Story*, 275-276.

251-252 Dark Echoes...The tide of war had once more been reversed: (Description of the destruction of Task Force Faith and the death of George E. Foster, class of 1950.) Blair, *The Forgotten War*, 514-515.

252-253 Blaik to MacArthur, 27 December 1950.

7 – 1951: West Point's Year of Adversity

255 Christmas leave schedule: Administrative Memorandum #143, Christmas Leave 1951, United States Corps of Cadets, West Point, New York, 15 November 1951; USMAA.

256 Harry Comeskey...was 'found' in academics: *Official Register of Officers and Cadets, United States Military Academy, West Point, N.Y., For the Academic Year Ending 30 June 1952*, 152.

258 Preparations For A Joyous Season...Eight men in the class of 1952 and eleven in the class of 1954 comprised the California bound revelers: "United States Air Force Aircraft Accident Report 12-30-51-1," Williams Air Force Base, Arizona, February 7, 1952; Air Force Historical Research Agency. (Hereinafter referred to as AFHRA).

260 On November 27 Guy McNeil, Sr. requested...two C-47 airplanes...The stated purpose was to ferry West Point cadets: Flight Orders No.'s 511 and 521, C-47 Serial Number 44-75266, "United States Air Force Inspector General Special Investigation Report, 17 January 1952"; AFHRA.

261 The 4th Air Force Flying Safety Officer...was Major Lester G. Carlson...He was released from active duty into reserve status in May 1948: Inspector General Special Investigation Report, 17 January 1952 and Special Investigation Report, September 1953; AFHRA.

261 Colonel McNeil believed a senior pilot rating and a green instrument card were necessary for...and 4th Air Force Flying Safety Officer: Statement of Colonel Guy McNeil to the Aircraft Accident Investigation Board, 3 January 1952; AFHRA.

261 Minimum Individual Training was...routine flying to maintain proficiency and build experience. Major Carlson...had almost two hundred hours...the preceding three months.: Aircraft Accident Report, Williams Air Force Base, Arizona, 7 February 1952, AFHRA.

261 There were no academics, except for men "found" deficient in academics...they lived in their own rooms...until they were released to go on leave...they...formed up in front of Washington Hall...rather than march from company areas with the plebes. Administrative Memorandum Number 157, Instructions for Cadets Present Dur-

ing Christmas Leave, United States Corps of Cadets, West Point, New York, 19 December 1951; USMAA.

262 Preparing for Christmas...We received a three hour course... from the first class and the Tactical Department...instructional principles and methods were the same...Review and prepare for writ – a test...the last hour was a series of skits...Tingle...was the instructor for the last hour.: Training Schedule Number 4A-4, Fourth Class Military Customs and Courtesy, Headquarters United States Corps of Cadets, West Point, New York, 13-21 December 1951; USMAA.

262-263 Hop schedule and dress for Hops: Administrative Memorandum Number 155, Christmas Hop Schedule, United States Corps of Cadets, West Point, New York, 18 December 1951; USMAA.

263 There were Ice Follies on Thursday...at Smith Rink...broom hockey,...and the southern sweepstakes...for cadets who had never worn a pair of ice skates, Administrative Memorandum Number 150, Ice Follies, 27 December 1951, United States Corps of Cadets, West Point, New York, 12 December 1951; USMAA.

264 There were "open houses" for two days beginning Christmas...between eleven and twelve noon....cadets brought visitors to their rooms...rooms in inspection order for two hours, the first hour to have an inspection...Additional Christmas Week Activities,...Four days there was no reveille...on all other days an hour later...meal formations an hour later: Memorandum to: All Fourth Classmen, United States Military Academy, 12 December 1951, USMAA; "Orientation for Parents of Fourth Classmen During the Christmas Holidays," Department of Tactics, United States Military Academy, West Point, New York, 21 November 1951; USMAA.

265-272 THE RETURN FLIGHT...Colonel Guy McNeil had been ill. He decided not to fly: "Report of Special Investigation of Aircraft Accident Involving C-47 Serial Number 44-75266, near Phoenix, Arizona, on 30 December 1951," 17 January 1952, AFHRA; "Proceedings of the Aircraft Accident Investigation Board," Williams Air Force Base, Arizona, 7 February 1952; AFHRA.

272 "I'm an old hand...My head says no, but my heart says yes": *San Francisco Chronicle*, January 1, 1952, 11.

272-274 A MASSIVE SEARCH...watches found among the wreckage marked the time: *Carmel Pine Cone*, January 4, 1952; *Monterey Peninsula Herald*, December 31, 1951 and January 1, 2, 3, 1952; *Palo Alto Times*, January 2, 3, 4, 7, 1952; *Phoenix Arizona Republic*, December 31, 1951 and January 1, 2, 3, 4, 1952; *New York Times*, January 2, 1952; *San Francisco Chronicle*, January 1, 2, 3, 4, 1952;

San Franciso Examiner, December 31, 1951 and January 1, 2, 1952.

275-280 TRAGIC IRONIES OF AIR FORCE 266...Bill Sharp's...funeral services in the Yale Ward of the Church of the Latter Day Saints: "Report of Special Investigation of Aircraft Accident Involving C-47 Serial Number 44-75266, near Phoenix, Arizona, on 30 December 1951, 17 January 1952, AFHRA; "Proceedings of the Aircraft Accident Investigation Board," Williams Air Force Base, Arizona, 7 February 1952, AFHRA; *Carmel Pine Cone*, January 4, 1952; *Monterey Peninsula Herald*, December 31, 1951 and January 1,2,3, 1952; *Palo Alto Times*, January 2, 3, 4, 7, 1952; *Phoenix Arizona Republic*, December 31, 1951 and January 1, 2, 3, 4, 1952; *New York Times*, January 2, 1952; *San Francisco Chronicle*, January 1, 2, 3, 4, 1952; *San Franciso Examiner*, December 31, 1951 and January 1, 2, 1952.

280 Lieutenant Colonel Jeff J. Irvin...to the west coast to assist bereaved families...making arrangements...and officer in charge of 21 cadets who attended funerals or escorted bodies back to West Point for burial: Letter from Major General Frederick A. Irving to Lieutenant General Joseph M. Swing, Headquarters Sixth Army, United States Military Academy, West Point, New York, 17 January 1952, USMAA; Message from Commanding General, Sixth Army to superintendent, USMA, Presidio of San Francisco, California, 7 January 1952, USMAA.

281 Letters and messages of condolence poured in to the Academy...General of the Army Dwight D. Eisenhower...and the New York Chapter of the Daughters of the American Revolution: Memorandums to the Corps of Cadets, Headquarters United States Corps of Cadets, West Point, New York, 4 and 14 January 1952; USMAA.

282 President Hoover's letter to Irving, and Irving's reply: Herbert Hoover Library and USMAA.

283-286 AN END TO THE BEGINNING...Both lines faced toward the center of the roads...At 1:13, a train pulled into the station...the West Point Provost Marshall led the...procession...cadet company commanders called their companies to "attention"...each cadet rendered a hand salute...companies in columns of platoons...march back to the cadet area...normal afternoon class schedule...life continued at West Point: Order No. 2, "Return to West Point of the Remains of the Late Cadets Glasbrenner, Karl F., Jr., Class of 1952, Co C-2; Mastelotto, Maurice J., Class of 1954, Co G-2; Wilson, Hugh R., Jr., Class of 1954, Co I-2." United States Corps of Cadets, West Point, New York, 8 January 1952; USMAA.

8 – REBUILDING BEGINS

290-293　MUTINY ON THE 'WHISKEY'...Apparently...cadets in such numbers never joined another Middie cruise...": Movement Order Number 10, U S Naval Academy Cruise for Selected Cadets, Class of 1954, Headquarters, United States Corps of Cadets, 10 July 1951, USMAA; "Mutiny on the 'Whiskey'," Robert H. Downey, Jr. and Walter F. Evans; U S Military Academy Class of 1954 40th yearbook, *'54-40...and still in The Fight*, 1994, 208; "Reflections on a Mid-Summer Cruise." Unpublished paper, Robert H. Downey, Jr., 1994.

293-295　REVEILLE GUN CAPER...Ed Moses garnered 20 demerits, 44 punishment tours, and two months special confinement: "The Reveille Gun Caper," Edward M. Moses; U S Military Academy Class of 1954 40th yearbook, *'54-40...and still in The Fight*, 1994, 209; Edward M. Moses, author's interview.

296　As a plebe the fall of 1950, Ed Moses tried out for the powerful, nationally ranked Army football team...The Emperor of Ethiopia didn't arrive early enough to grant Ed amnesty: Edward M. Moses, author's interview.

296　Ed Moses' interest in armor: *Ibid*.

297　The Collins Board's June 8 report had been read by General Maxwell Taylor...and General J. Lawton Collins: Academy Staff Diary, Army Staff papers, National Archives.

297-298　In addition to concluding ninety-four men total...had been participants...Art Collins, Jeff Irvin, and Tracy Harrington also concluded...There are academic and personnel matters beyond the scope of this Board: Collins Board, 69.

298　The Collins Board's first recommendation was...cadets determined to be guilty...resign...a Board of Senior Officers be appointed to study the background which contributed: *Ibid*.

299　On Tuesday, August 7...General Irving called a meeting of the Academic Board...read General Taylor's August 3 letter...Considerable discussion ensued: Minutes of Academic Board meeting, 7 August 1951, USMAA.

299-300　Before the meeting adjourned at twelve twenty, General Irving gave his decision...three members on the board of senior officers...one...from the Academic Board...a member of the Tactical Department...and General Irving would select a member: *Ibid*.

300　On Monday, August 13, 1951, the Bartlett Board convened. The letter...was brief and to the point:...Report will be made...in detail with a view to material change: Bartlett Board, Exhibit A.

300   The Board's mission evolved out of informal discussions between its president and General Irving...Consider means of...periodic check that would show up an incipient repetition of a similar incident in its early stages: Bartlett Board, 1.

300-301 Selection of Colonels Bartlett, Greene, and Miles as Board members; summaries of their backgrounds and experiences: Minutes of Academic Board meeting, 7 August 1951; "Boyd Wheeler Bartlett," *Assembly*, Fall 1965, 83-85; "Boyd Wheeler Bartlett," *1921 Howitzer* (class of 1919), 47; "Francis Martin Greene," *Assembly*, June 1975; "Francis Martin Greene," *1922 Howitzer*, 134; "Charles Harlow Miles, Jr.," *Assembly*, September 1985, 153; "Charles Harlow Miles, Jr.," *1933 Howitzer*, 189; *1996 Register of Graduates and Former Cadets*; Association of Graduates, USMA, West Point, NY.

305   There were only two senior, active duty officers within the Department of Athletics...The Graduate Manager of Athletics worked under Earl Blaik's supervision...The position gave Draper access to the superintendent...two Assistant Graduate Managers of Athletics...Russell P. "Red" Reeder...and Elliot W. Amick: Bartlett Board, Exhibit K (General Orders Number 33, "Athletic Policy U.S.M.A.," 8 July 1948); USMAA.

305-307 Earl Blaik, the Director of Athletics in addition to head football coach, was given the Director's responsibilities in 1946... He was 49 years old in 1946... There were no retirement or pension plans... Though Blaik was Director of Athletics...he was a civilian employee... he completed his second five year contract in the spring of 1951... direct access to the superintendent...Blaik insisted on reporting directly to the superintendent or college president...What's more...Blaik's success...garnered him...celebrity in his own right, whether he wanted such status...Thus, by the spring of 1951, Blaik was both well respected and established at the Academy...signed another contract...the first days of April: Blaik, *The "Red" Blaik Story*, 124, 173, 261.

307-308 THE NATURE OF INVESTIGATION REQUIRED...Analysis of the report of the board which originally investigated the honor violations...Many a boy who would not cheat would not inform on another boy who did...To obtain information about beliefs and opinions...we have made an effort to compare current beliefs and opinions...to determine if significant changes or trends are evident: Bartlett Board, 2.

308   The Board members first...selected eleven areas of investigation...six involved intercollegiate athletics. The remainder included... privileges and extra-curricular activities... The proceedings were deliberately informal... to obtain frank and uninhibited discussion of the problems at issue: Bartlett Board, 32.

308-309   The Board elected not to investigate in depth...matters... Collins Board testimony... needed to be considered. Instead those responsibilities were assigned follow-on boards and committees: Collins Board, 1-69; Bartlett Board, 32.

309   Boyd Bartlett and his two Board members worked at a hectic pace... People interviewed... came from every Academy organization... Among the cadets interviewed were the entire Cadet Duty Committee... The Board also interviewed...Jones...Barrett, J. W. Green, Jr.: Bartlett Board, Exhibit C ("Chronological Record"), 1-9.

309-310   Backgrounds and experiences of Colonels Charles J. Barrett, H. Crampton Jones, and James W. Green, Jr.: "Charles J. Barrett," *Assembly*, Fall 1963, 94-95; *1922 Howitzer*, 109; "James W. Green, Jr.," 1927 Howitzer; 177; *1996 Register of Graduates and Former Cadets*; "H. Crampton Jones," *1996 Register of Graduates and Former Cadets*; *Official Register of the Officers and Cadets, United States Military Academy for the Academic Year Ending 30 June 1952*, 11.

310   To each individual and group interviewed the Board explained its purpose... described the barest of essentials about the cheating incident... cautioned each witness... proceedings... confidential...no intent to obtain additional names of cadets... nor... information used... for disciplinary action: Bartlett Board, 3-6.

311   The Board's days were long and arduous from August 13...eight to nine in the mornings... until ten thirty or eleven at night...The morning of the first day... six cadets found guilty by the Collins Board... They worked every day until the report was complete on September 7... : Bartlett Board, Exhibit C, 1-9.

313-314   Summary and analysis of Bartlett Board conduct and its effects: Collins Board Report; Bartlett Board Report, 1-32; and Bartlett Board testimony transcripts of: Paul D. Harkins, Earl H. Blaik, H. Crampton Jones, Charles J. Barrett, Arthur S. Collins, Jr., Jefferson J. Irvin, Tracy B. Harrington, James W. Green, Jr.

314   Bartlett Board assumptions about recruited football players: Bartlett Board, 18-19.

318   "Specific Conclusions"...of the Bartlett Board: *Ibid.*, 25-29.

318-320   Board recommendations: *Ibid.*, 32.

323-324   "Summarizing Remarks" of the Bartlett Board: *Ibid.*, 30-31.

324   "...that they '...would go out with their heads held high...'": Y. Arnold Tucker, author's interview.

9 – PLEBE YEAR: HARD LESSONS AND LAUGHTER

326   Bob Mischak... recruited by Vince Lombardi and Captain Johnny Green... wondered, "What have I gotten myself into?": Robert Mischak, author's interview.

326   "To train new cadets...to march in company formation to and from the...ceremony...": Training Memorandum Number 25, <u>TRAINING PROGRAM, NEW CADETS, CLASS OF 1955, 3 July – 27 August 1951</u>, Headquarters United States Corps of Cadets, 26 April 1951; USMAA.

342   Schofield's Definition of Discipline: *1951 Bugle Notes*, 202.

348   Training Memorandum Number 24, USMAA.

349-350   Another more detailed memorandum...Officers and cadets in charge... From June 25 through July 2,...As a result, Beast Barracks ran like clockwork... There were 317 scheduled hours... including visits to the cadet store...and "free time": Training Memorandum Number 25, Headquarters United States Corps of Cadets, 26 April 1951; USMAA.

350   Attendance at chapel is part of a cadet's training; no cadet will be exempted. Each cadet... in one of the three principal faiths: Catholic, Protestant, or Jewish: Regulations, United States Corps of Cadets, 1949, 30.

352   "...the democracy of the Corps assure[s] every individual cadet...starts always without handicap in the same competition.": Douglas MacArthur, *Reminiscences*, 80.

357   Dick Shea's cartoon and the Special Swimming Squad story of Cadet Bill Grugin, class of '51: "The Corps in Column," *Pointer*, United States Corps of Cadets, 10 November 1950, 17.

359   "...Colonel Sylvanus Thayer...in 1824 reorganized the Corps of Cadets according to height...": Pappas, *To the Point*, 142.

362   More than ninety reporters...descended on the Academy...": Joseph F. H. Cutrona, author's interview.

363   ...Amounts offered rumored to be...from $20 to $200.": *George L. Hendricks*, author's interview.

363   Early the morning of August 4, of the soon-to-be-discharged cadets... attempted... a personal long distance call... to... Walter Winchell... prohibited... claimed... his phone calls monitored... triggered an investigation...proved his complaint groundless: Lieutenant Colonel Winfield L. Martin, Memorandum to the Chief of Staff, USMA, 14 August 1951; Academy Staff Diary, Army Staff papers, National Archives.

364   "Empty beer cans thrown our direction...": Willard Robinson, author's interview.

364   On August 9...Roy Thorsen...wrote in his journal... "There's been quite a bit of talk here... one takes great pride... we are able to... stand by our beliefs and convictions.": From the journal of Roy T. Thorsen, provided during author's interview.

365   Art Collins said years later, "... there has to be a nucleus... of honor and integrity the Army must have... if West Point doesn't provide that... do you need a West Point?": Collins' Oral History, 202.

366   Early that summer one of our luckless classmates... deception uncovered... confronted with an honor violation.: Gerald J. Samos, conversation with author.

10 – WHAT'S IT ALL ABOUT?

379-380 When Colonel John K. Waters...addressed the new cadets... Waters,...had strong feelings about honor...spoke of honor... "Cherish the honor system...Measure yourself by this standard...cannot go far wrong.": Proceedings of a Board of Officers Appointed by Letter Order No. 56, Headquarters, United States Corps of Cadets, 17 October 1951, Exhibit S. (Hereinafter referred to as the Greene Board).

380-382 Description of Colonel John K. Waters' background and experience: John K. Waters, Jr. and George P. Waters, "John Knight Waters," *Assembly*, July 1993, 164-165.

382   We heard nothing more...until Saturday morning, July 21, General Irving spoke to our class... "I think that everyone familiar with West Point would instantly agree...you are the instrument by which the honor system must be and can be made effective": Greene Board, Exhibit T, USMAA.

385-390 THE HONOR CODE AND SYSTEM: EVOLUTIONARY HISTORIES: "The USMA Honor System – a Due Process Hybrid," Major John H. Beasley, *Military Law Review*, Vol. 118, 1987, 187-191; Fleming, *West Point, The Men and Times of the United States Military Academy*; 51, 139-141, 317, 336; Pappas, *To the Point;* 139, 363, 365-366, 415.

390-392 Testimony from Company K-1 yearling: Collins Board, 2.

392   Testimony from Company H-2 yearling: Collins Board, 4.

392   Testimony from Company D-1 third classman: Collins Board, 10.

393   Testimony from another Company D-1 yearling: Collins Board, 11.

393-396 Descriptions of how ring members became involved and remained involved in organized cheating: Collins Board, 1-73.

396   Also contained in the Collins Board report... "The Honor Orientation...was adequate and complete": Collins Board, 68.

397   5th GENERAL PRINCIPLE on which the honor code was founded: 1923 *Bugle Notes,* 49; 1951-52 *Bugle Notes,* 13.

397   "...If the ideal of the honor system is strong... Beliefs and attitudes of mind will govern action...If the belief is firmly held the action will result...": Bartlett Board, 2.

397   ...forty-four other cases: "Survey of Honor Violations During the Past Five Years," Thomas S. Wertz, Letter to the Commandant of Cadets, 16 September 1951; USMAA.

399   Bartlett Board recommendation that a cadet committee review its honor system: Bartlett Board, 32.

399-403   Frank Greene's Board found fault lines in the administration of academics...The Greene Board did examine in detail the honor orientation given new cadets that summer: Greene Board, 4-36.

401-404   The August 4 lecture touched only briefly on the 5th principle... "We have considered honor offenses and the results...Nothing is bigger than the honor of the Corps...Board recommended several specific changes related to honor training for cadets"...training in honor be studied...overall character and moral training at West Point be reviewed...fifty-five specific recommended changes...an offense against society: Greene Board, 4-36, Exhibit R.

403-404   Louis Tomasetti also spoke of the future to the class of '55 in his lecture about honor... "when you graduate you will be in command of a unit and your word concerning the unit will be accepted as truth – be sure it is": Greene Board, Exhibit R.

405   ...outgrowth... was the Military Psychology and Leadership Course... Colonel Russell P. "Red" Reeder...was selected to develop the course: *Assembly*, "1997 Distinguished Graduates," July/August 1997, 6.

406   "Commencing with the first day that he enters... The teaching of leadership by example and instruction is ... the daily responsibility and privilege of each": "A Syllabus for Military Psychology and Leadership," Department of Military Psychology and Leadership, 1947, a; USMAA.

408-409   On August 6 Senator William Benton...urged the Senate investigate the cheating incident...curriculum...academic standards... discontinue football at West Point and Annapolis... strangle hold on...college life: Senate Congressional Record No. 143-5; August 6, 1951, 9704-9706; USMAA.

411   Reply from the Army's Department Counsel... cautioned against precipitate action,... calm deliberation... reviews already in place... constant interest to the Army: Letter, F. Shackelford to Senator William Benton; August 27, 1951; Army Staff papers, National Archives.

412   "Army football...from the rest of the Corps...": Public statement by Major General Frederick A. Irving, 18 August 1951; *Ibid*.

413 Answers to the Army counsel's questions...proved seldom, if ever, necessary...contained words few in the American public had heard: Army Staff papers, National Archives.

414-415 Secretary of War Newton D. Baker's comments about honor: "West Point Honor System and Its Objectives and Procedures," Major General Maxwell D. Taylor; July 23, 1947, 2; Staff Summary titled "Reappointment of Cadets Separated from West Point for Honor Violations," Brigadier General B. M. McFayden, 20 November 1951, Army Staff papers, National Archives.

11 – COACH: "THE COLONEL"

416-417 The Army of 1920-22...was essentially the Old Army...in which officers were spoken to in the third person...Sir, Private Carson has the first sergeant's permission to speak to the lieutenant.: Reeder, *Memoirs of an American Soldier*, 2.

418 ...it would not gain them monetary reward, but would gain them lifelong education, travel, associations with the finest of men, and a personal satisfaction that has no equal in any civilian pursuit: Blaik, *The "Red" Blaik Story*, 393.

418 He would say to Bob Mischak two years later, "Don't ever give up": Godwin Ordway III, author's interview.

418 "...Born to Coach,,,": Gerald A. Lodge, author's interview.

419 "I was fascinated by ... football... was ten when I formed... the... Riverdale Rovers": Blaik, *The "Red" Blaik Story*, 6.

419-420 The Riverdale Rovers played with a round, black, soccer-type ball... The... schedule of games... usually erupted into a free-for-all... only concession to the forward pass... When young Earl was an eighth grader in 1909... they didn't lecture him too severely, for fear of destroying the holiday spirit.: *Ibid.*, 6-7

420-425 Description of Earl Blaik's younger years: *Ibid.*, 7-28.

428-431 THE 1919 FOOTBALL EDUCATION: Blaik, *Ibid.*, 39-43, 79-83.

432 After he hired his assistants...careful watch of their behavior: Godwin Ordway III, author's interview.

433 Players referred to the camera as "cyclops," and couldn't "goof off": Donald Fuqua, author's interview.

434 If a manager blew his whistle indicating it was time for the next phase of practice...he would courteously ask the manager... "May we have another five minutes, please?": Freddie A. D. Attaya, author's interview.

435 Paul Lasley in the class of '56...estimated...annual expenditure for tape...$80,000 a year, stunned him: Paul Lasley, author's interview.

437-438 The Academy's Public Information Office emphasis on Army football didn't begin to change until 1949...The intent was to broaden and shift its emphasis, and throw the public spotlight on the quality of the Academy's curriculum and the performance of cadets on Graduate Record Examinations...Moore gave those instructions to Colonel Billy Leer, and he discussed the matter with Joe Cahill, enlisting Cahill's...understanding and cooperation: Testimony of Colonel Billy Leer, Bartlett Board, 334-335.

438 Following the scandal, when Army football came under attack from all directions... the Association supported seventeen varsity sports... built $2.5 million in ... athletic plants...and $460,000 addition to the gymnasium: Minutes of February 15, 1952 Athletic Board meeting, West Point, New York; USMAL.

438 "Blaik's natural instincts and Army training had molded him into a polished leader... one could not play for him or coach with him without feeling the same urge to overcome all obstacles on the path to victory...": O'Brien, *Vince*, 89.

440 Murray Warmath...said "He was the hardest working man I ever knew": *Ibid.*, 95.

440-441 Vince Lombardi remembered each hour and a half as, meticulously... organized and intense... Typically, the assistants... had fifteen minutes to give the cadets five new plays... Lombardi called it "organization at its highest level... and all those pictures would be developed and waiting": Lombardi, *Run to Daylight*, 71.

441 One summer day Vince and other coaches...intended playing some golf... Blaik... ordered reshowing the old Navy film... "Look at that. The fullback missed the block on that end": Lombardi, *Run to Daylight*, 14.

441 ...in their five years together...3500 hours of film... "brain lobes... worn as...sprocket holes looked." O'Brien, *Vince*, 95.

441 "I always drove my assistants hard...part of the price of our mission and the steepest part of all." *Ibid.*, 94.

442 He disliked public appearances and speeches, tending to stay out of the limelight... aroused storms of protest...when he publicly clashed... over the handling of the cheating scandal: *Ibid.*, 90.

442 "At the Academy or anyplace else...the importance of the game and the will to win": *Ibid.*, 90.

442 "Colonel Blaik was always on a pedestal": Ralph J. Chesnauskas, author's interview.

443 Indoctrinating coaches on the Blaik way of doing things... responsibility to work with players: Godwin Ordway III, author's interview.

443   Blaik constantly studied all his players: Joseph P. Franklin, author's interview.

443   Bob Farris, '56, saying, "He never swore...he never used curse words": Robert G. Farris, Godwin Ordway III, Patrick N. Uebel, and John R. Krobock, author's interviews.

444   Leroy Lunn...got into a scrap with a teammate during a pile-up. Blaik...said, "Don't you ever use words like that again on this field...you'll be off the team": John E. Krause, author's interview.

444   Freddie Attaya's lesson about "dirty play": Freddie A. D. Attaya, author's interview.

445   "...he had a pair of eyes that could rivet and hold you": Norman F. Stephen, author's interview.

445   "His eyes were piercing": Frank S. Wilkerson, Jr., author's interview.

445   Tommy Bell was particularly awed by Blaik... "It was always... Colonel Blaik, is this for real?"...a frequent source of subdued humor... but responded differently than Tom did: Thomas J. Bell, Patrick N. Uebel, Norman F. Stephen, author's interviews.

445   Earl Blaik was a teacher: Frank S. Wilkerson, Jr., John R. Krobock, author's interviews.

445   "Listen, learn, do better next time out": Lowell E. Sisson, author's interview.

445   He said to tackle Howard "Knobby" Glock...one day... "I'd hang my head in shame..." Howard Glock was a terror on the field the rest of the season: Joseph P. Franklin, author's interview.

446   "If you are prepared you are going to win. You will lose if you make a mistake." Al Paulekas carried Blaik's words into intercollegiate wrestling": Alfred E. Paulekas, author's interview.

446   "...Colonel Blaik made everyone feel so well prepared they couldn't lose... He knew how to make the team mentally prepare": Robert J. Farris, author's interview.

446   Blaik had "the best of the best in his coaching staff": Lowell E. Sisson, author's interview.

446-447 Howard Glock's recollections of Carney Laslie: Howard G. Glock, author's interview.

447   Leroy Lunn, '54, noticed Blaik analyzed "the physics" of the game. He insisted linemen use a three fingered stance ...ensure proper weight distribution...better mobility: Leroy T. Lunn, author's interview.

448   ..."coach the passing game and do some scouting.": Blaik, The "Red" Blaik Story, 77.

448   Garrison H. "Gar" Davidson's clash with Earl Blaik: Davidson, *Grandpa Gar, The Saga of One Soldier as Told to His Grandchildren* (unpublished), 20; USMAL.

450   Blaik to Dr. Upham: Special Collections, University of Miami, Ohio.

450   Introduced his assistants...outlined the system of attack...then told the... team... "We'll be as successful...Because we're going to bring home the bacon.": Blaik, *The "Red" Blaik Story*, 126.

451   "Our major problem at Dartmouth was to replace the spirit of good fellowship...victory more than justifies all the...work and sacrifice": *Ibid.*, 127.

451   "I suspect my players...considered me ...severe...a few may have hated me for it...sympathy was out of key...to regain...lost football respect": *Ibid.*, 128.

452   Don Fuqua...believed Army scrimmaged more than any other team in the nation...: Donald G. Fuqua, author's interview.

452   Godwin "Ski" Ordway... said scrimmages were brutal... pressure, hard hitting, fatigue... lot of fights... "The toughening wasn't just physical, it was mental." : Godwin Ordway III, author's interview.

452-453 Bob Mischak recalled Blaik saying, "Never kneel down. That's a sign of weakness"... "Some things were really brutal,"...a particular drill... that rankled... "bull in the ring": Robert M. Mischak, author's interview.

453   From team captain Leroy Lunn's perspective, Blaik's philosophy included physically wearing opponents down: Leroy T. Lunn, author's interview.

453   Don Fuqua...remembers... Doug Kenna intervening... with Blaik... to end what seemed to Kenna an excessively long period of wind sprints... He ignored Kenna's intercession and continued: Donald G. Fuqua, author's interview."

453   Krause... remembers wind sprints vividly... "He ran and ran and ran us,"...Blaik said, "If you're going to lose, you're going to lose hard.": John E. Krause, author's interview.

454   "The custom had grown up at Dartmouth of holding something of a wake over every fallen gladiator...We had to put a stop to that... Games are not won on the rubbing table": Blaik, *The "Red" Blaik Story*, 129.

454   ...the Army coach was deeply concerned for the welfare of injured players and always ensured they had the best of care: Y. Arnold Tucker, author's interview.

455   Bob Mischak's sprained shoulder in scrimmage with Syracuse, with Blaik's reaction to the injury: Robert M. Mischak, author's interview.

455   "Blaik *never* talked about injuries": Lowell E. Sisson, author's interview.

455   Gerry Lodge saw Blaik's reaction to injuries as a "blind spot"... recalled Blaik saying, "If you don't want to go back in, you don't want to play": Gerald A. Lodge, author's interview.

455-456   Gerry Lodge's encounter with Blaik after a home football game, escorting a roommate's date to a dance, then Blaik's visit with him in the West Point hospital: *Ibid.*

456   Howard Glock ran afoul of Blaik's injury policies: Howard G. Glock, author's interview.

456   Blaik's attitude toward injuries... "Spartan, too Spartan..." too many players played with injuries when they shouldn't.: Donald G. Fuqua, author's interview.

457-458   The disagreement between Earl Blaik and Rollie Bevan, and the senior physicians at the West Point Hospital, and how the issue was resolved: Colonel C. L. Kirkpatrick's testimony to the Bartlett Board, 27 August 1951, USMAA; Letter from the Bartlett Board to Major General Frederick A. Irving, August 1951, USMAA; Note from Irving to the Academy Chief of Staff, USMAA.

458   "I would walk through a brick wall for that man": Joseph D. Lapchick, Jr., author's interview.

458   ..."a giant of a man. I've never met one like him before or since. He was a man of his word": John E. Krause, author's interview.

458   ..."a man's man.": Frank S. Wilkerson, Jr., author's interview.

458   ..."a leader among young men preparing to be leaders. He set an impeccable example. He wasn't perfect, but he rose above his mistakes, and didn't fixate on them": Joseph P. Franklin, author's interview.

458   ...Blaik's teaching and coaching his assistants and his players... delegating responsibilities, delegating well...: Norman F. Stephen, author's interview.

458   ..."Blaik managed, and let the assistant coaches do their jobs": John R. Krobock, author's interview.

458   ..."Blaik was a gentleman on and off the field": Freddie A. D. Attaya, author's interview.

459   ...the Army coach saved many players ready to give up on their performance, themselves, or the Academy...: Ralph J. Chesnauskas, author's interview.

459 ..."leave West Point with their heads held high": Y. Arnold Tucker, author's interview.

460 "Football is secondary to the purpose for... college... Championship football and good scholarship are entirely compatible...The purpose of the game of football is to win, and to dilute the will to win is to destroy the purpose of the game": Blaik, *The "Red" Blaik Story*, 312-313.

461 "The tragedy of the expulsion brought "Red" Blaik closer than ever to the Corps of Cadets": Reeder, *Heroes and Leaders of West Point*, 130.

462 "Blaik talked mostly to quarterbacks and team captains": Paul A. Lasley, author's interview.

463 "It isn't what you don't know that gets you in trouble, but rather what you know – for sure – that isn't so": Blaik, 13.

12 – THE SEASON OF '51

467 Resignations in progress in August: United States Military Academy, *Official Register of the Officers and Cadets for the Period Ending 30 June 1952*, 69-94.

467-469 Blaik to MacArthur, September 4, 1951; USMAL.

468 MacArthur had not attended an Army game since 1934: Allison Danzig; *New York Times*, September 30, 1951.

468 "silhouettes": Blaik, *The "Red" Blaik Story*, 315.

468 ...never been noted for looking for too many dragons to slay... just a few choice dragons": Marshall Smith; *Life*, October 9, 1950.

468-469 Irving to MacArthur, September 5, 1951, USMAA.

469 ...emotionally exhausted and depressed staff: *Ibid.*, 437.

470 ...second only to Davis-Blanchard teams": *Ibid.*, 314.

470 ...ten years to build back...MacArthur": *Ibid.*, 315.

470-471 The Vince Lombardi-Bob Blaik-Blaik family relationship: O'Brien, *Vince: A Personal Biography of Vince Lombardi*, 87-106.

473 "We will make up in fight what we lack in talent": *Ibid.*, 103.

473 "I don't condone what they did": Blaik, *The "Red" Blaik Story*, 282.

473 Communicated with sportswriters through taped question and answer sessions... Blaik-Cutrona relationship: Joseph F.H. Cutrona; author's interview.

474 Drummed integrity into his players: Freddie A. D. Attaya; author's interview.

474 The word honor had been tarnished: Blaik, *The "Red" Blaik Story*, 279-300.

474 "There isn't a man among you with a single minute of varsity playing time": Norman F. Stephen; author's interview.

475 Two returning lettermen: United States Military Academy, 1952 *Howitzer*, 123.

474 Decision to stay with two platoons: United States Military Academy, 1952 *Howitzer*, 123.

478 Yearling group... less promising than any since he arrived at Army: Blaik, *The "Red" Blaik Story*, 315.

478 "Good fellows... an aggressive leader is priceless...": *Ibid.*, 251.

479 MacArthur's visit: Allison Danzig, *New York Times*, September 30, 1951.

479 MacArthur at the cadet review on the Plain: Mrs. Elizabeth Irving Maish; author's interview.

480 Televised football rally and game highlights, movies at the cadet theater: United States Corps of Cadets Daily Bulletin, September 28, 1951; USMAA.

486-489 Description of the Villanova game: Allison Danzig, Special to the *New York Times*, September 30, 1951.

487 ...Lombardi... "I'm going to make Tommy an All-American": Thomas J. Bell; author's interview.

489-490 Blaik's football season Sunday routines with the Army team: Paul A. Lasley; author's interview.

490-491 Description of the Northwestern game: *New York Times*, October 7, 1951.

491 ...gallant try... and "...this was the Lombardi I knew...": Blaik, 438.

491-494 Description of the Dartmouth game: Louis Effrat, Special to the *New York Times*, October 14, 1951.

492 Blaik let quarterbacks call 50 percent of plays: Peter J. Vann; author's interview.

492 Blaik walked and talked with Army teams as part of pre-game routine: Army players, author's interviews.

492 Blaik called three plays of first offensive series...first pass...out and down...dropped pass: Peter J. Vann and Freddie A. D. Attaya; author's interviews.

492-493 What are we going to do?...dropped pass...lesson learned: Freddie A. D. Attaya; author's interview.

494 Blaik to MacArthur, 18 October 1951, USMAL.

494-495 Description of the Harvard game: Louis Effrat, Special to The *New York Times*, October 21, 1951.

495-498 Description of the Columbia game: Joseph M. Sheehan, Special to the *New York Times*, October 28, 1951.

495 We play Columbia on the 27th of October and I hope that you will make plans to be with us that day... your presence will be an inspiration... I suppose in the end the decision will be made by General Collins...": Blaik to MacArthur, 18 October 1951; USMAL.

497 "Golden Mullets" weekly schedule...among toughest in the nation: Ernest F. Condina; author's interview.

498-499 Sat next to Mrs. Condina...Rosary beads: Richard J. Miller; author's interview.

498-499 Final Columbia play: Alfred E. Paulekas and Ernest F. Condina; author's interviews.

500 Credit Corps spirit...two touchdowns: United States Military Academy, 1954 *Howitzer*, 132.

500 MacArthur cabled...his congratulations: Blaik to MacArthur, 1 November 1951; USMAL.

501-503 Description of USC game: Allison Danzig, Special to the *New York Times*, November 4, 1951.

501 "Our only hope of launching an offense against [Southern California] is with the forward pass... The quick kick as an offensive weapon is included in our plans": Blaik to MacArthur, 1 November 1951; USMAL.

501 Blaik's pre-game quick-kick call against USC: Freddie A. D. Attaya; author's interview.

504 Lombardi...Blaik...Fred Attaya's memorable lesson...": *Ibid.*

504-505 Description of Citadel game: *New York Times*, November 11, 1951.

505-506 Description of Pennsylvania game: *Ibid.*, November 18, 1951.

506-511 Description of Navy Game; *Ibid.*, December 2, 1951.

510 Freddie Meyers... "What was the play, Coach?": Freddie A. D. Attaya; author's interview.

511 "I did not dream... that they were exploiting adversity...beginning of a football miracle.": Blaik, 315-316.

13 – 150 YEARS OF HISTORY AND TRADITION

513-590 Description of Sesquicentennial planning and celebration: *The Sesquicentennial of the United States Military Academy*, Baker, Jones, Hausauer and Savage, Inc., 1953; 16-106.

14 – TO BUILD A TEAM

594   Continue the two platoon system...a small number of skilled players ...going both ways: "The Rabble," Al Hamblin, *Pointer*, 19 September 1952, 18.

594-595 Blaik to MacArthur, 3 June 1952: Blaik letters, USMAL.

595   MacArthur to Blaik, 9 June 1952: *Ibid.*

595-598 Blaik to MacArthur, 24 June 1952: *Ibid.*

598   Al Paulekas' background and election as team captain: Alfred Paulekas, author's interview; "And Its Captain...," John P. Lovell, *Pointer*, 19 September 1952, 18.

600   Ralph Chesnauskas' background and experiences on the 1952 plebe football team: Ralph Chesnauskas, author's interview.

601   A SPECTACULAR BEGINNING: "Cadet Eleven Tops Gamecocks 28 to 7," Michael Strauss, *New York Times*, September 28, 1952.

602-603 John "Moose" Krause, the sled, and preparations for the USC Trojans: John E. Krause, author's interview.

603   TROJAN HORSES – AGAIN: "Trojans Conquer Cadet Squad 22-0: *New York Times*, October 5, 1952.

604   Ordway's recollections of blocked punt: Godwin Ordway III, letter to author.

605   Krause's recollections of "...Good job": John E. Krause, author's interview.

605-608 BACK HOME TO FACE THE DARTMOUTH GREEN: "Cadets Score, 37-7," Joseph M. Sheehan, *New York Times*, October 12, 1952.

609-613 DRAGONS, PANTHERS AND ROLLER COASTERS: "Panther Team Too Fast," Allison Danzig, *New York Times*, October 19,1952.

609-610 Walk past Trophy Point, across the Plain: John E. Krause, author's interview.

614-619 THE LION NEMESIS: "Pass Saves the Lions," *New York Times*, October 26, 1952.

619-620 VIRGINIA MILITARY INSTITUTE COMES TO MICHIE STADIUM: "West Point Team Triumphs by 42-14," *New York Times*, November 2, 1952.

621-622 ONE WAY TO PAY THE PRICE: Godwin Ordway III and Frank S. Wilkerson, Jr., author's interviews.

622-624 INTO THE DEEP SOUTH, AGAINST "THE RAMBLIN' WRECK": "Ga. Tech Conquers Cadet Eleven, 45-6," *New York Times*, November 9, 1952.

624   "I have long since forgotten...but I shall always recall...Dean Masson.": Blaik, *The "Red" Blaik Story*, 298.

624-628 QUAKERS IN THE MUD: "Cadets' Last Minute Tally by Bill Purdue Wins Game," Louis Effrat, *New York Times*, November 16, 1952.

627 628 18 November Blaik telegram to MacArthur: Blaik letters, USMAL.

628-635 "Operation Paintbrush": William R. Schulz, author's interview; Willis C. Tomsen, author's interview; 1954 *Howitzer;* 243, 365, 382, 396, 409, 419; Willis C. Tomsen's Explanation of Report, courtesy of Willis C. Tomsen.

635-640 NOT THIS TIME EITHER: "Monahan Tallies," Allison Danzig, *New York Times,* November 30, 1952.

640 John Green's good-bye after Army-Navy '52: Frank S. Wilkerson, Jr., author's interview.

640 Navy, Georgia Tech, Southern California, and Pittsburgh standing in the Board of Coaches' Poll, *New York Times,* December 2, 1952.

641 The Bob Guidera story: Willis C. Tomsen, correspondence to author, 7 July 1999.

641 "Our football rebound...to reach the peak in 1953": 1953 *Howitzer,* 118.

642-643 In the spring of 1952,...something else was afoot...a tradition was stirring... much improved: Robert H. Downey and Benjamin F. Schemmer, author's interviews and correspondence exchanges.

644 "A GOOD START EARL...: Blaik, *The "Red" Blaik Story,* 322.

644 Blaik reply to MacArthur, 29 September 1952: Blaik letters, USMAL.

644 "That wire may have...implied...more...than was possible...": Blaik, *The "Red" Blaik Story,* 323.

15 – BELLS OF LOVE, SOUNDS OF WAR

646 In mid-April 1953...The three lieutenants were going off to war: William J. Ryan, author's interview; Barbara (Inman) Colby, author's interview; Joyce (Shea) Himka, author's interview.

647 ...repeatedly won the mile and two mile events at track meets...in the northeastern United States: John P. Lovell, "Dick Shea – The Winning Habit"; *Pointer;* May 2, 1952, 14-15.

648 ...during the preceding Christmas holidays he and Joyce had made their plans: Joyce (Shea) Himka, author's interview.

648 ...he learned he was no match for America's strongest prospects for the Olympic team...literally passed out on his feet: Louis M. Davis, author's interview.

649 ...saw a circular explaining how a soldier could obtain an appointment to West Point: "Dick Shea – The Winning Habit," Pointer; May 2, 1952; 15.

649 ...married graduation day in 1952, at four thirty in the afternoon: Joyce (Shea) Himka, author's interview.

649 Dick Inman's fumble of Navy's kick off: Mary Jo Inman Vermillion, author's interview; *New York Times*, December 2, 1951.

650 ...left the field after the game with tears streaming down his face: Newspaper clipping and note in personal memorabilia from Richard G. Inman, courtesy of Mary Jo Inman Vermillion.

650-651 Dick Inman had no prospects of romance when he graduated...Love and marriage came later: Mary Jo Inman Vermillion, author's interview; Barbara (Inman) Colby, author's interview.

651 In the spring of 1948 Pat Ryan...and there would be a third time in his future: William J. Ryan, author's interviews.

653 At Randolph Air Force Base... 'sit-down strikes' affected Pat Ryan's family and future: *American Airpower Strategy in Korea, 1950-53*; Conrad C. Crane, Kansas University Press, 1999, 169.

656-670 In the spring of 1948,...Joyce Elaine Riemann met Dick Shea... he went to A Company, 1st Battalion... His company commander was First Lieutenant William S. 'Bill' Roberts, Jr.: Joyce Himka, author's interview; Letters, dated March 6 and 11, 1998, from Joyce Himka to the author; Personal correspondence from Richard T. Shea, Jr. to Joyce R. Shea, courtesy of Joyce Himka; Ronald M. Obach, author's interview; Louis M. Davis, author's interview; Walter G. Parks, author's interview; Command Reports, 17th Infantry Regiment, June and July 1953.

670-680 Dick Inman was a hurdler and high jumper in high school and at Army, but above all the sports he enjoyed, he loved football...In a short time Dick Inman would learn he was assigned to B Company, 1st Battalion of the 17th, as Platoon Leader, 2nd Platoon: Mary Jo Inman Vermillion, author's interview; Letters, dated March 6 and 17, 1998, from Mary Jo Vermillion to the author; Letter from Barbara (Inman) Colby to Mrs. "Becky" Inman and Bonnie Ruth Inman, Christmas c. 1963, with copy of Richard G. Inman's poem, "To My Sister Bonnie"; Richard G. Inman's papers, poems and letters courtesy of Mary Jo Vermillion; Lawrence H. Putnam, Sr., author's interview; George B. Bartel, author's interview; Richard C. Coleman, author's interview. John R. Aker, author's interview; *Vincennes-Sun Commercial*, March 8, 1953; 13 and 18; *Stars and Stripes*, August 6, 1951.

680-691 The 19th was operating entirely at night over Korea. Their airplanes were painted black...Welcome to the war. More missions to come: William J. Ryan, author's interviews; Air Force Form 5, Record of Flying Time, for William J. Ryan.

16 – THE SUMMER OF '53: NEW HEROES

692-693 May 27, 1953 letter from Dick Inman to his parents: Courtesy of Mary Jo Inman Vermillion.

693 In mid-April...there had been a fierce battle on Arsenal: S.L.A. Marshall; *Pork Chop Hill*, 31-32.

693 There had been no major Chinese assault on Arsenal or Erie... while Dick Inman and Dick Shea were there: Letters from Dick Inman and Dick Shea, courtesy of Mary Jo Inman Vermillion and Joyce (Shea) Himka.

693-694 June 4, 1953 letter from Dick Inman to his parents: Courtesy of Mary Jo Inman Vermillion.

694 In the beginning of June...The 1st Battalion was given new responsibilities which included Pork Chop Hill: 17th Infantry Command Report for June 1953.

694-695 June 9, 1953 letter from Dick Shea to Joyce: Courtesy of Joyce Himka.

695 The Chinese had tried three times to take Pork Chop: December of 1952, and the following March and April: Colonel William R. Kintner; "Pork Chop," *The Army Combat Forces Journal*, 40.

696 In the aftermath of April's battle on Arsenal...detailed, accurate map: Marshall, *Pork Chop Hill*, 31-32.

698-699 When it came to major offensive operations along the front, the Chinese had a predilection to... their manner of handling... battle plans... to the point of recklessness...it made them think the plan was the best possible...At least, that was the American explanation of Chinese reasoning...The Americans on Arsenal and Erie were ready... and it came slightly ahead of schedule: *Ibid.*

700 Descriptions of the purposes, organization, operations, and communications of an ambush patrol and outpost defense: *Ibid.*, 133-142.

701 Description of fighting typical of Pork Chop battles: *Ibid.*, 133-166.

701-704 June 14 and 22, 1953 letters from Dick Shea to Joyce: Courtesy of Joyce Himka.

704-706 June 25, 1953 letter from Dick Inman to his parents: Courtesy of Mary Jo Inman Vermillion.

706-709 June 29, 1953 letter from Dick Shea to Joyce: Courtesy of Joyce Himka.

709-711 June 30, 1953 letter from Barbara Kipp Inman to the Inman family: Courtesy of Mary Jo Inman Vermillion and Barbara Colby.

711   July 2, 1953 letter from Barbara Kipp Inman to the Inman family: *Ibid.*

711-713 July 2, 1953 letter from Dick Inman to his parents: Courtesy of Mary Jo Inman Vermillion.

713-715 Dick Inman's poem, "My Toys": Courtesy of Mary Jo Vermillion.

715   One night in May...the Americans dubbed the network the Rat's Nest: Kintner, "Pork Chop," *The Army Forces Combat Journal*, 41.

715   In mid-June plans were set in motion...aggressive ground action prohibited...in the period 18-29 June: 17th Regiment Command Report for July 1953.

715   ...throughout the second half of the month, increased activity... all along the 7th Division front...: 7th Infantry Division Command Report for July 1953.

715   The Americans fired numerous artillery barrages...Three airstrikes failed to hit the complex: Kintner, *The Army Forces Combat Journal*, 41.

715   July 3...Lt. Trudeau's B-29... released at an altitude of three thousand feet... over the area of "Old Baldy": William J. Ryan, author's interview.

716   The plan of maneuver...There were few enemy encountered...mortar and artillery began finding their mark: 17th Regiment Command Report for July 1953.

717   Returning raiders reported counting three enemy killed in action, and fourteen wounded...raiding party carried back twenty wounded, and two were missing in action: *Ibid.*

717   When Read...pleaded with Trudeau...he brushed tears from his cheeks...: "The Last Patrol," *Argosy*, December 1953, 35.

718   Names like Elko, Vegas, Charley, Harry, Christmas, and Pork Chop marked the bloodiest spots along the Korean battle line: Kintner, *The Army Combat Forces Journal*, 40.

718   After the April battle "The Chop" was a shambles: *Ibid.*, 41.

718   Major General Arthur Trudeau, 7th Division commander, decided to rebuild Pork Chop, make it impregnable...The task fell to the 17th Buffaloes: *Ibid.*

719   At the beginning of July the 17th's 2nd Battalion anchored the left sector... E and I Companies...moved on line to replace the Columbian soldiers... Arrayed against the 17th were...199th, 200th, and 201st Regiments...: 17th and 32nd Infantry Command Reports for July 1953.

719   The 17th Buffaloes...found the going rough...the Chinese...could observe every beam nailed into place: Kintner, *The Army Combat Forces Journal*, 41.

719   One company defended the hill day to day: *Ibid.*

719-720   Description of "guard duty": Dale W. Cain, Sr., correspondence and author's interview.

720-721   Dale Cain's call to return to A Company and his meeting, working with Dick Shea, Cain's two enlistments and return to Korea: *Ibid.*

721   The night of July 6 Dick Inman was to lead an ambush patrol of twenty men...in support of A Company: Depositions given by Pfc.'s Irwin Greenberg and Clarence R. Mouser, 27 September 1953: Courtesy of Mary Jo Inman Vermillion.

721   It was routine to move ambush patrols...and attempt to take a prisoner: S.L.A. Marshall, *Pork Chop Hill*, 133-134.

721   The night of July 5 Dick Inman rehearsed his...patrol: Deposition given by Pfc. Clarence R. Mouser, 27 September 1953: Courtesy of Mary Jo Vermillion.

722   The patrol included two elements...Finger 21: Depositions given by Pfc.'s Greenberg and Mouser, 27 September 1953: Courtesy of. Mary Jo Inman Vermillion.

722   From the hillsides to the north, Pork Chop's defenders heard demands they surrender... "We will take Pork Chop if we have to wade through blood": Lieutenant Colonel John A. Coulter II, unpublished manuscript, Chapter VI, "Richard Thomas Shea"; Dale W. Cain, Sr., correspondence and author's interviews.

723   Dale Cain had heard it all before... "We're in good shape": Dale W. Cain, Sr., author's interviews.

723   At eight thirty in the evening July 6, Dick Inman's ambush patrol...took up their positions out front of Finger 21: Depositions from members of ambush patrol.

723   A heavy rainstorm began: 17th Regiment Command Report for July 1953.

723   Private Miller asleep, preparing for guard duty: Robert E. Miller, author's interview.

723   Private Cain and Lieutenant Shea in their bunker, Cain monitoring the radio: Dale W. Cain, Sr., correspondence and interview with the author; Dale W. Cain, Sr., *Korea, the Longest War*, 67-74.

723   Under cover of the storm two battalions of Chinese infantry began moving...: 17th Infantry Command Report for July 1953.

723-724   At the checkpoint behind Hill 200, Private Emmett "Johnny" Gladwell...: Emmett D. Gladwell, author's interview.

724   Battalion aid station "...The place has just been blown all to hell": Lee H. Johnson, author's interview.

724-725 The first round a direct hit...no one was hurt: Gladwell, author's interview.

725   Private Bob Miller awoke with a start... "All hell's broken loose.": Robert E. Miller, author's interview.

725   Corporal Charlie Brooks' actions when Chinese artillery opened fire...: Charles Brooks, author's interview.

725   Dale Cain's radio came alive...last time Dale saw Dick Shea: Dale W. Cain, Sr., author's interview.

725   Another A Company soldier...said the attack looked like a "...moving carpet of yelling, howling men...driving their men up the hill...": Lieutenant Colonel John A. Coulter II; unpublished manuscript

725   At 10:38...small arms fire...: 17th Infantry Command Report, July 1953.

725   Among the lead elements were boys twelve to sixteen years of age: David R. Willcox, author's interviews.

725-726 As soon as the artillery began Lieutenant Willcox called his outguards and Dick Inman's patrol back...Mouser...saw Dick reenter the trench: Depositions given by Lt. David R. Willcox and Pfc. Clarence R. Mouser, 27 September 1953: Courtesy of Mary Jo Inman Vermillion; David R. Willcox, author's interviews.

726   When the artillery and mortar barrages began, MacPherson Conner, Dick Inman's and Dick Shea's West Point classmate...decided to get some sleep...By this time, E Company had already been attached to the 17th Regiment...on Pork Chop...: MacPherson Conner, author's interview; 17th Regiment Command Report for July 1953.

726-728 Descriptions of the start of the battle for Pork Chop, Dick Inman's actions attempting to bring his patrol into contact with 2nd Platoon, to support them, and providing covering fire as patrol survivors made a dash for the 2nd Platoon area: 17th Infantry Command Report for July 1953; Kintner, *The Army Combat Forces Journal*, 40-45; Depositions given by Pfc.'s Greenberg and Mouser, Cpl. Harm J. Tipton, and Lt. David R. Willcox, 27 September 1953: Courtesy of Mary Jo Inman Vermillion.

727   Suddenly, without warning...alive with Chinese infantrymen: Charles Brooks, author's interview.

727   On the right flank of the hill...In the melee David Willcox...fighting off repeated Chinese assaults: David R. Willcox, author's interview.

731   At 10:41...Division artillery reported... 'Flash Pork Chop in progress': 7th Division Command Report, July 1953.

731   "Differing little from the curtain barrage of World War I days...twelve to six rounds per tube per minute while maintaining the fire for six minutes": S.L.A. Marshall, *Pork Chop Hill*, 47.

731-734 Description of battle for Pork Chop: 17th Infantry Command Report for July 1953; Kintner, *The Army Combat Forces Journal*, 40-45.

732   "[B] Company retook...central sector near A Company CP.": *Ibid.*

728   Description of fighting on Pork Chop and MacPherson Conner's second encounter with Dick Shea: 17th Infantry Command Report for July 1953; MacPherson Conner, author's interview.

734-735 Description of Dick Shea's actions: Letter from Lieutenant Colonel Beverly M. Read to Mrs. Joyce R. Shea, 31 August 1953, Courtesy of Joyce Himka; Citation accompanying award of the Congressional Medal of Honor to Richard Thomas Shea, Jr., May 1954; Lieutenant Colonel John A. Coulter II; unpublished manuscript.

735   Easy Company Commander in Bunker 45: MacPherson Conner, author's interview.

735-737 MacPherson Conner was immediately awakened from his sleep... One distinctive voice he hadn't heard for a long while... "Dick Inman called in and said he had a broken leg": MacPherson Conner, author's interview.

737-739 Mac Conner's platoon and his actions: MacPherson Conner, correspondence and author's interview.

739-740 Description of the continuing battle for Pork Chop: 17th Regiment Command Report for July 1953; Kintner, *The Army Combat Forces Journal*, 40-45.

740   Charles Brooks and Bob Miller and the survival of five others: Brooks and Miller, author's interviews.

740   Corporal Jim McKenzie, 17th Medical Company, on Pork Chop Hill: James E. McKenzie, author's interview.

741-742 Approximately noon on July 7...General Trudeau... "A night-time single-file march...make a right-angle turn, and regroup for attack...": Kintner, *The Army Combat Forces Journal*, 42.

743-744 Mac Conner's platoon actions, and Conner and Dick Shea cross paths a second time in the tunnel near the A Company CP: MacPherson Conner, author's interview.

744-746 Continuing description of Pork Chop battle: 17th Infantry Command Report; Kintner, *The Army Combat Forces Journal*, 41-45.

745-746 At a conference at the 17th Infantry CP...Special instructions were issued to the troops to stay out of the trenches until they were on their objective: Kintner, *The Army Combat Forces Journal*, 43.

746-748 Bob Miller and Dick Shea, and Bob Miller's evacuation from Pork Chop; Robert E. Miller, author's interview.

750 Mac Conner's actions and first wounds: MacPherson Conner, correspondence exchange and author's interviews.

751 Corporal Bob Northcutt's actions and recollections of the G Company counterattack: Robert Northcutt, author's interview.

751-754 General description of the latter phases of the battle for Pork Chop: 17th Infantry Command Report; Kintner, *The Army Combat Forces Command Journal*, 43-45.

752-753 Corporal Dan Schoonover's actions in the G Company counterattack: Citation accompanying the award of the Congressional Medal of Honor, July 1955.

753-754 Dick Shea's actions in concert with G Company soldiers: *Ibid.*

753-754 Dick Shea's actions the afternoon of July 8, 1953: Letter from Lieutenant Colonel Read to Joyce Shea, 31 August 1953; Lieutenant Colonel John A. Coulter II, unpublished manuscript; Citation accompanying the award of the Congressional Medal of Honor to Richard Thomas Shea, Jr., May 1954.

755-756 Mac Conner's platoon's July 9 assault, his second wounding, and evacuation: MacPherson Conner, correspondence exchange and author's interviews.

757-758 Charlie Company counterattack and actions of Dan Peters, Frederick Tanaka, Richard Beacher, and Donald Swope: 17th Infantry Command Report, July 1953; Kintner, The Army Combat Forces Journal, 44; Danny L. Peters, author's interviews and exchange of correspondence.

758-759 Jim McKenzie's experiences, evacuation from Pork Chop: James E. McKenzie, author's interviews.

760 David Willcox's experiences and evacuation from Pork Chop: David R. Willcox, author's interviews.

762 Corporal Dan Schoonover's actions and death: Citation to accompany the award of the Congressional Medal of Honor, July 1955.

763 Dale Cain leaves Pork Chop on July 11: Dale W. Cain, Sr., correspondence and interviews with author; Dale W. Cain, *Korea, the Longest War*, 67-74.

764 Among those moved to a Norwegian mobile army surgical hospital was...MacPherson Conner...Mac received a visitor...General Trudeau told him...The outpost would be obliterated...made unusable by the Chinese: MacPherson Conner, author's interview; Kintner, *The Army Combat Forces Journal*, 44-45.

764 Ten battalions of American artillery opened fire at a prearranged time...VT fused rounds: Letter from Herbert C. Hollander to Louis Davis, 6 November 1953, courtesy of Joyce Himka.

765 Barbara Inman a...dream... She woke up screaming and crying: Barbara Colby, author's interview; Mary Jo Inman Vermillion, letter to author.

765 Notifications of Joyce Shea and Barbara Inman that their husbands were missing in action: Joyce Himka, author's interviews; Barbara Colby, author's interview.

766 Doug Halbert's recollections of the cease fire and Robert Hall, a combat marine... "The cease-fire was announced for ten o'clock that night...That's when we began to realize that it really was over.": Donald Knox, *The Korean War, Uncertain Victory*; 504-505; Douglas Halbert, correspondence with the author.

766 Korean War casualties: Knox, *The Korean War, Uncertain Victory*, 506.

768-781 ONE MORE MISSION: William J. Ryan, author's interview; B-29A 44-61920 Aircraft Accident Investigation Report for 24 August 1953, crash of 920, September 1953.

781-786 MISSING IN ACTION: Joyce Himka, author's interviews; Barbara Colby, author's interviews; Letter from the Adjutant General of the Army to Mrs. Joyce Shea, courtesy of Joyce Himka; Letter from Lieutenant Colonel Beverly M. Read to Mrs. Joyce Shea, Courtesy of Joyce Himka.

17 – ACADEMIC YEAR BEGINS

789-790 Description of contents of Beast Barracks honor lecture to classes of 1956 and 1957, differences between contents of lectures given the summers of '51, '52, and '53, and changes in honor investigation procedures: Unpublished scripts of Beast Barracks honor lectures, USMAA.

790-791 Descriptions of Academy class rings, their origin, evolving tradition, and the 1953 presentation ceremonies: *The Pointer*; September 11, 1953; 1-6; 1954 *Howitzer*, 81- 82; 1997 *Register of Graduates and Former Cadets*, 223.

791-797 Summer Military Training for the classes of 1954-57: *The Pointer*; September 11, 1953; "Military Training Program for Summer 1953," Training Memorandum Number 5, 9 April 1953; USMAA.

797 Reorganization Week honor conferences and Earl Blaik's athletic policy briefing: "Reorganization Week," Training Memorandum Number 12; 15 August 1953; USMAA.

797   Blaik to MacArthur, 3 September 1953: USMAL.

798-799   Earl Blaik's comments to *The Pointer* interviewer, about football players and "privileges": Dexter H. Shaler, "Blaik Looks at the Army Team"; *The Pointer*; September 25, 1953; 14.

797-801   Description of the birth of the Army "victory cannon": Unpublished, undated paper titled "The Army Victory Cannon" by Edward Moses and a graduate who requested his name not be used in this work.

802   Academic subjects for yearling and cow years: Admissions Directorate, "1950-51 United States Military Academy Catalog," 1950, 35-37; USMAA.

802   Numbers of cadets from the class of 1955 who left the Academy during the 1952-53 academic year: *Official Register of Officers and Cadets*, United States Military Academy, 30 June 1953; 119-120.

802-803   Description of 1821 travel and march of the Corps of Cadets from West Point to Boston; "...They marched through Lenox... 'you are responsible to your country.'": Pappas, *To the Point*, 146-147.

803   ...the smallest Army team in forty-five years: Dexter H. Shaler, "Blaik Looks at the Army Team, *"The Pointer*; September 25, 1953; 14.

804-807   National Collegiate Athletic Association rule changes for 1941 and 1953: Nelson, *The Anatomy of a Game*, 218 and 254-255; Stan Harvill, "The Army Team is Back," *The Pointer*; September 25, 1953; 21.

805   Bob Farris, a yearling in '56, became a linebacker on defense, a tackle on offense. Howard Glock, a tackle on defense, moved over to guard on offense: Blaik, 326.

805   Lowell Sisson's 1953 spring practice and preseason changes in position: Lowell E. Sisson, author's interview.

805   Gerry Lodge's 1952-'53 change in positions, and spring practice coaching by "Doc" Blanchard: Gerald A. Lodge, author's interview.

805-806   Freddie Attaya's 1951 spring practice change from defense to offense, and actions, comments by Vince Lombardi: Freddie A. D. Attaya, author's interview.

806   Neil Chamberlin's serious injury in a jeep accident: *The Pointer*; September 25, 1953; 1.

806   Bob Guidera's debilitating knee injuries: Willis C. Tomsen, author's interview.

807   "Players of the classes of '54 and '55 had been to the wars...against Pennsylvania": Blaik, 324 and 325.

808-809   Description of Army-Furman football game, and Associated Press post game national rankings: *New York Times*; September 27 and 29, 1953; 1954 *Howitzer*, 138.

809-811 Description of Army-Northwestern football game, and Blaik's thoughts, actions, and comments afterward: *New York Times*; October 4 and 11, 1953; 1954 *Howitzer*, 139; Blaik, 326-327.

811 Taking advantage of an airplane delay...he accepted he... hadn't adequately prepared... players for the game: Blaik, 327.

811-814 Description of Army-Dartmouth football game: William J. Briordy, *New York Times*; October 11, 1953; 1954 *Howitzer*, 140.

813 Paul Lasley's background and football experience before coming to the Academy: Paul A. Lasley, author's interview.

814-815 Earl Blaik's half-time tongue lashing of the Army team, and the team's second half break from the huddle: Norman F. Stephen, author's interview.

18 – ARMY VS DUKE: 17 OCTOBER 1953

816 "This game linked Army football past, present and future...this had to be the day of the beginning.": Blaik, *The "Red" Blaik Story*, 328.

816-817 Duke football team's record, national ranking, injury status, and strengths: *New York Times*, October 4 and 11, 1953.

817 Pete Vann's injury and ability to practice before the Duke game: Peter J. Vann, author's interview.

817 Earl Blaik's sixth football axiom, "Games are not won on the training table": Blaik, *The "Red" Blaik Story*, 251.

818 Army scouting reports on Duke and Blaik's pregame defensive preparations: Norman F. Stephen, author's interview.

818 Inside the 10-yard line Duke ran 95 percent...between tackles. Near goal lines... "quarterback sneaks": Gerald A. Lodge, author's interview.

819 When an opponent's fullback went in motion...the Blue Devil defense failed to cover him: Peter J. Vann and Patrick N. Uebel, author's interviews.

819 "...200 to 300...daily...": Peter J. Vann, author's interview.

820 He stirred them as he never had before... "The Corps and the football team are one unit. You cannot separate them": Reeder, *Heroes of West Point*, 130.

820 "Silence" imposed by cheerleaders: Edward M. Moses, author's interview.

820-821 "Men, after the team send-off...we're to remain silent... until the last man double time's off the field... We'll remain on our feet the entire afternoon... Our voices will be heard as they haven't been heard in years.": Author's recollections.

822   Pete Vann's Friday afternoon passing practice at the Polo Grounds, with Earl Blaik and Vince Lombardi protecting against press and opponent intelligence gathering, and possible ankle reinjury: Peter J. Vann, author's interview.

822-828 ALL QUIET AND READY, Description and timing of the Corps of Cadets' trip and march-on for the Duke football game: United States Corps of Cadets, "Operations Order 1," 8 October 1953; USMAA.

824-825 A MISSION TO REMEMBER...for fear midshipmen would discover the weapon and... "borrow" it: Willard L. Robinson, author's interview.

825-828 GO! GO! GO! GO! To the applause of a small crowd gathering in the Polo Grounds, the Academy Band...Company A-1's left flank on the 10-yard line...brigade staff on the 50-yard line..., and description of the pregame ceremony: "Operations Order 1," 8 October 1953; Edward M. Moses, author's interview; author's recollections.

828   "...small crowd of 21,284...": *New York Times*; October 18, 1953.

829   "a game never to be forgotten": Blaik, 328.

829-831 Description of first half's play: *New York Times*; October 18, 1953; New York *Sunday News*; October 18, 1953; Jim Fleming, "Army Varsity," *Pointer*; 23 October 1953, 14.

832-834 Description of half time cheerleader exchange and tradition of "pass 'im up.": Willard L. Robinson, author's interview; author's recollections.

834   Bill Schulz's post game arrangements to assist the plebe on emergency leave and join his friends at the Roosevelt Hotel: William R. Schulz III, author's interview.

834-842 Description of second half action against Duke: *New York Times*; October 18, 1953; New York *Sunday News*; October 18, 1953; Jim Fleming, "Army Varsity," Pointer; 23 October 1953, 14.

837   With fourth down and five yards to go...Blaik ordered...a field goal... On the bench...was Lieutenant Martin "Tiger" Howell.... Howell unknowingly countermanded Blaik's instructions... Duke went over to the offensive... "Take this. You'll need it in Korea"... Blaik remembered, "The suffering... Tiger went through that afternoon...": Blaik, 329.

838   Blaik, Howell, and the Army bench turned their attention back to the game...The prior 56 1/2 minutes had been enough to wring any crowd dry of emotion... "New York could not comprehend...they were being rewarded by one of the classics of...Army history...": Blaik, 328-329.

838   Description of Duke's double reverse and game saving tackle by Bob Mischak: Blaik, 329-330; *New York Times*; October 18, 1953;

New York *Sunday News*; October 18, 1953; Robert M. Mischak, Lowell E. Sisson, and Thomas J. Bell, exchanges of correspondence and author's interviews.

838-842 Description of the last three and one half minutes of the Army-Duke game, and the after game scene: Blaik, 329-330; *New York Times*; October 18, 1953; New York *Sunday News*; October 18, 1953; "A United 'Army' Wins Its Greatest Victory," *Pointer*; 23 October 1953, 22; Preston S. Harvill, Jr., e-mail exchange, April 1998; Peter J. Vann, author's interview; Norman F. Stephen, author's interview; Gerald A. Lodge, author's interview; Freddie A. D. Attaya, author's interview; Robert G. Farris, author's interview; Lowell E. Sisson, author's interview; Edward M. Moses, author's interview.

843 Scissors had to be used: Lowell E. Sisson, author's interview.

843 "Don't ever give up...": Godwin Ordway III, author's interview.

844 For nearly fifteen years afterward...Bob Mischak's play against Duke University was...an example of the power of motivation: Reeder, *Heroes of West Point*, 130.

844 Lowell Sisson's experience typified what was different that day...for the first time he clearly heard the Corps...In Lowell's mind, the Corps won for the team: Lowell E. Sisson, author's interview.

844 Pat Uebel heard the sounds...he had never seen the Corps so aroused: Patrick N. Uebel, author's interview.

844 Bob Farris...knew "the 12th man" was there: Robert G. Farris, author's interview.

845 Godwin "Ski" Ordway... "felt at one with the Corps like never before": Godwin Ordway III, author's interview.

845 Bob Mischak...went through...intense preparation for Duke...But what struck him...was...excitement and interest...his roommates... The Corps "had the fever": Robert M. Mischak, author's interview.

845 ...happy men lifted...Tommy Bell... onto their shoulders... His little Irish mother attended the game... she too was "...one of his fans": Thomas J. Bell, author's interview.

845 Freddie Attaya heard Vince Lombardi's shout to him... "Run the gauntlet!" yelled Lombardi...Freddie ripped the...Duke line for twelve yards and a first down. Lombardi rushed to hug him, telling him he had done a great job: Freddie A. D. Attaya, author's interview.

846 In 1951, following the Northwestern game, Pete had missed the team's flight home... was a plebe that year...tardiness and behavior were noted by Blaik... belief...Pete was immature, "a mere boy,"... wouldn't grow up: Blaik, 323.

846   "[Earl Blaik] pounded three characteristics into his quarterbacks:... When you demonstrated on the field...your knowledge of the game... the team will follow your leadership and produce touchdowns.": Reeder, *Heroes of West Point*, 127.

846   Pete was no longer "a mere boy." He was a man, and indispensable to the team: Blaik, 323.

846   Pete Vann's encounter with Vince Lombardi, in the hotel elevator after the game: Peter J. Vann, author's interview.

847   Ralph Sasse's visit to the Army dressing room after the game: Blaik, 331.

847-848 MISSION ACCOMPLISHED. The sequestering of the Army victory cannon: Willard L. Robinson, author's interview.

848-849 Bill Schulz's post game evening with Lelia McGill: William R. Schulz III, author's interview.

849   Allison Danzig wrote:"...many valiant performers on the playing field": *New York Times*, 18 October 1953.

850   Gene Ward...wrote: "...a victory jamboree the like of which had never been seen": *New York Sunday News*, 18 October 1953.

19 – THE GREAT AND GOOD TEAM

853-857 Description of Columbia game: *New York Times*; October 25, 1953; 1954 *Howitzer*, 142.

857   The Tulane team was better prepared for the changed substitution rules: *The Pointer*; September 25, 1953; 24.

859   1947 Army Athletic Policy: Bartlett Board, Exhibit F.

860   Description of the drunk in the New Orleans Hotel lobby: John E. Krause, author's interview.

860-862 Description of Tulane game: *New York Times*; November 1, 1953; 1954 *Howitzer*, 143.

862-865 Description of the North Carolina State game: *New York Times*; November 8, 1953; 1954 *Howitzer*, 144.

867-868 Description of the class of '54 march-on at Franklin Field: Operations Order 2: USMAA.

869-872 Description of the Pennsylvania game: *New York Times*; November 15, 1953; 1954 *Howitzer*, 145; Blaik, 332-333.

872   "He made several... key tackles, two preventing touchdowns.": Blaik, 333.

873   ..."evoked stentorian cheers...Peter Vann became a man": *Ibid*.

874-895 A SHAGGY GOAT STORY AND THE GOAT REBELLION: "A Shaggy Goat Story," Schemmer, *Assembly*; January 1995, 17; "More to the Point,"

Pappas, *Assembly*; January 1995, 45; Benjamin F. Schemmer, Emery S. Wetzel, Jr., and John C. Bard, author's interviews; Preston S. Harvill, Jr., e-mail exchange with author; *Diamondback*, University of Maryland, 14 October and 13 November, 1952; "The Tale of a Goat," McIntyre, *Assembly*, January 1954.

895   "WHO THE HELL ARE THEY, ANYWAY?": Army players, author's interviews.

896   Blaik told them he would sing a song... "The small part of me that was *Buttons the Bellboy* had not been subordinated": Blaik, 333.

898-899 "The Last for '54," naming the men in the class of 1954 who were playing their last game for Army: *The Pointer*; November 27, 1953.

899   Paul Lasley's encounter with Earl Blaik prior to Army-Navy 1953: Paul Lasley, author's interview.

899-900 Cadet and midshipmen pregame hijinks: *Philadelphia Inquirer*; November 29, 1953.

901   Names of attending dignitaries and description of Army-Navy game and published reports by coaches, sports writers, and sports broadcasters: *Philadelphia Inquirer*; November 29, 1953; *The Pittsburgh Press*; November 29, 1953; *Boston Sunday Advertiser*; November 29, 1953; *Pittsburgh Post-Gazette*; November 29, 1953; *New York Times*; November 29, 1953; 1954 *Howitzer*, 146-147.

901   Former prisoners of war at Army-Navy: Al Laney, *New York Times*, 29 November 1953.

912   Bob Mischak's award of All-American certificate: Robert M. Mischak, author's interview.

913   On December 15, John Bard...wrote... "The Lambert Trophy is going to be awarded...the first time to a football team and its student body": John C. Bard, author's interview.

914   "Lambert Trophy Presentation" memorandum: USMAA.

914   "...FOOTBALL TROPHY ACCEPTED BY ARMY": *New York Times*; December 21, 1953.

914   Irving to Lambert brothers' letter, December 1953: USMAA.

915   "I never have coached a team that gave me more than you did...you have done more for football at West Point...": Blaik, 335.

20 – HONOR AND VICTORY REVISITED

917   Thompson Cummings, the last to fall from the class of 1952: *1998 Register of Graduates and Former Cadets*, 3-217; 1952 *Howitzer*, 260.

925   Barbara Inman's search for Dick Inman: Barbara Colby, author's interview; Mary Jo Inman Vermillion, author's interview.

925-926   Congressman Bray to Howard Greenlee, 6 October 1953: Courtesy of Mary Jo Inman Vermillion.

927   Barbara Inman's attempts to publish Dick Inman's book: Barbara Colby, author's interview; Mary Jo Inman Vermillion, correspondence and author's interview.

931   Pat Ryan's return home: William J. Ryan, author's interview.

931   David Willcox's January 1954 visit to Joyce Shea's home in New Milford, New Jersey: Joyce Himka, author's interview; David R. Willcox, author's interview; Read to Willcox letter, 23 December 1953, courtesy of David R. Willcox.

932   October 28 work party in the DMZ to recover remains of American soldiers killed on Pork Chop Hill: Danny L. Peters, author's interview.

933-934   Jeremiah Ford to Mr. and Mrs. Inman, February 23, 1954: Courtesy of Mary Jo Inman Vermillion.

934   Major General John A. Klein to Mrs. Shea, 21 April 1954: Courtesy of Joyce Himka.

935   Colonel John D. Martz to Mrs. Shea: *Ibid.*

935-937   Lieutenant Colonel B. M. Read to Mrs. Shea: *Ibid.*

937   Dick Shea's funeral: Joyce Himka, author's interview.

937   Posthumous presentation of Dick Inman's Silver Star to Barbara Inman: *Vincennes Sun Commercial*, July 7, 1954; Barbara Colby, author's interview.

937   Posthumous presentation of Dick Shea's Medal of Honor to Joyce Shea: Joyce Himka, author's interview; Citation accompanying Dick Shea's award.

950-952   During the Korean war... "...Where did we get such men?": Michener, *The Bridges of Toko-Ri*, 93.

953-955   *This Is My Song*, excerpts from poem by Richard G. Inman: Courtesy of Mary Jo Inman Vermillion.

1064   USMA graduate casualties in the Korean War: *1973 Register of Graduates and Former Cadets*, 821-823.

APPENDIX I

HISTORY OF UNITED STATES MILITARY ACADEMY CLASSES 1950-1957

AND

GRADUATES WHO WERE KILLED OR MISSING IN ACTION OR DIED OF BATTLE WOUNDS IN THE KOREAN WAR

UNITED STATES MILITARY ACADEMY CLASSES 1950-57
Class Entry, Graduation, and Commissioning Data

CLASSES	50	51	52	53	54	55	56	57
TOTAL ENTERED*	915	639	678	697	797	649	670	730
GRADUATED	670	475	527	512	633	470	479	546
COMMISSIONED								
•INFANTRY	218	156	174	166	187	143	158	146
•ARMOR	36	39	37	46	33	36	40	
•CAVALRY	50							
•FIELD ARTILLERY	83	87	98	98	119	85	91	131
•COAST ARTILLERY	43							
•AIR FORCE	167	116	126	127	177	140	119	135
•ENGINEERS	68	47	55	49	60	41	46	51
•SIGNAL	35	25	28	26	33	22	25	38
TOTALS								
COMMISSIONED	664	468	520	503	622	464	475	541
•Foreign Cadets			4	5	3	2	1	4
•Not Commissioned (Honorably)		3	2	3	5	4	2	1
•Discharged (Physical Disability)								
•Not Comm. (Misc. Reasons)	6	4	1	1	3		1	

*Includes men turned back from prior classes
SOURCES: SUPERINTENDENTS' ANNUAL REPORTS FOR 1950-1957

CLASSES	50	51	52	53	54	55	56	57	TOTALS
GENERAL	7	3		3	4	1	3	2	23
LT. GENERAL	11	8	4	4	10	4	9	5	55
MAJOR GENERAL	23	15	17	12	16	14	14	13	124
BRIGADIER GENERAL	23	9	16	10	9	6	14	10	97
TOTALS	64	35	37	29	39	25	40	30	299

GENERALS	CLASS	NOTES
Fidel V. Ramos	1950	Chief of Staff, Philippine Armed Forces, 1986-92 President, Philippines, 1992-98
Paul F. Gorman, Jr.	1950	
John A. Wickham, Jr.	1950	Chief of Staff, United States Army, 1983-87
Wallace H. Nutting	1950	
Charles A. Gabriel	1950	Chief of Staff, United States Air Force, 1982-86
Bennie L. Davis	1950	
Volney F. Warner	1950	
Edward C. Meyer	1951	Chief of Staff, United States Army, 1979-83
William R. Richardson	1951	
Roscoe Robinson, Jr.	1951	
Glenn K. Otis	1953	
Arthur E. Brown, Jr.	1953	
Jerome F. O'Malley	1953	
Joseph P. Palastra, Jr.	1954	
Louis C. Wagner, Jr.	1954	
James E. Dalton	1954	
John R. Galvin	1954	
Frederick F. Woerner, Jr.	1955	
H. Norman Schwarzkopf	1956	
John W. Foss, 2d	1956	
John A. Shaud	1956	
Donald J. Kutyna	1957	
Carl E. Vuono	1957	Chief of Staff, United States Army, 1987-91

SOURCE: ASSOCIATION OF GRADUATES, 28 NOVEMBER 1995.

ASTRONAUTS	CLASS
Frank Borman	1950
Edwin E. Aldrin, Jr.	1951
Edward H. White II	1952
Michael Collins	1952
David R. Scott	1954
Alfred M. Worden	1955
Donald H. Peterson	1955

SOURCE: "1990 REGISTER OF GRADUATES AND FORMER CADETS"

RHODES SCHOLARS CLASS

James M. Thompson	1950
Andrew C. Remson, Jr.	1951
Charles R. Wallis	1952
Ames S. Albro, Jr.	1954
John C. Bard	1954
Dale A. Vesser	1954
Lee D. Olvey	1955
Harvey A. Garn	1955
John T. Hamilton	1955
Martin C. McGuire	1955
B. Conn Anderson, Jr.	1956
Richard D. Sylvester	1956
James R. Murphy	1957

SOURCE: "1990 REGISTER OF GRADUATES AND FORMER CADETS"

CONGRESSIONAL MEDAL OF HONOR CLASS

Richard T. Shea, Jr.	Korea	1952
Andre C. Lucas	Vietnam	1954

SOURCE: "1974 REGISTER OF GRADUATES AND FORMER CADETS"

KOREAN WAR CASUALTIES

CLASSES	50	51	52	53	54	55	56	57	TOTALS
WOUNDED	84	47	13						144
BATTLE DEATHS	34	9	5						48
NON-BATTLE DEATHS	17	4	3						24
TOTALS	135	60	21						216

NOTE: In Korea there were 157 total battle deaths from 21 graduated classes.

VIETNAM WAR CASUALTIES

CLASSES	50	51	52	53	54	55	56	57	TOTALS
BATTLE DEATHS*	3	2	5	9	9	3	9	10	50**

KOREA AND VIETNAM

CLASSES	50	51	52	53	54	55	56	57	TOTALS
BATTLE DEATHS*	37	11	10	9	9	3	9	10	98**

*NOTE: Vietnam data on wounded and non-battle deaths not available. Battle deaths include killed in action or died of wounds.

**Note: In Vietnam there were 278 total battle deaths from 29 graduated classes.

SOURCE: "1973 REGISTER OF GRADUATES AND FORMER CADETS"

UNITED STATES MILITARY ACADEMY GRADUATES WHO WERE KILLED OR MISSING IN ACTION OR DIED OF BATTLE WOUNDS IN THE KOREAN WAR

They shall not grow old as we that are left grow old Age shall not weary them nor the years condemn.

At the going down of the sun and in the morning, We will remember them.
From the tomb of the Unknown Soldier in Edinburg, Scotland

NAME	CLASS	NAME	CLASS
GEN Walton H. Walker	1912	1LT Jared W. Morrow	1945
MG Bryant E. Moore	Aug '17	CPT Donald E. Myers	"
COL John R. Lovell	1927	CPT Edmund D. Poston	"
COL Frank L. Forney	1929	CPT Charles W. Pratt	"
COL Allan D. MacClean	1930	1LT Clarence V. Slack, Jr.	"
LTC William H. Isbell, Jr.	1931	CPT Robert E. Spragins	"
COL Thomas B. Hall	1933	CPT Robert I. Starr	"
LTC James L. McBride, Jr.	1939	CPT James L. Treester	"
MAJ Donald L. Driscoll	1941	CPT Arthur H. Truxes, Jr.	"
MAJ Frank B. Howze	"	1LT Dirck deR. Westervelt	"
MAJ William T. McDaniel	"	CPT Peter J. Arend	1946
MAJ John E. Roberts	Jan '43	1LT James M. Becker	"
MAJ Boone Seegers	"	CPT Frank A. Doyle	"
MAJ William J. Greene	June '43	CPT Loren G. Dubois	"
MAJ William P. Hunt, Jr.	"	1LT John N. Munkres	"
CPT John H. Nelson	"	CPT Fred B. Rountree	"
LTC Lewis F. Webster	"	1LT Carl P. Schmidt	"
CPT Dean G. Crowell	1944	1LT Roland W. Skilton	"
CAPT George A. Davis, Jr.	"	1LT David B. Spellman	"
CPT Louis W. Howe	"	CPT Frank B. Tucker	"
CPT James T. Milam	"	CPT Marshall McD. Williams III	"
CPT John A. Bruckner	1945	1LT Jerome B. Christine	1947
1LT Taylor K. Castlen	"	1LT Robert B. Coleman	"
1LT William B. Crary	"	1LT Stanley W. Crosby, Jr.	"
1LT Milton H. De Vault	"	1LT Robert M. Garvin	"
1LT Ralph A. Ellis, Jr.	"	1LT David W. Gibson	"
CPT William J. Glunz	"	1LT Frederick G. Hudson III	"
1LT Alfred H. Herman, II	"	1LT Leon J. Jacques, Jr.	"
1LT Robert M. Horan	"	1LT Henry T. MacGill	"
1LT John H. Jones	"	CPT LeRoy E. Majeske	"
1LT Thomas A. Lombardo	"	CPT Lee G. Schlegel	"
CPT Raymond J. McCarrell	"	1LT Gordon M. Strong	"
CPT Edward R. McElroy	"	1LT Louis L. Anthis	1948
CPT Harry R. Middleton	"	1LT David W. Armstrong	"

NAME	CLASS	NAME	CLASS
1LT Raymond U. Bloom	1948	1LT Medon A. Bitzer	1949
CPT Charles E. Coons	"	1LT Warner T. Bonfoey, Jr.	"
1LT Raymond C. Drury, Jr.	"	1LT Thomas W. Boydston	1950
1LT Patteson Gilliam	"	2LT Howard G. Brown	"
1LT Rufus J. Hyman	"	2LT Willard H. Coates	"
1LT Charles F. McGee	"	2LT Frank P. Christensen, Jr.	"
1LT John M. Nelson	"	1LT Gene A. Dennis	"
1LT William T. O'Connell, Jr.	"	1LT George B. Eichelberger, Jr.	"
1LT Tenney K. Ross	"	1LT Charles K. Farabaugh	"
1LT Richard J. Sequin	"	1LT George E. Foster	"
1LT James A. Van Fleet, Jr.	"	1LT John H. Green	"
1LT Richard L. Warren	"	2LT Carter B. Hagler	"
1LT John E. Watkins	"	2LT George E. Hannan	"
1LT Edward A. White	"	2LT Edmund J. Lilly, III	"
1LT David P. Barnes	1949	2LT Warren C. Littlefield	"
1LT Ralph M. Buffington	"	2LT Frank R. Lloyd, Jr.	"
1LT William D. Bush, Jr.	"	1LT John M. McAlpine	"
1LT Samuel S. Coursen	"	1LT James D. Michel	"
1LT Bernard Cummings, Jr.	"	1LT Peter H. Mon fore	"
2LT Courtenay C. Davis, Jr.	"	1LT William F. Nelson	"
2LT Frederic N. Eaton	"	1LT Stanley D. Osborne	"
2LT Roger L. Fife	"	2LT William E. Otis, Jr.	"
1LT Joseph A. Giddings, Jr.	"	1LT James R. Pierce, Jr.	"
1LT Thomas G. Hardaway	"	1LT Robert W. Robinson	"
1LT William S. Kempen, Jr.	"	2LT Harry E. Rushing	"
1LT Leslie W. Kirkpatrick	"	1LT William B. Slade	"
2LT Roger R. Kuhlman	"	1LT Kenneth A. Tackus	"
1LT Munro Magruder	"	2LT John C. Trent	"
2LT Herbert E. Marshburn, Jr.	"	1LT John L. Weaver	"
1LT Wilbur J. Mueller	"	1LT Warren Webster III	"
2LT Cecil E. Newman, Jr.	"	1LT Roland E. Cooper	1951
2LT Fenton McG. Odell	"	1LT Maynard B. Johnson	"
1LT Jerome J. Paden	"	1LT Samuel A. Lutterloh	"
1LT William R. Penington	"	2LT Richard R. McCullough	"
1LT John J. Ragucci	"	1LT Edward J. Mueller, Jr.	"
1LT Robert B. Ritchie	"	1LT Robert F. Niemann	"
1LT Floyd A. Stephenson, Jr.	"	1LT Kenneth V. Riley, Jr.	"
2LT George W. Tow	"	1LT Louis J. Storck	"
1LT William McC. Wadsworth	"	1LT John R. Wasson	"
2LT Harry W. Ware, Jr.	"	2LT Thompson Cummings	1952
2LT William H. Wilbur, Jr.	"	2LT Richard G. Inman	"
2LT Courtenay L. Barrett, Jr.	1950	1LT Karl G. Koenig, Jr.	"
1LT John O. Bates, Jr.	"	2LT Kennis E. Lockard, Jr.	"
2LT Thurston R. Baxter	"	1LT Richard T. Shea, Jr.	"

APPENDIX II

PRINCIPLES OF WAR

AND

MAXIMS OF LEADERSHIP

STUDIED AT THE UNITED STATES MILITARY ACADEMY

1950-1953

PRINCIPLES OF WAR

The following is an excerpt from a pamphlet titled, *Notes for the Course in the History of Military Art*, used by cadets in the course taught at the United States Military Academy in the period of this story.

THE EXERCISE OF COMMAND

1. The Principles of War. – The principles of war are fundamental truths governing the prosecution of war. The application of those principles to the planning for and direction of war is called *strategy*; their application on the battlefield is called *tactics*. They constitute a guide, and their use in specific circumstances demands sound judgment and common sense. Successive developments in weapons and techniques of waging war have continually influenced the application of principles, but basically they are as true today as they have been throughout the history of the art of war.

    a. *The Objective. – Direct all efforts toward a decisive, obtainable goal.* The ultimate objective of all military operations is the destruction of the enemy's armed forces and his will to fight. The selection of intermediate objectives whose attainment contributes most decisively and quickly to the accomplishment of the ultimate objective at the least cost, human and material, must be based on as complete knowledge of the enemy and theater of operations as is possible for the commander to gain by the exploitation of all sources and means of information available to him.

    b. *Simplicity. – Prepare uncomplicated plans and concise orders to ensure thorough understanding and execution.* Plans should be as simple

1067

and direct as the attainment of the objective will permit. Simplicity of plans must be emphasized, for in operations even the most simple plan is usually difficult to execute. The final test of a plan is its execution; this must be borne constantly in mind during planning.

c. Unity of Command. – For every task there should be unity of effort under one responsible commander. Unity of command obtains the unity of effort which is essential to the decisive application of the full combat power of the available forces. Unity of effort is furthered by full cooperation between elements of the command. Command of a force of joint or combined arms is vested in the senior officer present eligible to exercise command unless another is specifically designated to command.

d. The Offensive. – Seize, retain, and exploit the initiative. Through offensive action, a commander preserves his freedom of action and imposes his will on the enemy. The selection by the commander of the right time and the right place for offensive action is a decisive factor in the success of the operation. A defensive attitude may be forced on a commander by many situations; but a defensive attitude should be deliberately adopted only as a temporary expedient while awaiting an opportunity for counteroffensive action, or for the purpose of economizing forces on a front where a decision is not sought.

e. Maneuver. – Position your military resources to favor the accomplishment of your mission. Maneuver in itself can produce no decisive results, but if properly employed it makes decisive results possible through the application of the principles of the offensive, mass, economy of forces, and surprise. Better armament and equipment, more effective fire, higher morale, and better leadership, coupled with skillful maneuver, will frequently overcome hostile superior numbers.

f. Mass. – Achieve military superiority at the decisive place and time. Mass, or the concentration of superior forces, on the ground, at sea, and in the air, at the decisive place and time, and their employment in a decisive direction, creates conditions essential to victory. Such concentration requires strict economy in the strength of forces assigned secondary missions. Detachments during combat are justifiable only when the execution of tasks assigned them contribute directly to success in the main battle.

g. Economy of Force. – Allocate to secondary efforts minimum essential combat power. The principle of economy of force is a corollary to the principle of mass. In order to concentrate superior combat strength in one place, economy of forces must be exercised in other places. The situation will frequently permit a strategically defensive mission to be effectively executed through offensive action.

h. Surprise. – Accomplish your purpose before the enemy can effectively react. Surprise must be sought throughout the action by every means and by every echelon of command. Surprise may be produced by measures which deny information to the enemy or deceive him as to our dispositions, movements, and plans; by variation in the means and methods employed in combat; by rapidity and power of execution; and by the utilization of terrain which appears to impose great difficulties. Surprise may compensate for numerical inferiority.

i. Security. – Never permit the enemy to acquire an unpredicted advantage. Adequate security against surprise requires a correct estimate of enemy capabilities, resultant security measures, effective reconnaissance, and readiness for action. Every unit takes the necessary measures for its own local ground and air security. Provision for security of flanks and rear is of special importance.

2. Leadership.– *Leadership is the art of influencing and directing people to an assigned goal in such a manner as to command their obedience, confidence, respect, and loyal cooperation.*

The first demand in war is decisive action. Commanders inspire confidence in their subordinates by their decisive conduct and their ability to gain material advantage over the enemy.

Man is a fundamental instrument in war; other instruments may change, but he remains relatively constant. War severely tests the physical endurance and moral stamina of the individual soldier. Strong men, inoculated with a proper sense of duty, a conscious pride in unit, and a feeling of mutual obligation to their comrades in the group, can dominate the demoralizing influences of battle far better than those imbued only with fear of punishment or disgrace. Patriotism and loyalty coupled with knowledge of and a firm belief in the principles for which war is being fought are essential.

A leader must have superior knowledge, will power, moral and physical courage, self-confidence, initiative, resourcefulness, force, and selflessness. A bold and determined leader will carry his troops with him no matter how difficult the enterprise, aware always of the great responsibility imposed upon him.

The combat value of a unit is determined in great measure by the soldierly qualities of its leaders and members, and by its will to fight. Superior combat value will offset numerical inferiority. Superior leadership combined with superior combat value of troops equipped with superior combat weapons constitutes a sure basis for success in battle.

MAXIMS OF LEADERSHIP

The following is an excerpt from a 1950 instructional aid titled, *An Instructor's Guide to Leadership Standards and Methods in the Service,* used in a leadership course taught at the United States Military Academy in the period of this story.

Maxim A   "TAKE RESPONSIBILITY FOR YOUR ACTIONS, REGARDLESS OF THEIR OUTCOME"

Maxim B   "SET THE EXAMPLE"

Maxim C   "KNOW YOURSELF AND SEEK SELF IMPROVEMENT"

Maxim D   "DEVELOP A SENSE OF RESPONSIBILITY AMONG SUBORDINATES"

Maxim E   "SEE THAT THE TASK IS UNDERSTOOD, SUPERVISE AND FOLLOW THROUGH TO SEE THAT IT IS CARRIED OUT"

Maxim F   "KNOW YOUR MEN AND LOOK OUT FOR THEIR WELFARE"

Maxim G   "KEEP YOUR MEN INFORMED"

Maxim H   "TRAIN YOUR MEN AS A TEAM"

Maxim I   "EMPLOY YOUR COMMAND IN ACCORDANCE WITH ITS CAPABILITIES"

Maxim J   "KNOW YOUR JOB"

NOTE: During the Korean War the majority of soldiers in the United States Army were draftees.

APPENDIX III

ACADEMIC CURRICULUM

INTERCOLLEGIATE ATHLETICS

INTRAMURAL ATHLETICS

AND ACTIVITIES

AT THE UNITED STATES MILITARY ACADEMY – 1950-1953

PROGRAM OF INSTRUCTION FOR ACADEMIC YEAR 1950-51

Class	Subject	Attendance	Length of period (minutes)
FOURTH (Freshman year).	Mathematics.................	Whole class daily...........	80
	Military topography and graphics.	One-half class daily except Saturday.	120
	Physical Education.......	One-half class daily except Saturday.	45
		Whole class Saturday....	45
	English........................	One-half class daily except Saturday.	60
	Languages...................	One-half class daily except Saturday.	60
THIRD (Sophomore year).	Mathematics.................	One-half class daily.......	80
	Physics........................	One-half class daily.......	80
	Chemistry....................	One-half class daily (91 periods).	80
	Languages...................	One-half class daily.......	70
	English........................	One-half class daily except Saturday (63 periods).	60
	Military psychology and leadership.	One-half class daily except Saturday (27 periods).	60 or 120
	Military topography and graphics.	One-half class daily except Saturday.	60 or 120
	Military Hygiene............	One-half class daily except Saturday	60

Class	Subject	Attendance	Length of period (minutes)
SECOND (Junior year).	Mechanics of fluids.......	One-half class daily.......	80
	Mechanics of solids......	One-half class daily.......	80
	Electricity....................	Whole class daily (158 periods).	80
		One-half class daily (27 periods).	80
	Military instructor training.	One-half class daily (23 periods).	80
	Military correspond- ence.	One-half class daily (4 periods).	80
	Social sciences (geog- graphy, government, and history).	Whole class daily except Saturday.	60
FIRST (Senior year).	Military Engineering	One-half class daily.......	80
	History of Military Art..........................	One-half class daily.......	80
	Social Sciences (economics and international relations)................	One-half class daily.......	70
	Ordnance....................	One-half class daily (94 periods)	70
	English.......................	One-half class daily (27 periods)	60
	Military hygiene.............	One-half class daily (12 periods)	60
	Tactics.......................	One-half class daily (10 periods)	60
	Law............................	One-half class daily except Saturday.	60
	Military psychology and leadership...............	One-half class daily except Saturday.	60
After 3 p. m. (all classes)	Tactics.......................	2 attendances a week...	60
	Physical education........	2 attendances a week... (36 periods)	75

NOTE: The average number of periods available for the courses prescribed in this table are as follows:

Whole class daily 212
Half class daily 106
Half class daily (except Sat.) 90
Two attendances a week.....72

SOURCE: Admissions Catalog for 1950-51

INTERCOLLEGIATE ATHLETICS

FALL	WINTER	SPRING
Football	Basketball	Baseball
Soccer	Indoor Track	Lacrosse
Cross Country	Boxing	Track
	Wrestling	Tennis
	Swimming	Golf
	Gymnastics	
	Fencing	
	Hockey	
	Rifle	
	Squash	

Teams representing West Point take part during the academic year in about 300 "at home" and 70 "away" contests.

Intercollegiate athletics at West Point are supported entirely by the Army Athletic Association. No appropriated funds were used. The Army Athletic Association contributed to the support of other cadet activities and organizations.

The athletic plant includes the Field House, Michie Football Stadium, Smith Rink, a golf course, four gymnasium buildings, 23 tennis courts, and several athletic fields.

SOURCE: 1950-51 ADMISSIONS CATALOG

INTRAMURAL ATHLETICS

Intramural Athletics: Intramural athletics at West Point are a specific part of the physical education program. With the exception of the voluntary winter intramural program they are compulsory for all cadets not currently members of intercollegiate squads. Their purpose is to provide a broad experience in sports competition and, in addition, for First Classmen experience in organization, coaching, and officiating in competitive sports.

Each cadet company provides a team in each sport, the company program being organized by a cadet athletic director. Teams compete twice weekly during each season under the leadership of trained First Classmen. Strict eligibility requirements permit cadets to participate only one season in a sport, prohibit cadets with intercollegiate squad experience from playing the sport in which they have been so trained, and require that each cadet on a squad participate for a specified length of time in each team contest or in a specified number of events or matches in individual contests.

All sports equipment, including uniforms and team supplies, is furnished to cadets from the Intramural Storeroom.

The following is the annual schedule of intramural athletics:

Fall: 20 attendances (compulsory) in football, golf, lacrosse, or track;

Winter: 16 attendances (voluntary) in basketball, boxing, handball, squash, swimming, wrestling, or volleyball;

Spring: 16 attendances (compulsory) in cross country, golf, softball, tennis, soccer, or water polo;

Summer: 20 attendances (compulsory to Third Class only) in basketball, softball, skeet, swimming, tennis, football, volleyball, golf, or canoeing.

SOURCE: 1950-51 ADMISSIONS CATALOG.

PHYSICAL EDUCATION
Fourth Class

a. *Summer.*

(1) Conditioning exercises. *7 hours.*

(2) Athletics, including water polo, speedball, softball, swimming, touch football, and volleyball. *30 hours.*

b. *Academic Year.* Instructional classes for developing basic physical and recreational skills. Twenty-four lessons in each subject: boxing, gymnastics, swimming, and wrestling. *92 hours (one hundred and twenty-three 45-minute periods.)*

In the spring, instruction is held out of doors. Eleven attendances each are required in golf and tennis.

Third Class

The development of advanced physical skills and the enlargement of the repertory of individual sports. Volleyball and basketball; swimming or squash and handball. *30 hours.*

Second Class

Instructor training in preparation for leading an army physical education program. The command voice, physiology of exercise, leadership of conditioning exercises; and coaching techniques in the following sports: basketball, boxing, water polo, cross country, golf, football, lacrosse, track, swimming, softball, soccer, and wrestling. The study of coaching techniques prepares the second classmen for their duties as intramural coaches and officials during their First Class year. *19 hours.*

First Class

Conferences on programs and schedules, tournaments and meets, physiology of exercise, procurement and care of athletic equipment, physical education, testing administration, principles and methods of athletic leadership, and field problems. Designed to prepare for administration of army athletic programs. *9 hours.*

SOURCE: 1950-51 ADMISSIONS CATALOG.

ACTIVITIES AND SOCIAL LIFE

Contrary to popular opinion, the cadets do not spend all their time parading, shining shoes, and studying; they also enjoy practically the same recreational activities as any college students.

During the summer there is swimming in Delafield Pond. Picturesque Flirtation Walk, winding for three quarters of a mile along the majestic Hudson offers a peaceful and shady retreat from the walls of barracks. Cadets stationed at Camp Buckner enjoy swimming, canoeing, fishing, and sailing on Lake Popolopen. There are two or more football trips to New York City and Philadelphia, where the bright lights of the cities are a welcome diversion. During the winter months, ice skating at Smith Rink and skiing on the Constant Slope are extremely popular Weekly hops are held in either the gymnasium or Cullum Hall, with music furnished by the Cadet Dance Band or one of the two Post orchestras. Cadets may attend movies in the Army Theater on Saturday nights, Sundays, and holidays. Frequently, outside talent, sponsored by the cadet special program committee, is brought to the Post for Sunday evening performances.

Aside from the general recreation activities, there are many organized extracurricular activities. Standing cadet committees performing functions of great importance include the Honor Committee, whose job it is to interpret and apply the honor system; and the Hop Committee, which plans and schedules social functions.

Young men with a musical bent are encouraged to indulge their talents. Cadet Protestant, Catholic, and Jewish choirs sing at religious services on the Post and usually make several trips each year to sing in New York and Washington. For those who prefer a more formal type of music, there are the Cadet Glee Club and the Cadet Dance Band.

As members of the Radio, Model Airplane, Model Railroad, and Camera clubs, technically minded hobbyists will find pleasure and opportunities to test their skill.

Those to whom literary activities appeal may seek outlets for their talents in The Howitzer, yearbook of the Corps of Cadets; The Pointer, official magazine of the Corps of Cadets; and Bugle Notes, the cadet handbook, more commonly known as the "Plebe Bible". Cadet press representatives conduct interviews and prepare hundreds of releases for hometown newspapers.

Recently organized clubs corresponding to each foreign language taught in the regular curriculum: Portuguese, Spanish, German, Russian, and French, where members, in addition to devoting further study to the language, seek to broaden their knowledge of the various countries and peoples.

The One Hundredth Night Show, the time-honored dramatic highlight presented annually by the Dialectic Society, is written, produced, and enacted solely by cadets.

The most active organization is the West Point debate council. In the course of an academic year its members engage in some 50 intercollegiate debates, traveling to all parts of the United States. In the spring of 1947 the efforts of the West Point debaters to organize and stage the first truly national intercollegiate debating tournament culminated with their playing host to representatives of 29 colleges, victors in regional tournaments.

Organized extracurricular activities are directed and administered almost entirely by the cadets themselves subject to the approval of the Superintendent. There is an officer in charge of each activity, who acts in an advisory capacity. From these activities cadets acquire a wealth of specialized knowledge or develop latent talent, which subsequently will serve them well and be a source of pleasure and relaxation in their careers as officers.

SOURCE: 1950-51 ADMISSIONS CATALOG.

APPENDIX IV

ARMY FOOTBALL

1950-1953

AND

EARL BLAIK'S ANNUAL GAME RECORDS

EARL H. BLAIK'S ANNUAL GAME RECORDS

From 1927 through 1933 Earl H. Blaik was assistant coach at West Point; from 1934 through 1940 head coach at Dartmouth; from 1941 through 1958 head coach at West Point.

ARMY		OPPONENT			ARMY	OPPONENT
					1929	
				26	Boston U.	0
	1927			33	Gettysburg	7
13	Boston U.	0		23	Davidson	7
6	Detroit	12		20	Harvard	20
21	Marquette	12		13	Yale	21
27	Davis & Elkins	6		33	South Dakota	6
6	Yale	10		7	Illinois	17
34	Bucknell	0		89	Dickinson	7
45	Franklin & Marshall	0		19	Ohio Wesleyan	6
18	Notre Dame	0		0	Notre Dame	7
13	Ursinus	0		13	Stanford	34
14	Navy	9			**1930**	
	1928			39	Boston U.	0
35	Boston U.	0		54	Furman	0
14	Southern Methodist	13		39	Swarthmore	0
44	Providence	0		6	Harvard	0
15	Harvard	0		7	Yale	7
18	Yale	6		33	North Dakota	6
38	De Pauw	12		13	Illinois	0
6	Notre Dame	12		47	Kentucky Wesleyan	2
32	Carleton	7		18	Ursinus	0
13	Nebraska	3		6	Notre Dame	7
0	Stanford	26		6	Navy	0

ARMY	OPPONENT	
1931		
60	Ohio Northern	0
67	Knox College	6
20	Michigan State	7
13	Harvard	14
6	Yale	6
27	Colorado College	0
20	LSU	0
0	Pittsburgh	26
54	Ursinus	0
12	Notre Dame	0
17	Navy	7
1932		
13	Furman U.	0
57	Carleton	0
13	Pittsburgh	18
20	Yale	0
33	William & Mary	0
46	Harvard	0
52	North Dakota State	0
7	West Virginia Wesleyan	0
0	Notre Dame	21
20	Navy	0
1933		
19	Mercer	6
32	VMI	0
52	Delaware	0
6	Illinois	0
21	Yale	0
34	Coe College	0
27	Harvard	0
12	Penn. Military College	0
12	Navy	7
12	Notre Dame	13

DARTMOUTH	OPPONENT	
1934		
39	Norwich	0
32	Vermont	0
27	Maine	0
27	Virginia	0
10	Harvard	0
2	Yale	7
21	New Hampshire	7

6	Cornell	21
13	Princeton	38
1935		
39	Norwich	0
47	Vermont	0
59	Bates	7
41	Brown	0
14	Harvard	6
14	Yale	6
34	William & Mary	0
41	Cornell	6
6	Princeton	26
7	Columbia	13
1936		
58	Norwich	0
56	Vermont	0
0	Holy Cross	7
34	Brown	0
26	Harvard	7
11	Yale	7
20	Columbia	13
20	Cornell	6
13	Princeton	13
Ivy League Champions		
1937		
39	Bates	0
31	Springfield	0
42	Holy Cross	7
41	Brown	0
20	Harvard	2
9	Yale	9
33	Princeton	9
6	Cornell	6
27	Columbia	0
Ivy League Champions		
Ranked No. 7 nationally		
1938		
46	Bates	0
51	St. Lawrence	0
22	Princeton	0
34	Brown	13
13	Harvard	7
24	Yale	6
44	Dickinson	6
7	Cornell	14

DARTMOUTH		OPPONENT
13	Stanford	23

1939

41	St. Lawrence	9
34	Hamp. Sydney	6
0	Navy	0
14	Lafayette	0
16	Harvard	0
33	Yale	0
7	Princeton	9
6	Cornell	35
3	Stanford	14

1940

35	St. Lawrence	0
21	Franklin & Marshall	23
6	Columbia	20
7	Yale	13
7	Harvard	6
26	Sewanee	0
9	Princeton	14
3	Cornell	0
20	Brown	6

ARMY		OPPONENT

1941

19	Citadel	6
27	VMI	20
20	Yale	7
13	Columbia	0
0	Notre Dame	0
6	Harvard	20
7	Pennsylvania	14
7	West Virginia	6
6	Navy	14

1942

14	Lafayette	0
28	Cornell	8
34	Columbia	6
14	Harvard	0
0	Pennsylvania	19
0	Notre Dame	13
19	VPI	7
40	Princeton	7
0	Navy	14

1943

27	Villanova	0
42	Colgate	0
51	Temple	0
52	Columbia	0
39	Yale	7
13	Pennsylvania	13
0	Notre Dame	26
16	USNTS Sampson	7
59	Brown	0
0	Navy	13

1944

46	North Carolina	0
59	Brown	7
69	Pittsburgh	7
76	Coast Guard Academy	0
27	Duke	7
83	Villanova	0
59	Notre Dame	0
62	Pennsylvania	7
23	Navy	7

National Champions

1945

32	PDC AAF, Louisville, Ky.	0
54	Wake Forest	0
28	Michigan	7
55	MTBS, Melv'e, RI	13
48	Duke	13
54	Villanova	0
48	Notre Dame	0
61	Pennsylvania	7
32	Navy	13

National Champions

1946

35	Villanova	0
21	Oklahoma U.	7
46	Cornell	21
20	Michigan	13
48	Columbia	14
19	Duke	0
19	West Virginia	0
0	Notre Dame	0
34	Pennsylvania	7
21	Navy	18

Eastern Champions

Army & Notre Dame National Champions
Blaik chosen Coach of the Year by the Football Coaches Association of America

ARMY		OPPONENT
1947		
13	Villanova	0
47	Colorado U.	0
0	Illinois	0
40	VPI	0
20	Columbia	21
65	Washington & Lee	13
7	Notre Dame	27
7	Pennsylvania	7
21	Navy	0
1948		
28	Villanova	0
54	Lafayette	7
26	Illinois	21
20	Harvard	7
27	Cornell	6
49	VPI	7
43	Stanford	0
26	Pennsylvania	20
21	Navy	21

Ranked No. 6 nationally

	1949	
47	Davidson	7
42	Penn State	7
21	Michigan	7
54	Harvard	14
63	Columbia	6
40	VMI	14
35	Fordham	0
14	Pennsylvania	13
38	Navy	0

Eastern Champions
Ranked No. 4 nationally

	1950	
28	Colgate	0
41	Penn State	7
27	Michigan	6
49	Harvard	0

34	Columbia	0
28	Pennsylvania	13
51	New Mexico	0
7	Stanford	0
2	Navy	14

Ranked No. 2 nationally

	1951	
7	Villanova	21
14	Northwestern	20
14	Dartmouth	28
21	Harvard	22
14	Columbia	9
6	Southern California	28
27	Citadel	6
6	Pennsylvania	7
7	Navy	42
1952		
28	South Carolina	7
0	Southern California	22
37	Dartmouth	7
14	Pittsburgh	22
14	Columbia	14
42	VMI	14
6	Georgia Tech	45
14	Pennsylvania	13
0	Navy	7
1953		
41	Furman	0
20	Northwestern	33
27	Dartmouth	0
14	Duke	13
40	Columbia	7
0	Tulane	0
27	North Carolina State	7
21	Pennsylvania	14
20	Navy	7

Eastern Champions
Ranked No. 14 nationally
Blaik chosen Coach of the Year by the Touchdown Club of Washington, D.C.

	1954	
20	South Carolina	34
26	Michigan	7
60	Dartmouth	6
28	Duke	14

ARMY		OPPONENT
67	Columbia	12
21	Virginia	20
48	Yale	7
35	Pennsylvania	0
20	Navy	27

Ranked No. 7 nationally

1955

81	Furman	0
35	Penn State	6
2	Michigan	26
0	Syracuse	13
45	Columbia	0
27	Colgate	7
12	Yale	14
40	Pennsylvania	0
14	Navy	6

1956

32	VMI	12
14	Penn State	7
14	Michigan	48
0	Syracuse	7
60	Columbia	0
55	Colgate	46
34	William & Mary	6
7	Pittsburgh	20
7	Navy	7

1957

42	Nebraska	0
27	Penn State	13
21	Notre Dame	23
29	Pittsburgh	13
20	Virginia	12
53	Colgate	7
39	Utah	33
20	Tulane	14
0	Navy	14

1958

45	South Carolina	8
26	Penn State	0
14	Notre Dame	2
35	Virginia	6
14	Pittsburgh	14
68	Colgate	6
14	Rice	7

26	Villanova	0
22	Navy	6

Eastern Champions
Ranked No. 3 nationally

Overall Dartmouth record,
1934 through 1940:

Won	45
Lost	15
Tied	4
Percentage	*.750*

Overall Army record,
1941 through 1958:

Won	121
Lost	32
Tied	10
Percentage	*.790*

Overall totals, Blaik teams,
1934 through 1958:

Won	166
Lost	47
Tied	14
Percentage	*.784*

43 first team All-American players

ARMY FOOTBALL LETTERMEN AND ASSISTANT COACHES
1950-53

Number in parenthesis indicates selection to 1st, 2nd, or 3rd team All-American.
x - Did not graduate.

SEASON 1950

NAME	CLASS	YRS LTRD
Bruce A. Ackerson	1950	'48,'49,'50
Raymond M. Bara	x1952	'50
Donald A. Beck	x1952	'49, '50
Robert M. Blaik	x1952	'49, '50
Ben F. Brian	x1952	'50
James W. Cain	1952	'48, '49 (3), '50
Bruce E. Emblad	1951	'48, '49, '50
Eugene C. Filipski	x1953	'50
Frank R. Fischl	1951	'49, '50
John D. Foldberg	1951	'48 (2), '49 (1), '50 (Capt) (1)
Eugene P. Gribble	x1952	'50
Robert J. Haas	x1952	'49, '50
Gerald E. Hart	x1953	'50
Herbert L. Johnson	x1952	'49, '50
J. D. Kimmel	x1952	'49, '50 (1)
John R. Krobock	1953	'50, '52
Harold J. Loehlein	x1952	'49, '50
Raymond J. Malavasi	x1953	'50
Jack W. Martin	1951	'49, '50
John E. McShulskis	x1953	'50
Alfred L. Pollard	x1953	'50 (1)
Victor J. Pollock	x1952	'49, '50
Gilbert M. Reich	x1953	'50
John Ritchie	1951	'50 (Team Mgr)
Richard J. Roberts	x1952	'49, '50
William H. Rowekamp	x1953	'50
Charles N. Shira	x1952	'49, '50
Edward J. Stahura	x1953	'50
Elmore E. Stout	x1952	49, '50 (1)
Robert L. Volonnino	x1953	'50
John E. Weaver	1954	'50, '52
Lewis R. Zeigler	x1952	'50

Total: 32

ASSISTANT COACHES

Paul J. Amen	John F. Green
Douglas E. Kenna, Jr.	Vincent T. Lombardi
Murray Warmath	

SEASON 1951

NAME	CLASS	YRS LTRD
Freddie A.D. Attaya	1954	'51, '52, '53
Thomas J. Bell	1955	'51, '52, '53,'54
Raymond O. Bergeson	1952	'51
Neil A. Chamberlin	1954	'51, '52
Donald G. Fuqua	1953	'51, '52
Theodore O. Gregory	1952	'51
Carl B. Guess	1952	'51
Robert F. Guidera	1954	'51, '52
Wallace K. Haff	1954	'51
Richard G. Inman	1952	'51
Kenneth R. Kramer	1954	'51, '52
John E. Krause	1954	'51, '52, '53
Stanley J. Kuick	1952	'51
Ronald H. Lincoln	x1954	'52, '52
Gerald A. Lodge	1954	'51, '52, '53
Leroy T. Lunn	1954	'51, '52, '53 (Capt)
William MacPhail	x1954	'51
Frederic D. Meyers	x1955	'51
Robert M. Mischak	1954	'51, '52, '53 (1)
Alfred E. Paulekas	1953	'51, '52 (Capt)
Richard J. Reich	x1955	'51
John C. Rogers	1954	'51
Myron W. Rose	1953	'51, '52
Robert L. Rutte	1952	'51 (Team Mgr)
Lowell E. Sisson	1954	'51, '52, '53
Norman F. Stephen	1954	'51, '52, '53
Frank S. Wilkerson	1953	'51, '52

Lewis A. Williams 1952 '51
John R. Wing 1955 '51, '52
Total: 29
ASSISTANT COACHES
Paul J. Amen   John F. Green
Douglas E. Kenna, Jr.   Vincent T. Lombardi
Robert J. St. Onge   Murray Warmath

SEASON 1952

NAME	CLASS	YRS LTRD
Freddie A.D. Attaya	1954	'51, '52, '53
Thomas J. Bell	1955	'51, '52, '53,'54
Richard D. Boyle	1953	'52
Neil A. Chamberlin	1954	'51, '52
Earl L. Chambers	1953	'52 (Team Mgr)
Mario L. DeLucia	x1954	'52
William A. Doremus	x1955	'52
Donald G. Fuqua	1953	'51, '52
Robert F. Guidera	1954	'51, '52
James H. Harris	1953	'52
Kenneth R. Kramer	1954	'51, '52
John E. Krause	1954	'51, '52, '53
John R. Krobock	1953	'50, '52
Ronald H. Lincoln	x1954	'52, '52
Gerald A. Lodge	1954	'51, '52, '53
Peter C. Manus	1954	'52
Leroy T. Lunn	1954	'51, '52, '53 (Capt)
John D. Meglen	1953	'52
Robert M. Mischak	1954	'51, '52, '53 (1)
Godwin Ordway III	1955	'52, '53', '54
Alfred E. Paulekas	1953	'51, '52 (Capt)
Myron W. Rose	1953	'51, '52
Paul Schweikert	1954	'52
Lowell E. Sisson	1954	'51, '52, '53
Norman F. Stephen	1954	'51, '52, '53
Peter J. Vann	1956	'52, '53, '54 (2)
John E. Weaver	1954	'50, '52
Frank S. Wilkerson	1953	'51, '52
John R. Wing	1955	'51, '52

Richard G. Ziegler 1954 '52, '53
Total: 30
ASSISTANT COACHES
Paul J. Amen   John F. Green
Douglas E. Kenna, Jr.   Laslie G. Carney
Vincent T. Lombardi

SEASON 1953

NAME	CLASS	YRS LTRD
Freddie A.D. Attaya	1954	'51, '52, '53
Thomas J. Bell	1955	'51,'52, '53, '54 (1)
Ralph J. Chesnauskas	1956	'53, '54, '55
Robert G. Farris	1956	'53
Howard G Glock	1956	'53, '54
Jerome F. Hagan	1955	'53
Donald W. Holleder	1956	'53, '54 (1), '55
John E. Krause	1954	'51, '52, '53
Joseph D. Lapchick	1954	'53
Paul A. Lasley	1956	'53
Gerald A. Lodge	1954	'51, '52, '53
Leroy T. Lunn	1954	'51, '52, '53 (Capt)
Marion F. Meador	1954	'53 (Team Mgr)
Ronald P. Melnik	x1956	'53
Robert M. Mischak	1954	'51, '52, '53 (1)
Godwin Ordway III	1955	'52, '53', '54
William P. Purdue	1954	'53
Elwin R. Shain	1954	'53
Lowell E. Sisson	1954	'51, '52, '53
Norman F. Stephen	1954	'51, '52, '53
Patrick N. Uebel	1956	'53, '54, '55 (Capt)
Peter J. Vann	1956	'52, '53, '54 (2)
Michael G. Zeigler	1956	'53, '54, '55
Richard G. Ziegler	1954	'52, '53

Total: 23
ASSISTANT COACHES
Paul J. Amen   Paul F. Dietzel
Robert L. Dobbs   Laslie G. Carney
Vincent T. Lombardi

FOOTBALL OFFENSE AND DEFENSE FORMATIONS COMMON IN COLLEGIATE FOOTBALL 1950-1953

6-3-2 DEFENSE

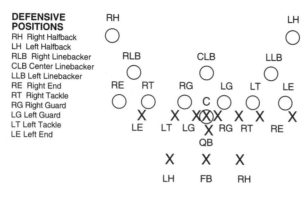

DEFENSIVE POSITIONS
RH Right Halfback
LH Left Halfback
RLB Right Linebacker
CLB Center Linebacker
LLB Left Linebacker
RE Right End
RT Right Tackle
RG Right Guard
LG Left Guard
LT Left Tackle
LE Left End

OFFENSIVE POSITIONS
RH Right Halfback
FB Fullback
LH Left Halfback
QB Quarterback
LE Left End
LT Left Tackle
LG Left Guard
C Center
RG Right Guard
RT Right Tackle
RE Right End

T-FORMATION

OFFENSE

6-3-2 DEFENSE

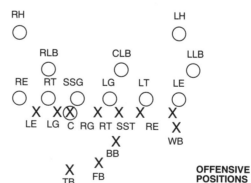

DEFENSIVE POSITIONS
RH Right Halfback
LH Left Halfback
RLB Right Linebacker
CLB Center Linebacker
LLB Left Linebacker
RE Right End
RT Right Tackle
SSG Strong Side Guard
LG Left Guard
LT Left Tackle
LE Left End

OFFENSIVE POSITIONS
WB Wingback
BB Blockingback
FB Fullback
TB Tailback
LE Left End
LG Left Guard
C Center
RG Right Guard
RT Right Tackle
SST Strong Side Tackle
RE Right End

SINGLE WING RIGHT FORMATION UNBALANCED LINE

OFFENSE

APPENDIX IV

6-3-2 DEFENSE

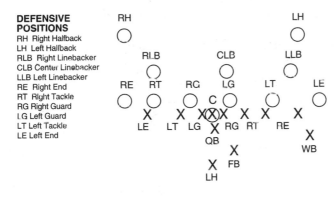

DEFENSIVE POSITIONS
RH Right Halfback
LH Left Halfback
RLB Right Linebacker
CLB Center Linebacker
LLB Left Linebacker
RE Right End
RT Right Tackle
RG Right Guard
LG Left Guard
LT Left Tackle
LE Left End

OFFENSIVE POSITIONS
WB Wingback
FB Fullback
LH Left Halfback
QB Quarterback
LE Left End
LT Left Tackle
LG Left Guard
C Center
RG Right Guard
RT Right Tackle
RE Right End

WING-T FORMATION

OFFENSE

6-2-2-1 DEFENSE

DEFENSIVE POSITIONS
RH Right Halfback
S Safety
LH Left Halfback
LLB Left Linebacker
RLB Right Linebacker
RE Right End
RT Right Tackle
RG Right Guard
LG Left Guard
LT Left Tackle
LE Left End

OFFENSIVE POSITIONS
RH Right Halfback
FB Fullback
LH Left Halfback
QB Quarterback
LE Left End
LT Left Tackle
LG Left Guard
C Center
RG Right Guard
RT Right Tackle
RE Right End

T-FORMATION

OFFENSE

1085

APPENDIX V

A GLOSSARY OF CADET SLANG

AREA BIRD, n. A cadet who is serving punishment by being obliged to walk punishment on the area.

ARMY BRAT, n. Son or daughter of an Army Officer

B-ACHE, n. An explanation of a report. A complaint.
> v. To explain a report; to complain about something.

BEAST BARRACKS, n. First summer of military training – for new cadets entering West Point.

BENO, n. A cancellation, negative report derived from the official phrase, "There will be no...." Often comes in the form of a letter from a femme, i.e. "Sorry, can't come."

B. J., a. Fresh, lacking in respect. Bold before June.

BOARD FIGHT, n. A recitation in which Cadets of a section are sent to the blackboard, and are assigned questions or problems, being given a limited time in which to answer them.

BONE, v. To study; to strive for something.
> B. CHECK BOOK, To practice economy.
> B. FILE, To strive for class standing.
> B. MAKE, To work for chevrons.
> B. MUCK, To exercise in the gymnasium for the purpose of increasing one's physical strength.
> B. REVERSE, To be regarded with disfavor.
> B. TENTHS, To work for good marks in academic duties.

BOODLE, n. Cake, candy, ice cream, etc.

BOODLE FIGHT, n. A gathering at which boodle is served.

BOODLER'S, n. The Cadet Restaurant.

BRACE, n. The correct military carriage for a plebe.
> v. To take up a military position; to correct a plebe's posture.

BUCK, n. A cadet private.
> v. To work against, to oppose.

BUST, v. To revoke the appointment of a Cadet commissioned or non-commissioned officer.

BUTT, n. The remains of anything, as the butt of a month, the butt of a cigarette.

CIVVIES, n. Civilian clothing.

COLD, a. Absolutely without error, as "a cold max."

COLD JUG, n. One who has a sober air.

COM, n. The Commandant of Cadets.

CON, n. Confinement to quarters, as punishment for breaches of discipline.

COW, n. A member of the second class, a junior.

CRAWL, v. To correct a fourth classman; to rebuke.

D., adj. Deficient; below average, as in academics.

DEADBEAT, n. An easy time. One who believes in as little work as possible.
  v To get out of a disagreeable duty.

DEMO, n. A demerit.

DIV., n. A division of barracks.

D. P., n. A dining permit.

DRAG, n. A young lady whom a cadet is escorting. Pull with a superior.
  v. To escort a young lady. To apply water, pomade, and shoe blacking to
    any deserving cadet, such as newly made cadet officer.

D. T., v. Double time.

ENGINEER, n. One well up in his studies. A cadet in the upper sections in
  academic work.

F. D., n. Full dress uniform.

FEMME, or FEM, n. A young lady.

FIFTY-FIFTY, n. Uniform composed of dress gray coat and white trousers.

FILE, n. A person, male usually, ordinarily in the military service. A grade in
  class or military rank.

FILEBONE, v. Any act to gain class or academic standing.

FILEBONER, n. One who strives to get ahead.

FIND, v. To discharge a cadet for deficiency in studies or conduct.

FIRST CLASSMAN, n. A member of the first class, a senior.

FIRSTIE, n. A first classman.

FLANKER, n. A tall person.

FORE, interj. A warning signal, as in golf.

FOUNDATION, n. The day when the list of cadets dismissed for academic
  deficiency is published.

FOUNDLING, n. A cadet who has been dismissed.

FOURTH CLASSMAN, n. A member of the fourth class, a freshman, a plebe.

FRIED EGG, n. Insignia of the U.S.M.A. worn on the headpiece.

G.I., a. General issue, or government issue.

GIG, n. See quill.

  v. See quill.

GNOME (G-nomie), n. Member of a runt company. See "runt."

GOAT, n. A man in the lower academic sections. A man near the bottom of his class.

GOOSE, n. The Portuguese language.

GRIND, n. A joke.

  v. To laugh or smirk.

GROSS, a. Disreputable. Unacceptable. Offensive.

HELL CATS, n. Orderlies; musicians who sound reveille and the calls.

HIVE, n. An intelligent person or one who learns quickly.

  v. To understand, to comprehend.

HIVEY, a. Bright in academic work. Quick to learn.

JUICE, n. Electricity.

KAYDET, n. A Cadet.

LIMITS, n. The limits on the reservation, to which cadets are restricted.

MAKE, n. Cadet officer or non-commissioned officer.

  v. To appoint a Cadet as officer or non-commissioned officer in the Corps of Cadets.

MAX, n. A complete success; a maximum.

  v. To make a 3.0 in academic recitation; to do a thing perfectly.

MEAT WAGON, n. The ambulance.

MISSOURI NATIONAL, n. A tune supposed to bring rain.

MUCK, n. Muscle, brawn, physical strength.

  v. To strain at physical work.

O.A.O., n. One and only. HER.

O.C., n. Officer in charge. Overcoat.

O.D., n. Officer of the Day (Cadet).

O.G., n. Officer of the Guard (Cadet).

OID, Suffix denoting agent or doer, as; sluggoid, hopoid, spoonoid, specoid, etc.

P., n. A professor or instructor.

P.C.S., n. Previous conditions of servitude. Occupation before entering the U.S.M.A.

P.D., n. Police detail.

PLEBE, n. A cadet of the fourth class, a freshman.

PLEBE BIBLE, n. "Bugle Notes." The Handbook of the Corps of Cadets.

PLEBE SKINS, n. First issue gray flannel trousers; gymnasium trousers.

PODUNK, n. A cadet's home town. The newspaper thereof.

POLICE, v. To throw away, to discard. To be thrown from a horse. To clean up.

POLICING, n. A general transfer in academic sections. A spill at riding.

POOP, n. Information to be memorized.

    v. To become tired from physical exertion.

POOP-DECK, n. The balconies on cadet headquarters, where the O.C. watches formations. Also the balcony in the dining hall where the O.C. eats and from which orders are published.

POOP-SHEET, n. Page of information.

P-RADE, n. A parade.

PRO, adj. Proficient, above passing in studies or looks.

QUILL, n. A report for a delinquency.

    v. To report a cadet for a breach of regulations.

QUILL BOOK, n. Company delinquency book.

QUILL SHEET, n. Company delinquency sheet published daily for each company.

RATRACE, n. Violent or noisy activity, as "ratracing" in the halls of barracks.

RECOGNIZE, n. To place a fourth classman on upper class status.

REVERSE, n. Disfavor.

R. H. I. P., Rank Hath Its Privileges (as well as obligations).

ROGER, I understand.

RUNT, n. Opposite of "flanker." A short person.

SACK, n. Cadet bed.

SAMMY, n. Syrup.

S. C. C., A telegram saying "Sorry can't come," usually received from a femme Saturday noon.

SECOND CLASSMAN, n. A member of the second class, a junior, a cow.

SHAVE TAIL, n. A new second lieutenant.

S. I., n. Saturday Inspection.

SKIN, n. See "quill."

    v. See "quill."

SKIN SHEET, n. Same as "quill sheet."

SLUG, n. A special punishment for a serious offense. A disagreeable duty.

    v. To impose a disagreeable duty on someone.

SLUM, n. Stew, a dish in the dining hall.

SMALL MILK, n. A small pitcher of milk, for coffee, etc.

SMALL DISH, n. A dessert dish in the dining hall.

SNAKE, n. One who will cut-in at hops.
    v. To cut-in.

SOIREE, n. A task requiring begrudged effort.
    v. To inconvenience.

SOUND OFF, n. A powerful voice.
    v. To use the voice so as to be heard, shout.

S. O. P., Standard operating procedure.

SPEC (speck), v. To memorize verbatim, as; "to spec blind."

SPIC, n. The Spanish language.

SPOON UP, v. To put in order, to clean up.

SPOONY, a. Neat in personal appearance.

STEP OUT, v. Hurry, to increase one's gait.

STORM, n. A disordered condition. (Said of things.) A nervous haste. (Said of persons.)

SUPE, n. The superintendent.

TAC, n. An officer in the Department of Tactics.

TARBUCKET, n. A full dress hat.

T. D., n. Department of Tactics.

TENTH, n. A tenth of a unit, one thirtieth of the max; the smallest division of the system of marking.

TENTH AVENUE, n. The street (there aren't nine others) running between the East and West academic buildings. Part of Thayer Road.

THIRD CLASSMAN, n. A sophomore, a yearling.

TIE UP, v. To make a gross error.

TOUR, n. One hour's walk on the area (punishment); a tour of duty, as a "guard tour."

TROU, n. Trousers.

TURKEY, n. Hash served in cadet dining hall.

TURNBACK, n. A re-admitted cadet.

WHEEL, n. High ranking cadet officer, as member of Brigade Staff, etc.

WHITE ELEPHANT, n. South Barrack's cupboard.

WIFE, n. A roommate.

WRIT, n. A written recitation, and examination.

YEARLING, n. A member of the third class, a sophomore.

SOURCE: *1951-52 Bugle Notes*

BIBLIOGRAPHY

AUTHOR'S INTERVIEWS

John R. Aker ('52)
Michael G. Arrison ('54)
Freddie A.D. Attaya ('54)
John C. Bard ('54)
George B. Bartel ('52)
Thomas J. Bell ('55)
Charles Brooks
Dale W. Cain, Sr.
Ralph J. Chesnauskas ('56)
Barbara Kipp Inman Colby
Richard C. Coleman ('52)
Ernest F. Condina ('52)
MacPherson Conner ('52)
Thomas E. Courant ('52)
John H. Craigie ('51)
Joseph F. H. Cutrona ('44)
Howard H. Danford ('52)
Louis M. Davis ('52)
Robert G. Farris ('56)
Joseph P. Franklin ('55)
Donald G. Fuqua ('53)
Emmett D. Gladwell
Howard G. Glock ('56)
Preston S. Harvill, Jr.
George L. Hendricks (X'53)
Joyce E. Shea Himka
David R. Hughes ('50)
Jefferson J. Irvin ('38)
Frederick F. Irving ('51)
Lee H. Johnson
John E. Krause ('54)
John R. Krobock ('53)
Joseph D. Lapchick, Jr. ('54)
Paul A. Lasley ('56)
Gerald A. Lodge ('54)
Leroy T. Lunn ('54)

Elizabeth Irving Maish
Thomas P. McKenna ('53)
James E. McKenzie
Richard J. Miller ('52)
Robert E. Miller
Robert M. Mischak ('54)
Edward M. Moses ('54)
Daniel J. Myers, Jr. ('51)
Brian C. Nolan ('53)
Robert Northcutt
Ronald M. Obach ('52)
William Van D. Ochs, Jr. ('45)
Godwin Ordway III ('55)
George S. Pappas ('44)
Walter G. Parks ('52)
Alfred E. Paulekas ('53)
Danny L. Peters
Lawrence A. Putnam, Sr. ('52)
Willard L. Robinson ('55)
Robert L. Royem, Jr. ('44)
William J. Ryan ('51)
Gerald J. Samos ('55)
Benjamin F. Schemmer ('54)
William R. Schulz III ('54)
Lowell E. Sisson ('54)
Norman F. Stephen ('54)
William K. Stockdale ('51)
Willis C. Tomsen ('54)
Roy T. Thorsen ('55)
Y. Arnold Tucker ('47)
Patrick N. Uebel ('56)
Peter J. Vann ('56)
Mary Jo Inman Vermillion
Emery S. Wetzel, Jr. ('54)
Frank S. Wilkerson, Jr. ('53)
David R. Willcox

BOOKS

Blaik, Earl H. *The "Red" Blaik Story*. New Rochelle, New York: Arlington House Publishers, 1974.

Blair, Clay. *The Forgotten War*. New York, New York: Doubleday, 1987.

1951-52 Bugle Notes. West Point, New York: United States Military Academy, 1951.

1922-23 Bugle Notes. West Point, New York: United States Military Academy, 1951.

Campbell, Richard M.; Painter, John D.; Straziscar, Sean W. *NCAA Football, The Official 1996 College Football Record Books*. Overland Park, Kansas: National Collegiate Athletic Association, 1996.

Cain, Dale W., Jr. Korea, *The Longest War*. Seoul, Korea: Ministry of Patriots and Veterans Affairs, 1997.

Fleming, Thomas J. *West Point, The Men and Times of the United States Military Academy*. New York, New York: William Morrow and Company, Inc., 1969.

Halberstam, David. *The Fifties*. New York, New York: Fawcett Columbine, 1993.

Knox, Donald. *The Korean War, Uncertain Victory*. San Diego, New York, London: Harcourt Brace Jovanovich, Publishers, 1988.

Lombardi, Vince. *Run to Daylight!* Englewood Cliffs, New Jersey: Prentice-Hall, Inc., 1963.

MacArthur, Douglas. *Reminiscences*. New York, New York: McGraw- Hill Book Company, 1964.

Marshall, S.L.A. *Pork Chop Hill*. Nashville, Tennessee: The Battery Press, 1986.

Michener, James A. *The Bridges at Toko-Ri*. Garden City, New Jersey: Nelson Doubleday, Inc., 1953.

Nelson, David M. *Anatomy of a Game*. Cranbury, New Jersey: Associated University Presses, 1994.

O'Brien, Michael. *Vince, A Personal Biography of Vince Lombardi*. New York, New York: William Morrow and Company, 1987.

Pappas, George S. *To the Point, The United States Military Academy, 1802-1902*. Westport, Connecticut: Praeger, 1993.

Porter, David L. (editor). *Biographical Dictionary of American Sports - Football*. Westport, Connecticut: Greenwood Press, Inc., 1987.

Reeder, Russell P. *Heroes and Leaders of West Point*. New York and Camden, New Jersey: Thomas Nelson, Inc., 1970.

Reeder, Russell P. *Memoirs of an American Soldier*. Quechee, Vermont: Vermont Heritage Press, 1996.

Sesquicentennial of the United States Military Academy. Buffalo, New York: Baker, Jones, Hausuer and Savage, Inc., 1953.

Twombly, Wells. *Shake Down the Thunder! The Official Biography of Notre Dame's Frank Leahy*. Radnor, Pennsylvania: Chilton Book Company, 1974.

54-40...and Still in the Fight. United States of America: Class of 1954, United States Military Academy, 1994.

1917 Howitzer. West Point, NY: Class of 1917, United States Corps of Cadets.

1921 Howitzer. West Point, NY: Class of 1919, United States Corps of Cadets.

1922 Howitzer. West Point, NY: Class of 1922, United States Corps of Cadets.

1927 Howitzer. West Point, NY: Class of 1927, United States Corps of Cadets.

1929 Howitzer. West Point, NY: Class of 1929, United States Corps of Cadets.

1933 Howitzer. West Point, NY: Class of 1933, United States Corps of Cadets.

1938 Howitzer. West Point, NY: Class of 1938, United States Corps of Cadets.

1950 Howitzer. West Point, NY: Class of 1950, United States Corps of Cadets.

1951 Howitzer. West Point, NY: Class of 1951, United States Corps of Cadets.

1952 Howitzer. West Point, NY: Class of 1952, United States Corps of Cadets.

1953 Howitzer. West Point, NY: Class of 1953, United States Corps of Cadets.

1954 Howitzer. West Point, NY: Class of 1954, United States Corps of Cadets.

1955 Howitzer. West Point, NY: Class of 1955, United States Corps of Cadets.

1991 Guide to Army Football. West Point, New York: USMA Sports Information Office, 1991.

1973 Register of Graduates and Former Cadets, United States Military Academy. West Point, New York: Association of Graduates, USMA.

1980 Register of Graduates and Former Cadets, United States Military Academy. West Point, New York: Association of Graduates, USMA.

1990 Register of Graduates and Former Cadets, United States Military Academy. West Point, New York: Association of Graduates, USMA.

1996 Register of Graduates and Former Cadets, United States Military Academy. West Point, New York: Association of Graduates, USMA.

1997 Register of Graduates and Former Cadets, United States Military Academy. West Point, New York: Association of Graduates, USMA.

1998 Register of Graduates and Former Cadets, United States Military Academy. West Point, New York: Association of Graduates, USMA.

Operations in Korea. West Point, New York: United States Military Academy, 1955.

Official Register of Officers and Cadets for the Academic Year Ending 30 June 1950. West Point, New York: United States Military Academy, 1950.

Official Register of Officers and Cadets for the Academic Year Ending 30 June 1951. West Point, New York: United States Military Academy, 1951.

Official Register of Officers and Cadets for the Academic Year Ending 30 June 1952. West Point, New York: United States Military Academy, 1952.

Official Register of Officers and Cadets for the Academic Year Ending 30 June 1953. West Point, New York: United States Military Academy, 1953.

Official Register of Officers and Cadets for the Academic Year Ending 30 June 1954. West Point, New York: United States Military Academy, 1954.

ARTICLES

Barber, Ransom E. "Big Rabble Shaping Up for the 1950 Campaign..." *The Pointer*, September 15, 1950.

Barber, Ransom E. "Clippings." *The Pointer*, September 15, 1950.

Barber, Ransom E. "Clippings." *The Pointer*, September 29, 1950.

Beasley, John H. "The USMA Honor System – a Due Process Hybrid." *Military Law Review*, Volume 118, 1987.

"Boyd Wheeler Bartlett." *Assembly,* Fall 1965.

"Charles J. Barrett." *Assembly*, Fall 1963.

"Charles Harlow Miles, Jr." *Assembly,* September 1985.

"John Knight Waters." *Assembly,* July 1993.

Downey, Robert H., Jr.; Evans, Walter F. "Mutiny on the 'Whiskey,'" *54-40...and Still in the Fight.* United States of America: Class of 1954, United States Military Academy, 1994.

Fleming, Jim. "Army Varsity." *The Pointer*, 23 October 1953.

"Francis Martin Greene." *Assembly,* June 1975.

Hamblin, Al. "The Rabble." *The Pointer*, 19 September 1952.

Harvill, Preston S., Jr. "The Army Team is Back." *The Pointer*, 25 September 1953.

Hughes, David R. "Report...from Korea." *The Pointer*, 2 November 1951.

Kintner, William R. "Pork Chop." *The Army Combat Forces Journal*, March 1955.

Love, J.B. "Remember the Price They Paid." *Assembly*, March/April 1997.

Lovell, John P. "And Its Captain...." *The Pointer*, 19 September 1952.

Lovell, John P. "Dick Shea – The Winning Habit." *The Pointer*, 2 May 1952.

Moses, Edward M. "The Reveille Gun Caper." *54-40...and Still in the Fight.* United States of America: Class of 1954, United States Military Academy, 1994.

Newman, Major General A.S. "The Forward Edge." *Army*, December 1980.

Pappas, George S. "More to the Point." *Assembly*, January 1995.

Shaler, Dexter H. "Blaik Looks at the Army Team." *The Pointer*, 25 September 1953.

Schemmer, Benjamin F. "A Shaggy Goat Story" and "The Goat Rebellion." *Assembly*, January 1995.

Swensson, John K. "A Race Well Run...." *Assembly*, Fall 1962.

"The Corps in Column." *The Pointer*, 10 November 1950.

"The Last for '54." *The Pointer*, 27 November 1953.

"A United 'Army' Wins Its Greatest Victory." *The Pointer*, 23 October 1953.

MAGAZINES

Argosy. New York, New York: Argosy, Inc. December 1953.

Assembly. West Point, New York: Association of Graduates, USMA, October 1951.

Assembly. West Point, New York: Association of Graduates, USMA, January 1954.

Assembly. West Point, New York: Association of Graduates, USMA, October 1954.

Assembly. West Point, New York: Association of Graduates, USMA, July 1955.

Assembly. West Point, New York: Association of Graduates, USMA, Winter 1959.

Assembly. West Point, New York: Association of Graduates, USMA, June 1985.

Assembly. West Point, New York: Association of Graduates, USMA, March 1987.

Assembly. West Point, New York: Association of Graduates, USMA, July/August 1997.

Life. New York, New York: October 9, 1950.

Newsweek. New York, New York: June 18, 1951.

The Pointer. West Point, New York: United States Corps of Cadets, USMA, November 30, 1950.

The Pointer. West Point, New York: United States Corps of Cadets, USMA, September 11, 1953.

The Pointer. West Point, New York: United States Corps of Cadets, USMA, September 25, 1953.

United States Military Academy Catalog, 1950-51. West Point, New York. Directorate of Admissions, 1950.

NEWSPAPERS

Boston Sunday Advertiser, November 29, 1953 • Carmel, California *Pine Cone*, January 4, 1952 •*New York Times*, October 11, 1953 •*New York Times*, October 18, 1953 • *Buffalo Evening News*, October 14, 1954 • *Diamondback*, University of Maryland, October 14, 1952 • *Diamondback*, University of Maryland, November 13, 1952 • Monterey, California *Monterey Penisula Herald*, December 31, 1951 • Monterey, California *Monterey Penisula Herald*, January 1, 1952 • Monterey, California *Monterey Penisula Herald*, January 2, 1952 • Monterey, California *Monterey Penisula Herald*, January 3, 1952 • Norfolk, Virginia *Pilot,* October 7, 1944 • *New York Sunday News*, October 18, 1953 • *New York Times*, November 12, 1950 • *New York Times*, November 14, 1950 • *New York Times*, November 19, 1950 • *New York Times*, November 21, 1950 • *New York Times*, December 3, 1950 • *New York Times*, September 30, 1951 • *New York Times*, October 7, 1951 •*New York Times*, October 14, 1951 • *New York Times*, October 21, 1951 • *New York Times*, October 28, 1951 • *New York Times*, November 4, 1951 • *New York Times*, November 11, 1951 • *New York Times*, November 18, 1951 • *New York Times*, December 2, 1951 • *New York Times*, January 2, 1952 • *New York Times*, September 28, 1952 • *New York Times*, October 5, 1952 • *New York Times*, October 12, 1952 •*New York Times*, October 19, 1952 • *New York Times*, October 26, 1952 • *New York Times*, November 2, 1952 • *New York Times*, November 9, 1952 • *New York Times*, November 16, 1952 • *New York Times*, November 30, 1952 • *New York Times*, December 2, 1952 • *New York Times*, September 27, 1953 • *New York Times*, September 29,1953 • *New York Times*, October 4, 1953 • *New York Times*, October 11, 1953 • *New York Times*, October 18, 1953 • *New York Times*, October 25, 1953 • *New York Times*, November 1, 1953 • *New York Times*, November 8, 1953 • *New York Times*, November 15, 1953 • *New York Times,* November 29, 1953 • *New York Times,* December 21, 1953 • *Pacific Stars and Stripes,* August 6, 1951 • *Pacific Stars and Stripes,* August 25, 1951 • Palo Alto, California *Palo Alto Times,* January 2, 1952 • Palo Alto, California *Palo Alto Times,* January 3, 1952 •

Palo Alto, California *Palo Alto Times,* January 4, 1952 • Palo Alto, California *Palo Alto Times,* January 7, 1952 • *Philadelphia Inquirer,* November 29, 1950 • *Phoenix Arizona Republic,* December 31, 1951 • *Phoenix Arizona Republic,* January 1, 1952 • *Phoenix Arizona Republic,* January 2, 1952 • *Phoenix Arizona Republic,* January 3, 1952 • *Phoenix Arizona Republic,* January 4, 1951 • *Pittsburgh Post-Gazette,* November 29, 1953 • San Francisco *Chronicle,* January 1, 1952 • San Francisco *Chronicle,* January 2, 1952 • San Francisco *Chronicle,* January 3,1952 • San Francisco *Chronicle,* January 4, 1952 • San Francisco *Examiner,* December 31, 1951 • San Francisco *Examiner,* January 1, 1952 • San Francisco *Examiner,* January 2, 1952 • *The Pittsburgh Press,* November 29, 1953 • Vincennes, Indiana *Vincennes-Sun Commercial,* March 8, 1953 • Vincennes, Indiana *Vincennes-Sun Commercial,* July 7, 1954.

VIDEOS

Army Football Office. 1953 *Football Highlights.*

Sports Information Office. *Field of Honor, 100 Years of Army Football.* National Football League, 1990.

UNPUBLISHED SOURCES

Administrative Memorandum #143. "Christmas Leave 1951": West Point, New York, United States Corps of Cadets, 15 November 1951.

Administrative Memorandum #150. "Ice Follies, 27 December 1951": West Point, New York, United States Corps of Cadets, 12 December 1951.

Administrative Memorandum # 155. "Christmas Hop Schedule": West Point, New York, United States Corps of Cadets, 18 December 1951.

Administrative Memorandum #157. "Instructions for Cadets Present During Christmas Leave": West Point, New York, United States Corps of Cadets, 19 December 1951.

Air Force Form 5. Record of Air Force Flying Time for William J. Ryan.

Bartlett, Boyd W.; Counts, Gerald A.; Stamps, Thomas D. Study of " Section 2 of the Act of June 3,1942": United States Military Academy, August 1, 1951.

Collins, Arthur S. "Senior Officers' Oral History Program": United States Army Military History Institute, Carlisle, Pennsylvania, 1981.

Collins, Arthur S. Comments recorded in minutes of Army Athletic Board Meeting: United States Military Academy, 15 February 1952.

Collins, Arthur S. Memorandum for Record, 28 July 1951.

Collins, J. Lawton. Memorandum for Record, 9 August 1951.

Coulter, John A. Chapter IV, "Richard Thomas Shea." Unpublished manuscript.

Daily Bulletin. United States Corps of Cadets, 19 February 1951.

Daily Bulletin. United States Corps of Cadets, 28 September 1951.

Davidson, Garrison H. *Grandpa Gar, The Saga of One Soldier as Told to His Grandchildren.* Unpublished book, 1974.

Downey, Robert H., Jr. "Reflections on a Mid-Summer Cruise," Unpublished paper, 1994.

Citation accompanying the posthumous award of the Congressional Medal of Honor to Richard Thomas Shea, Jr., given to Joyce Riemann Shea, May 1954.

Citation accompanying the posthumous award of the Silver Star to Richard George Inman, given to Barbara Kipp Inman, 6 July 1954.

Command Report. 17th Infantry Regiment, June 1953.

Command Report. 17th Infantry Regiment, July 1953.

Command Report. 32nd Infantry Regiment, July 1953.

Directive from Colonel R.S. Nourse to C.J. Barrett: United States Military Academy, West Point, New York, 9 August 1951

Directive from Colonel R.S. Nourse to H.C. Jones: United States Military Academy, West Point, New York, 4 August 1951.

Escalante, Edward J. "Last Patrol," paper of undated recollections.

Flight Order Nos. 511 and 521, C-47 Serial Number 44-75266, "United States Air Force Inspector General Special Investigation Report," 17 January 1952.

Greenberg, Pfc. Irwin. Deposition given Congressman William Bray of Indiana, 27 September 1953. (Courtesy of Mary Jo Inman Vermillion)

Harkins, Paul D. "Senior Officers' Oral History Program": United States Army Military History Institute, Carlisle, PA, 1974.

Hartle, Anthony C. "1990 Case Study," United States Military Academy, March 1990.

Harvill, Preston S., Jr., exchange of e-mails, November 1997.

Hughes, Lieutenant David R. Deposition accompanying recommendation for decoration of Edward J. Escalante for "...extraordinary courage and gallant determination" on the 23 September 1951 combat patrol, 10 December 1951.

Inman, Richard G. Personal papers, correspondence, and poems, courtesy of Mary Jo Inman Vermillion.

McFayden, Brigadier General B.M. Staff summary sheet titled "Reappointment of Cadets Separated from West Point for Honor Violations": Washington, D.C., United States Army, 20 November 1951.

Mauser, Pfc. Clarence R. Deposition given Congressman William Bray of Indiana, 27 September 1953. (Courtesy of Mary Jo Inman Vermillion)

Moses, Edward M. "The Army Victory Cannon." Unpublished recollection.

Taylor, Major General Maxwell D. Pamphlet titled "West Point Honor System and Its Objectives and Procedures": West Point, New York. United States Military Academy, 23 July 1947.

Thorsen, Roy T. Journal entry of 8 August 1951.

Tipton, Cpl. Harm J. Deposition given Congressman William Bray of Indiana, 27 September 1953. (Courtesy of Mary Jo Inman Vermillion)

Willcox, Lt. David R. Deposition given Congressman William Bray of Indiana, 27 September 1953. (Courtesy of Mary Jo Inman Vermillion)

Letter, David R. Hughes to Mrs. Helen I. Hughes, 1951.

Letter, David R. Hughes to Captain John R. Flynn, undated.

Letter, F. Shackelford to Senator William Benton, 27 August 1951.

Letter, Earl H. Blaik to Douglas MacArthur, 4 September 1951.

Letter, Earl H. Blaik to Douglas MacArthur, 18 October 1951.

Letter, Earl H. Blaik to Douglas MacArthur, 1 November 1951.

Letter, Earl H. Blaik to Douglas MacArthur, 3 June 1952.

Letter, Douglas MacArthur to Earl H. Blaik, 9 June 1952.

Letter, Earl H. Blaik to Douglas MacArthur, 24 June 1952.

Letter, Herbert C. Hollander to Louis M. Davis, 6 November 1953.

Letter, Barbara K. Inman to Mr. and Mrs. Inman, 30 June 1953.

Letter, Barbara K. Inman to Mr. and Mrs. Inman, 2 July 1953.

Letter, Richard G. Inman to Mr. and Mrs. Inman, 27 May 1953.

Letter, Richard G. Inman to Mr. and Mrs. Inman, 4 June 1953.

Letter, Richard G. Inman to Mr. and Mrs. Inman, 25 June 1953.

Letter, Richard G. Inman to Mr. and Mrs. Inman, 2 July 1953.

Letter, Barbara Kipp (Inman) Colby to Mrs. 'Becky' Inman and Bonnie Ruth Inman, circa Christmas 1963, with copy of Richard G. Inman's poem, "To My Sister Bonnie."

Letter, Joyce Riemann (Shea) Himka to the author, 6 March 1998.

Letter, Joyce Riemann (Shea) Himka to the author, 11 March 1998.

Letter, Major General Frederick A. Irving to General Douglas MacArthur.

Letter, Major General Frederick A. Irving to Lieutenant General Joseph M. Swing, Headquarters, Sixth Army: West Point, New York. United States Military Academy, 17 January 1952.

Letter, with enclosures, from Colonel R.P. Eaton, Adjutant General, USMA, to Superintendent, *USMA, Subject: Summary of Actions Taken on Mass Honor Violations* April - December 1951, 21 April, 1952.

Letter, Lt. Col. Beverly M. Read to Joyce Riemann Shea, 31 August 1953. (Courtesy of Joyce (Shea) Himka)

Letter, Lt. Col. Beverly M. Read to Joyce Riemann Shea, Easter 1954. (Courtesy of Joyce (Shea) Himka)

Letter, Richard T. Shea, Jr. to Joyce Riemann Shea, 9 June 1953.

Letter, Richard T. Shea, Jr. to Joyce Riemann Shea, 14 June 1953.

Letter, Richard T. Shea, Jr. to Joyce Riemann Shea, 23 June 1953.

Letter, Richard T. Shea, Jr. to Joyce Riemann Shea, 29 June 1953.

Letter, Warrant Officer Junior Grade Thomas E. Wertz to Commandant of Cadets: "Survey of Honor Violations During Past Five Years," United States Military Academy, September 16, 1951.

Letter accompanying the Collins Board Report from Colonel John K. Waters to the Superintendent: 10 June 1951.

Letter, Mary Jo Inman Vermillion to the author, 6 March 1998.

Letter, Mary Jo Inman Vermillion to the author, 17 March 1998.

Memorandum to the Corps of Cadets: "Letters and Messages of Condolence." West Point, New York. United States Corps of Cadets, 4 January 1952.

Memorandum to the Corps of Cadets: "Letters and Messages of Condolence." West Point, New York. United States Corps of Cadets, 14 January 1952.

Memorandum, Collins Board to the Commandant of Cadets: "Additional List of Cadets Involved in Investigation," 24 June 1951.

Memorandum from Colonel John K. Waters to the Superintendent, 7 July 1951.

Memorandum from Major General E.M. Brannon: "Court-Martial Jurisdiction to Try Officer Offense Committed as a Cadet," 20 July 1951.

Memorandum from Lieutenant Winfield L. Martin to the Chief of Staff, United States Military Academy. 14 August 1951.

Memorandum from Lieutenant General Maxwell D. Taylor to the Army Chief of Staff: "Dishonorable Practices of Cadets at USMA," 27 July 1951.

Memorandum to: All Fourth Classmen. West Point, New York, United States Corps of Cadets, 12 December 1951.

Message from Colonel R.P. Eaton to Commanding Officer, Williams Air Force Base, Arizona: West Point, New York. United States Military Academy, 3 January 1952.

Message from Commanding General, Sixth Army to Superintendent, United States Military Academy: Presidio of San Francisco, California, 7 January 1952.

Minutes from the United States Military Academy Academic Board. West Point, New York. 7 August 1951.

Movement Order Number 10: "U.S. Naval Academy Cruise for Selected Cadets." West Point, New York. Headquarters, United States Corps of Cadets, 10 July 1951.

Operations Order 1 (Corps of Cadets Travel to New York City's Polo Grounds): West Point, New York. Headquarters, United States Corps of Cadets, 8 October 1953.

Order No. 2: "Return to West Point of the Remains of the Late Cadets Glassbrenner, Karl F. Jr., Class of 1952, Co C-2; Mastelotto, Maurice J., Class of 1954, Co G-2; Wilson, Hugh R. Jr., Class of 1954, Co I-2." West Point, New York. United States Corps of Cadets, 8 January 1952.

Order No. 5: Headquarters, United States Corps of Cadets, West Point, New York, 29 March 1951.

"Orientation for Parents of Fourth Classmen During the Christmas Holidays": United States Military Academy. West Point, New York, Department of Tactics, 21 November 1951.

Personal correspondence from Lieutenant Richard T. Shea, Jr. to Joyce Riemann Shea, 1953.

Proceedings of a Board of Officers convened pursuant to Letter Orders: Headquarters, United States Corps of Cadets, West Point, New York, 28 May 1951. (Collins Board)

Proceedings of a Board of Officers Appointed by Letter Orders: Headquarters, United States Military Academy, 13 August 1951. (Bartlett Board)

Proceedings of a Board of Officers Appointed by Letter Orders: Headquarters, United States Military Academy, 17 October 1951. (Greene Board)

"Recommendation for Decoration for Heroism" (Distinguished Service Cross). United States Army, 7th Cavalry Regiment, 11 January 1952.

Regulations. West Point, New York. United States Corps of Cadets, 1949. (Blue Book)

Report of Board Appointed by Confidential Order of the Secretary of the Army, 17 July 1951. (Hand Board)

Ryan, William J. Record of Interview: "1990 Case Study," United States Military Academy, March 1990.

Scripts for Honor lectures given to Fourth Class Cadets during "Beast Barracks. West Point, New York. United States Military Academy; 1951, 1952, and 1953.

Senate Congressional Record No. 143-5: Washington, D.C., 6 August 1951.

Statement of Colonel Guy McNeil to the Aircraft Accident Investigation Board, for C-47 Serial Number 44-75266, 3 January 1952.

"Syllabus for Military Psychology and Leadership:" West Point, New York. Department of Military Psychology and Leadership, 1947.

Telegram, Earl H. Blaik to Douglas MacArthur, 29 September 1952.

Telegram, Earl H. Blaik to Douglas MacArthur, 18 November 1952.

Training Memorandum Number 25: "Training Program, New Cadets, Class of 1955, 3 July-27 August 1951." West Point, New York. Headquarters, United States Corps of Cadets, 26 April 1951.

Training Memorandum Number 5: "Military Training Program for Summer 1953." West Point, New York. Headquarters, United States Corps of Cadets, 9 April 1953.

Training Memorandum Number 12: "Reorganization Week." West Point, New York. Headquarters, United States Corps of Cadets, 15 August 1953.

Training Schedule Number 4A-4: "Fourth Class Military Customs and Courtesy." Headquarters, United States Corps of Cadets. West Point, New York, 13-21 December 1951.

"United States Air Force Aircraft Accident Report 12-30-51-1." Williams Air Force Base, Arizona, 7 February 1952.

"United States Air Force Aircraft Accident Report on Crash of B-29A 44-61920, on 24 August 1953."

United States Military Academy Staff Diary, Part II, Entry dated 10 August 1951.

United States Military Academy Staff Diary, Parts I and II, and related Army Staff Papers, from 1 April to 10 December 1951.

Vann, Peter J., A book proposal titled *Vince Lombardi's West Point, 1949-1953*.

INDEX

Doremus (football player) 615, 642, 912
Dorr (football player) 511, 642
Downey, Robert H., Jr. 642, 799
Dozier, Wayne 27
Draper, Philip H., Jr. 305
Drews, Barry 537
Duke University 133, 215, 231, 489, 499, 624, 800, 801, 816, 817, 818, 819, 820, 821, 822, 823, 824, 825, 826, 827, 828, 829, 830, 831, 832, 833, 834, 835, 836, 837, 838, 839, 840, 841, 842, 843, 844, 845, 846, 847, 848, 849, 850, 851, 853, 855, 860, 862, 868, 872, 874, 897, 922, 947, 996, 997, 999
Dulles, John Foster 901

E

Edwards, Jay 821
Eglin Air Force Base 794
Eichelberger, Robert 229, 236, 306, 417, 475, 593, 919, 962
Eisenhauer, Steve 635, 897, 902, 906
Eisenhower, Dwight D. 10, 127, 281, 383, 384, 396, 404, 405, 406, 411, 553, 555, 614, 619, 620, 635, 645, 853, 884, 887, 901, 902, 949, 967, 971, 984, 989, 1004
Eisenhower, John S. D. 902
Eisenhower, Milton S. 547, 619
Elko (Korea) 718
Ellinger, Harry 449
Elliot, James H. 570
Ellis, Ross 608
Ely, Anne Carter Lee 540
Ely, Hanson Edward, III 540, 542
Emmanuel, Victor 967
Erdlatz, Eddie 507, 508, 636, 896, 910
Erie (Korea) 692, 693, 694, 695, 699, 719, 754
Ersfeld, Ronald A. 771, 773
Escalante, Edward J. 183, 185, 186
Evans, Bob 505

F

Faith, Don 251, 252
Farris, Bob 443, 446, 599, 805, 814, 832, 841, 844, 861, 907, 911, 912, 998
Favole, "Cheech" 900
Fechteler, William M. 522, 554
Felver, Bob 870
Fenton, Chauncey L. 514, 549, 572
Finch, Dick 711
Finletter, Thomas K. 281, 522, 619
Fischl, Frank 655
Fisher, Don 509
Flanagan, L.J. 866
Flynn, John Robert 164, 169, 170, 173, 177, 196, 544, 792
Folberg, Dan 250
Fontaine, Richard A. 329
Forbes Air Force Base 653
Ford, Jeremiah, II 934
Fort Belvoir 796
Fort Benning 24, 34, 156, 157, 163, 169, 232, 251, 267, 381, 547, 651, 667, 677, 792, 985, 993
Fort Bliss 34, 199, 242, 257, 416, 547
Fort Bragg 157, 206, 795
Fort Dix 792
Fort Eustis 794
Fort Hamilton 847
Fort Jackson 792
Fort Knox 156, 791, 792, 893
Fort Lee 793
Fort Monmouth 797
Fort Myer 34, 986
Fort Ord 220, 991
Fort Riley 34, 156, 163, 164, 380, 547
Fort Sill 156, 206, 792
Foster, George E. 252
Franco, Fred 509, 636, 637
Franklin, Joe 912
Freeman, Douglas Southall 538
Frey, Roger 623
Fry, Virginia K. 1008
Fullam, Don 905
Fullbright, J. William 16, 408

I

W

Wade, James O. 213
Wagner, Hayden W. 514
Wagner, Louis C., Jr. 570
Wainwright, Jonathan M. 68, 241
Walker, Lieutenant 186
Walker, Walton H., Jr., "Johnnie" 20, 171, 253, 980
Wallace, Bob 617
Wallis, Charles R. 982
Walsh, Ellard A. 519, 525, 529, 536
Walthour, John B. 572
Walton (football player) 642
Ward, Al 496, 615, 616, 617
Ward, Gene 850
Warmath, Murray 210, 244, 440, 446, 469, 968
Washington, George 9, 385, 517, 524, 548, 567, 570, 588
Waters, John K. 11, 12, 25, 32, 33, 37, 42, 52, 59, 61, 92, 116, 127, 128, 130, 133, 248, 300, 301, 311, 338, 348, 362, 363, 379, 380, 381, 382, 411, 467, 480, 525, 536, 549, 677, 800, 944, 947, 976, 984
Watson, Numa A., Jr. 982
Weathers, A. R. 783, 925
Weaver, Ed 511, 607, 615, 616, 638, 642
Weaver, John 637, 907, 908
Weaver, Joseph U. 878, 879
Weems, Wharton 538
Welsh, George 897, 904, 908, 999
West, Eddie 864
Wetzel, Emery Scott 68, 876, 878, 879
Wheeler, Boyd 300, 302
White, Edward H., II 982, 983
Whitney, Courtney 972
Wickham, John A., Jr. 981
Wilkerson, Frank 445, 458, 511, 613, 621, 641, 642
Wilkinson, Bud 968

Willcox, David R. 722, 725, 727, 746, 748, 752, 758, 760, 926, 931, 933, 938, 941, 992
Williams (football player) 511
Williams Air Force Base 268, 270
Wilner, Jack 636
Wilson, Charles E. 901
Wilson, Gail 241
Wilson, Hugh R., Jr. 258, 283, 284
Wilson, "Woody" 834, 848
Winchell, Walter 363
Wing, Johnny 477, 487, 488, 490, 496, 504, 509, 511, 600, 601, 608, 612, 616, 617, 642, 912
Winger, Robert F. C. 570
"Whiskey," USS *Wisconsin* 132, 290, 292, 793, 982
Wodeschick, Gene 616
Wolf, Coach Bear 857
Wonsan, Korea 171, 245, 252, 686, 773
Woodward (football player) 511
Worden, Alfred M. 821, 982
Wright-Patterson Air Force Base 791
Wynn (football player) 912

Y

Yakimowicz 809
Yale University 229, 635, 996, 997, 998
Yalu River, Korea 160, 245, 252, 669, 697
Yoju, Korea 21
Yonch'on, Korea 682
Young, Archie 421
Young, Jim 969
Young, Jimmy 422

Z

Zaborowski (football player) 842, 912
Zastrow, Bob 250
Zeigler, Mike 642, 812, 819, 831, 853, 857, 864
Ziegler, Dick 511, 640, 832, 840, 898, 912